DMU 0316336 01 4

AF606290
WITHDRAWN
DE MONTFORT UNIVERSITY
LIBRARY

Advances in ___________________

Pharmacology

Volume 31

Advances in 

Pharmacology

Volume 31

Anesthesia and Cardiovascular Disease

Edited by

Zeljko J. Bosnjak
Departments of Anesthesiology and Physiology
The Medical College of Wisconsin
Milwaukee, Wisconsin

John P. Kampine
Departments of Anesthesiology and Physiology
The Medical College of Wisconsin
Milwaukee, Wisconsin

Academic Press
San Diego New York Boston London Sydney Tokyo Toronto

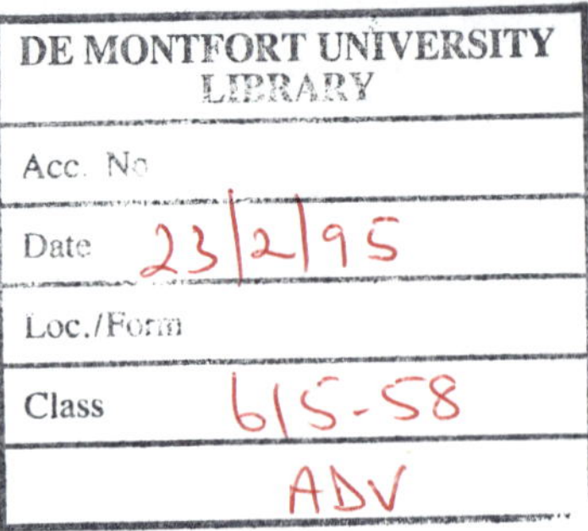

This book is printed on acid-free paper. ∞

Academic Press, Inc.
A Division of Harcourt Brace & Company
525 B Street, Suite 1900, San Diego, California 92101-4495

United Kingdom Edition published by
Academic Press Limited
24-28 Oval Road, London NW1 7DX

International Standard Serial Number: 1054-3589

International Standard Book Number: 0-12-032932-8 (case)

International Standard Book Number: 0-12-118860-4 (paper)

PRINTED IN THE UNITED STATES OF AMERICA
94 95 96 97 98 99 QW 9 8 7 6 5 4 3 2 1

Contents

Part I
Cardiac Muscle

Regulation of the Calcium Slow Channels of Heart by Cyclic Nucleotides and Effects of Ischemia

Nicholas Sperelakis

Functional Adaptation to Myocardial Ischemia: Interaction with Volatile Anesthetics in Chronically Instrumented Dogs

Patrick F. Wouters, Hugo Van Aken, Marc Van de Velde, Marco A. E. Marcus, and Willem Flameng

Excitation–Contraction Uncoupling and Vasodilators for Long-Term Cold Preservation of Isolated Hearts

David F. Stowe

Troponin T as a Marker of Perioperative Myocardial Cell Damage

H. Mächler, H. Gombotz, K. Sabin, and H. Metzler

Silent Myocardial Ischemia: Pathophysiology and Perioperative Management

Anders G. Hedman

Interaction of Anesthetics and Catecholamines on Conduction in the Canine His–Purkinje System

Lawrence A. Turner, Sanja Vodanovic, and Zeljko J. Bosnjak

Anesthetics, Catecholamines, and Ouabain on Automaticity of Primary and Secondary Pacemakers

John L. Atlee III, Martin N. Vicenzi, Harvey J. Woehlck, and Zeljko J. Bosnjak

The Role of L-Type Voltage-Dependent Calcium Channels in Anesthetic Depression of Contractility

Thomas J. J. Blanck, D. L. Lee, S. Yasukochi, C. Hollmann, and J. Zhang

Effects of Inhibition of Transsarcolemmal Calcium Influx on Content and Releasability of Calcium Stored in Sarcoplasmic Reticulum of Intact Myocardium

Hirochika Komai and Ben F. Rusy

Arrhythmogenic Effect of Inhalation Anesthetics: Biochemical Heterogeneity between Conduction and Contractile Systems and Protein Unfolding

Issaku Ueda and Jang-Shing Chiou

Part II
Coronary Circulation

Potassium Channel Current and Coronary Vasodilatation by Volatile Anesthetics

Nediljka Buljubasic, Jure Marijic, and Zeljko J. Bosnjak

Potassium Channel Opening and Coronary Vasodilation by Halothane

D. R. Larach, H. G. Schuler, K. A. Zangari, and R. L. McCann

Volatile Anesthetics and Coronary Collateral Circulation

Judy R. Kersten, J. Craig Hartman, Paul S. Pagel, and David C. Warltier

Myocardial Oxygen Supply–Demand Relations during Isovolemic Hemodilution

George J. Crystal

Part III
Cellular Targets

Plasma Membrane Ca^{2+}-ATPase as a Target for Volatile Anesthetics
Danuta Kosk-Kosicka

Enhancement of Halothane Action at the Ryanodine Receptor by Unsaturated Fatty Acids
Jeffrey E. Fletcher and Vincent E. Welter

Adrenergic Receptors: Unique Localization in Human Tissues
Debra A. Schwinn

Volatile Anesthetic Effects on Inositol Triphosphate-Gated Intracellular Calcium Stores in GH_3 Cells
Alex S. Evers and M. Delawar Hossain

Part IV
Reflex Regulation

Differential Control of Blood Pressure by Two Subtypes of Carotid Baroreceptors
Jeanne L. Seagard

Sympathetic Activation with Desflurane in Humans
Thomas J. Ebert and Michael Muzi

Randomized, Prospective Comparison of Halothane, Isoflurane, and Enflurane on Baroreflex Control of Heart Rate in Humans

Michael Muzi and Thomas J. Ebert

Baroreflex Modulation by Isoflurane Anesthesia in Normotensive and Chronically Hypertensive Rabbits

Leonard B. Bell

Part V
Peripheral Circulation

Effects of Isoflurane on Regulation of Capacitance Vessels under Normotensive and Chronically Hypertensive Conditions

Thomas A. Stekiel, Leonard B. Bell, Zeljko J. Bosnjak, and John P. Kampine

Effect of Volatile Anesthetics on Baroreflex Control of Mesenteric Venous Capacitance

J. Bruce McCallum, Thomas A. Stekiel, Anna Stadnicka, Zeljko J. Bosnjak, and John P. Kampine

Effect of General Anesthesia on Modulation of Sympathetic Nervous System Function

Margaret Wood

First Pass Uptake in the Human Lung of Drugs Used during Anesthesia

David L. Roerig, Susan B. Ahlf, Christopher A. Dawson, John H. Linehan, and John P. Kampine

Lactic Acidosis and pH on the Cardiovascular System

Yuguang Huang, James B. Yee, Wang-Hin Yip, and K. C. Wong

Part VI
Cerebral Circulation

Role of Oxygen Free Radicals and Lipid Peroxidation in Cerebral Reperfusion Injury

Laurel E. Moore and Richard J. Traystman

Effects of Volatile Anesthetics on Cerebrocortical Laser Doppler Flow: Hyperemia, Autoregulation, Carbon Dioxide Response, Flow Oscillations, and Role of Nitric Oxide

Antal G. Hudetz, Joseph G. Lee, Jeremy J. Smith, Zeljko J. Bosnjak, and John P. Kampine

Cerebral Blood Flow during Isovolemic Hemodilution: Mechanistic Observations

Michael M. Todd

Cerebral Physiology during Cardiopulmonary Bypass: Pulsatile versus Nonpulsatile Flow

Brad Hindman

Anesthetic Actions of Cardiovascular Control Mechanisms in the Central Nervous System

William T. Schmeling and Neil E. Farber

Contributors

Numbers in parentheses indicate the pages on which the authors' contributions begin.

Susan B. Ahlf (531), Department of Veterans Affairs, Zablocki Veterans Affairs Medical Center, Milwaukee, Wisconsin 53295

John L. Atlee III (185), Department of Anesthesiology, The Medical College of Wisconsin, Milwaukee, Wisconsin 53226

James B. Yee (551), Department of Anesthesiology, University of Utah School of Medicine, Salt Lake City, Utah 84132

Leonard B. Bell (389, 409), Departments of Anesthesiology and Physiology, The Medical College of Wisconsin, and Zablocki Veterans Affairs Medical Center, Milwaukee, Wisconsin 53226

Saiid Bina (459), Department of Anesthesiology, Uniformed Services University of the Health Sciences, F. Edward Hébert School of Medicine, Bethesda, Maryland 20814

Thomas J. J. Blanck (207), Department of Anesthesiology, Cornell University Medical College, New York, New York 10021

Zeljko J. Bosnjak (167, 185, 235, 409, 431, 577), Departments of Anesthesiology and Physiology, The Medical College of Wisconsin, Milwaukee, Wisconsin 53226

Nediljka Buljubasic (235), Department of Anesthesiology, The Medical College of Wisconsin, Milwaukee, Wisconsin 53226

Grace L. Chien (99), Department of Anesthesiology, Oregon Health Sciences University, and Portland Veterans Affairs Medical Center, Portland, Oregon 97201

Jang-Shing Chiou (223), Department of Anesthesia, University of Utah School of Medicine and Veterans Affairs Medical Center, Salt Lake City, Utah 84148

George J. Crystal (285), Departments of Anesthesiology and Physiology and Biophysics, University of Illinois College of Medicine, Illinois Masonic Medical Center, Chicago, Illinois 60680

Bruce A. Davidson (109), Department of Anesthesiology, The State University of New York at Buffalo, Buffalo General Hospital, Buffalo, New York 14203

Richard F. Davis (99), Department of Anesthesiology, Oregon Health Sciences University, and Portland Veterans Affairs Medical Center, Portland, Oregon 97201

Christopher A. Dawson (531), Department of Physiology, The Medical College of Wisconsin, Milwaukee, Wisconsin 53226

Benjamin Drenger (89), Department of Anesthesiology and Critical Care Medicine, Hadassah University Hospital, 91120 Jerusalem, Israel

Ron Dueck (505), Department of Anesthesiology, Veterans Affairs Medical Center, and University of California at San Diego, La Jolla, California 92161

Thomas J. Ebert (369, 379), Department of Anesthesiology, Veterans Affairs Medical Center, and The Medical College of Wisconsin, Milwaukee, Wisconsin 53226

Alex S. Evers (343), Departments of Anesthesiology and Pharmacology, Washington University School of Medicine, St. Louis, Missouri 63110

Neil E. Farber (617), Departments of Anesthesiology and Pharmacology and Toxicology, The Medical College of Wisconsin, and Zablocki Veterans Affairs Medical Center, Milwaukee, Wisconsin 53226

Willem Flameng (25), Department of Anesthesiology, University Hospitals Gasthuisberg, Katholieke Universiteit, Leuven, Belgium

Jeffrey E. Fletcher (323), Department of Anesthesiology, Hahnemann University, Philadelphia, Pennsylvania 19102

Martha J. Frazer (145), Department of Anesthesiology, University of Virginia Health Sciences Center, Charlottesville, Virginia 22908

Yehuda Ginosar (89), Department of Anesthesiology and Critical Care Medicine, Hadassah University Hospital, 91120 Jerusalem, Israel

H. Gombotz (63), University Department of Anesthesiology, A-8036 LKH Graz, Austria

Yaacov Gozal (89), Department of Anesthesiology and Critical Care Medicine, Hadassah University Hospital, 91120 Jerusalem, Israel

Jayne L. Hart (459), Department of Biology, George Mason University, Fairfax, Virginia 22030

J. Craig Hartman (269), The Upjohn Company, Kalamazoo, Michigan 49001

Anders G. Hedman (75), Department of Medicine, Ludvika Hospital, Ludvika S-77181, Sweden

Brad Hindman (607), Department of Anesthesiology, University of Iowa College of Medicine, Iowa City, Iowa 52242

Quinn H. Hogan (471), Department of Anesthesiology, The Medical College of Wisconsin, Milwaukee, Wisconsin 53226

C. Hollmann (207), Department of Anesthesiology, Cornell University Medical College, New York, New York 10021

M. Delawar Hossain (343), Department of Anesthesiology, Washington University School of Medicine, St. Louis, Missouri 63110

Yuguang Huang (551), Peking Union Medical College, Peking, China

Antal G. Hudetz (577), Departments of Anesthesiology and Physiology, The Medical College of Wisconsin, Wisconsin 53226

William E. Hurford (531), Department of Anesthesia, Massachusetts General Hospital, Boston, Massachusetts 02114

Ming Jing (459), Department of Anesthesiology, Uniformed Services University of the Health Sciences, F. Edward Hébert School of Medicine, Bethesda, Maryland 20814

John P. Kampine (409, 431, 471, 531, 577), Departments of Anesthesiology and Physiology, The Medical College of Wisconsin, Milwaukee, Wisconsin 53226

Judy R. Kersten (269), Department of Anesthesiology, The Medical College of Wisconsin, Milwaukee, Wisconsin 53226

Paul R. Knight (109), Department of Anesthesiology, The State University of New York at Buffalo, Buffalo General Hospital, Buffalo, New York 14203

Hirochika Komai (215), Department of Anesthesiology, University of Wisconsin—Madison, Madison, Wisconsin 53792

Danuta Kosk-Kosicka (313), The Johns Hopkins University School of Medicine, Department of Anesthesiology and Critical Care Medicine, Baltimore, Maryland 21287

D. R. Larach (253), Department of Anesthesia, College of Medicine, The Pennsylvania State University, Hershey, Pennsylvania 17033

D. L. Lee (207), Department of Anesthesiology, Cornell University Medical College, New York, New York 10021

Joseph G. Lee (577), Departments of Anesthesiology and Physiology, The Medical College of Wisconsin, Milwaukee, Wisconsin 53226

John H. Linehan (531), Department of Biomedical Engineering, Marquette University, Milwaukee, Wisconsin 53223

Carl Lynch III (145), Department of Anesthesiology, University of Virginia Health Sciences Center, Charlottesville, Virginia 22908

H. Mächler (63), University Department of Anesthesiology, A-8036 LHK Graz, Austria

Marco A. E. Marcus (25), Department of Anesthesiology, University Hospitals Gasthuisberg, Katholieke Universiteit, Leuven, Belgium

Jure Marijic (235), Department of Anesthesiology, The Medical College of Wisconsin, Milwaukee, Wisconsin 53226

J. Bruce McCallum (431), Department of Anesthesiology, The Medical College of Wisconsin, Milwaukee, Wisconsin 53226

R. L. McCann (253, 269), Department of Anesthesia, College of Medicine, The Pennsylvania State University, Hershey, Pennsylvania 17033

H. Metzler (63), University Department of Anesthesiology, A-8036 LHK Graz, Austria

Ning Miao (145), Department of Anesthesiology, University of Virginia Health Sciences Center, Charlottesville, Virginia 22908

Laurel E. Moore (565), Department of Anesthesiology and Critical Care Medicine, Johns Hopkins Medical Institutions, Baltimore, Maryland 21287

Sheila M. Muldoon (459), Department of Anesthesiology, Uniformed Services University of the Health Sciences, F. Edward Hébert School of Medicine, Bethesda, Maryland 20814

Paul A. Murray (485, 505), Department of Anesthesiology and Critical Care Medicine, The Johns Hopkins University School of Medicine, Baltimore, Maryland 21205

Michael Muzi (369, 379), Department of Anesthesiology, Veterans Affairs Medical Center, and The Medical Collge of Wisconsin, Milwaukee, Wisconsin 53226

Paul S. Pagel (125, 269), Department of Anesthesiology, The Medical College of Wisconsin, and the Zablocki Veterans Affairs Medical Center, Milwaukee, Wisconsin 53226

David L. Roerig (531), Departments of Anesthesiology and Pharmacology and Toxicology, The Medical College of Wisconsin, Milwaukee, Wisconsin 53226

Ben F. Rusy (215), Department of Anesthesiology, University of Wisconsin—Madison, Madison, Wisconsin 53792

K. Sabin (63), University Department of Anesthesiology, A-8036 LKH Graz, Austria

William T. Schmeling (617), Departments of Anesthesiology and Pharmacology and Toxicology, The Medical College of Wisconsin, and Zablocki Veterans Affairs Medical Center, Milwaukee, Wisconsin 53226

H. G. Schuler (253), Department of Anesthesia, College of Medicine, The Pennsylvania State University, Hershey, Pennsylvania 17033

Debra A. Schwinn (333), Department of Anesthesiology, Duke University Medical Center, Durham, North Carolina 27710

Jeanne L. Seagard (351), Department of Anesthesiology, and Zablocki Veterans Affairs Medical Center, The Medical College of Wisconsin, Milwaukee, Wisconsin 53226

Jeremy J. Smith (577), Departments of Anesthesiology and Physiology, The Medical College of Wisconsin, Milwaukee, Wisconsin 53226

Mitchell D. Smith (109), Department of Anesthesiology, The State University of New York at Buffalo, Buffalo General Hospital, Buffalo, New York 14203

Nicholas Sperelakis (1), Department of Physiology and Biophysics, University of Cincinnati, Cincinnati, Ohio 45267

Anna Stadnicka (431, 471), Department of Anesthesiology, The Medical College of Wisconsin, Milwaukee, Wisconsin 53226

Thomas A. Stekiel (409, 431), Departments of Anesthesiology and Physiology, The Medical College of Wisconsin, and Zablocki Veterans Affairs Medical Center, Milwaukee, Wisconsin 53226

David F. Stowe (39), Departments of Anesthesiology and Physiology, and Zablocki Veterans Affairs Medical Center, The Medical College of Wisconsin, Milwaukee, Wisconsin 53226

Michael M. Todd (595), Department of Anesthesiology, University of Iowa College of Medicine, Iowa City, Iowa 52242

Richard J. Traystman (565), Department of Anesthesiology and Critical Care Medicine, Johns Hopkins Medical Institutions, Baltimore, Maryland 21287

Lawrence A. Turner (167), Departments of Anesthesiology and Physiology, The Medical College of Wisconsin, Milwaukee, Wisconsin 53226

Issaku Ueda (223), Department of Anesthesia, University of Utah School of Medicine and Veterans Affairs Medical Center, Salt Lake City, Utah 84148

Hugo Van Aken (25), Department of Anesthesiology, University Hospitals Gasthuisberg, Katholieke Universiteit, Leuven, Belgium

Donna M. Van Winkle (99), Department of Anesthesiology, Oregon Health Sciences University, and Portland Veterans Affairs Medical Center, Portland, Oregon 97201

Marc Van de Velde (25), Department of Anesthesiology, University Hospitals Gasthuisberg, Katholieke Universiteit, Leuven, Belgium

Martin N. Vicenzi (185), Department of Anesthesiology, The Medical College of Wisconsin, Milwaukee, Wisconsin 53226

Sanja Vodanovic (167), Departments of Anesthesiology and Physiology, The Medical College of Wisconsin, Milwaukee, Wisconsin 53226

David C. Warltier (125, 269), Departments of Anesthesiology, Pharmacology, and Medicine, The Medical College of Wisconsin, Milwaukee, Wisconsin 53226

Vincent E. Welter (323), Department of Anesthesiology, Hahnemann University, Philadelphia, Pennsylvania 19102

Harvey J. Woehlck (185), Department of Anesthesiology, The Medical College of Wisconsin, Milwaukee, Wisconsin 53226

K. C. Wong (551), Department of Anesthesiology, University of Utah School of Medicine, Salt Lake City, Utah 84132

Margaret Wood (449), Department of Anesthesiology, Vanderbilt University Medical Center, Vanderbilt University School of Medicine, Nashville, Tennessee 37232

Patrick F. Wouters (25), Department of Anesthesiology, University Hospitals Gasthuisberg, Katholieke Universiteit, Leuven, Belgium

S. Yasukochi (207), Department of Anesthesiology, Cornell University Medical College, New York, New York 10021

James B. Yee (551), Department of Anesthesiology, University of Utah School of Medicine, Salt Lake City, Utah 84132

Wang-Hin Yip (551), Kaohsiung Medical College, Kaohsiung, Taiwan, Republic of China

K. A. Zangari (253), Department of Anesthesia, College of Medicine, The Pennsylvania State University, Hershey, Pennsylvania 17033

Warren M. Zapol (513), Department of Anesthesia, Massachusetts General Hospital, Boston, Massachusetts 02114

J. Zhang (207), Department of Anesthesiology, Cornell University Medical College, New York, New York 100211

Preface

The purpose of this book is to emphasize the close relationship and interdependence between anesthesiology and physiology, especially in the area of cardiovascular function. This book represents some of the most recent experimental observations and hypotheses with emphasis on the cardiovascular effects of anesthetics and the mechanisms that may be involved in normal and pathophysiological states.

Many patients anesthetized for surgical procedures under general anesthesia have some degree of cardiovascular disease. Sadly, knowledge of how anesthetics and drugs used during anesthesia contribute and/or interact with altered physiological states that are responsible for cardiovascular instability is far from complete. Although there are a number of likely causes for perioperative cardiovascular morbidity, noteworthy among these are the involvement of altered physiological states, autonomic imbalance, altered stress responses, concurrent disease, and factors that directly relate to the quality or type of care provided in the perioperative setting. Many of the drugs commonly used during general anesthesia, including the inhalational anesthetics, alter cardiovascular regulation as a side effect of their primary purposes to produce surgical anesthesia.

The cardiovascular depressant properties of potent volatile anesthetics are due not only to their direct effects on the heart and peripheral circulation but also to their depressant effects on reflex regulation of the cardiovascular system. There is a considerable body of evidence that the inhibitory effects of potent volatile anesthetics occur at multiple sites. These include central nervous system (CNS) regulatory sites, sympathetic preganglionic and postganglionic sites, sympathetic and parasympathetic ganglionic transmission sites, and the nerve endings, as well as direct end organ effects. Although the depression of the cardiovascular system at anesthetic levels required for surgical anesthesia is usually moderate and reversible in healthy patients, it may be much more extensive in patients with either chronic or acute disorders of the cardiovascular system, as well as associated diseases which may impair cardiovascular function. Many of the topics in this book deal with effects of anesthetics in the presence of altered physiological states including myocardial ischemia, diastolic dysfunction, hypertension, respiratory distress syndrome, acidosis, cerebral injury, and isovolemic hemodilution. It is not surprising that the anesthesiologist is faced with increased challenges to provide safe anesthesia for complex surgical procedures especially in these groups of patients. It is imperative, therefore, that we gain a better understanding of not only normal cardiovascular function but also the mechanisms by which the volatile anesthetics interfere with disorders of the cardiovascular system and associated diseases.

It is hoped that this book will in part contribute to a better understanding of the ways in which the potent anesthetics alter cardiac function in order to provide a more rational basis for therapeutic interventions and as a guide in choosing anesthetic approaches which are more likely to provide the level of anesthesia required for surgery with the least disturbance in the cardiovascular regulation. The 42 chapters in this book are organized into six parts covering (1) cardiac muscle, (2) coronary circulation, (3) cellular targets, (4) reflex regulation, (5) peripheral circulation, and (6) cerebral circulation.

The book begins with a section that deals primarily with the basic cardiac effects of anesthetics and provides information on the normal and ischemic processes controlling or affecting cardiac function. Part II focuses on the effects of volatile anesthetics on coronary circulation and presents data obtained from *in situ* experiments, isolated coronary rings, and isolated single smooth muscle cells including single-channel recordings. Part III addresses the various effects of anesthetics on specific cellular targets. Part IV examines the effects of anesthesia on reflex regulation along with the major circulatory effects of chronic hypertension. Part V focuses on the effects of anesthetics on the peripheral circulation including pulmonary vascular regulation. The effects of nitric oxide in the adult respiratory distress syndrome and other lung diseases are also reviewed. Part VI deals with anesthetic effects on cerebral circulation along with cerebral reperfusion injury.

It is our hope that future work will lead to further understanding of basic mechanisms of the interaction between anesthetics and cardiovascular disorders and a greater appreciation of the therapeutic and pathophysiological implications of these mechanisms.

The editors thank all the contributors for providing their own original presentation and interpretation of the data. We gratefully acknowledge the excellent secretarial support of Anita Tredeau and her superb care in coordinating the editorial phases of publication. We also thank Criticare Systems, Inc., for generous support. The editors also express profound gratitude to Jasna Markovac, the Biomedical Sciences Editor of Academic Press, for help in overseeing production of this book.

Zeljko J. Bosnjak
John P. Kampine

Regulation of Calcium Slow Channels of Heart by Cyclic Nucleotides and Effects of Ischemia

Nicholas Sperelakis
Department of Physiology and Biophysics
University of Cincinnati
Cincinnati, Ohio 45267

I. Introduction

Phosphorylation of ion channels is a means of regulating or modulating the activity of the channels. There is evidence for such regulation or modulation of function of Ca^{2+}, K^+, and Na^+ channels by phosphorylation, and biochemical evidence shows that one or a few sites on the channel proteins can be phosphorylated by various protein kinases. Most of the physiological evidence for modulation of ion channel function by cyclic-nucleotide-dependent phosphorylation is for slow (L-type) Ca^{2+} channels of cardiac muscle, vascular smooth muscle (VSM), skeletal muscle, and nerve, and for K^+ channels (delayed rectifier type) of cardiac muscle and nerve. This article focuses primarily on the slow Ca^{2+} channel of cardiac muscle and VSM. These examples will illustrate the important principles that are involved.

The voltage- and time-dependent slow Ca^{2+} channels in the myocardial cell membrane are the major pathway by which calcium ions enter the cell during excitation for initiation and regulation of the force of contraction of cardiac muscle. The Ca^{2+} that enters the cell via the Ca^{2+} channels triggers the release of more Ca^{2+} stored in the sarcoplasmic reticulum (SR) by a Ca^{2+}-induced Ca^{2+} release that involves the Ca^{2+}-release channels of the

Advances in Pharmacology, Volume 31

SR. It is estimated that, of the Ca^{2+} necessary to activate the contractile proteins, about 10–20% enters the cell through the Ca^{2+} channels, and the remainder is released into the myoplasm from the SR. However, the Ca^{2+} influx through the Ca^{2+} channels is the key determinant of the level of myoplasmic $[Ca]_i$ and hence the force of contraction of the heart.

In addition to its prime role in excitation–contraction coupling, Ca^{2+} also acts as a second messenger in several other ways, including activation of some enzymes and activation of two or three types of ion channels, for example, a Ca^{2+}-activated K^+ conductance, $g_{K(Ca)}$, a Ca^{2+}-activated mixed cation conductance, $g_{Na,K(Ca)}$, and a Ca^{2+}-activated Cl^- conductance, $g_{Cl(Ca)}$. These Ca^{2+}-activated channels are important in normal physiology and pathophysiology (e.g., arrhythmias) of the heart.

The slow Ca^{2+} channels have some special properties, including functional dependence on metabolic energy, selective blockade by acidosis, and regulation by intracellular cyclic nucleotide levels. Because of these special properties of the slow channels, Ca^{2+} influx into the myocardial cell can be controlled by extrinsic factors (such as autonomic nerve stimulation or circulating hormones) and by intrinsic factors (such as cellular pH or ATP level). The special properties also serve to protect the myocardial cells during periods of ischemia.

The Ca^{2+} influx that occurs during each cardiac cycle is regulated by cyclic nucleotides. This regulation is presumably mediated by phosphorylation(s) of the Ca^{2+} slow channel protein (L-type). Phosphorylation of the slow Ca^{2+} channels (or of an associated regulatory protein) by cAMP-dependent protein kinase (PK-A) (Fig. 1) presumably (a) increases the number of Ca^{2+} slow channels available for voltage activation during the action potential (AP); (b) increases the probability of their opening, and (c) increases the mean open time. A greater density of available Ca^{2+} channels increases Ca^{2+} influx and inward Ca^{2+} slow current (I_{Ca}) during the AP, and so increases the force of contraction of the heart. Phosphorylation by cGMP-dependent protein kinase (PK-G) depresses the activity of the slow Ca^{2+} channels (1).

II. Types of Calcium Channels

Five different subtypes of voltage-dependent Ca^{2+} channels have been described for nerve and muscle cells (Table I). Two of these subtypes are known as L-type (or long-lasting or kinetically slow) and T-type (or transient or kinetically fast) (2). Muscle fibers, in general, appear to possess only the L-type and T-type, with the T-type channels being very sparse or absent in some types of muscles. In other words, in muscle cells, the

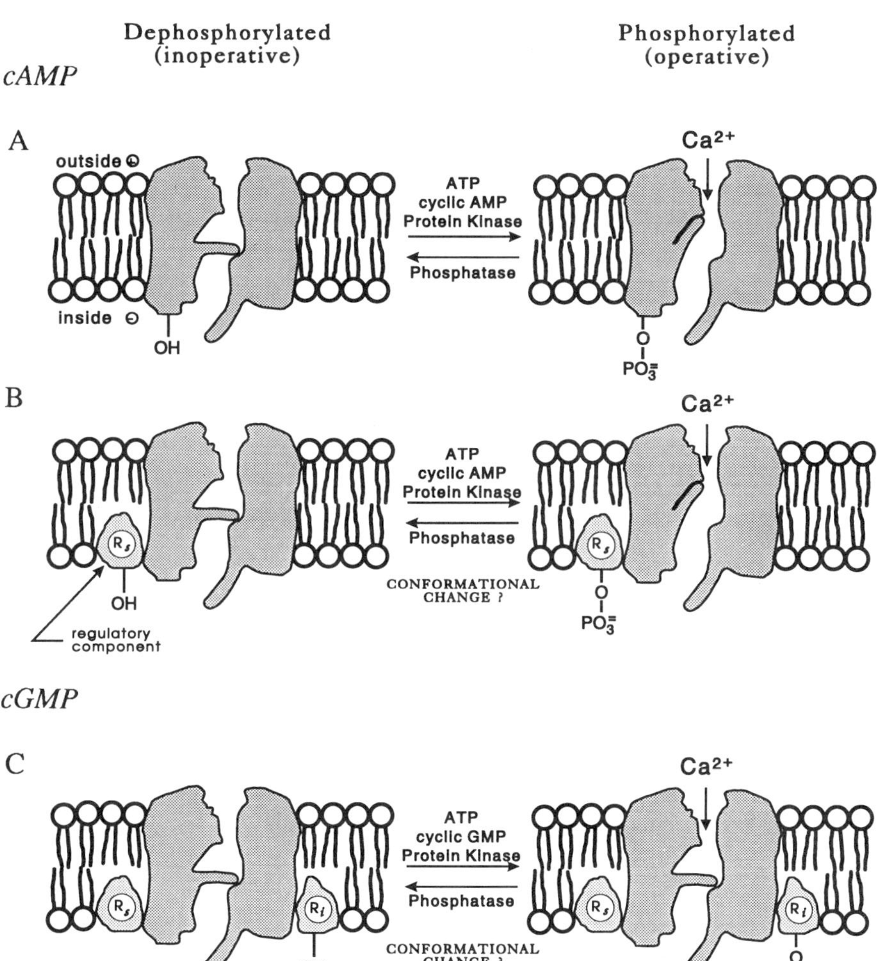

Fig. 1 Schematic model for a Ca^{2+} slow channel in myocardial cell membrane in two hypothetical forms: dephosphorylated (or electrically silent) form (left diagrams) and phosphorylated form (right diagrams). The two gates associated with the channel are an activation gate and an inactivation gate. The phosphorylation hypothesis states that a protein constituent of the slow channel itself (A) or a regulatory protein associated with the slow channel (B) must be phosphorylated in order for the channel to be in a state available for voltage activation. Phosphorylation of a serine or threonine residue occurs by a cAMP-dependent protein kinase (PK-A) in the presence of ATP. Phosphorylation may produce a conformational change that effectively allows the channel gates to operate. The slow channel (or an associated regulatory protein) may also be phosphorylated by a cGMP-dependent enzyme (C), thus mediating the inhibitory effects of cGMP on the slow Ca^{2+} channel. Modified from Sperelakis and Schneider (1976). A metabolic control mechanism for calcium ion influx that may protect the ventricular myocardial cell. *Am. J. Cardiol.* **37,** 1079–1085.

major inward Ca^{2+} current (involved in excitation–contraction coupling) is through the L-type slow Ca^{2+} channels. This is true of cardiac muscle, skeletal muscle, and most types of smooth muscles.

The L-type and T-type Ca^{2+} channels have a single-channel conductance of about 18–26 pS (L-type) and 8–12 pS (T-type) (Tables I and II). The L-type Ca^{2+} channel is high threshold, having an activation voltage of −45 to −35 mV, whereas the T-type is low threshold, with an activation voltage of −60 to −50 mV (Tables I and II).

Table II summarizes the major differences between the slow (L-type) Ca^{2+} channels and the fast (T-type) Ca^{2+} channels. As indicated, the kinetics of activation and inactivation are slower for the L-type. That is, the slow $I_{Ca(L)}$ turns on (activates) more slowly and turns off (inactivates) more slowly. In addition, the voltage ranges over which these channels operate are different, the threshold potential and inactivation potential being higher (more positive or less negative) for the slow Ca^{2+} channels. Therefore, the L-type channels are high threshold, and the T-type channels are low threshold. The single-channel conductance is greater for the slow Ca^{2+} channel: 18–26 versus 8–10 pS. The slow Ca^{2+} channels are regulated by cyclic nucleotides and phosphorylation, whereas the fast Ca^{2+} channels are not. Finally, the slow Ca^{2+} channels are blocked by Ca^{2+} antagonist drugs (such as verapamil, diltiazem, and nifedipine) and opened by Ca^{2+} agonist drugs (such as Bay k 8644, a dihydropyridine which is chemically very close to nifedipine), whereas the fast Ca^{2+} channels are not. In some respects, the fast Ca^{2+} channels behave like fast Na^{+} channels, except that they are Ca^{2+} selective (rather than Na^{+} selective) and are not blocked by tetrodotoxin (TTX). The T-type channels are relatively selectively blocked by tetramethrine and by low concentrations of Ni^{2+} (e.g., 30 M) (Table I). Higher concentrations of Ni^{2+} block the L-type channels as well.

In addition to the two Ca^{2+} channel subtypes described above, a new subtype was discovered in 18-day-old fetal rat ventricular (heart) cells (3). A substantial fraction (e.g., 30%) of the total I_{Ca} remained in the presence of a high concentration (3 M) of nifedipine (nifedipine-resistant I_{Ca}), and it was not blocked by diltiazem (another L-type channel blocker) or by ω-conotoxin (N-type channel blocker). This Ca^{2+} current had a half-inactivation potential about 20 mV more negative than the L-type I_{Ca}, in this respect being like a T-type Ca^{2+} current. It was called F-type (or fetal-type) Ca^{2+} current [$I_{Ca(F)}$]. The single-channel conductance (γ) is not known.

Another difference discovered during the embryonic or fetal period, first observed in chick and later in rat, is that the slow Ca^{2+} channels (L-

Table I

Summary of Various Subtypes of Calcium Channels and Drugs or Toxins That Block or Open Them

Channel subtype	Single-channel conductance (PS)	Threshold	Blockers	Openers
L-type (slow)	18–26	High	Verapamil diltiazem, nifedipine	Bay k 8644
T-type (fast)	8–12	Low	Tetramethrine, Ni^{2+} (30 *M*)	—
N-type	14–18	Low	ω-Conotoxin	—
P-type[a]	9–20	High	ω-Agatoxin IVA[b] polyamine (FTX)[c]	—
F-type (fetal)	—	Low	—	—

[a] One subtype was initially found in Purkinje neurons of the cerebellum, and hence called P-type [Llinas, R., Sugimori, D., Hillman, E., and Cherksey, B. (1992). Distribution and functional significance of the P-type, voltage-dependent Ca^{2+} channels in the mammalian central nervous system. *Trends Neurosci.* **15,** 351–355]. The P-type channel has now been identified at the nerve terminals of neuromuscular junctions of both vertebrates and invertebrates, and they are involved in presynaptic Ca^{2+} influx and associated neurotransmitter release.

[b] ω-Agatoxin IVA, a polypeptide (5202 Da) from funnel-web spider venom (*Agelenopsis aperta*), blocks P-type Ca^{2+} channels of rat Purkinje neurons with a K_D of about 2 n*M*. A smaller (~254 Da) polyamine (FTX) extracted from the venom of this spider also blocks the P-type channels; it contains 7 N atoms and 11 C atoms [Mintz, I. M., Venema, V. J., Swiderek, K. M., Lee, T. D., Bean, B. P., and Adams, M. E. (1992). P-type calcium channels blocked by the spider toxin omega-Aga-IVA. *Nature (London)* **355,** 827–829].

[c] FTX block is antagonized by Ba^{2+}.

type) exhibit an unusually high incidence of very long openings, as observed in single-channel recordings (cell-attached patch) (4); (H. Masuda and N. Sperelakis, unpublished observations, 1994). The incidence of long openings diminished during development and approached the adult channel behavior. The adult behavior consists primarily of bursting patterns, namely, rapid openings and closings of short duration (flickerings) that persist for a short period.

Table II

Major Differences between Slow (L-Type) and Fast (T-Type) Calcium Channels

	Calcium channels	
Property	Slow (L-type)	Fast (T-type)
Duration of current	Long-lasting (sustained)	Transient
Inactivation kinetics	Slower	Faster
Activation kinetics	Slower	Faster
Threshold	High (~ −35 mV)	Low (~ −60 mV)
Half-inactivation potential	~ −20 mV	~ −50 mV
Single-channel conductance	High (18–26 pS)	Low (8–12 pS)
Regulated by cAMP and cGMP	Yes	No
Regulated by phosphorylation	Yes	No
Blocked by Ca^{2+} antagonist drugs	Yes	No (slight)
Opened by Ca^{2+} agonist drugs	Yes	No
Permeation by M^{2+}	Ba > Ca	Ba ≃ Ca
Inactivation by $[Ca]_i$	Yes	Slight (?)
Recordings in isolated patches	Runs down	Relatively stable

III. Cyclic AMP Stimulation of Slow Calcium Channels

Cyclic AMP (cAMP) modulates the functioning of the Ca^{2+} slow channels (5–7). β-Adrenergic agonists and histamine, after binding to their specific receptors, lead to rapid stimulation of adenylate cyclase with resultant elevation of cAMP levels. Methylxanthines enter the myocardial cells and inhibit the phosphodiesterase, thus causing an elevation of cAMP. These positive inotropic agents also concomitantly stimulate Ca^{2+}-dependent slow APs by increasing I_{Ca}.

Additional evidence for the regulatory role of cAMP in heart cells includes the following. (a) The GTP analog GPP(NH)P and forskolin, which directly activate adenylate cyclase, induce Ca^{2+}-dependent slow APs (8). (b) cAMP microinjection into ventricular muscle cells (by iontophoresis, pressure, or liposomes) induces slow APs in the injected cells within seconds (9–11). (c) The I_{Ca} of isolated single cardiac cells is enhanced by injection of cAMP (12,13). (d) Single-channel analysis suggests that cAMP increases the number of functional slow channels available in the sarcolemma and/or the probability of channel opening (14–16). Isoproterenol (Iso) increases the mean open time of single Ca^{2+} channels and decreases the intervals between bursts; the conductance of the single channel, however, is not changed (17).

Therefore, the increase in the slow Ca^{2+} current caused by Iso could be produced by the observed increase in mean open time of each channel and probability of opening, as well as by an increase in the number of available channels. Consistent with the latter mechanism, it is well known that the number of specific dihydropyridine (DHP) binding sites (indicative of the total number of slow Ca^{2+} channels) is much greater than the number of functioning Ca^{2+} channels estimated from electrophysiological measurements of $I_{Ca(L)}$. In other words, there is a very large fraction of silent channels or surplus channels. In agreement with this concept, it was found that the DHP Ca^{2+} channel agonist Bay k 8644 recruited previously silent (nonfunctioning) channels (18).

IV. Phosphorylation Hypothesis

Because of the relationship between cAMP and the number of available slow Ca^{2+} channels, and because of the dependence of the functioning of these channels on metabolic energy, it was postulated that the slow channel protein must be phosphorylated in order for it to become available for voltage activation (5,6). Elevation of cAMP by a positive inotropic agent activates a cAMP-dependent protein kinase (PK-A), which phosphorylates a variety of proteins in the presence of ATP. One protein that is phosphorylated may be the slow Ca^{2+} channel protein itself or a contiguous regulatory type of protein (Fig. 1). Agents that elevate cAMP increase the fraction of channels that are in the phosphorylated form, and hence readily available for voltage activation. Phosphorylation could make the slow Ca^{2+} channel available for activation either by a conformational change that allows the activation gate to be opened on depolarization or by an increase in the pore diameter.

In this phosphorylation model, the phosphorylated form of the slow Ca^{2+} channel is the active (operational) form, and the dephosphorylated form is the inactive (inoperative) form. The dephosphorylated channels are either electrically silent or have a very low probability of opening. Thus, phosphorylation markedly increases the probability of channel opening with depolarization. An equilibrium would exist between the phosphorylated and dephosphorylated forms of the channel for a given set of conditions. For example, some agents like fluoride ion (<1 mM) increase the force of contraction of the heart and potentiate the Ca^{2+}-dependent slow APs and Ca^{2+} influx (I_{Ca}), without increasing the level of cyclic AMP (19). Fluoride may act by inhibiting the phosphatase which dephosphorylates the channel protein, thus prolonging the life span of the phosphorylated channel. Thus, channel stimulation can be produced either by increasing the rate of phosphorylation (by PK-A activation) or by decreasing the rate of dephosphorylation (inhibition of the phosphatase). Some negative inotropic agents or drugs could depress the rate of phosphorylation or stimulate the rate of phosphorylation. It might be difficult to distinguish between a drug that inhibited phosphorylation or stimulated dephosphorylation of the slow Ca^{2+} channel and one that physically blocked the channel.

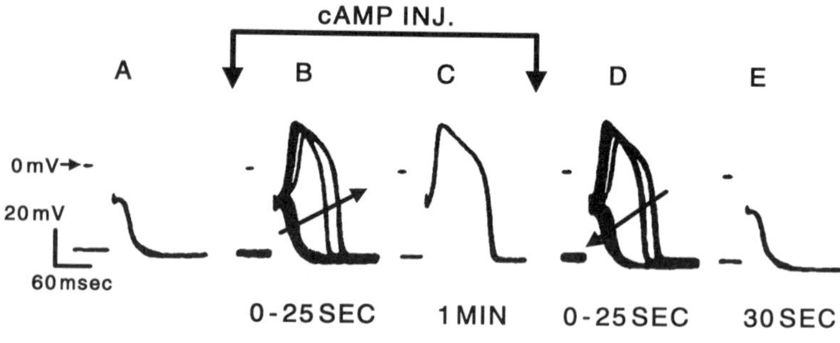

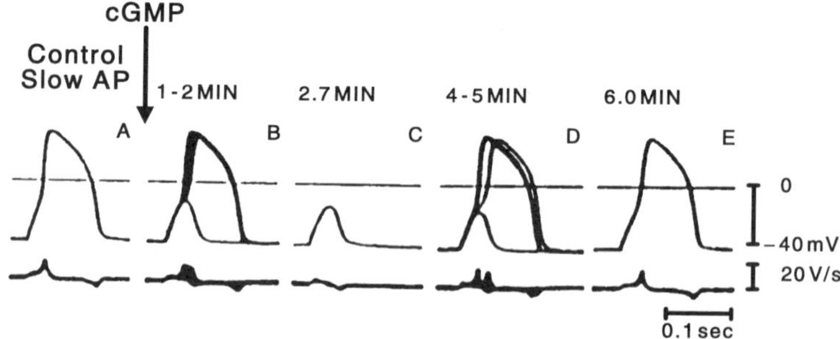

Fig. 2 *Top row:* Induction of Ca^{2+}-dependent slow action potentials (APs) in guinea pig papillary muscle by intracellular pressure injection of cyclic AMP. The muscle was depolarized in 22 m*M* $[K]_o$ to voltage inactivate fast Na^+ channels. (A) Small graded response (stimulation rate 30/min). (B) Superimposed records showing the gradual appearance of slow APs on cAMP injection over a 25-sec period. (C) Presence of stable slow APs after injection for 1 min. (D) Gradual depression of slow APs over a period of 25 sec after stopping injection. (E) Complete decay of slow APs 30 sec after cessation of cAMP injection. All records are from one impaled cell. (Data from Ref. 10.) *Bottom row:* Transient abolition of Ca^{2+}-dependent slow APs by pressure injection of cGMP. (A) Control slow AP induced by 10 m*M* tetra-ethyl-ammonium (TEA) plus 4.0 m*M* $[Ca]_o$ in 25 m*M* K^+ to inactivate fast Na^+ channels (B, C) Approximately 1–2 min following the onset of cGMP injection (10 sec duration), the slow APs were depressed and then abolished. (D, E) At 4–6 min, the slow APs recovered spontaneously to control levels. All records are from the same cell. [From Wahler, G. M., and Sperelakis, N. (1985). Intracellular injection of cyclic GMP depresses cardiac slow action potentials. *J. Cyclic Nucleotide Protein Phosphoryl. Res.* **10,** 83–95].

Based on the rapid decay of the response to microinjected cAMP (Fig. 2, top), the mean life span of a phosphorylated channel is likely to be only a few seconds at most, and it is possible that the channels are phosphorylated and dephosphorylated with every cardiac cycle (9,10). Hence, agents which affect or regulate the phosphatase that dephosphorylates the channel would affect the life span of the phosphorylated channel.

V. Protein Kinase A Stimulation

To test whether the regulatory effect of cAMP is exerted by means of PK-A and phosphorylation, intracellular injection of the catalytic subunit of PK-A was done. Such injections induced and enhanced the slow Ca^{2+}-dependent APs and potentiated I_{Ca} (20,21). Another test of the phosphorylation hypothesis was done by injection of a protein inhibitor of PK-A into heart cells, which showed that it inhibited the spontaneous slow Ca^{2+}-dependent APs and I_{Ca} (21,22). Phosphatases have been shown to decrease the Ca^{2+} current in neurons (23) and ventricular myocardial cells (24). The catalytic subunit of protein phosphatases type 1 and type 2A inhibited I_{Ca} prestimulated by β-adrenergic agents. Okadaic acid, a protein phosphatase inhibitor, stimulated I_{Ca}.

Consistent with the phosphorylation hypothesis, the slow Ca^{2+} channel activity disappears within 90 sec in isolated membrane inside-out patches (25), but it can be restored (in neurons) by applying the catalytic subunit of PK-A and MgATP (26). This is consistent with the washing away of regulatory components of the slow Ca^{2+} channels or of the enzymes necessary to phosphorylate the channel. Even in whole-cell voltage clamp, there is a progressive rundown of the slow Ca^{2+} current, which is slowed or partially reversed by conditions that enhance PK-A phosphorylation.

VI. Cyclic GMP Inhibition of Slow Calcium Current

The physiological role played by cyclic GMP on cardiac function is still controversial. If has been proposed that cGMP plays an antagonistic role to that of cAMP, namely, that there was a yin–yang relationship between cAMP and cGMP (27). 8-Bromo-cyclic GMP (8Br-cGMP) (10^{-4} *M*) shortened the AP duration in rat atria accompanied by a negative inotropic effect, and it was suggested that cGMP might decrease the Ca^{2+} conductance (28). Acetylcholine (ACh) and 8Br-cGMP reduced upstroke velocity and duration of the Ca^{2+}-dependent slow APs in guinea pig atria (29). The abbreviation of AP duration was also observed following pressure injection of cGMP into isolated guinea pig cardiomyocytes (30). However, elevation of intracellular cGMP by photoactivation of a derivative had no effect on the L-type Ca^{2+} current [$I_{Ca(L)}$] in isolated rat ventricular cells (13).

Superfusion of isolated ventricular muscle with 8Br-cGMP abolished the Ca^{2+}-dependent slow APs and accompanying contractions within 7–20 min (31). A similar inhibition by 8Br-cGMP was shown for the slow APs of atrial muscle and Purkinje fibers (32). Introduction of cGMP into heart cells by the liposome method also abolished the slow APs (11). Pressure injection of cGMP intracellularly into ventricular cells transiently depressed or abolished the slow APs much more quickly (e.g., 1–2 min) (31) (Fig. 2, bottom). It was also demonstrated that 8Br-cGMP markedly inhibits the basal I_{Ca} (unstimulated by cAMP) in voltage-clamped single ventricular myocytes of embryonic chick (17-day) (1) (G. Haddad, N. Sperelakis, and G. Bkaily, unpublished, 1994) (Figs. 3 and 4). The I–V curves showed that cGMP did not alter the voltage for maximum current (+10 mV) or the reversal potential (V_{rev} of +80 mV).

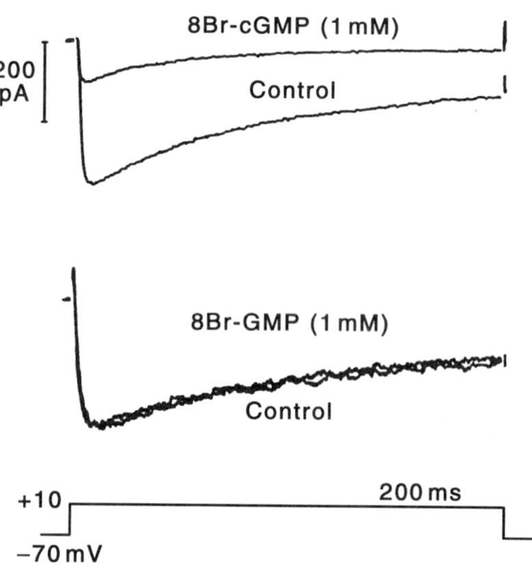

Fig. 3 Effect of 8Br-cGMP on the slow inward Ca^{2+} current in two cultured embryonic chick ventricular myocytes. *Top traces:* Currents elicited by depolarizing pulses from −70 to +10 mV in the control bath solution and following 10 min of superfusion with a solution containing 1 m*M* 8Br-cGMP. Note the large inhibition of $I_{Ca(s)}$. *Bottom traces:* Currents elicited by depolarizing pulses in the control bath solution and following 10 min of superfusion with a solution containing 1 m*M* 8Br-GMP, the noncyclic analog of 8Br-cGMP. The bath solution (20–22°C) included 10 m*M* $BaCl_2$ and 135 m*M* TEA chloride; the pipette solution included 150 m*M* Cesium glutamate, 5 m*M* MgATP, and 1 m*M* EGTA. (Reprinted from Ref. 1.)

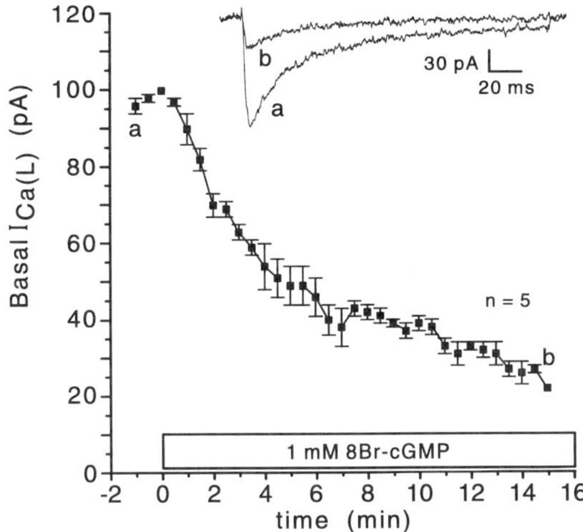

Fig. 4 Time course of inhibition of basal $I_{Ca(L)}$ by 8Br-cGMP (1 m*M*) in 17-day-old embryonic chick heart cells. Plotted are the means ± Se for five cells. Tracings in the inset at top show the original current recordings of $I_{Ca(L)}$ in one cell taken at the time points indicated by the corresponding letters (*a*, *b*) in the graph. $I_{Ca(L)}$ was elicited by 200 msec depolarizing pulses to +10 mV from a holding potential of −60 mV. Temperature was 20°C; $[Ca]_o$ was 2.5 m*M*.

Inhibition by cGMP of slow Ca^{2+} channel activity of embryonic chick heart cells at the single-channel level was also demonstrated (Fig. 5) (33). Cyclic GMP did not change unit amplitude and slope conductance of the Ca^{2+} channel, but it prolonged the closed times and shortened the open times. Similar observations were made on isolated rabbit ventricular myocytes (3). Because 8Br-cGMP is a potent activator of PK-G (G-kinase) and does not stimulate cAMP hydrolysis, the inhibition of the basal activity of the Ca^{2+} channels (not prestimulated by cAMP) by cGMP is likely to be mediated by PK-G.

The Ca^{2+} slow channels of young (3-day-old) embryonic chick heart cells often exhibit long-lasting openings (e.g., for 300 msec) under normal conditions, especially at the more positive command potentials (4,33). Long-lasting openings were much less frequently observed in 17-day-old embryonic cells. In other words, the Ca^{2+} slow channels in early development naturally possess some mode-2 behavior. Generally, mode-2 behavior is induced pharmacologically by Ca^{2+} channel agonists such as the dihydropyridine Bay k 8644. Addition of Bay k 8644 to cells exhibiting mode-2 behavior naturally did not further prolong the already long open times, but recruited silent Ca^{2+} channels (18). Addition of 8Br-cGMP to the bath of cells (3-day) exhibiting long openings completely inhibited Ca^{2+} slow channel activity (Fig. 4).

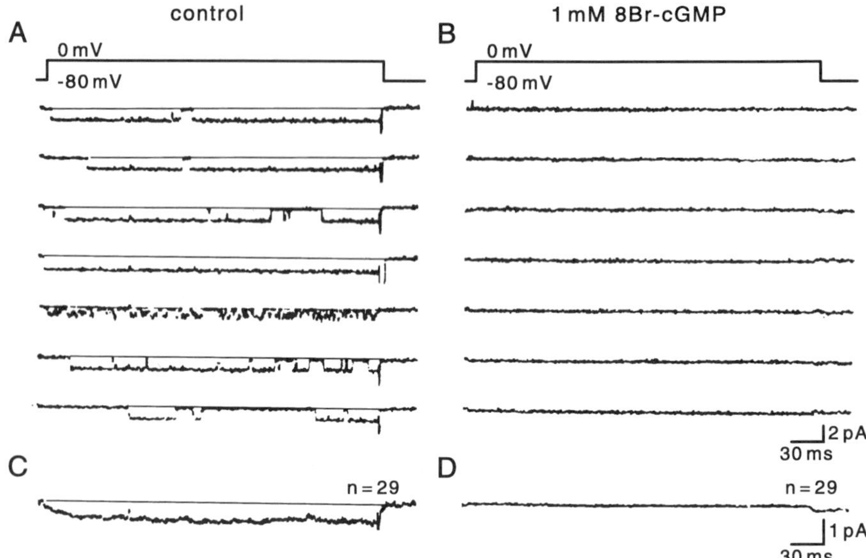

Fig. 5 Current recordings from a cell-attached patch showing effect of 8Br-cGMP on the Ca^{2+} slow channel activity in a single myocardial cell isolated from a 3-day-old embryonic chick heart. Single-channel currents were evoked by depolarizing voltage pulses to 0 mV from a holding potential of −80 mV, at a duration of 300 msec and repetition rate of 0.5 Hz. (A, B) Examples of original current recordings from the same patch, before (A) and after (B) superfusion of 1.0 m*M* 8Br-cGMP. (C, D) Ensemble-averaged currents calculated from the current recordings ($n = 29$). The current tracings were low-pass filtered at 1 kHz and corrected for leakage and capacitive currents (Data from Ref. 33).

In whole-cell voltage clamp experiments on single ventricular cardiomyocytes from early neonatal (2-day) rats, it was demonstrated that the stimulated $I_{Ca(L)}$ produced by 8Br-cAMP added to the bath also could be markedly inhibited by the addition of 8Br-cGMP (Fig. 6) (H. Masuda and N. Sperelakis, unpublished, 1994). Therefore, the ratio of cAMP to cGMP should determine the degree of stimulation of $I_{Ca(L)}$. Clearly, even the basal I_{Ca} is inhibited by cGMP, at least in the case of chick, and perhaps in rat (see section on PK-G inhibition).

The results demonstrate that cGMP regulates the functioning of the myocardial Ca^{2+} slow channels of chick and rat in a manner that is antagonistic to that of cAMP (Figs. 1–6, and 9). It is possible that the slow Ca^{2+} channel protein has a second site that can be phosphorylated by PK-G and which, when phosphorylated, inhibits the slow channel. Another possibility is that there is a second type of regulatory protein that is inhibitory when phosphorylated (Fig. 1). A 47-kDa membrane protein was found to be phosphorylated by PK-G (34).

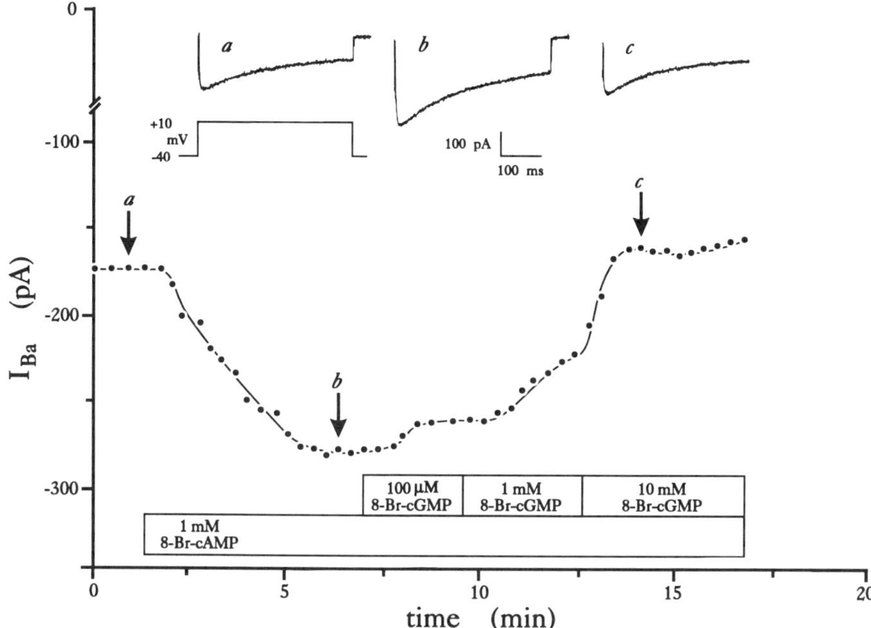

Fig. 6 Antagonism of the stimulating effect of 8Br-cAMP (1 m*M*) on $I_{Ca(L)}$ by 8Br-cGMP in a single young (2-day) neonatal rat ventricular myocyte. As shown, 8Br-cAMP almost doubled $I_{Ba(L)}$, and stepwide elevation of 8Br-cGMP (100 μM, 1 m*M*, and 10 m*M*) reversed the stimulation. Tracings at top show original current recordings of I_{Ca} corresponding to the three time points labeled in the lower graph. $I_{Ca(L)}$ was elicited by 300 msec depolarizing pulses to +10 mV from a holding potential of −40 mV. Ba^{2+} (2.0 m*M*) was used as the charge carrier. Experiments were carried out at room temperature (25°C). (H. Masuda and N. Sperelakis, unpublished, 1994).

Another mechanism proposed for cGMP inhibition is based on cGMP depression of the cAMP level. Intracellular application of cGMP inhibited I_{Ca} of frog ventricular myocytes, but only after the cAMP levels had been increased; that is, there was no effect of cGMP on the basal I_{Ca} (35,36). It was concluded that cGMP inhibited I_{Ca} by stimulating one of the phosphodiesterase isozymes, resulting in increased degradation of the elevated cAMP. However, in a later study on guinea pig and rat cardiomyocytes, the same group reported a direct inhibition of I_{Ca} by cGMP and PK-G (37,38). In addition, 8Br-cAMP inhibition of the Ca^{2+}-dependent slow APs in mammalian cardiac muscle occurs without a decrease in cAMP levels (39). Thus, it appears that in avian and mammalian ventricular muscle, cGMP inhibits I_{Ca} directly through a cGMP-mediated phosphorylation (8Br-cGMP is a potent activator of PK-G) of a protein involved in the functioning of the slow Ca^{2+} channels (Fig. 1).

VII. Protein Kinase G Inhibition

When PK-G (G-kinase, 25 nM) is added to the patch pipette for diffusion into the cell during whole-cell voltage clamp, it was found that basal I_{Ca} is inhibited markedly and rapidly, maximum inhibition being reached within 2–3 min (Figs. 7A and 8). Data from 17-day chick cardiomyocytes are illustrated in Fig. 7 (H. Haddad, N. Sperelakis, and G. Bkaily, unpub-

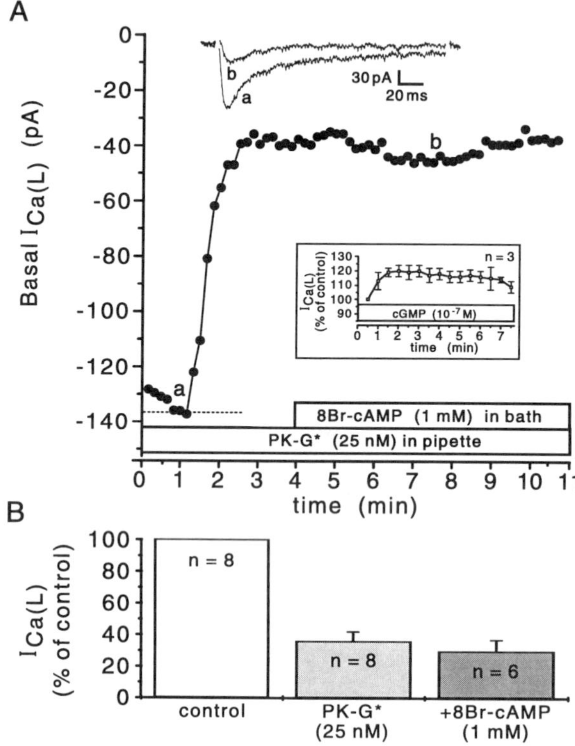

Fig. 7 Inhibition of basal $I_{Ca(L)}$ by protein kinase G (25 nM) present in the patch pipette in a 17-day embryonic chick cardiomyocyte. The PK-G diffused into the cell during the whole-cell voltage clamp. (A, inset) Control experiment in cells with no PK-G present in the pipette. There was very little rundown of the basal I_{Ca} over the 7-min period. (A) Experiment in a cell with PK-G in the pipette. Inhibition of basal I_{Ca} began within 1.5 min after breaking into the cell and reached maximum at 3 min. The two current tracings illustrated at top correspond to the time points labeled *a* and *b* on the graph. (B) Summary of data from eight cells. PK-G inhibited $I_{Ca(L)}$ to about 31% of the control level. The addition of 8Br-cAMP (1 mM) to the bath was not able to reverse the effect of the PK-G. Experiments were done at 20°C, and $[Ca]_o$ was 2.5 mM. (G. Haddad, N. Sperelakis, and G. Bkaily, unpublished, 1994).

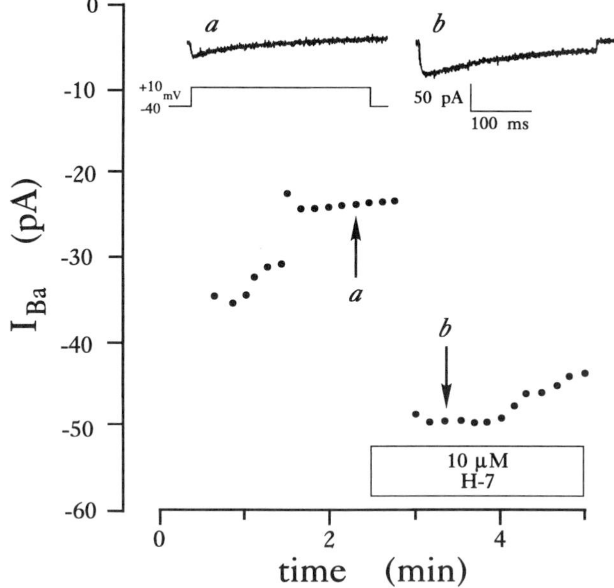

Fig. 8 Inhibition of basal I_{Ca} by G-kinase (25 n*M*) present in the patch pipette in a ventricular myocyte from an early (2-day) neonatal rat heart. Very shortly after breaking into the cell, there was a marked decrease in I_{Ca}. This inhibition by G-kinase was reversed by application of the kinase inhibitor H-7 to the bath. This action could be reversed by washout of H-7. 8Br-cAMP (1 m*M*) added to the bath produced a small stimulation of I_{Ca} (in the continued presence of G-kinase). This action of the cyclic nucleotide was also reversed by washout. The current tracings illustrated correspond to the two time points labeled *a* and *b* in the graph. Experiments were done at room temperature (25°C), with 2.0 m*M* Ba^{2+} as the charge carrier. (H. Masuda and N. Sperelakis, 1994).

lished, 1994). Note that inhibition of basal I_{Ca} began within about 90 sec after breaking into the cell. At steady-state inhibition produced by PK-G, $I_{Ca(L)}$ was reduced to about 31% of control (Fig. 7). The control with no PK-G to demonstrate the lack of significant I_{Ca} rundown in 7 min is shown in the inset of Fig. 7A. Addition of 8Br-cGMP (1 m*M*) to the bath was unable to reverse the inhibition of basal I_{Ca} produced by PK-G (Fig. 7).

Similar effects of PK-G infusion were observed in early neonatal rat ventricular myocytes, as illustrated in Fig. 8 (H. Masuda and N. Sperelakis, unpublished, 1994). There was a very rapid and prominent inhibition of the basal I_{Ca} by G-kinase (24 n*M*) following breaking into the cell. Addition of H-7 (a blocker of protein kinases including PK-G) to the bath caused a rapid restoration of I_{Ca} to about the original basal level (Fig. 8). Therefore, these findings indicate that the inhibitory effects of cGMP on basal I_{Ca} are mediated by activation of PK-G (G-kinase) and resultant phosphorylation.

A single protein of approximately 47 kDa has been found to be specifically phosphorylated by PK-G (in the presence of 10^{-5} M cGMP) in guinea pig sarcolemmal preparations (34). Thus, this protein may be a possible mediator involved in regulation of Ca^{2+} channels of the heart by the cGMP/PK-G pathway.

VIII. Inhibition by Muscarinic Agonists

The parasympathetic neurotransmitter acetylcholine exerts a negative inotropic effect on ventricular myocardium prestimulated by β-adrenergic agonists. This action is produced by at least four different mechanisms (Fig. 9). First, activation of the muscarinic receptor by ACh exerts an inhibitory effect on adenylate cyclase (and thereby lowers cAMP levels), via the G_i (inhibitory) coupling protein, to reverse the stimulation of adenylate cyclase produced by means of the G_S coupling protein due to, for example, activation of the β-adrenoceptor (Fig. 9, top left). Thus, ACh may depress Ca^{2+} influx and contraction by reversing cAMP elevation produced by various agonists.

Second, ACh may act by stimulating guanylyl cyclase (Fig. 9, top right). Muscarinic agonists are known to elevate cGMP (40). The elevated cGMP would activate PK-G which would phosphorylate the slow Ca^{2+} channel and thereby inhibit channel activity, as described previously in Sections VI and VII.

Third, ACh may activate a G_k protein that directly activates ("gates") a K^+ channel (Fig. 9, bottom right). In other words, an ACh-activated K^+ current [$I_{K(ACh)}$] is turned on, which causes the cardiac action potential to repolarize prematurely. This, in turn, causes the slow Ca^{2+} channels to turn off (deactivate) prematurely, thereby lowering the Ca^{2+} influx.

Fourth, it has been proposed that muscarinic agonists also act to inhibit the Ca^{2+} slow channel by PK-G stimulation of a phosphatase (Type 1) that dephosphorylates the channel (41) (Fig. 7, bottom center). This would have the effect of lowering the fraction of channels in the phosphorylated form, and therefore the Ca^{2+} influx. In this mechanism, the rate of phosphorylation is unaffected, but the rate of dephosphorylation is increased.

Acetylcholine depressed the I_{Ca} of cultured chick ventricular cells that had been prestimulated by isoproterenol (8). ACh also reverses the electrophysiological effects of direct adenylate cyclase stimulation by forskolin (42). Additionally, in ventricular cells, ACh may inhibit the slow Ca^{2+}

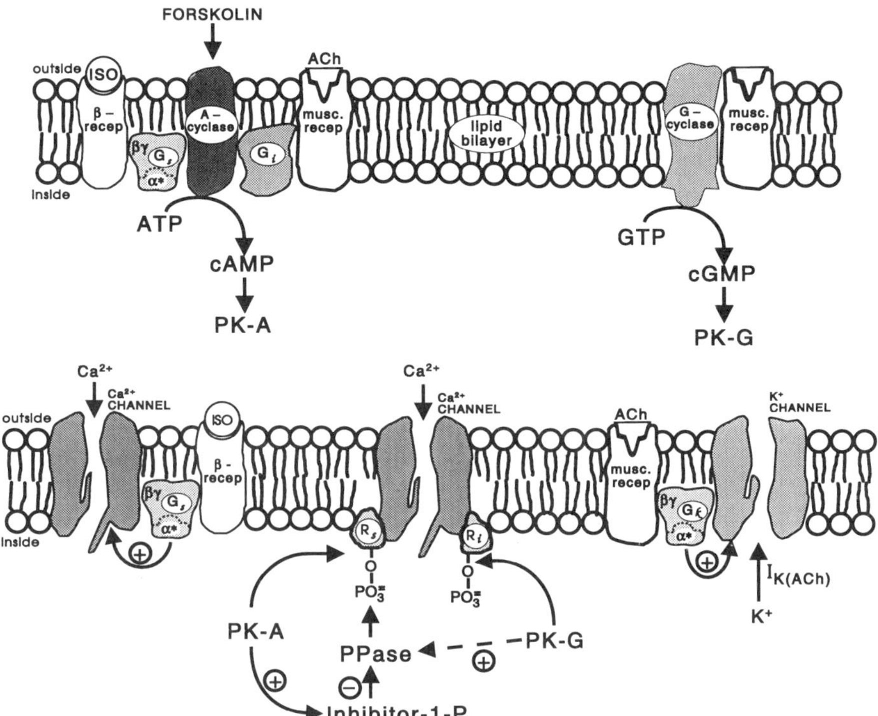

Fig. 9 Diagrammatic summary of regulation of Ca^{2+} slow channels in the myocardial cell membrane and mechanisms of action of some inotropic agents. The β-adrenergic agonists and histamine H_2 agonists act via their receptors on a GTP-binding protein (G_s) to stimulate adenylate cyclase and cAMP production. The voltage-dependent myocardial slow Ca^{2+} channels are stimulated by cAMP, presumably because the channel (or an associated regulatory protein) must be phosphorylated in order for it to be in a form that is available for voltage activation. cGMP-dependent phosphorylation also regulates the slow channel in a manner antagonistic to cAMP, namely, producing inhibition. Thus, the muscarinic receptor activated by ACh can produce inhibition of Ca^{2+} influx by at least the four mechanisms depicted: (a) reversal of adenylate cyclase stimulation produced by β-agonists or H_2 agonists; (b) stimulation of guanylate cyclase and production of cGMP; (c) activation of a K^+ channel [$I_{K(ACh)}$] that produces an outward K^+ current which depresses excitability and terminates the AP earlier, and thereby voltage deactivates the Ca^{2+} slow channels earlier; and (d) stimulation of a phosphatase (PPase) by cGMP and PK-G. Mechanism (c) may be absent in ventricular myocardial cells.

channels due, in part, to elevation of cGMP levels (43). Adenosine exerts effects similar to those of ACh. However, it was reported that ACh was ineffective in reducing the basal I_{Ca} (44), and that cGMP actually potentiated the stimulating effect of ISO on I_{Ca}, perhaps mediated by phosphodiesterase inhibition (45).

IX. Protein Kinase C and Calmodulin Protein Kinase

Protein kinase C (PK-C) also appears to be involved in regulation of the myocardial slow Ca^{2+} channels. Angiotensin II and a high concentration of the α-adrenergic agonist phenylephrine (variable findings on elevation of cAMP) cause a positive inotropic effect in cardiac muscle (46,47). These agonists stimulate the phosphatidylinositol cycle and generation of inositol trisphosphate (IP_3) and diacylglycerol (DAG). The IP_3 acts as a second messenger to release stored Ca^{2+} from the SR. DAG and Ca^{2+} activate PK-C, which phosphorylates a number of proteins, including presumably the slow Ca^{2+} channel protein (or an associated regulatory protein). Phorbol ester (direct activator of PK-C) and angiotensin II stimulate $I_{Ca(L)}$ in rat and chick hearts (46,48), but not in guinea pig heart (46,49).

Inhibitors of calmodulin (e.g., calmidazolium) inhibit the Ca^{2+}-dependent slow APs of heart cells, and subsequent injection of calmodulin reverses the inhibition (50). Therefore, it appears that maximal activation of the slow Ca^{2+} channels requires two separate phosphorylation steps (calmodulin-dependent and cAMP-dependent). These may be on the same protein or on two separate proteins.

X. Comparison with Vascular Smooth Muscle and Skeletal Muscle

Table III summarizes the effects of the cyclic nucleotides on $I_{Ca(L)}$ in cardiac muscle, vascular smooth muscle (VSM), and skeletal muscle (frog). As indicated, in myocardial cells, cAMP and cGMP act in an antagonistic manner, cAMP stimulating and cGMP inhibiting. In contrast, in VSM cells, cAMP and cGMP act in the same direction, both inhibiting

Table III

Effect of Cyclic Nucleotides on Calcium Slow Channels in Muscle[a]

Cyclic nucleotide	Cardiac muscle	Vascular smooth muscle cells	Skeletal muscle
cAMP	Stimulation	Inhibition	Stimulation
cGMP	Inhibition	Inhibition	Stimulation

[a] From bullfrog (Kokate, T. G., Heiny, J. A., and Sperelakis, N. (1993). Stimulation of the slow calcium current in bullfrog skeletal muscle fibers by cAMP and cGMP. *Am. J. Physiol.* **265,** C47–C53).

$I_{Ca(L)}$. Similarly, in skeletal muscle fibers, both cyclic nucleotides act in the same direction, but here cAMP and cGMP both stimulate $I_{Ca(L)}$. In uterine smooth muscle, neither cyclic nucleotide has any effect on $I_{Ca(L)}$. Therefore, there are a variety of effects concerning the regulation exerted by cyclic nucleotides, depending on the tissue.

Cyclic AMP also has effects on other types of ion channels (Table IV). For example, the following channels of heart are stimulated by cAMP: (a) delayed rectifier K^+ channel (30,51), (b) hyperpolarization-activated Na^+, K^+ I_f (I_h) channel (52), and (c) adrenaline-activated Cl^- channel (53). The fast Na^+ channel is almost unaffected by cAMP. The cAMP stimulation of $I_{K(del)}$ in myocardial cells, coupled with its stimulation of $I_{Ca(L)}$, serves to shorten APD_{50} of the cardiac AP when heart rate is increased. For example, β-adenergic agonists, such as isoproterenol or epinephrine, raise cAMP levels, and thereby increase automaticity and force of contraction.

Cyclic GMP was reported to stimulate, presumably by phosphodiesterase inhibition, the delayed rectifier K^+ current (45) and the Cl^- current activated by β-adrenergic agonists (54).

Table IV

Listing of Some Channels Modulated by Cyclic AMP and Protein Kinase A Phosphorylation

Channel type	Tissue	Action
$I_{Ca(L)}$	Heart, nerve, skeletal muscle	Stimulated
$I_{Ca(L)}$	Vascular smooth muscle	Inhibited
$I_{Ca(L)}$	Uterine smooth muscle	No effect
$I_{K(del)}$	Heart	Stimulated
$I_f(I_h)$	Heart	Stimulated
$I_{Cl(ag)}$	Heart	Stimulated
$I_{Na(f)}$	Heart	No effect[a]

[a] Variable subtle effects have been reported for $I_{Na(f)}$, including slight inhibition [Ono, K., Kiyosue, T., and Arita, M. (1989). Isoproterenol, DBcAMP, and forskolin inhibit cardiac sodium current. *Am. J. Physiol.* **256,** C1131–C1137], slight stimulation [Ono, K., Fozzard, H. A., and Nanck, D. A. (1993). Mechanism of cAMP-dependent modulation of cardiac sodium channel current kinetics. *Circ. Res.* **72,** 807–815], and shifting of the voltage dependency.

XI. Direct Stimulation of Slow Calcium Channels by G_S Proteins and β-Receptors

There is evidence (55) that the β-adrenergic receptors can also stimulate the slow (L-type) Ca^{2+} channels by a faster and more direct pathway, via the α subunit of the G_S protein, as depicted in Fig. 9 (bottom left). The $\alpha_S{}^*$ subunit must be activated by GTP binding, as denoted by the superscript asterisk. Thus, in heart cells, both pathways—the faster, more direct one mediated by $\alpha_S{}^*$ and the slower, indirect one mediated by cAMP/PK-A—cause stimulation of slow Ca^{2+} channel activity.

Preliminary evidence has been obtained for a similar G_S protein gating of the slow Ca^{2+} channel of vascular smooth muscle cells (from rat and rabbit portal vein) on activation of the β-adrenergic receptor by isoproterenol (Z. Xiong and N. Sperelakis, unpublished, 1994). However, in this case, the slower, indirect pathway (via cAMP/PK-A) leads to channel inhibition, whereas the faster, more direct pathway leads to channel stimulation.

XII. Summary

The slow Ca^{2+} channels (L-type) of the heart are stimulated by cAMP. Elevation of cAMP produces a very rapid increase in the number of slow channels available for voltage activation during excitation. The probability of a Ca^{2+} channel opening and the mean open time of the channel are increased. Therefore, any agent that increases the cAMP level of the myocardial cell will tend to potentiate I_{Ca}, Ca^{2+} influx, and contraction. The action of cAMP is mediated by PK-A and phosphorylation of the slow Ca^{2+} channel protein or an associated regulatory protein (stimulatory type).

The myocardial slow Ca^{2+} channels are also regulated by cGMP, in a manner that is opposite or antagonistic of that of cAMP. This has been demonstrated at both the macroscopic level (whole-cell voltage clamp) and the single-channel level. The effect of cGMP is mediated by PK-G and phosphorylation of a protein, for example, a regulatory protein (inhibitory type) associated with the Ca^{2+} channel. It has been demonstrated that introduction of PK-G intracellularly causes a relatively rapid inhibition of $I_{Ca(L)}$ in both chick and rat heart cells. In addition, cGMP/PK-G act to stimulate a phosphatase that dephosphorylates the Ca^{2+} channel.

In addition to the slower, indirect pathway–exerted via cAMP/PK-A—there is a faster, more direct pathway for $I_{Ca(L)}$ stimulation by the β-

adrenergic receptor. The latter pathway involves direct modulation of the channel activity by the α subunit (α_S*) of the G_S protein.

PK-C and calmodulin-PK also may play roles in the regulation of the myocardial slow Ca^{2+} channels, possibly mediated by phosphorylation of some regulatory type of protein. Both protein kinases stimulate the activity of the slow Ca^{2+} channels. Thus, it appears that the slow Ca^{2+} channel is a complex structure, including perhaps several associated regulatory proteins, which can be regulated by a number of factors intrinsic and extrinsic to the cell (Fig. 9).

The cyclic nucleotides also have effects on the slow Ca^{2+} channels in cells other than cardiac muscle, including neurons, smooth muscle, and skeletal muscle fibers (Tables III and IV). In cardiac muscle, the two cyclic nucleotides have opposing effects, cAMP stimulating and cGMP inhibiting. In some smooth muscles (e.g., vascular), both cyclic nucleotides act in the same direction, namely, both inhibit $I_{Ca(L)}$. In skeletal muscle, both cAMP and cGMP act in the same direction on $I_{Ca(L)}$, but to stimulate. The cyclic nucleotides and phosphorylation may also modulate the activity of several other types of ion channels, including K^+ channels (delayed rectifier type), Cl^- channels (β-agonist-activated type), and I_f pacemaker channels (Table IV).

References

1. Wahler, G. M., Rusch, N. J., and Sperelakis, N. (1990). 8-Bromo-cyclic GMP inhibits the calcium channel current in embryonic chick ventricular myocytes. *Can. J. Physiol. Pharmacol.* **68,** 531–534.
2. Nowycky, M. C., Fox, A. P., and Tsien, R. W. (1985). Three types of neuronal calcium channels with different calcium agonist sensitivity. *Nature* (*London*) **316,** 440–443.
3. Tohse, N., Nakaya, H., Takeda, Y., and Kanno, M. (1992). Inhibitory effect of human atrial natriuretic peptide on cardiac L-type Ca channels. *Jpn. J. Pharmacol.* **58,** 184P.
4. Tohse, N., and Sperelakis, N. (1990). Long-lasting openings of single slow (L-type) Ca^{2+} channels in chick embryonic heart cells. *Am. J. Physiol.* **259,** H639–H642.
5. Shigenobu, K., and Sperelakis, N. (1972). Ca^{2+} current channels induced by catecholamines in chick embryonic hearts whose fast Na^+ channels are blocked by tetrodotoxin or elevated K^+. *Circ. Res.* **31,** 932–952.
6. Tsien, R. W., Giles, W., and Greengard, P. (1972). Cyclic AMP mediates the action of adrenaline on the action potential plateau of cardiac Purkinje fibers. *Nature* (*London*) **240,** 181–183.
7. Reuter, H., and Scholz, H. (1977). The regulation of the calcium conductance of cardiac muscle by adrenaline. *J. Physiol.* (*London*) **264,** 49–62.
8. Josephson, I., and Sperelakis, N. (1978). 5′-Guanylimidodiphosphate stimulation of slow Ca^{2+} current in myocardial cells. *J. Mol. Cell. Cardiol.* **10,** 1157–1166.
9. Vogel, S., and Sperelakis, N. (1981). Induction of slow action potentials microiontophoresis of cyclic AMP into heart cells. *J. Mol. Cell. Cardiol.* **13,** 51–64.
10. Li, T., and Sperelakis, N. (1983). Stimulation of slow action potentials in guinea pig

papillary muscle cells by intracellular injection of cAMP, Gpp(NH)p, and cholera toxin. *Circ. Res.* **52,** 111–117.

11. Bkaily, G., and Sperelakis, N. (1985). Injection of cyclic GMP into heart cells blocks the slow action potentials. *Am. J. Physiol* **248,** H745–H749.
12. Irisawa, H., and Kokobun, S. (1983). Modulation by intracellular ATP and cyclic AMP of the slow inward current in isolated single ventricular cells of the guinea pig. *J. Physiol. (London)* **338,** 321–327.
13. Nargeot, J., Nerbonne, J. M., Engels, J., and Lester, H. A. (1983). Time course of the increase in the myocardial slow inward current after a photochemically generated concentration jump of intracellular cAMP. *Proc. Natl. Acad. Sci. U.S.A.* **80,** 2395–2399.
14. Cachelin, A. B., dePeyer, J. E., Kokubun, S., and Reuter, H. (1983). Ca^{2+} channel modulation by 8-bromo-cyclic AMP in cultured heart cells. *Nature (London)* **304,** 462–464.
15. Trautwein, W., and Hoffman, F. (1983). Activation of calcium current by injection of cAMP and catalytic subunit of cAMP-dependent protein kinase. *Proc. Int. Union Physiol. Sci.* **15,** 75–83.
16. Bean, B. P., Nowycky, M. C., and Tsien, R. W. (1984). β-Adrenergic modulation of calcium channels in frog ventricular heart cells. *Nature (London)* **307,** 371–375.
17. Reuter, H., Stevens, C.-F., Tsien, R. W., and Yellen, G. (1982). Properties of single calcium channels in cardiac cell culture. *Nature (London)* **297,** 501–504.
18. Tohse, N., Conforti, L., and Sperelakis, N. (1991). Bay k 8644 enhances Ca^{2+} channel activities in embryonic chick heart cells without prolongation of open times. *Eur. J. Pharmacol.* **203,** 307–310.
19. Vogel, S., Sperelakis, N., Josephson, J., and Brooker, G. (1977). Fluoride stimulation of slow Ca^{2+} current in cardiac muscle. *J. Mol. Cell. Cardiol.* **9,** 461–475.
20. Osterrieder, W., Brum, G., Hescheler, J., Trautwein, W., Flockerzi, V., and Hofmann, F. (1982). Injection of subunits of cyclic AMP-dependent protein kinase into cardiac myocytes modulates Ca^{2+} current. *Nature (London)* **298,** 576–578.
21. Bkaily, G., and Sperelakis, N. (1984). Injection of protein kinase inhibitor into cultured heart cells blocks calcium slow channels. *Am. J. Physiol.* **246,** H630–H634.
22. Kameyama, M., Hofmann, F., and Trautwein, W. (1986). On the mechanism of β-adrenergic regulation of the Ca^{2+} channel in the guinea pig heart. *Pfluegers Arch.* **405,** 285–293.
23. Chad, J. E., and Eckert, R. J. (1986). An enzymatic mechanism for calcium current inactivation in dialysed *Helix* neurons. *J. Physiol. (London)* **378,** 31–51.
24. Hescheler, J., Kameyama, M., Trautwein, W., Mieskes, G., and Soling, H. D. (1987). Regulation of the cardiac calcium channel by protein phosphatases. *Eur. J. Biochem.* **165,** 261–266.
25. Reuter, H. (1983). Calcium channel modulation by neurotransmitters, enzymes, and drugs. *Nature (London)* **301,** 569–574.
26. Armstrong, D., and Eckert, R. (1987). Voltage-activated calcium channels that must be phosphorylated to respond to membrane depolarization. *Proc. Natl. Acad. Sci. U.S.A.* **84,** 2518–2522.
27. Goldberg, N. D., Haddox, M. K., Nicol, S. E., Glass, D. B., Sanford, C. H., Kuehl, F. A., Jr., and Estensen, R. (1975). Biological regulation through opposing influences of cyclic GMP and cyclic AMP: The Yin Yang hypothesis. *Adv. Cyclic Nucleotide Res.* **5,** 307–330.
28. Nawrath, H. (1977). Does cyclic GMP mediate the negative inotropic effect of acetylcholine in the heart? *Nature (London)* **267,** 72–74.

29. Kohlhardt, M., and Haap, K. (1978). 8-Bromo-guanosine 3′,5′-monophosphate mimics the effect of acetylcholine on slow response action potential and contractile force in mammalian atrial myocardium. *J. Mol. Cell. Cardiol.* **10,** 573–578.
30. Trautwein, W., Taniguchi, J., and Noma, A. (1982). The effect of intracellular cyclic nucleotides and calcium on the action potential and acetylcholine response of isolated cardiac cells. *Pfluegers Arch.* **392,** 307–314.
31. Wahler, G. M., and Sperelakis, N. (1985). Intracellular injection of cyclic GMP depresses cardiac slow action potentials. *J. Cyclic Nucleotide Protein Phosphorylation Res.* **10,** 83–95.
32. Mehegan, J. P., Muir, W. W., Unverferth, D. V. Fertel, R. H., and McGuirk, S. M. (1985). Electrophysiological effects of cyclic GMP on canine cardiac Purkinje fibers. *J. Cardiovasc. Pharmacol.* **7,** 30–35.
33. Tohse, N., and Sperelakis, N. (1991). Cyclic GMP inhibits the activity of single calcium channels in embryonic chick heart cells. *Circ. Res.* **69,** 325–331.
34. Cuppoletti, J., Thakkar, J., Sperelakis, N., and Wahler, G. (1988). Cardiac sarcolemmal substrate of the cGMP-dependent protein kinase. *Membr. Biochem.* **7,** 135–142.
35. Hartzell, H. C., and Fischmeister, R. (1986). Opposite effects of cyclic GMP and cyclic AMP on Ca^{2+} current in single heart cells. *Nature (London)* **323,** 273–275.
36. Fischmeister, R., and Hartzell, R. C. (1987). Cyclic guanosine 3′,5′-monophosphate regulates the calcium current in single cells from frog ventride. *J. Physiol. (London)* **387,** 455–472.
37. Levi, R. C., Alloatti, G., and Fischmeister, R. (1989). Cyclic GMP regulates the Ca-channel current in guinea pig ventricular myocytes. *Pfluegers Arch.* **413,** 685–687.
38. Mery, P. F., Lohmann, S. M., Walter, U., and Fischmeister, R. (1991). Ca^{2+} current is regulated by cyclic GMP-dependent protein kinase in mammalian cardiac myocytes. *Proc. Natl. Acad. Sci. U.S.A.* **88,** 1197–1201.
39. Thakkar, J., Tang, S., Sperelakis, N., and Wahler, G. (1988). Inhibition of cardiac slow action potentials by 8-bromo-cyclic GMP occurs independent of changes in cyclic AMP levels. *Can. J. Physiol. Pharmacol.* **66,** 1092–1095.
40. George, W. J., Polson, J. B., O'Toole, A. G., and Goldberg, N. D. (1970). Elevation of guanosine 3′,5′-cyclic phosphate in rat heart after perfusion with acetylcholine. *Proc. Natl. Acad. Sci. U.S.A.* **66,** 398–403.
41. Ahmad, Z., Green, F. J., Subuhi, H. S., and Watanabe, A. M. (1989). Purification and characterization of a α-1,2-mannosidase involved in processing asparagine linked oligosaccharides. *J. Biol. Chem.* **264,** 3859–3863.
42. Wahler, G. M., and Sperelakis, N. (1986). Cholinergic attenuation of the electrophysiological effects of forskolin. *J. Cyclic Nucleotide Protein Phosphorylation Res.* **11,** 1–10.
43. MacLeod, K. M., and Diamond, J. (1986). Effects of the cyclic GMP lowering agent LY83583 on the interaction of carbachol with forskolin in rabbit isolated cardiac preparations. *J. Pharmacol. Exp. Ther.* **238,** 313–318.
44. Hescheler, J., Kameyama, M., and Trautwein, W. (1986). On the mechanism of muscarinic inhibition of the cardiac Ca current. *Pfluegers Arch.* **407,** 182–189.
45. Ono, K., and Trautwein, W. (1991). Potentiation by cyclic GMP of β-adrenergic effect on Ca^{2+} current in guinea pig ventricular cells. *J. Physiol. (London)* **443,** 387–404.
46. Freer, R. J., Pappano, A. J., Peach, M. J., Bing, K. T., McLean, M. J., Vogel, S. M., and Sperelakis, N. (1976). Mechanism of the positive inotropic effect of angiotensin II on isolated cardiac muscle. *Circ. Res.* **39,** 178–183.
47. Bruckner, R., and Scholz, H. (1984). Effects of α-adrenoceptor stimulation with phenylephrine in the presence of propranolol on force of contraction, slow inward current and cyclic AMP content in the bovine heart. *Br. J. Pharmacol.* **82,** 223–232.

48. Dosemeci, A., Dhalla, R. S., Cohen, N. M., Lederer, W. J., and Rogers, T. B. (1988). Phorbol ester increases calcium current and stimulates the effects of angiotensin II on cultured neonatal rat heart myocytes. *Circ. Res.* **62,** 347.
49. Tohse, N., Kameyama, M., Sakiguchi, K., Shearman, M. S., and Kanno, M. (1990). Protein kinase C activation enhances the delayed rectifier K^+ current in guinea pig heart cells. *J. Mol. Cell. Cardiol.* **22,** 725–734.
50. Bkaily, G., and Sperelakis, N. (1986). Calmodulin is required for a full activation of the calcium slow channels in heart cells. *J. Cyclic Nucleotide Protein Phosphorylation Res.* **11,** 25–34.
51. Yazawa, K., and Kameyama, M. (1990). Mechanism of receptor-mediated modulation of the delayed outward potassium current in guinea pig ventricular myocytes. *J. Physiol. (London)* **421,** 135–150.
52. DiFrancesco, D., and Tromba, C. (1988). Muscarinic control of the hyperpolarization-activated current (I_f) in rabbit sino-atrial node myocytes. *J. Physiol. (London)* **405,** 493–510.
53. Ehara, T., and Ishihara, K. (1990). Anion channels activated by adrenaline in cardiac myocytes. *Nature (London)* **347,** 284–286.
54. Tareen, F. M., Ono, K., Noma, A., and Ehara, T. (1991). β-Adrenergic and muscarinic regulation of the chloride current in guinea pig ventricular cells. *J. Physiol. (London)* **440,** 225–241.
55. Yatani, A., Codina, J., Imoto, Y., Reeves, J. P., Birnbaumer, L., and Brown, A. M. (1987). Direct regulation of mammalian cardiac calcium channels by a G protein. *Science* **238,** 1288–1292.

Functional Adaptation to Myocardial Ischemia: Interaction with Volatile Anesthetics in Chronically Instrumented Dogs

Patrick F. Wouters, Hugo Van Aken, Marc Van de Velde, Marco A. E. Marcus, and Willem Flameng

Department of Anesthesiology
University Hospitals Gasthuisberg
Katholieke Universiteit,
Leuven, Belgium

I. Introduction

It has been demonstrated previously that volatile anesthetics attenuate myocardial stunning following a single brief coronary artery occlusion in chronically instrumented dogs (1). Proposed mechanisms for this protective effect include a reduction of metabolic requirements during the ischemic intervention and inhibition of intracellular calcium overload during subsequent reperfusion. However, Mattheussen *et al.* reported that volatile anesthetics did not improve cardiac recovery from postischemic contractile dysfunction in an experimental model of global myocardial ischemia (2). It is now clear that stunning may result from a number of different experimental and clinical conditions which induce a temporary imbalance between myocardial metabolic demand and supply (3). As it is unknown whether these different conditions have a common underlying mechanism leading to contractile dysfunction, findings obtained in one particular experimental model might not be applicable to other models of stunning.

Advances in Pharmacology, Volume 31

In this respect, it should also be recognized that myocardial injury following repetitive ischemic cycles may differ from damage secondary to a single insult. Murray *et al.* demonstrated that after a first exposure to myocardial ischemia and reperfusion, canine myocytes temporarily become more resistant to subsequent ischemic events (4). This state of enhanced tolerance to ischemia is characterized by a reduced rate of ATP utilization and lactate production. It can be induced in a variety of species and has been termed ischemic preconditioning (IP) (5–7). Whether volatile anesthetics have protective effects on stunning during conditions of repetitive ischemia remains to be established. Volatile anesthetics and IP might share similar pathways through which myocardial protection is provided. Alternatively, ATP-sensitive potassium channels have been shown to play an important role in IP (8,9). Because volatile anesthetics suppress potassium channel currents (10), they might interfere with the mechanism of IP. The goal of this study was to determine whether an interaction between the myocardial protective effects of IP and those offered by volatile anesthetics exists.

Although it is generally accepted that IP protects the myocardium from irreversible ultrastructural damage and major arrhythmias, its effect on recovery from stunning following subsequent brief ischemic insults remains controversial (11,12). For this reason, we first examined the effects of IP on myocardial stunning. Functional recovery from a brief coronary artery occlusion was assessed in dogs that were previously exposed to a preconditioning episode and compared to control animals with untreated "virgin myocardium." Subsequently, the effects of isoflurane on recovery from repetitive ischemia were studied and compared to the conscious condition. All studies were performed in chronically instrumented dogs in order to avoid the potentially adverse effects of acute surgical instrumentation and baseline anesthetic drugs.

II. Chronic Instrumentation and Monitoring

Mongrel dogs (either sex, weighing between 19 and 24 kg) were premedicated intramuscularly with piritramide at 1 mg/kg and Hypnorm at 0.25 ml/kg (10 mg fluanisone plus 0.2 mg fentanyl per milliliter), and anesthetized with intravenous pentobarbital at 20 mg/kg. The trachea of the animals was intubated, and the lungs were mechanically ventilated. Anesthesia was maintained with halothane at 0.8–1.2% in 50% oxygen. Under aseptic conditions, a left thoracotomy was performed through the fourth intercostal space. Tygon catheters were inserted in the left atrium and ascending aorta, and a micromanometer (Janssen Pharmaceutics,

Beerse, Belgium) was positioned in the left ventricle through an apical stab wound. The 20 MHz pulsed Doppler flow transducers (Baylor College of Medicine, Houston, TX) were fitted around the left anterior descending (LAD) and circumflex coronary artery. A cuffed hydraulic occluder was positioned around the LAD, proximally to the flow probe (Fig. 1). Finally, a 10 MHz pulsed Doppler wall thickening probe (Baylor College of Medicine) was sutured to the epicardium of the LAD coronary artery perfusion area. A brief coronary artery occlusion resulted in hypokinesia within seconds, confirming the correct position of the wall thickening probe prior to closure of the pericardium. The thorax was closed in layers, and all leads were tunneled to the dorsum of the neck of the animal and exteriorized. The dogs were fitted with a jacket to protect the probe leads from damage. The animals were treated with ampicillin anhydricum (400 mg/kg) for 5 days during recovery from surgery and were trained to get accustomed to the laboratory.

Pressure, flow, and wall thickening signals were processed through a 6-channel pulsed Doppler system (Baylor College of Medicine). The left ventricular microtransducer was calibrated against pressures measured through fluid-filled catheters in the aorta and left atrium. The left ventricular rate of pressure rise was obtained by electronic differentiation of the pressure wave signal (Gould Inc., Cleveland, OH). The systolic wall thickening fraction was measured using the pulsed Doppler method. This tech-

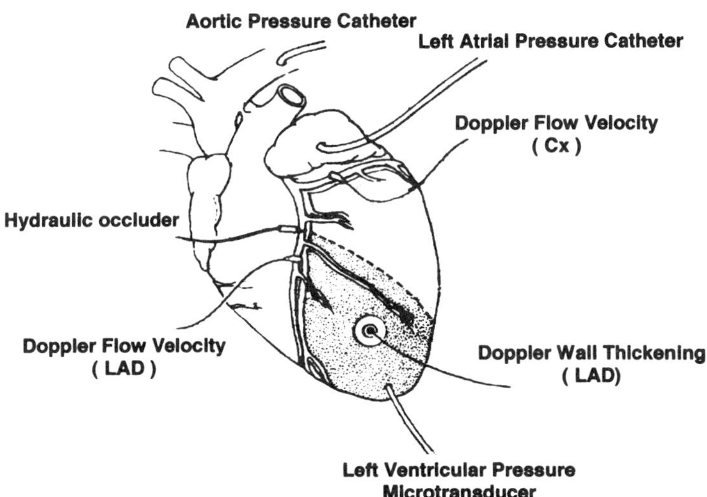

Fig. 1 Cardiovascular instrumentation technique. Cx, Left circumflex coronary artery equivalent; LAD, left anterior descending coronary artery.

nique has been previously validated in a variety of pharmacological and physiological interventions including acute myocardial ischemia (13). It requires the implantation of a single 10 MHz piezoelectric crystal which is sutured to the epicardium and serves both as a transmitter and receiver of pulsed Doppler signals. Wall thickening is obtained by integrating the velocity of myocardial layers passing through the sample volume at a preselected depth. All signals were simultaneously recorded on an 8-channel thermal-writing polygraph (Gould Inc.) and on a Macintosh CI personal computer using LABVIEW software for data storage and off-line analysis.

Stunning was induced in six animals by a 10-min coronary artery occlusion followed by 120 min of reperfusion. Each animal underwent two studies separated by at least 7 days. On one occasion, the 10-min ischemic event was preceeded by a shorter episode of acute myocardial ischemia (2 min followed by 30 min reperfusion, namely, ischemic preconditioning), whereas the control experiment consisted of a single 10-min coronary artery occlusion. The sequence of studies was randomized, and a comparison was made between recovery from stunning in control and preconditioned animals. Hemodynamics were recorded continuously prior to and following coronary artery occlusions in unsedated animals lying on their right side. No analgesics or antiarrhythmic drugs were administered. Prior to coronary artery occlusions, heparin (1 mg/kg) was administered intravenously to avoid thrombus formation and allow immediate and complete reperfusion.

In a second experiment which included five dogs, the technique of IP was modified. Ischemic pretreatment was extended to cumulative intermittent coronary artery occlusions up to a total ischemic duration of 5 min (90 + 90 + 120 sec separated by 5 min of reperfusion). Fifteen minutes later, a sustained 5-min ischemic event was induced. Each dog underwent two experiments (with or without IP preceeding the sustained 5-min occlusions) in a randomized fashion, separated by at least 3 days.

A similar experimental sequence was repeated during isoflurane anesthesia. Dogs were anesthetized by mask with isoflurane in 100% oxygen. The trachea was intubated, and the lungs were mechanically ventilated with a tidal volume of 15 ml/kg, the rate being adjusted to maintain normocarbia. Inspired oxygen concentration was titrated to keep arterial oxygen pressure between 100 and 150 mmHg. Anesthesia was maintained with doses required to lower mean arterial blood pressure to 60 mmHg. The end-tidal anesthetic concentration was measured using an infrared absorption technique (Irina, Dräger, Lübeck, Germany). After a 45-min equilibration period, baseline hemodynamic measurements were performed and experimental coronary artery occlusions were initiated.

Data were analyzed using a two-way analysis of variance for repeated measure (time and group) design. Paired post hoc comparisons were performed using Students *t*-test with modified Bonferroni correction. A probability level *p* below 0.05 was considered statistically significant. Data were presented as means ± SEM.

III. Effects of Ischemic Preconditioning on Functional Recovery from Stunning in Conscious Dogs

As indicated in Table I, global hemodynamic performance prior to, during, and following a 10-min coronary artery occlusion in preconditioned and nonpreconditioned hearts were essentially similar. Acute coronary artery occlusion in conscious animals resulted in a significant increase in heart rate. Coronary blood flow through the patent circumflex coronary artery increased by 25% and returned toward baseline immediately following reperfusion. Recovery from regional contractile dysfunction in reperfused myocardium did not differ between preconditioned and untreated animals (Fig. 2). In both groups, the systolic wall thickening fraction (WTF) recovered to 75% of preocclusion values within 2 hr following reperfusion.

Brief intermittent episodes of cumulative ischemia up to a total of 5 min also failed to affect recovery from a subsequent sustained 5-min ischemic episode in conscious animals (Table II). In contrast, early recovery of systolic wall thickening tended to be worse in preconditioned animals, suggesting cumulative functional deterioration rather than a protective effect of IP on stunning.

Compared to the awake condition, isoflurane significantly decreased wall thickening fraction (from 23 ± 1 to 14 ± 1%), mean arterial blood pressure (from 90 ± 10 to 61 ± 3 mmHg), LV DP/*dt* MAX (from 2077 ± 126 to 1452 ± 102 mmHg/sec) and LV DP/*dt* MIN (from −1978 ± 182 to −1481 ± 57 mmHg/sec). Heart rate and left anterior descending coronary blood flow velocity increased [from 69 ± 4 to 91 ± 5 beats/min (bpm) and from 2.6 ± 0.3 to 4 ± 0.4 kHz, respectively]. Coronary artery occlusion resulted in moderate tachycardia and a mild increase in circumfex blood flow velocity (Table III). IP resulted in significant residual contractile dysfunction after 15 min of reperfusion. Temporal progression of recovery from stunning following the subsequent 5-min sustained coronary artery occlusion, however, showed little difference between isoflurane-anesthetized and conscious animals (Fig. 3). However, a significantly more pronounced transient hypercontractile response was noted immediately on reperfusion.

Table I

Hemodynamic Data in Conscious Dogs before, during, and after 10-Minute Coronary Artery Occlusion as Single Event (Control) or Subsequent to 2-Minute Ischemic Episode[a]

		Baseline	Occlusion 1	Reperfusion 1	Occlusion 2	Reperfusion 2					
						5 min	10 min	30 min	60 min	90 min	120 min
HR	Control	88 ± 10	—	—	126 ± 9[b]	89 ± 11	94 ± 8	93 ± 11	87 ± 8	83 ± 9	84 ± 8
(bpm)	IP	85 ± 10	108 ± 13	80 ± 12	116 ± 13[b]	99 ± 13	89 ± 15	84 ± 13	92 ± 14	85 ± 14	92 ± 17
ABPM	Control	91 ± 8	—	—	101 ± 8	89 ± 8	94 ± 7	89 ± 9	91 ± 10	86 ± 8	91 ± 5
(mmHg)	IP	90 ± 10	91 ± 10	89 ± 10	95 ± 9	93 ± 11	84 ± 11	88 ± 11	92 ± 10	88 ± 11	92 ± 8
EDLVP	Control	3.8 ± 0.5	—	—	5.3 ± 0.8	4.1 ± 0.8	4.1 ± 1.0	3.8 ± 0.9	3.4 ± 1.3	4.0 ± 1.1	4.0 ± 1.6
(mmHg)	IP	4.3 ± 0.6	5.4 ± 1.2	2.5 ± 1.1	4.6 ± 1.4	3.7 ± 0.9	4.0 ± 0.6	4.7 ± 1.0	5.9 ± 1.6	6.4 ± 1.8	6.0 ± 1.5
DP/*dt* MAX	Control	2099 ± 126	—	—	2229 ± 211	2069 ± 209	2091 ± 146	1922 ± 135	1928 ± 214	1873 ± 155	1914 ± 165
(mmHg/sec)	IP	2077 ± 126	1958 ± 114	1966 ± 141	2161 ± 235	1926 ± 159	1741 ± 129	1854 ± 150	1927 ± 194	1780 ± 193	1927 ± 209
DP/*dt* MIN	Control	−2064 ± 134	—	—	−2035 ± 240	−1980 ± 251	−1976 ± 167	−1776 ± 154	−1762 ± 178	−1750 ± 194	−1832 ± 169
(mmHg/sec)	IP	−1978 ± 182	−1889 ± 111	−1930 ± 194	−1985 ± 218	−1937 ± 191	−1747 ± 154	−1817 ± 156	−1754 ± 174	−1619 ± 184	−1723 ± 154
LAD BFv	Control	4.3 ± 0.8	—	—	0[b]	11.2 ± 2.0[b]	3.9 ± 0.6	3.1 ± 0.7	3.7 ± 0.7	3.5 ± 0.7	3.9 ± 0.7
(kHz)	IP	4.6 ± 1.2	0[b]	5.1 ± 1.2	0[b]	10.5 ± 1.6[b]	5.1 ± 1.1	5.1 ± 1.2	5.0 ± 1.2	4.7 ± 1.2	4.6 ± 1.0
Cx BFv	Control	4.0 ± 0.7	—	—	5.6 ± 0.9[b]	3.8 ± 0.6	3.9 ± 0.6	2.8 ± 0.5	3.6 ± 0.5	3.3 ± 0.4	3.6 ± 0.5
(kHz)	IP	3.1 ± 0.4	4.0 ± 0.6	3.2 ± 0.5	4.5 ± 0.7[b]	3.5 ± 0.6	3.3 ± 0.6	3.2 ± 0.5	3.3 ± 0.5	3.0 ± 0.4	3.0 ± 0.4

[a]Data are presented as means ± SEM. IP, Ischemic preconditioning; ABPM, mean arterial blood pressure; HR, heart rate; LAD and Cx BFv, blood flow velocity in left LAD and Cx coronary arteries; DP/*dt* MAX and MIN, maximum rate of left ventricular pressure rise and decline; EDLVP, left ventricular end-diastolic pressure.

[b]$p < 0.05$ versus baseline.

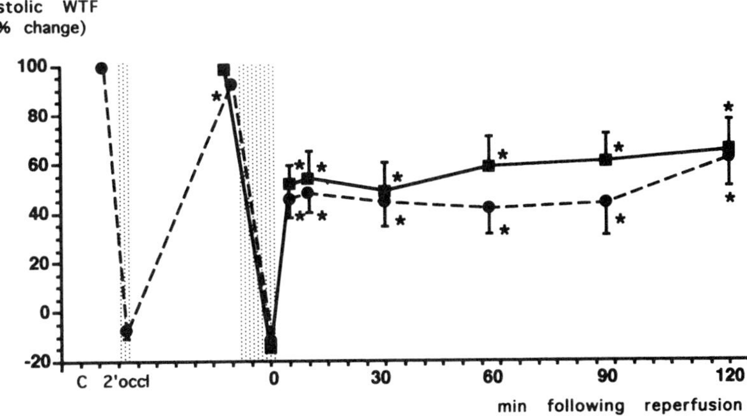

Fig. 2 Percent recovery of systolic wall thickening fraction (WTF) following a 10-min coronary artery occlusion in conscious dogs. Circles (●) and dashed lines represent dogs pretreated with a 2-min preconditioning event followed by 30 min of reperfusion. Squares (■) and solid lines represent control animals with no exposure to ischemia prior to the 10-min occlusion. The dashed areas indicate coronary artery occlusions. C, Control; *, $p < 0.05$ versus control. Data are means ± SEM.

IV. Discussion

This report indicates that prior exposure of myocardium to brief coronary artery occlusions does not enhance its ability to functionally recover from stunning following subsequent ischemic insults. Our observations in conscious chronically instrumented dogs appear to contradict the data of Cohen *et al.* (5). Those authors demonstrated a markedly better contractile recovery following prolonged ischemia in rabbit hearts after prior exposure to short ischemic episodes. However, it was not clear whether improved systolic recovery was due to smaller infarct size or reduced stunning. By using short coronary artery occlusions of insufficient duration to result in myocardial necrosis, we could attribute contractile dysfunction of post-ischemic myocardium entirely to stunning.

It has become apparent that brief sublethal ischemic challenges rapidly lead to an increased resistance against lethal myocardial injury following a subsequent prolonged coronary artery occlusion. This phenomenon of endogenous myocardial protection, termed ischemic preconditioning, is the most powerful means currently available to reduce infarct size *in vivo* (14). The lack of any effect of IP on recovery from stunning contrasts sharply with the protection offered against ultrastructural damage, suggesting that the pathogenesis of myocardial necrosis and stunning involves different mechanisms.

Table II

Hemodynamic Data in Conscious Dogs before, during, and after 5-Minute Coronary Artery Occlusion as Single Event (control) or Subsequent to Brief Repetitive Ischemic Episodes[a]

					Reperfusion 2					
		Baseline	Reperfusion 1	Occlusion 2	1 min	5 min	30 min	60 min	90 min	120 min
WTF (%)	Control	25 ± 1	—	-6 ± 3^b	20 ± 1	21 ± 2	20 ± 2^b	18 ± 2^b	20 ± 2^b	19 ± 3
	IP	23 ± 1	19 ± 2	-5 ± 2^b	17 ± 2^b	16 ± 2^b	14 ± 3^b	15 ± 3^b	17 ± 2^b	17 ± 3*
ABPM (mmHg)	Control	83 ± 5	—	94 ± 7	85 ± 5	79 ± 6	82 ± 5	86 ± 5	81 ± 6	98 ± 5
	IP	89 ± 3	81 ± 4^b	93 ± 6	92 ± 4	86 ± 4	88 ± 6	82 ± 4	83 ± 3	90 ± 5
HR (bpm)	Control	79 ± 7	—	108 ± 12^b	83 ± 8	80 ± 8	81 ± 13	71 ± 9	66 ± 8	79 ± 10
	IP	69 ± 4	66 ± 6	102 ± 7^b	81 ± 6	73 ± 6	71 ± 8	72 ± 7	71 ± 5	76 ± 11
LAD BFv (kHz)	Control	3.2 ± 0.6	—	0.0 ± 0.0^b	15.0 ± 2.7^b	6.1 ± 1.8	3.9 ± 1.0	3.6 ± 0.9	3.9 ± 1.1	4.6 ± 1.1
	IP	2.6 ± 0.3	2.8 ± 0.5	0.0 ± 0.0^b	19.5 ± 1.9^b	4.8 ± 1.1	3.0 ± 0.3	3.1 ± 0.3	2.9 ± 0.5	2.8 ± 0.2
Cx BFv (kHz)	Control	3.4 ± 0.4	—	5.4 ± 0.6^b	4.0 ± 0.5	3.6 ± 0.5	3.8 ± 0.6	3.3 ± 0.5	3.5 ± 0.5	3.4 ± 0.5
	IP	2.6 ± 0.4	2.9 ± 0.6	4.6 ± 0.7^b	3.3 ± 0.5	2.8 ± 0.5	3.0 ± 0.6	3.1 ± 0.5	2.8 ± 0.5	3.1 ± 0.6
EDLVP (mmHg)	Control	7.0 ± 0.6	—	11.5 ± 1.1	7.3 ± 1.6	7.8 ± 1.4	7.6 ± 1.5	6.9 ± 1.6	7.4 ± 1.5	8.1 ± 1.9
	IP	8.0 ± 0.5	7.6 ± 1.0	9.5 ± 2.0	8.1 ± 1.2	8.5 ± 1.0	7.5 ± 1.6	8.5 ± 1.9	10.3 ± 1.4	11.2 ± 1.3
DP/*dt* MAX (mmHg/sec)	Control	2579 ± 128	—	2813 ± 207	2751 ± 175	2397 ± 132	2399 ± 194	2515 ± 154	2381 ± 149	2832 ± 231
	IP	2195 ± 165	2070 ± 141	2441 ± 153	2397 ± 152	2053 ± 128	2154 ± 165	2053 ± 125	2035 ± 94	2096 ± 233
DP/*dt* MIN (mmHg/sec)	Control	−2193 ± 161	—	−2340 ± 148	−2297 ± 138	−2148 ± 128	−2156 ± 122	−2273 ± 131	−2147 ± 124	−2779 ± 320
	IP	−2208 ± 207	−2134 ± 172	−2401 ± 214	−2332 ± 202	−2135 ± 147	−2294 ± 209	−2216 ± 179	−2188 ± 111	−2250 ± 295

[a]Data are presented as means ± SEM. IP, Ischemic preconditioning; WTF, systolic wall thickening fraction; ABPM, mean arterial blood pressure; HR, heart rate; LAD and Cx BFv, blood flow velocity in left LAD and Cx coronary arteries; DP/*dt* MAX and MIN, maximum rate of left ventricular pressure rise and decline; EDLVP, left ventricular end-diastolic pressure. Values at reperfusion 1 were obtained 15 min following IP, that is, immediately prior to the final 5-min occlusion (occlusion 2).

[b]$p < 0.05$ versus baseline.

Table III

Hemodynamic Data in Isoflurane-Anesthetized Dogs before, during, and after 5-Minute Coronary Artery Occlusion Preceeded by Brief Repetitive Ischemic Events[a]

				Reperfusion 2					
	Baseline	Reperfusion 1	Occlusion 2	1 min	5 min	30 min	60 min	90 min	120 min
WTF (%)	14 ± 1^{b}	$10 \pm 2^{b,c}$	-6 ± 2^{c}	16 ± 1	15 ± 1	$9 \pm 2^{b,c}$	$8 \pm 2^{b,c}$	$9 \pm 1^{b,c}$	$10 \pm 2^{b,c}$
ABPM (mmHg)	61 ± 3^{b}	65 ± 4^{b}	69 ± 5^{b}	70 ± 3^{b}	69 ± 3^{b}	71 ± 3^{b}	71 ± 3^{b}	69 ± 3^{b}	67 ± 2^{b}
HR (bpm)	91 ± 5^{b}	92 ± 6^{b}	103 ± 5^{c}	98 ± 4^{b}	95 ± 5^{b}	98 ± 4^{b}	$99 \pm 4^{b,c}$	97 ± 5^{b}	95 ± 4^{b}
LAD BFv (kHz)	4 ± 0.4^{b}	4.1 ± 0.4^{b}	0 ± 0^{c}	15.5 ± 1.6^{c}	5.7 ± 0.7^{c}	4.4 ± 0.4^{b}	4.4 ± 0.4^{b}	4.5 ± 0.5^{b}	4.3 ± 0.4^{b}
Cx BFv (kHz)	2.9 ± 0.3	3.1 ± 0.4	3.6 ± 0.4^{c}	3.4 ± 0.4^{c}	3.2 ± 0.4	3 ± 0.5	3.4 ± 0.4^{c}	3.5 ± 0.4^{c}	3.4 ± 0.3^{c}
EDLVP (mmHg)	11.3 ± 1.4	10.8 ± 1.2	11.1 ± 2	9.9 ± 1.2	9.5 ± 1.4	9.6 ± 1.2	9.4 ± 2	9.4 ± 2.3	8.7 ± 2.3
LV DP/*dt* MAX (mmHg/sec)	1452 ± 102^{b}	1457 ± 93^{b}	1540 ± 95^{b}	1653 ± 108^{b}	1519 ± 87^{b}	1466 ± 82^{b}	1489 ± 98^{b}	1489 ± 97^{b}	1464 ± 91^{b}
LV DP (mm Hg/sec)	-1481 ± 57^{b}	-1449 ± 69^{b}	-1442 ± 89^{b}	-1585 ± 81^{b}	-1554 ± 48^{b}	-1499 ± 44^{b}	-1549 ± 63^{b}	-1577 ± 79^{b}	-1520 ± 72^{b}

[a]Data are presented as means ± SEM. IP, Ischemic preconditioning; WTF, systolic wall thickening fraction; ABPM, mean arterial blood pressure; HR, heart rate; LAD and Cx BFv, blood flow velocity in left LAD and Cx coronary arteries; DP/*dt* MAX and MIN, maximum rate of left ventricular pressure rise and decline respectively; EDLVP, left ventricular end-diastolic pressure. Values at reperfusion 1 were obtained 15 min following IP, that is, immediately prior to the final 5-min occlusion (occlusion 2).

[b]$p < 0.05$ versus corresponding condition in conscious dogs.

[c]$p < 0.05$ versus baseline.

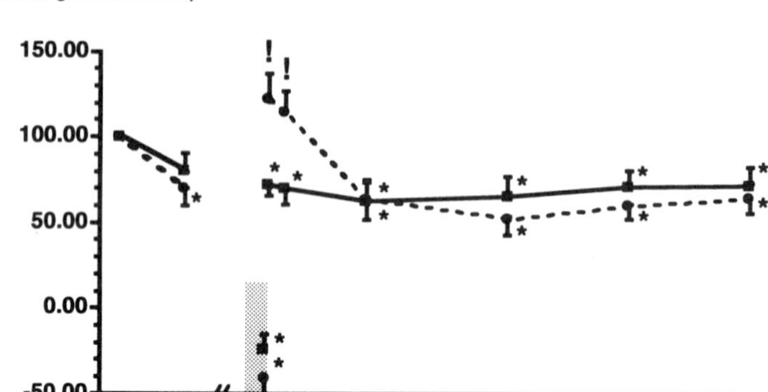

Fig. 3 Percent recovery of systolic wall thickening fraction (WTF) following a 5-min coronary artery occlusion preceeded by intermittent ischemic episodes (preconditioning of 90 + 90 + 120 sec) in conscious (B) and isoflurane-anesthetized (J) dogs. C, Control; IP, contractile performance 15 min following ischemic preconditioning and immediately prior to the 5-min ischemic event; *, $p < 0.05$ versus control; !, $p < 0.05$ isoflurane versus conscious. Data are means ± SEM.

Isoflurance failed to attenuate stunning from repetitive coronary artery occlusions. In contrast to our results, Warltier *et al.* clearly demonstrated that both halothane and isoflurane improved functional recovery from a single 15-min coronary artery occlusion in dogs when compared to conscious controls (1). Because the use of repetitive ischemia versus single coronary artery occlusion in our study did not affect the ultimate functional recovery from stunning, other factors must be considered to explain our discrepant findings. First, although a nearly similar experimental model of regional myocardial ischemia was used, methodological details during ischemia and reperfusion were different. Warltier *et al.* reported the administration of lidocaine during ischemia and employed a gradual reperfusion technique to reduce the incidence of ventricular arrhythmias on relief of coronary artery occlusion. In contrast, we have allowed an immediate restoration of blood flow following coronary artery occlusions and avoided the use of antiarrhythmic drugs. Lidocaine decreases influx through fast Na^+ channels so that less Na^+ is available for Na^+/Ca^{2+} exchange during subsequent reperfusion (15). It has been clearly shown that experimental conditions which reduce sarcolemmal Ca^{2+} influx during reperfusion attenuate stunning (16,17). Okamoto *et al.* demonstrated that gentle reperfusion is also associated with improved contractile function after coronary artery occlusion when compared to sudden reperfusion techniques (18). From these observations it appears that the extent of reperfusion injury was relatively more severe in our experimental model.

Although stunning is generally considered to be a form of "reperfusion injury" the abnormalities which occur in the postischemic myocardium are governed primarily by the extent of damage during ischemia (19). Consequently, because volatile anesthetics reduce the severity of ischemia during coronary artery occlusion by decreasing myocardial oxygen demand, they are expected to affect subsequent postischemic recovery beneficially. A possible explanation for the discrepancy between the data of Warltier *et al.* and our data is then hypothetized as follows. The beneficial effects of volatile anesthetics during ischemia results in attenuation of stunning provided that reperfusion injury is tempered, as reported in the former study. These potentially beneficial effects of volatile anesthetics become overshadowed when reperfusion damage occurs to its full extent. It is obvious that such a hypothesis requires further investigations.

The temporary hypercontractile state that occurred during reactive hyperemia was significantly more pronounced in isoflurane-anesthetized animals, when compared to conscious dogs. The response has been well documented in postischemic myocardium and has been attributed to the Gregg phenomenon (20), namely, increases in coronary blood flow directly stimulate myocardial performance and oxygen consumption. The consequences of such a direct hyperemia-induced increase in energy expenditure during reperfusion are unknown. In addition, it is not clear why the extent of temporary hypercontractile performance was more pronounced during isoflurane anesthesia. It is known that isoflurane lowers ventricular afterload, which also improves considerably the performance of stunning myocardium (21). However, whereas the effect of isoflurane on afterload persisted throughout the experiment, the hypercontractile condition lasted only for the duration of postischemic hyperemia. In addition, the extent of postischemic hyperemia was equal in isoflurane-anesthetized and conscious animals. Furthermore, preliminary data obtained in halothane-anesthetized dogs have shown essentially the same response. Consequently, the phenomenon appears unrelated to drug-specific effects on the coronary and systemic vascular system.

It should be noticed that the extent of paradoxical wall motion during coronary artery occlusion was equal in conscious and isoflurane-anesthetized animals. The degree of wall motion abnormalities which result from acute myocardial ischemia have previously been shown to correlate directly with the severity and duration of stunning (22). This may also explain why no differences in recovery from stunning were observed between isoflurane-anesthetized and conscious dogs. Finally, it remains possible that the degree of stunning, as present in our experimental model, was too modest to reveal any protective effect of volatile anesthetics.

In summary, our data demonstrate that ischemic preconditioning offers no protection against reversible contractile dysfunction following subsequent episodes of myocardial ischemia. Consequently, when developing future therapies based on IP to reduce myocardial infarct size, it should be realized that preservation of contractile function is not guaranteed. Second, we failed to demonstrate that isoflurane attenuates stunning in an experimental model of repetitive ischemic insults. We postulated that variations in experimental reperfusion techniques may modify the effects of volatile anesthetics on stunning and offer an explanation for the current controversial findings. Further studies are necessary to understand the interaction between volatile anesthetics and the complex pathophysiology of myocardial ischemia and reperfusion.

References

1. Warltier, D. C., Al-Wathiqui, M. H., Kampine, J. P., and Schmeling, W. T. (1988). Recovery of contractile function of stunned myocardium in chronically instrumented dogs is enhanced by halothane and isoflurane. *Anesthesiology* **69,** 552–565.
2. Mattheussen, M., Rusy, B. F., Van Aken, H., and Flameng, W. (1993). Recovery of function and adenosine triphosphate metabolism following myocardial ischemia induced in the presence of volatile anesthetics. *Anesth. Analg.* (*N.Y.*) **76,** 69–75.
3. Bolli, R. (1992). Myocardial stunning in man. *Circulation* **86,** 1671–1691.
4. Murry, C. E., Jennings, R. B., and Reimer, K. A. (1986). Preconditioning with ischemia: A delay of lethal cell injury in ischemic myocardium. *Circulation* **74,** 1124–1136.
5. Cohen, M. V., Liu, G. S., and Downey, J. M. (1991). Preconditioning causes improved wall motion as well as smaller infarcts after transient coronary occlusion in rabbits. *Circulation* **84,** 341–349.
6. Asimakis, G. K., Inners-McBride, K., Medellin, G., and Conti, V. R. (1992). Ischemic preconditioning attenuates acidosis and postischemic dysfunction in isolated rat heart. *Am. J. Physiol.* **263,** H887–H894.
7. Schott, R. J., Rohmann, S., Braun, E. R., and Schaper, W. (1990). Ischemic preconditioning reduces infarct size in swine myocardium. *Circ. Res.* **66,** 1133–1142.
8. Gross, G. J., and Auchampach, J. A. (1992). Blockade of ATP-sensitive potassium channels prevents myocardial preconditioning in dogs. *Circ. Res.* **70,** 223–233.
9. Grover, G. J., Sleph, P. G., and Dzwonczyk, S. (1992). Role of myocardial ATP-sensitive potassium channels in mediating preconditioning in the dog heart and their possible interaction with adenosine A1 receptors. *Circulation* **86,** 1310–1316.
10. Buljabasic, N., Flynn, N. M., Marijic, J., Rusch, N. J., Kampine, J. P., and Bosnjak, Z. J. (1992). Effects of halothane and isoflurane on calcium and potassium channel currents in canine coronary arterial cells. *Anesthesiology* **76,** 990–998.
11. Ovize, M., Pryzklenk, K., Hale, S. L., and Kloner, R. A. (1992). Ischemic preconditioning does not attenuate myocardial stunning. *Circulation* **85,** 2247–2254.
12. Shizukuda, Y., Iwamoto, T., Mallet, R. T., and Downey, H. F. (1993). Hypoxic preconditioning attenuates stunning caused by repeated coronary artery occlusions. *Cardiovasc. Res.* **27,** 559–564.

13. Hartley, C. J., Latson, L. A., Michael, L. H., Seidel, C. L., Lewis, R. M., and Entman, M. L. (1983). Doppler measurement of myocardial thickening with a single epicardial transducer. *Am. J. Physiol.* **245,** H1066–H1072.
14. Lawson, C. S., and Downey, J. M. (1993). Preconditioning: State of the art myocardial protection. *Cardiovasc. Res.* **27,** 542–550.
15. Ritchie, J. M., and Greene, N. M. (1990). Local anesthetics. *In* "The Pharmacological Basics of Therapeutics" (A. G. Gilman, T. W. Rall, A. S. Nies, and P. Taylor, eds.), 8th Ed. pp. 311–331. Pergamon, New York.
16. Kusuoka, H., Porterfield, J. K., Weisman, H. F., Weisfeldt, M. L., and Marban, E. (1987). Pathophysiology and pathogenesis of stunned myocardium: Depressed Ca^{2+} activation of contraction as a consequence of reperfusion-induced cellular calcium overload in ferret hearts. *J. Clin. Invest.* **79,** 950.
17. Kitakaze, M., Weisfeldt, M. L., and Marban, E. (1988). Acidosis during early reperfusion prevents stunning in perfused hearts. *J. Clin. Invest.* **82,** 920.
18. Okamoto, F., Allen, B. S., Buckberg, G. D., Bugyi, H., and Leaf, J. (1986). Studies of controlled reperfusion after ischemia–reperfusion conditions: Importance of ensuring gentle versus sudden reperfusion during relief of coronary occlusion. *J. Thorac. Cardiovasc. Surg.* **92,** 613.
19. Bolli, R. (1990). Mechanisms of myocardial stunning. *Circulation* **82,** 723–738.
20. Heusch, G. (1991). The relationship between regional blood flow and contractile function in normal, ischemic, and reperfused myocardium. *Basic Res. Cardiol.* **86,** 197–218.
21. Stahl, L. D., Aversano, T. R., and Becker, L. C. (1986). Selective enhancement of function of stunned myocardium by increased flow. *Circulation* **74,** 843–851.
22. Przyklenk, K., and Kloner, R. A. (1989). What factors predict recovery of contractile function in the canine model of the "stunned" myocardium? *Am. J. Cardiol.* **64**(Suppl.), 18F.

Excitation–Contraction Uncoupling and Vasodilators for Long-Term Cold Preservation of Isolated Hearts

David F. Stowe

Departments of Anesthesiology and Physiology
The Medical College of Wisconsin and
Zablocki Veterans Affairs Medical Center
Milwaukee, Wisconsin 53226

I. Introduction

Improved methods to preserve donor hearts for longer periods is a crucial goal in cardiac transplantation research. Several potential experimental approaches are listed in Table I. The usual clinical approach to reduce myocardial reperfusion damage is to flush the clamped aortic root with a cold, high-potassium (KCl cardioplegic) solution to arrest the heart prior to harvest. However, because high KCl produces arrest by depolarizing the cell membrane, this activates voltage-sensitive slow Ca^{2+} channels and Na^{+}/Ca^{2+} exchangers to promote intracellular calcium loading. On normothermic reperfusion with normal KCl, this could lead to diastolic contracture, ventricular dysrhythmias, decreased contractility, and increased energy expenditure.

Most experimental models have focused on return on mechanical function as an indicator of the quality of preservation following cold storage or cold perfusion (1). Suboptimal global myocardial perfusion, however, may be the most important factor limiting contractility on normothermic reperfusion; that is, supply may not meet demand (2). For example, KCl-induced vasoconstriction might result in an indaequate coronary flow

Advances in Pharmacology, Volume 31

Table I

Potential Approaches to Preserve Cardiac Function and Vascular Tone during Long-term Hypothermic Hypoperfusion and Reperfusion[a]

Decrease O_2 free radical damage	Decrease calcium influx
Decrease O_2 delivery	Decrease $[Ca^{2+}]_e$[b]
Allopurinol	Volatile anesthetics
Glutathione, ascorbic acid	Ca^{2+} blockers
Desferroxamine	BDM (higher levels)
SOD, catalase, vitamin E	
Decrease ion pump work	Decrease substrate loss
Increase $[K^+]_e$[b]	NBTI,[b] acadesine
Increase $[H^+]_e$	Dipyridamole
Hypothermia[b]	Pentostatin
Block Na^+/K^+ pump (ouabain)	
Block Na^+/H^+ pump (amiloride)	
K^+ agonists (nicorandil, cromakalim)	
Na^+ antagonists (tetrodotoxin, lidocaine)	
Counteract edema	Counteract increased resistance to flow
Mannitol,[b] sorbitol	Mannitol[b]
Raffinose	Nitroso vasodilators (NP)[b]
Lactobionate	Decrease $[K^+]_e$
Hyodroxyethylcellulose	Ca^{2+} antagonists
Dextran	Adenosine,[b] BDM[b]
Supply energy substrates	Decrease metabolic demands
Adenosine[b]	Hypothermia
Aspartamate, glutamate[b]	Decrease $[Ca^{2+}]_e$[b]
Pyruvate[b] (TCA cycle)	Raise $[K^+]_e$[b] (arrest heart)
	BDM[b] (uncouple $[Ca^{2+}]_i$ excitation from myofibrillar contraction)

[a] BDM, Butanedione monoxime; NBTI, nitrobenzylthioinosine; SOD, superoxide dismutase; TCA, tricarboxylic acid.

[b] Approaches used in our laboratory (Refs. 2, 6, 8, 9, 14).

despite cardiac arrest so that oxygen delivery may not match the decrease in metabolic demand. In addition, restoration of mechanical function may be dependent not only on restoration of basal coronary flow, but also on adequacy of coronary flow reserve. Vasodilation can be mediated through specific agents producing effects directly on vascular smooth muscle cells or indirectly through vascular endothelial cells (3,4). Prolonged cold storage has been shown to abolish endothelium-dependent relaxing responses in porcine coronary arteries (5). This has also been shown after hypothermic perfusion of isolated hearts (6).

Another experimental approach for preserving hearts is to perfuse isolated hearts for one day with a cold, but normal, extracellular ionic solution containing metabolic inhibitors and vasodilators. Several potential advantages of a normal ionic solution for preservation are (i) better maintenance of ionic gradients and ion pump function, (ii) prevention or attenuation of resting membrane depolarization and resulting intracellular Ca^{2+} overload, which can depress cardiac function and cause dysrhythmias with normothermic reperfusion, and (iii) shortening of the time to reestablish ionic equilibration. Hypothermia alone is inadequate for long-term cardiac protection because all metabolism is not halted. Hypothermia itself can impair ion pump activities so that intracellular Na^{+} and Ca^{2+} rise over time (7). Additional metabolic methods are required to protect hearts over long periods.

Studies from our laboratory show that 2,3-butanedione monoxime (BDM, also called diacetyl monoxime) protects isolated guinea pig hearts when BDM is added to a normal ionic Krebs–Ringer solution before, during, and initially after either 30 min of coronary occlusion (8) or 1 day of hypothermic perfusion (9). BDM is a small hydrophobic molecule that has little effect on cardiac electrical activity but has a marked effect on contractility when infused at concentrations of 10 m*M* of lower (10,11). BDM appears to greatly decrease myofibrillar Ca^{2+} sensitivity with a lesser effect to alter the uptake or release of Ca^{2+} from the sarcoplasmic reticulum (10,12,13). BDM is also a vasodilator (8,9). BDM may afford protection not only by reducing metabolic demand and by increasing coronary flow during prolonged hypothermic perfusion, but also by unknown direct intracellular effects. In models of both ischemia (8) and cold preservation (9), BDM enhanced functional recovery and decreased the incidence and severity of dysrhythmias. Additional studies showed that the improvement in recovery following perfusion with added BDM is enhanced even more when adenosine (9) or nitrobenzylthioinosine (6) is given with nitroprusside just before, and initially after, normothermic reperfusion during rewarming. This approach to cardiac preservation, using metabolic inhibitors and vasodilators, may minimize vasoconstriction and effectively suppress contractile function both during low-flow, cold perfusion and during the initial period of normothermic reperfusion.

The purpose of this article is to compare several treatments for preserving hearts after long-term hypothermic perfusion. Complete preservation could be defined as elimination of ventricular dysrhythmias and restoration of oxygen extraction, contractility, contractile response to epinephrine, coronary flow, efficiency of oxygen utilization, and responses to endothelium-dependent and endothelium-independent vasodilators on normothermic reperfusion. Four hypothermia groups of isolated guinea pig hearts were perfused at about one-third normal coronary flow at 3.8°C for 23 hr. Three of the groups were perfused continuously with a normal extracellular solution containing 5 m*M* KCl: one group served as the treatment control; another group was perfused with BDM before, during, and after hypothermic perfusion; and a third group was perfused before, during, and after hypothermic perfusion not only with BDM, but also with adenosine, and additionally with nitroprusside given only during the initial reperfusion period along with BDM and adenosine. A fourth hypothermia group was perfused before, during, and after hypothermic perfusion with 20 m*M* KCl ("cardioplegic solution"). A fifth normothermia group was not hypothermically perfused and served as the time control group. Inotropic and chronotropic responsiveness to epinephrine, as well as coronary vasodilatory responsiveness to acetylcholine (endothelium-dependent vasodilator), nitroprusside (endothelium-independent vasodilator), and adenosine (a mixed effect vasodilator), were tested at the beginning and end of each experiment in all groups).

II. Long-Term Cardiac Perfusion

Albino English short-haired guinea pigs (400–600 g) were injected intraperitoneally with 10 mg ketamine and 1000 units heparin, and decapitated when unresponsive to noxious stimulation. Isolation and preparation of hearts as used for this study have been detailed in other reports (2,6,8,9,14). The inferior and superior venae cavae were cut after thoracotomy, and the aorta was cannulated distal to the aortic valve. Each heart was immediately perfused retrogradely through the aorta and excised. All hearts were perfused initially at a constant aortic root perfusion pressure of 55 mmHg. The perfusate, a modified Krebs–Ringer solution, was filtered (5 μm pore size) in-line (Astrodisc, Gelman Scientific, Ann Arbor, MI) and had the following control composition (in m*M*): Na^+ 137, K^+ 5, Mg^{2+} 1.2, Ca^{2+} 2.5, Cl^- 134, HCO_3^- 15.5, $H_2PO_4^-$ 1.2, pyruvate 2, glucose 11.5, mannitol 16, glutamate 0.05, EDTA (ethylenediaminetetraacetic acid) 0.05, and insulin (5 units/liter).

Left ventricular pressure (LVP) was measured isovolumetrically with a transducer connected to a thin saline-filled latex balloon that was inserted into the left ventricle through the mitral valve from a cut in the left atrium. Balloon volume was adjusted to maintain a diastolic LVP of 0 mmHg during the initial control period. Two pairs of bipolar electrodes (Teflon-coated silver, diameter 125 μm) were placed in each heart to monitor intracardiac electrograms from which spontaneous sinoatrial rate and atrial–ventricular (AV) conduction time were measured as noted previously (15). Coronary sinus effluent was collected by placing a cannula into the right ventricle through the pulmonic artery after ligating the venae cavae. Coronary inflow (aortic) was measured at constant temperature by an electromagnetic flowmeter (1.5 mm ID) which was calibrated daily by four timed collections into a volumetric cylinder over the flow range of 0 to 24 ml/min. Zero inflow was periodically established by temporarily bypassing the flow transducer.

Coronary inflow and outflow (coronary sinus) O_2 tensions were measured continuously on-line (Instech 203B, Instech Laboratories, Plymouth Meeting, PA) and verified simultaneously off-line with an intermittently self-calibrating analyzer system (Instrumentation Labs 813, Chicago, IL) as described previously (2,6,8,9,14). Oxygen consumption and percent oxygen extraction were measured in all studies to assess the direct vasodilatory response of drugs apart from the response due to autoregulatory factors, for example, a decrease in coronary flow and oxygen delivery secondary to decreased contractility and oxygen consumption. Use of this measurement assumes that local metabolites are produced in proportion to myocardial oxygen consumption, and that local metabolites are major factors controlling autoregulation of coronary flow. The percentages of O_2 extraction, O_2 consumption, and cardiac efficiency were calculated as reported previously (9).

Perfusate and bath temperature were maintained at 37.2 ± 0.1°C before and after hypothermia using a thermostatically controlled water circulator. During the 23-hr hypothermic perfusion period, perfusate and bath temperature were maintained at 3.8 ± 0.1°C. A switch to hypoperfusion at 3.8°C was accomplished by use of a separate refrigerated jacket and perfusion circuit (VWR Scientific 1160, Chicago, IL) placed in parallel with the warm perfusion circuit. Normothermic perfusion for 4 hr at 37.2 ± 0.1°C following cold perfusion was reinstated by switching back to the warm circuit. Warm and cold perfusion circulation circuits were temperature equilibrated in advance. The time to reach half the temperature fall from 37.2 to 3.8°C was 5 min. On lowering the temperature, at 15°C, cardiac perfusion was switched from constant pressure to a low constant flow (1.8 ml/g-min) of about one-third the baseline normothermic flow during constant pressure perfusion. On raising the temperature, at 25°C following hypothermia, cardiac perfusion was returned to the constant pressure mode. Coronary vascular resistance was measured at constant perfusion pressure before and after hypothermia, and at constant flow during hypothermia, and was defined as coronary perfusion pressure divided by coronary flow normalized to wet heart weight. The time to reach half the temperature rise from 3.8 to 37°C was 3 min. Warm and cold perfusate solutions were equilibrated with a gas mixture of 96% O_2 and 4% CO_2.

For hearts in all groups during the initial normothermic period, mean coronary arterial (inflow) pH was 7.44 ± 0.02 (mean ± SEM), p_{CO_2} was 27 ± 1 mmHg, and p_{O_2} was 560 ± 13 mmHg; samples, collected at 3.8°C during the hypothermic period and measured at 37°C, had values of 7.15 ± 0.03, 46 ± 2 mmHg, and 786 ± 18 mmHg, respectively. There were no significant differences for these values among the groups at each the two temperatures.

Electrograms, heart rate, AV conduction time, outflow O_2 tension, coronary flow, LVP and coronary perfusion pressure were tape recorded (model D1, A. R. Vetter, Rebersburg, PA) for later detailed analysis. All measured variables were displayed on a fast writing (3 kHz), thermal array 8-channel recorder (Astro-Med, model MT9500, West Warwick, RI). Calculated variables were computed using a software program (Microsoft Excel, Microsoft Corp, Redmond, WA). Hearts were weighed immediately after each experiment (28 hr for hypothermia groups and 6 hr for the time control group), and dehydrated weights determined to calculate dry heart weight were expressed as percentage of wet heart weight.

Peak coronary flow responsiveness was tested before and after hypothermic perfusion by maximally vasodilating the vasculature with adenosine (0.2 ml of a 200 μM solution) injected into the aortic (coronary perfusion) cannula. Percent oxygen extraction was measured to assess direct vasodilatory responses as differentiated from those due to an autoregulatory response. Interpretation of the data relies on the assumption that production of local metabolites is proportional to myocardial oxygen consumption, and that local metabolites are major factors controlling autoregulation of coronary flow (16). Thus, an imbalance of oxygen consumption to oxygen delivery reflects a change in coronary vascular tone. A submaximal concentration of epinephrine hydrochloride, 0.5 μM, was infused for 2 min before and after the hypothermic period to test chronotropic and inotropic responsiveness. Coronary responsiveness to 1 μM acetylcholine (Ach) and 100 μM sodium nitroprusside (NP) (Nipride, Abbott Labs, Chicago, IL), indicators of endothelium-dependent and -independent vasodilation, respectively, were tested during 2-min infusions 30 min before cold perfusion and 60 min after rewarming (i.e., 30 min after reperfusion) with standard Krebs–Ringer solution). For Ach, flow changes were measured during initial (m, maximal) exposure, during the steady-state response at spontaneously (s) slowed heart rates (146 ± 14 beats/min), and during pacing (p) at 240 beats/min to approximate intrinsic heart rates. At the end of each experiment wet and dry hearts were obtained to determine relative changes in cardiac water.

Once isolated, hearts (n = 65) were assinged randomly to one of five groups of 11–15 hearts each: Group 1, time control (normothermia, normal ionic solution), Group 2, 10 m*M* 2,3-butanedione monoxime (BDM) (Sigma, St. Louis, MO, or Aldrich, Milwaukee, WI) with added adenosine and NP; Group 3, 10 m*M* BDM (hypothermia, normal ionic solution); Group 4, treatment control (hypothermia, normal ionic solution) and; Group 5, 20 m*M* KCl (hypothermia, high-KCl solution). For Groups 2, 3 and 5, changes in perfusate were made 20 min before hypothermic perfusion, during 22 hr of hypothermic perfusion, and for 30 min after restoring normothermic perfusion. For Group 2, 10 μM adenosine (ADE) (Sigma or Aldrich) was infused with BDM before, during, and after hypothermic perfusion, and 100 μM NP was infused only during the 30-min reperfusion period. Higher concentrations of NP and ADE than those reported in this study did not additionally increase coronary flow.

All variables were measured during the last minute of (a) a 30-min initial control period, (b) an initial 2-min epinephrine infusion period 30 min later, (c) initial test infusions of Ach and NP, (d) the prehypothermic control period (C or C1), (e) controls or onset of treatment with BDM, BDM plus ADE, of high KCl before hypothermic perfusion, (f) infusions of BDM, BDM plus ADE, and NP initially after hypothermic perfusion, (g) every 30-min period during normothermic reperfusion for 3.5 hr, (h) repeat test infusions of ACH and NP, and (i) a posthypothermia, repeated epinephrine infusion period 2 hr after normothermic reperfusion. Because rewarming provoked ventricular dysrhythmias in many hearts at about 25°C, all hearts in each group received prophylactically one bolus injection of 0.1 ml of 10 mg lidocaine hydrochloride during rewarming at 25°C to reduce the occurrence of dysrhythmias. The time control group underwent the same drug protocol as the hypothermia treatment control group except that there was no period of hypothermia.

All data are expressed as means plus or minus standard error of the means (SEM). Mean values were considered significant at $p < 0.05$. Significance was determined by analysis of variance with repeated measures; if the *F*-test was significant, Fisher's least significant difference test was used to compare means (Statview, Abacus Concepts, Berkeley, CA). Derived variables and changes from controls were computed (Microsoft, Excel). Among-group (treatment effects) comparisons were used to assess differences among the five groups for a given variable at a specific time or test interval. For each variable the following groups were statistically compared: BDM, ADE, NP (Group 2) versus time control (Group 1), denoted in Figs. 1–8 as "†"; BDM (Group 3) versus BDM, ADE, NP (Group 2), "§"; treatment control (Group 4) versus BDM (Group 3), "¶"; and 20 m*M* KCl (Group 5 versus treatment control, Group 4), "‡." Within-group comparisons (time or test effects) were used to assess differences within a given group from the initial, prehypothermic control (C) "*", or from the control before (C1) or after (C2) "●." Software programs were run on a personal computer. Significance for the incidence and duration of dysrhythmias was determined by chi-squared (χ^2) and t-tests, respectively. The expected incidence of dysrhythmias was assumed to be zero.

III. Preservation of Isolated Hearts

A. Atrial Rate, Atrial–Ventricular Conduction, and Cardiac Rhythm

Atrial rate was similar in each group initially (C) and increased similarly in each group with epinephrine (EPI) infusion (Fig. 1). After initiating treatment before hypothermic perfusion (at 1 hr), atrial rate ceased in the KCl group. Between approximately 17 and 12°C, during cooling to 3.8°C, atrial and ventricular electrical and mechanical activity ceased in each hypothermia group without fibrillation following a slowing of pacing and development of AV dissociation (data not displayed). Hearts remained quiescent during 22 hr of hypothermia, and slow atrial activity began to occur earlier in the cold treatment control group than in the two BDM

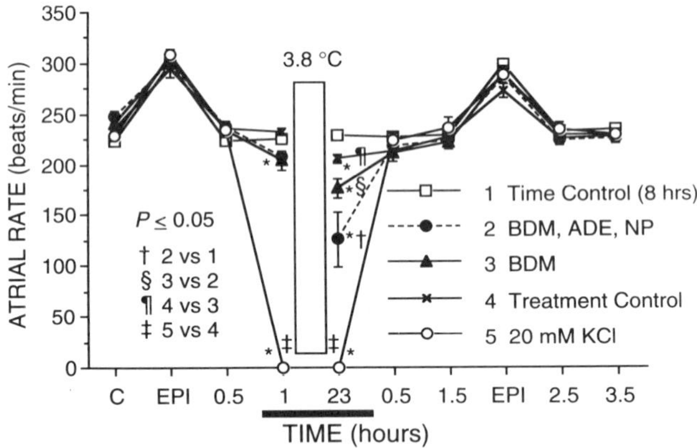

Fig. 1 Spontaneous atrial rate in five groups of hearts before and after long-term (1 to 23 hr) hypothermic perfusion. Time control hearts (Group 1) were perfused without treatment or hypothermia for 8 hr. Group 2 hearts were infused (1 to 23 hr) with BDM (2,3-butanedione monoxime) and ADE (adenosine) before, during, and, with NP (nitroprusside), initially after hypothermic perfusion. Group 3 hearts were similarly infused (1 to 23 hr) with BDM alone. Group 4 hearts served as hypothermia alone treatment controls. Group 5 hearts were infused (1 to 23 hr) with 20 mM KCl (normal 5 mM KCl) before, during, and after hypothermic perfusion. EPI denotes the response to epinephrine infusion. Data are expressed as means ± SEM. Not all statistical symbols are noted. See text for additional details.

groups at approximately 20°C during rewarming. At 25°C lidocaine was given prophylactically to all hearts to temporarily arrest hearts not arrested. During the initial phase of rewarming (Fig. 1) with continued treatments (at 23 hr), atrial rate was lower in the BDM, ADE, NP group than in the BDM group, and lower in the treatment control group than in the BDM group; in the KCl group hearts were arrested. After discontinuation of drug treatment, atrial rate was similar in all groups. Responses to epinephrine infusion were similar among all groups after hypothermia.

Control AV time was similar for all groups (65 ± 2 msec). High-KCl solution caused AV dissociation initially on reperfusion with normal ionic solution (data not displayed). In all groups except the time control and BDM groups, there were severe cardiac ventricular dysrhythmias (ventricular fibrillation, tachycardia, and electromechanical dissociation) not converting to sinus rhythm with repeated lidocaine injection on rewarming. The incidence of ventricular fibrillation compared to sinus rhythm was significant ($p < 0.05$) in the treatment control (36%) group but not in the KCl group (22%). All hearts remained in or reverted to sinus rhythm by 2 hr of warm reperfusion. The AV dissociation occured in most hearts in the BDM, ADE, NP group before, during, and initially after hypothermic perfusion; otherwise, hearts were in sinus rhythm.

B. Left Ventricular Pressure and Cardiac Efficiency

Developed (systolic minus diastolic) LVP (Fig. 2), initially similar in each group, increased about 20% with epinephrine infusion in each group, decreased similarly by 82 to 85 ± 1% in the two BDM groups, respectively, and decreased by 100% (arrested) in the KCl group. After reperfusion, diastolic LVP was continuously elevated in cold treatment control (8 ± 3 mmHg) and KCl (9 ± 3 mmHg) groups but not in the BDM groups or in the time control group (data not displayed). Developed LVP was initially greatly depressed on normothermic reperfusion during treatment in all hypothermia groups. Over time with discontinuation of treatment at 23 hr, LVP was similarly and most depressed in the treatment control and KCl groups. LVP was significantly improved after treatment in the BDM alone group, and LVP approached the time control values after treatment in the BDM, ADE, NP group. LVP decreased 15% in the time control group over 8 hr ($p < 0.05$). Compared to the prehypothermia perfusion period, the responsiveness to epinephrine was blunted in all groups including the time control group.

Relative cardiac efficiency (Fig. 3), an index of oxygen consumed per unit of developed pressure per beat, was initially similar (C) in all groups. Cardiac efficiency increased by approximately 15% with infusion of epi-

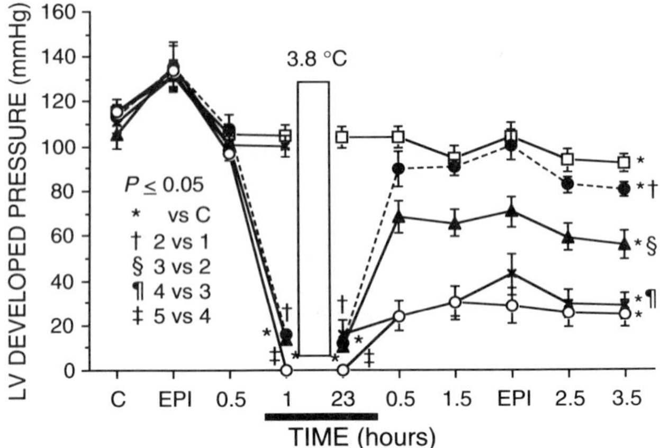

Fig. 2 Left ventricular (LV) developed (systolic − diastolic) pressure in five groups of hearts before and after long-term (1 to 23 hr) hypothermic perfusion. Note highest LV pressure in the BDM, ADE, NP treated group after hypothermia. See Fig. 1 and text for other details.

nephrine, decreased in the three treated groups before hypothermia (at 1 hr), and remained low in all hypothermia groups with initial warm reperfusion (at 23 hr). After ending treatments, relative cardiac efficiency remained lowest in the treatment control and KCL groups, increased in the BDM alone group, and was equivalent to that of the time control in the BDM, ADE, NP group.

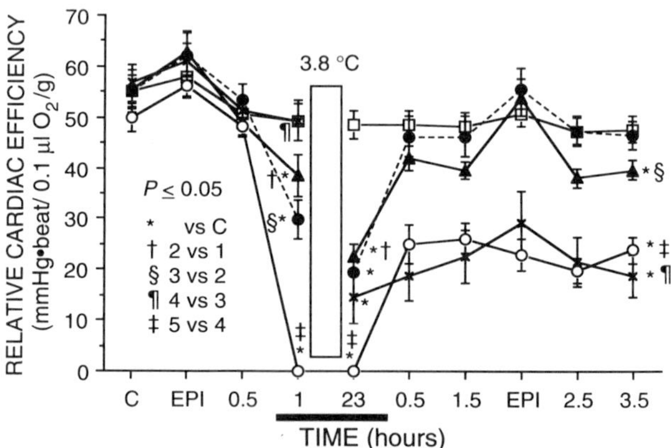

Fig. 3 Relative cardiac efficiency in five groups of hearts before and after long-term (1 to 23 hr) hypothermic perfusion. Note highest utilization of oxygen in the BDM, ADE, NP group after hypothermia. See Fig. 1 and text for other details.

C. Coronary Flow and Oxygen Extraction

Coronary flow (Fig. 4) was similar in each group initially (C). Initial adenosine injection (ADE) increased flow maximally about 2-fold in each group. On initial treatment with high-KCl perfusate (at 1 hr), coronary flow markedly decreased in the arrested hearts. Flow increased significantly only in the BDM, ADE, NP group (statistic not shown) on changing to the treatment solution (at 1 hr). On normothermic reperfusion, flow was reduced in all hypothermia groups both initially (at 23 hr), and throughout the reperfusion period with normal perfusate. Of the hypothermic groups, flow remained lowest in the high-KCl-treated group and was highest in the BDM, ADE, NP group throughout the reperfusion period.

Coronary vascular resistance (Fig. 5) was initially similar for all groups (C) and increased only with KCl treatment (at 1 hr). During and after hypothermic perfusion vascular resistance remained greatly increased only in the KCL group. Resistance was lowest in the BDM, ADE, NP group during hypothermic perfusion and approached values obtained in the time control group after reperfusion.

Percent oxygen extraction (Fig. 6) generally changed in opposite direction to coronary flow. Before hypothermia (at 1 hr), percent oxygen extraction markedly decreased only in the two BDM groups. Initially after hypothermia with continued treatment (at 23 hr), percent oxygen extraction remained lower than initial controls (C) in both BDM groups, whereas it was elevated above control in the KCl group. During reperfusion with normal ionic perfusate, percent oxygen extraction was increased most in

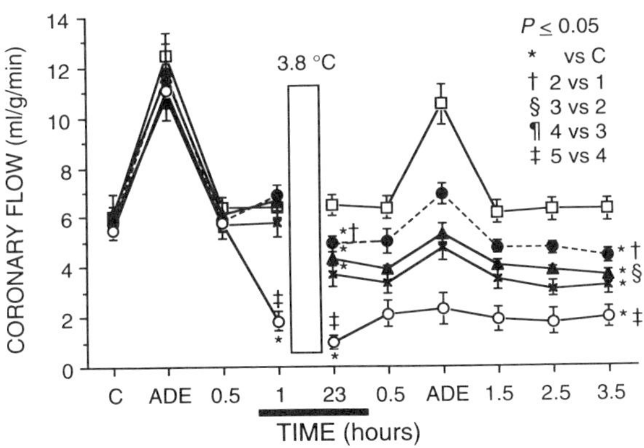

Fig. 4 Coronary flow in five groups of hearts before and after long-term (1 to 23 hr) hypothermic perfusion. ADE denotes the transient flow response to adenosine. Note lowest flow in the KCl group and highest flow (of hypothermic groups) in the BDM, ADE, NP group after hypothermia. See Fig. 1 and text for other details.

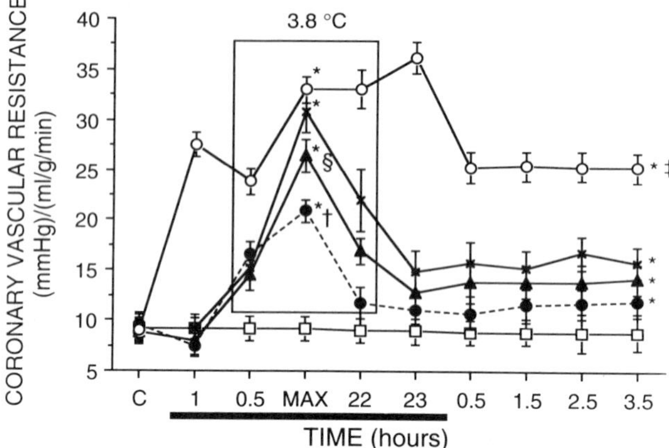

Fig. 5 Coronary vascular resistance in five groups of hearts before, during, and after long-term (1 to 23 hr) hypothermic perfusion. During hypothermia hearts were perfused at constant flow equivalent to about one-third normothermic natural (constant perfusion pressure) flow. MAX denotes the highest resistance recorded during the 23-hr hypothermic perfusion period. Note highest resistance to flow in the KCl group and lowest resistance to flow in the BDM, ADE, NP group after hypothermia. See Fig. 1 for other details.

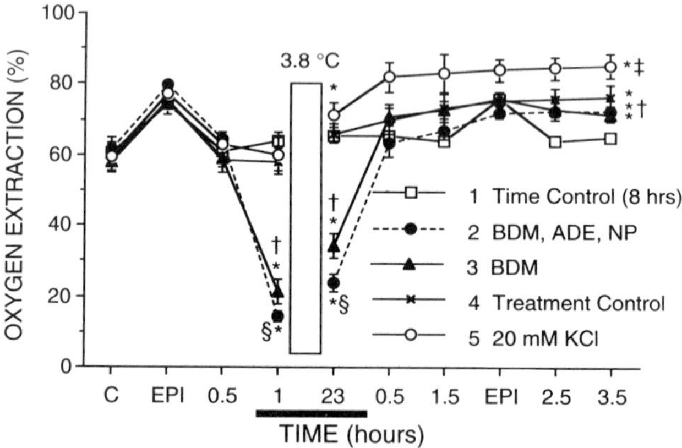

Fig. 6 Percent oxygen extraction in five groups of hearts before and after long-term (1 to 23 hr) hypothermic perfusion. Note highest oxygen extraction in the KCL-treated group after hypothermia. See Fig. 1 for other details.

the KCL groups compared to all other groups; oxygen extraction was not additionally increased by infusion of epinephrine except in the time control group.

D. Responses to Vasodilators

Coronary flow and percent oxygen extraction responses to brief infusions of three vasodilators, namely adenosine (Ade), acetylcholine (Ach), and nitroprusside (NP), given before and after hypothermic perfusion, are shown in Figs. 7 and 8, respectively. Coronary flow was initially similar in all groups (Cl), and each group responded similarly to infusion of vasodilators. Averaged for all groups, adenosine increased flow about 100% before hypothermic perfusion (Fig. 7, top). Continuous infusion of acetylcholine initially increased flow by 68% (maximally, m), next increased steady-state (s) flow by 28% at a lowered atrial rate of about 150 beats/min, and then increased flow by 19% during ventricular pacing (p) at 240/min (intrinsic rate 230/min). Infusion of nitroprusside increased flow 35% at a mean atrial rate of 260 beats/min.

One hour (C2) after reperfusion with normal ionic perfusate (Fig. 7, bottom), flow was reduced in all hypothermia groups (most in the KCl group and equivalently so in the treatment control and BDM alone groups (and was reduced least in the BDM, ADE, NP group. Similar results are shown in Fig. 4 at 30 min after hypothermia. In the nonhypothermia time control group, responses to these vasodilators persisted, but were attenuated, 5 hr later. Posthypothermia responses to vasodilators were absent in KCl and treatment control groups. Responses to Ade, Ach_m, Ach_s, and NP were greatly attenuated but significantly different from controls (C2) after hypothermia in the BDM alone group. In the BDM, ADE, NP group, flow increased significantly with Ade and NP testing by 48 and 21%, respectively, compared to the posthypothermia control (C2); these responses were attenuated compared to time control values.

Percent oxygen extraction, average for all groups before cold perfusion (Fig. 8, top), similarly decreased by about 48 and 27% with acetylcholine before and during pacing, respectively, and similarly decreased by about 26% with nitroprusside. Oxygen extraction was not stable during bolus injection of adenosine or during the initial maximal flow response to acetylcholine and was not recorded.

After hypothermia (Fig. 8, bottom), basal percent oxygen extraction (C2) was significantly elevated only in the KCl group. Similar results are shown in Fig. 6 at 30 min after hypothermia. For all groups, percent oxygen extraction did not decrease as much after hypothermia as before

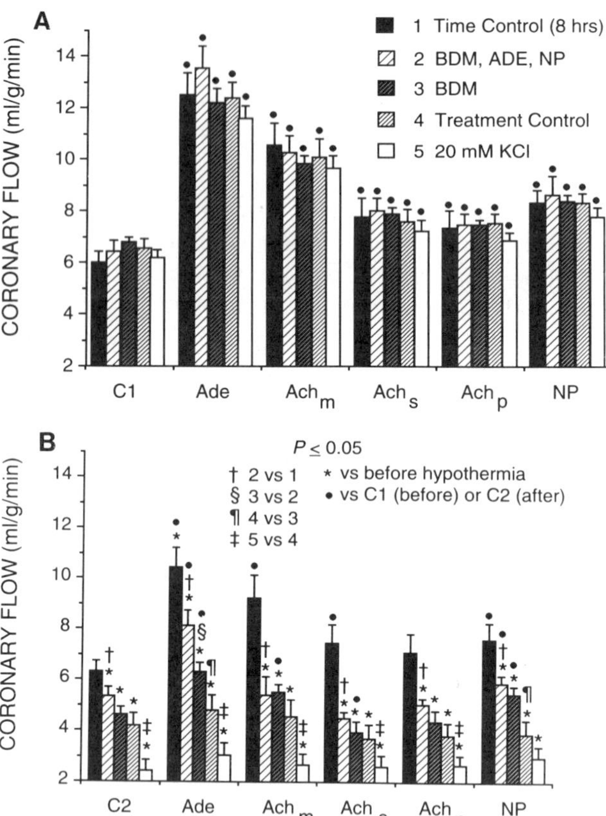

Fig. 7 Coronary flow responses to bolus intraaortic (coronary) injection of adenosine (Ade), infusion of acetylcholine (Ach) (m, s, and p represent maximal, spontaneous rate, and paced—240 beats/min—rate, respectively), and infusion of nitroprusside (NP) before and after long-term hypothermic perfusion in five groups of hearts. Time control hearts (8 hr) were not subjected to hypothermia. Note equivalent responses to endothelium-dependent (Ach), endothelium-independent (NP), and mixed (Ade) vasodilators among the hypothermic groups before (A) hypothermia and the attenuated flow responses after (B) hypothermic perfusion. See text for additional details.

hypothermia in response to acetylcholine and nitroprusside. However, percent oxygen extraction decreased significantly by 17 and 12% compared to posthypothermia controls (C2) during spontaneous atrial rate with acetylcholine in the BDM, ADE, NP group and BDM alone group, respectively. In treatment control and KCl groups, oxygen extraction did not change significantly during testing with acetylcholine or nitroprusside. In

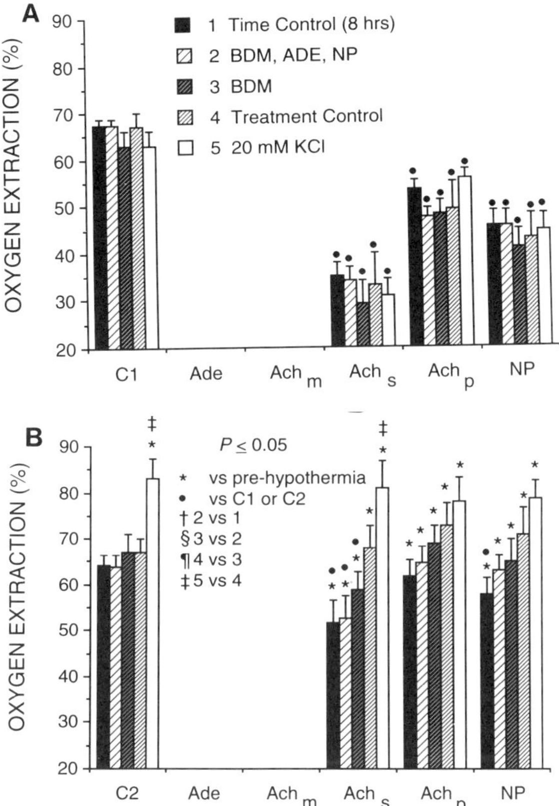

Fig. 8 Percent oxygen extraction responses to infusions of Ach and NP before (A) and after (B) long-term hypothermic perfusion in five groups of hearts. Responses were not measured during bolus injection of adenosine or during the initial response to Ach. Time control hearts (8 hr) were not subjected to hypothermia. Note equivalent responses to endothelium-dependent (Ach) and endothelium-independent (NP) vasodilators among the hypothermic groups before hypothermia and the attenuated responses after hypothermic perfusion. See text for additional details.

the nonhypothermia, time control group, the decreases in percent oxygen extraction were attenuated, but persisted, with acetylcholine and nitroprusside testing 5 hr later.

Dry heart weight as a percentage of wet heart weight was not significantly different in the time control (13.4 ± 0.5%), BDM, ADE, NP (13.6 ± 0.5%), BDM alone (13.6 ± 0.4%), treatment control (13.4 ± 0.4%), or high KCl (13.1 ± 0.3%) groups.

IV. Discussion

The major objective was to determine which of four cold perfusate solutions best preserves myocardial perfusion and function. The results indicate that low-flow cardiac perfusion with a normal ionic solution containing BDM, ADE, and NP gives better long-term protection in *in vitro* hearts than solutions containing BDM alone or high KCl; moreover, high-KCl solution protects no better than extracellular solutions containing normal KCl. This assessment is based on a reduced incidence of dysrhythmias, on improved left ventricular pressure development and flow responses, and on other metabolic indices measured during normothermic reperfusion after 23 hr of hypothermic perfusion. In general, the following results were obtained: atrial rate and AV conduction returned similarly in all groups; ventricular dysrhythmias were frequent in each cold group except the two BDM groups; developed LVP and cardiac efficiency approached the time control only in BDM groups and remained equivalently low in the cold treatment control and high KCl groups; coronary flow was most nearly restored to time control levels in the BDM, ADE, NP group and was farthest removed from time control levels in the high-KCl group; and percent oxygen extraction increased in all groups but was highest in the KCl group. Moreover, flow responses to endothelium-dependent and endothelium-independent vasodilators, Ach and NP, respectively, and the mixed vasodilator Ade disappeared in all except the two BDM groups wherein responses, although present, were blunted. There was no relative uptake or loss of cardiac water among the groups.

One initial aim was to produce equivalent levels of cardiac depression with BDM or KCl before inducing long-term hypothermic perfusion. Treatment with BDM plus ADE or BDM alone decreased contractility approxiamtely 84%, whereas high KCl decreased contractility 100% (cardiac arrest) before hypothermic perfusion. However, high-KCl treatment caused a marked fall in coronary flow with no net reduction in oxygen extraction. On the other hand, perfusion with BDM alone, or with BDM plus ADE plus NP, caused a relative vasodilation before and after hypothermic perfusion. This is shown by similar decreases in percent oxygen extraction and, for the BDM, ADE, NP group, a demonstrable increase in flow. Therefore, the oxygen supply to demand ratio was about three times higher for BDM groups than for the KCl group, despite cardiac arrest, or for the untreated cold group. This could explain why the high-KCl solution did not protect hearts any better than the control ionic solution. It seems likely, then, that the marked improvement in contractile function after cold preservation with BDM is due to more than an induced reduction of contractility and oxygen consumption before, during, and after cold perfusion. Indeed, a low-$CaCl_2$ solution, which reduces contractility and oxygen extraction the same as BDM, does not restore function much better than a control solution (2). Moreover, it may be that the vasoconstriction and higher oxygen extraction ratio caused by the high-KCl solution may be detrimental during reperfusion, particularly if intracellular Ca^{2+} is elevated.

Many approaches have been taken to preserve hearts for long periods. The most common approach is to arrest hearts with a high potassium "cardioplegic" solution. Continuous perfusion with a high-KCl solution, however, causes calcium loading of cardiac cells (17). This is thought to occur because prolonged membrane depolarization promotes calcium influx by activating voltage-sensitive slow calcium channels and voltage sensitive sodium–calcium exchangers (17–19). When the sarcoplasmic reticulum (SR) is saturated with Ca^{2+}, Ca^{2+} from extracellular fluid may bypass the SR to potentiate contraction and cause diastolic contracture (18). Indeed, high-KCl-arrested hearts have a higher basal oxygen consumption, which is likely calcium dependent, than do very low-$CaCl_2$-arrested hearts (20,21). In the present study a very low level of cardiac efficiency was associated with the low flow after KCl treatment. This indicates that the amount of oxygen consumed for a given amount of LV isovolumetric pressure developed per beat is higher after KCl treatment. Enhanced contracture is not observed because of cardiac cell depolarization with KCl, but may be observed on washout of the high-KCl solution as an elevation of diastolic pressure as observed in this study and in another (20) study. If intracellular Ca^{2+} is high on reperfusion, abnormal repolarization potentials result, and this can allow random reentrant pathways to develop into ventricular fibrillation (22).

Another approach is to decrease intrinsic metabolic activity before and during cold perfusion with a reversible intracellular metabolic inhibitor, without changing the extracellular ionic composition. Butanedione monoxime (BDM), also called diacetyl monoxine (DAM), was chosen because it is a potent negative iontropic agent in papillary muscle (10) and isolated hearts (2,8–10,14) that causes little change in electrical activity at concentations up to 10 m*M*. BDM appears to have little effect on the slow calcium inward current but to have prominent effects on other "downstream" factors involved in excitation–contraction coupling.

Although the specific site(s) of action for BDM has not been fully elucidated (10,13,23), it now appears that BDM exerts a major effect on contractile system proteins. Specifically, BDM may decrease responsiveness of troponin C to Ca^{2+} (13). BDM may also depress mobilization of calcium from the SR and reduce sarcolemmal Ca^{2+} influx at higher concentrations (10,13). Along with its negative inotropic effect, BDM causes vasodilation (2,6,8,9,14). Thus, BDM appears to offer protective effects by enhancing oxygen supply as well as by reducing oxygen demand.

Earlier studies from our laboratory indicated that BDM, given before, during, and initially after cold perfusion to depress cardiac function, greatly attenuated the deleterious functional and metabolic effects of prolonged hypothermic perfusion (9) and the BDM better preserved cardiac function and coronary flow responses than either low-$CaCl_2$ or high-KCl solutions (2). The present study demonstrates additionally that ADE and NP, when infused with BDM, offer much better protection than high KCl against the deleterious functional and metabolic effects of prolonged hypothermic perfusion. A dissociation between oxidative metabolism and contractile function after postischemic reperfusion of the myocardium is well documented (24). The relative amount of potential work (isovolumetric developed LVP times heart rate) for a given rate of oxygen consumption was considerably improved in both BDM groups compared to high-KCl or treatment control groups. This may be accounted for by the improved basal coronary flow with BDM alone, and with ADE and NP, and by the absence of diastolic contracture after hypothermic perfusion in these groups.

The greatest insult attenuating the return of myocardial function could occur during the initial period of normothermic reperfusion, as has been demonstrated for reperfusion following coronary occlusion (24,25), Damage to the vascular system may preclude damage to myocardial contractile tissue. A major factor retarding full return of contractile function could be inadequate coronary vasodilator capacity. Indeed, the present study shows that both contractility and coronary flow during reperfusion after hypothermic perfusion were improved best after treatment with BDM, ADE, and NP. Moreover, this improvement in flow likely underlies the observed improvement in relative cardiac efficiency, that is, potential cardiac work per rate of oxygen consumption. Although coronary responses to Ach, Ade, and NP were severely blunted in all hypothermia groups, they were improved most in the BDM, ADE, NP group. Although the responses to Ach were small, this suggests a continued release of NO by vascular endothelium to induce vasodilation. In the endothelium Ach generates inositol 1,4,5-trisphosphate (IP_3), which elicits Ca^{2+} release from intracellular stores (4). The increase in Ca^{2+}, which bind to calmodulin, forms one of several cofactors required to activate nitric oxide synthase (NOS) to produce NO from L-arginine. NO diffuses to adjacent vascular smooth muscle and binds to iron in the heme at the active site of guanylyl cyclase which generates cGMP, causing relaxation. NO is constitutively expressed normally and accounts for a portion of basal coronary flow. This was demonstrated in another study by the decrease in coronary flow in the presence of nitro-L-arginine methylester (L-NAME) (14) and by similar results from other laboratories (26,27).

Adenosine may offer beneficial effects during cardiac preservation techniques. In the present study ADE was infused with BDM before and during hypothermia as well as after hypothermia. Although not detailed here statistically, this resulted in a better return of LV pressure after hypothermia than when ADE was infused only during the rewarming period (9). Moreover, coronary vascular resistance was lower during and after hypothermia in the BDM, ADE, NP group than in the BDM alone group. ADE exerts a negative chronotropic effect, which is mediated through A_1 receptors coupled to K^+ channels through a pertussis toxin-sensitive G protein to cause membrane hyperpolarization, and a negative iontropic effect, also mediated through A_1 receptors but coupled to adenylyl cyclase activity (28). ADE has been shown to improve cardiac contractile function during reperfusion following ischemia (29,30) and continuous cold perfusion (31) as well as to prevent free radical generation by neutrophils through A_2 receptors, thereby protecting vascular endothelium from damage by neutrophils (32). In an earlier series of studies from our laboratory, equivalent results were obtained by substituting nitrobenzylthioinosine, a nucleoside transport inhibitor, for ADE when either drug was given with BDM and NP during the initial normothermic reperfusion period (6). It can be inferred that the most important beneficial effect of ADE during preservation in this model, with hearts already metabolically depressed by hypothermia and BDM, is its vasodilatory effect. Primarily, ADE relaxes vascular smooth muscle through G_s-coupled A_2 receptor stimulation of adenylyl cyclase to form cyclic AMP; this effect ultimately leads to a reduced Ca^{2+} effect on contractile proteins (28). The effect of ADE may be at least partially endothelium dependent, as noted by the reduced flow with ADE in the presence of L-NAME in another study from our laboratory (14) and as observed by others (24).

Nitroprusside may furnish additional protection of coronary vascular tone in a manner different from that of ADE. *In vitro,* NP, unlike ADE, has no inotropic effect and only a small positive chronotropic effect, so its major effect is relaxation of vascular smooth muscle. NP is really an exogenous NO donor and could improve flow after reperfusion by enhancing vasorelaxation. Evidence indicates that endogenous vasodilation by endothelial-dependent relaxing factor (EDRF), which is now known as NO, is attenuated during early reperfusion following ischemia (33–35). It is notable that myocardial ischemia–reperfusion injury is decreased by administration of NO donors (36). The small direct vasodilator response to NP following prolonged hypothermia in the two BDM groups of the present study show that smooth muscle can be direcly relaxed. However, because the Ade and NP responses were greatly attenuated, it could not be ascertained whether the defect was greater in vascular muscle or in vascular endothelium. Finally, because there was no relative uptake of water in the treated groups compared to the time control group, global myocardial edema is unlikely to be a cause for the meager vascular responses. However, the distribution of water may have been altered so that cellular versus interstitial edema cannot be ruled out.

A limitation of this study is use of a small isolated heart preparation to examine the mechanisms of cardiac preservation. Although the *in vitro,* crystalloid perfused heart has adequate coronary and mechanical reserves, these are less than in the *in vivo,* blood-perfused heart. In addition, blood-borne factors, such as platelets, may play a significant role in reperfusion injury following hypothermia. Moreover, only potential (pressure) work was measured, and kinetic (volume) work may not be so well preserved. It will be important to examine such approaches for long-term cardiac preservation using *in vivo* animal models to validate the potential success of transplanting the human donor heart harvested many hours earlier.

In summary, the best return of cardiac mechanical function, cardiac efficiency,and coronary flow following hypothermic perfusion was found with normal extracellular solution containing BDM, ADE, and NP. A high-KCl solution afforded no better protection against mechanical dysfunction in *in vitro* hearts after long-term hypothermic perfusion relative to a normal extracellular solution. Moreoever, the vasoconstriction induced by high KCl may be additionally detrimental to restoring function. Because responses to Ade, Ach, and NP were blunted after reperfusion in all hypothermia groups, it appears that both the endothelium and vascular smooth muscle are damaged. It follows that if coronary flow is sufficiently reduced because of damage to resistance vessels or to their endothelial lining on reperfusion, myocardial contractility would likely be depressed as oxygen supply decreases relative to demand.

Aknowledgments

Other principal participants in this research project are Drs. Mladen Boban, Berhard M. Graf, Helmut Habazettl, Barbara Palmisano, Zeljko J. Bosnjak, and John P. Kampine. We are especially grateful to James S. Heisner for help in conducting these studies, and to Susan Lawrence, David Chang, and Jolene Andryk for laboratory assistance. Research was supported in part by grants from the American Heart Association of Wisconsin (88-GA-06), the National Institutes of Health (HL 34708), and an Anesthesiology Research Training Grant (GM 08377).

References

1. Hausen, B., Fieguth, H. G., Schafers, H. J., Winter, A., Spring, E. A., and Haverich, A. (1988). Distant heart procurement: Impacts of storage solution composition on cardiac performance following transplantation. *Transplant. Int.* **1,** 140–145.
2. Stowe, D. F., Boban, M., Graf, B. M., Kampine, J. P., and Bosnjak, Z. J. (1994). Contraction uncoupling vs low calcium or high potassium for long-term cold preservation of isolated hearts. *Circulation* **89,** 2412–2420.
3. Furchgott, R. F., and Vanhoutte, P. M. (1989). Endothelium-derived relaxing and contracting factors. *FASEB J.* **3,** 2007–2018.
4. Dinerman, J. L., Lowenstein, C. J., and Snyder, S. H. (1993). Molecular mechanisms of nitric oxide regulation: Potential relevance to cardiovascular disease. *Circ. Res.* **73,** 217–222.
5. Hashimoto, M., Ishida, Y., Naruse, I., and Paul, R. J. (1992). Prolonged cold storage abolishes endothelium-dependent relaxing responses to A23187 and substance P in porcine coronary arteries. *J. Vasc. Res.* **29,** 64–70.
6. Stowe, D. F., Boban, M., Palmisano, B. W., Kampine, J. P., and Bosnjak, Z. J. (1994).Coronary flow response to vasodilators in isolated hearts cold perfused for one day with butanedione monoxime. *Endothelium* **2,** 87–98.
7. Bridge, J. H. B. (1986). Relationship between the sarcoplasmic reticulum and transsarcolemmal Ca transport revealed by rapidly cooling rabbit ventricular muscle. *J. Gen. Physiol.* **88,** 437–473.
8. Boban, M., Stowe, D. F., Kampine, J. P., Goldberg, A. H., and Bosnjak, Z. J. (1993). Effects of 2,3-butanedione monoxime in isolated hearts: Protection during reperfusion following global ischemia. *J. Thorac. Cardiovasc. Surg.* **105,** 532–540.
9. Stowe, D. F., Boban, M., Kampine, J. P., and Bosnjak, Z. J. (1993). Reperfusion with adenosine and nitroprusside improves preservation of isolated guinea pig hearts after 22 hours of cold perfusion with 2,3-butanedione monoxime. *J. Cardiovasc. Pharmacol.* **21,** 578–586.
10. Marijic, J., Buljubasic, N., Stowe, D. F., Turner, L. A., Kampine, J. P., and Bosnjak, Z. J. (1991). Opposing effects of diacetyl monoxime on contractility and calcium transients in isolated myocardium. *Am. J. Phyisol.* **260,** H1153–H1159.
11. Stowe, D. F., Boban, M., Bosnjak, Z. J., and Kampine, J. P. (1991). Protective effects of 2,3-butanedione monoxime following hypothermia of isolated hearts for 22 hours. *Anesthesiology* **75,** A523 (abstract).
12. Li, T., Sperelakis, N., Teneick, R. E., and Solaro, R. J. (1985). Effects of diacetyl monoxime on cardiac excitation–contraction coupling. *J. Pharmacol. Exp. Ther.* **232,** 688–695.
13. Gwathmey, J. K., Hajjar, R. H., and Solaro, R. J. (1991). Contractile deactivation and uncoupling of crossbridges. Effects of 2,3-butanedione monoxime on mammalian myocardium. *Circ. Res.* **69,** 1280–1292.
14. Stowe, D. F., Boban, M., Andryk, J., Kampine, J. P., and Bosnjak, Z. J. (1993). Endothelium dependent and independent coronary flow responses to L-arginine and L-NAME after cold perfusion of isolated hearts for one day. *FASEB J.* **7,** A718.
15. Stowe, D. F., Bosnjak, Z. J., Marijic, J., and Kampine, J. P. (1988). Effects of halothane with and without histamine and/or epinephrine on automaticity, intracardiac conduction times and development of dysrhythmias in the isolated guinea pig heart. *Anesthesiology* **68,** 695–706.

16. Hoffman, J. I. E. (1984). Maximal coronary flow and the concept of coronary vascular reserve. *Circulation* **70,** 153–158.
17. Rich, T. L., and Brady, A. J. (1974). Potassium contracture and utilization of high-energy phosphates in rabbit heart. *Am. J. Physiol.* **226,** 105–113.
18. Jynge, P. (1985). Calcium control with special reference to myocardial protection in cardiac surgery. *In* "Control and Manipulation of Calcium Movement" (J. R. Parratt, ed.), pp. 253–271. Raven, New York.
19. Somlyo, A. P., and Himpens, B. (1989). Cell calcium and its regulation in smooth muscle. *FASEB J.* **3,** 2266–2276.
20. Burkhoff, D., Kalil-Filho, R., and Gerstenblith, G. (1990). Oxygen consumption is less in rat hearts arrested by low calcium than by high potassium at fixed flow. *Am. J. Physiol.* **259,** H1142–H1147.
21. Pernot, A. C., Ingwall, J. S., Menasche, P., Grousset, C., Bercot, M., Piwnica, A., and Fossel, E. T. (1983). Evaluation of high-energy phosphate metabolism during cardioplegic arrest and reperfusion: A phosphorus-31 nuclear magnetic resonance study. *Circ. Res.* **67,** 1296–1303.
22. Billman, G. E. (1992). Cellular mechanisms for ventricular fibrillation. *Int. Union Physiol. Sci.* **7,** 254–259.
23. Blanchard, E. M., Smith, G. L., Allen, D. G., and Alpert, N. R. (1990). The effects of 2,3-butanedione monoxime on initial heat, tension, and light output of ferret papillary muscles. *Eur. J. Physiol.* **416,** 219–221.
24. Gœrge, G., Chatelain, P., Schaper, J., and Lerch, R. (1991). Effect of increasing degrees of ischemic injury on myocardial oxidative metabolism early after reperfusion in isolated rat hearts. *Circ. Res.* **68,** 1681–1692.
25. Jeremy, R. W., Stahl, L., Gillinov, M., Litt, M., Aversano, T. R., and Becker, L. C. (1989). Preservation of coronary flow reserve in stunned myocardium. *Am. J. Physiol.* **256,** H1303–H1310.
26. Kelm, M., and Schrader, J. (1990). Control of coronary vascular tone by nitric oxide. *Circ. Res.* **66,** 1561–1575.
27. Ueda, M., Silvia, S. K., and Olsson, R. A. (1992). Nitric oxide modulates coronary autoregulation in the guinea pig. *Circ. Res.* **70,** 1296–1303.
28. van Galen, P. J. M., Stiles, G. L., Michaels, G., and Jacobson, K. A. (1992). Adensoine A_1 and A_2 receptors: Structure–function relationships. *Med. Res. Rev.* **12,** 423–471.
29. Schubert, T., Vetter, H., Owne, P., Reichart, B., and Opie, L. H. (1989). Adenosine cardioplegia. Adenosine versus potassium cardioplegia: Effects of cardiac arrest and postischemic recovery in the isolated rat heart. *J. Thorac. Cardiovasc. Surg.* **98,** 1057–1065.
30. Ledingham, S., Katayama, O., Lachno, D., Patel, N., and Yacoub, M. (1990). Beneficial effect of adenosine during reperfusion following prolonged cardioplegic arrest. *Cardiovasc. Res.* **24,** 247–253.
31. Petsikas, D., Ricci, M. A., Baffour, R., de Varennes, B., Symes, J., and Guerraty, A. (1990). Enhanced 24 hours *in vitro* heart preservation with adenosine and adenosine monophosphate. *J. Heart Transplant.* **9,** 114–118.
32. Cronstein, B. H., Levine, R. I., Belanoff, J., Weissman, G., and Hirschhorn, R. (1986). Adenosine: An endogenous inhibitor of neutrophil mediated injury to endothelial cells. *J. Clin. Invest.* **78,** 760–770.
33. Rubanyi, G., and Vanhoutte, P. M. (1985). Endothelium-removal decreases relaxations of canine coronary arteries caused by β-adrenergic agonists and adenosine. *J. Cardiovasc. Pharmacol.* **7,** 139–144.

34. Mehta, J. L., Nichols, W. W., Donnelly, W. H., Lawson, D. L., and Saldeen, T. G. P. (1989). Impaired canine coronary vasodilator response to acetylcholine and bradykinin after occlusion–reperfusion. *Cir. Res.* **64,** 43–54.
35. Lefer, A. M., Tsao, P. S., Lefer, D. J., and Ma, X. L. (1991). Role of endothelial dysfunction in the pathogenesis of reperfusion injury after myocardial ischemia. *FASEB J.* **5,** 2029–2034.
36. Siegfried, M. R., Erhardt, J., Rider, T., Ma, X. L., and Lefer, A. M. (1992). Cardioprotection of organic nitric oxide donors in myocardial ischemia reperfusion. *J. Pharmacol. Exp. Ther.* **260,** 668–675.

Troponin T as a Marker of Perioperative Myocardial Cell Damage

H. Mächler, H. Gombotz, K. Sabin, and H. Metzler
University Department of Anesthesiology,
A-8036 LHK Graz, Austria

I. Introduction

Coronary artery bypass (CABG) surgery in patients with unstable angina is associated with higher morbidity and mortality when compared to patients with stable angina. The extent and contribution of the preoperative ischemic status and myocardial cell injury, however, could not be quantified unless acute myocardial infarction occurred. If myocardial ischemia results in reversible or irreversible cell damage, the structurally bound protein troponin T may be released into the circulation. With the enzyme-linked immunosorbent assay (ELISA)–troponin T assay (Boehringer Mannheim, Mannheim, Germany), plasma troponin T levels can now be determined in routine clinical settings.

Twenty-four hours before anesthesia 22 of 26 CABG patients with unstable angina had significantly elevated levels of troponin T (>0.2 μg/liter), whereas in 37 of 38 CABG patients with stable angina normal values were found. This statistically highly significant difference did not change before onset of cardiopulmonary bypass. After bypass there was a significant elevation of troponin T in patients with unstable as well as in patients with stable angina when compared to the prebypass period, and there was no longer a significant difference between both groups. Except for 3 patients with unstable angina, who had elevated creatine kinase isoenzyme

Advances in Pharmacology, Volume 31

MB (CK-MB) mass values preoperatively, no patient had increased CK-MB isoenzyme activity before extracorporal circulation.

Although this phenomenon was not reflected by a significant difference in adverse outcome in the unstable angina group, the results of this study would strongly suggest that aggressive antianginal therapy be undertaken until anesthesia has to be continued or even completed in patients who seem to be medically stabilized. The troponin T assay may be highly useful for routine diagnosis and treatment of minor myocardial cell damage in patients undergoing CABG surgery and may also serve as a sensitive marker for the effectiveness of different cardioprotective measures.

II. Cardiac Troponin

The troponin complex is part of the thin filament of the myocardial muscle cell and comprises three protein subunits. Troponin C binds calcium and is responsible for regulating the process of thin filament activation during heart muscle contraction. Troponin I prevents contraction in the absence of calcium and troponin C. Troponin T is the tropomyosin-binding protein of the troponin regulatory complex (Fig. 1). Different genes encode troponin T in cardiac, skeletal, and smooth muscle, resulting in proteins with differing amino acid compositions and molecular weights. Cardiac troponin T, therefore, can be differentiated by immunological techniques. Furthermore, the high concentration of troponin T in cardiac muscle also results in high sensitivity (1–3). The measurement of cardiac troponin T may, therefore, be an alternative and superior technique to measuring CK-MB for diagnosis of myocardial ischemia (1,4–6).

According to an intracellular compartmentation model, circulating troponin T may have two origins: a smaller free cytosolic pool and a larger

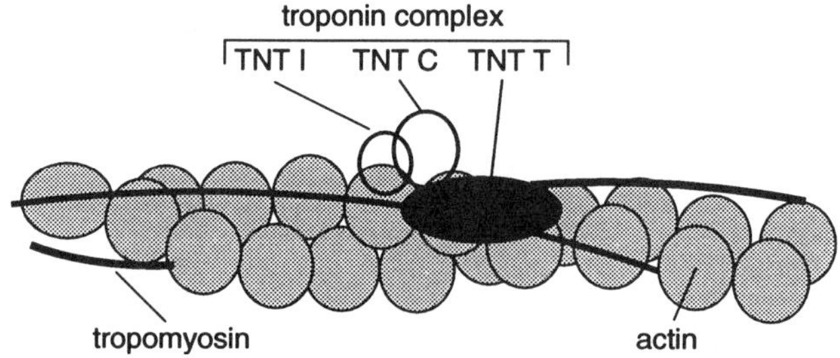

Fig. 1 Illustration of part of the thin filament presenting the troponin complex (TNT I, troponin I; TNT C, troponin C; TNT T, troponin T), actin helix, and tropomyosin coil [Mair, J., Dienstl, F., and Puschendorf, B. (1992). Cardiac troponin T in the diagnosis of myocardial injury. *Crit. Rev. Clin. Lab. Sci.* **29**, 31–57].

structurally bound fraction. In patients with acute myocardial infarction, therefore, a biphasic peak of serum troponin T has been found (7,8) (Fig. 2). The first peak reflects the characteristics of cytosolic proteins like CK beginning about 3.5 hr and ending about 32 hr after the onset of pain; the second peak indicates a constant liberation of the bound troponin T fraction. Because of its short half-life of about 120 min a constant elevation of serum troponin T may indicate a continuous release into the circulation, probably from an immediately preceding and continuing ischemic process.

Determinations of cardiac troponin T have been used not only to detect acute myocardial infarction, but also to check the effect of thrombolytic therapy and percutaneous transluminal angioplasty (7,9). Furthermore, in a multicenter study, Hamm *et al.* demonstrated that troponin T measured in the serum of patients with unstable angina appears to be a more sensitive prognostic indicator of myocardial cell injury than conventional diagnostic markers (10).

III. Troponin T in Coronary Artery Bypass Patients with Unstable Angina

Because of its heterogeneous nature, unstable angina covers a broad spectrum of syndromes ranging from stable angina to acute myocardial in-

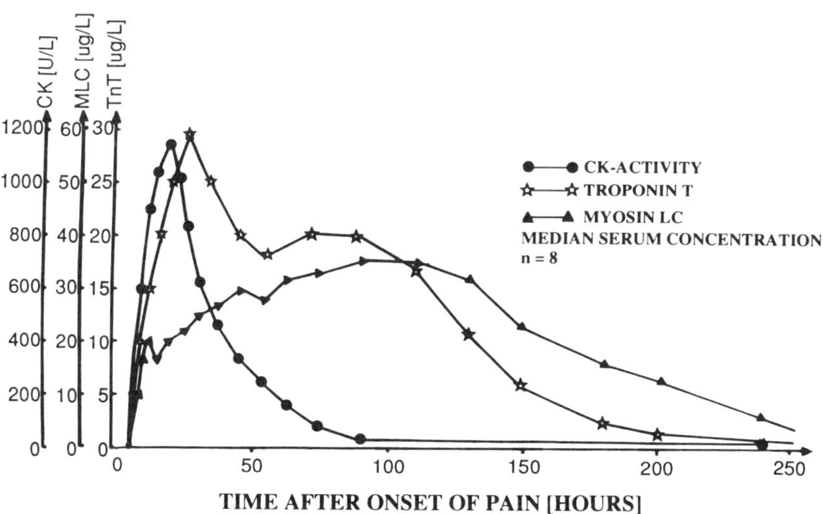

Fig. 2 Changes in median serum concentration of myosin light chains (MLC) and troponin T (TnT) and total CK activity in patients with Q-wave infarction (8).

farction, but the condition does not meet the standard enzymatic and electrocardiographic (ECG) criteria for myocardial infarction. A basic mechanism in patients with unstable angina may be recurrent episodes of mural thrombosis that slowly leads to vascular occlusion accompanied by intermittent embolization rather than a single abrupt thrombotic event (11). In patients undergoing CABG surgery, unstable angina is reported to be associated with a higher morbidity and mortality compared to patients with stable angina (12). The extent of the preoperative ischemic status and myocardial cell damage, however, could not be quantified unless acute myocardial infarction occurred. Corresponding to the severity of transient episodes of myocardial ischemia, only severe episodes resulting in cell necrosis could be assessed by increased creatine kinase activity (8).

Using the new troponin T assay, which has been developed by H. A. Katus in association with Boehringer Mannheim (4), troponin T plasma levels were determined in 64 consecutive patients scheduled for elective coronary artery bypass surgery and compared to conventional markers of myocardial ischemia. The patients were assigned to two groups according to the clinical presentation of angina pectoris. Unstable angina was diagnosed in patients with chest pains lasting for 30 min or longer. This group was further divided according to the classification of Braunwald (13). Patients with acute myocardial infarction in the previous 6 weeks as assessed by electrocardiographic, biochemical, or angiographic diagnosis as well as those with failed coronary angioplasty (PTCA) and those with angina unresponsive to medical therapy and therefore scheduled for emergency operation were excluded.

Medical pretreatment of unstable angina included heparin and/or aspirin. Aspirin was stopped at least 5 days before surgery. Heparin (300–800 IU/hr) was applied continuously in patients assigned to class II and III of Braunwald; intravenous nitrates were given continuously in patients unresponsive to heparin therapy. β-Blockers, calcium channel antagonists, and oral nitrates were given until the day of surgery. Conventional 12-lead ECG was performed 24 hr before surgery and immediately before induction of anesthesia. Perioperative clinical monitoring included 7-lead ECG, arterial pressure, central venous pressure, pulmonary artery catheter, as well as capnography and pulse oximetry.

After oral premedication with 2 mg flunitrazepam all patients received etomidate at 300 mg/kg, fentanyl at 10 μg/kg for induction of anesthesia, and 0.3 mg/kg pancuronium for relaxation intravenously. Anesthesia was supplemented with isoflurane up to 0.5% and bolus doses of fentanyl at 50–100 μg intravenously in the presence of light anesthesia. Cardiopulmonary bypass was performed with a bubble oxygenator using hemodilution and systemic hypothermia of 25°C. St. Thomas Hospital II solution was used for cardioplegia.

The biochemical assays used are summarized in Table I. Measurements were performed 24 hr before anesthesia and surgery, immediately before induction of anesthesia, before and after cardiopulmonary bypass, at the end of surgery, and 24 hr after surgery. Six criteria were used for the definition of major cardiac events resulting in adverse outcome: perioperative myocardial infarction based on electrocardiographic and enzymatic diagnosis, high inotropic support (dopamine or dobutamin > 10 g/kg/min), malignant arrhythmias, the need for intraaortic balloon pumping, death in the first 24 hr postoperatively, and death during the hospital stay.

Using the criteria above, of 64 CABG patients 38 patients were considered to have stable and 26 to have unstable angina. Of 26 patients with unstable angina 22 had already elevated levels of troponin T 24 hr before anesthesia and surgery (median 0.32 μg/liter, range 0.15–5.13), whereas 37 of 38 patients with stable angina had values below 0.2 μg/liter at the same observation time (median 0.0 μg/liter, range 0.0–0.53) (Fig. 3). At the time of induction of anesthesia again 20 of 26 patients with unstable angina had elevated troponin T levels (median 0.27 μg/liter, range 0.1–1.2)

Table I
Biochemical Assays

Substance analyzed	Assay
Troponin T	TNT immunoassay (ELISA TNT) using EnzymTest System ES 300 analyzer (Boehringer Mannheim, Germany)
Creatine kinase	
CPK Biotrol	Ultraviolet test (Biotrol, Paris, France)
CPK Kodak	Spectrometric determination on immunometric basis (Kodak, Ektachem, Rochester, NY)
Creatine kinase isoenzyme MB	
CK-MB Merck	Photometric determination on immunometric basis (Merck, Darmstadt, Germany)
CK-MB Kodak	Spectrometric determination (Kodak, Ektachem)
CK-MB mass concentration	Immunoassay using Abbott IMX automated analyzer (Abbott, Chicago, IL)

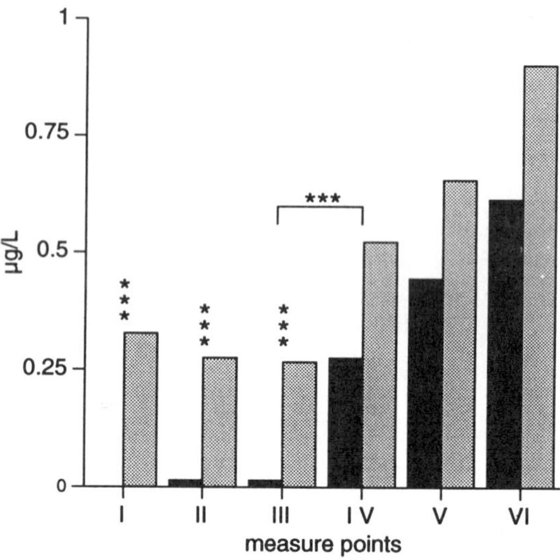

Fig. 3 Troponin T values (median) in two groups of patients (solid bars, stable angina; shaded bars, unstable angina pectoris) at six different measure points: I, 24 hr preoperatively; II, before induction of anesthesia; III, before cardiopulmonary bypass; IV, after cardiopulmonary bypass; V, at the end of the operation; VI, 24 hr postoperatively. Troponin T was compared between the two groups at each measure point (*** $p < 0.001$, no significance) and between measure point III and IV of each group (** $p < 0.05$ in both groups). The Mann–Whitney test and the Wilcoxon test for paired differences were used for statistics.

compared to 37 of 38 patients with stable angina showing normal values (median 0.0 μg/liter, range 0.0–0.18) (Table II). This highly significant difference did not change before onset of cardiopulmonary bypass. After termination of extracorporeal circulation there was a permanent significant elevation of troponin T in both groups compared to the prebypass values. There was, however, no longer a statistically significant difference between both groups. No patient had elevated CK-MB activity (absolute value and percentage of total activity) 24 hr and immediately before induction of anesthesia. In the group of patients with unstable angina 3 of 26 patients had elevated CK-MB mass activity either 24 hr or immediately before induction of anesthesia (Tables II and III).

Preoperatively three patients in the stable angina group showed negative T-waves and one patient a Q-wave. Six patients in the unstable angina group were presented with T-flattening or isoelectric T, and three patients with negative T-waves. The ECG phenomena did not change immediately before induction of anesthesia.

Table II

Number of Patients above Normal Range in Patients with Stable[a] and in Patients with Unstable Angina Pectoris[b]

		Measurement point[c]					
Assay	Patient group	I	II	III	IV	V	VI
CPK[d] Biotrol	Stable AP	0	2	1	7	15	35
	Unstable AP	1	2	3	6	13	22
CPK Kodak	Stable AP	0	1	0	0	3	24
	Unstable AP	0	1	1	2	3	15
CK-MB[e] Merck	Stable AP	0	0	0	10	16	12
	Unstable AP	0	0	0	6	10	7
CK-MB Kodak	Stable AP	0	0	0	3	7	7
	Unstable AP	0	0	0	1	2	2
CK-MB mass	Stable AP	0	1	4	34	35	34
	Unstable AP	3	3	4	21	25	25
TNT[f]	Stable AP	1	1	0	22	32	32
	Unstable AP	22	20	15	22	26	25

[a] $n = 38$.
[b] $n = 26$.
[c] Measurement points: I, 24 hr preoperatively; II, before induction of anesthesia; III, before cardiopulmonary bypass; IV, after cardiopulmonary bypass; V, at end of operation; VI, 24 hr postoperatively.
[d] CPK, Creatine kinase.
[e] CK-MB, Creatine kinase isoenzyme MB.
[f] TNT, Troponin T.

Until cardiopulmonary bypass no major hemodynamic events occurred in the stable angina group, whereas in the unstable angina group there was one patient with ventricular fibrillation before going on bypass. After cardiopulmonary bypass four patients in the stable angina group and eight patients in the unstable angina group had major cardiac events. One patient with stable angina preoperatively died on the day of surgery in the intensive care unit (ICU). He suffered from malignant arrhythmia, perioperative myocardial infarction, low cardiac output, and intraaortic balloon pumping. He had a poor, incomplete revascularization, a long bypass time, and severe problems coming off bypass. One patient from the unstable angina group died 1 week after surgery suffering from multiorgan failure.

Table III

Comparison of Median Values of Each Assay in Patients with Stable[a] and in Patients with Unstable Angina Pectoris[b]

Assay	Patient group	Measurement point[c]					
		I	II	III	IV	V	VI
CPK Kodak (U/l)	Stable AP	10	12	17	56	99	448
	Unstable AP	15	15	22	55	95	396
CPK Biotrol (U/l)	Stable AP	23	21	24	77	129	618
	Unstable AP	23	18	26	77	133	531
CK-MB Kodak (U/l)	Stable AP	0.4	0.4	0.4	8.7	11.1	8.1
	Unstable AP	1.0	0.4	0.4	10.4	11.6	4.5
CK-MB Merck (U/l)	Stable AP	5	4.5	4.5	18.5	21	18
	Unstable AP	4.5	4	4	19.5	22	20.5
CK-MB mass (μg/l)	Stable AP	0.5	1.2	2.0	12.6	23.9	18
	Unstable AP	1.4	1.3	2.2	15.0	30.1	20.6

[a] $n = 38$.
[b] $n = 26$.
[c] Measurement points: I, 24 hr preoperatively; II, before induction of anesthesia; III, before cardiopulmonary bypass; IV, after cardiopulmonary bypass; V, at end of operation; VI, 24 hr postoperatively.

IV. Troponin T versus Conventional Markers of Myocardial Cell Damage in Perioperative Settings

Debate continues on the individual laboratory tests for the detection of myocardial ischemic damage. Following a myocardial ischemic event, the loss of integrity of myocardial cell membranes results in a release of proteins of the cardiac contractile apparatus into the circulation (4,8). Among these proteins troponin T and troponin I appear to be unique cardiac antigens. In addition, they are not detectable in the serum of healthy people.

Beside higher sensitivity and specifity, the advantage of troponin T compared to conventional CK and CK isoenzyme MB assays is the detection and quantification of myocardial cell damage in cases where CK and CK isoenzyme MB are still in a normal range (1). Increased troponin T concentrations in a patient with coronary artery disease indicate a recently developed severe coronary artery narrowing (8). On the other hand, testing of troponin T cannot be used to screen severe coronary artery disease but may be useful in detecting patients with angina pectoris at risk or with poor prognosis.

The time for determination of troponin T limits its use in the triage. A bedside test which is now under investigation may be more helpful in the future. Nevertheless Collinson concluded that at present troponin T measurement meets the criteria for the best biochemical test for differential diagnosis of cardiac damage (14). The cutoff of 0.2 μg/liter used in this investigation exceeds the value quoted by the manufacturer, who claims that any detectable (>0.1 μg/liter) troponin T indicates myocardial damage. A multicenter study by Gerhardt, however, revealed a cutoff of 0.2 μg/liter (3).

Many of the patients with unstable angina, although clinically and medically stabilized, undergo anesthesia and CABG surgery with already elevated troponin T levels. Increased creatine kinase activity is found in only a small percentage of these patients (15). Additionally CK measurements still have false-positive rates as high as 15–25% (16). On the other hand, histological studies revealed myocardial cell necrosis in high-risk patients even if serum creatine kinase activity was in a normal range (17). Technetium-99*m* scans also detected subendocardial necrosis in one-third of patients with unstable angina (18).

Reversible as well as irreversible cell injury may occur in unstable angina (10). This may be explained by an intermittent critical flow reduction as a result of intracoronary thrombus formation and also by minor local cell necrosis due to thrombotic microembolism. Currently, it remains unclear whether elevated troponin T levels in patients with unstable angina indicate only ischemic cell damage or irreversible microinfarctions which are not detectable by conventional diagnostic methods. Nevertheless, we have to assume that most of the patients with unstable angina present pathophysiological changes resulting in constant release of troponin T into the circulation, although they seem to be clinically stabilized (19).

V. Summary

Unstable angina in patients undergoing CABG surgery is associated with a higher morbidity and mortality compared to patients with stable angina. Mortality ranges between 2 and 10% (20,21). The importance of the preoperative status is only clear and well documented for patients with unstable angina who are unresponsive to medical treatment, patients who undergo emergency revascularization, and for patients with failed angioplasty. The adverse outcome in elective patients with unstable angina was statistically not significantly different from those with stable angina. Therefore, we may assume that in stabilized patients with unstable angina and minor myocardial cell damage intraoperative determinants like the duration of the aortic clamping period or the degree of revascularization are more relevant than the preoperative ones. These determinants may also be reflected by a marked and significant increase of troponin T in both groups during and after surgery. As for other cardiac enzymes, this increase of troponin T beginning immediately after reperfusion of the cardioplegic heart may limit its diagnostic value after cardiac surgery (6,22). On the other hand, troponin T may serve as a marker in assessing the effectiveness of different cardioprotective measures.

Nevertheless, preoperatively elevated troponin T levels may indicate a jeopardized myocardium with an ongoing process of myocardial cell damage and may be of prognostic value. Antianginal and antiischemic therapy, therefore, has to be continued and completed until the day of surgery in these high-risk patients.

References

1. Katus, H. A., Remppis, A., Neumann, F. J., Scheffold, Th., Diederich, K. W., Vinar, G., Noe, A., Matern, G., and Kuebler, W. (1991). Diagnostic efficiency of troponin T measurements in acute myocardial infarction. *Circulation* **83,** 902–912.
2. Katus, H. A., Looser, S., Hallermayer, K., Remppis, A., Scheffold, Th., Borgya, A., Essig, U., and Geu, U. (1992). Development and *in vitro* characterization of a new immunoassay of cardiac troponin T. *Clin. Chem.* **38,** 386–393.
3. Gerhardt, W., Katus, H., Ravkilde, J., Hamm, C., Jergenson, P. J., Peheim, E., Ljungdahl, L., and Lîfdahl, P. (1991). S-Troponin T in suspected ischemic myocardial injury compared with mass and catalytic concentrations of S-creatine kinase isoenzyme MB. *Clin. Chem.* **37,** 1405–1411.
4. Katus, H. A. (1989). Enzyme linked immunoassay of cardiac troponin T for the detection of acute myocardial infarction in patients. *J. Mol. Cell. Cardiol.* **21,** 1349–1353.
5. Fox, R. (Editorial) (1991). Troponin T and myocardial damage. *Lancet* **338**(6), 23–24.
6. Mair, J., Wieser, Ch., Seibt, I., Artner-Dworzak, E., Furtwängler, W., Waldenberger, F., Balogh, D., and Puschendorf, B. (1991). Troponin T to diagnose myocardial infarction in bypass surgery. *Lancet* **337,** 434–435.
7. Katus, H. A., Remppis, A., Scheffold, Th., Diederich, K. W., and Kübler, W. (1991). Intracellular compartmentation of cardiac troponin T and its release kinetics in patients with reperfused and nonreperfused myocardial infarction. *Am. J. Cardiol.* **67,** 1360–1367.
8. Katus, H. A., and Kübler, W. (1990). Detection of myocardial cell damage in patients with unstable angina by serodiagnostic tools. *In* "Unstable Angina" (W. Bleifeld, eds.), pp. 92–100. Springer-Verlag, Berlin and Heidelberg.

9. Talasz, H., Genser, N., and Mair, J. (1992). Side Branch occlusion during percutaneous transluminal coronary angioplasty. *Lancet* **228,** 639.
10. Hamm, C. W., Ravkilde, J., Gerhardt, W., Jorgensen, P., Peheim, M. E., Ljungdahl, L., Goldmann, B., and Katus, H. A. (1992). The prognostic value of serum troponin T in unstable angina. *N. Engl. J. Med.* **327**(3), 146–150.
11. Falk, E. (1985). Unstable angina with fatal outcome: Dynamic coronary thrombosis leading to infarction and/or sudden death. *Circulation* **71,** 699–708.
12. Jain, U. (1992). Myocardial infarction during coronary artery bypass surgery. *J. Cardiothorac. Vasc. Anesth.* **6,** 612–623.
13. Braunwald, E. (1989). Unstable angina—A classification. *Circulation* **80,** 410–414.
14. Collinson, P. O., and Path, M. R. C. (1992). The prognostic value of serum troponin T in unstable angina. *N. Engl. J. Med.* **327,** 1760–1761.
15. Armstrong, P. W., Chiong, M. A., and Parker, J. O. (1982). The spectrum of unstable angina: Prognostic role of serum creatine kinase determination. *Am. J. Cardiol.* **49,** 1849–1852.
16. Lee, T. H. H., Rousan, C. W., and Weiberg, M. C. (1987). Sensitivity of routine clinical criteria for diagnosing myocardial infarction within 24 hours of hospitalization. *Ann. Intern. Med.* **106,** 181–186.
17. Davies, M. J., Thomas, A. C., Knapman, P. A., and Hangartner, J. R. (1986). Intramyocardial platelet aggregation in patients with unstable angina suffering sudden ischemic cardiac death. *Circulation* **73,** 418–427.
18. Olson, H. G., Lyons, K. P., Aronow, W. S., Stinson, P. J., Kuperus, J., and Waters, H. J. (1981). The high-risk angina patient: Identification by clinical features, hospital course, electrocardiography and technetium-99*m* stannous pyrophosphate scintigraphy. *Circulation* **64,** 674–684.
19. Hendricks, G. R., Amano, J., Kenna, T., Fallon, J. T., Patrick, T. A., Manders, W. T., Rogers, G. G., Rosendorff, C., and Vatner, S. F. (1985). Creatine kinase release is not associated with mycardial necrosis after short periods of coronary occlusion in conscious baboons. *J. Am. Coll. Cardiol.* **6,** 1299–1303.
20. Kaiser, G. C., Schaff, H. V., and Kilip, Th. (1989). Myocardial revascularization for unstable angina pectoris. *Circulation* **79,** I-60–I-67.
21. Sharma, G. V. R. K., Deupree, R. H., Khuri, S. F., Parisi, A. F., Luchi, R. J., and Scott, S. M. (1991). Coronary bypass surgery improves survival in high-risk unstable angina. *Circulation* **84,** III-260–III-267.
22. Katus, H. A., Schoeppenthau, M., and Tanzem, A. (1991). Non-invasive assessment of perioperative myocardial cell damage by circulating troponin T. *Br. Heart J.* **65,** 259–264.

Silent Myocardial Ischemia: Pathophysiology and Perioperative Management

Anders G. Hedman
Department of Medicine
Ludvika Hospital
Ludvika, S-77181, Sweden

I. Introduction

Silent myocardial ischemia should not be regarded as a separate disease entity, but rather as one of several possible manifestations of myocardial ischemia and coronary heart disease. It may be defined as objective evidence of transient myocardial ischemia (by direct or indirect measurements of left ventricular function, perfusion, metabolism, or electrical activity) without chest pain or other equivalents of angina (1). Other manifestations of silent ischemia include unrecognized (silent) myocardial infarction, ischemic cardiomyopathy, and sudden cardiac death. Common to all these forms is evidence of ischemia or its secondary effects.

Cohn has identified three different types of silent ischemia (2). Such a classification is important as prevalence, management, and prognosis vary among the three types.

A. Type I Silent Myocardial Ischemia

Type I silent myocardial ischemia occurs in patients with totally asymptomatic coronary artery disease and is usually detected by a screening exercise test. Based on autopsy studies, the prevalence of coronary artery disease in the asymptomatic United States population has been conserva-

Advances in Pharmacology, Volume 31

tively estimated to be in the range of 4 to 5% (3). In a Norwegian study of 2014 apparently healthy male office workers aged 40 to 59 years who underwent an exercise test, 69 patients had angiographically verified significant coronary artery disease with at least 50% stenosis in one coronary artery. Fifty of these patients were completely asymptomatic (4).

B. Type II Silent Myocardial Ischemia

Type II silent myocardial ischemia is seen in patients who are asymptomatic after a myocardial infarction, and it can be identified by early postinfarction exercise stress testing or Holter monitoring. In a well-known Canadian study by Theroux and co-workers (5), 18% of asymptomatic postinfarction patients had silent ischemia on early exercise testing. Of all the patients with a positive exercise electrocardiogram (ECG), silent ischemia was found in 58%. Thus, in the United States, about 50,000 asymptomatic postinfarction patients per year would be expected to have silent ischemia, at least in the initial 30-day postinfarction period.

C. Type III Silent Myocardial Ischemia

Included in the type III class are patients with angina who also have episodes of silent myocardial ischemia. A review of several different studies using Holter monitoring found that about 40% of all patients with clinical angina had episodes of silent ischemia (6), but considerably higher figures have been reported. Furthermore, in this group of patients, 60 to 75% of all episodes of ischemia are silent. The proportion of 60 to 75% painless episodes remains remarkably constant in several studies from many different centers.

II. Pathophysiology of Silent Myocardial Ischemia

The neurophysiology of cardiac pain involves pain receptors, afferent nerve pathways, and spinal cord and supraspinal regulation. Despite this, our understanding of cardiac pain is still incomplete, and the pathophysiology of silent myocardial ischemia remains elusive. Three separate theories listed below have been widely proposed in an effort to explain why significant myocardial ischemia sometimes fails to elicit symptoms of angina pectoris (7).

A. Global Deficiency in Pain Perception

A generalized alteration in the sensitivity of a patient to pain, somatic as well as visceral, could explain the total absence of symptoms in type I patients and the incomplete pain awareness of the type III patients. As early as 1983, Droste and Roskamm in Germany investigated 42 patients who had ST segment depressions on exercise stress testing and angiographically confirmed significant coronary artery disease. Of these patients, 22 had typical angina and 20 had silent myocardial ischemia. Three different types of pain-receptive modalities including electric pain threshold test, cold pressor stimulation, and tourniquet-induced forearm ischemia were studied. Patients with silent ischemia demonstrated a higher pain threshold and, in particular, a greater tolerance for all types of pain (8). As the groups were similar in terms of frequency of prior infarction, extent of multivessel coronary disease, and prevalence of alcoholism and diabetes, it is very unlikely that the altered pain perception was due to destruction of specific myocardial nociceptive pathways.

Considerable research efforts have been devoted to the search for a consistent biochemical difference in patients with symptomatic versus silent ischemia. Available data include measurements of plasma levels of β-endorphin, metenkephalin, norepinephrine, and epinephrine. At the present time, the exact role of the endorphins remains inconclusive, and most studies have failed to link endorphins to silent myocardial ischemia.

B. Anatomic Changes in Pain Receptors and Nerves

A specific deficiency in cardiac nociception has been proposed as an explanation for silent myocardial ischemia. Autopsy studies of diabetic patients with previous painless myocardial infarction show clearly identifiable sympathetic and parasympathetic neuropathy (9). It is generally claimed that diabetic patients have an increased prevalence of silent myocardial ischemia, although this has been questioned by Finnish investigators who found no statistically convincing evidence in a metaanalysis of several studies (10). This theory might account for the type II silent ischemia following myocardial infarction with myocardial pain receptors damaged at the time of the myocardial infarction.

C. Quantitative Theory of Silent Myocardial Ischemia

There is considerable evidence that myocardial ischemia resulting in angina is quantitatively greater than that associated with silent ischemia. For example, continuous hemodynamic monitoring of patients with angina at rest shows that silent episodes are, on average, shorter in duration and produce less impairment of left ventricular function (11). Another means of ischemia detection, ambulatory electrocardiographic monitoring of type III patients with both painful and painless ischemia, has shown a longer mean duration and greater degree of ST segment depression in symptomatic episodes (12). It should be kept in mind, however, that the interindividual and intraindividual variation is considerable, and a study by Deanfield and associates established that painless ischemic episodes detected during Holter monitoring could not be distinguished from painful episodes merely on the basis of the severity of ST segment depression (13).

In our own institution 48-hr Holter monitoring was performed on 21 patients with clinical angina and angiographically verified significant coronary heart disease. A total of 274 episodes of significant ST segment depression were recorded. Of these episodes, 61% were silent. Also consistent with the quantitative theory, the duration of the painful episodes was longer [11.6 ± 9.9 versus 7.1 ± 8.5 min (mean ± SD; $p < 0.001$)] (Fig. 1) and the magnitude of ST segment depression was greater [2.6 ± 1.2 versus 1.7 ± 0.7 mm ($p < 0.001$)] (Fig. 2).

Similarly, radionuclear ventriculography stress testing has shown a significantly greater reduction in global left ventricular ejection fraction in angina patients compared with silent ischemia patients (14).

Transient coronary occlusion during balloon angioplasty provides a useful tool to the study of the sequence of ischemic events. The balloon obstruction allows the onset of ischemia to be precisely timed. The first changes are abnormalities of left ventricular relaxation and contraction followed by elevated left ventricular filling pressures and later by electrocardiographic evidence of ischemia. Angina, when it occurs, appears later than 25 sec after balloon occlusion and is usually preceded by electrocardiographic changes. Thus, ischemia in conscious man is always characterized by a transition period during which it remains silent, and this transition to a symptomatic stage does not necessarily occur in many instances (15). A puzzling phenomenon is that, within a given patient, some balloon inflations may be painless, whereas others of equal duration and with a similar degree of myocardium at jeopardy produce clinical angina. This suggests that factors other than pain threshold and degree of myocardial

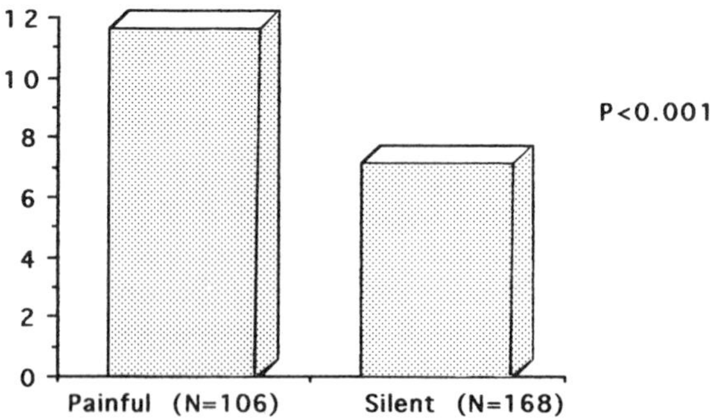

Fig. 1 Mean duration of episodes of ST segment depression (minutes).

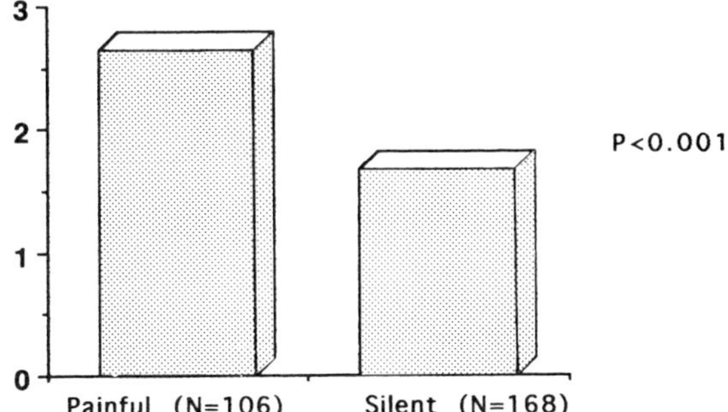

Fig. 2 Mean amplitude of episodes of ST segment depression (mm), measured at J-point plus 60 msec.

ischemia are operative in producing silent ischemia. It remains to be determined which other factors are involved.

In addition, there may be different pathophysiological characteristics resulting in different patterns of myocardial blood flow distribution during silent, as opposed to painful, ischemia. Some data support that contention, whereas other studies have failed to find any consistent difference.

III. Detection of Silent Myocardial Ischemia

For the anesthesiologist, the detection of silent ischemia in patients who are about to undergo surgery is the mainstay of the management of this condition. The first task is to determine which individuals are in jeopardy and which methods should be used in course of the workup.

The likelihood that a coronary event will occur during general surgery can be predicted partly on the basis of simple clinical information. Not surprisingly, patients without angina or myocardial infarction, those who are young, and women have the least risk. Moderate risk occurs in older men with chest pain syndromes or in patients with a number of coronary risk factors. Patients at high risk are more typically older men with known coronary disease, for example, typical angina or previous myocardial infarction, or earlier symptoms or signs of left ventricular dysfunction (3). A high index of suspicion is necessary to detect asymptomatic episodes, which appear to carry similar adverse prognosis as painful episodes.

Preoperative testing is primarily useful in stratifying risk in patients at medium or high clinical risk, and a decision to proceed with laboratory testing should be made only after careful consideration of the clinical risk profile and the risk of the planned surgical procedure. Eagle and coworkers at the Massachusetts General Hospital consider expected event rates above 10 to 15%, but probably not below 5%, high enough to justify screening all patients in the category by further laboratory testing. It is still a matter of personal opinion whether patients with no evidence of coronary disease but with multiple coronary risk factors should also be tested. Testing may be especially important in patients with diabetes. Patients with known coronary heart disease or heart failure probably warrant periodic cardiological evaluation whether they are to undergo surgery or not (16).

IV. Which Laboratory Methods Should Be Used for Screening Purposes?

It remains to be determined which tests are the safest and most cost effective as well as in which sequence they should be performed. Many of the available studies have been carried out in patients with peripheral vascular disease because of the high prevalence of associated coronary heart disease, in one study estimated to be 33% (17). Noninvasive tests appear to be sufficiently sensitive to identify the majority of patients likely to suffer perioperative or postoperative cardiac events. Thus, according to present data it is unnecessary to make coronary angiography the initial test in the majority of patients (16).

A. Exercise Stress Test

Exercise testing is the standard means to detect myocardial ischemia of all types, but it may be less useful before noncardiac surgery because of associated medical problems. Patients with vascular, orthopedic, thoracic and other problems are often elderly, out of condition, and unable to exercise adequately, and it should be kept in mind that submaximal effort may lead to false-negative results. Exercise stress test results are also less reliable in asymptomatic populations with a low prevalence of coronary artery disease because of the frequency of false-positive test results. This follows from Bayes' theorem which states that test results cannot be adequately interpreted without knowing the prevalence of the disease in the population under study (18).

For patients who are unable to undergo a maximal stress test or in whom the test results are inconclusive, two alternative diagnostic methods are available.

B. Thallium-201 Dipyridamole Myocardial Imaging

Dipyridamole is a potent vasodilator which is used to increase coronary blood flow without producing myocardial ischemia. Although test protocols may vary, the drug is usually given as an intravenous infusion at a rate of 0.15 mg/kg/min over 4 min. After the infusion, the patient walks in place for 2–3 min. Then 2.0 mCi of thallium-201 is injected, and the patient continues to walk in place for another 2 min. Because highly stenotic coronary vessels cannot dilate normally, areas of myocardium supplied by them will take up less thallium during scanning than areas supplied by normal vessels. Later, after the vasodilation has resolved, these underperfused areas will "fill in" as the thallium redistributes. Thus, what might be labeled a "cold" area on the scan that fills in during the redistribution phase of the procedure implies the presence of a highly stenotic coronary artery. Eagle and co-workers found that preoperative dipyridamole–thallium imaging was most useful to stratify vascular patients determined to be at intermediate risk by clinical evaluation (19).

C. Ambulatory Long-Term Electrocardiographic Monitoring of ST Segment

The validity of long-term ECG monitoring of the ST segment, also known as Holter monitoring, has been extensively investigated. ST segment depressions during Holter monitoring have not always been accepted as firm evidence of myocardial ischemia. However, a strong association between ambulatory ST depression and other simultaneous objective evidence of ischemia has been demonstrated by means of perfusion scintigraphy, radionuclide cardioangiography, and hemodynamic monitoring in patients with known coronary heart disease, but the clinical relevance of ST segment changes in normal individuals remains doubtful. In addition, Holter data can be used to measure ischemia duration and to some extent also the severity of ischemia, based on the amplitude of the ST segment depression.

Holter monitoring is far less expensive and more widely available than dipyridamile–thallium scanning. On the other hand, it has several limitations. Ten percent or more of patients have electrocardiographic abnormalities that limit or even preclude the interpretation of ST segment depression, such as bundle branch blocks or left ventricular hypertrophy due to hypertension or digoxin treatment. Patients unable to undergo monitoring for silent ischemia because of such electrocardiographic abnormalities are well suited for dipyridamole–thallium testing (16). There are also important technical considerations concerning signal recording. The American Heart Association has published valuable recommendations regarding frequency responses and bandwidth (20). Many older amplitude-modulated systems amplify the low-frequency information present in the ST segment. This might cause an overestimation of the degree of ST depression. Current amplitude-modulated systems, however, are as reliable as frequency-modulated recorders, which have been considered the "golden standard" (1).

Another problem with Holter monitoring are artifacts caused by patient motion, loose leads, etc. Monitoring is usually performed for 24 hr. Studies from our institution indicate that longer periods of monitoring do not significantly increase the sensitivity of detection, even if the variation from day to day is considerable.

Slogoff and co-workers have evaluated an on-line operating room monitor (Spacelabs Cardule) for its accuracy of detection of ST segment shifts. It was found to have a higher sensitivity than Holter equipment in diagnosing myocardial ischemia with no apparent loss of specificity (21).

To date, no direct comparison of the prognostic value of Holter monitoring of the ST segment with that of thallium redistribution in a large number of preoperative patients is available. Such data are needed to clarify the correct sequence of these tests in the preoperative assessment. That would be a very important study, suitable for a cooperative effort by anesthesiologists and cardiologists.

V. Perioperative Medical Management of Silent Myocardial Ischemia

According to official statistics from the late 1980s, 25 million patients in the United States undergo noncardiac surgery each year. Three million are considered having or being at risk of having coronary artery disease. Some 50,000 of these patients sustain a perioperative myocardial infarction, and it has been estimated that more than half the 40,000 deaths after surgery each year are caused by cardiac events (22). Perioperative myocardial ischemia is strongly associated with postoperative cardiac events, and painless myocardial ischemia seems to be just as dangerous as clinical angina. The importance of detecting silent ischemia in patients who will undergo anesthesia is clear.

Several groups of anesthesiology researchers, among them Mangano and co-workers in the Study of Perioperative Ischemia Research Group of the University of California at San Francisco, have investigated in considerable detail the mechanisms of perioperative ischemia of which silent ischemia is such an important and integrated part. At least three studies have been published clearly demonstrating that preoperative silent ischemia significantly predicts postoperative coronary events in patients undergoing vascular surgery (23–25).

Pasternack and associates (23) performed real-time electrocardiographic monitoring for silent ischemia on 200 patients undergoing peripheral vascular surgery. The population was predominantly male and elderly, mean age 70 years, and had a high prevalence of coronary artery disease. During the preoperative, intraoperative, and postoperative periods, silent ischemia was detected in 127 patients (63.5%). Nine patients sustained acute perioperative myocardial infarctions, all having had evidence of silent ischemia. There were no perioperative myocardial infarctions among patients without perioperative silent ischemia. The authors concluded that patients with preoperative silent myocardial ischemia had a 5-fold increased risk of a perioperative infarction.

Furthermore, Mangano *et al.* have established that early postoperative myocardial ischemia is an important correlate of adverse cardiac outcome. They studied 243 men with known coronary artery disease who were undergoing noncardiac surgery. Eighteen percent of the patients had postoperative cardiac events, whereas postoperative myocardial ischemia occurred in 41%. There was a 9-fold increase in the risk of an ischemic event among those shown to have ischemia by Holter monitoring. Virtually all ischemic episodes were silent (22).

All interventions—medications, angioplasty, coronary bypass surgery—proven to be effective in the relief of symptomatic ischemia are also effective against silent ischemia, and the treatment principles are essentially the same. It is important to remember, however, that therapy aimed at control of symptoms does not necessarily eliminate silent ischemic episodes, and all antianginal therapies must also be shown to be antiischemic if the problem of silent ischemia is to be dealt with in a satisfactory manner (26). Furthermore, in evaluating any drug response, the natural variability of ischemic events must be considered.

Agents that reduce myocardial oxygen demand as well as those that primarily improve supply have been effective. In the longer perspective, the Swedish RISK study has proved that treatment with low-dose aspirin, 75 mg daily, in patients with unstable coronary heart disease significantly reduces the risk of subsequent myocardial infarction at least as well in silent as in symptomatic myocardial ischemia (27). Because improvement in prognosis is the main treatment objective in symptom-free patients, aspirin should be a mainstay of the treatment unless contraindications are present.

So far no study has proved that treatment of type I silent ischemia apart from risk factor intervention has a beneficial effect on the long-term prognosis. The U.S. National Heart, Lung, and Blood Institute has completed an international pilot study, called ACIP for Asymptomatic Cardiac Ischemia Pilot Study, designed to determine if a clinical trial testing this hypothesis is feasible. It involved 600 patients with coronary arteries suitable for revascularization, a positive stress test, and at least one episode of silent ischemia on 48-hr Holter monitoring. The medical regimens were atenolol, for patients needing a β-blocker, or diltiazem. In addition, all patients were advised to take 325 mg of aspirin per day. The results are expected in the near future.

In the perioperative situation several types of rapidly acting drugs are available, among which nitrates, β-adrenergic blockers, and calcium antagonists are the most important.

A. Nitrates

The effect of nitrates, regardless of the mode of administration, has been demonstrated very clearly, and it is even more striking if β-blockers and calcium antagonists are used simultaneously. Intravenous nitroglycerin has been shown to normalize regional wall motion abnormalities during an ischemic episode (28).

B. α-Adrenergic Blockers

There are no reports describing convincing effects of α-adrenergic blockers against silent myocardial ischemia.

C. β-Adrenergic Blockers

β-Adrenergic blockers reduce asymptomatic ST depressions during Holter monitoring, normalize exercise-induced reduction in left ventricular ejection fraction (29), and eliminate the usual circadian variation of silent myocardial ischemia (30). They appear to be the most effective drugs in eliminating the morning surge of ischemic activity. Drugs with combined α- and β-adrenergic blocking effects (e.g., labetalol) have demonstrated similar efficacy.

D. Calcium Antagonists

Calcium antagonists reduce the frequency and duration of silent ischemia. No single calcium channel blocker has proved clearly superior, although rate-accelerating agents might be less effective when used alone (31). Combinations of calcium antagonists and β-adrenergic blockers are particularly effective (32).

E. Surgery and Angioplasty

Both coronary angioplasty (PTCA) and coronary bypass surgery (CABG) are effective against silent ischemia, but they fall outside the scope of this article.

VI. Summary

Silent ischemia has been called the silent killer. Pain does not kill patients with coronary heart disease—ischemia does, whether it happens to be painful or silent. An increased awareness of this still puzzling phenomenon may be of great importance in the pre- and perioperative management of patients with coronary heart disease, and improved detection and management of silent ischemia are likely to reduce the risk of perioperative cardiac events.

References

1. Cohn, P. F. (1989). "Silent Myocardial Ischemia and Infarction," 2nd Ed. Dekker, New York.
2. Cohn, P. F. (1985). Silent myocardial ischemia: Classification, prevalence and prognosis. *Am. J. Med.* **79,** 2–6.
3. Diamond, G. A., and Forrester, J. S. (1979). Analysis of probablity as an aid in the clinical diagnosis of coronary artery disease. *N. Engl. J. Med.* **300,** 1350–1358.
4. Eriksen, J., and Thaulow, E. (1984). Follow-up of patients with asymptomatic myocardial ischemia. *In* "Silent Myocardial Ischemia" (W. Rutishauser and H. Roskamm, eds.), pp. 156–164. Springer-Verlag, Berlin.
5. Theroux, P., Waters, D. D., Halphen, C., Debaisieux, J. C., and Mizgala, H. F. (1979). Prognostic value of exercise testing soon after myocardial infarction. *N. Engl. J. Med.* **301,** 341–345.
6. Cohn, P. F. (1988). Silent myocardial ischemia. *Ann. Intern. Med.* **109,** 312–317.
7. Assey, M. E. (1990). The recognition and treatment of silent myocardial ischemia. *In* "The Heart" (J. W. Hurst, R. C. Schlant, C. E. Rackley, E. H. Sonnenblick, and N. K. Wenger, eds.), 7th Ed., pp. 1079–1086. McGraw-Hill, New York.
8. Droste, C., and Roskamm, H. (1983). Experimental pain measurement in patients with asymptomatic myocardial ischemia. *J. Am. Coll. Cardiol.* **1,** 940–945.
9. Faerman, I., Faccio, E., Milei, J, Nunez, R., Jadzinsky, M., Fox, D., and Rapaport, M. (1977. Autonomic neuropathy and painless myocardial infarction in diabetic patients. *Diabetes* **26,** 1147–1158.
10. Airaksinen, K. E. L., and Koistinen, M. J. (1992). Association between silent coronary artery disease, diabetes and autonomic neuropathy—Fact or fallacy? *Diabetes Care* **15,** 288–292.
11. Chierchia, S., Lazzari, M., Freedman, B., Brunelli, C., and Maseri, A. (1983). Impairment of myocardial perfusion and function during painless myocardial ischemia. *J. Am. Coll. Cardiol.* **1,** 924–930.
12. Cecchi, A. C., Dovellini, E. V., Marchi, F., Pucci, P., Santoro, C. M., and Fazzini,

P. F. (1983). Silent myocardial ischemia during ambulatory electrocardiographic monitoring in patients with effort angina. *J. Am. Coll. Cardiol.* **1,** 934–939.

13. Deanfield, J. E., Maseri, A., Selwyn, A. P., Chierchia, S., Ribiero, P., Krikler, S., and Morgan, M. (1983). Myocardial ischemia during daily life in patients with stable angina: Its relations to symptoms and heart rate changes. *Lancet* **2** 753–758.
14. Bonow, R. O., Bacharach, S. L., Green, M. V., La Freniere, R. L., and Epstein, S. E. (1987). Prognostic implications of symptomatic versus asymptomatic (silent) myocardial ischemia induced by exercise in mildly symptomatic and in asymptomatic patients with angiographically documented coronary artery disease. *Am. J. Cardiol.* **60,** 778–783.
15. Sigwart, U., Grbic, M., Payot, M., Goy, J.-J., Essinger, A., and Fischer, A. (1984). Ischemic events during coronary artery balloon occlusion. *In* "Silent Myocardial Ischemia" (W. Rutishauser and H. Roskamm, eds.), pp. 29–36. Springer-Verlag, Berlin.
16. Eagle, K. A., and Boucher, C. A. (1989). Cardiac risk of non-cardiac surgery. *N Engl. J. Med.* **321,** 1330–1332.
17. Hertzer, N. R., Beven, E. G., Young, J. R., O'Hara, P. J., Ruschhaupt, W. F., Graor, R. A., Dewolfe, V. G., and Maljovec, L. C. (1984). Coronary artery disease in peripheral vascular patients. A classification of 1000 coronary angiograms and results of surgical management. *Ann. Surg.* **199,** 223–233.
18. Rifkin, R. D., and Hood, W. B. (1977). Bayesian analysis of electrocardiographic exercise stress testing. *N. Engl. J. Med.* **297,** 681–686.
19. Eagle, K. A., Coley, C. M., Newell, J. B., Brewster, D. C., Darling, R. C., Strauss, H. W., Guiney, T. E., and Boucher, C. A. (1989). Combining clinical and thallium data optimizes preoperative assessment before major vascular surgery. *Ann. Intern. Med.* **110,** 859–866.
20. Pipberger, H. V., Arzbaecher, R. C., Berson, A. S., and the American Heart Association Committee on Electrocardiography (1975). Recommendations for standardization of leads and of specifications for instruments in electrocardiography and vectorcardiography. *Circulation* **52,** 11–13.
21. Slogoff, S., Keats, A. S., David, Y., and Igo, S. R. (1990). Incidence of perioperceptive myocardial ischemia detected by different electrocardiographic systems. *Anesthesiology* **73,** 1074–1081.
22. Mangano, D. T., Browner, W. S., Hollenberg, M., London, M. J., Tubau, J. F., Tateo, I. M., and the Study of Perioperative Ischemia Research Group (1990). Association of perioperative myocardial ischemia with cardiac morbidity and mortlity in men undergoing noncardiac surgery. *N. Engl. J. Med.* 323, 1781–1788.
23. Pasternack, P. F., Grossi, E. A., Baumann, F. G., Riles, T. S., Lamparello, P. J., Giangola, G., Primis, L. K., Mintzer, R., and Imparato, A. M. (1989). The value of silent myocardial ischemia monitoring in the prediction of perioperative myocardial infarction in patients undergoing peripheral vascular surgery. *J. Vasc. Surg.* **10,** 617–625.
24. McCann, R. L., and Clements, F. M. (1989). Silent myocardial ischemia in patients undergoing peripheral vascular surgery: Incidence and association with perioperative cardiac morbidity and mortality. *J. Vasc. Surg.* **9,** 583–587.
25. Raby, K. E., Goldman, L., Creager, M. A., Cook, E. F., Weisberg, M. C., Whittemore, A. D., and Selwyn, A. P. (1989). Correlation between preoperative ischemia and major cardiac events after peripheral vascular surgery. *N. Engl. J. Med.* **321,** 1296–1300.
26. Frishman, W. H., and Teicher, M. (1987). Antianginal drug therapy for silent myocardial ischemia. *Am. Heart J.* **114,** 140–147.
27. Nyman, I., Larsson, H., Wallentin, L., and the Research Group on Instability in Coro-

nary Artery Disease in Southwest Sweden (1992). Prevention of serious cardiac events by low-dose aspirin in silent myocardial ischemia. *Lancet* **340,** 497–501.

28. Pepine, C. I., Feldman, F. L., Ludbrook, P., Holland, P., Lambert, C. R., Conti, C. R., and McGrath, P. D. (1986). Left ventricular dyskinesia reversed by intravenous nitroglycerin: A manifestation of silent myocardial ischemia. *Am. J. Cardiol.* **58,** 38B–42B.
29. Cohn, P. F., Brown, E. I., Swinford, R., and Atkins, H. L. (1986). Effect of beta blockade on silent regional left ventricular wall motion abnormalities. *Am. J. Cardiol.* **57,** 521–526.
30. Imperi, G. A., Lambert, C. R., Coy, K., Lopez, L., and Pepine, C. J. (1987). Effects of titrated beta blockade (metoprolol) on silent myocardial ischemia in ambulatory patients with coronary artery disease. *Am. J. Cardiol.* **60,** 519–524.
31. Bala Subramanian, V., Bowles, M. J., Khurmi, N. S., Davies, A. B., and Raftery, E. B. (1982). Rationale for the choice of calcium antagonists in chronic stable angina: An objective double-blind, placebo-controlled comparison of nifedipine and verapamil. *Am. J. Cardiol.* **50,** 1173–1179.

Effect of Halothane on Sarcolemmal Calcium Channels during Myocardial Ischemia and Reperfusion

Benjamin Drenger, Yehuda Ginosar, and Yaacov Gozal

Department of Anesthesiology and Critical Care Medicine
Hadassah University Hospital
91120 Jerusalem, Israel

I. Introduction

Myocardial ischemia, even for a short period, results in significant depletion of high-energy stores, followed by defective calcium transport activities in the subcellular organelles, such as the sarcolemma, sarcoplasmic reticulum, and mitochondria. The development of myocardial dysfunction on reperfusion, described as myocardial "stunning" is dependent on a massive increase in calcium ion influx, which is related to the severity and duration of ischemia. The damage to the cell is magnified by the generated free radicals, probably acting through lipid peroxidation mechanisms (1–3).

The sequence of events when the oxygen concentration approaches zero, with rapid depletion of ATP stores, is primarily attributed to impairment in membrane ion transport mechanisms with the occurrence of abnormal ion shifts. The consequent decrease in cytoplasmic pH, owing to lactate and CO_2 accumulation, leads to acidification of the cell membrane and protonation of the channels (4). As a result, a progressive failure of the intracellular ionic pumps is caused, and the myocardial cell loses control over cytosolic calcium movements and calcium concentration. Loss of K^+ and Mg^{2+} is evident (5) with intracellular accumulation of Na^+ and Ca^{2+}. If ischemia persists, it will lead eventually to calcium overload, persisting contracture of the contractile proteins, and cell death. The possibility that volatile anesthetics (VA) may slow down or interrupt this process before irreversible damage occurs has already attracted great interest (6–10).

These studies strongly suggest that the VA and particularly halothane have a beneficial effect on the ischemic myocardium. Our experimental approach in studying the anesthetic effect on calcium channels during ischemia is based on the hypothesis that VA attenuate voltage-dependent Ca^{2+} currents, probably by interacting with dihydropyridine binding sites in the sarcolemma (11–16). It has been known for some time that halothane increases the tolerance of myocardium to ischemia (17). This effect was originally attributed to enhanced preservation of the high-energy phosphate compounds. These observations have been confirmed when halothane administration produced a decrease in the extent of myocardial infarction in a canine model of coronary artery ligation (7). Kroll and Knight (9) showed a reduced incidence of ventricular fibrillation in a canine ischemia–reperfusion model, and Warltier *et al.* (10) demonstrated better recovery of contractile function of stunned canine myocardium if halothane was administered during acute ischemia. Another important aspect of VA is their ability to prevent reduction in myocardial contractility and coronary blood flow produced by oxygen-free radicals generated in isolated rabbit heart preparations (8).

The aim of the present study was to identify, at the cellular and organelle level, the functional derangements which occur in ischemia. The preinfarction phase of ischemia is particularly of interest as assessment of the degree of injury by histological methods is often not fruitful. Specific investigations to identify biochemical markers of reversible myocardial organelle dysfunction are therefore indicated.

In the present study we have evaluated the effect of a short period of ischemia and reperfusion on the voltage-sensitive calcium channels (VSCC). We first focused on changes occurring in binding capacity during the ischemic injury, followed by additional changes associated with reperfusion. Only then we defined the contribution of VA during ischemia–reperfusion toward altering or preventing the changes previously observed in binding capacity.

II. Canine Model for Myocardial Ischemia and Reperfusion

To simulate the clinical conditions of surgical coronary revascularization or evolving myocardial infarction, we have chosen a canine "region-at-risk" model for our studies. Equilibrium binding studies of the dihydropyridine calcium channel blocker [^{3}H]isradipine were used in order to evaluate the influence of ischemia and reperfusion on the VSCC, as well as the ischemia effect in the presence or absence of halothane anesthesia. Isradipine binds to the VSCC in a specific, saturable, and reversible manner (18), and it can be used as a probe of these channels. Previous studies have demonstrated that a decrease in binding correlates with a reduction of calcium influx associated with a decrease in contractile force (19).

Twenty-four experiments were carried out on 13 mongrel dogs weighing 20–28 kg. After induction of general anesthesia using intravenous pentobarbital and mechanical ventilation with nitrous oxide and oxygen, anesthesia was maintained by muscle relaxation and additional doses of pentobarbital or 1.6% (v/v) halothane in the inhalation mixture, according to the experimental protocol. Electrocardiogram, temperature, and femoral arterial blood pressure were continuously monitored. When halothane was used, the end-tidal concentration was maintained steady using an anesthetic gas monitor (Drager Iris, Lubeck, Germany). It was administered from a calibrated vaporizer (Fluotec 3, Cyprane Keighley, Yorkshire, England) starting 10 min prior to the coronary occlusion and continued during the 10-min ischemia period, constantly maintaining a steady end-tidal concentration of 1.6 (v/v). The heart was exposed in left thoracotomy, and an atraumatic elastic vascular ligature was applied to the left anterior descending (LAD) artery, distal to the first diagonal artery, for 10 min to create a reversible regional ischemia.

The LAD ischemic region was separated from the rest of the myocardial muscle, which served as control, nonischemic tissue, and was perfused by the left circumflex artery (LCX) and right coronary artery (RCA). Ischemic and control regions were determined by simultaneous injection of 20 ml brillant blue dye into the LAD artery and normal saline into the aortic root, immediately after aortic cross-clamping and removal of the heart (20). The ventricular tissue was immediately excised according to the color landmarks and put into separate ice containers.

The regional myocardial ischemia animal experiments were divided into four groups: I, control studies, in which a nonischemic myocardial tissue was obtained from the LAD, LCX, and RCA perfusion territory and the [^{3}H]isradipine binding characteristics of each of the regions were analyzed by radioligand binding studies (n = 3 binding studies); II, ischemia studies of 10 min ischemia duration (n = 8); III, ischemia and reperfusion studies involving 10 min ischemia duration and 20 min of reperfusion (n = 6); and IV, ischemia experiments in the presence of *in vivo* halothane inhalation (n = 6).

III. Isolation of Sarcolemma-Enriched Preparation

Sarcolemmal membranes were isolated in a series of extraction and centrifugation steps. The homogenized tissue was centrifuged in 10 m*M* histidine, 0.75 *M* NaCl, pH 7.5 (10,400 *g*, 30 min), 25°C, and the pellet received was homogenized again and reextracted in 10 m*M* $NaHCO_3$, 5 m*M* histidine (10,400 *g*, 20 min). The supernatant, which was now free of contractile proteins, nuclear debris, and mitochondria, was centrifuged (17,000 *g*, 40 min), and the pellet obtained was homogenized in 0.25 *M* sucrose, 10-m*M* histidine and stored in a −80°C freezer.

[^{3}H]Isradipine (PN200-110, Amersham, Buckinghamshire, England), a dihydropyridine, was used to label the VSCC. The assay was carried out in triplicate at 25°C in 50 m*M* Tris-HCl at pH 7.5. Sarcolemmal membranes (80–100 μg) were incubated for 60 min with 0.05 to 1.0 n*M* [^{3}H]isradipine in the absence or presence of 1 μM nitrendipine as the displacing agent, to define total and nonspecific binding, respectively; specific binding is calculated from the difference between these values. The reaction was terminated by filtration through Whatman (Clifton, NJ) GF/B filters, and binding of the radioligand was quantitated by scintillation counting. The data were fitted by nonlinear regression analysis with explicit weighing (Scatchard analysis) using the Enzfitter (Leatherbarrow AJ, Elsevier, Amsterdam) program to obtain estimates of the dissociation constant (K_d) and density of binding sites (B_{max}). Data obtained in each experiment from the control sarcolemmal membranes were compared to those achieved from the ischemic (LAD) membranes. Statistical analysis was performed using the paired *t*-test, in which the effect of ischemia was compared to that of control value for each isradipine concentration. In the same way data of ischemia–reperfusion experiments and ischemia–halothane experiments were analyzed.

IV. Binding Capacity of Isradipine

Three control studies were performed in order to evaluate the binding capacity of [^{3}H]isradipine to different areas of the ventricular muscle. [^{3}H]Isradipine specific binding to the sarcolemmal membranes was similar in all three types of membranes, obtained separately from the regions perfused by the three arteries (LAD, LCX, and RCA). Mean [^{3}H]isradipine specific binding values at 1 n*M* concentration, in each preparation, were 82.8 ± 14, 78 ± 12, and 78.2 ± 15 fmol/mg, respectively (mean ± SEM). These control studies indicate that the binding capacity of the dihydropyridine to the VSCC is comparable in the different regions of the myocardium.

Ischemia caused a significant rise of 50 to 95% in [^{3}H]isradipine specific binding to sarcolemma membranes ($p < 0.01$; Fig. 1), which resulted from an increase in the number of available binding sites in the cell membrane (B_{max}) and not to a change in K_d, the dissociation constant (Table I). When the myocardium was exposed to ischemia followed by reperfusion, [^{3}H]isradipine specific binding tended to increase, but the effect did not reach significance. The present findings indicate that ischemia increases the number of available VSCC in the sarcolemma. However, this increase does not appear to be irreversible. Indeed, following 20 min of reperfusion the change in [^{3}H]isradipine specific binding was already partially reversed.

At present, the literature does not directly support a linkage between an increased number of VSCC and an increase in Ca^{2+} influx to the myocardial cell. Schmidt *et al.* (21) described a direct correlation between a decrease in dihydropyridine binding to the calcium channels and a parallel decrease in force of contraction of human ventricular muscle strips. However, experiments which used radioligand binding studies in brain tissue in order to evaluate ischemia-induced changes in cerebral calcium mechanisms showed an increase in [^{3}H]isradipine specific binding to the

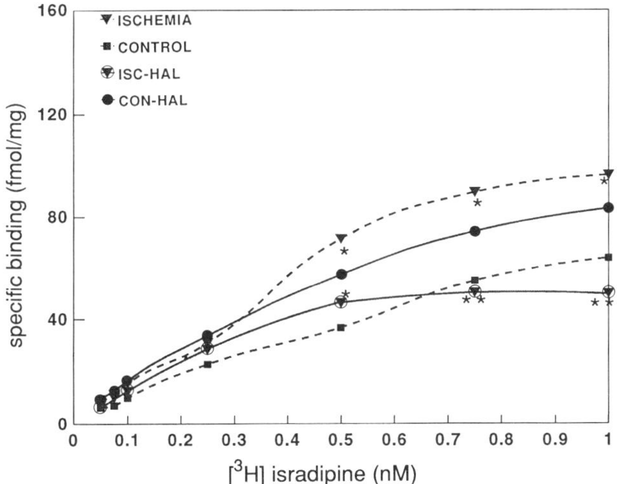

Fig. 1 [^{3}H]Isradipine specific binding to canine heart sarcolemma. Sarcolemmal membranes were prepared from hearts exposed to a 10-min ischemic event, in the presence or absence of *in vivo* administration of 1.6% (v/v) halothane. Specific binding data are presented as femtomoles per milligram protein; each point represents a triplicate measurement of 8 ischemia or 6 ischemia–halothane experiments, for each of the [^{3}H]isradipine concentrations. Asterisks mark values significantly different from controls.

Table I
Binding Properties of [^{3}H]Isradipine[a]

	B_{max}[b]		K_d[c]	
	Control	Ischemia	Control	Ischemia
Ischemia ($n = 8$)	98.1 ± 15	181.1 ± 34	0.84 ± 0.10	0.99 ± 0.20
Ischemia–reperfusion ($n = 6$)	209 ± 32	288 ± 46	0.92 ± 0.30	0.93 ± 0.30
Ischemia–halothane ($n = 6$)	251 ± 69	173 ± 51	0.93 ± 0.40	0.89 ± 0.35

[a] Values are means ± SEM.
[b] B_{max} is the maximal binding capacity (fmol/mg protein).
[c] K_d is the dissociation constant (n*M*).

ischemic neural tissue (22,23). Similar experiment further demonstrated an increase in dopamine release from the striatum brain area, a cellular function which appears to be mediated by an increase in inward calcium current (24).

When halothane was administered before the ischemic challenge, the effect of ischemia on [^{3}H]isradipine binding was reversed (Fig. 1). In each of the experiments a clear decrease of up to 42% in specific binding was observed with significance of $p < 0.02$ and $p < 0.01$ in the 0.75 and 1.0 n*M* concentration, respectively. The results are explained by a decrease of 31% in B_{max}, with no change in K_d (Table I). These changes in VSCC binding kinetics are identical to those described in experiments which used *in vitro* direct halothane application to sarcolemmal membranes isolated from nonischemic bovine heart (14). The similarity of action in ischemic and nonischemic membranes suggests that halothane has a direct effect on sarcolemmal calcium channels. We propose that the effect on the VSCC during ischemia is not mediated by conservation of ATP stores, but rather results from a direct effect of halothane on the general structure of the channel. This effect may be influencing the high-affinity, inactivated state of the channel, thus reducing its favorable binding capacity to the calcium channel blockers (25,26). The mechanism responsible for the increase in VSCC binding capacity during ischemia is uncertain. It might be explained by methylation of the membrane phospholipids (27,28), which leads to recruitment or unmasking of latent channels.

V. Summary

The present results provide indirect support for other studies which showed that halothane inhibited the Ca^{2+} accumulation associated with myocardial ischemia in isolated guinea pig hearts (6), demonstrating a potentially beneficial effect of the anesthetic on the ischemic heart. The role of halothane in preventing ischemia-induced dysrhythmias and attenuation of free radical generation on reperfusion offers a new potential use during open heart surgery. The method of continuous perfusion of oxygenated blood cardioplegia, retrogradely, through the coronary sinus, enables a concomitant administration of the VA before and during the ischemic period of the cardiopulmonary bypass. Further studies may promote the use of the volatile anesthetic when myocardial ischemia and reperfusion are present during open heart surgery.

Acknowledgments

This work was supported in part by the Chief Scientist's Fund, Ministry of Health, State of Israel, and by the Joint Research Fund of the Hebrew University Medical School and Hadassah University Hospital.

References

1. Opie, L. H. (1989). Reperfusion injury and its pharmacologic modification. *Circulation* **80,** 1049–1062.
2. Dhalla, N. S., Panagia, V., Singal, P. K., Makino, N., Dixon, I. M., and Eyolfson, D. A. (1988). Alterations in heart membrane calcium transport during the development of ischemia–reperfusion injury. *J. Mol. Cell. Cardiol.* **20**(Suppl. 2), 3–13.
3. Bolli, R. (1990). Mechanisms of myocardial "stunning." *Circulation* **82,** 723–738.
4. Sperelakis, N. (1988). Regulation of calcium slow channels of cardiac muscle by cyclic and nucleotides and phosphorylation. *J. Mol. Cell. Cardiol.* **20**(Suppl. 2), 75–105.
5. Trump, B. E., Mergner, W. J., Won Kahng, M., and Saladino, A. J. (1976). Studies on the subcellular pathophysiology of ischemia. *Circulation* **52**(Suppl. 1), 17–26.
6. Hoka, S., Bosnjak, Z. J., and Kampine, J. P. (1987). Halothane inhibits calcium accumulation following myocardial ischemia and calcium paradox in guinea pig hearts. *Anesthesiology* **67,** 197–202.
7. Davis, R. F., DeBoer, L. W. V., Rude, R. E., Lowenstein, E., and Maroko, P. R. (1983). The effect of halothane anesthesia on myocardial necrosis, hemodynamic performance, and regional myocardial blood flow in dogs following coronary artery occlusion. *Anesthesiology* **59,** 402–411.
8. Tanguay, M., Blaise, G., Dumont, L., Beique, G., and Hollmann, C. (1991). Beneficial effects of volatile anesthetics on decrease in coronary flow and myocardial contractility induced by oxygen-derived free radicals in isolated rabbit hearts. *J. Cardiovasc. Pharmacol.* **18,** 863–870.
9. Kroll, D. A., and Knight, P. R. (1984). Antifibrillatory effects of volatile anesthetics in acute occlusion/reperfusion arrhythmias. *Anesthesiology* **61,** 657–661.

10. Warltier, D. C., Al-Wathiqui, M. H., Kampine, J. P., and Schmeling, W. T. (1988). Recovery of contractile function of stunned myocardium in chronically instrumented dogs is enhanced by halothane or isoflurane. *Anesthesiology* **69,** 552–565.
11. Lynch III, C. (1986). Differential depression of myocardial contractility by halothane and isoflurane *in vitro*. *Anesthesiology* **64,** 620–631.
12. Bosnjak, Z. I., and Kampine, J. P. (1986). Effects of halothane on transmembrane potentials, Ca^{2+} transients, and papillary muscle tension in the cat. *Am. J. Physiol.* **251,** H374–H381.
13. Blanck, T. J. J., Runge, S., and Stevenson, R. L. (1988). Halothane decreases calcium channel antagonist binding to cardiac membranes. *Anesth. Analg. (N.Y.)* **67,** 1032–1035.
14. Drenger, B., Quigg, M., and Blanck, T. J. J. (1991). Volatile anesthetics depress calcium channel blocker binding to bovine cardiac sarcolemma. *Anesthesiology* **74,** 155–165.
15. Bosnjak, Z. J., Supan, F. D., and Rusch, N. J. (1991). The effects of halothane, enflurane, and isoflurane on calcium current in isolated canine ventricular cells. *Anesthesiology* **74,** 340–345.
16. Bosnjak, Z. J., Aggarwal, A., Turner, L. A., Kampine, J. M., and Kampine, J. P. (1992). Differential effects of halothane, enflurane, and isoflurane on Ca^{2+} transients and papillary muscle tension in guinea pig. *Anesthesiology* **76,** 123–131.
17. Spieckermann, Von P. G., Bruckner, J., Kubler, W., Lohr, B, and Bretschneider, H. J. (1969). Preischemic stress and resuscitation time of the heart. *Verh. Dtsch. Ges. Kreislaufforsch.* **35,** 358–364.
18. Miller, R. J. (1987). Calcium channels in neurones: Biochemical and molecular properties of neuronal calcium channels. Structure and physiology of the slow inward calcium channel. *Recept. Biochem. Methodol.* **9,** 189–246.
19. Janis, R. A., and Scriabine, A. (1983). Sites of action of Ca^{2+} channel inhibitors. *Biochem. Pharmacol.* **32,** 3499–3507.
20. Bolli, R., Kuo, L. C., and Roberts, R. (1984). Influence of acute arterial hypertension on myocardial infarct size in dogs without left ventricular hypertrophy. *J. Am. Coll. Cardiol.* **4,** 522–528.
21. Schmidt, U., Schwinger, R. H. G., Bohm, S., Uberfuhr, P., Kruezer, E., Reichart, B., Meyer, L. V., Erdmann, E., and Bohm, E. (1993). Evidence for an interaction of halothane with the L-type Ca^{2+} channel in human myocardium. *Anesthesiology* **79,** 332–339.
22. Magnoni, M. S., Govoni, S., Battaini, F., and Trabucchi, M. (1988). L-type calcium channels are modified in rat hippocampus by short-term experimental ischemia. *J. Cereb. Blood Flow Metab.* **8,** 96–99.
23. Hoehner, P. J., Blanck, T. J. J., Roy, R., Rosenthal, R. E., and Fiskum, G. (1992). Alteration of voltage-dependent calcium channels in canine brain during global ischemia and reperfusion. *J. Cereb. Blood Flow Metab.* **12,** 418–424.
24. Hoehner, P. J., Werling, L. L., Fox, L., Blanck, T. J. J., Rosenthal, R. E., and Fiskum, G. (1993). Increased ^{3}HPN200-110 binding is associated with nitrendipine depression of NMDA-stimulated ^{3}H-dopamine release in ischemic and reperfused dog striatum. *SCA 15th Annu. Meeting,* 343.
25. Drenger, B., Runge, S. R., Hoehner, P., Quigg, M., and Blanck, T. J. J. (1991). Halothane inhibits binding of calcium channel blockers to cardiac sarcolemma. Mechanisms of anesthetic action in skeletal cardiac and smooth muscle. *Adv. Exp. Med. Biol.* **301,** 109–114.
26. Hoehner, P. J., Quigg, M. C., and Blanck, T. J. J. (1991). Halothane depresses D600 binding to bovine heart sarcolemma. *Anesthesiology* **75,** 1019–1024.

27. Sobel, B. E., Corr, P. B., and Robinson, R. A. (1978). Accumulation of lysophosphoglycerides and arrhythmogenic properties in ischemic myocardium. *J. Clin. Invest.* **62,** 546–553.
28. Corr, P. B., Shayman, J. A., Kramer, J. B., and Kipnis, R. J. (1981). Increased α-adrenergic receptors in ischemic cat myocardium. *J. Clin. Invest.* **67,** 1232–1236.

Myocardial Ischemic Preconditioning

Donna M. Van Winkle, Grace L. Chien, and Richard F. Davis
Department of Anesthesiology
Oregon Health Sciences University
and Portland Veterans Affairs Medical Center
Portland, Oregon 97201

I. Introduction

Myocardial ischemic preconditioning is defined as the protective effect (from cardiac myocyte death produced by prolonged ischemia) that is generated by an antecedent exposure to a sublethal ischemic interval. Myocardium that has been exposed to a short episode(s) of ischemia has a decreased rate of high-energy phosphate loss during a subsequent prolonged ischemic period when compared to myocardium without the antecedent ischemic challenge (1). Based on this observation, Murry *et al*. examined the effect of four 5-min ischemic periods (each separated by a 5-min reperfusion interval) on the myocardial infarct size resulting from a subsequent 40-min period of ischemia (2). The results showed an approximately 75% reduction in infarct size, expressed as a percentage of risk area, in the animals that received the antecedent ischemic periods as compared to controls. If the prolonged ischemia was increased to 3 hr, the protective effect was lost. Regional blood flow studies did not show a difference in collateral flow into the risk area in the two groups. Subsequently the profound protective effect from ischemia-mediated cardiac myocyte death has been reported by other research groups using a variety of mammalian experimental models (3–7). During percutaneous transluminal coronary angioplasty, anginal symptoms and electrocardiographic ST

Advances in Pharmacology, Volume 31

segment changes have been reported to be less during subsequent balloon inflation periods than during the first inflation period (9). Also, in patients undergoing surgical coronary artery revascularization, transient aortic cross-clamping (and release) prior to subsequent cardiopulmonary bypass resulted in preservation of myocardial ATP levels (10). These studies indicate that the preconditioning effect may also occur in humans.

Two major hypotheses have emerged to explain the initiation and subsequent subcellular events leading to the preconditioning effect. Based on data showing attenuation of preconditioning by blockade of the ATP-sensitive potassium channel (K_{ATP}) with either glibenclamide or 5-hydroxydecanoate (5-HD) and protection from ischemic myocyte death with pretreatment with the K_{ATP} channel opener aprikalim, Gross and colleagues have hypothesized that activation of the K_{ATP} channel mediates ischemic preconditioning (11–12). On the other hand, blockade of cell surface adenosine (A_1) receptors with 8-(*p*-sulfophenyl theophylline attenuates the preconditioning effect, whereas A_1 receptor activation with $R(-)N^6$-(2-phenylisopropyl)adenosine (R-PIA) or 2-chloro-N^6-cyclopentyladenosine (CCPA) produces protection from ischemic injury that is similar to that seen with preconditioning (6–8, 11–12). These results lead Downey and colleagues to postulate that activation of the A_1 receptor mediates ischemic preconditioning.

The experiments described in this article were performed to determine whether antecedent intracoronary infusion of an adenosine A_1 receptor agonist or adenosine itself produces a cytoprotective effect during ischemia and whether blockage of the K_{ATP} channel interferes with the postulated protective effect of the A_1 adenosine receptor agonist. The data reported here have previously been published (14–15). Hence, this article is a redaction of those previously reported studies. The porcine model was selected in order to provide a heart of sufficient size to permit coronary instrumentation for drug infusion and to obviate the need for regional blood flow measurements that are necessary for proper interpretation of infarct size data when significant collateral flow exists, as with canine preparations.

II. Myocardial Ischemia

Domestic swine (30–46 kg) acquired through the institutional animal care facility (AAALAC accredited since 1973, last site visit 1991) were anesthetized with ketamine, 25 mg/kg i.m., followed by α-chloralose, 100 mg/kg i.v. Maintenance anesthesia was α-chloralose at 25 mg/kg i.v. administered approximately every 30 min. Additionally morphine sulfate, 1 mg/kg i.v., was administered prior to opening the thorax (see below) and 0.5 mg/kg approximately every 30–45 min thereafter. After orotracheal intubation, mechanical ventilation was adjusted by monitoring end-tidal CO_2 tension to achieve an arterial blood pH of 7.38–7.42. The inspired oxygen fraction was adjusted to maintain arterial oxygen tension between 100 and 150 mm Hg. Core temperature was measured from the esophagus and maintained 37–39°C by means of a servo-controlled heating pad. If additional heating was necessary a servo-controlled heated humidifier was placed in line in the breathing circuit.

Polyethylene catheters were placed into the right femoral artery or the right common carotid artery to allow measurement of aortic pressure and blood sampling. Venous access was similarly achieved via the femoral or jugular vein in addition to an 18-gauge catheter placed into an ear vein after the initial ketamine dose. A left fourth interspace thoracotomy was performed, and 1–3 ribs were resected to provide optimum cardiac exposure. The pericardium was widely opened longitudinally, and the heart was suspended via pericardial cradle. The left anterior descending coronary artery (LAD) was dissected free of surrounding tissue, and a silicone ligature was passed around the LAD immediately distal to the first major diagonal branch. A 24-gauge catheter was placed into the artery proximally so that the tip of the catheter was proximal to the future occlusion site. Coronary blood flow (LAD) was measured using an ultrasound transit time flow probe placed around the LAD distal to the occlusion site. However, in some animals LAD anatomy made placement proximal to the first diagonal branch necessary, in which case the metered flow included both flow to the ischemic zone and flow to the first diagonal branch perfusion territory. Flowmeters were calibrated at the conclusion of each experiment by timed volumetric collection of blood.

The experimental time line is illustrated in Fig. 1. Briefly, after the surgical preparation was completed the animals were randomly assigned to experimental subgroups (control, preconditioned, adenosine, R-PIA,

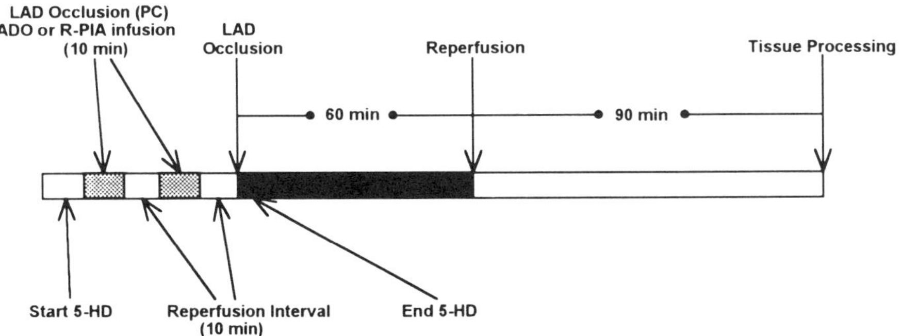

Fig. 1 Schematic representation of the experimental time line.

R-PIA plus 5-HD, and 5-HD alone). In the R-PIA plus 5-HD and the 5-HD alone groups, intracoronary (LAD) infusion of 5-HD began at $T = -45$ min, where T = 0 indicates the start of the 60 min LAD occlusion (see Fig. 1). In the preconditioned group (PC) the LAD was occluded for 10 min at $T = -40$ and $T = -20$ with each 10-min occlusion interval followed by 10 min of reperfusion. In the adenosine group (ADO), the R-PIA group, and the R-PIA plus 5-HD group, the respective drug was infused into the LAD for 10 min followed by a 10-min period of no infusion to mimic the timing of the LAD occlusion in the PC group. Adenosine was infused at 0.04 or 0.12 mg/kg/min (infusion volume approximately equal to 1 ml/min), and R-PIA was infused at 2.5 μg/kg/min (also approximating an infused volume of 1 ml/min). In the R-PIA plus 5-HD and the 5-HD alone groups, the 5-HD dose was 0.2 mg/kg/min delivered in a volume of approximately 1 ml/min.

At $T = 0$ the LAD was occluded and remained so for the next 60 min in all experimental groups. The 5-HD infusion was stopped (in the R-PIA plus 5-HD and the 5-HD alone groups) at $T = 5$ min, so that the K_{ATP} blocker was administered only during the time interval corresponding to the temporary occlusion–reperfusion cycle in the PC group plus 5 min on either end of that interval. Heparin (750 U/kg IV) was administered at $T = 55$ min. At $T = 60$ min the LAD occlusion was released, and the myocardium was reperfused for the next 90 min. The LAD was reoccluded after 90 min of reperfusion ($T = 150$ min). The ascending aorta was then immediately occluded and the heart was fibrillated. Zinc–cadmium sulfide particles (ZnCDS, 1–10 μm diameter, 75 ml of suspension containing 1.2 mg particles/ml) were injected into the aortic root to delineate the region of myocardial perfusion.

After the ZnCDS infusion the heart was excised, and the left ventricle plus the interventricular septum (LV) was dissected free of atrial, valvular, and right ventricular tissue. The LV was weighed and, after first being frozen, was cut into 8 mm thick transverse (parallel to the atrioventricular groove) slices, from the base to the apex. The LV slices were incubated in a solution of triphenyltetrazolium chloride [TTC, 1% (w/v) in sodium phosphate buffer] at 37°C for 20 min. After TTC staining the slices were viewed under ultraviolet light and the myocardial areas with and without ZnCDS fluorescence were measured using a digital planimeter for each tissue slice. After the initial planimetry, the LV tissue slices were incubated in 10% neutral buffered formalin for 15 min to enhance the TTC contrast. Subsequently, the distributions of TTC staining and of ZnCDS fluorescence were digitally planimetered. The ratio of the ZnCDS fluorescence area before formalin to after formalin was used to correct to the TTC stained area for tissue shrinkage owing to dehydration from formalin.

Planimetered LV tissue areas (cm^2) were converted to volume (cm^3) using slice thickness (0.8 cm). The area of myocardium stained by TTC was taken to indicate the area of viable myocardium, whereas the TTC-negative (unstained) area was accepted as indicating infarcted myocardium. The distribution of ZnCDS fluorescence indicated the area of perfused myocardium, and the area without ZnCDS fluorescence indicated the nonperfused (therefore at risk of infarction) zone of the LV. The primary measure of infarct size in this report is as the ratio of infarct zone (TTC negative) volume to risk zone (ZnCDS fluorescence negative) volume.

Physiological signals (electrocardiogram, heart rate, aortic pressure, and LAD flow) were acquired and recorded continuously. Data acquired at $T = -41, -31, -21, -11, -1, 5, 30, 59, 90, 120$, and 150 min (Fig. 1) were transcribed for subsequent off-line analysis. Statistical comparisons of data utilized analysis of variance techniques with post hoc testing by two-tailed t-test for between group comparisons and the Dunnett test for within group comparisons.

III. Effects on Infarct Size

Of 63 animals prepared for this study, 49 completed the protocol. The primary cause of attrition was refractory ventricular fibrillation and cardiogenic shock. The distribution of this complication was not different among the treatment groups. Hemodynamic and infarct size data were similar in the two ADO subgroups; therefore, the data were pooled for analysis of infarct size data into a single group. Similarly, in the R-PIA group two subsets were studied because of the apparent residual negative chronotrophic effect of the drug. In one subset the native heart rate was allowed ($n = 4$), and in the other subset demand atrio–ventricular sequential pacing was used to control the heart rate if the observed rate became less than the baseline ($T = -41$) value. Infarct size was not different in the two R-PIA subgroups so that the data were pooled to form a single R-PIA group. Heart rate, mean aortic pressure, and LAD flow values over the entire experimental time line are shown in Fig. 2, 3, and 4, respectively, for all six experimental groups. Infarct size determination showed the expected decrease in the PC group (Fig. 5). Both adenosine and R-PIA provided protection statistically equivalent to preconditioning, although the variance of the data in both groups is visibly greater than that seen in the PC group (Fig. 5). Addition of 5-HD to R-PIA abolished the protective effect seen with R-PIA alone, but 5-HD alone did not increase infarct size from the control value.

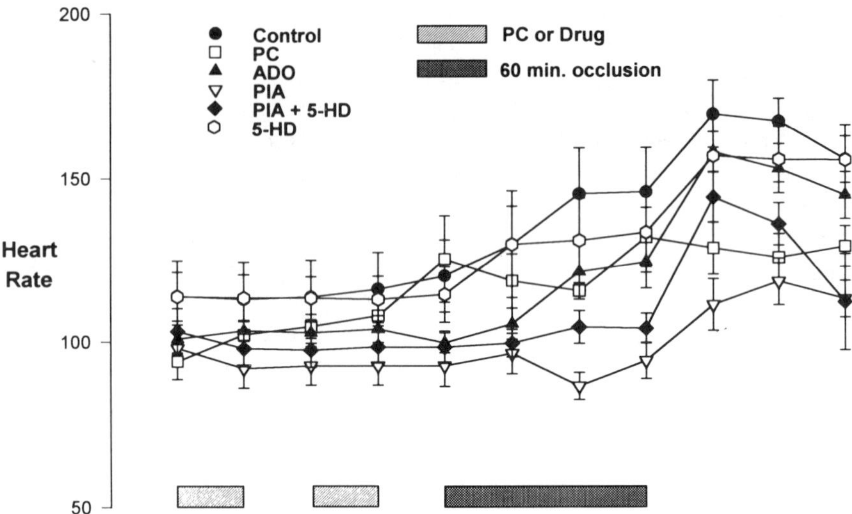

Fig. 2 Heart rate is plotted for 11 time points during the experiment in each of the six treatment groups. Single hatched bars indicate periods of temporary coronary occlusion or drug infusion. The double hatched bar indicates the 1-hr occlusion period. Note that the abscissa is not drawn to scale according to the experimental timeline (refer to Fig. 1). The two treatment intervals were 10 min in duration with an intervening 10-min intertreatment interval. During the 1 hr occlusion data were collected at 5, 30, and 59 min occlusion. Values are mean ± SEM.

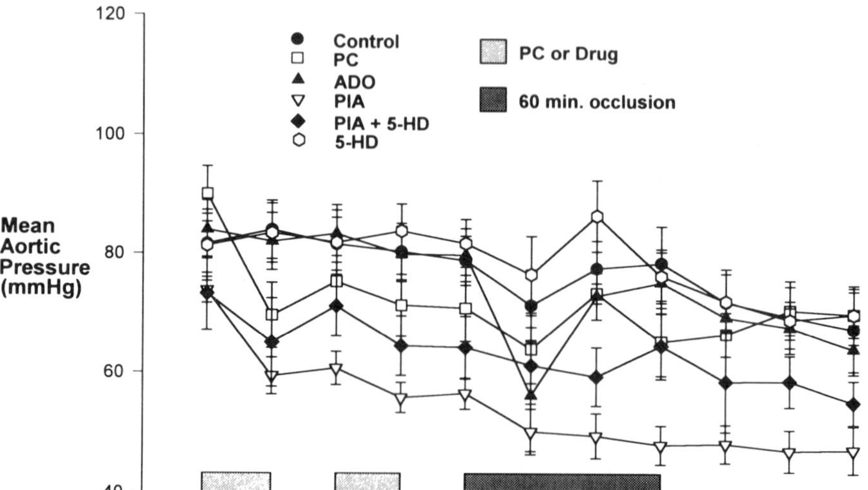

Fig. 3 Mean aortic pressure plotted over time for the six treatment groups. See legend to Fig. 2.

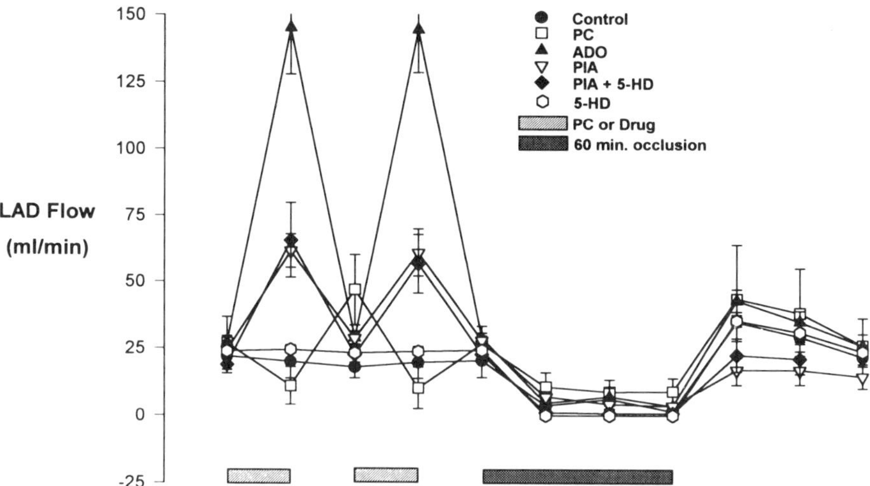

Fig. 4 Mean LAD flow plotted over time for each of the six treatment groups. See legend to Fig. 2.

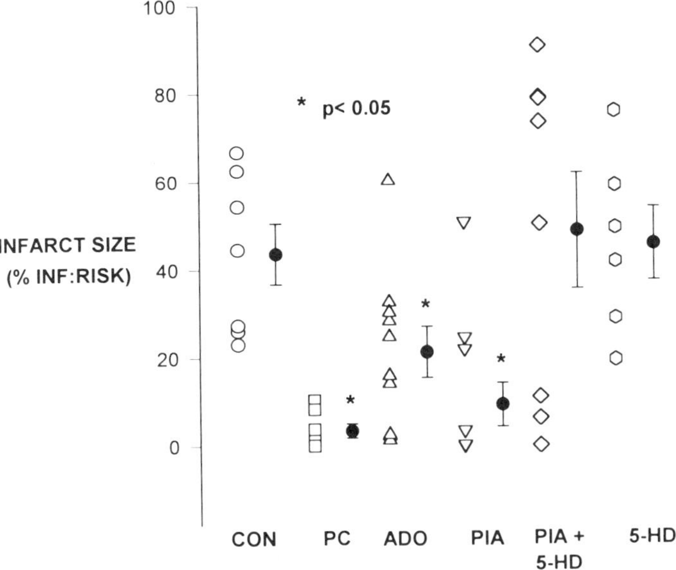

Fig. 5 Infarct size for individual animals in each group. Filled circles show the corresponding mean value ± SEM. Some individual data points may be obscured because of overlapping symbols. CON, $n = 7$; PC, $n = 7$; ADO, $n = 10$; R-PIA, $n = 11$; R-PIA + 5-HD, $n = 8$; 5-HD, $n = 6$.

IV. Discussion

The principal finding of this study is that localized activation of the adenosine A_1 receptor by intracoronary infusion of either adenosine or the A_1 receptor specific agonist R-PIA produces a decreased myocardial infarct size of a magnitude similar to that seen with ischemic preconditioning. Also, concomitant administration of the K_{ATP} channel blocker 5-HD results in a loss of the protective effect produced by R-PIA infusion. Because the 5-HD infusion alone did not increase infarct size above the control value, this apparent abolition of the R-PIA protective effect was not likely to have been due to an increased infarct size from the 5-HD that was then partially obscured by the protective effect of R-PIA.

The data intimate an interplay between the A_1 receptor and the K_{ATP} channel in providing protection from ischemic injury. Although our data do not definitely establish the sequence of such an interaction, careful analysis may provide some preliminary insight in this regard. The process(es) leading to cardiac myocyte protection could, conceptually, take the form of serial or parallel sequences of events (Fig. 6). If the K_{ATP} channel and the A_1 receptor events are seen as elements of separate arms of a parallel series of events that either independently or via a common element produce protection, then inhibition of either limb of the sequence should have little influence on the protective effect. This would be true

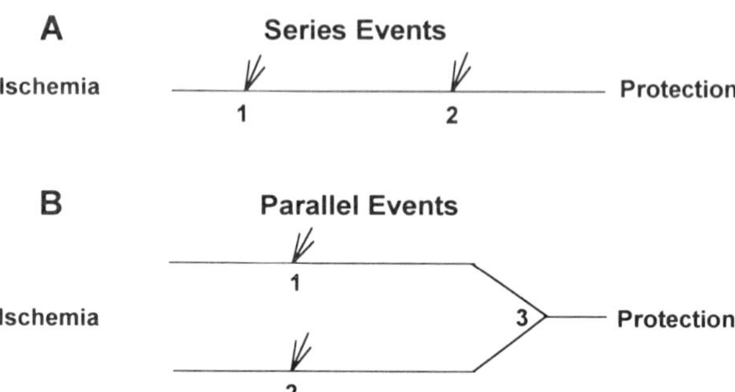

Fig. 6 Sequence of events leading from transient ischemia to protection schematically depicted as either a single series of events (A) or as parallel series of events (B). In a single series, blockade of a downstream event (A.2) would be expected to inhibit the effect of any activation of an upstream event (A.1), but the reverse would not be true. However, the protection produced by activation of the downstream event would not be affected by blockade of upstream activation. Two events on separate limbs of a parallel system (B.1 and B.2) would be less likely to be interactive unless a common intermediate is present (B.3), and a summation of the effects of the two limbs of the system is necessry in order to move the sequence past the point of confluence.

unless there were a requirement for a summation of the effects of the separate arms at the confluence of the system. Previous studies have shown that both K_{ATP} channel opening and A_1 receptor activation produce cardioprotection, an observation that is consistent with either the parallel or the series construct. Our data show that K_{ATP} channel blockade with 5-HD completely abolishes the protective effect of A_1 receptor activation with R-PIA, without itself increasing infarct size. If the hypothesized interaction of the A_1 receptor and K_{ATP} channel events is as elements in a single series of events, then inhibition of the downstream element would inhibit the protective effect despite activation of the upstream element; however, the reverse would not be true. For these reasons we speculate that both A_1 receptor activation and K_{ATP} channel opening are components of a series of events leading to protection in which K_{ATP} channel opening occurs subsequent to A_1 receptor activation. Further experiments are in progress to help clarify this point and to help elucidate the further subcellular events involved in the ischemic preconditioning phenomenon.

References

1. Reimer, K. A., Murry, C. E., Yamasawa, I., Hell, M. L., and Jennings, R. B. (1986). Four brief periods of myocardial ischemia cause no cumulative ATP loss or necrosis. *Am. J. Physiol.* **251,** H1306–H1315.
2. Murry, C. E., Jennings, R. B., and Reimer, K. A. (1986). Preconditioning with ischemia: A delay of lethal cell injury in ischemic myocardium. *Circulation* **74,** 1124–1136.
3. Schott, R. J., Rohmann, S., Braun, E. R., and Schaper, W. (1990). Ischemic preconditioning reduces infarct size in swine myocardium. *Circ. Res.* **66,** 1133–1142.
4. Tsuchida, A., Miura, T., Miki, T., Shimanoto, K., and Iimura, O. (1992). Role of adenosine receptor activation in myocardial infarct size limitation by ischemic preconditioning. *Cardiovasc. Res.* **26,** 456–461.
5. Velasco, C. E., Turner, M., Cobb, M. A., Virmani, R., and Forman, M. B. (1991). Myocardial reperfusion injury in the canine model after 40 minutes of ischemia: Effect of intracoronary adenosine. *Am. Heart J.* **122,** 1561–1570.
6. Liu, G. S., Thornton, J., Van Winkle, D. M., Stanley, A. W. H., Olsson, R. A., and Downey, J. M. (1988). Promotion against infarction afforded by preconditioning is mediated by A1 adenosine receptors. *Naunyn-Schmiedeberg's Arch. Pharmacol.* **337,** 687–689.
7. Gross, G. J., and Auchampach, J. A. (1992). Blockade of ATP-sensitive potassium channels prevents myocardial preconditioning in dogs. *Circ. Res.* **70,** 223–233.
8. Deutsch, E., Birger, M., Kussmaul, W. G., Hirshfeld, J. R., Jr., Herrmann, H., and Laskey, W. K. (1990). Adaptation to ischemia during PTCA. *Circulation* **82,** 2044–2051.
9. Jain, A. and Gettes, L. S. (1991). Patterns of ST-segment change during acute no-flow myocardial ischemia produced by balloon occlusion during angioplasty of the left anterior descending coronary artery. *Am. J. Cardiol.* **67,** 305–307.
10. Yellon, D. M., Alkhulaifi, A. M., and Pugsley, W. B. (1993). Preconditioning the human myocardium. *Lancet* **342,** 276–277.
11. Auchampach, J. A., Grover, G. J., and Gross, G. J. (1991). The ischemia-selective

ATP-sensitive potassium (K_{ATP}) channel antagonist, sodium 5-hydroxydecanoate (5-HD), blocks myocardial preconditioning in dogs. *Circulation* **84**(4), II-432 (abstract).

12. Auchampach, J. A., Maruyama, M., Cavero, I., and Gross, G. J. (1991). The new K^+ channel opener Aprikalim (RP 52891) reduces experimental infarct size in dogs in the absence of hemodynamic changes. *J. Pharmacol. Exp. Ther.* **259,** 961–967.
13. Thornton, J. D., Liu, G. S., Olsson, R. A., and Downey, J. M. (1992). Intravenous pretreatment with A1-selective adenosine analogues protects the heart against infarction. *Circulation* **85,** 659–665.
14. Van Winkle, D. M., Chien, G. L., Wolff, R. A., and Davis, R. F. (1992). Intracoronary infusion of *R*-phenylisopropyl adenosine prior to ischemia/reperfusion reduces myocardial infarct size in swine. *Circulation* **86**(4), I-213. (abstract).
15. Van Winkle, D. M., Chien, G. L., Wolff, R. A., Soifer, B. E., Kozume, K., and Davis, R. F. (1994). Cardioprotection provided by adenosine receptor activation is abolished by blockade of the ATP sensitive potassium channel. *Am. J. Physiol.* **266,** H829–H839.

Effects Hypoxia/Reoxygenation on Intracellular Calcium Ion Homeostasis in Ventricular Myocytes during Halothane Exposure

Paul R. Knight, Mitchell D. Smith, and Bruce A. Davidson

Department of Anesthesiology
The State University of New York at Buffalo
Buffalo General Hospital
Buffalo, New York 14203

I. Introduction

The pathogenesis of acute myocardial ischemia involves the interaction of a complex series of responses to hypoxia, diminished availability of substrates, and the accumulation of toxic cellular waste products. Although interruption of the oxygen supply to the heart represents only one insult, understanding the pathogenesis of hypoxic heart injury is important to provide a basis for determining how other factors, associated with a reduction in coronary perfusion, may modify the injury response. Furthermore, oxygen deprivation-induced myocardial cell injury may occur during depressed states of consciousness (e.g., anesthesia) where respiration is unintentionally interrupted while circulation to the heart is normal. During the anesthetic state the presence of potent myocardial depressant agents may also modify the response to hypoxia with or without myocardial hypoperfusion.

Previously, we have demonstrated that, in canines, 1% halothane is protective against the development of associated lethal ventricular tachy-

Advances in Pharmacology, Volume 31

arrhythmias produced by a brief period (20 min) of myocardial ischemia followed by reperfusion (1). We and others have demonstrated that the only other agents which offer protection in these types of ischemia–reperfusion experiments are the Ca^{2+} channel blockers (2). Because Ca^{2+} flux and the intracellular accumulation of Ca^{2+} are believed to be important in mediating the development of arrhythmias following coronary ischemia–reperfusion (3), we hypothesize that halothane prevents the occurrence of these tachyarrhythmias by either inhibiting the acute $[Ca^{2+}]_i$ overload by blocking Ca^{2+} channels or by depressing the direct effects of the high $[Ca^{2+}]_i$ on myocyte excitability. Because other drugs, known to decrease myocardial irritability, do not prevent the ischemia–reperfusion arrhythmias, we feel that the effects of halothane on voltage-gated Ca^{2+} channels is the most likely explanation for these findings (1). This hypothesis is also supported by evidence that halothane can decrease binding of Ca^{2+} channel blockers to voltage dependent Ca^{2+} channels and depress myocyte slow action potentials (4,5).

Possible causes for the $[Ca^{2+}]_i$ overload in the myocyte during ischemia include (1) sarcolemmal influx of extracellular Ca^{2+} through Ca^{2+} channels, (2) oxidant damage to the sarcolemma, and/or (3) reversal of ionic flow through the Na^+/Ca^{2+} antiport following acute cytosolic Na^{2+} overload (6). Mechanisms by which myocytes attempt to preserve a normal $[Ca^{2+}]_i$ in the presence of hypoxia-induced- Ca^{2+} leakage into the cytosolic compartment include extrusion by the Ca^{2+}-ATPase pump of the sarcolemma and sequestration of the ion by energy-requiring pumps into internal organelles, primarily the mitochondria and sarcoplasmic reticulum (SR). Ischemia may inhibit the Ca^{2+}-ATPase pumps owing to decreased ATP stores and inhibit the mitochondrial Ca^{2+} electrophoretic uniporter by preventing oxidative phosphorylation (6,7). The SR plays an extremely important function in this regard. Prevention of the ability of this organelle to take up Ca^{2+} during myocyte hypoxia *in vitro* by caffeine is associated with increased cytotoxicity (8).

In the experiments presented in this article, we test the hypothesis that exposure of myocytes to halothane decreases the severity of hypoxia–reoxygenation injury to the myocardium by preventing the accumulation of Ca^{2+} in the intracellular organelles, primarily the SR, during the hypoxic period. We also propose that this decreases the likelihood that the organelles will be injured because of high Ca^{2+} levels. Furthermore, owing to halothane-induced inhibition of SR Ca^{2+} accumulation during hypoxia and subsequent release following reoxygenation, possible injury to the contractile mechanism as a result of $[Ca^{2+}]_i$ overload (e.g., hypercontraction) as well as increased irritability of the ventricular myocyte (and myocardium as a whole) is prevented. We hypothesize that the decrease in hypoxia-induced releasable SR Ca^{2+} caused by halothane can explain the inhibition of ventricular fibrillation that usually occurs following the occlusion/release of regional coronary blood flow in canines receiving volatile anesthetics (1).

In these experiments, we have examined the role of altered Ca^{2+} homeostasis in the pathogenesis of acute cellular hypoxia and how halothane modifies these changes to decrease myocardial cell injury. The level of available Ca^{2+} in the SR is determined by using caffeine-induced release of these stores. We have also ascertained the viability and retention of function of the individual myocytes by morphology and the ability to electrically pace the cells during and after the hypoxic period. It is our hope that new insights into the effects of anesthetic agents on the pathogenesis of hypoxia–reoxygenation injury to single ventricular myocytes may serve as a foundation for developing strategies to decrease myocardial injury to patients.

II. Intracellular Calcium Measurements

A. Preparation of Ventricular Myocytes

Our method for the isolation of individual myocytes has been previously described (9,10) and is based on a modification of the procedure of Mitra and Morad (11). Briefly, female Sprague-Dawley rats (Charles River, Portage, MI) weighing 250–300 g, were anesthetized with 1–1.6% halothane and the heart mounted by the aorta on a modified Langendorff apparatus. The isolated hearts were perfused at 35°C with nominally Ca^{2+}-free ($[Ca^{2+}] < 50\ \mu M$) Tyrode's solution containing Type I collagenase (Sigma, St. Louis, MO). The bicarbonate-buffered Tyrode's solution also contained 0.1% bovine serum albumin (Sigma) and 15 mM taurine (Sigma). Following collagenase perfusion, the hearts were minced, placed in the collagenase solution again, and shaken in a water bath at 35°C. Serial samples were examined microscopically every 5 min until healthy single myocytes began appearing. The cells were then washed in bicarbonate-buffered Dulbecco's modified essential medium (DMEM; Gibco, Grand Island, NY) containing 1.8 mM Ca^{2+} and kept at 22°C. For determination of $[Ca^{2+}]_i$, cell suspensions were loaded with Fura-2 by bathing the cells in DMEM containing 4 μM Fura-2AM (Molecular Probes, Eugene, OR) for 10 min at 22°C. The myocytes were washed in DMEM and not used for at least 20 min to allow for the formation of the membrane-impermeant potassium salt of Fura-2 from deesterification of the permeant acetoxymethyl ester.

Suspensions of isolated Fura-2-loaded cells were placed in chambers in which the atmosphere was maintained with a flow of fresh gas containing 95% O_2/5% CO_2 or 95% N_2/5% CO_2 (by volume) with or without 1% halothane. The cells were also continuously superfused (~3 ml/min) with solutions aerated with the same gases listed above at 35°C. Myocytes were exposed to the anesthetic by adding halothane, via a Drager vaporizer, to the gas aerating the superfusion solution, as well as the gas flowing through the air space of the atmosphere chamber. The gas concentrations were analyzed periodically by using a Rascal II gas analyzer (Ohmeda, Salt Lake City, UT). The O_2 content of the superfusion solution was analyzed by periodic sampling of the effluent. During the experiments myocytes were maintained at 35°C.

B. Measurement of Intracellular Calcium-Ion Concentration

For each experiment, a single quiescent myocyte with normal morphology was selected. In all single-cell experiments, the myocyte had to initially elicit a Ca^{2+} transient following electrical impulses derived from field stimulation or exposure to 10 m*M* caffeine. The $[Ca^{2+}]_i$ was measured as described previously (10). Briefly, the intensity of the epifluorescence emission at 510 nm, resulting from the alternating excitation of intracellular Fura-2 by 340 and 380 nm light, was measured using a Leitz Diavert inverted microscope fitted with quartz optics and a Leitz MVP photomultiplier. Alternating 340 and 380 nm excitation light was produced by passing the beam from a xenon lamp through band-pass filters mounted in a computer-controlled filter wheel. Continuous recordings of $[Ca^{2+}]_i$ trends were expressed as a simple ratio of fluorescence intensity between the 340 and 380 nm excitation; however, at individual times (e.g., every 5 min), determinations of $[Ca^{2+}]_i$ were made from these measurements according to the formula of Grynkiewicz *et al.* using measurement of fluorescence intensities of Fura-2, unbound (EGTA-buffered) and maximally bound (Ca^{2+}-saturated), at the end of each individual experiment (12).

Myocytes were placed in the atmosphere chamber and allowed to attach loosely to the cover slide. Following an initial period of observation, during which the fluorescent Fura-2 signal stablized, the selected cell was stimulated either by electrical field stimulation, using periodic voltage pulses applied to platinum electrodes situated at both ends of the atmosphere chamber, or by superfusion of the atmosphere chamber with 10 m*M* caffeine. Myocytes were then superfused with a normoxic solution with or without 1% halothane for 5 min after which time the superfusate was changed to one aerated with N_2/CO_2 with or without 1% halothane. The myocytes were exposed to the deoxygenated solution for 20 min followed by superfusion with normoxic Tyrodes solution without anesthetic. The continuous flow of gases through the air space of the atmosphere chamber was also changed to be consistent with that which was aerating the superfusate. Stimulation of intracellular Ca^{2+} transients was attempted every 10 min by electrical field stimulation, or after 5, 10, and 20 min of hypoxia or 15 min following reoxygenation with 10 m*M* caffeine. Intracellular Ca^{2+} was continuously monitored during normoxia, hypoxia, and reoxygenation and the presence of spontaneous contractions noted. In addition, the cells were observed periodically ($\sim$1 min^{-1}) for changes in cell morphology. After visual observation, the myocytes were scored as being either (1) a normal rod with striations (normal), (2) a hypercontracted myocyte with striations (hypercontracted), or (3) an amorphous, rounded cell without striations (rounded). The rounded cells were also found to take up trypan blue dye. In cultures not loaded with Fura-2 ($n = 3$), aliquots of cells were periodically examined during and following hypoxia and reoxygenation at 4°C and scored for morphological type.

For statistical analysis, a minimum of 10 myocytes was used for each condition ($n = 10$) when individual cells were examined. When aliquots of hypoxic myocyte preparations were assayed for morphology, 250 cells were counted from each of four quadrants of the culture dish. Statistical inferences for parametric variables were determined using Student's, unpaired *t*-test corrected for multiple comparisons by analysis of variance (ANOVA) and the Scheffe *f*-test. Values are expressed as means ± SEM. Non parametric data were analyzed using chi-square (χ^2) and maximum likelihood test. Statistical significance was accepted at the 5% level ($p < 0.05$).

III. Effects of Hypoxia, with or without Halothane, on Myocyte Morphology

The morphological integrity of myocytes exposed to halothane during hypoxia and reoxygenation was better maintained than that of cells which were not exposed to the anesthetic during the hypoxic insult (controls). During the hypoxic period, most of the observed myocytes (80%) underwent a slight, but recognizable, shortening after approximately 10 min of hypoxia but maintained otherwise normal morphology. The first hypercontraction of an individual control myocyte was observed following 10 min of hypoxia, and two additional control cells hypercontracted between 10 and 20 min of hypoxia. Within 5 min following reoxygenation, there was a second increase in the number of hypercontracted cells in both control and halothane-exposed preparations. Of the individual cells examined,

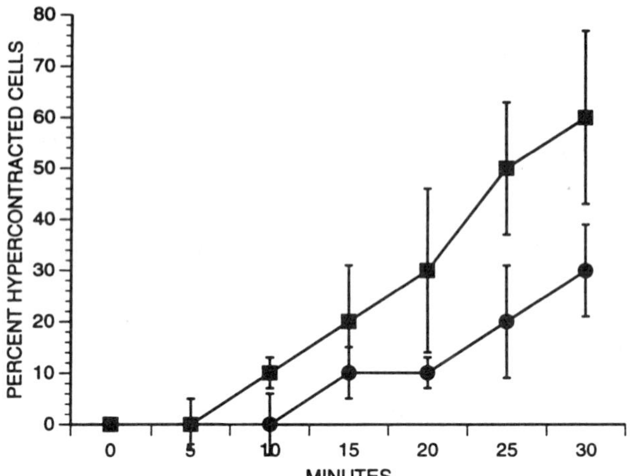

Fig. 1 Comparison of the effects of halothane and hypoxia–reoxygenation on cell morphology. Preparations of individual myocytes were incubated under hypoxic conditions, in the presence (●) or absence (■) of 1% halothane followed by reoxygenation. Aliquots of cells were removed at 5-min intervals and scored as to the increase in hypercontracted (and rounded) cells in the population over prehypoxic levels. Each point represents the average of three separate experiments with 1000 cells per experiment (250 × 4).

only 30% of those receiving halothane underwent hypercontraction and subsequent rounding by the end of the hypoxia and reoxygenation observation period. However, 60% of control cells hypercontracted during this period ($p < 0.05$). Similarly, when large populations of myocytes were examined morphometrically, fewer had hypercontracted or rounded up in the preparations exposed to halothane compared to the controls during hypoxia and following reoxygenation (Fig. 1).

IV. Effects of Hypoxia, with or without Halothane, on Intracellular Calcium

In preliminary studies, approximately 20% of the control cells underwent spontaneous Ca^{2+} transients, with associated contractions, within 2 min of the cultures becoming hypoxic. These hypoxia-induced Ca^{2+} transients tended to last for 3–4 min and then stopped. The cells that did develop spontaneous contractions eventually underwent hypercontraction and rounded up. None of the myocytes that received halothane prior to and during hypoxia developed spontaneous contractions. In subsequent experiments, only myocytes which did not develop spontaneous Ca^{2+} transients following hypoxia were studied. Interestingly, exposure of the cells to 1%

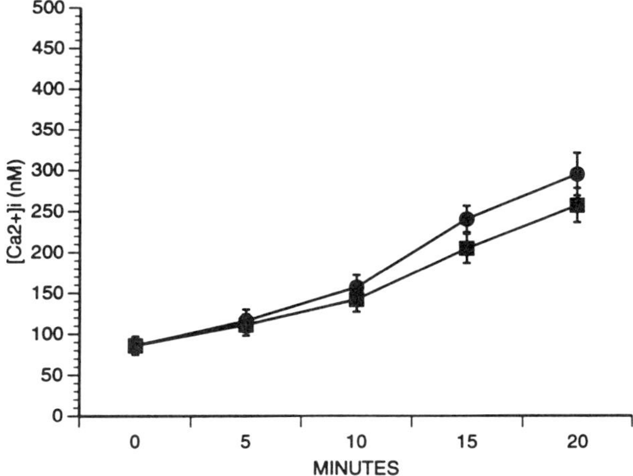

Fig. 2 Comparison of the effects of hypoxia in the presence (●) or absence (■) of 1% halothane on basal $[Ca^{2+}]_i$. Isolated, normal appearing, ventricular myocytes were electrically stimulated and $[Ca^{2+}]_i$ determined every 5 min as described in the text, during a 20-min period of hypoxia. Data are presented as means ± SEM (n = 10/group).

halothane or 1 μM nitrendopine after the initiation of hypoxia-induced spontaneous contractions resulted in an inhibition of the spontaneous myocyte activity (data not shown). This suggests that leakage of Ca^{2+} through voltage-gated calcium channels in the sarcolemma was the etiology of these contractions.

The effects of hypoxia on $[Ca^{2+}]_i$ in control myocytes and those exposed to 1% halothane are demonstrated in Fig. 2. Prior to hypoxia the resting

Fig. 3 Trace of changes in $[Ca^{2+}]_i$ recorded from a single rat ventricular myocyte that underwent 20 min of hypoxia, without halothane exposure, followed by reoxygenation. Hypoxia was initiated at the first arrow, and reoxygenation began at the second arrow. Note the voltage-stimulated Ca^{2+} transients prior to hypoxia. Morphologically, this cell appeared normal until shortly after reoxygenation, following which the $[Ca^{2+}]_i$ increased dramatically and the cell underwent hypercontraction.

$[Ca^{2+}]_i$ in the myocytes was 86 ± 11 nM (n = 20). The myocytes exposed to halothane had a slightly greater increase (not statistically significant) in $[Ca^{2+}]_i$ by the end of 20 min of hypoxia when compared to controls: 298 ± 26 nM (n = 9) and 256 ± 21 nM (n = 7), respectively. Interestingly, 70% of the myocytes exposed to halothane during hypoxia returned to prehypoxic $[Ca^{2+}]_i$, whereas in 60% of the control cells the $[Ca^{2+}]_i$ kept increasing during the reoxygenation period until the Fura-2 signal was identical to the maximal Ca^{2+}-bound state of the dye (i.e., after treating the cell with ionomycin). Figure 3 is a trace of a single cell that demonstrates this observation. This increase in $[Ca^{2+}]_i$ was also associated with hypercontraction of the myocyte followed by loss of structural integrity and rounding. The rounded cells did not exclude trypan blue dye.

V. Effects of Hypoxia, with or without Halothane, on Electrically Induced Calcium Transients

All myocytes (n = 20) could be paced prior to hypoxia. After 5 min of hypoxia, with or without halothane, none of the cells could be electrically stimulated. None of the the control myocytes, which were able to maintain normal morphology in the posthypoxic period, could be stimulated to elicit a Ca^{2+} transient using electrical field stimulation (n = 4). In comparison, 3 of the 7 cells in the group exposed to halothane during hypoxia, which demonstrated normal morphology, could be paced in the posthypoxic period (Fig. 4).

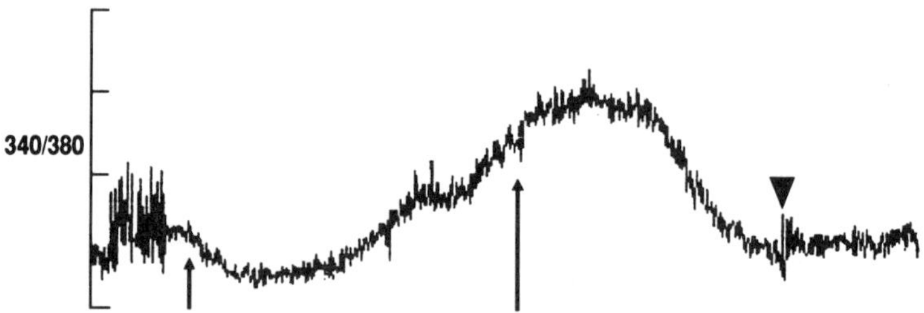

Fig. 4 Trace of changes in $[Ca^{2+}]_i$ recorded from a single rat ventricular myocyte that underwent 20 min of hypoxia, in the presence of 1% halothane, followed by reoxygenation. Hypoxia was initiated at the first arrow, and reoxygenation began at the second arrow. Note the voltage-stimulated Ca^{2+} transients prior to hypoxia and following reoxygenation (▼). These transients were also associated with contraction–relaxation of the myocyte, which could be concurrently observed visually. Morphologically, this cell appeared normal for the entire observation period. The ability to electrically pace myocytes following hypoxia–reoxygenation only occurred in the myocytes exposed to 1% halothane during the hypoxic insult (30% of the time, n = 10).

VI. Effects of Hypoxia, with or without Halothane, on Caffeine-Induced Calcium Transients

Halothane-exposed myocytes demonstrated a different response to 10 m*M* caffeine when compared to controls. No Ca^{2+} transients could be elicited from the anesthetized cells with 10 m*M* caffeine during the hypoxic period ($n = 10$). However, following hypoxic exposure, a normal caffeine transient (e.g, a rapid rise in $[Ca^{2+}]_i$ followed by a return to basal levels) was produced 15 min following reoxygenation in 8 of the cells. Conversely, in control myocytes, a reduced Ca^{2+} transient could be elicited in the first 5 min of hypoxia, in 50% of the cells. In 50% of the caffeine-stimulated control cells, with normal morphology after 10 or 15 min of hypoxia ($n = 10$), or in 40% of the control myocytes 15 min after reoxygenation ($n = 10$), the $[Ca^{2+}]_i$ increased rapidly, followed by hypercontraction of the cell and subsequent rounding (Fig. 5). The control myocytes that did not hypercontract following superfusion with Tyrodes solution, containing 10 m*M* caffeine, failed to produce any Ca^{2+} transient.

VII. Discussion

The object of the present study has been to explore the interaction of hypoxia and halothane on intracellular Ca^{2+} homeostasis, in order to examine possible mechanisms by which the anesthetic may protect the myocardium during hypoxic insult. These experiments demonstrate that concurrent exposure of individual myocytes to halothane during hypoxia

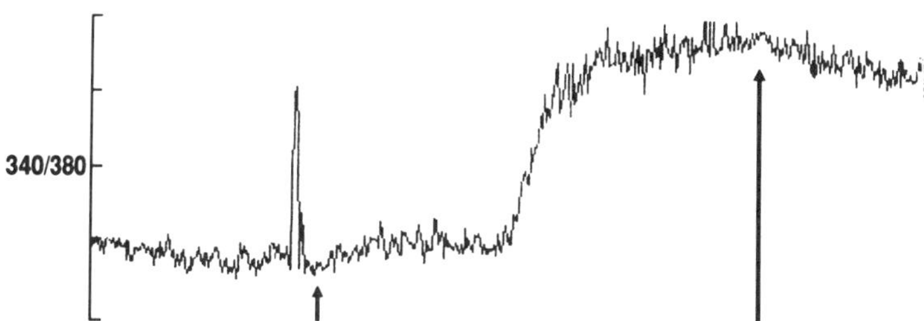

Fig. 5 Trace of changes in $[Ca^{2+}]_i$ recorded from a single rat ventricular myocyte that underwent 10 min of hypoxia, without halothane, and was then exposed to 10 m*M* caffeine. Hypoxia was initiated at the first arrow, and reoxygenation began at the second arrow. Note the caffeine-elicited Ca^{2+} transient prior to hypoxia. Morphologically, this cell appeared normal until shortly after adding 10 m*M* caffeine to the superfusate, following which the $[Ca^{2+}]_i$ increased dramatically and the cell underwent hypercontraction. This change in $[Ca^{2+}]_i$ following 10 m*M* caffeine only occurred in cells not exposed to halothane (50% of the time, $n = 10$/group).

is associated with a beneficial outcome based on indicies of structure and function. Greater numbers of anesthetic-exposed cells remain rod shaped (with striations) than control cells following hypoxia. Furthermore, approximately 40% of the halothane-treated myocytes could be induced to undergo normal appearing Ca^{2+} transients, and associated contraction–relaxation, in response to electrical field pacing, and 80% of the cells had transitory Ca^{2+} increases in response to 10 m*M* caffeine pulse stimulation following reoxygenation. In addition, the differences in Ca^{2+} release from internal stores, primarily the SR, by caffeine in the halothane-treated and control myocytes also suggest a possible mechanism by which the anesthetic may exert its beneficial effects during hypoxia. At the end of the hypoxic period, no Ca^{2+} transients could be elicited by 10 m*M* caffeine exposure in the group exposed to the anesthetic, whereas, in 50% of the control group, rapid $[Ca^{2+}]_i$ increases occurred after caffeine stimulation, which was ultimately associated with hypercontraction of the cell. Assuming that caffeine-releasable internal stores represent, primarily, SR-sequestered Ca^{2+} that has leaked into the cell as a result of hypoxia, these findings suggest that halothane prevents accumulation of Ca^{2+} in this organelle during hypoxia. Prevention of Ca^{2+} accumulation during hypoxia represents a possible mechanism by which the volatile anesthetic may protect the myocardium during hypoxia. Although somewhat limited in scope, these findings are consistent with the hypothesis that halothane protects the myocardium at the cell level by beneficially altering changes in Ca^{2+} handling during the hypoxic period.

In previous studies in canines, we have demonstrated that halothane can prevent the occurrence of fatal ventricular arrhythmias associated with temporary occlusion of the coronary artery for 20 min followed by reestablishment of flow (occlusion–release) (1). We have hypothesized that this advantageous anesthetic action is a result of either inhibition of the acute $[Ca^{2+}]_i$ overload associated with myocardial ischemia–reperfusion or depression of the direct effects of elevated $[Ca^{2+}]_i$ on myocyte excitability. Although caution must be exercised when extrapolating findings from one experimental model to another, the large quantity of Ca^{2+} released from control myocytes by caffeine following 20 min of hypoxia offers an attractive cause and effect for the increase in ventricular irritability associated with myocardial reperfusion following ischemia. If this is indeed true, the lack of an increase in caffeine-releasable Ca^{2+} in the anesthetized cells following reoxygenation suggests a possible cellular mechanism for halothane-induced prevention of occlusion–release ventricular arrhythmias in the anesthetized canine model.

In another report also using isolated rat ventricular myocytes, we have demonstrated that exposure of myocytes to halothane results in a decrease in caffeine-releasable intracellular Ca^{2+} during normoxia (9). In addition, the experiments performed in this study support the contention that the decrease in intracellular Ca^{2+} associated with anesthetic exposure is a result of leakage out of and/or decreased uptake by the SR. The increased leakage of Ca^{2+} out of the SR is also supported by data from other investigators that there is increased binding of ryanodine to isolated SR vesicles in the presence of halothane (13). Increased binding of this compound to the internal Ca^{2+} release channel is associated with an increase in the open state of the channel and leakage of Ca^{2+} from the SR. Because caffeine is unable to produce a Ca^{2+} transient in the anesthetic-treated, hypoxic cells, anesthetic-induced depletion of SR Ca^{2+} would appear to be an attractive hypothesis to account for these findings. However, if this hypothesis is correct, the hypoxia-induced leakage of Ca^{2+} into the cell (6) would also tend to increase $[Ca^{2+}]_i$ in the halothane-exposed hypoxic myocytes as compared to controls, since, during halothane exposure, the SR does not have the capacity to take up the increased Ca^{2+} leak associated with hypoxia. Furthermore, inhibition of the ability of this organelle to take up Ca^{2+} during myocyte hypoxia *in vitro* by caffeine is associated with increases in $[Ca^{2+}]_i$ and increased cytotoxicity (8). Because this increase $[Ca^{2+}]_i$ over controls does not occur, even though there is an obvious decrease in SR releasable Ca^{2+}, these findings suggest that halothane may work by blocking leakage of Ca^{2+} into the myocyte at one or more sites.

Organelles other than the SR also exert control over $[Ca^{2+}]_i$. Involvement of the sarcolemma in control of $[Ca^{2+}]_i$ occurs primarily via the Ca^{2+}-ATPase pump, voltage-gated Ca^{2+} channels, the Na^+/Ca^{2+} antiport, and Ca^{2+}–calcium exchange mechanisms. Hypoxia-induced leakage of extracellular Ca^{2+} into the myocyte may occur through the sarcolemma at several sites. These include the voltage-gated Ca^{2+} channels, oxidant-induced sites of sarcolemmal damage, or the Na^+/Ca^{2+} antiport activated as a result of hypoxia-associated increases in $[Na^+]_i$. Mitochondria also control $[Ca^{2+}]_i$ by controlling Ca^{2+} influx and efflux between the sarcoplasm and the organelles via the Ca^{2+} electrophoretic uniporter, Ca^{2+} calcium release channels, and a Ca^{2+}-ATPase pump (13). The mitochondria, like the SR, can sequester a large amount of Ca^{2+} internally in both readily releasable and complexed forms and, therefore, are potentially capable of protecting against injurious increases in $[Ca^{2+}]_i$, although this process is quite slower when compared to the SR and sacrolemmal responses.

Mitochondria represent a high-capacity Ca^{2+} buffering system which can avidly take up Ca^{2+} during periods of elevated $[Ca^{2+}]_i$ (14). Some investigators have demonstrated a role for mitochondrial dysfunction in the pathogenesis of ischemia–reperfusion injury to the myocardium (15,16). After 30 min of anoxia, reoxygenation is associated with increased leakiness of the mitochondrial membrane to Ca^{2+}, and there is a net increase in the cycling of Ca^{2+} across the mitochondrial membrane which can lead to an increase $[Ca^{2+}]_i$ (16). Halothane inhibits oxidative phosphorylation, thereby inhibiting the negative intramitochondrial potential which is required to drive the Ca^{2+} uniporter (17). This would tend to inhibit the accumulation of Ca^{2+} within the mitochondria and its subsequent release during reoxygenation. However, as with reducing SR accumulation of Ca^{2+}, an increase in $[Ca^{2+}]_i$ over controls would be predicted if halothane does inhibit Ca^{2+} uptake by the mitochondria. As previously stated this does not occur. Furthermore, several investigators have reported no increase in mitochondrial Ca^{2+} following short-term (30 min) hypoxia (16). Based on these studies it would appear unlikely that the mitochondrion is the site of the beneficial actions of halothane.

A number of sites on the sarcolemma have been implicated for cytosolic leakage of Ca^{2+} during and following hypoxia. As previously stated, these include the voltage-gated Ca^{2+} channels, the Na^+/Ca^{2+} antiport, and damage to the sarcolemmal membrane by oxygen free radicals. Intracellular Na^+ can accumulate as a result of cellular hypoxia (6). The increase in $[Na^+]_i$ would make available additional levels of this ion to drive the Na^+/Ca^{2+} exchange. Inhibition of the Na^+/Ca^{2+} antiport activity by halothane represents a possible mechanism by which the anesthetic may inhibit the increase in intracellular Ca^{2+} during hypoxia; however, data supporting a major role in protecting the myocyte are not very compelling (18).

Oxidant-mediated damage of the sarcolemma is associated with hypoxia, and reperfusion and causes increased Ca^{2+} influx (19). Antioxidants used during the hypoxic period can decrease the Ca^{2+} influx and improve ventricular performance following hypoxia and reoxygenation (20). However, this source of increased Ca^{2+} entry into the myocyte is not likely to be inhibited by halothane exposure, since the anesthetic increases cellular injury during concurrent oxidant treatment (21).

Voltage-gated Ca^{2+} channels are a potential source of increased Ca^{2+} influx during hypoxia and reoxygenation and may serve as a site of anesthetic action to inhibit these increases in $[Ca^{2+}]_i$. Treatment of myocytes with verapamil prior to hypoxia decreases the accumulation of Ca^{2+} during and following reoxygenation (15). Interestingly, the Ca^{2+} channel blocker must be given prior to the institution of hypoxia in order to produce a beneficial effect. As previously stated, changes in myocyte Ca^{2+} flux are believed to be intricately involved with the pathogenesis of arrhythmias during acute coronary occlusion (1,2,3). In this regard, the voltage-gated Ca^{2+} channels are thought to play a major role in the alterations in Ca^{2+} flux since verapamil, as well as other Ca^{2+} channel blockers, exerts a protective effect against the generation of experimental ischemia–reperfusion ventricular fibrillation (1,2,22,23,24). A similar protective effect is not afforded by other antiarrhythmic drugs which do not work at the Ca^{2+} channel (1). Interestingly, in addition to Ca^{2+} channel blockers, halothane, and to a lesser degree enflurane and isoflurane, can also protect the myocardium against ischemia–reperfusion ventricular fibrillation (1). Clearly, voltage-gated Ca^{2+} channels represent a major site of action of volatile anesthetics in the nonischemic heart (4,25,26). Halothane inhibition of hypoxia-induced increases in voltage-gated Ca^{2+} channel activity, and the resultant Ca^{2+} influx through the sarcolemma, in combination with an anesthetic-induced decrease in SR Ca^{2+} levels, via inhibition of Ca^{2+} uptake and/or increased leakage, would provide an attractive mechanism to explain the lack of accumulation of Ca^{2+} in the organelles of the hypoxic myocyte during and following hypoxia in the anesthetic-exposed cells. In addition, the ability of halothane to prevent Ca^{2+} influx through the sarcolemma could also explain how halothane protects against occlusion–release ventricular arrhythmias.

In summary, hypoxia–reoxygenation of ventricular myocytes is associated with pathological changes in the way that myocytes handle Ca^{2+}. Even apparently normal cells (normal morphology and resting $[Ca^{2+}]_i$) are impaired (i.e., have an inability to be electrically paced following hypoxia and difficulty in handling large sudden increases in the $[Ca^{2+}]_i$ produced by 10 m*M* caffeine). Exposure to halothane during hypoxia and reoxygenation appears to give some protection to the ventricular myocyte as demonstrated by a greater propensity for maintenance of normal morphology and resting $[Ca^{2+}]_i$ in the postischemia reoxygenation period and the ability to handle pacing-induced and caffeine-induced sudden increases in $[Ca^{2+}]_i$. Furthermore, exposure of the myocytes to the anesthetic prevents the large increase in caffeine-releasable Ca^{2+} associated with hypoxia and reperfusion. These findings support the hypothesis that exposure of myocytes to halothane decreases the hypoxia–reoxygenation injury to the myocardium by preventing the accumulation of Ca^{2+} in the intracellular organelles, primarily the SR, during the hypoxic period. This inhibition of Ca^{2+} overload can be expected to decrease the likelihood that cytosolic organelles will be injured because of high Ca^{2+} levels, and that reoxygenation-induced increases in $[Ca^{2+}]_i$ as a result of Ca^{2+} accumulated during hypoxia may injure the contractile mechanism. These anesthetic-induced changes in how the hypoxic myocytes handle Ca^{2+} can help explain the decrease in myocardial damage and irritability seen *in vivo*.

Acknowledgments

This work has been supported in part by National Institutes of Health Grant GM 39227.

References

1. Kroll, D. A., and Knight, P. R. (1984). Antifibrillatory effects of volatile anesthetics in acute occulusion/reperfusion arrhytmias. *Anesthesiology* **61,** 657–661.
2. Ribeiro, L. G. T., Brandon, T. A., Debauche, T. L., Maroko, P. R., and Miller, R. R. (1981). Antiarrhythmic effects of calcium channel blocking agents during coronary artery reperfusion. Comparative effects of verapimil and nifedipine. *Am. J. Cardiol.* **48,** 69–74.
3. Clusin, W. T., Bristow, M. R., Karaguezian, H. S., Katzung, B. G. and Schroeder, J. S. (1982). Do calcium-dependent ionic currents mediate ischemic ventricular fibrillation? *Am. J. Cardiol.* **49,** 606–612.
4. Drenger, B., Quigg, M., and Blanck, T. J. J. (1991). Volatile anesthetics depress calcium channel blocker binding to cardiac membranes. *Anesthesiology* **74,** 155–165.
5. Lynch, C., Vogel, S., and Sperelakis, N. (1981). Halothane depression of myocardial slow action potentials. *Anesthesiology* **55,** 360–368.
6. Murphy, J. G., Marsh, J. D., and Smith, T. W. (1987). The role of calcium in ischemic myocardial injury. *Circulation* **75,** v15–v24.
7. Altschuld, R. A., Hostetler, J. R., and Brierley, G. P. (1981). Response of isolated rat heart cells to hypoxia, reoxygenation, and acidosis. *Circ. Res.* **49,** 307–316.
8. Allshire, A., Piper, H. M., Cuthbertson, K. S. R., and Cobbold, P. H. (1987). Cytosolic free Ca^{2+} in single rat heart cells during anoxia and reoxygenation. *Biochem. J.* **244,** 381–385.
9. Wilde, D. W., Knight, P. R., Sheth, N., and Williams, B. A. (1991). Halothane alters control of intracellular Ca^{+2} mobilization in single rat ventricular myocytes. *Anesthesiology* **75,** 1075–1086.
10. Wilde, D. W., Davison, B. A., Smith, M. D., and Knight, P. R. (1993). The effects of isoflurane and enflurane on intracellular Ca^{+2} mobilization in isolated cardiac myocytes. *Anesthesiology* **79,** 73–82.
11. Mitra, R., and Morad, M. (1985). A uniform enzymatic method for dissociation of myocytes from hearts and stomachs of vertebrates. *Am. J. Physiol.* **249,** H1056–H1160.
12. Grynkiewicz, G., Poenie, M., and Tsien, R. Y. (1985). A new generation of Ca^{2+} indicators with greatly improved fluorescence properties. *J. Biol. Chem.* **260,** 3440–3448.
13. Frazer, M. J., and Lynch, C. Ca release channels of cardiac (SR) are activated by halothane, but not by isoflurane. Submitted for publication.
14. Carafoli, E. (1985). The homeostasis of calcium in heart cells. *J. Mol. Cell. Cardiol.* **17,** 203.
15. Miller, L. S., Barrett, J. N., Cameron, J. S., and Bassett, A. L. (1984). Differential effects of calcium antagonists on viability of adult rat ventricular myocytes. *J. Mol. Cell Cardiol.* **16,** 427.
16. Cheung, J. Y., Leaf, A., and Bonventre, J. V. (1986). Mitochondrial function and intracellular calcium in anoxic cardiac myocytes. *Am. J. Physiol.* **250,** C18.
17. Nahrwold, M. L., and Cohen, P. J. (1975). Anesthetics and mitochondrial respiration, *In* "Molecular Aspects of Anesthesia" (P. J. Cohen, ed.), pp. 25–44. Davis, Philadelphia, Pennsylvania.
18. Hayworth, R. A., and Goknur, A. B. (1991). Inhibition of Na/Ca exchange in heart cells by enflurane, isoflurane, and halothane. *Anesthesiology* **75,** A576 (abstract).
19. Jolly, S. R., Kane, W. J., Baille, M. B., Abrams, G. D., and Lucchesi, B. R. (1984). Canine myocardial reperfusion injury. *Circ. Res.* **54,** 277.
20. Otani, H., Engelman, R. M., Roussou, J. A., Breyer, R. H., and Lemeshow, S. das D. K. (1986). Cardiac performance during reperfusion improved by pretreatment with oxygen free radical scavengers. *J Thorac Cardiovasc. Surg.* **91,** 290.

21. Shayevitz, J. R., Varani, J., Ward, P. A., and Knight, P. R. (1991). Halothane and isoflurane increase pulmonary artery endothelial cell sensitivity to oxidant mediated injury. *Anesthesiology* **74,** 1067–1077.
22. Brooks, W. W., Verrier, R. L., and Lown, B. (1980). Protective effect of verapamil on vulnerability to ventricular fibrillation during myocardial ischemia and reperfusion. *Cardiovasc. Res.* **14,** 295–302.
23. Fagbemi, O., and Parratt, J. R. (1981). Calcium antagonists prevent early post infarction ventricular fibrillation. *Eur. J. Pharmacol.* **75,** 179–185.
24. Fagbemi, O., and Parratt, J. R. (1981). Suppression by oral-administered nifedipine, nisoldipine, and niludipine of early life-threatening ventricular arrhythmias resulting from acute myocardial ischeamia. *Br. J. Pharmacol.* **74,** 12–14.
25. Blanck, T. J. J., Runge, S., and Stevenson, R. L. (1988). Halothane decreases calcium channel antagonist binding to cardiac membranes. *Anesth. Analg. (N.Y.)* **67,** 1032–1035.
26. Bosnjak, Z. I., Supan, F. D., and Rusch, N. J. (1988). The effects of halothane, enflurane, and isoflurane on calcium currents in isolated canine ventricular cells. *Anesthesiology* **74,** 340–345.

Mechanical Consequences of Calcium Channel Modulation during Volatile Anesthetic-Induced Left Ventricular Systolic and Diastolic Dysfunction

Paul S. Pagel* and David C. Warltier†,‡

*Departments of *Anesthesiology, †Pharmacology, and ‡Medicine*
The Medical College of Wisconsin
Milwaukee, Wisconsin 53226

I. Introduction

A major goal of recent work from our laboratory has been directed toward a comprehensive understanding of the effects of volatile anesthetics on left ventricular mechanical function during systole and diastole. Potent inhalational agents, including isoflurane and halothane, depress myocardial contractility, (1–3), prolong isovolumic relaxation (4–6), and impair rapid ventricular filling (7,8) during early diastole in a dose-related manner *in vivo*. Halothane may also decrease global ventricular chamber compliance, though that conclusion remains controversial (4,7,9,10). The negative inotropic and lusitropic effects produced by volatile anesthetics in the intact heart occur as a result of modulation of calcium ion (Ca^{2+}) homeostasis within the cardiac myocyte (11). Isoflurane and halothane differentially alter transsarcolemmal Ca^{2+} flux by affecting the function, reducing the number, or interfering with the dihydropyridine binding sites of voltage-dependent Ca^{2+} channels (12–16). This action precipitates a sequence of events which ultimately lead to a decline in the intracellular availability of Ca^{2+} for contractile activation. Volatile anesthetics also

Advances in Pharmacology, Volume 31

provoke enhanced Ca^{2+} leak from the sarcoplasmic reticulum (17), contributing to decreases in total Ca^{2+} accumulation and release by this organelle during systole and, possibly, to delays in the removal of Ca^{2+} from the environment of the contractile apparatus during diastole. These intracellular actions of volatile anesthetics cause direct declines in contractile force and may also result in concomitant delays in isovolumic relaxation and impairment of rapid ventricular filling *in vivo*.

Previous investigations from this (7,8) and other laboratories (18–22) have demonstrated that administration of positive inotropes such as exogenous calcium chloride ($CaCl_2$), β_1-adrenoceptor agonists, or inhibitors of cardiac phosphodiesterase fraction III (PDE III) reverse volatile anesthetic-induced depression of myocardial contractility by increasing the intracellular concentration of Ca^{2+} available for contraction. This occurs by direct augmentation of Ca^{2+} supply or via the intracellular actions of cyclic adenosine monophosphate (cAMP). Reversal of depression of left ventricular function during early diastole caused by halothane and isoflurane has also been observed with $CaCl_2$ (7) and amrinone (8), a PDE III inhibitor. The positive lusitropic effects of $CaCl_2$ probably result directly from enhanced ventricular performance during systole. Although PDE III inhibitors may also enhance diastolic function through direct positive inotropic effects and optimized ventricular loading conditions, these agents also stimulate protein kinase-mediated phosphorylation of phospholamban, the regulatory protein of the sarcoplasmic reticulum Ca^{2+}-ATPase, through increases in the concentration of intracellular cAMP, thereby augmenting the diastolic Ca^{2+} uptake by this organelle (23). In addition, cAMP-induced phosphorylation of the troponin I subunit of the troponin–tropomyosin complex decreases the affinity of troponin C for Ca^{2+}, enhancing the dissociation of Ca^{2+} from this regulatory protein during diastole (24). Thus, the positive lusitropic properties produced by PDE III inhibitors in the intact heart can also be attributed to enhanced release of Ca^{2+} from the contractile apparatus coupled with improved sequestration of Ca^{2+} into the sarcoplasmic reticulum during diastole.

The interaction between volatile anesthetics and the voltage-dependent Ca^{2+} channel *in vitro* provides another avenue to explore the impact of intracellular Ca^{2+} modulation on left ventricular function in the presence of volatile anesthetics *in vivo*. Calcium channel antagonists, including nifedipine, nicardipine, and verapamil, have been shown to exacerbate depression of left ventricular systolic function induced by halothane and isoflurane (21,22,25–29); however, the impact of Ca^{2+} channel antagonists on volatile anesthetic-induced diastolic dysfunction have yet to be described. The dihydropyridine Ca^{2+} channel agonist Bay k 8644 (Bay k) has been shown partially reverse halothane-induced depression of transsarcolemmal Ca^{2+} flux *in vitro* (30–32), an action which should translate directly into improved systolic ventricular performance in the intact heart (33,34). The impact of Bay k on specific indicators of left ventricular systolic and diastolic function *in vivo,* however, remains unexplored. The work presented in this chapter represents preliminary data from an ongoing series of experiments aimed at further understanding of the functional interactions between dihydropyridine Ca^{2+} channel modulators and volatile anesthetic agents in the chronically instrumented dog.

II. Materials and Methods

Surgical implantation of instruments has been described in detail elsewhere (1,4,7,8). Briefly, under general anesthesia and aseptic conditions, dogs were instrumented for measurement of aortic and left ventricular pressure, left ventricular *dP/dt,* subendocardial segment length, intrathoracic pressure, and cardiac output. All instrumentation was firmly secured, tunneled subcutaneously between the scapulae, and exteriorized via several small incisions. The chest wall was closed in layers, and the pneumothorax was evacuated by a chest tube. After surgery, each dog was treated with analgesics as needed [Innovar-Vet (fentanyl–droperidol)]. Antibiotic prophylaxis consisted of cephalothin (40 mg/kg) and gentamicin (4.5 mg/kg). Dogs were allowed to recover for a minimum of 7 days before experimentation.

Segment length signals were driven and monitored by ultrasonic amplifiers. End-systolic segment length (ESL) was determined at maximum negative left ventricular *dP/dt,* and end-diastolic segment length (EDL) was determined just before the onset of left ventricular isovolumic contraction. An estimate of myocardial oxygen consumption, the pressure work index, was determined using the formula of Rooke and Feigl (35). All hemodynamic data were continuously monitored on a polygraph and digitized via a computer interfaced with an analog-to-digital converter. Ventricular pressure and segment length data were also transmitted to a digital storage oscilloscope for recording of left ventricular pressure–segment length waveforms and loops.

A. Nifedipine

Dogs were assigned to receive nifedipine in the conscious state or during isoflurane or halothane anesthesia in a random fashion on separate days. After instrumentation was calibrated and baseline hemodynamic data were recorded, the autonomic nervous system was pharmacologically blocked with intravenous propranolol (2 mg/kg), atropine methyl nitrate (3 mg/kg), and hexamethonium (20 mg/kg). Continuous left ventricular pressure, intrathoracic pressure, and segment length waveforms were recorded for later off-line analysis of diastolic function. Left ventricular preload was altered to generate a series of left ventricular pressure–segment length loops used to evaluate myocardial contractility in the conscious and anesthetized states. The inferior vena cava was abruptly occluded to reduce left ventricular pressure approximately 30 mmHg over 10–20 cardiac cycles. Respiratory variation in ventricular pressure in the conscious state was later eliminated by electronic subtraction of the continuous intrathoracic pressure waveform from the left ventricular pressure waveform as previously described (7). The resultant left ventricular transmural pressure–segment length loops were used to evaluate myocardial contractility in the conscious state.

In one group of experiments, nifedipine was administered in the conscious state after systemic hemodynamics and left ventricular pressure–segment length waveforms and loops had been recorded. Intravenous infusions of nifedipine at 0.5, 1, 2, or 4 mcg/kg/min were administered in a random fashion. Hemodynamics were recorded, and ventricular pressure–segment length waveforms and loops were obtained using the techniques described above after 10 min of equilibration at each dose of nifedipine. The infusion rate of nifedipine was then changed, and measurements were repeated after a similar period of equilibration.

Nifedipine was also administered to autonomically blocked dogs anesthetized with isoflurane or halothane in two other groups of experiments. After autonomic blockade, inhalation induction, and tracheal intubation, anesthesia was maintained with 0.75 MAC (end-tidal) isoflurane or halothane in a mixture of nitrogen and oxygen (79 and 21% by volume). The canine MAC values for isoflurane and halothane used in this investigation were 1.28 and 0.86%, respectively. Systemic hemodynamics were recorded, and ventricular pressure–segment length waveforms and loops were generated and stored on the digital oscilloscope after 30 min of equilibration in the anesthetized state. Intravenous infusions of nifedipine were administered and data recorded as described above. Each dog was allowed to recover from anesthesia and autonomic nervous system blockade for 3 days prior to subsequent experimentation. A total of 21 experiments in 3 separate groups (nifedipine administered in the conscious state and during isoflurane or halothane anesthesia) were completed in which the same 7 dogs were used.

B. Bay k 8644

Dogs were assigned to receive Bay k in the conscious state or during isoflurane or halothane anesthesia in a random fashion on separate days. Autonomic nervous system blockade was accomplished using the drugs and dosages mentioned above. In one group of experiments, intravenous infusions of Bay k at 0.5, 1, 2, and 4 mcg/kg/min were administered for 10 min each in a random fashion in the conscious state. In two additional groups of experiments, Bay k was administered after each dog had been anesthetized with isoflurane or halothane. In this series of experiments, anesthesia was maintained at 1.2 MAC (end-tidal) isoflurane or halothane. Data were collected after 30 min of equilibration at this anesthetic concentration. Continuous intravenous infusions of Bay k at 0.5, 1, 2, or 4 mcg/kg/min were then administered in a random fashion in the anesthetized state. Hemodynamics and pressure–length waveforms and loops were recorded during the conscious and/or anesthetized status after autonomic blockade and at each dose of Bay k as described above. Dogs were allowed to recover from anesthesia and autonomic nervous system blockade for at least 3 days prior to subsequent experimentation. A total of 24 experiments were used in 3 separate groups (Bay k administered in the conscious state and during isoflurane or halothane anesthesia) in which the same 8 dogs were used.

Solutions of all drugs were prepared on the day of each experiment. Propranolol, atropine methyl nitrate, and hexamethonium were dissolved in 0.9% normal saline. A racemic mixture of optical isomers of Bay k was used in this investigation (Research Biochemicals, Natick, MA). The drug vehicle for nifedipine and Bay k consisted of 5 ml of 95% ethanol, 5 ml of 5% polyethylene glycol, and 10 ml of 0.9% normal saline. No hemodynamic effects were observed with the administration of the drug vehicle alone as previously described (33). Because dihydropyridines such as nifedipine and Bay k may be unstable during normal lighting conditions, all experiments were performed under light emitted by a sodium vapor lamp.

C. Calculation of Indices of Systolic and Diastolic Left Ventricular Function

Myocardial contractility was evaluated using the slope (M_w) of the preload recruitable stroke work (PRSW) relationship as previously described (1,7,8,36). A series of left ventricular pressure–segment length loops were obtained by transient occlusion of the inferior vena cava in the conscious or anesthetized state and during each dose of nifedipine or Bay k. The area of each loop, corresponding to segmental stroke work (SW), was plotted against the corresponding EDL for each loop, and a linear regression analysis was used to describe the PRSW relationship slope (M_w) and length intercept (L_w): $SW = M_w(EDL - L_w)$. The time constant of isovolumic relaxation (τ) was calculated assuming a nonzero asymptote of left ventricular pressure decay using the method of Raff and Glantz

(37). Maximum segment lengthening velocity during rapid ventricular filling (dL/dt) was determined by differentiation of the continuous segment length waveform as previously characterized (7). The regional chamber stiffness constant (K_p) was derived from ventricular pressure–segment length data between minimum ventricular pressure and the onset of atrial systole using a simple monoexponential relationship assuming an elastic model (4,7,8,38).

D. Statistics

Statistical analysis of data within and between groups in the conscious state with and without autonomic nervous system blockade, during anesthetic interventions, and during nifedipine or Bay k infusions was performed with multiple analysis of variance with repeated measures followed by application of the Bonferroni modification of Student's *t*-test. Changes within and between groups were considered statistically significant when the *p* value was below 0.05.

III. Effects of Anesthetics and Calcium Channel Modulation

Autonomic nervous system blockade produced alterations in systemic hemodynamics which were similar to those described previously (1,7,8). Briefly, autonomic nervous system blockade caused an increase in heart rate and decreases in mean arterial pressure, left ventricular systolic pressure, and systemic vascular resistance. Administration of nifedipine to conscious dogs in the presence of autonomic nervous system blockade produced significant ($p < 0.05$) and dose-related decreases in mean arterial pressure, left ventricular systolic pressure, systemic vascular resistance, rate–pressure product, and pressure work index. Nifedipine also increased left ventricular end-diastolic pressure and cardiac output at higher doses. Nifedipine decreased myocardial contractility as assessed by M_w in a dose-dependent fashion in conscious dogs (Fig. 1). Nifedipine also caused dose-related increases in τ consistent with prolongation of the isovolumic relaxation phase of diastole. However, no changes in maximum segment lengthening velocity (dL/dt) or regional chamber stiffness (K_p) were observed when nifedipine was administered to conscious dogs (Fig. 2).

In the presence of autonomic blockade, isoflurane anesthesia (0.75 MAC) resulted in decreases in heart rate, mean arterial pressure, left ventricular systolic pressure, cardiac output, and pressure work index.

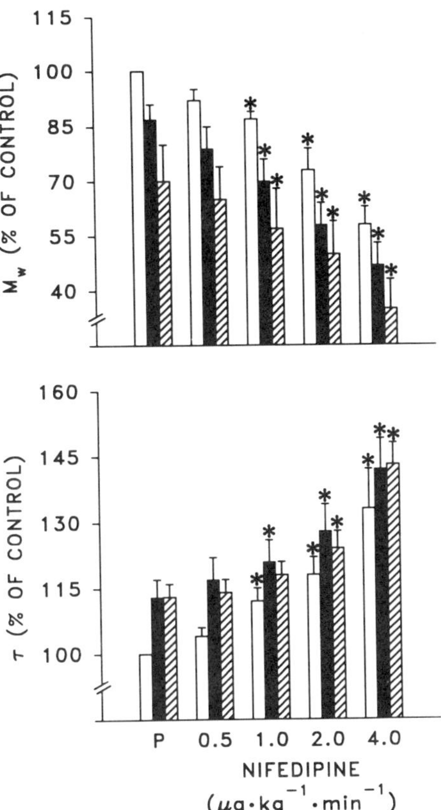

Fig. 1 Effect of nifedipine on preload recruitable stroke work slope (M_w) and the time constant of isovolumic relaxation (τ) in conscious (open bars), isoflurane-anesthetized (solid bars), and halothane-anesthetized (hatched bars) dogs as expressed as a percentage of control. * Significantly ($p < 0.05$) different from prenifedipine infusion (P) in the conscious or anesthetized state.

No changes in systemic vascular resistance or left ventricular end-diastolic pressure was observed with isoflurane. In the presence of isoflurane, nifedipine decreased M_w and increased τ in a dose-related manner (Fig. 1). In contrast to the findings in conscious dogs, nifedipine caused declines in dL/dt and increases in K_p in isoflurane-anesthetized dogs consistent with decreases in early ventricular filling and increases in regional chamber stiffness, respectively (Fig. 2).

Halothane anesthesia (0.75 MAC) produced alterations in systemic hemodynamics which were similar to those produced by isoflurane in autonomically blocked dogs. In the presence of halothane anesthesia, nifedi-

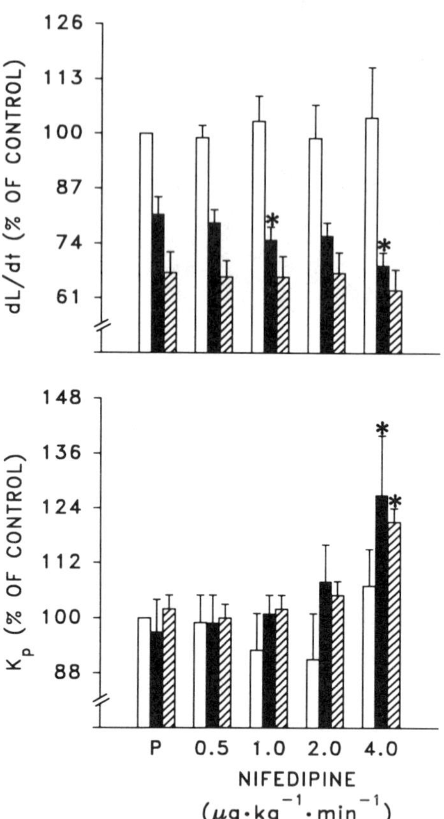

Fig. 2 Effect of nifedipine on the maximum segment lengthening velocity during rapid ventricular filling (*dL*/*dt*) and regional chamber stiffness (K_p) in conscious (open bars), isoflurane-anesthetized (solid bars), and halothane-anesthetized (hatched bars) dogs as expressed as a percentage of control. * Significantly ($p < 0.05$) different from prenifedipine infusion (P) in the conscious or anesthetized state.

pine caused dose-related decreases in mean arterial pressure, left ventricular systolic pressure, and indices of myocardial oxygen consumption (rate–pressure product and pressure–work index) and increases in left ventricular end-diastolic pressure. Heart rate and systemic vascular resistance also decreased at the 4 mcg/kg/min dose of nifedipine in halothane-anesthetized dogs. Nifedipine produced dose-related declines in myocardial contractility (M_w) and early ventricular filling (*dL*/*dt*), prolongation of isovolumic relaxation (τ), and increases in regional chamber stiffness (K_p) in halothane-anesthetized dogs, consistent with further depression of indices of left ventricular systolic and diastolic function (Figs. 1 and 2).

The dihydropyridine calcium channel agonist Bay k increased heart rate, mean arterial pressure, left ventricular systolic and end diastolic pressures, derived indices of myocardial oxygen consumption (rate–pressure product and pressure–work index), and systemic vascular resistance in conscious dogs. No change in cardiac output was observed in conscious dogs. Bay k caused dose-dependent increases in M_w consistent with augmentation of myocardial contractile state (Fig. 3). Bay k also enhanced indices of performance during several phases of diastole. Dose-related increases in dL/dt as well as decreases in τ and K_p consistent with improved early ventricular filling, shortened isovolumic relaxation, and increased

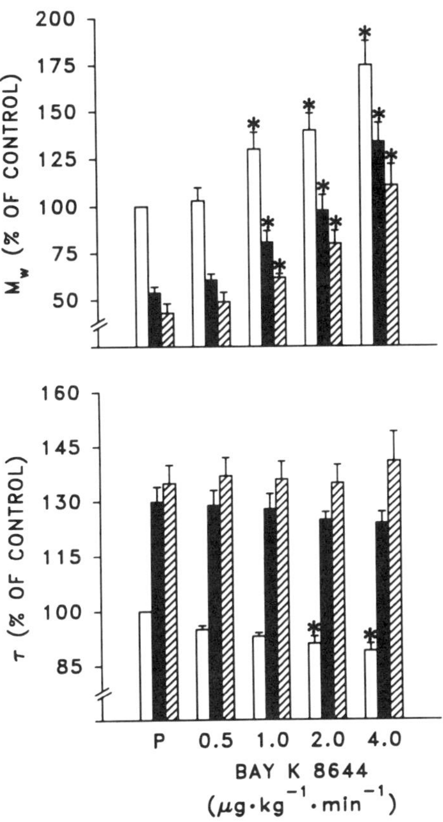

Fig. 3 Effects of Bay k 8644 on preload recruitable stroke work slope (M_w) and the time constant of isovolumic relaxation (τ) in conscious (open bars), isoflurane-anesthetized (solid bars), and halothane-anesthetized (hatched bars) dogs as expressed as a percentage of control. * Significantly ($p < 0.05$) different from pre-Bay k infusion (P) in the conscious and anesthetized state.

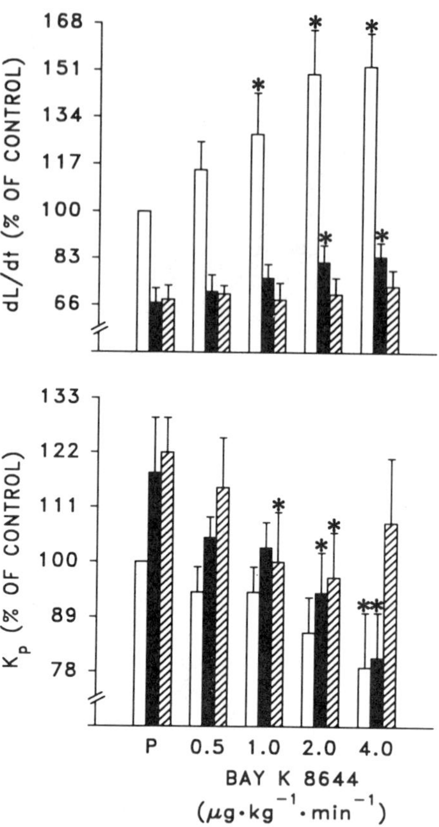

Fig. 4 Effect of Bay k 8644 on maximum segment lengthening velocity during rapid ventricular filling (*dL/dt*) and regional chamber stiffness (K_p) in conscious (open bars), isoflurane-anesthetized (solid bars), and halothane-anesthetized (hatched bars) dogs as expressed as a percentage of control. * Significantly ($p < 0.05$) different from pre-Bay k infusion (P) in the conscious and anesthetized state.

regional chamber compliance, respectively, were produced by Bay k in conscious dogs (Figs. 3 and 4).

In the presence of isoflurane or halothane anesthesia, Bay k produced dose-related increases in mean arterial and left ventricular systolic pressure, rate–pressure product, and pressure–work index. A significant increase in heart rate was also observed at the 4 mcg/kg/min infusion rate of Bay k. In contrast to the findings in the conscious state, Bay k administration to isoflurane-anesthetized dogs resulted in a dose-related increase in cardiac output without change in systemic vascular resistance. In halothane-anesthetized dogs, however, Bay k caused dose-related increases in systemic vascular resistance and no change in cardiac output.

Bay k caused a dose-dependent increase in myocardial contractility (M_w) in isoflurane- and halothane-anesthetized dogs (Fig. 3). During isoflurane anesthesia, significant increases in dL/dt and decreases in K_p consistent with enhanced rapid ventricular filling and decreases in regional chamber stiffness, respectively, were produced by Bay k (Fig. 4). No changes in τ were observed during administration of Bay k in isoflurane-anesthetized dogs. During halothane anesthesia, no changes in dL/dt or τ were observed during administration of Bay k. In fact, a paradoxical increase in τ occurred at the highest dose of Bay k. In addition, although declines in K_p were observed at the 1 and 2 mcg/kg/min doses of Bay k, regional chamber stiffness increased to the pre-Bay k control at the 4 mcg/kg/min dose in halothane-anesthetized dogs.

IV. Discussion

Preliminary results of these experiments indicate that modulators of the cardiac slow Ca^{2+} channel have important functional consequences on left ventricular mechanical performance during systole and diastole *in vivo*. Nifedipine, a dihydropyridine Ca^{2+} channel antagonist, caused dose-dependent decreases in myocardial contractility in conscious and anesthetized dogs as evaluated using the regional PRSW slope (M_w), a relatively heart rate- and load-independent index of contractile state in canine myocardium (36). Nifedipine also produced increases in the time constant of isovolumic relaxation (τ) in the conscious and anesthetized states, indicating a direct prolongation of this phase of diastole. However, no changes in maximum segment lengthening velocity during rapid ventricular filling (dL/dt) or regional chamber stiffness (K_p) were observed in conscious dogs. In contrast, decreases in dL/dt and increases in K_p were produced by higher doses of nifedipine in anesthetized dogs, suggesting that diastolic function may be more adversely affected by nifedipine in the presence of volatile anesthetics.

Slow Ca^{2+} channel blocking agents, including nifedipine, nicardipine, diltiazem, and verapamil, inhibit transsarcolemmal Ca^{2+} flux through the voltage-dependent Ca^{2+} channel, decreasing the availability of Ca^{2+} for contractile activation and causing depression of myocardial contractility (39). This direct negative inotropic effect has been shown to be additively affected by volatile anesthetics (22,25–29), agents which are also known to depress transsarcolemmal Ca^{2+}influx during cardiac myocyte depolarization (13–16). The myocardial depressant actions of Ca^{2+} channel antagonists are accompanied by dose-related prolongation of isovolumic relaxation when these drugs are administered systemically in the absence of autonomic nervous system reflexes or via an intracoronary route (39–41). Direct Ca^{2+} channel blockade-induced delays of isovolumic relaxation are attenuated or even reversed in the presence of intact autonomic nervous system function by reflex sympathetic stimulation, optimization of heart rate and ventricular loading conditions, improvement of myocardial oxygen supply–demand relations, and normalization of regional left ventricular uniformity (42–44). Because of these advantageous indirect effects, Ca^{2+} channel antagonists have been used successfully in the clinical treatment of left ventricular diastolic dysfunction associated with myocardial ischemia or ventricular hypertrophy despite direct actions blocking transsarcolemmal Ca^{2+} entry (44).

The negative lusitropic effects of nifedipine observed in conscious dogs in this investigation probably occur as a direct consequence of depression of myocardial contractility (44). Nifedipine-induced declines in intracellular Ca^{2+} concentration may result in relative decreases in Ca^{2+}–calmodulin kinase II-mediated phosphorylation of the regulatory protein of the sarcoplasmic reticulum Ca^{2+}-ATPase, phospholamban. This leads to subsequent declines in Ca^{2+} uptake by the sarcoplasmic reticulum during diastole and delays global isovolumic relaxation in the intact heart (45). The changes in relaxation produced by nifedipine did not influence early ventricular filling characteristics or regional chamber compliance in the conscious state, however, confirming the findings of Walsh *et al.* (41) and Kurnik *et al.* (38) in conscious dogs and humans, respectively. In contrast, higher doses of nifedipine exacerbated volatile anesthetic-induced depression of dL/dt and caused increases in K_p in the presence of isoflurane and halothane, suggesting that nifedipine may produce a differentially greater effect on ventricular filling and compliance in the presence of volatile anesthetics. The mechanism by which nifedipine accentuates of diastolic dysfunction in isoflurane- and halothane-anesthetized dogs is unknown but presumably occurs as a consequence of further decreases in transsarcolemmal Ca^{2+} influx.

Bay k 8644 has been shown to produce positive inotropic effects and systemic vasoconstriction *in vivo* (33) by enhancing transsarcolemmal Ca^{2+} entry into myocardial (30,31) and vascular smooth muscle cells *in vitro* (34), respectively. These actions can be inhibited in a competitive fashion by nifedipine and other dihydropyridine Ca^{2+} channel antagonists (34). Baum (30) recently demonstrated that Bay k-induced increases in transsarcolemmal Ca^{2+} entry can ameliorate the myocardial depressant actions of halothane in isolated guinea pig ventricular myocytes. A major mechanism by which volatile anesthetics depress myocardial contractility also involves the actions of these agents on the transsarcolemmal Ca^{2+} current. Volatile anesthetics inhibit the influx of Ca^{2+} into myocytes in a nonspecific fashion by suppressing both L- (slow) and T-type (fast) Ca^{2+} currents (12–15). Bay k enhances the L-type Ca^{2+} current, increasing the magnitude and duration of Ca^{2+} influx in response to depolarization (30,31). This increase in Ca^{2+} current in the myocyte can be attributed to Bay k-induced prolongation of Ca^{2+} channel open time and increases in open-channel probability observed in patch clamp experiments *in vitro* (46–48).

The preliminary results of this investigation confirm and extend the findings of previous studies (32,33) demonstrating that Bay k produces dose-related increases in arterial blood pressure, systemic vascular resistance, and myocardial contractility in conscious dogs. Bay k also improved indices of diastolic performance in several phases of diastole in the conscious state as indicated by decreases in τ, increases in dL/dt, and declines in K_p. These positive lusitropic effects probably resulted directly from the positive inotropic actions of Bay k. During isoflurane and halothane anesthesia, Bay k also increased myocardial contractility as indicated by the PRSW slope (M_w). The magnitude of increase in contractility was similar in conscious and anesthetized dogs. Administration of Bay k to isoflurane-anesthetized dogs improved indices of rapid ventricular filling and regional compliance; however, the magnitude of increases of dL/dt was less than that observed in the conscious state. In addition, no changes in the time constant of isovolumic relaxation were noted when Bay k was administered to dogs during isoflurane anesthesia. No changes in either τ or dL/dt were observed with the administration of Bay k to halothane-anesthetized dogs. In fact, τ paradoxically increased at the highest dose of Bay k in the presence of halothane, indicating further prolongation of this phase of diastole. In addition, although lower doses of Bay k caused decreases in regional chamber stiffness, K_p increased to control values at the highest dose of Bay k. These findings suggest that Bay k did not enhance but, instead, actually worsened measures of left ventricular diastolic performance during halothane anesthesia despite producing concomitant increases in systolic function.

Differences between the systemic hemodynamic effects produced by Bay k in the conscious and anesthetized states may partially account for the observed differences in diastolic mechanics between groups. Bay k caused increases in cardiac output with no change in systemic vascular resistance in isoflurane-anesthetized dogs. In contrast, Bay k caused dose-related increases in systemic vascular resistance in conscious and halothane-anesthetized dogs, a qualitative index of left ventricular afterload. Increases in afterload are associated with concomitant increases in τ (49–51), an effect which occurs as a result of the length dependence of Ca^{2+} affinity for the contractile apparatus (52–54). Increases in end-systolic length of individual sarcomeres are associated with concomitant increases in the affinity of Ca^{2+} for the troponin C subunit of the troponin–tropomyosin regulatory complex. This enhanced affinity for activator Ca^{2+} leads to delays in Ca^{2+} dissociation at greater muscle lengths, prolonging the relaxation phase of diastole. Thus, the failure of Bay k in low doses to decrease and paradoxically increase τ at the highest dose in the presence of increases in M_w in halothane-anesthetized dogs may be directly related to increases in afterload produced by the Ca^{2+} channel agonist. However, similar changes in systemic vascular resistance were produced by Bay k in conscious dogs, and, under these conditions, Bay k caused dose-related declines in τ consistent with direct, positive lusitropic effects. In addition, no change in τ occurred during administration of Bay k in isoflurane-anesthetized dogs despite maintenance of constant systemic vascular resistance and increases in M_w. Thus, differences in isovolumic relaxation caused by Bay k in conscious or anesthetized dogs cannot be completely attributed to differential effects on left ventricular afterload.

Volatile anesthetics have been shown to impair isovolumic relaxation *in vivo* (4–8). These negative lusitropic actions may be attributed to decreases in the clearance of Ca^{2+} from the myoplasm during diastole resulting from volatile anesthetic-induced alterations in storage and mobilization functions of the sarcoplasm reticulum. Because Bay k prolongs the transsarcolemmal Ca^{2+} transient (30,31), relative increases in intracellular Ca^{2+} derived from this source may contribute to further delays in Ca^{2+} extrusion and sequestration during diastole in the presence of volatile anesthetics. Bay k may also directly enhance Ca^{2+} loss from the sarcoplasmic reticulum during diastole (55), an effect which should contribute to further exacerbation of volatile anesthetic-induced Ca^{2+} leak from this organelle. These actions may precipitate relative prolongation of isovolumic relaxation despite concomitant increases in myocardial contractility in anesthetized dogs.

The preliminary results of this investigation must be interpreted within the constraints of several other limitations. Bay k caused an increase in heart rate and preload (as indicated by left ventricular end-diastolic pressure) in conscious and anesthetized dogs which was not dose-related. The time constant of isovolumic relaxation is inversely related to heart rate and may also be partially dependent on preload (49–51). Bay k-induced alterations in these variables may have contributed to the observed changes in isovolumic relaxation. The rate of rapid ventricular filling (as evaluated by dL/dt) is partially dependent on the gradient between left atrial and left ventricular pressure during this period of the cardiac cycle

(56), which was not specifically measured in the present study. Changes in ventricular loading conditions and increases in myocardial contractility produced by Bay k may have also influenced passive ventricular elastic properties (49) and subsequent interpretations of decreases in regional chamber stiffness.

In summary, preliminary results of these experiments indicate that the dihydropyridine Ca^{2+} channel agonist Bay k 8644 and antagonist nifedipine produce dose-related positive and negative inotropic effects, respectively, as evaluted using M_w in conscious and anesthetized, chronically instrumented dogs with autonomic nervous system blockade. Nifedipine also prolonged isovolumic relaxation, decreased early ventricular filling, and increased regional chamber stiffness in both the conscious and anesthetized states. These effects were more pronounced in anesthetized dogs, consistent with the additive actions of Ca^{2+} channel blockers and volatile agents on transsarcolemmal Ca^{2+} influx. Thus, nifedipine-induced depression of left ventricular performance may be related not only to negative inotropic actions but also to negative lusitropic effects as well. Bay k caused enhancement of several indices of diastolic function in conscious dogs. In contrast, the drug caused minimal changes in indices of diastolic performance and did not change these parameters during isoflurane and halothane anesthesia, respectively, despite causing concomitant augmentation of the myocardial contractile state. Bay k-induced decreases in diastolic Ca^{2+} clearance resulting from persistent Ca^{2+} channel activation and enhanced Ca^{2+} release from the sarcoplasmic reticulum combined with volatile anesthetic-induced disruption of normal sarcoplasmic reticulum Ca^{2+} sequestration and storage functions represent potential mechanisms by which Bay k may produce diastolic dysfunction while increasing myocardial contractility during halothane and isoflurane anesthesia.

Acknowledgments

The authors thank Doug Hettrick, John Tessmer, and Dave Schwabe for technical assistance and Angela M. Barnes for help in preparation of this article. This work was supported in part by U.S. Public Health Service Grant HL 32911, Anesthesiology Research Training Grant GM 08377, and Veterans Administration Medical Research Funds.

References

1. Pagel, P. S., Kampine, J. P., Schmeling, W. T., and Warltier, D. C. (1990). Comparison of end-systolic pressure–length relations and preload recruitable stroke work as indices of myocardial contractility in the conscious and anesthetized, chronically instrumented dog. *Anesthesiology* **73,** 278–290.
2. Horan, B. F., Prys-Roberts, C., Roberts, J. G., Bennett, M. J., and Foex, P. (1977).

Haemodynamic responses to isoflurane anaesthesia and hypovolaemia in the dog, and their modification by propanolo. *Br. J. Anaesth.* **49,** 1179–1187.
3. Merin, R. G. (1981). Are the myocardial functional and metabolic effects of isoflurane really different from those of halothane and enflurane? *Anesthesiology* **55,** 398–408.
4. Pagel, P. S., Kampine, J. P., Schmeling, W. T., and Warltier, D. C. (1991). Alteration of left ventricular diastolic function by desflurane, isoflurane, and halothane in the chronically instrumented dog with autonomic nervous system blockade. *Anesthesiology* **74,** 1103–1114.
5. Humphrey, L. S., Stinson, D. C., Humphrey, M. J., Finney, R. S., Zeller, P. A., Judd, M. R., and Blanck, T. I. (1990). Volatile anesthetic effects on left ventricular relaxation in swine. *Anesthesiology* **73,** 731–738.
6. Doyle, R. L., Foex, P., Ryder, W. A., and Jones, L. A. (1989). Effects of halothane on left ventricular relaxation and early diastolic coronary blood flow in the dog. *Anesthesiology* **70,** 660–666.
7. Pagel, P. S., Kampine, J. P., Schmeling, W. T., and Warltier, D. C (1993). Reversal of volatile anesthetic-induced depression of myocardial contractility by extracellular calcium also enhances left ventricular diastolic function. *Anesthesiology* **78,** 141–154.
8. Pagel, P. S., Hettrick, D. A., and Warltier, D. C. (1993). Amrinone enhances myocardial contractility and improves left ventricular diastolic function in conscious and anesthetized chronically instrumented dogs. *Anesthesiology* **79,** 753–765.
9. Rusy, B. F., Moran, J. E., Vongvises, P., Lattanand, S., MacNab, M., Much, D. R., and Lynch, P. R. (1972). The effects of halothane and cyclopropane on left ventricular volume determined by high-speed biplane cineradiography in dogs. *Anesthesiology* **36,** 369–373.
10. Moores, W. Y., Weiskopf, R. B., Baysinger, M., and Utley, J. R. (1981). Effects of halothane and morphine sulfate on myocardial compliance following total cardiopulmonary bypass. *J. Thorac. Cardiovasc. Surg.* **81,** 163–170.
11. Rusy, B. F., and Komai, H. (1987). Anesthetic depression of myocardial contractility: A review of possible mechanisms. *Anesthesiology* **67,** 745–766.
12. Bosnjak, Z, J., and Kampine, J. P. (1986). Effects of halothane on transmembrane potentials, Ca^{2+} transients, and papillary muscle tension in the cat. *Am. J. Physiol.* **251,** H374–H381.
13. Bosnjak, Z. J., Aggarwal, A., Turner, L. A., Kampine, J. M., and Kampine, J. P. (1992). Differential effects of halothane, enflurane, and isoflurane on Ca^{2+} transients and papillary muscle tension in guinea pigs. *Anesthesiology* **76** 123–131.
14. Lynch III, C., Vogel, S., and Sperelakis, N. (1981). Halothane depression of myocardial slow action potentials. *Anesthesiology* **55** 360–368.
15. Eskinder, H., Rusch, N. J., Supan, F. D., Kampine, J. P., and Bosnjak, Z. J. (1991). The effects of volatile anesthetics of L- and T-type calcium channel currents in canine cardiac Purkinje cells. *Anesthesiology* **74,** 919–926.
16. Drenger, B., Heitmiller, E. S., Quigg, M., and Blanck, T. J. J. (1992). Depression of calcium channel blocker binding to rat brain membranes by halothane. *Anesth. Analg. (N.Y.)* **74,** 758–761.
17. Blanck, T. J. J., Peterson, C. V., Baroody, B., Tegazzin, V., and Lou, J. (1992). Halothane, enflurane, and isoflurane stimulate calcium leakage from rabbit sarcoplasmic reticulum. *Anesthesiology* **76,** 813–821.
18. Denlinger, J. K., Kaplan, J. A., Lecky, J. H., and Wollman, H. (1975). Cardiovascular responses to calcium administered intravenously to man during halothane anesthesia. *Anesthesiology* **42,** 390–397.
19. Hysing, E. S., Chelly, J. E., Jacobsen, L., Doursout, M. F., and Merin, R. G. (1990).

Cardiovascular effects of acute changes in extracellular ionized calcium concentration induced by citrate and $CaCl_2$ infusions in chronically instrumented dogs, conscious and during enflurane, halothane, and isoflurane anesthesia. *Anesthesiology* **72,** 100–104.

20. Rooney, R. T., Stowe, D. F., Marijic, J., Bosnjak, Z. J., and Kampine, J. P. (1991). Amrinone reverses cardiac depression and augments coronary vasodilation with isoflurane in the isolated heart. *Anesthesiology* **74,** 559–567.
21. Makela, V. H. M., and Kapur, P. A. (1987). Amrinone blunts cardiac depression caused by enflurane or isoflurane anesthesia in the dog. *Anesth. Analg. (N.Y.)* **66,** 215–221.
22. Makela, V. H. M., and Kapur, P. A. (1987). Amrinone and verapamil–propanolol induced cardiac depression during isoflurane anesthesia in dogs. *Anesthesiology* **66,** 792–797.
23. Hicks, M. J., Shigekawa, M., and Katz, A. M. (1979). Mechanism by which cyclic adenosine 3′ : 5′-monophosphate-dependent protein kinase stimulates calcium transport in cardiac sarcoplasmic reticulum. *Circ. Res.* **44,** 384–391.
24. Katz, A. M. (1983). Cyclic adenosine monophosphate effects on the myocardium: A man who blows hot and cold with one breath. *J. Am. Coll. Cardiol.* **2,** 143–149.
25. Chelly, J. E., Rogers, K., Hysing, E. S., Taylor, A., Hartley, C., and Merin, R. G. (1986). Cardiovascular effects of and interaction between calcium blocking drugs and anesthetics in chronically instrumented dogs. I. Verapamil and halothane. *Anesthesiology* **64,** 560–567.
26. Rogers, K., Hysing, E. S., Merin, R. G., Taylor, A., Hartley, C., and Chelly, J. E. (1986). Cardiovascular effects of and interaction between calcium blocking drugs and anesthetics in chronically instrumented dogs. II. Verapamil, enflurane, and isoflurane. *Anesthesiology* **64,** 568–575.
27. Hysing, E. S., Chelly, J. E., Doursout, M.-F., Hartley, C., and Merin, R. G. (1986). Cardiovascular effects of and interaction between calcium blocking drugs and anesthetics in chronically instrumented dogs. III. Nicardipine and isoflurane. *Anesthesiology* **65,** 385–391.
28. Merin, R. G., Chelly, J. E., Hysing, E. S., Rogers, K., Dlewati, A., Hartley, C. J., Abernethy, D. R., and Doursout, M.-F. (1987). Cardiovascular effects of and interaction between calcium blocking drugs and anesthetics in chronically instrumented dogs. IV. Chronically administered oral verapamil and halothane, enflurane, and isoflurane. *Anesthesiology* **66,** 140–146.
29. Marshall, A. G., Kissin, I., Reves, J. G., Bradley, E. L., Jr., and Blackstone, E. H. (1983). Interaction between negative inotropic effects of halothane and nifedipine in the isolated rat heart. *J. Cardiovasc. Pharmacol.* **5,** 592–597.
30. Baum, V. C. (1992). Will calcium channel agonist Bay k 8644 inhibit halothane-induced impairment of calcium current? *Anesth. Analg. (N.Y.)* **74,** 865–869.
31. Thomas, G., Chung, M., and Cohen, C. J. (1985). A dihydropyridine (Bay k 8644) that enhances calcium currents in guinea pig and calf myocardial cells. A new type of positive inotropic agent. *Cir. Res.* **56,** 87–96.
32. Adachi, H., Shoji, T., and Goto, K. (1991). Comparison of the effects of endothelin and Bay k 8644 on cardiohemodynamics in anesthetized pigs. *Eur. J. Pharmacol.* **193,** 57–65.
33. Wynsen, J. C., Gross, G. J., Brooks, H. L., and Warltier, D. C. (1987). Changes in adrenergic pressor responses by calcium channel modulation in conscious dogs. *Am. J. Physiol.* **253,** H531–H539.
34. Preuss, K. C., Gross, G. J., Brooks, H. L., and Warltier, D. C. (1985). Slow channel calcium activators, a new group of pharmacological agents. *Life Sci.* **37,** 1271–1278.
35. Rooke, G. A., and Feigl, F. O. (1982). Work as a correlate of canine left ventricular oxygen consumption, and the problem of catecholamine oxygen wasting. *Circ. Res.* **50,** 273–286.

36. Glower, D. D., Spratt, J. A., Snow, N. D., Kabas, J. S., Davis, J. W., Olsen, C. O., Tyson, G. S., Sabiston, D. C., Jr., and Rankin, J. S. (1985). Linearity of the Frank–Starling relationship in the intact heart: The concept of preload recruitable stroke work. *Circulation* **71** 994–1009.
37. Raff, G. L., and Glantz, S. A. (1981). Volume loading slows left ventricular isovolumic relaxation rate. Evidence of load-dependent relaxation in the intact dog heart. *Circ. Res.* **48,** 813–824.
38. Kurnik, P. B., Courtois, M. R., and Ludbrook, P. A. (1986). Effects of nifedipine on intrinsic myocardial stiffness in man. *Circulation* **74,** 126–134.
39. Gelpi, R. J., Mosca, S. M., Rinaldi, G. J., Kosoglov, A., and Cingolani, H. E. (1983). Effect of calcium antagonism on contractile behavior of canine hearts. *Am. J. Physiol.* **244,** H378–H386.
40. Walsh, R. A., and O'Rourke, R. A. (1985). Direct and indirect effects of calcium entry blocking agents on isovolumic left ventricular relaxation in conscious dogs. *J. Clin. Invest.* **75,** 1426–1434.
41. Walsh, R. A., Badke, F. R., and O'Rourke, R. A. (1981). Differential effects of systemic and intracoronary calcium channel blocking agents on global and regional left ventricular function in conscious dogs. *Am. Heart J.* **102,** 341–350.
42. Ehring, T., and Heusch, G. (1990) Left ventricular asynchrony: An indicator of regional myocardial dysfunction. *Am. Heart J.* **120,** 1047–1057.
43. Bonow, R. O. (1990). Regional left ventricular nonuniformity: Effects on left ventricular diastolic function in ischemic heart disease, hypertrophic cardiomyopathy, and the normal heart. *Circulation* **81**(Suppl. 3), 54–65.
44. Walsh, R. A. (1989). The effects of calcium entry blockade on normal and ischemic ventricular diastolic function. *Circulation* **80**(Suppl. 4), 52–58.
45. Vittone, L., Cingolani, H. E., and Mattiazzi, R. A. (1985). The link between myocardial contraction and relaxation: The effects of calcium antagonists. *J. Mol. Cell. Cardiol.* **17,** 255–263.
46. Ochi, R., Hino, N., and Niimi, Y. (1984). Prolongation of calcium channel open time by the dihydropyridine derivative Bay k 8644 in cardiac myocytes. *Proc. Jpn. Acad.* **60B,** 153–156.
47. Kass, R. S. (1987). Voltage-dependent modulation of cardiac calcium channel current by optical isomers of Bay k 8644: Implications for channel gating. *Circ. Res.* **61,**(Suppl. 1), 1–5.
48. Brown, A. M., Kunze, D. L., and Yatani, A. (1984). The agonist effect of dihydropyridines on Ca channels. *Nature (London)* **311,** 570–572.
49. Gilbert, J. C., and Glantz, S. A. (1989). Determinants of left ventricular filling and of the diastolic pressure–volume relation. *Circ. Res.* **64,** 827–852.
50. Weiss, J. L., Frederiksen, J. W., and Weisfeldt, M. L. Hemodynamic determinants of the time-course of fall in canine left ventricular pressure. *J. Clin. Invest.* **58,** 751–760.
51. Gaasch, W. H., Blaustein, A. S., Andrias, C. W., Donahue, R. P., and Avitall, B. (1980). Myocardial relaxation: II. Hemodynamic determinants of rate of left ventricular isovolumic pressure decline. *Am. J. Physiol.* **239,** H1–H6.
52. Brutsaert, D. L., Housmans, P. R., and Goethals, M. A. (1980). Dual control of relaxation: Its role in the ventricular function in the mammalian heart. *Circ. Res.* **47,** 637–652.
53. Kentish, J. C., ter Keurs, H. E. D. J., Ricciardi, L., Bucx, J. J. J., and Noble, M. I. M. (1986). Comparison between the sarcomere length–force relations of intact and skinned trabeculae from rat right ventricle: Influence of calcium concentrations on these relations. *Circ. Res.* **58,** 755–768.

54. Housmans, P. R., Lee, N. K. M., and Blinks, J. R. (1983). Active shortening retards the decline of the intracellular calcium transient in mammalian heart muscle. *Science* **221,** 159–161.
55. Hryshko, L. V., Kobayashi, T., and Bose, D. (1989). Possible inhibition of canine ventricular sarcoplasmic reticulum by Bay k 8644. *Am. J. Physiol.* **257,** H407–H414.
56. Zile, M. R., Blaustein, A. S., and Gaasch, W. H. (1989). The effect of acute alterations in left ventricular afterload and β-adrenergic tone on indices of early diastolic filling rate. *Circ. Res.* **65,** 406–416.

Anesthetic Actions on Calcium Uptake and Calcium-Dependent Adenosine Triphosphatase Activity of Cardiac Sarcoplasmic Reticulum

Ning Miao, Martha J. Frazer, and Carl Lynch III

Department of Anesthesiology
University of Virginia Health Sciences Center
Charlottesville, Virginia 22908

I. Introduction

The inhalational anesthetics have long been known to depress myocardial contractility. These effects appear to result from a number of mechanisms involving sarcolemma Ca^{2+} entry, sarcoplasmic reticulum (SR) uptake and release of Ca^{2+}, as well as direct effects on contractile proteins (1). The latter effect appears to be relatively minor, whereas substantial evidence suggests that multiple effects are present with regard to Ca^{2+} uptake and release by the SR. Su *et al.* (2) had demonstrated that in the presence of halothane there is decreased Ca^{2+} available for subsequent caffeine-stimulated release in mechanically skinned myocardial fibers from the rabbit, an effect which was less prominent with isoflurane (3). A number of mechanical studies have suggested that on intact tissue halothane causes decreased accumulation of calcium within the SR (4–7), although other studies have suggested specific effects of isoflurane on certain pools of Ca^{2+} which may transit the SR (8–9). In previous studies on isolated SR vesicles, decreased retention of Ca^{2+} has been demonstrated in both actively loaded and passively loaded vesicles (10,11). The presence of a

Advances in Pharmacology, Volume 31

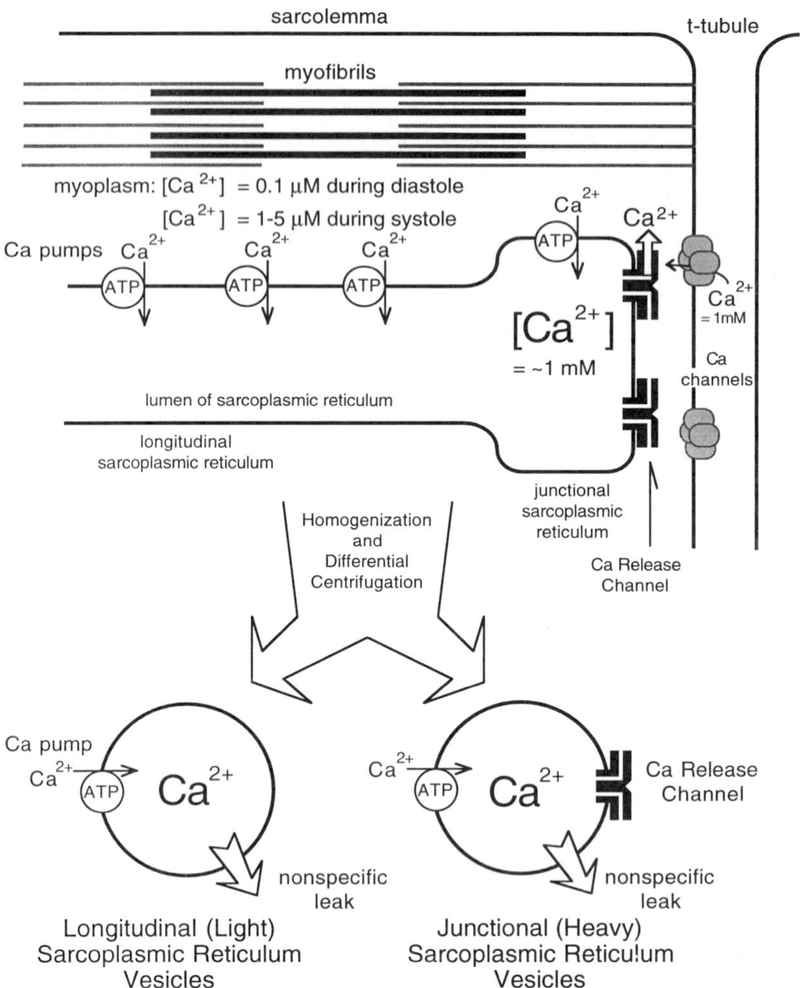

Fig. 1 Schematic of a partial myocardial sarcomere showing myofibrils, a T-tubule, and the sarcoplasmic reticulum (SR). By orchestration of sequential release (from the junctional SR or terminal cisternae), uptake, and retention (both longitudinal and junctional SR) of Ca^{2+}, the SR provides the major control of Ca^{2+} fluxes which activate tension. When homogenized and centrifuged on a sucrose gradient, the SR membrane can be subfractionated into longitudinal SR vesicles (LSR), which contain primarily Ca^{2+}-ATPase (the calcium pump), and junctional SR vesicles (JSR), the latter retaining in addition the release channels through which Ca^{2+} can rapidly efflux, and which also bind ryanodine. Both vesicles also possess some component of nonspecific leak. [Adapted from Frazer and Lynch (11).]

nonspecific leak which was not blocked by ruthenium red obscured potential anesthetic effects on the calcium release channel (11).

The present studies were undertaken to better define which specific aspects of myocardial SR function were altered by anesthetics. The distinct functions of Ca^{2+} uptake and release have distinct morphological correlates at the subcellular level, as indicated in Fig. 1. The longitudinal SR (LSR) located along the length of the myofibrils is responsible for Ca^{2+} uptake from the myofibrils to bring about relaxation. On homogenizing cardiac tissue, vesicles of the LSR containing almost exclusively Ca^{2+}-ATPase can be separated by centrifugation from heavier vesicles derived from junctional SR (JSR) (11). In intact tissue at the junction of the SR and the T-tubules, the JSR contains calcium release channels which regulate release of activator Ca^{2+} from the SR. Calcium release channels, which are present in three isoforms in various tissues, are large homotetrameric proteins (~5000 amino acids/tetramer) which have been termed ryanodine receptors, based on the high affinity for the plant alkaloid ryanodine (12). High-affinity ryanodine binding occurs with Ca^{2+}-dependent channel activation and opening. Connelly *et al.* (13) and Frazer and Lynch (14) have reported that halothane, isoflurane, and enflurane differentially enhance ryanodine binding by the skeletal and cardiac ryanodine receptors.

To assess how anesthetic effects on the ATPase function may influence the amount of releasable SR Ca^{2+} as well as ATP energy consumption, the effects of these agents on Ca^{2+}-ATPase function and SR vesicle Ca^{2+} retention were determined under identical conditions. We compared Ca^{2+} uptake with Ca^{2+}-ATPase activity in the presence of halothane, isoflurane, and enflurane to determine if the anesthetics differentially alter the ability of the vesicles to accumulate and retain Ca^{2+}, thus contributing to their differing depressant actions on myocardial contractility.

II. Sarcoplasmic Reticulum

A. Preparation of Cardiac Sarcoplasmic Reticulum

Following a protocol approved by the Animal Research Committee, dogs were anesthetized and the heart excised. Cardiac SR was isolated from canine ventricle according to a modification of the method of Alderson and Feher (15). Ventricles were combined with three times their weight of 10 m*M* imidazole at pH 7.0. All solutions were maintained at 4°C and contained the following protease inhibitors: 200 μM phenylmethylsulfonyl flouride (PMSF), 77 n*M* aprotinin, 0.83 m*M* benzamidine, 1 m*M* iodocetamide, 1.1 μM leupeptin, and 0.70 μM pepstatin A. The supernatant solutions from two separate centrifugations (20 min, 6000 g) were combined and centrifuged for 2 hr at 100,000 g, which resulted in a crude microsomal pellet. The pellet was subsequently homogenized in 1 M KCl and 10 mM imidazole, layered on a discontinuous sucrose gradient consisting of 0.6, 0.8, 1.0, 1.1, and 1.6 mM sucrose, and centrifuged for 10 hr at 150,000 g. Aliquots of the resulting fractions were separated by sodium dodecyl sulfate–polyacrylamide gel electrophores is (SDS-PAGE) on a gel constructed with a linear 6–12% polyacrylamide gradient and stained with Coomassie blue; protein concentrations were determined using the method of Bradford (16). In the fractions from the 1.0–1.1 and 1.1–1.6 M sucrose interfaces and from the pellet, the most prominently staining proteins were at 105 and 45 kDa, characteristic for cardiac SR Ca^{2+}-ATPase and calsequestrin, respectively. The lighter of the fractions (2.84 mg/ml protein) showed no visible band at 500–450 kDa corresponding to the Ca^{2+} release channel (CaRC, ryanodine receptor) and was employed for the studies described.

B. Calcium Uptake Measurement

The assays were performed in a constantly stirred, temperature-controlled Plexiglas chamber which accepted a sample volume of 1–2.5 ml. An enclosure around the electrodes and chamber permitted control of the surrounding atmosphere. Extravesicular Ca^{2+} was measured with a Ca^{2+} ion-selective electrode (Orion Model 93-20, Boston, MA). The uptake buffer (1 ml) contained 100 mM KCl, 5 mM $MgCl_2$, 10 mM sodium azide, 10 mM HEPES, 10 mM potassium oxalate, and 10 μM EGTA (pH 7.4). Following incremental addition of 50 μM $CaCl_2$ into the 1-ml reaction mixture containing 5 μl SR (14.2 μg protein), Ca^{2+} uptake into the vesicles was initiated by the addition of 5 mM ATP (Fig. 2). Owing to the complex Ca^{2+} buffering by the combined oxalate and EGTA, the voltage reading of the Ca^{2+}-sensitive electrode was calibrated for each assay by preceding incremental addition of five 10-nmol $CaCl_2$ aliquots. The resulting voltage at 10, 20, 30, 40, and 50 μM total added Ca^{2+} could be fit by a linear regression (r values 0.90–0.99, Fig. 2 inset), from which the decline in total Ca^{2+} by uptake could be determined. The linear range for the Ca^{2+} electrode response to total Ca^{2+} corresponded to a free [Ca^{2+}] range of approximately 2 μM up to 25 μM. The Ca^{2+} electrode gave results similar to those obtained previously employing a spectrophotometric assay employing antipyrylazo III (11). To estimate the level of CaRC contamination, 10 μM ruthenium red (RR) was added to block any specific vesicular efflux of Ca^{2+} via the CaRC.

C. Phosphate Production Measurement

As an indirect measurement of ATPase activity, production of inorganic phosphate (P_i) was measured by a colorimetric assay with a sensitivity to less than 0.2 nmol P_i (17). At 0.25, 1, 1.5, 2, 4, 7, and 10 min after SR addition, 200-μl aliquots of the reaction mixture were removed from the chamber and added to a cuvette containing 2.5 ml of 7.7 m*M* ammonium molybdate, 46 μM polyvinyl alcohol, and 0.77 m*M* malachite green in 1.8 *M* HCl solution. After 2 min of incubation, the absorbance was measured at 630 nm. Total P_i was calibrated from a standard curve which employed varying concentrations of K_2HPO_4. To permit a sufficient quantity for repeated sampling, 2 ml instead of 1 ml of reaction mixture (with 28.4 μg SR protein) was used in the same chamber and under identical conditions as employed for Ca^{2+} uptake studies. Although simultaneous determinations of Ca^{2+} uptake and P_i production were not performed routinely, no difference in Ca^{2+} uptake rate was observed with the 2-ml volume and aliquot removal for P_i determination. Background P_i arising from non-Ca^{2+}-ATPase activity was measured in the absence of added Ca^{2+}; these values (typically 5% of the total) were subtracted from the total P_i to obtain P_i attributable to Ca^{2+}-ATPase activity. The Ca^{2+}-ATPase activity for each time point was calculated as the Ca^{2+}-dependent P_i production divided by the preceding time interval over which it occurred.

To evaluate the response of the SR used to test anesthetic effects, Ca^{2+} uptake and P_i production rate were measured with various control conditions. Altered conditions examined were as follows: (1) ATP concentrations (m*M*) of 0.25, 0.5, 1, 2, and 5; (2) ruthenium red concentrations (μM) of 1, 3, 10, 30, and 100; (3) pH values of 6.6, 7.0, 7.2, 7.4, 7.6, and 7.8; and (4) temperatures (°C) of 25, 30, 35, 37, and 40.

D. Reagents and Anesthetic Administration

Reagents and protease inhibitors were obtained from Sigma Chemical Co. (St. Louis). Halothane was supplied by Halocarbon Industries (Newark, NJ); isoflurane and enflurane by Anaquest (Madison, WI). For a minimum equilibration period of 20 min, buffer solution at the experimental temperature was bubbled via a fritted glass tube with 0.2 μm filtered air that passed through an appropriate calibrated, temperature-compensated anesthetic vaporizer, providing approximately equipotent anesthetic concentrations (18): 0.75 and 1.5% (of atmosphere, by volume) halothane, 1.25 and 2.5% isoflurane, and 1.75 and 3.5% enflurane. To assure minimal anesthetic loss from solution during experiments, filtered air containing anesthetics was delivered over the top of the preequilibrated solution in the enclosed chamber. The presence of anesthetic concentrations in solution appropriate for the known buffer/gas partition coefficient (19) were verified by gas chromatography. Halothane at 0.75 and 1.5% yielded 0.25 and 0.5 m*M* in solution; 1.25 and 2.5% isoflurane yielded 0.3 and 0.6 m*M*; 1.75 and 3.5% enflurane yielded 0.5 and 1 m*M*.

Differences among controls and anesthetics were compared by analysis of variance (ANOVA) employing Fisher's PLSD test to verify differences at $p < 0.05$ (Statview, Abacus Concepts, Berkeley, CA). Values are plotted and tabulated as means ± SEM.

III. Calcium Uptake and ATPase Activity

The Ca^{2+} electrode potential shown in Fig. 2 demonstrates an initial acceleration in Ca^{2+} uptake into SR vesicles over the first minute, which was sustained for up to 2 min. The uptake rate subsequently declined, so that after 5–7 min net Ca^{2+} uptake virtually ceased. Figure 3A shows the mean Ca^{2+} uptake rate and Ca^{2+}-dependent P_i production rate observed at various time points with 5 m*M* at pH 7.4 and 35°C, both in the absence or presence of 10 μ*M* RR. It is evident that the Ca^{2+} uptake rate was initially highest at 1–1.5 min and subsequently declined thereafter (Fig. 3). The amount of added Ca^{2+} was 3.52 mmol/μg (50 nmol Ca^{2+}/14.2 μg

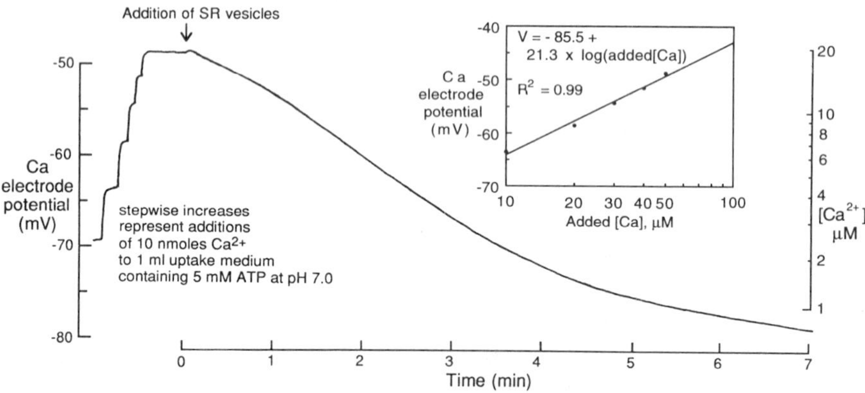

Fig. 2 Calcium electrode potential with stepwise addition of Ca^{2+} to 10 m*M* potassium oxalate-containing buffer medium, and subsequent cardiac SR vesicle uptake of Ca^{2+}. Ca^{2+} was added as five 10-μl aliquots of 1 m*M* $CaCl_2$ solution, which provied a calibration of the Ca^{2+} content of the medium based on the electrode potential, as shown in the inset. Such a calibration was performed for each uptake measurement. Experiments were performed with 14.2 μg of SR vesicle protein in 1 ml of reaction solution, at pH 7.0 and 35°C, containing 5 m*M* ATP.

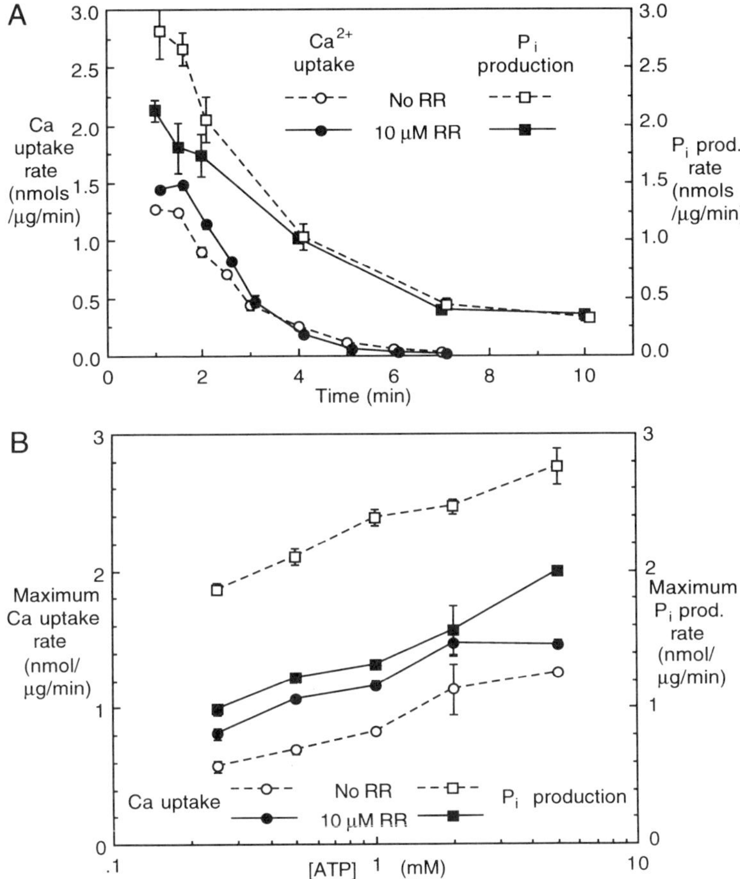

Fig. 3 (A) Rate of Ca^{2+} uptake and Ca^{2+}-dependent P_i liberation by canine cardiac SR at the indicated points in time. The 1 ml of medium (pH 7.4, 35°C) contained 5 m*M* ATP and 50 nmol added Ca^{2+}. The control Ca^{2+} uptake rate was determined from the Ca^{2+} electrode potential as indicated in Fig. 2 in the absence (○) or presence (●) of 10 μ*M* RR ($n = 16$), and P_i liberation was determined under identical conditions except that 2 ml of medium was used (□, no RR, $n = 15$; ■, 10 μ*M* RR, $n = 8$). (B) Effect of ATP concentration on the maximum Ca^{2+} uptake and Ca^{2+}-dependent P_i production rate by cardiac SR vesicles at pH 7.4 and 35°C, in the presence (○) or absence (●) of 10 μ*M* RR. For each experiment the mean of the values measured at 1 and 1.5 min was employed as a measure of maximum activity (see A). For 5 m*M* ATP, the number of experiments (n) is indicated above (see A); $n = 3$ for 0.25 to 2 m*M* ATP.

SR protein), which for the example in Fig. 3A is slightly less than the integral of Ca^{2+} taken up over time, which was 3.8 and 4.0 nmol Ca^{2+}/μg protein in the absence and presence of RR, respectively. The difference probably represents uptake of the contaminating 3–7 nmol Ca^{2+} typically present in the 1-ml reaction mixture. The decline with time in P_i production, presumably reflecting ATPase activity, was concurrent with the decrease in Ca^{2+} uptake. The presence of 10 μM RR increased the initial Ca^{2+} uptake rate by 16% and decreased the Ca^{2+}-dependent P_i production by 27%. By 3–4 min the rates were similar with or without 10 μM RR. When the net Ca^{2+} uptake had virtually ceased after 7 min, Ca^{2+}-dependent ATPase activity reached a stable level which was 15–20% of the maximum value. This pattern was consistently observed in all subsequent studies and may represent Ca^{2+}-ATPase activity not coupled to Ca^{2+} accumulation (membrane fragments or nonretentive vesicles). The mean values at 1 and 1.5 min for Ca^{2+} uptake or P_i production rate, which represented the maximum rates, were used for comparison among various conditions and in the presence of anesthetics. As expected Ca^{2+} uptake rate and ATPase activity was dependent on [ATP], with a significant decrease in both uptake and ATPase activity observed for 0.25–1.0 mM ATP (Fig. 3B).

Dependences of maximum Ca^{2+} uptake rate on the RR concentration, pH, and temperature are indicated in Fig. 4. RR caused a dose-dependent enhancement of Ca^{2+} uptake (Fig. 4A) at both pH 7.0 and 7.4, with a maximal increase of 27% (pH 7.4) and 31% (pH 7.0) at 30 μM RR. More alkaline conditions result in a lower rate of Ca^{2+} uptake, with a marked decline occurring over pH 7.4 (Fig. 4B). At pH 7.8 the maximal Ca^{2+} uptake rate was only 29% of that at the optimal rate (pH 6.6), and the absolute effect of 10 μM RR in accelerating Ca^{2+} uptake was diminished at alkaline pH. Under conditions identical to those of the indicated points in Fig. 3, rates of Ca^{2+}-dependent P_i production (rP_i) are listed with the calculated coupling ratio (CR, Ca^{2+} uptake/Ca^{2+}-dependent P_i production). Changes in the rate of P_i production were typically reciprocal to those of Ca^{2+} uptake; that is, when Ca^{2+} uptake was enhanced (e.g., RR present, decreased pH) the rate of P_i production typically was decreased. In the absence (−RR) or presence of RR (+RR), the maximum Ca^{2+} uptake rate was 20% (−RR) or 31% (+RR) higher at pH 7.0 versus pH 7.4, whereas ATPase activity at pH 7.0 was decreased to 82% (−RR) or 92% (+RR) of the pH 7.4 value. This resulted in the coupling ratio at pH 7.0 being 45% (−RR) to 41% (+RR) higher than that observed at pH 7.4. As a function of temperature, Ca^{2+} uptake was highest at 35–37°C with a significant decrease evident at temperatures less than or equal to 30°C and at 40°C (Fig. 4C). Comparing rates at 25 versus 35° C, the Ca^{2+}-dependent ATPase activity was stimulated at the higher temperature by

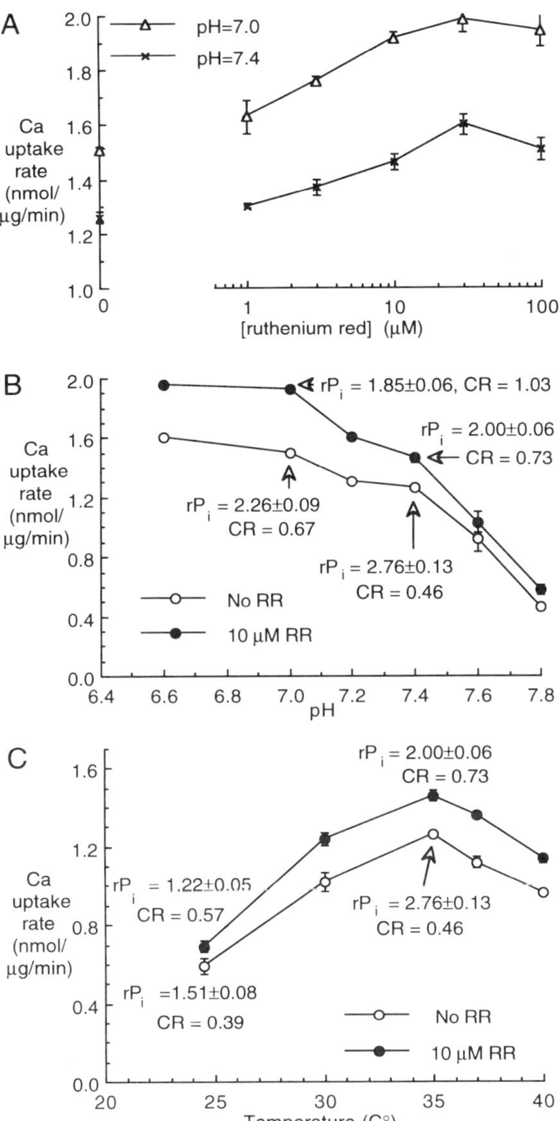

Fig. 4 Effects of varied ruthenium red concentration, pH, and temperature on peak Ca^{2+} uptake rate by cardiac SR vesicles. Maximum Ca^{2+} uptake rate was the mean of the values measured at 1 and 1.5 min, as shown in Fig. 3A. For the specific variable values indicated, the P_i production rate (rP_i) was measured under identical conditions, and the resulting coupling ratio (CR, Ca^{2+} uptake rate/P_i production rate) is also noted. (A) Effect of RR concentration on the maximum Ca^{2+} uptake rate at pH 7.4 (x) or pH 7.0 (△). For 1, 3, 30, and 100 μM RR, $n = 3$; for 0 and 10 μM RR, $n = 16$. (B) Effect of pH on maximum Ca^{2+} uptake rate and rP_i in the absence (○) or presence (●) of 10 μM RR. Except for pH 7.0 (Ca^{2+} uptake, $n = 5$; rP_i, $n = 4$) and pH 7.4 (see Fig. 3A), $n = 3$. (C) Effect of temperature on the Ca^{2+} uptake rate and rP_i in the absence (○) or presence (●) of 10 μM RR; $n = 3$, except for 35°C (see Fig. 3A).

64% (+RR) or 82% (−RR). However, the increase in Ca^{2+} uptake was even greater (112–114%), resulting in a higher coupling ratio at 35°C.

Table I lists the Ca^{2+} uptake rate and ATPase activity measured following exposure to isoflurance, eflurane, and halothane. At pH 7.4 and 5 m*M* ATP, the rates of net Ca^{2+} uptake and ATPase activity were increased by either 1.7 and 3.5% enflurane. In contrast, 1.5% halothane caused a marked depression in the net Ca^{2+} uptake rate while enhancing the ATPase activity. Although the anesthetic-equivalent 2.5% isoflurane caused slight depression of Ca^{2+} uptake, the rate was still greater than in the presence of 1.5% halothane ($p < 0.05$). Furthermore, there was no increase in P_i production seen with isoflurane as with halothane.

The enhancement of Ca^{2+} uptake and the reduced ATPase activity induced by the presence of 10 μM RR persisted in the presence of the

Table I

Anesthetic Effects on Maximum Calcium Uptake Rate and Ca^{2+}-ATPase Activity [a]

Conditions	Maximum Ca^{2+} uptake rate (nmol/μg/min)		Maximum Ca^{2+}-ATPase activity (nmol/μg/min)	
	No RR	10 μM RR	No RR	10 μM RR
pH 7.4, 5 m*M* ATP				
Control	1.26 ± 0.02	1.46 ± 0.03	2.76 ± 0.13	2.00 ± 0.08
0.75% Halothane	1.21 ± 0.04	1.34 ± 0.04	2.68 ± 0.09	2.18 ± 0.05
1.5% Halothane	0.92 ± 0.02[b]	1.13 ± 0.03[b]	2.97 ± 0.12	2.79 ± 0.09[c]
1.3% Isoflurane	1.36 ± 0.03	1.48 ± 0.03	2.40 ± 0.20	2.20 ± 0.21
2.5% Isoflurane	1.10 ± 0.03[b]	1.28 ± 0.02[b]	2.54 ± 0.23	2.11 ± 0.07
1.7% Enflurane	1.54 ± 0.02[c]	1.73 ± 0.03[c]	3.51 ± 0.17[c]	2.93 ± 0.11[c]
3.5% Enflurane	1.38 ± 0.03[c]	1.57 ± 0.01[c]	3.25 ± 0.09	2.71 ± 0.22[c]
	(Control n = 16, other n = 6)	(Control n = 16, other n = 6)	(Control n = 15, other n = 6)	(Control n = 8, other n = 6)
pH 7.0, 5m*M* ATP				
Control	1.51 ± 0.01	1.91 ± 0.01	2.26 ± 0.09	1.85 ± 0.06
1.5% Halothane	1.46 ± 0.01[b]	1.61 ± 0.05[b]	2.69 ± 0.03[c]	2.03 ± 0.02[c]
2.5% Isoflurane	1.50 ± 0.02	1.86 ± 0.03	2.50 ± 0.13	1.82 ± 0.03
3.5% Enflurane	1.56 ± 0.01[c]	1.96 ± 0.01	2.82 ± 0.05[c]	2.33 ± 0.04[c]
	(n = 5)	(n = 5)	(n = 4)	(n = 4)

[a] Values given are means ± SEM. Ca^{2+}-ATPase activity represents the Ca^{2+}-dependent component of P_i production rate. Ruthenium red (RR) at 10 μM was employed to block the SR Ca^{2+} release channel. Statistical comparisons were made by ANOVA and Fisher's test.

[b] Significant decrease ($p < 0.05$) compared to control state.

[c] Significant increase ($p < 0.05$) compared to control state.

anesthetics, whereas the specific effects for each anesthetic were also still apparent. Likewise, at pH 7.0 and 5 m*M* ATP, conditions under which the Ca^{2+} uptake rate was increased compared to pH 7.4, the addition of anesthetics caused effects similar to those seen at 7.4. At the lower pH, enflurane still stimulated Ca^{2+} uptake rate and ATPase activity. The depression by 1.5% halothane of net Ca^{2+} uptake and the increase in ATPase activity was less prominent but still significant, whereas 2.5% isoflurane caused no change in either measure.

Coupling ratios calculated from Table I are plotted graphically in Fig. 5. The depressant effects of the three anesthetics on the coupling ratio

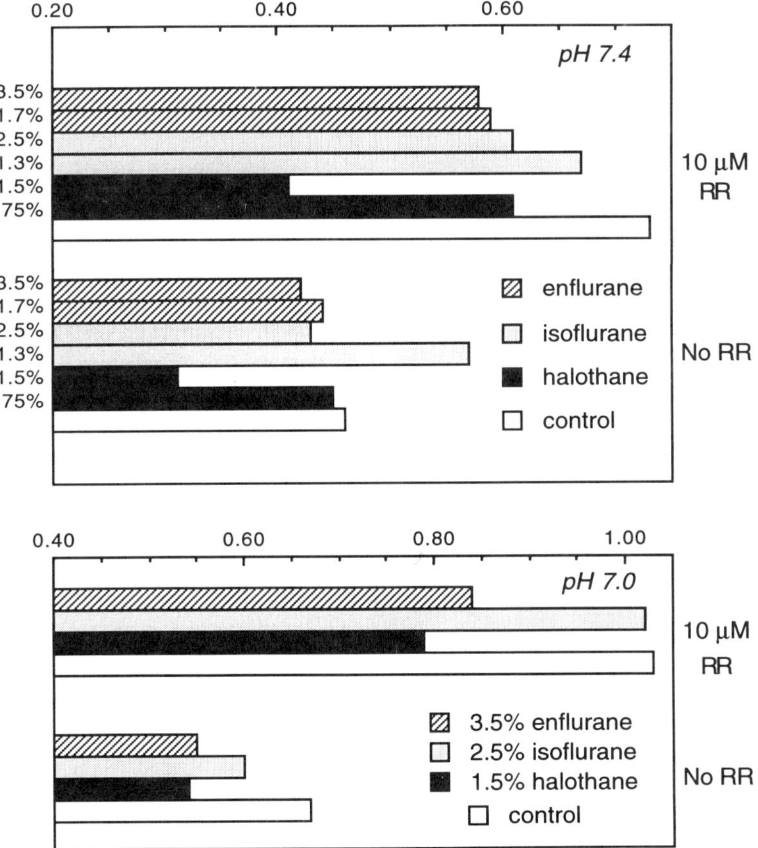

Fig. 5 Alterations in coupling ratio caused by equianesthetic concentrations of halothane, isoflurane, and halothane. Values listed are calculated as the Ca^{2+} uptake rate divided by the Ca^{2+}-dependent ATP hydrolysis rate presented in Table I.

Table II

Maximum Calcium Uptake Rate with 0.5 m*M* ATP at pH 7.4[a]

	Maximum Ca^{2+} uptake rate (nmol/μg/min)	
Conditions	No RR	10 μ*M* RR
Control	0.69 ± 0.02	1.06 ± 0.01
0.75% Halothane	0.49 ± 0.02[b]	0.66 ± 0.01[b]
1.3% Isoflurane	0.58 ± 0.02[b]	0.73 ± 0.02[b]
1.7% Enflurane	0.82 ± 0.08	1.08 ± 0.01

[a] Values given are means ± SEM ($n = 3$). Statistical comparisons were made by ANOVA and Fisher's test.

[b] Significant decrease ($p < 0.05$) compared to control state or in the presence of 1.7% enflurane.

appeared to be greater with RR than those without RR, particularly for 1.5% halothane. At pH 7.0 and 7.4 the effects of the anesthetics seemed similar: isoflurane had minimal apparent action, whereas halothane, especially at 1.5%, caused marked depression of the coupling ratio. Although enflurane increased Ca^{2+} uptake rate by up to 18–22%, the Ca^{2+}-dependent ATPase activity was increased even further (27–46%); thus, the result was a modest decrease in coupling ratio.

To explore the influence of substrate limitation on anesthetic actions, Ca^{2+} uptake was examined with 0.5 mM ATP at pH 7.4 and 35°C (Table II). In both the absence or presence of 10 μ*M* RR, Ca^{2+} uptake was significantly reduced by 0.75% halothane and 1.3% isoflurane. In contrast, enflurane maintained Ca^{2+} uptake in this setting, although there was no significant enhancement as seen in 5 m*M* ATP.

IV. Discussion

The major result of the present study concerns the modest but distinct effects of the halogenated volatile anesthetics on Ca^{2+} uptake and retention of isolated myocardial SR function at clinically relevant concentrations. The differing functional changes appear to arise from agent-specific actions on the ATPase activity as well as the ability of vesicles to retain accumulated Ca^{2+}.

A. Control Behavior

After addition of Ca^{2+} to the SR vesicles in the presence of ATP, a reproducible pattern of Ca^{2+} accumulation occurred: initially, a high net uptake rate was achieved which subsequently decreased after 1.5 min, eventually declining to zero, where any Ca^{2+} uptake must have been offset by an equal Ca^{2+} efflux (Fig. 3A). The decrease in the rate of Ca^{2+} uptake by SR vesicles with time is due to a gradual decrease in Ca^{2+} influx and/or an increase in Ca^{2+} efflux as the intra- versus extraluminal $[Ca^{2+}]$ gradient increases. Because the experiments were performed with oxalate present in the vesicle lumen, the rise in intraluminal free Ca^{2+} was blunted by binding to oxalate. Consequently, the gradient for Ca^{2+} efflux should have been modest, so that any feedback depression of ATPase activity by the rising free Ca^{2+} in the lumen should also have been minor. To avoid the potentially complicating factors of intraluminal Ca^{2+}, comparisons were made of the initial maximal rates at 1–1.5 min. The decrease in Ca^{2+} uptake, as well as the decrease in ATPase activity, became most prominent after 2 min, when the $[Ca^{2+}]_o$ typically decreased to below 5 μM (see Fig. 2). This decrease in extravesicular Ca^{2+} may be primarily responsible for the declining Ca^{2+} uptake and ATP hydrolysis rates. The presence of an ongoing Ca^{2+}-dependent ATP hydrolysis when net Ca^{2+} uptake was zero suggests that, in spite of intraluminal oxalate binding of Ca^{2+} in most vesicles, a significant fraction of the ATPase activity was not linked to Ca^{2+} uptake, or there was continuing Ca^{2+} loss from certain vesicles.

The preparation employed for this study was derived largely from longitudinal SR as evident from the lack of detectable Ca^{2+} release channels (CaRC) on electrophoresis. Nevertheless, some contamination with junctional SR was present based on the changes induced by the presence of RR, which enhanced Ca^{2+} uptake and coincidentally decreased Ca^{2+}-ATPase activity. Because RR blocks CaRC (20), the increased uptake suggests that CaRC were present in at least some of the vesicles. Furthermore, the decrease in ATPase activity coincident with the increased Ca^{2+} uptake supports the notion that faster depletion of the extravesicular Ca^{2+} (or possibly intraluminal Ca^{2+} accumulation) leads to a decreased ATP hydrolysis rate. Although some depression of ATPase activity by RR has been reported in isolated skeletal SR, it was not observed in the potassium oxalate buffer employed here (21). Nevertheless, since a trend toward decreased Ca^{2+} uptake was noted above 30 μM RR in the studies of pH, [ATP], and anesthetic effects, 10 μM RR was employed to block the majority of contaminating CaRC. As anticipated from prior work (22,23), the Ca^{2+} uptake and Ca^{2+}-dependent ATPase activity were dependent on

[ATP], with increased uptake apparent with concentrations from 0.25 up to 2 m*M* ATP and a greater effect evident at 5 m*M* ATP (Fig. 3B).

In a fashion similar to RR, acidification of the medium also enhanced Ca^{2+} uptake, and alkalinization above pH 7.4 markedly decreased Ca^{2+} uptake, a result reported previously both for heart (10) and for isolated skeletal muscle SR (24). Compared to rates at pH 7.4, Ca^{2+}-dependent ATPase activity was decreased at pH 7.0, coincident with the enhanced Ca^{2+} uptake, an effect which was additive with that of RR. The decrease in ATPase activity was associated with increased Ca^{2+} uptake and likely represents reduced pump activity secondary to local extravesicular Ca^{2+} depletion (possibly luminal Ca^{2+} accumulation). In isolated skeletal muscle SR vesicles, alkalinization of the medium activated a marked release of Ca^{2+} and decreased Ca^{2+} uptake rates, secondary to increased efflux via a CaRC (RR-sensitive) and via a nonspecific (RR-insensitive) pathway (24). An acidic medium appears to reduce the open probability and conductance of the CaRC (25,26). In the present case, the very large decrease in Ca^{2+} uptake rate was identical in the presence and absence of RR (to 29% of the value at pH 6.6), suggesting that the primary effect was unrelated to the CaRC and was similar to effects in "leaky light SR" (24). Unlike the case in skeletal muscle, however, Ca^{2+} uptake was reduced even in the presence of oxalate, which should improve vesicle Ca^{2+} retention. It is therefore unclear whether alkalinization inhibits SR Ca^{2+} capacity, as suggested for skeletal muscle, or whether an increased Ca^{2+} leak is induced via the ATPase or some other pathway.

With 10 μM at pH 7.0, the maximum coupling ratio observed was 1.03, similar to that reported by others for similar cardiac SR preparations (15,27). This ratio does not approach the theoretically ideal ratio for the Ca^{2+}-ATPase of 2 Ca^{2+} translocated for 1 ATP hydrolyzed (28,29), which is more readily observed in skeletal muscle SR vesicles. Because the coupling ratio improved when Ca^{2+} efflux through the Ca^{2+} channel was blocked by RR, a component of the low coupling ratio appeared to be the result of increased efflux through this channel rather than an intrinsic uncoupling of the ATPase. A higher RR concentration and greater acidosis might have improved the ratio, but these were not explored owing to potential direct actions on ATPase activity. Isolated cardiac SR vesicles possess more Ca^{2+} leakage pathways, or more vesicles or membrane fragments possess Ca^{2+}-dependent ATPase activity which is uncoupled from Ca^{2+} uptake. If the steady-state (10 min) Ca^{2+}-ATPase activity is subtracted, the coupling ratios listed are increased by 20–25%.

Although hypothermia decreased the ATPase activity of the SR membrane as expected (30), Ca^{2+} uptake was reduced even more, leading to a decreased coupling ratio at 25 versus 35°C. This effect occurred even in the presence of RR, suggesting that it could not be explained by a hypothermia-induced increase in open probability of the CaRC (31), unless hypothermia reduced RR binding. The Ca^{2+} uptake rate decreased beyond 35–37°C, suggesting that hyperthermia may reduce the ability of isolated SR membranes to retain Ca^{2+}, which is consistent with evidence of uncoupling of Ca^{2+} uptake from ATPase activity in isolated skeletal SR (32).

B. Anesthetic Effects

Although all volatile anesthetics are myocardial depressants, these commonly employed agents differ in potency. Isoflurane is significantly less depressant than halothane, whereas enflurane frequently has intermediate effects (6,7,9,33–36). The negative inotropic effects of volatile anesthetics have been attributed to intracellular mechanisms of action that involve excitation–contraction coupling in ventricular myocardium, especially in regard to Ca^{2+} uptake, retention, and release by cardiac SR (3,9,33,37,38). Depending on pH, ATP concentration, and anesthetic concentration, volatile anesthetics have produced varying results on isolated cardiac SR: increased Ca^{2+} uptake (10,22,23,39), increased Ca^{2+} efflux (22,23,39,40,41), decreased Ca^{2+} uptake (22,23), decreased ATPase activity (42), or little to no effect (22,23).

The present study in which Ca^{2+} uptake and P_i production were determined under identical conditions permits greater clarification of anesthetic actions. In a previous study in isolated cardiac SR vesicles, Frazer and Lynch (11) found that both isoflurane and halothane increased nonspecific (RR-insensitive) Ca^{2+} efflux, although halothane had an initially greater effect. The present results more clearly define the greater propensity for halothane to depress Ca^{2+} uptake to a greater extent than isoflurane. The depression in Ca^{2+} uptake is associated with a compensatory rise in ATPase activity, similar to the effect of an increase from pH 7.0 to pH 7.4 attributable to enhanced efflux (24). Although direct uncoupling of the enzymatic cycle could contribute, it seems most likely that the decrease in coupling ratio is associated with halothane-enhanced efflux (11). In intact myocardium, such an effect could result in increased energy utilization, but the depression of sarcolemmal Ca^{2+} entry by halothane (43–45) will prevent cellular Ca^{2+} loading and the occurrence of excessive Ca^{2+} cycling. Additional effects of halothane in promoting Ca^{2+} loss from the SR (2,5,38) may very well include its activation of the CaRC (13,14), which will contribute to loss of SR Ca^{2+} and decreased Ca^{2+} available to activate contractions. In contrast, Ca^{2+} uptake is only modestly decreased by isoflurane, a finding consistent with other studies in myocardial tissue which suggest that isoflurane has less effect on Ca^{2+} stores (3,5,38).

The differential effects of the agents persisted at pH 7.0. Although the SR vesicles had a higher coupling ratio, the anesthetic-induced changes with or without RR were similar to those at pH 7.4. None of the anesthetics depressed the ATPase activity. Isoflurane (2.5%) caused no change; 1.5% halothane decreased Ca^{2+} uptake while increasing ATP hydrolysis so that coupling ratio decreased. Enflurane (3.5%) enhanced both Ca^{2+} uptake and ATPase activity, with a somewhat greater effect on the latter, so that coupling ratio also decreased.

The most curious effect is the increase in ATPase activity and Ca^{2+} uptake caused by enflurane. These results confirm previous work by Blanck and co-workers that in rabbit cardiac SR (10,23) and also in skeletal muscle SR (46) enflurane was the least depressant or, under certain circumstances (i.e., low [ATP]), actually enhanced Ca^{2+} uptake. Although we found that enflurane did not significantly enhance Ca^{2+} uptake with 0.5 m*M* ATP, enflurane, unlike halothane and isoflurane, did not depress uptake. Whereas the greater increase in ATPase activity compared to Ca^{2+} uptake rate suggests a slight degree of uncoupling, the predominate effect appears to be a primary increase in the enzyme cycling rate. Interestingly, isoflurane which is a chemical isomer of enflurane with similar physical properties, demonstrated no ability to enhance the Ca^{2+}-ATPase.

In addition to the aforementioned anesthetic actions, there are in junctional SR additional differential effects of the anesthetics on the Ca^{2+} release channel (CaRC) of cardiac SR (see Fig. 1). Ryanodine binding to the CaRC occurs when the channel opens, so that ryanodine binding can serve as a measure of CaRC activation (47–50). For example, ryanodine binding is enhanced by increasing [Ca^{2+}] above 1 μM or by caffeine, which are also known to activate CaRC opening. Both Connelly *et al.* (13) and Frazer and Lynch (14) have found that halothane enhances ryanodine binding to cardiac CaRCs, whereas isoflurane does not. Figure 6 demonstrates the effects of anesthetics on specific ryanodine binding to isolated junctional SR containing CaRC subjected to submaximal stimulation with 5 μM Ca^{2+} (14). Halothane causes dose-dependent enhancement, which is consistent with an increased number of high-affinity (open channel) sites. Such opening of CaRC by halothane, combined with enhancement of nonspecific leak when CaRC are apparently blocked by RR, would explain the transient enhancement by halothane of tension on sudden application to isolated tissue (41,43,51) and subsequent contractile depression associated with decreased SR Ca^{2+} (2,41). The increased myocardial tension is all the more surprising since halothane has been clearly documented to depress sarcolemmal Ca^{2+} current (43,45,52), which triggers Ca^{2+}-induced Ca^{2+} release. Thus the Ca^{2+} sensitization of the CaRC by halothane overrides depression of Ca^{2+} current. However, if such an action is present during rest at diastolic levels of Ca^{2+} ($\sim$100 n*M*), then

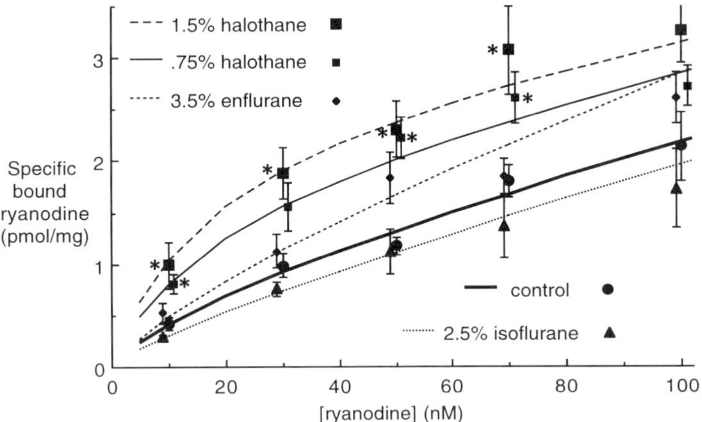

Fig. 6 Mean specific binding of ryanodine in the presence of 5 μM Ca^{2+} and with the anesthetics indicated. * Significant difference from control ($p < 0.05$ by ANOVA). The quantity of total bound ryanodine (bRy_{total}) could be described by a model of ligand-binding in which ryanodine could bind to a high- or low-affinity site (1 and 2, respectively) with maximum capacity B_{max} and dissociation constant K_d where

$$K_d = \frac{[Ry][B_{max} - bRy]}{[bRy]}$$

so that

$$bRy_{total} = \left[\frac{(B_{max1} - bRy_1)}{k_{d1}} + \frac{(B_{max2} - bRy_2)}{K_{d2}}\right]$$

As an added constraint on the system, it was assumed that for the ryanodine binding sites total $B_{max} = B_{max1} + B_{max2}$. Control values for K_{d1}, K_{d2}, B_{max1}, and B_{max2} were obtained using a least squares fitting program (Sigma Plot, Jandel Scientific, San Rafael, CA) for a single-ligand–two-site model as described by Feldman [Feldman, H. A. (1972). Mathematical theory of complex ligand-binding systems at equilibrium: Some methods for parameter fitting. *Anal. Biochem.* **48,** 317–338]. All curves employed the following values: high-affinity binding K_{d1} = 17 nM; total maximum number of binding sites B_{max} = 48.5 pmol/mg. The values for the number of high-affinity sites (B_{max1}) and the low-affinity binding constant (K_{d2}) were as follows:

Conditions	B_{max1} (pmol/mg)	K_{d2} (μM)
Control	0.66	2.86
0.75% Halothane	1.80	3.5
1.5% Halothane	2.5	4.5
3.5% Enflurane	0.66	2.0
2.5% Isoflurane	0.35	2.86

[Data adapted from Frazer and Lynch[14] (full report in press).]

Table III

Effects of Anesthetics on Sarcoplasmic Reticulum Function

Anesthetic	Aspect of SR function		
	ATPase activity	Nonspecific leak	CaRC activation
Halothane	↑ (secondary to →)	↑ ↑ ↑	↑ ↑ ↑
Enflurane	↑ ↑	↑	? ↑ ↑
Isoflurane	? ↓	? ↓	—

halothane will induce leak of accumulated Ca^{2+} from SR (2,38,46,53), reducing Ca^{2+} available for subsequent contraction. This action, which will cause ongoing depletion of SR Ca^{2+} stores, particularly when combined with Ca^{2+} current depression, thereby provides a mechanism for myocardial contractile depression if halothane administration is sustained for more than a few beats. Isoflurane had no such comparable effect on activation of CaRC and has been shown not to cause the transient tension increase seen with halothane (51). Enflurane appeared to enhance low-affinity binding, and interpretation of this effect is not clear.

Table III summarizes the actions of the various anesthetics on SR. In isolated myocardium, enflurane has depressant effects on contractily which frequently fall between those of halothane and isoflurane, but in the absence of a detailed model of myocyte Ca^{2+} cycling it is unclear how the various effects on SR will combine with other anesthetic effects, for example, depression of Ca^{2+} current (45) to depress contractility. Although certain observations correlate with findings in whole tissue, we cannot exclude the possibility that the present data result from an increased susceptibility of isolated SR vesicles; these anesthetic actions may not occur in the *in situ* cellular organelle. If such anesthetic-induced alterations in subcellular function are present in intact tissue, these distinct effects are likely to contribute to the known differential effects of the anesthetic agents on myocardial function.

Acknowledgments

This work was supported by National Institutes of Health Grant R01 GM 31144. The authors thank Dr. Joseph Pancrazio for helpful discussion through the course of the work.

References

1. Rusy, B. F., and Komai, H. (1987). Anesthetic depression of myocardial contractility: A review of possible mechanisms. *Anesthesiology* **67,** 745–766.

2. Su, J. Y., and Kerrick, W. G. L. (1979). Effects of halothane on caffeine-induced tension transients in functionally skinned myocardial fibers. *Pfluegers Arch.* **380,** 29–34.
3. Su, J. Y., and Bell, J. G. (1986). Intracellular mechanism of action of isoflurane and halothane on striated muscle of the rabbit. *Anesth. Analg.* (*N.Y.*) **65,** 457–462.
4. Komai, H., and Rusy, B. F. (1984). Differences in the myocardial depressant action of thiopental and halothane. *Anesth. Analg.* **63,** 313–318.
5. Komai, H., and Rusy, B. F. (1987). Negative inotropic effects of isoflurane and halothane in rabbit papillary muscles. *Anesth. Analg.* (*N.Y.*) **66,** 29–33.
6. DeTraglia, M. C., Komai, H., and Rusy, B. F. (1988). Differential effects of inhalational anesthetics on myocardial potentiated-state contractions *in vitro*. *Anesthesiology* **68,** 534–540.
7. Lynch III, C., and Frazer, M. J. (1989). Depressant effects of volatile anesthetics upon rat and amphibian ventricular myocardium: Insights into mechanisms of action. *Anesthesiology* **70,** 511–522.
8. Lynch III, C. (1986). Depression of myocardial contractility *in vitro* by bupivacaine, etidocaine, and lidocaine. *Anesth. Analg.* (*N.Y.*) **65,** 551–559.
9. Lynch III, C. (1990). Differential depression of myocardial contractility by volatile anesthetics *in vitro:* Comparison with uncouplers of excitation–contraction coupling. *J. Cardiovasc. Pharmacol.* **15,** 655–665.
10. Casella, E. S., Suite, D. A., Fisher, Y. L., and Blanck, T. J. J. (1987). The effect of volatile anesthetics on the pH dependence of calcium uptake by cardiac sarcoplasmic reticulum. *Anesthesiology* **67,** 386–390.
11. Frazer, M. J., and Lynch III, C. (1992). Halothane and isoflurane effects on Ca^{2+} fluxes of isolated myocardial sarcoplasmic reticulum. *Anesthesiology* **77,** 316–323.
12. McPherson, P. S., and Campbell, K. P. (1993). The ryanodine receptor/Ca^{2+} release channel. *J. Biol. Chem.* **268,** 13765–13768.
13. Connelly, T. J., Hayek, R.-E., Rusy, B. F., and Coronado, R. (1992). Volatile anesthetics selectively alter [^{3}H]ryanodine binding to skeletal and cardiac ryanodine receptors. *Biochem. Biophys. Res. Commun.* **186,** 595–600.
14. Frazer, M. J., and Lynch III, C. (1992). Ca release channels of cardiac sarcoplasmic reticulum are activated by halothane, but not by isoflurane. *Anesthesiology* **77,** A630 (abstract).
15. Alderson, B. H., and Feher, J. J. (1987). The interaction of calcium and ryanodine with cardiac sarcoplasmic reticulum. *Biochim. Biophys. Acta* **900,** 221–229.
16. Bradford, M. (1976). New method for protein quantification. *Anal. Biochem.* **72,** 248–250.
17. Chan, K.-M., Delfert, D., and Junger, K. D. (1986). A direct colorimetric assay for Ca-stimulated ATPase activity. *Anal. Biochem.* **157,** 375–380.
18. Firestone, L. L., Miller, J. C., and Miller, K. W. (1986). Appendix: Tables of physical and pharmacological properties of anesthetics. *In* "Molecular and Cellular Mechanisms of Anesthetics" (S. H. Roth and K. W. Miller, eds.), pp. 455–470. Plenum, New York.
19. Renzi, E., and Waud, B. E. (1977). Partition coefficients of volatile anesthetics in Krebs' solution. *Anesthesiology* **47,** 62–63.
20. Chamberlain, B. K., Volpe, P., and Fleischer, S. (1984). Inhibition of calcium-induced Ca^{++} release from purified cardiac sarcoplasmic reticulum vesicles. *J. Biol. Chem.* **259,** 7547–7553.
21. Vale, M. G. P., and Carvalho, A. P. (1973). Effects of ruthenium red on Ca^{2+} uptake and ATPase of sarcoplasmic reticulum of rabbit skeletal muscle. *Biochim. Biophys. Acta* **325,** 29–37.
22. Blanck, T. J. J., and Thompson, M. (1981). Calcium transport by cardiac sarcoplasmic

reticulum: Modulation of halothane action by substrate concentration and pH. *Anesth. Analg.* (*N.Y.*) **60,** 390–394.

23. Blanck, T. J. J., and Thompson, M. (1982). Enflurane and isoflurane stimulate calcium transport by cardiac sarcoplasmic reticulum. *Anesth. Analg.* (*N.Y.*) **61,** 142–145.
24. Dettbarn, C., and Palade, P. (1991). Effect of alkaline pH on sarcoplasmic reticulum Ca^{2+} release and Ca^{2+} uptake. *J. Biol. Chem.* **266,** 8993–9001.
25. Meissner, G., and Henderson, J. S. (1987). Rapid calcium release from cardiac sarcoplasmic reticulum vesicles is dependent on Ca^{2+} and is modulated by Mg^{2+}, adenine nucleotide, and calmodulin. *J. Biol. Chem.* **262,** 3065–3073.
26. Rousseau, E., and Pinkos, J. (1990). pH modulates conductance and gating behaviour of single calcium release channels. *Pfluegers Arch.* **415,** 645–647.
27. Lain, R. F., Hess, M. L., Gertz, E. W., and Briggs, E. N. (1968). Calcium uptake activity of canine myocardial sarcoplasmic reticulum in the presence of anesthetic agents. *Circ. Res.* **23,** 597–604.
28. MacLennan, D. H. (1990). Molecular tools to elucidate problems in excitation–contraction coupling. *Biophys. J.* **58,** 1355–1365.
29. Langer, G. A. (1992). Calcium and the heart: Exchange at the tissue, cell, and organelle levels. *FASEB J* **6,** 893–902.
30. Kurihara, S., and Sakai, T. (1985). Effects of rapid cooling on mechanical and electrical responses in ventricular muscle of guinea pig. *J. Physiol.* (*London*) **361,** 361–378.
31. Sitsapesan, R., Montgomery, R. A. P., MacLeod, K. T., and Williams, A. J. (1991). Sheep cardiac sarcoplasmic reticulum calcium-release channels: Modulation of conductance and gating by temperature. *J. Physiol.* (*London*) **434,** 469–488.
32. Lepock, J. R., Rodahl, A. M., Zhang, C., Heynen, M. L., Waters, B., and Cheng, K. H. (1990). Thermal denaturation of the Ca^{2+}-ATPase of sarcoplasmic reticulum reveals two thermodynamically independent domains. *Biochemistry* **29,** 681–689.
33. Lynch III, C. (1986). Differential depression of myocardial contractility by halothane and isoflurane *in vitro*. *Anesthesiology* **64,** 620–631.
34. Housmans, P. E., and Murat, I. (1988). Comparative effects of halothane, enflurane, and isoflurane at equipotent anesthetic concentrations on isolated ventricular myocardium of the ferret. I. Contractility. *Anesthesiology* **69,** 451–463.
35. Herland, J. S., Julian, F. J., and Stephenson, D. G. (1990). Halothane increases Ca^{2+} efflux via Ca^{2+} channels of sarcoplasmic reticulum in chemically skinned rat myocardium. *J. Physiol.* (*London*) **426,** 1–18.
36. Housmans, P. E. (1990). Negative inotropy of halogenated anesthetics in ferret ventricular myocardium. *Am. J. Physiol.* **259,** H827–H834.
37. Katsuoka, M., and Ohnishi, S. T. (1989). Inhalation anaesthetics decrease calcium content of cardiac sarcoplasmic reticulum. *Br. J. Anaesth.* **62,** 669–673.
38. Komai, H., and Rusy, B. F. (1990). Direct effect of halothane and isoflurane on the function of the sarcoplasmic reticulum in intact rabbit atria. *Anesthesiology* **72,** 694–698.
39. Housmans, P. E., and Murat, I. (1988). Comparative effects of halothane, enflurane, and isoflurane at equipotent anesthetic concentrations on isolated ventricular myocardium of the ferret. II. Relaxation. *Anesthesiology* **69,** 464–471.
40. Palade, P. (1987). Drug-induced Ca^{2+} release from isolated sarcoplasmic reticulum: II. Releases involving Ca^{2+} induced Ca^{2+} release channel. *J. Biol. Chem.* **262,** 6142–6148.
41. Katsuoka, M., Kobayashi, K., and Ohnishi, T. (1989). Volatile anesthetics decrease calcium content of isolated myocytes. *Anesthesiology* **70,** 954–960.
42. Malinconico, S. M., and McCarl, R. L. (1982). Effect of halothane on cardiac sarcoplasmic reticulum Ca-ATPase at low calcium concentrations. *Mol. Pharmacol.* **22,** 8–10.

43. Lynch III, C., Vogel, S., and Sperelakis, N. (1981). Halothane depression of myocardial slow action potentials. *Anesthesiology* **55,** 360–368.
44. Ikemoto, Y., Yatani, A., Arimura, H., and Yoshitake, J. (1985). Reduction of the slow inward current of isolated rat ventricular cells by thiamylal and halothane. *Acta Anaesthesiol. Scand.* **29,** 583–586.
45. Bosnjak, Z. L., Supan, F. D., and Rusch, N. J. (1991). The effects of halothane, enflurane and isoflurane on calcium currents in isolated canine ventricular cells. *Anesthesiology* **74,** 340–345.
46. Blanck, T. J. J., Peterson, C. V., Baroody, B., Tegazzin, V., and Lou, J. (1992). Halothane, enflurane, and isoflurane stimulate calcium leakage from rabbit sarcoplasmic reticulum. *Anesthesiology* **76,** 813–821.
47. Rardon, D. P., Cefali, D. C., Mitchell, R. D., Seiler, S. M., and Jones, L. R. (1989). High molecular weight proteins purified from cardiac junctional sarcoplasmic reticulum vesicles are ryanodine-sensitive calcium channels. *Circ. Res.* **64,** 779–789.
48. Chu, A., Díaz-Muñoz, M., Hawkes, M. J., Brush, K., and Hamilton, S. L. (1990). Ryanodine as a probe for the functional state of the skeletal muscle sarcoplasmic reticulum calcium release channel. *Mol. Pharmacol.* **37,** 735–741.
49. Pessah, I. N., Durie, E. M., Schiedt, M. J., and Zimanyi, I. (1990). Anthraquinone-sensitized Ca^{2+} release channel from rat cardiac sarcoplasmic reticulum: Possible receptor-mediated mechanism of doxorubicin cardiomyopathy. *Mol. Pharmacol.* **37,** 503–514.
50. Sitsapesan, R., and Williams, A. J. (1990). Mechanisms of caffeine activation of single calcium-release channels of sheep cardiac sarcoplasmic reticulum. *J. Physiol.* (*London*) **423,** 425–439.
51. Luk, H. N., Lin, C. I., Chang, C. L., and Lee, A. R. (1987). Differential inotropic effects of halothane and isoflurane in dog ventricular tissues. *Eur. J. Pharmacol.* **136,** 409–413.
52. Terrar, D. A., and Victory, J. G. G. (1988). Influence of halothane on membrane currents associated with contraction in single myocytes isolated from guinea pig ventricle. *Br. J. Pharmacol.* **94,** 500–508.
53. Rock, E., Mammar, M. S., Thomas, M. A., Viret, J., and Vignon, X. (1990). Halothane-induced functional and structural modifications in sarcoplasmic reticulum membranes from pig skeletal muscle. *Biochimie* **72,** 245–250.

Interaction of Anesthetics and Catecholamines on Conduction in the Canine His–Purkinje System

Lawrence A. Turner, Sanja Vodanovic, and Zeljko J. Bosnjak
Departments of Anesthesiology and Physiology
The Medical College of Wisconsin
Milwaukee, Wisconsin 53226

I. Introduction

The electrophysiological basis underlying the adverse interaction between low doses of epinephrine and halothane, which sensitizes the heart to the arrhythmogenic effects of catecholamines, is not known. The adrenergic receptor mechanisms mediating this interaction in dogs anesthetized with halothane involve activation of both cardiac α_1- and β_1-adrenoceptors because the so-called threshold arrhythmogenic dose of epinephrine is elevated by pretreatment with either prazosin or metoprolol (1), two relatively selective adrenoceptor antagonists. In addition, Hayashi *et al.* (2) have shown that the induction of dysrhythmias with halothane involves a synergistic interaction between α_1- and β-adrenergic-mediated effects as assessed by measurements of the arrhythmogenic doses of phenylephrine and isoproterenol administered in combination. Although *in vitro* studies have demonstrated antiarrhythmic actions of halothane on delayed afterdepolarizations and triggered activity associated with β-adrenoceptor activation and abnormal intracellular calcium transients (3,4), relatively little is known about the role of α_1-adrenoceptor activation in the arrhythmogenic interaction between anesthetics and catecholamines. A potential contribution to abnormal conduction was suggested by Reynolds and Chiz

Advances in Pharmacology, Volume 31

(5) who originally reported that high concentrations of epinephrine (4.5 μM), by an action sensitive to phentolamine but not propranolol, markedly potentiated the depression of conduction produced by halothane in Purkinje fibers. However, one subsequent study failed to find evidence for α-mediated negative dromotropic effects of epinephrine or methoxamine with halothane (6).

Studies in our laboratory have been directed toward examining the actions of catecholamines on conduction in canine Purkinje fibers in the absence and presence of volatile anesthetics. To study these actions we utilized Purkinje fibers from the free-running false tendons superfused in a low-volume (2 ml), high-flow (5–7 ml/min) tissue chamber. The tendons were stimulated at 150 bpm in the orthodromic direction, and action potentials were recorded utilizing standard electrophysiological techniques from two fibers located about 5 mm apart and several millimeters away from the stimulating electrodes. The signals were monitored continuously on oscilloscopes for measurement of interelectrode conduction time and velocity at each minute of drug trials which were conducted by rapidly switching between solutions equilibrated with the anesthetics and test drugs. Additional action potential characteristics, including the rate of phase 0 depolarization (electronically differentiated and held with a peak detector), were measured by analog-to-digital conversion and computerized analysis. The values obtained were evaluated by repeated measures analysis of variance and the means compared utilizing least significant difference methods.

II. Catecholamine–Anesthetic Interaction

Figure 1 illustrates the conduction velocities obtained over time on exposure to 5 μM epinephrine alone and epinephrine in combination with halothane or isoflurane in a single group of preparations (7). Epinephrine alone (Fig. 1, circles) did not influence conduction in the absence of anesthetics. Neither halothane nor isoflurane alone (0.4 mM, about 1.6 MAC) decreased conduction velocity ($0.05 \leq p \leq 0.10$) over an interval of 20 min, whereas epinephrine in the presence of either anesthetic depressed velocity by 10–15% within 3–5 min, with return toward control values by 10 min despite continuing the anesthetics and epinephrine. In another group similar transient depression of conduction occurred on exposure to the neurotransmitter norepinephrine (5 μM) with either anesthetic, and in both groups the mean conduction velocities at the times of maximum depression were lower ($p \leq 0.01$) with halothane than with isoflurane.

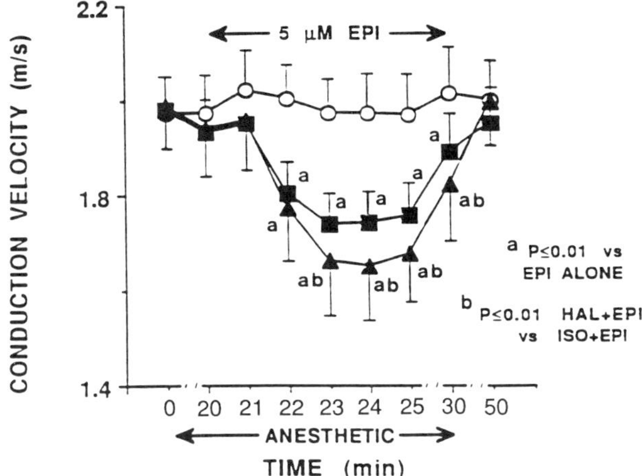

Fig. 1 Effects of epinephrine alone (circles) and in combination with halothane (HAL, triangles) or isoflurane (ISO, squares) on conduction velocity (mean ± SEM) in Purkinje fibers ($n = 12$). On the x axis, time 0 corresponds to controls before exposure to anesthetics for 20 min; times 20–30 cover the period of epinephrine (EPI) exposure; and times 50 denotes washout of both anesthetic and epinephrine. [Reprinted with permission from Vodanovic *et al.* (7).]

Figure 2 illustrates the accompanying changes of action potential characteristics, treated as a percentage of drug-free control values, associated with the sequence of exposure to halothane and then epinephrine plus halothane. The actions of epinephrine with halothane, which decreased conduction velocity by an average of 17% ($p \leq 0.01$) at 3–4 min, were not accompanied by significant changes of the action potential amplitude or rate of phase 0 depolarization (V_{max}), major determinants of conduction velocity. The results not only confirm the report of Reynolds and Chiz (5) that epinephrine abnormally depresses conduction in the presence of halothane, but also demonstrate the remarkable transient time course of this interaction and indicate that the degree of depression is greater with the more sensitizing anesthetic (8) halothane than with isoflurane.

The cellular electrophysiological mechanisms responsible for the depression of conduction by catecholamines in Purkinje fibers exposed to volatile anesthetics are not known but presumably involve separate or combined effects on active (phase 0 ionic channel conductances, largely peak Na^+ current) or passive (membrane capacitance, internal longitudinal resistance, gap junctional conductance) properties of the fibers. Anesthetic actions that may contribute to depression of conduction (halothane > isoflurane) (9) include inhibition of inward Na^+ current

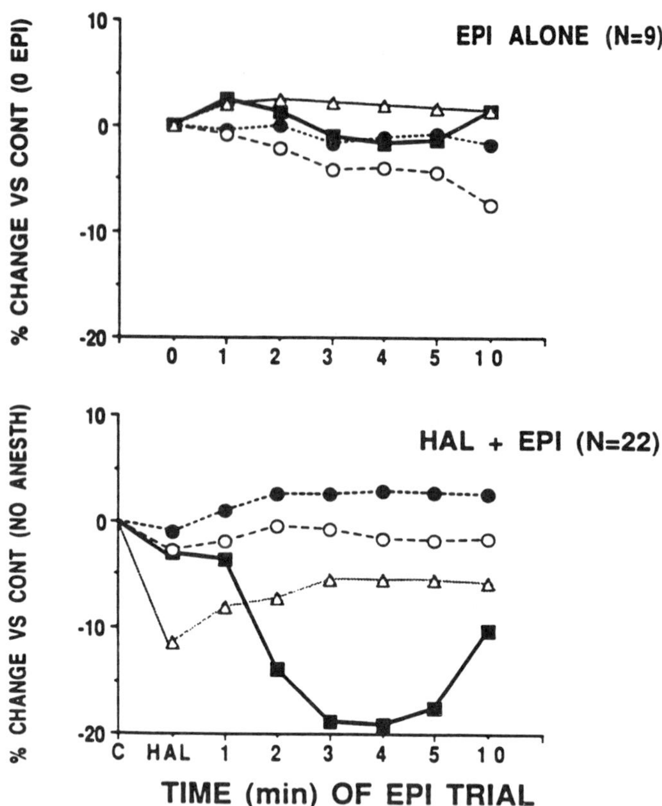

Fig. 2 Simultaneous changes of Purkinje fiber conduction velocity (squares) and action potential characteristics during trials of 5 μM epinephrine in the absence (top) and presence (bottom) of halothane. Filled circles denote action potential amplitude, empty circles V_{max}, and triangles action potential duration at 50% repolarization. Time C corresponds to drug-free control values. [Reprinted with permission from Vodanovic *et al.* (7).]

(10,11) increased membrane capacitance and internal longitudinal resistance (12), and either direct reduction of gap junctional conductance (13–16) as a consequence of disordering effects on the sarcolemmal membrane or transsarcolemmal proteins or indirect depression secondary to anesthetic-induced changes in the levels of intracellular regulatory substances (Ca^{2+}, cAMP, and cGMP) known to modulate gap junctional conductance. (17–19).

Although catecholamines were earlier thought to have little influence on propagation in Purkinje fibers except to improve conduction in depressed fibers (20), more recent studies indicate that β_1-receptor activation and elevation of cAMP depress Na^+ current in a voltage-dependent manner (21). However, other reports (19,22) suggest that both β- and α-adrenoceptor activation may influence cell-to-cell coupling, the former increasing and the latter decreasing coupling. There is increasing evidence that the various tissue-specific connexin proteins forming gap junctions exhibit distinct channel conductances and potential phosphorylation sites that may be regulated by intracellular enzymes including protein kinase C (23,24). Although the findings do not distinguish between the possible mechanisms leading to inhibition of Na^+ currents or cell-to-cell coupling, application of cable theory to propagation in linear Purkinje fiber strands (25,26) indicates that conduction velocity is directly related to the square root of the change of V_{max} and inversely to the root of the sum of external and internal longitudinal resistances, the latter largely determined by gap junctional resistance. Therefore, the relatively small decreases of V_{max} (only $\pm$ 5%) in association with a greater degree of depression of velocity (10–15% versus halothane) by epinephrine with halothane (Fig. 2) suggests that the electrophysiological mechanism underlying the depression of conduction involves inhibition of cell-to-cell coupling alone or in combination with reduction of inward Na^+ current during the upstroke of the action potential.

The time course of changes of conduction velocity on exposure to high doses of epinephrine (5 μM) and halothane (about 2.4%) alone and in combination, as well as the effect of changing the order of administration of the two drugs, (27), was evaluated as shown in Fig. 3. Epinephrine alone produced small biphasic changes, decreases and increases of about $\pm$5% ($p \leq 0.05$) from the control velocity of 2.1 m/sec, whereas halothane alone decreased velocity in a monophasic manner to about 1.9 m/sec. Halothane added to epinephrine decreased velocity, with a monophasic time course similar to that with halothane alone, to a lower steady-state value of 1.8 m/sec after 20 min of combined exposure. However, the usual clinical order of administration of epinephrine after halothane (EPI added to HAL in Fig. 3) produced a transient depression of velocity to about 1.3 m/sec within 2–3 min followed by return to about 1.7 m/sec after 20 min of combined exposure. The transient decrease of velocity, about -33% from preceding values with halothane alone, was again associated with a smaller (-7%) although significant ($p \leq 0.05$) decrease of V_{max}.

The findings *in vitro* clearly indicate that full manifestation of the negative dromotropic interaction between epinephrine and halothane is critically dependent on time and prior anesthetic exposure. This result suggests that the potential contribution of this self-limited proarrhythmic effect of epinephrine to abnormal conduction and arrhythmogenesis during halothane anesthesia may not be readily assessed under steady-state conditions

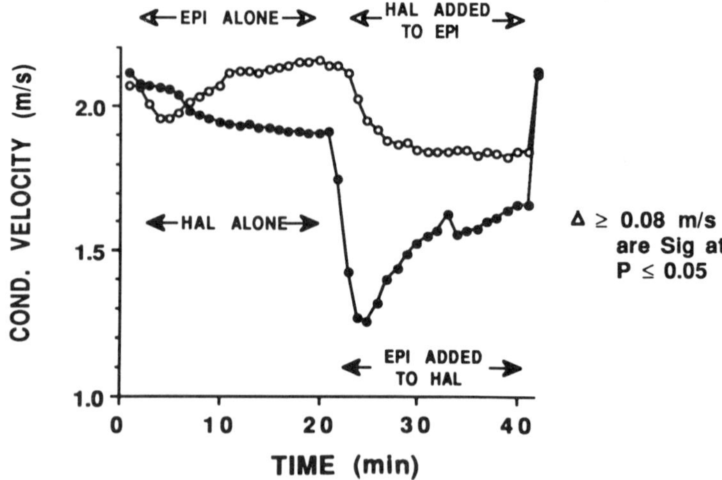

Fig. 3 Effects of changing the order of administration of epinephrine (5 μM) and halothane (0.75 μM) on Purkinje fiber conduction velocity in a single group of preparations (n = 8). The symbols represent mean velocity values in the two possible orders: epinephrine followed by halothane added to epinephrine (empty circles) and halothane followed by epinephrine added to halothane (filled circles). Only the latter order produced marked transient depression of conduction.

and may vary with the rate of change of the plasma epinephrine concentration and perhaps with the duration of preceding anesthetic exposure. The brief time course of conduction slowing by epinephrine in the presence of halothane may indicate that two separate but opposing processes are involved. One process rapidly slows conduction to a minimum value 2–3 min after introduction of epinephrine. This depressant effects occurs with a time course similar to that in which epinephrine precipitates dysrhythmias in halothane-anesthetized dogs given just threshold doses utilizing the Pace protocol (28) of epinephrine infusions. The second process, with a slower time constant, appears to gradually return velocity toward control despite continuing both halothane and epinephrine. Both processes modulating conduction velocity in fibers exposed first to halothane and then to epinephrine are either absent, minimal, or completely balanced in their opposing influences on conduction following exposure to epinephrine in the absence of anesthetic.

Various reports (29–32) indicate that "just threshold" arrhythmogenic doses of epinephrine in dogs anesthetized with halothane produce plasma concentrations of epinephrine in the range of 40–60 ng/ml, or about 0.2 to 0.3 μM. In contrast, the plasma concentrations of epinephrine with just threshold doses are about five times higher (200–300 ng/ml ~1 μM) during anesthesia with nonsensitizing agents (etomidate, pentobarbital), and presumably plasma epinephrine levels inducing ventricular tachycardia and fibrillation are proportionately higher both with and without halothane.

The epinephrine dose–response relationship for depression of conduction velocity (33) was determined at two halothane levels as illustrated in Fig. 4. This experiment utilized a randomized design in which the responses to each of 4 epinephrine concentrations were evaluated during 5-min trials at each anesthetic level in a single group of 8 preparations. The values of conduction velocity shown represent the nadir of epinephrine effect at 3 min of exposure. Compared to the drug-free control velocity of about 2.2 m/sec, halothane alone (0 epinephrine values in Fig. 4) decreased ($p \leq 0.05$) conduction velocity by 5 and 11% at the low (0.46 mM) and high (0.86 mM) halothane concentrations, respectively. Epinephrine at 5 μM with 0.86 mM halothane transiently decreased velocity by 33% (to 1.3 m/sec) of the value (~2 m/sec) with halothane alone. However, significant depression of conduction (−5% relative to halothane alone) was observed with 0.2 μM epinephrine in combination with the lower halothane dose (0.46 mM, about 1.5%, v/v); this represents an epinephrine concentration comparable to those reported to just induce

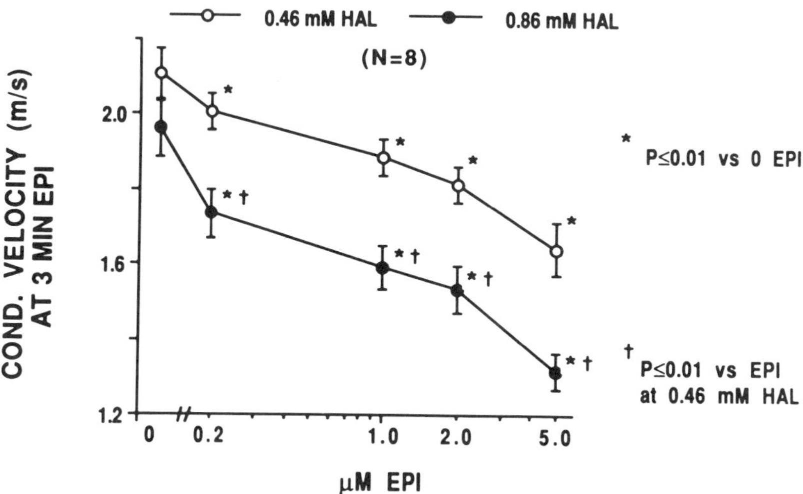

Fig. 4 Dose–response relationship for the peak depression of Purkinje fiber conduction velocity by epinephrine at two halothane concentrations. Values at 0 EPI represent controls with HAL alone; values with EPI represent those at 3 min of EPI exposure with HAL.

ventricular dysrhythmias in halothane-anesthetized dogs. The degree of depression of conduction was more dependent on the halothane than epinephrine doses because high concentrations of epinephrine have no (Fig. 1) or minimal (Fig. 3) effects on velocity in the absence of anesthetic, and about a 5- to 10-fold increase of the epinephrine concentration is required (Fig. 4) to transiently decrease velocity about the same as doubling the halothane level at the lowest (0.2 μM) epinephrine concentration. The results indicate that clinically relevant epinephrine concentations produce at least some degree of abnormal conduction in Purkinje fibers exposed to halothane but do not establish the lowest halothane concentration at which higher epinephrine concentrations slow conduction or any relationship between this interaction on conduction and the more complicated phenomena of unidirectional block and altered conduction leading to reentry.

The adrenergic receptor mechanisms underlying the actions of epinephrine on conduction were evaluated in different groups utilizing selected adrenergic antagonists and high anesthetic and agonist concentrations (5 μM)to permit statistical evaluation of different responses within and between treatment groups. Figure 5 shows control responses (top) to epinephrine in the presence of halothane and the responses in the same preparations (bottom) after treatment with either the β_1-adrenergic blocker metoprolol or the α_1-adrenergic antagonist prazosin. The depression of conduction was completely antagonized by prazosin but not by metoprolol. In another group of preparations (data not shown), the decreases of velocity due to epinephrine with halothane were not attenuated by 1 μM propranolol, which would exclude a role of simultaneous β_2 activation.

Figure 6 illustrates the actions of α_1- and α_2-agonists in combination with anesthetics on conduction velocity. In Fig. 6 (top), phenylephrine alone, which at 5 μM concentration slightly decreased velocity, produced larger decreases with halothane than isoflurane. The experiment shown in Fig. 6 (bottom) indicates that the α_2-agonist clonidine has no influence on Purkinje fiber conduction velocity in preparations that exhibit conduction slowing with epinephrine and halothane. Together the results clearly indicate that the depression of Purkinje fiber conduction velocity in the presence of halothane is mediated by activation of α_1-adrenoceptors. The findings of a proarrhythmic α_1-mediated effect of epinephrine on Purkinje fibers *in vitro* are consistent with studies in halothane-anesthetized dogs demonstrating (1) correlation between the arrhythmogenic dose of epinephrine in individual animals and their responsiveness to the pressor effects of phenylephrine, but not to the chronotropic effects of isoproterenol, (34), and (2) dose-related elevation of the epinephrine arrhythmogenic threshold dose by α_1-adrenergic blocking drugs (35).

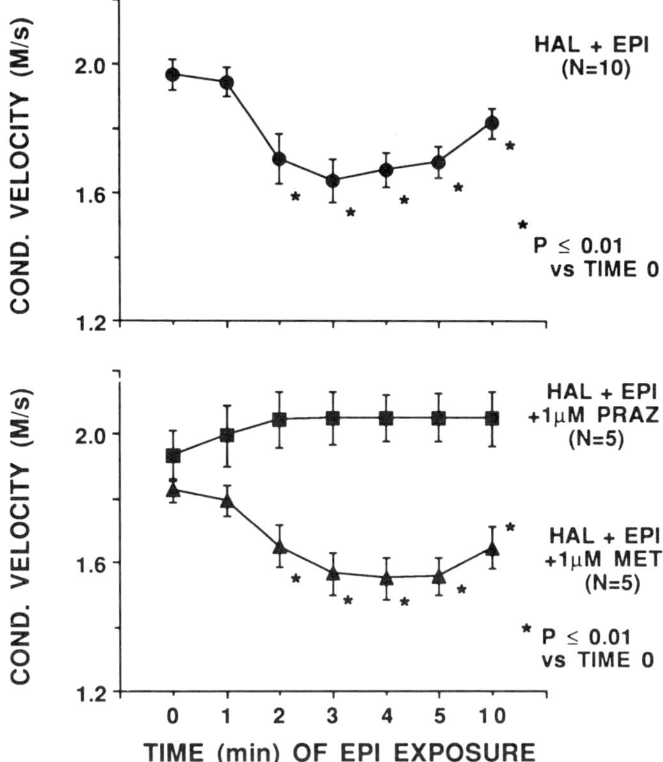

Fig. 5 Action of 5 μM epinephrine with 0.45 μM halothane on conduction in Purkinje fibers before (top) and after (bottom) α_1- and β_1-adrenergic receptor blockage. PRAZ, Prazosin; MET, metoprolol.

Adrenergic modulation of Purkinje fiber conduction by activation of α_1- rather than β-adrenoceptors is surprising and suggests that neither the depression or recovery of conduction velocity involve well-known β_1-receptor–effector mechanisms leading to increased cyclic AMP (cAMP) and intracellular Ca^{2+}. Thus, the depression and recovery are not consistent with cAMP-mediated depression of peak Na^+ current (21) followed by cAMP-enhanced junctional conductance (19) because neither process was altered by β_1-adrenergic blockage. In addition the insensitivity to propranolol and metoprolol would appear to exclude a sequence of changes involving transiently increased intracellular Ca^{2+} resulting from β_1-receptor activation, increased cAMP, and phosphorylation of L-type Ca^{2+} channels, which could reduce gap junctional conductance (17), followed

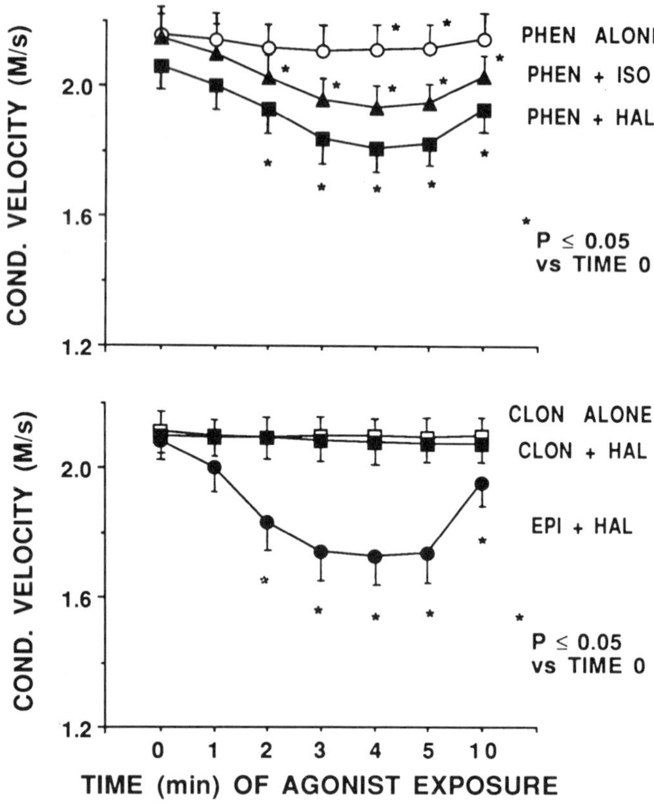

Fig. 6 Depression of Purkinje fiber conduction by the α_1-agonist phenylephrine (top, $n = 8$) but not by the α_2-agonist clonidine (bottom, $n = 12$). CLON denotes 5 μM clonidine, EPI 5 μM epinephrine, HAL 0.4 μM halothane, and ISO 0.4 μM isoflurane.

by recovery of velocity related to the effects of increasing cAMP on gap junctional conductance.

Studies have shown that at least two subtypes of α_1-receptors are involved in modulation of Purkinje fiber electrical activity (36,37). One subtype sensitive to the alkylating antagonist chloroethylclonidine (CEC) depresses automaticity at low catecholamine concentrations. This action may involve activation of the Na^+/K^+ electrogenic pump and is transduced by a pertussis toxin-sensitive G protein (38,39). The other subtype sensitive to competitive blockage by WB4101 mediates α_1-adrenoceptor-induced increases of automaticity (36) and action potential duration (40) at high catecholamine concentrations, probably by reducing outward K^+

currents. These actions are insensitive to pertussis toxin and may involve G-protein coupling to phospholipase C (36,40) resulting in generation of the intracellular second messengers inositol 1,4,5-trisphosphate (IP_3) and diacylglycerol (DAG). These studies clearly established that equimolar (0.1 μM) CEC and WB4101 antagonize the α_1-mediated effects of low concentrations (≤ 1 μM) of norepinephrine and phenylephrine in superfused Purkinje fiber preparations.

To examine the relative contribution of the α_1-receptor subtypes to conduction depression over a range of epinephrine concentrations up to 5 μM, we utilized a 5-fold higher concentration of CEC and WB4101 in the presence of 0.2 μM propranolol (41). The control epinephrine dose–response curve was determined in 12 preparations. The velocity was lowest between 2 and 5 min of epinephrine exposure, and the averaged values at these times are shown as the control curve in Fig. 7. Six of the preparations were treated with 0.5 μM WB4101 and six with 0.5 μM CEC, and the dose–response curve was repeated. As shown in Fig. 7, the depression of conduction velocity was substantially attenuated by WB4101 but only partially antagonized by CEC. Between groups the mean velocity was significantly higher after WB than CEC, indicating that the negative

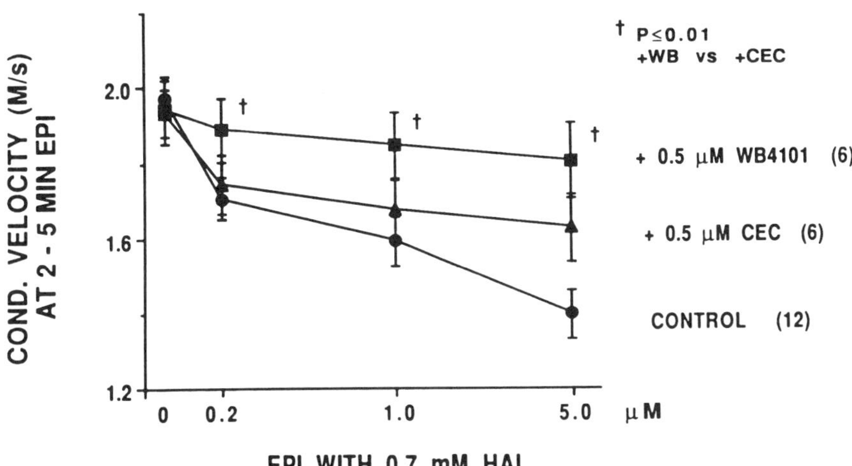

Fig. 7 Antagonism of the dose-related effects of epinephrine with halothane on conduction velocity (control responses in 12 preparations) by treatment with the α_1-receptor subtype antagonists WB4101 (WB, 0.5 μM) and chloroethylclonidine (CEC, 0.5 μM). The mean velocities averaged over the times of maximum depression (2–5 min) were higher between groups (6 each) after WB than after CEC. Propranolol (0.2 μM) was present throughout the experiments.

dromotropic effects of epinephrine with halothane are probably mediated by the α_1-receptor subtype sensitive to WB4101.

These studies, which are consistent with a potential contribution of α_1-adrenoceptor-mediated depression of conduction to halothane–epinephrine dysrhythmias, were conducted entirely in isolated Purkinje fiber false tendons at fixed stimulation rate. The interaction, although producing substantial conduction slowing (up to about 33% at a constant rate of 150 bpm) at high epinephrine (5 μM) and halothane (0.86 mM, ~2.75%) concentrations, was not observed to produce conduction block and would not appear to produce much conduction slowing with combinations of submicromolar concentrations of epinephrine ($\leq 0.3\ \mu M$) and subanesthetic levels of halothane ($\leq 0.5\%$, v/v) known to induce ventricular dysrhythmias *in vivo* (32). In theory the occurrence of unidirectional block and slow conduction leading to reentry is more likely to involve propagation at low membrane potentials during the relatively refractory period at ventricular sites manifesting disparate refractory periods and excitability characteristics or specific fiber geometries resulting in greater susceptibility to conduction depression (42). The Purkinje fiber–ventricular muscle junctions (PVJ) are sites known to be particularly susceptible to unidirec-

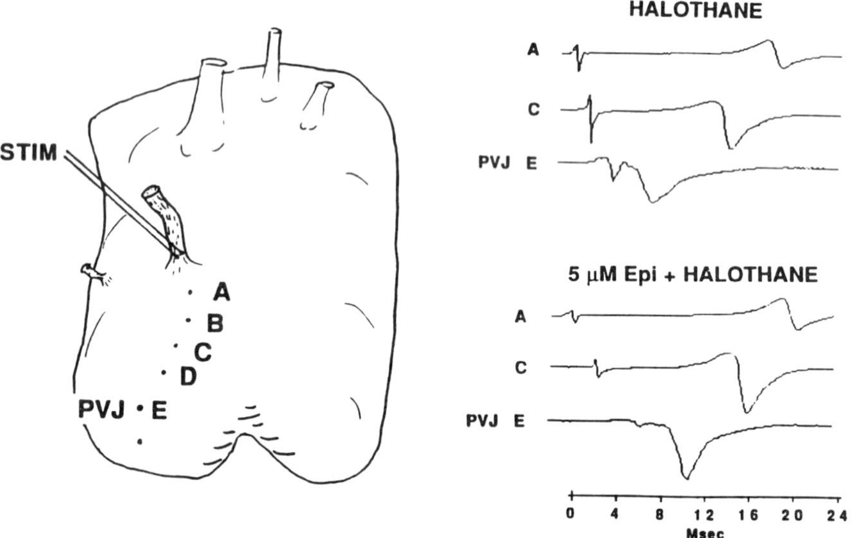

Fig. 8 Diagram of a typical canine left ventricular papillary muscle preparation and associated electrograms recorded at sites A and C between the false tendon stimulation site (STIM) and a Purkinje fiber–ventricular muscle junction (PVJ at site E), the site of earliest endocardial activation.

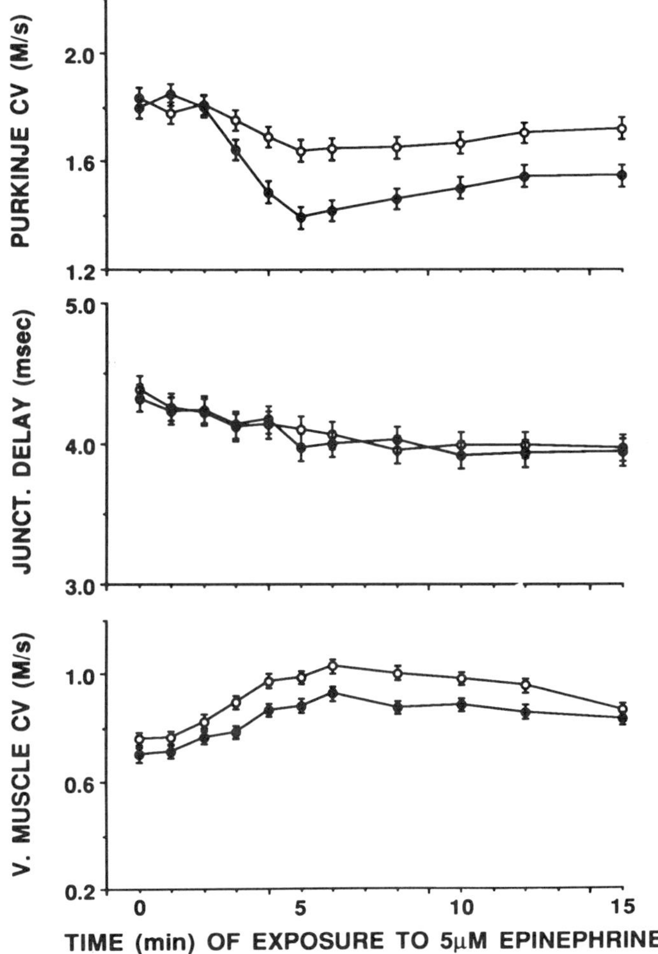

Fig. 9 Time-dependent effects of epinephrine alone (empty circles) and with halothane (filled circles) on Purkinje and ventricular muscle conduction velocities and delay at Purkinje fiber–ventricular muscle junctions in 10 preparations. The error bars represent approximate 95% confidence limits based on repeated measures analysis of variance and comparisons by least significant difference methods.

tional block (43) and to the depressant effects of lipophilic substances such as octanol and halothane on cell-to-cell coupling (16,44).

Additional experiments were performed on canine left ventricular papillary muscles to evaluate the time-dependent effects of epinephrine alone and epinephrine in combination with halothane on conduction from the

Purkinje fiber layer through Purkinje fiber–muscle junctions and the underlying endocardium. The muscles, as illustrated in Fig. 8, were superfused in a larger chamber permitting two exchanges per minute, and the false tendons were stimulated at 150 bpm. Activation mapping was performed utilizing a mobile bipolar electrode, in a manner similar to that used by Joyner and Overholt, (44) to locate a Purkinje–muscle junction, the site of earliest muscle activation (shown as site E in Fig. 8). Additional electrodes were then aligned between the stimulation site and the junction to record electrograms as shown to the right in Fig. 8. Propagation occurred through the Purkinje layer from the stimulation site to the PVJ and then, after a delay of about 4 msec, in a retrograde direction toward the muscle tip through the muscle layer. Purkinje conduction velocity, junctional delay, and muscle conduction velocity were measured from the electrograms and distance during randomized trials of exposure to 5 μM epinephrine alone and in combination with 0.4 mM halothane.

The preliminary results (45) are shown in Fig. 9. Epinephrine alone slightly (−10.5%) and transiently decreased velocity in the Purkinje layer, decreased junctional conduction delay about 10%, and increased muscle conduction velocity by 35%. The depression of velocity by 5 μM epinephrine in the Purkinje layer overlying the muscle was qualitatively different and larger in degree than the minimal changes (±5%) of velocity we usually observe with 5 μM epinephrine alone in false tendon fibers. This finding did not appear to be related to inadequate washout of anesthetic because the decrease was about the same (−12%) in the group of 5 muscles in which the trials of exposure to epinephrine alone were conducted before any exposure to halothane. In contrast, in the presence of 0.4 mM halothane, epinephrine produced a larger (−22%, $p \leq 0.01$) decrease of Purkinje layer velocity than with epinephrine alone, whereas junctional conduction delay decreased (−8%) and muscle velocity increased (+32%) about the same as in the absence of halothane.

III. Summary

The findings in papillary muscles that epinephrine facilitates conduction at Purkinje fiber–muscle junctions and in the endocardium are consistent with older observations that activation of myocardial β-adrenergic receptors speeds conduction and activation in the heart and thereby increases the synergy of contraction (46,47). The cellular mechanism underlying this action is probably increased cell-to-cell coupling between muscle fibers secondary to elevation of cyclic AMP (19,48). However, the findings that epinephrine alone or with halothane transiently slows conduction in the Purkinje layer while simultaneously improving conduction across Purkinje–muscle junctions and in the endocardium may represent proarrhythmic actions. These actions could facilitate arrhythmogenesis by transiently increasing regional differences of activation and repolarization times in the conduction system and myocardium and thereby increasing vulnerability to induction of reentry by premature impulses. Such a proarrhythmic effect could explain an older observation that low-dose norepinephrine infusions decrease the threshold for induction of fibrillation by two premature beats in pentobarbital-anesthetized animals (49).

The cellular basis underlying the different responses of Purkinje fibers and the endocardial muscle layer to catecholamines, in which velocity decreased and increased, respectively, is not known. Our working hypothesis to explain this action in canine Purkinje fibers is a mechanism involving activation of WB4101-sensitive α_1-adrenoceptor, G-protein coupling to phospholipase C and the generation of DAG and IP_3 leading to modulation of cell-to-cell coupling, which is potentiated in the presence of partial uncoupling by halothane. The different responses of Purkinje and mycardial fibers are speculated to result from differences in the relative density of this subtype of α_1-adrenoceptor, differences in the subcellular effector coupling mechanisms, or differences in the specific connexin proteins forming gap junctions between Purkinje and myocardial fibers (50).

References

1. Maze, M., and Smith, C. M. (1983). Identification of receptor mechanism mediating epinephrine-induced arrhythmias during halothane anesthesia in the dog. *Anesthesiology* **59,** 322–326.
2. Hayashi, Y., Sumikawa, K., Tashiro, C., and Yoshniya, I. (1988). Synergistic interaction of α_1- and β_1-adrenoceptor agonists on induction arrhythmias during halothane anesthesia in dogs. *Anesthesiology* **68,** 902–907.
3. Freeman, L. C., and Li, Q., (1991). Effects of halothane on delayed after depolarizations and calcium transients in dog ventricular myocytes exposed to isoproterenol. *Anesthesiology* **74,** 146–154.
4. Zukerman, R. L., and Wheeler, D. M. (1991). Effect of halothane on arrhythmogenic responses induced by sympathomimetic agents in single rat heart cells. *Anesth. Analg. (N.Y.)* **72,** 596–603.
5. Reynolds, A. K., and Chiz, J. F. (1974). Epinephrine-potentiated slowing of conduction in Purkinje fibers. *Res. Commun. Chem. Pathol. Pharmacol.* **9,** 633–645.
6. Freeman, L. C., and Muir, W. W. (1991). α-Adrenoceptor stimulation in the presence of halothane: Effects on impulse propagation in cardiac Purkinje fibers. *Anesth. Analg. (N.Y.)* **72,** 11–17.
7. Vodanovic, S., Turner, L. A., Hoffmann, R. G., Kampine, J. P., and Bosnjak, Z. J. (1993). Transient negative dromotropic effects of catecholamines on canine Purkinje fibers exposed to halothane and isoflurane. *Anesth. Analg. (N.Y.)* **76,** 592–597.
8. Johnston, R. R., Eger, E. I., and Wilson, C. (1976). A comparative interaction of epinephrine with enflurane, isoflurane and halothane in man. *Anesthesiology* **55,** 709–712.

9. Vodanovic, S., Turner, L. A., Kampine, J. P., and Bosnjak, Z. J. (1993). Comparative effects of halothane and isoflurane on Purkinje fiber conduction. *Anesthesiology* **79**(3A), A687.
10. Ikemoto, Y., Yatani, A., Imoto, Y., and Arimura, H. (1986). Reduction of the myocardial sodium current by halothane and thiamylal. *Jpn. J. Physiol.* **36,** 107–121.
11. Buljubasic, N., Berczi, V., Supan, D. F., Marijic, J., Turner, L. A., Kampine, J. P. and Bosnjak, Z. J. (1993). Depression of the Na^+ current by halothane and isoflurane in the rabbit Purkinje fibers. *Anesthesiology* **79,**(3A), A392.
12. Hauswirth, O. (1986). The influence of halothane on the electrical properties of cardiac Purkinje fibers. *J. Physiol.* (*London*) **201,** P42–43.
13. Terrar, D. A., and Victory, J. G. G. (1988). Influence of halothane on electrical coupling in cell pairs isolated from guinea pig ventricle. *Br. J. Pharmacol.* **94,** 509–514.
14. Niggli, E., Rudisuli, A., Maurer, P., and Weingart, R. (1989). Effects of general anesthetics on current flow across membranes in guinea pig myocytes. *Am. J. Physiol.* **256,** C273–281.
15. Burt, J. M., and Spray, D. C. (1989). Volatile anesthetics block intracellular communication between neonatal rat myocardial cells. *Circ. Res.* **65,** 829–837.
16. Freeman, L. C., and Muir, W. W. (1991). Effects of halothane on impulse propagation in Purkinje fibers and at Purkinje–muscle junctions: Relationship of V_{max} to conduction velocity. *Anesth. Analg.* (*N.Y.*) **72,** 5–10.
17. Noma, A., and Tsuboi, N. (1987). Dependence of junctional conductance on proton, calcium and magnesium ions in cardiac paired cells of guinea pig. *J. Physiol.* (*London*) **382,** 193–211.
18. Vulliemoz, Y., Verosky, M., and Triner, L. (1986). Effect of halothane on myocardial cyclic AMP and cyclic GMP content of mice. *J. Pharmacol. Exp. Ther.* **236,** 181.
19. Burt, J. M., and Spray, D. C. (1988). Inotropic agents modulate gap junctional conductance between cardiac myocytes. *Am. J. Physiol.* **254,** H1206–H1210.
20. Hoffman, B. F., and Singer, D. H. (1967). Appraisal of the effects of catecholamines on cardiac electrical activity. *Ann. N.Y. Acad. Sci.* **139,** 914–939.
21. Katsushige, O., Kiyosue, T., and Arita, M. (1989). Isoproterenol, DBcAMP, and forskolin inhibit cardiac sodium current. *Am. J. Physiol.* **256,** C1131–C1137.
22. Burt, J. M., and Spray, D. C. (1988). Adrenergic control of gap junction conductance in cardiac myocytes. *Circulation* **78**(Suppl. 2), 258.
23. Veenstra, R. D., Wang, H. Z., Westphale, E. M., and Beyer, E. C. (1992). Multiple connexins confer distinct regulatory and conductance properties of gap junctions in developing heart. *Circ. Res.* **71,** 1277–1283.
24. Munster, P. N., and Weingart, R. (1993). Effects of phorbol ester on gap junctions of neonatal rat heart cells. *Pfluegers Arch.* **423,** 181–188.
25. Arnsdorf, M. F. (1984). Cable properties and conduction of the action potential. *In* "Physiology and Pathophysiology of the Heart" (N. Sperelakis, ed.), pp. 109–140. Nijhoff, Boston.
26. Gettes, L. S. (1990). Effects of ionic changes on impulse propagation. *In* "Cardiac Electrophysiology: A Textbook" (M. R. Rosen, M. J. Janse, and A. L. Wit, eds.), pp. 459–479. Futura Publ. Mount Kisco, New York.
27. Turner, L. A., Vodanovic, S., Kampine, J. P., and Bosnjak, Z. J. (1993). Effects of order of administration of epinephrine and halothane on Purkinje fiber conduction. *Anesthesiology* **79**(3A), A688.
28. Pace, N. L., Ohmura, A., and Wong, K. C. (1979). Epinephrine-induced arrhythmias: Effects of exogenous prostaglandin synthesis inhibition during halothane–O_2 anesthesia in the dog. *Anesth. Analg.* (*N.Y.*) **58,** 401.

29. Metz, S., and Maze, M. (1985). Halothane concentration does not alter the threshold for epinephrine-induced arrhythmias in dogs. *Anesthesiology* **62,** 470–474.
30. Sumikawa, K., Ishizaka, N., and Suzaki, M. (1983). Arrhythmogenic plasma levels of epinephrine during halothane, enflurance and pentobarbital anesthesia in the dog. *Anesthesiology* **58,** 322–325.
31. Hayashi, Y., Sumikawa, K., Yamatodani, A., Tashiro, C., Wada, H., and Yoshiya, I. (1989). Myocardial sensitization by thiopental to arrhythmogenic action of epinephrine in dogs. *Anesthesiology* **71,** 929–935.
32. Hayashi, Y., Sumikawa, K., Yamatodani, A., Kamibayashi, T., Kuro, M., and Yoshiya, I. (1991). Myocardial epinephrine sensitization with subanesthetic concentrations of halothane in dogs. *Anesthesiology* **74,** 134–137.
33. Turner, L. A., Vodanovic, S., Hoffmann, R. G., Kampine, J. P., and Bosnjak, Z. J. (1994). A subtype of α_1-adrenoceptor mediates depression of conduction in Purkinje fibers exposed to halothane. *Anesthesiology* submitted.
34. Spiss, K., Maze, M., and Smith, C. M. (1984). α-Adrenergic responsiveness correlates with epinephrine dose for arrhythmias during halothane anesthesia in dogs. *Anesth. Analg.* (*N.Y.*) **63,** 297–300.
35. Maze, M., Hayward, E., and Gaba, D. M. (1985). α_1-adrenergic blockade raises epinephrine-arrhythmia threshold in halothane-anesthetized dogs in a dose-dependent fashion. *Anesthesiology* **63,** 611–615.
36. Del Balzo, U., Rosen, M. R., Malfatto, G., Kaplan, L. M., and Steinberg, S. F. (1990). Specific α_1-adrenergic receptor subtypes modulate catecholamine-induced increases and decreases in ventricular automaticity. *Circ. Res.* **67,** 1535–1551.
37. Robinson, R. B. (1990). α-Adrenergic receptor–effector coupling. *In* "Cardiac Electrophysiology: A Textbook" (M. R. Rosen, M. J. Janse, and A. L. Wit, eds.), pp 819–829. Future Publ., Mount Kisco, New York.
38. Shah, A., Cohen, I. S., and Rosen, M. R. (1988). Stimulation of cardiac alpha receptors increases Na/K pump current and decreases g_k via a pertussis toxin-sensitive pathway. *Biophys J.* **54,** 219–225.
39. Zaza, A., Kline, R. P., and Rosen, M. R. (1990). Effects of α-adrenergic stimulation on intracellular sodium activity and automaticity in canine Purkinje fibers. *Circ. Res.* **66,** 416–426.
40. Lee, J. H., Steinberg, S. F., and Rosen, M. R. (1991). A WB4104-sensitive α_1-adrenergic receptor subtype modulates repolarization in canine Purkinje fibers. *J. Pharmacol. Exp. Ther.* **259,** 681–687.
41. Turner, L. A., Vodanovic, S., Kampine, J. P., and Bosnjak, Z. J. (1993). WB-4101-sensitive α_1-adrenergic depression of conduction in canine Purkinje fibers exposed to halothane. *FASEB J.* **7**(3)1, A96.
42. Quan, W., and Rudy, Y. (1990). Unidirectional block and reentry of cardiac excitation; a model study. *Circ. Res.* **66,** 367–382.
43. Overholt, E. D., Joyner, R. W., Veenstra, R. D., Overholt, E. D., Joyner, R. W., Veenstra, R. D., Rawlings, D., and Wiedman, R. (1984). Unidirectional block between Purkinje and ventricular layers of papillary muscles. *Am. J. Physiol.* **247,** H584–595.
44. Joyner, R., and Overholt, E. D. (1985). Effects of octanol on subendocardial Purkinje-to-ventricular transmission. *Am. J. Physiol.* **249,** H1228–1231.
45. Turner, L. A., Vodanovic, S., Kampine, J. P., and Bosnjak, Z. J. (1993). Interaction of epinephrine with halothane on endocardial conduction and activation. *Anesthesiology* **79**(3A), A689.
46. DeMello W. C. (1984). Modulation of junctional permeability. *Fed. Proc.* **43,** L2692.

47. DeMello, W. C. (1989). Effects of isoproterenol and 3-isobutyl-1-methylxanthine on junctional conduction in heart cell pairs. *Biochim. Biophys. Acta* **1012,** 291.
48. Veenstra, R. D. (1991). Physiologic modulation of cardiac gap junctional channels. *J. Cardiovasc. Electrophysiol.* **2,** 168–189.
49. Rabinowitz, S. H., Verrier, R. L., and Lown, B. (1976). Muscarinic effects of vagosympathetic trunk stimulation on the repetitive extrasystole (RE) threshold. *Circulation* **53,** 622.
50. Kanter, H. L., Laing, J. G., Bean, S. L., Beyer, E. C., and Saffitz, J. E. (1993). Distinct pattern of connexin expression in canine Purkinje fibers and ventricular muscle. *Circ. Res.* **72,** 1124–1131.

Anesthetics, Catecholamines, and Ouabain on Automaticity of Primary and Secondary Pacemakers

John L. Atlee III, Martin N. Vicenzi, Harvey J. Woehlck, and Zeljko J. Bosnjak

Department of Anesthesiology
The Medical College of Wisconsin
Milwaukee, Wisconsin 53226

I. Introduction

Cardiac arrhythmias are detected in 60% or more of patients undergoing anesthesia and surgery when continuous methods are used for surveillance (1,2). Most arrhythmias (up to 90–95%) are relatively innocuous, since they do not jeopardize circulatory dynamics or predispose to lethal ventricular arrhythmias. The other 5–10% of arrhythmias are considered dangerous because they do produce circulatory compromise or predispose to lethal arrhythmias. As for causes, new or worse arrhythmias in perioperative settings must be the result of some imposed physiological imbalance (1,2). However, knowledge of how such imbalance contributes to arrhythmias is incomplete. Nevertheless, such knowledge is key to prevention, as well as to effective remedial or definitive treatment. Therefore, elucidation of mechanisms for arrhythmias during anesthesia has been the focus of investigation in our laboratory for over 20 years.

Cardiac arrhythmias are usually seen as disturbances of impulse generation, propagation, or both. Previous investigation in our laboratory focused more on anesthetic effects on properties pertaining to impulse propagation in the intact canine heart. This work, reviewed elsewhere (1,2),

Advances in Pharmacology, Volume 31

suggests that clinically useful concentrations of contemporary volatile inhalation anesthetics (halothane, enflurane, isoflurane) are expected to cause physiologically inconsequential prolongation of specialized atrioventricular (AV) conduction times or increased refractoriness in the normal heart. In other words, the amount of conduction prolongation is not sufficient to cause AV heart block, and increased refractoriness does not appear arrhythmogenic. In fact, increased refractoriness by anesthetic drugs might even oppose some reentrant tachyarrhythmias (3,4).

The foregoing suggests that ectopic supraventricular and ventricular beats or rhythms, commonly observed during anesthesia and surgery in presumably normal hearts (2), might be due to enhanced normal automaticity of secondary or latent pacemakers compared to the sinoatrial (SA) node. It is unlikely that anesthetic drugs can initiate arrhythmias due to triggered activity or abnormal automaticity (i.e., automaticity from a depolarized level of cell membrane potential) (1,2) in normal hearts. If so, increased endogenous catecholamines or enhanced efferent sympathetic discharge occasioned by perioperative stress might interact with anesthetics to increase automaticity of latent pacemakers, and toxic digitalis might also do the same. Thus, enhanced automaticity of subsidiary atrial pacemakers might explain earlier observations that ectopic atrial rhythms precede ventricular arrhythmias during epinephrine–anesthetic sensitization (5,6) and paroxysmal atrial tachycardia (PAT) with toxic digitalis (7).

We supposed that potent inhalation anesthetic agents, by direct or indirect (autonomic-mediated) effects, can alter automaticity of the SA node and latent pacemakers [subsidiary atrial (8,9), ectopic atrial, AV junctional] in a manner conducive to the formation of ectopic atrial or AV junctional rhythm disturbances. Further, catecholamines would enhance these effects of anesthetics. However, volatile anesthetics would oppose a possible effect of digitalis to enhance automaticity of subsidiary or ectopic atrial pacemakers to explain PAT with toxic digitalis. These hypotheses were tested in both *in vitro* and *in vivo* canine models. The findings to date can have relevance to the management of patients with SA node dysfunction, susceptibility to atrial tachyarrhythmias, or advanced AV heart block, or those receiving digitalis.

II. Isolated and Chronic Atrial Preparations

In vitro studies were carried out using a perfused canine right atrial preparation that included primary and subsidiary pacemakers within the distribution of the SA node artery, according to methods described by Rozanski *et al.* (10) and Woods *et al.* (11). Studies in intact heart were performed in chronically instrumented dogs according to methods previously described by Randall *et al.* (8), Tuna *et al.* (12), and Atlee *et al.* (13).

A. Canine Right Atrial Preparation

The canine right atrial preparation excluded left atrium, ventricular tissue, AV node, and coronary sinus. It was perfused with warmed, oxygenated Krebs solution via the SA node artery (Fig. 1) (14–16) and also superfused with the same Krebs solution. Bipolar, extracellular recording electrodes ($n = 4$) recorded the site of earliest activation (SEA), which could be the SA node or one of three potential sites of subsidiary atrial pacemakers (SAP) 1, 2, and 3 cm caudal to the SA node along the sulcus terminalis (Fig. 1). Preparations were exposed to halothane (1 or 2%), isoflurane (1.4 or 2.8%, % v/v), ouabain (low or mid therapeutic, 25 and 50 n*M*, respectively; borderline toxic, 0.1 μM), and epinephrine (1, 2, or 5 mg/liter) or norepinephrine (2, 5, and 10 mg/liter). The protocol used for these experiments is summarized in Fig. 2. Exposure to epinephrine or norepinephrine was chosen as a model for anesthetic sensitization and increased adrenergic input, respectively, and that to ouabain as a model for atrial rhythm disturbances with digitalis. Shifts in SEA from the SA node to sites of SAP were scored 1, 2, or 3 (SA node = 0) to provide a magnitude score. Magnitude scores for all preparations with shifts were

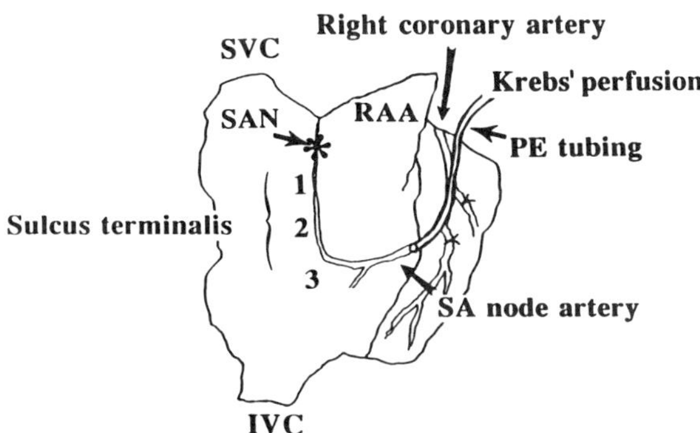

Fig. 1 Drawing of the perfused canine right atrial preparation with bipolar, extracellular recording electrodes at SA node (SAN) and three, increasingly remote sites of subsidiary atrial pacemakers along the sulcus terminalis (1, 2, and 3). The perfusion cannula (polyethylene tubing) was introdued via the right coronary artery and passed through to the SA node artery. The region indicated by the asterisk, namely, the SAN, is normally the site of earliest activation (SEA). However, potential subsidiary atrial pacemaker sites (1, 2, and 3) along the sulcus terminalis can become the SEA after exposure to drugs and other interventions. SVC, Superior vena cava; IVC, inferior vena cava; RAA, right atrial appendage. [From Polic *et al.* (14), with permission.]

CONTROL 1 (15 min)	L(H) AN (15 min)	CONTROL 2 (15 min)
E(NE) (5 min X 2 or 3)	L(H) AN + E(NE) (5 min X 2 or 3)	CONTROL 3 (15 min)
L(H) AN (15 min)	L(H) AN + E(NE) (5min X 2 or 3)	CONTROL 4 (15 min)

Fig. 2 Depiction of protocol for experiments with the perfused canine right atrial preparation. The sequence reads from left to right and top to bottom. Preparations were exposed in randomized order to low (L) or high (H) concentrations of anesthetics (AN, halothane or isoflurane), with observations recorded after 15 min for equilibration. Each preparation was also exposed for 5 min to each of two or three concentrations of epinephrine (E) or norepinephrine (NE), with or without anesthetics. Control measurements were performed after 15 min for washout. A similar procedure was used for experiments with ouabain.

summed for each test condition. In turn, summed magnitude scores were normalized by dividing by the total number of preparations studied for each test condition.

B. Chronically Instrumented Dogs

Conditioned mongrel dogs of either sex were prepared for chronic electrophysiological investigation by implantation of bipolar epicardial electrodes (Fig. 3) and an indwelling aortic catheter (17,18). Electrodes were sewn to the left and right atrial appendages and to the apex of the right ventricle. A patch electrode with five pairs of electrodes was sewn along the sulcus terminalis (Fig. 3). The most rostral pair of electrodes recorded from the anatomical region of the SA node, and more caudal pairs from previously described sites of SAP (19–21). The most caudal of these was situated at the junction of the right atrium with the inferior vena cava. A bipolar needle electrode was advanced into the interventricular septum from the aortic root to record the His bundle electrogram (12,13), and the inferior aorta was cannulated from the femoral artery for chronic pressure recording. After at least 2 weeks for recovery, dogs were ready for testing.

On weekly occasions, dogs were tested awake and during exposure to epinephrine and end-tidal concentrations of halothane, enflurane, or isoflurane equivalent to 1.2 and 2.0 minimum anesthetic concentration (MAC) with or without atropine methyl nitrate blockade (Fig. 4). Data for determination of atrial SEA during exposure to epinephrine were recorded for nine 5-sec intervals (i.e., total of 45 sec) during the second and third minutes of infusion. The electrocardiographic SEA was scored 1 to 6 depending on whether the SEA was the SA node (score 1) or some other atrial site (Table I). Atrioventricular junctional or ventricular ectopic beats were not scored. Instead, they were tabulated as a percentage of total beats. The fraction (as decimal) of beats for the 45-sec period from each

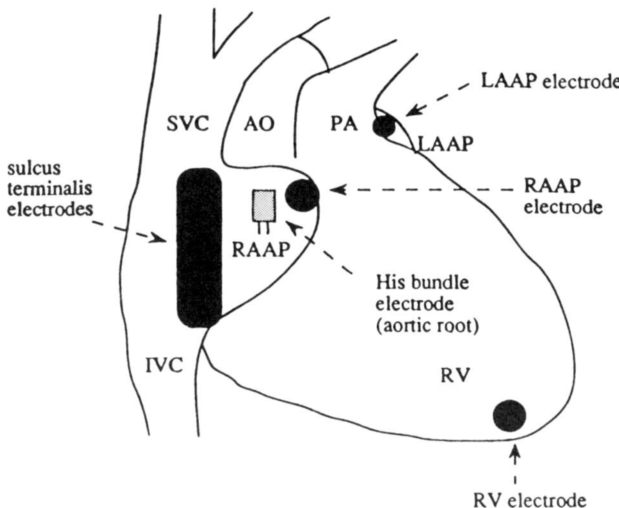

Fig. 3 Depiction of location of bipolar, epicardial recording electrodes in the chronically instrumented dog preparation. The patch electrode along the sulcus terminalis contained five evenly spaced electrode pairs. Other electrodes were located on the right (RAAP) and left atrial appendages (LAAP) and on the right ventricular apex (RV). A bipolar, His bundle needle electrode was advanced into the interventricular septum from the aortic root to record the His bundle electrogram. SVC, Superior vena cava; AO, aorta; PA, pulmonary artery; IVC, inferior vena cava. [From Woehlck *et al.* (17), with permission.]

CONTROL (no drugs)	±AMN (init) (3 mg/kg)	BASELINE (awake)	EPI-1 (awake)	EPI-2 (awake)
L(H) AN (anes-1)	±AMN (rep) (1.5 mg/kg)	BASELINE (anes-1)	EPI-1 (anes-1)	EPI-2 (anes-1)
L(H) AN (anes-2)	±AMN (rep) (1.5 mg/kg)	BASELINE (anes-2)	EPI-1 (anes-2)	EPI-2 (anes-2)

Fig. 4 Representation of protocol for experiments in chronically instrumented dogs. The sequence reads from left to right and top to bottom. Measurements were first made in conscious dogs without any drugs (control). If atropine methyl nitrate (AMN; int, initial dose) was used on the particular test occasion, awake measurements were repeated 30 min following its administration (baseline, awake). Dogs were then exposed to epinephrine (1.0 or 2.0 mg/kg/min for 3 min), with at least 30 min between infusions for return to baseline conditions. Next dogs were anesthetized with one or the other of two possible concentrations of anesthetic (L, low; H, high), with or without a repeat dose (rep) of AMN, and baseline anesthesia measurements (anes-1 or anes-2) made. This was followed by exposure to epinephrine (1.0 or 2.0 mg/kg/min for 3 min), with at least 30 min between infusions for return to baseline (anes-1 or anes-2) conditions. The same procedure was repeated with the second anesthetic concentration. Randomization was used for anesthetics (halothane, enflurane, or isoflurane), anesthetic concentrations, presence or absence of AMN, and epinephrine dose.

Table I

Scores for Site of Earliest Activation in Chronically Instrumented Dogs

Score	SEA
1	Sinoatrial node region (first patch electrode)
2	High rostral sulcus terminalis (second patch electrode)
3	Midrostral sulcus terminalis (third patch electrode)
4	Midcaudal sulcus terminalis (fourth patch electrode)
5	Low caudal sulcus terminalis (fifth patch electrode)
6	Remote atrial (left atrial appendage electrode)

of the six possible atrial SEAs was multiplied by its corresponding SEA score (1–6) and the resulting products summed. This sum was then divided by the fraction of all beats of atrial origin to give the atrial SEA value for the specified test condition.

III. Anesthetic Interactions with Ouabain and Catecholamines

A. Canine Right Atrial Preparation

1. Halothane and Catecholamines

Halothane was chosen as the model for sensitizing anesthetic agents. Under control conditions (no drugs), the site of earliest activation (SEA) in the canine right atrial preparation was always the SA node (Fig. 5). Alone, neither 1 nor 2% halothane produced a significant number of pacemaker shifts from the SA node to distal SAP sites along the sulcus terminalis (Fig. 1). Without halothane, increasing concentrations of epinephrine (EPI) or norepinephrine (NE) did produce shifts in SEA to SAP sites (Fig. 5) and increased normalized SEA magnitude scores. Although this effect of EPI or NE to increase normalized magnitude scores was not altered by exposure to 1 or 2% halothane, EPI or NE was required for pacemaker shifts away from the SA node to SAP with halothane.

2. Halothane and Ouabain

Tachycardic atrial rhythm disturbances with toxic digitalis might be explained by enhanced automaticity of the SA node or SAP (7,15), and halothane might be supposed effective against these arrthythmias (1,2). In fact, halothane has been suggested as ‘‘treatment’’ for atrial arrhythmias

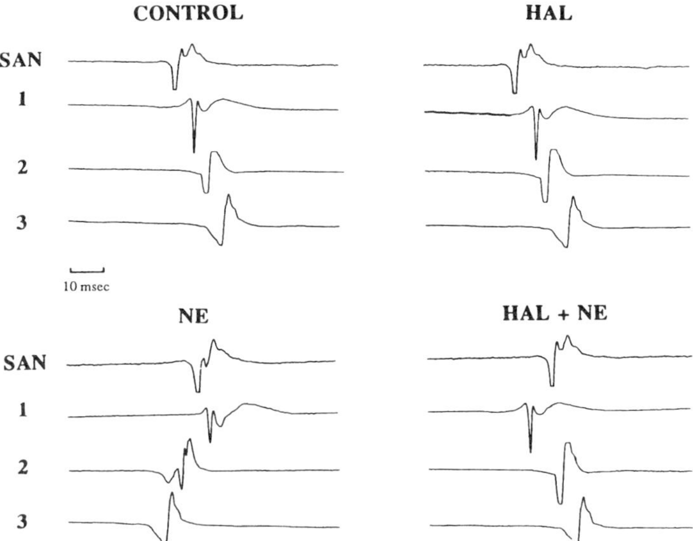

Fig. 5 Results of representative experiment with halothane (HAL) in the canine right atrial preparation. Under control conditions, the site of earliest activation (SEA) was the electrode in the SA node region (SAN), with activation proceeding in a caudal direction along the sulcus terminalis to sites 1, 2, and 3, in that order. During exposure to 1% HAL, the SEA was still the SAN, with activation of sites 1 to 3 as for control. With norepinephrine (NE), the nonlinear activation sequence shifted the SEA from the SAN to electrode site 3 (magnitude score of 3). Finally, with NE and HAL, the SEA shifted to electrode site 1 (magnitude score of 1). [From Polic *et al.* (14), with permission.]

in hypocapneic patients receiving digitalis (22). Therefore, we tested the hypotheses that (1) increasing ouabain would enhance automaticity of SAP relative to the SA node and (2) halothane would oppose ouabain-enhanced automaticity of SAP. Under control conditions (no drugs), the SA node was always the SEA (e.g., the normalized SEA magnitude score was 0). Normalized SEA magnitude scores were not significantly different from 0 during exposure to 1 or 2% halothane or any concentration of ouabain alone. However, normalized magnitude scores for 2% halothane with midtherapeutic or borderline toxic ouabain were significantly different from 0. An example of the effect of toxic ouabain and 2% halothane to increase the severity of pacemaker shifts to SAP is shown in Fig. 6. Regardless of pacemaker location, spontaneous heart rate tended to increase with ouabain and slow with halothane. This was most pronounced for borderline toxic ouabain with 2% halothane. Finally, total atrial arrest was observed in six preparations exposed to 1 or 2% halothane with borderline toxic ouabain (Fig. 7).

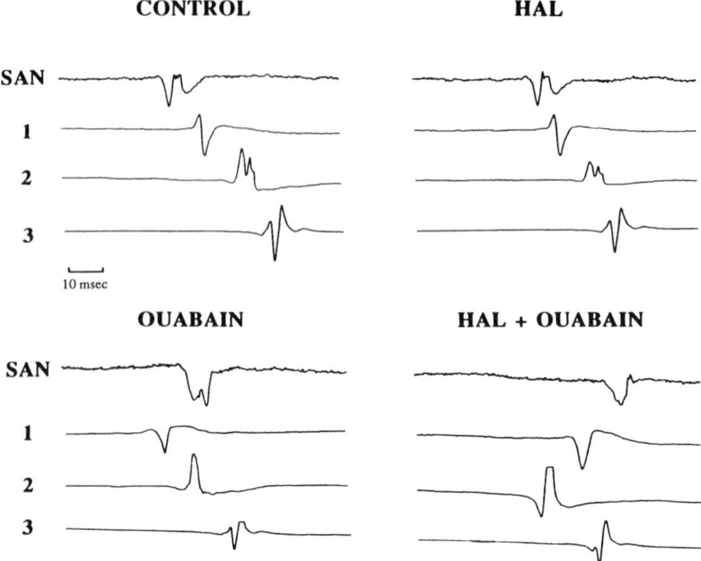

Fig. 6 Pacemaker shifts to subsidiary atrial pacemakers (SAP) with 2% halothane (HAL) and borderline toxic ouabain in the canine right atrial preparation. Note that both under control conditions and during exposure to 2% HAL, the SA node (SAN) was the site of earliest activation (SEA). With ouabain alone, the SEA shifted to SAP beneath electrode site 1, and with ouabain and HAL to SAP beneath electrode site 2.

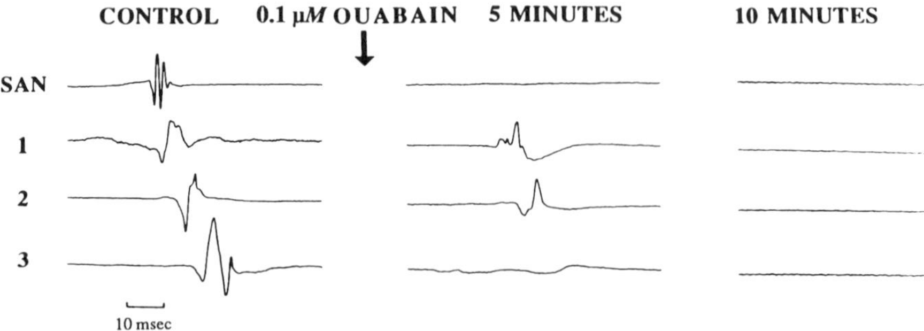

Fig. 7 Atrial quiesence during exposure to borderline toxic ouabain in the canine right atrial preparation. The SA node (SAN) was the pacemaker under control conditions. After 5 min of exposure to ouabain, the site of earliest activation (SEA) shifted to subsidiary atrial pacemaker site 1 (SAP-1). Activation, however, did not propagate to SAN or to SAP-3. Following 10 min of exposure to toxic ouabain, there was total atrial quiesence.

3. Isoflurane and Catecholamines

Isoflurane was selected as a model for less sensitizing anesthetics. As in previous experiments (14,15), the SA node was the pacemaker in the absence of drugs. Only the highest epinephrine (EPI) concentration, but none of the norepinephrine (NE) or isoflurane (ISO) concentrations tested, increased normalized SEA magnitude scores. However, all possible combinations of ISO with EPI or NE increased the incidence of pacemaker shifts and normalized SEA magnitude scores. Additionally, there was a tendency for EPI or NE alone to cause a larger increase in SEA magnitude scores following earlier exposure to isoflurane compared to conditions without such exposure (i.e., there was a possible effect of residual isoflurane). Finally, normalized SEA magnitude scores with the highest concentrations of ISO and EPI or NE were higher than earlier found with halothane and the same concentrations of EPI or NE (14).

B. Chronically Instrumented Dogs

1. Muscarinic Blockade and Epinephrine in Awake Dogs

We expected that pacemaker shifts away from the SA node to sites of SAP with epinephrine (EPI) would require participation of the vagus. Therefore, studies in both awake and anesthetized dogs were conducted with and without muscarinic blockade by atropine methyl nitrate (AMN). Awake dogs had respiratory sinus arrhythmia, with the SEA shifting between the first (SA node) and second (most rostral SAP) sulcus terminalis patch electodes. Along with respiratory sinus arrhythmia, there was variation in the morphology of surface electrocardiograph (ECG) (lead II) P-waves and bipolar atrial electrograms. AMN or panting (versus quiet breathing) abolished respiratory sinus arrhythmia, and the SEA returned to the SA node electrode (Fig. 8). In awake dogs, EPI increased site of earliest activation (SEA) values (Fig. 8) and blood pressure, and it decreased heart rate. With AMN and EPI, however, SEA values were unchanged from control (Fig. 8), although heart rate and blood pressure increased. Without AMN during exposure to 1 or 2 mg/kg/min EPI, slightly less than 1% of beats were of AV junctional (His bundle) origin (Fig. 9, Table II). With AMN, no junctional beats occurred (Table II). Further, without AMN during exposure to 2 mg/kg/min EPI, 2.9% of beats were of ventricular origin. This reduced to 0.2% following AMN (Table II).

2. Anesthetics and Epinephrine

a. Halothane With either concentration of halothane (HAL), SEA values were increased by both 1 and 2 mg/kg/min EPI (Fig. 10). However,

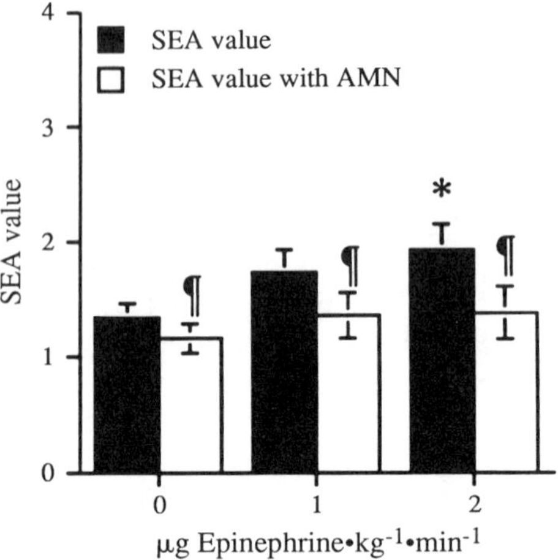

Fig. 8 Effects of epinephrine and atropine methyl nitrate (AMN) on SEA values in conscious dogs ($n = 8$). Data are shown as means ± SEM. *Significant difference ($p < 0.05$) versus control (awake, no drugs); ¶significant difference ($p < 0.05$) versus same test condition without AMN. See text for discussion. [From Woehlck *et al.* (17), with permission.]

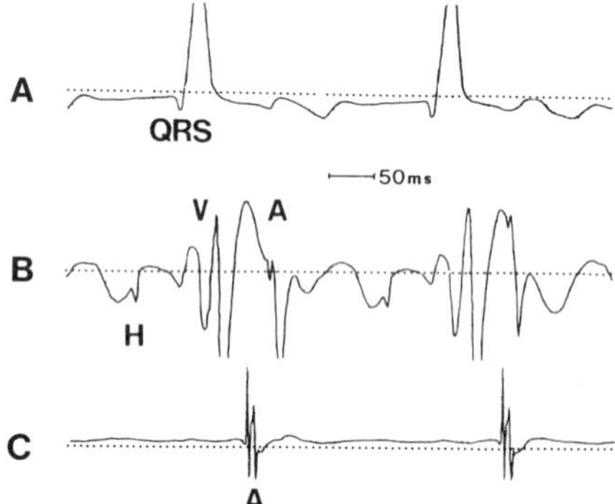

Fig. 9 Atrioventricular junctional beats originating at or above the His bundle. Simultaneous recordings were performed in chronically instrumented dogs of surface ECG lead II (A), His bundle electrogram (B), and the most proximal of the five patch electrode pairs, that is, electrogram from region of the SA node (C). Note the absence of P-waves preceding a normal QRS complex in lead II. However, a retrograde P-wave occurs in the ST segment of lead II, with atrial activity (A) corresponding to that retrograde P-wave apparent in both the His bundle and patch electrode electrograms. Ventricular beats were diagnosed by a widened, bizarre QRS, AV dissociation (±VA association), and no apparent His bundle activity preceding the QRS complex. H, His bundle electrogram; V, ventricular activity. The paper speed was 200 mm/sec. The adjustable gain was from 1 V = 1 cm to 25 mV = 1 cm. [From Woehlck *et al.* (17), with permission.]

Table II

Percentage of Atrioventricular Junctional-Origin and Ventricular-Origin Beats during Exposure to Epinephrine in Chronically Instrumented Dogs Anesthetized with Halothane[a]

	No AMN			With AMN		
Conditions	EPI (mg/kg/min)	AVJ (%)	VB (%)	EPI (mg/kg/min)	AVJ (%)	VB (%)
Conscious	0	0	0	0	0	0
	1	0.7	0	1	0	0
	2	0.2	2.9	2	0	0.2
HAL, 1.2 MAC	0	0	0	0	0	0
	1	0	3.4	1	16	0
	2	1.3	7.5	2	13.2	25.2
HAL, 2.0 MAC	0	0	0	0	0	0
	1	0	0.7	1	8.5	0.6
	2	25.5	2.4	2	22.2	8.2

[a] AMN, Atropine methyl nitrate; EPI, epinephrine; AVJ, AV junctional-origin beats; VB, ventricular-origin beats; HAL, halothane.

in dogs with AMN anesthetized with 1.2 MAC HAL (not 2.0 MAC HAL), there was no effect of 1 or 2 mg/kg/min EPI to increase SEA values (Fig. 10). Blood pressure increases with EPI were not as great with HAL compared to conscious dogs. Without AMN blockade, heart rate decreased in response to EPI with HAL but not in dogs with AMN. Also, in

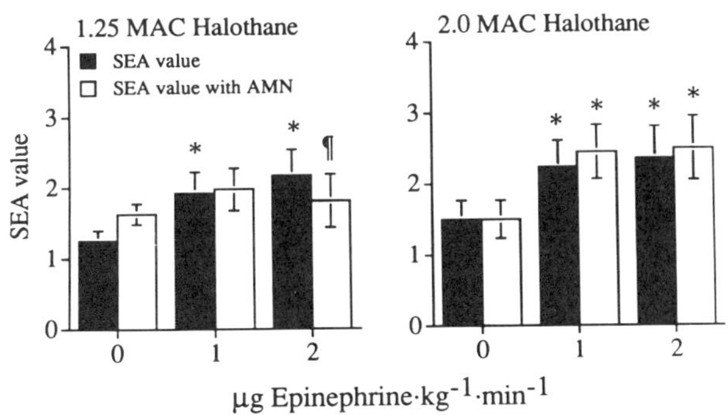

Fig. 10 Effect of epinephrine on site of earliest activation (SEA) values in dogs anesthetized with halothane, with and without atropine methyl nitrate (AMN) vagal blockade. Data are shown as means ± SEM. *Significant difference ($p < 0.05$) versus same test condition without epinephrine; ¶significant difference ($p < 0.05$) versus same test condition without AMN. See text for discussion. [From Woehlck *et al.* (17), with permission.]

dogs with HAL and EPI, the percentage of AV junctional- and ventricular-origin beats was greater compared to the conscious state (Table II). Further, in contrast to conscious dogs, AMN appeared to facilitate AV junctional- and ventricular-origin beats (Table II).

b. Enflurane Except with 2.0 MAC enflurane (ENF) and 2 mg/kg/min EPI, SEA values with ENF increased compared to baseline during exposure to EPI (Fig. 11). As for HAL, AMN had little effect on atrial pacemaker shifts with EPI, except with 2.0 MAC ENF and 1 mg/kg/min EPI (Fig. 11). Blood pressure increases during exposure to EPI were comparable to those in conscious dogs and were little affected by AMN blockade (18). With ENF (Table III) compared to HAL (Table II), there was a similar incidence of ventricular-origin beats during exposure to EPI

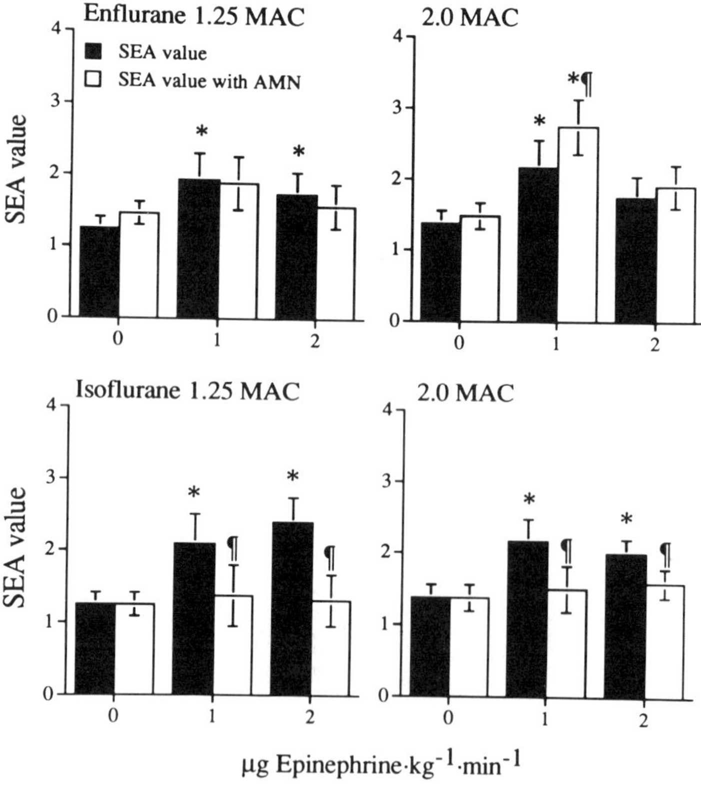

Fig. 11 Effect of epinephrine on site of earliest activation (SEA) values in dogs anesthetized with enflurane or isoflurane, with and without atropine methyl nitrate (AMN) vagal blockade. Data are shown as means ± SEM. *Significant difference ($p < 0.05$) versus same test condition without epinephrine; ¶significant difference ($p < 0.05$) versus same test condition without AMN. See text for discussion. [From Vicenzi *et al.* (18), with permission.]

Table III

Percentage of Atrioventricular Junctional-Origin and Ventricular-Origin Beats during Exposure to Epinephrine in Chronically Instrumented Dogs Anesthetized with Enflurane or Isoflurane[a]

	No AMN			With AMN		
Conditions	EPI (mg/kg/min)	AVJ (%)	VB (%)	EPI (mg/kg/min)	AVJ (%)	VB (%)
Conscious	0	0	0	0	0	0
	1	0.7	0	1	0	0
	2	0.2	2.9	2	0	0.2
ENF (SIO),	0 (0)	0 (0)	0 (0)	0 (0)	0 (0)	0 (0)
1.2 MAC	1	0 (10.1)	2.9 (4.3)	1	0 (6.1)	3 (0)
	2	4.1 (8.6)	7.8 (11)	2	1.8 (1.1)	9.5 (0)
ENF (ISO),	0	0	0	0	0	0
2.0 MAC	1	0 (0)	1.8 (1.0)	1	0 (0)	0.2 (0)
	2	0 (0.3)	2.3 (0)	2	1.1 (6.1)	0 (0.1)

[a] ENF, Enflurane; ISO, isoflurane; see Table II for other abbreviations. Percentages for ISO are given in parentheses.

but a lower incidence of AV junctional-origin beats. In contrast to HAL (Table II), AMN blockade did not facilitate AV junctional- or ventricular-origin beats with EPI and ENF (Table III).

c. Isoflurane The SEA values with isoflurane (ISO) were increased during exposure to EPI (Fig. 11). However, the effect of EPI with ISO to increase SEA values was opposed by AMN (Fig. 11). Blood pressure increases during exposure to EPI were comparable to those in conscious dogs, and little affected by AMN blockade (18). Compared to ENF, ISO-anesthetized dogs without AMN and exposed to EPI were more likely to have AV junctional-origin beats (Table III). However, the incidence of AV junctional beats was considerably higher in dogs anesthetized with 2.0 MAC HAL exposed to 2.0 mg/kg/min EPI, both with and without AMN (Table II). The AMN blockade tended to oppose the effect of ISO and EPI to facilitate either AV junctional- or ventricular-origin beats (Table III).

IV. Discussion

Until relatively recently, it was believed that atrioventricular (AV) junctional pacemakers would control the heart without a functioning sinoatrial (SA) node. This idea was supported by experiments of several groups in which arteries supplying the SA node or AV junctional region were selectively perfused with chronotropic substances [see Polic *et al.* (14) for discussion and references]. However, selective SA node artery perfusion with negative chronotropes to suppress SA node automaticity and cause emergence of AV junctional pacemakers assumes exclusive SA node distribution of the drug. Several investigators using different methods have shown that the SA node artery supplies other potential pacemaker sites within the atria [see Randall *et al.* (23) for discussion and references], including sites of subsidiary atrial pacemakers (SAP) along the sulcus terminalis. Thus, pharmacological interventions which depress SA node automaticity [e.g., acetylcholine (ACh)] are also expected to depress automaticity of SAP. This will permit pacemakers outside the distribution of the SA node artery, ones less affected by ACh (e.g., the AV junctional pacemakers), to emerge to control the rhythm of the heart.

Knowledge that SAP can control the heart with an absent or nonfunctioning SA node is physiologically significant. It means that there is an atrial mechanism available for control of cardiac rhythm with default of the SA node. Properly synchronized atrial contractions (i.e., the "atrial kick") and increased heart rate may be required for adequate cardiac output in patients with end-stage heart disease and systolic or diastolic dysfunction (24). The SAP are nearly as responsive as the SA node to exercise and changes in autonomic tone (8–10, 19–21,23). Compared to AV junctional or ventricular rhythms, SAP rhythms preserve atrial function. Thus, with SAP rhythms, adequate hemodynamics can be preserved in patients with advanced heart disease. Consequently, knowledge of anesthetic effects on SAP has importance to anesthesiologists, aside from furthering understanding of mechanisms of perioperative cardiac arrhythmias.

A. Experiments in Isolated Hearts

The effect of direct sympathetic (or vagal) stimulation could not be tested in the isolated heart preparation, so exposure to norepinephrine (NE) was used to simulate increased sympathetic neural input. Epinephrine (EPI) was used to test anesthetic sensitization. Responses to acetylcholine were not tested. Although the direct effect of both halothane (HAL) and isoflurance (ISO) was to decrease SA node rate, neither agent alone caused pacemaker shifts to sites of SAP. Whereas both EPI and NE alone did cause pacemaker shifts to SAP, with the addition of HAL there was no increase in the number of shifts. In contrast, with ISO and EPI or NE, there was an increase in pacemaker shifts to SAP. Thus, in the isolated heart, ISO does and HAL does not appear to sensitize the atrium to catecholamines. Further, with HAL, ectopic atrial rhythms caused by enhanced SAP automaticity would require augmented adrenergic tone or exogenous EPI. For both anesthetics, enhanced automaticity of SAP might be a mechanism for ectopic atrial rhythm disturbances with stress or exogenous catecholamines. Finally and importantly, neither anesthetic is expected to prevent emergence of SAP with SA node default.

Neither HAL nor ouabain (OUA), including borderline toxic OUA, caused pacemaker shifts away from the SA node to SAP sites. However, borderline toxic OUA with HAL did produce pacemaker shifts, and atrial quiescence was observed in some preparations. In preparations with pacemaker shifts, the spontaneous rate was not different from the control. Further, regardless of pacemaker location, the spontaneous rate tended to be slowed by halothane and increased by ouabain, a tendency which was nost pronounced for borderline toxic ouabain with increased HAL. In spite of the increased rate with borderline toxic OUA, under no test condition was the preparation rate sufficient to account for paroxysmal atrial tachycardia with block, commonly seen with digitalis toxicity (2). Perhaps this was because we tested only borderline toxic OUA. In preliminary experiments, however, higher OUA doses (1 and 2 μM) were tested. Either concentration produced total atrial quiescence within 60–100 min of exposure, preceded by a regular tachycardia (rate 200–300 beats/min) and then irregular beating.

Differential effects of HAL and OUA on automaticity of atrial pacemakers favoring shifts in SEA away from the SA node to SAP might have resulted from drug-induced SA node pacemaker desynchronization. Results of Jalife suggest that rhythm coordination of multiple pacemaker cells in the SA node depends on mutual entrainment, as well as on the degree of electrical coupling between cells (25). Both OUA and HAL have been shown to reduce electrical coupling between ventricular myocytes [see Polic *et al.* (15) for further discussion], and it is reasonable to suppose that this might extend to atrial myocytes or SA node cells as well.

If HAL or OUA do reduce cell-to-cell coupling in the SA node, or between SAP and atrial myocytes, our findings (canine right atrial preparation) could be explained as follows. First, reduction in cell-to-cell coupling is greatest with higher concentrations of OUA and HAL. This would explain more pacemaker shifts to SAP with higher combined drug concentrations. Second, preferential suppression of automaticity in the SA node compared to SAP may occur because cells of the SA node are more homogeneous, clustered, and mutually interdependent compared to SAP (9,26,27). Therefore, the SA node would be more vulnerable to uncoupling. In contrast, the pacemaker process in SAP is possibly not so dependent on mutual entrainment. Rather, as suggested by Rubenstein *et al.* (26), SAP automaticity is more affected by electrotonic interactions between adjacent, atrial myocytes. If so, and if OUA and HAL reduce cell-to-cell coupling between atrial myocytes and SAP, then SAP should become less restrained and available to control the atria. Further, it must be assumed that SAP automaticity is less affected by HAL and OUA than is SA node automaticity [arguments for this are given elsewhere (15)].

These *in vitro* data for comparative HAL and OUA effects on automaticity of the SA node and SAP may have clinical relevance but must be interpreted cautiously. Both HAL and OUA have significant autonomic effects (2). With this in mind, the following are suggested. First, enhanced automaticity of SAP as compared to SA node does not explain paroxysmal atrial tachycardia (PAT) with digitalis excess. Second, HAL should not be supposed to oppose PAT until more data are available, especially relevant *in vivo* testing. Third, it is possible that high concentrations of digitalis and HAL could interact to suppress atrial pacemakers, leading to atrial electrical quiescence. The latter would be most likely to occur in patients with sinus node dysfunction caused by intrinsic disease, drugs, or impaired autonomic regulation.

B. Experiments in Intact Hearts

Pacemaker shifts from the SA node to the most rostral sulcus terminalis patch electrode in conscious dogs with respiratory sinus arrhythmia exemplify the concept of a "physiologic" pacemaker complex confined to the rostral sulcus terminalis, as described by Boineau *et al.* in dogs (28,29). Shifts to more caudal SAP sites along the sulcus terminalis during respiratory sinus arrhythmia were not observed. Further, even when the SA node electrode was the site of earliest activation (SEA), the morphology, polarity, and amplitude of electrograms from sulcus terminalis patch electrodes frequently changed. This suggests that different, competing pacemaker foci were possibly located rostral to or adjacent to the SA node electrode pair. This could be determined by high-resolution electrocardiographic mapping techniques, for example, electrode patches with multiple, bipolar electrodes ($\geq 16/cm^2$) sewn over the SA node region. Regardless, muscarinic blockade with atropine methyl nitrate abolished pacemaker shifts from the SA node to the most rostral sulcus terminalis electrode.

It is known that vagal innervation of the atria declines progressively from the SA node to more remote atrial tissues, making the SA node potentially the most vagally inhibited pacemaker (30). Clinical observations support the contention that distinct cases of SA node dysfunction are related to altered vagal tone (31). In experiments with epinephrine, hypertension presumably activates the baroreceptor reflex arc. In turn, enhanced efferent vagal tone reduced heart rate. Because the SA node pacemaker cells were more restrained by the vagus, SEA could shift to

SAP sites less influenced by the vagus along the sulcus terminalis, evidenced by increased SEA values. With vagal blockade, SA node automaticity was either increased more by epinephrine (versus SAP) or sufficiently enhanced to overdrive suppress SAP automaticity. Regardless, SEA values were lower during exposure to epinephrine in dogs with vagal blockade. This suggests that there is no direct effect of epinephrine to enhance automaticity of SAP, in contrast to results in the isolated heart (see above). In part, the discrepancy might be related to "excess tachycardia" with pharmacological vagal blockade (32), possibly mediated by histamine or other receptors and mediators not present *in vitro* (33). Whether a similar phenomenon affects relations between automaticity of atrial pacemakers and the genesis of atrial arrhythmias is unknown.

The morphology of the P-wave in surface ECG lead II did not change so long as the SA node or most rostral SAP patch electrodes were the SEA. Thus, only pacemaker shifts to more caudal SAP or ectopic atrial sites result in morphologically distinct P-waves (i.e., flattened, notched, negative) in the surface ECG. Possibly, pacemaker shifts to caudal SAP sites along the sulcus terminalis explain wandering atrial pacemakers, commonly observed during anesthesia (2).

1. Inhalation Anesthetics

Bosnjak and Kampine (34) and Atlee *et al.* (35,36) have provided data in both the isolated and intact heart indicating that SA node automaticity is directly depressed by enflurane (ENF), HAL, and ISO. Although all agents similarly depress SA node automaticity, compensatory cardiovascular reflexes may variably mask such an effect in patients (37). Therefore, variable depression of the baroreflex arc by inhalation anesthetics [HAL most, ENF similar, ISO least (37)] must be considered when interpreting results of present experiments for anesthetic effects on automaticity of primary and secondary pacemakers.

2. Halothane

Epinephrine (EPI) increased SEA values during exposure to both 1.2 and 2.0 MAC HAL. Vagal blockade opposed atrial pacemaker shifts with 1.2 MAC HAL, which suggests that epinephrine-caused atrial pacemaker shifts require intact muscarinic transmission. However, vagal blockade did not oppose pacemaker shifts with 2.0 MAC HAL. This may be explained by relatively greater depression of SA node compared to SAP automaticity by HAL, allowing the SAP to emerge in response to stimulation by EPI. Regardless, results do support the concept of atrial sensitization by HAL; that is, increasing EPI with HAL compared to awake favors pacemaker shifts to SAP sites, independent of the vagus. Pacemaker shifts also occurred to AV junctional and ventricular pacemakers with HAL and EPI, and vagal blockade favored such shifts under most test conditions. This probably has clinical relevance, because AV junctional rhythms are fairly common in anesthetized patients and can respond unpredictably to atropine (2). Finally, AV junctional and ventricular beats and rhythms with EPI were more common with HAL than with the other two anesthetics.

3. Enflurane

Dogs anesthetized with 1.2 MAC enflurane (ENF) showed an EPI-induced increase in SEA values along with hypertension and baroreceptor-mediated reductions in heart rate. Atropine had no effect on pacemaker shifts with 1.2 MAC ENF and EPI, suggesting that a baroreflex-mediated increase in vagal tone is not required for EPI-induced pacemaker shifts away from the SA node. This suggests, in contrast to HAL, atrial sensitization by ENF. With 2.0 MAC ENF, SEA values actually increased during exposure to 1 mg/kg/min EPI. In addition to atrial sensitization, this suggests direct depression of SA node automaticity by ENF. Alternatively, there was an absolute increase in automaticity of SAP by ENF and EPI. The incidence of AV junctional and ventricular beats with ENF was noticably less than with HAL, and vagal blockade had little effect on this incidence. Thus, ENF appears less likely to sensitize lower latent pacemakers.

4. Isoflurane

Epinephrine produed hypertension and bradycardia in ISO-anesthetized dogs, indicating a functioning baroreceptor reflex arc. As in the conscious state, this was accompanied by pacemaker shifts to SAP. Vagal blockade prevented pacemaker shifts with both concentrations of ISO and EPI. Therefore, atrial pacemaker shifts with ISO appear more likely to be caused by reflex suppression of SA node automaticity than by atrial sensitization by ISO and EPI per se. In other words, reflex-enhanced vagal tone causes preferential inhibition of SA node compared to SAP automaticity, allowing emergence of SAP. Lack of atrial sensitization by ISO contrasts with *in vitro* findings (above), especially since plasma EPI concentrations measured in present experiments (random sampling) exceeded those *in vitro* by a factor of 5 to 20. That *in vitro* preparations appear more sensitive to atrial arrhythmic interaction between EPI and ISO suggests that unidentified neurohumoral factors present in the intact organism are protective. Finally, similar to ENF, the incidence of AV junctional and ventricular beats with ISO was noticeably less than with HAL, and vagal blockade had little effect on incidence. Thus, ISO too causes little sensitization of AV junctional or ventricular myocardium.

V. Conclusions

Enhanced automaticity of subsidiary atrial pacemakers (SAP) may explain atrial rhythm disturbances such as wandering atrial pacemaker and ectopic atrial beats or rhythm. Catecholamines may contribute by atrial sensitization or baroreflex-mediated suppression of SA node automaticity, and anesthetics by direct depression of SA node automaticity. The relative contribution of baroreflex-mediated suppression varies among inhalation anesthetics, depending on their ability to depress the baroreflex arc. Finally, rate responses to ouabain or catecholamines make it unlikely that enhanced automaticity of SAP can explain paroxysmal atrial tachycardia.

VI. Future Directions

Work to date has not compared the effects of equivalent concentrations of the volatile anesthetics on spontaneous phase 4 depolarization (automaticity) in SAP cells, as has been done for SA node (34) and Purkinje fibers (38). Although halothane (HAL), enflurane (ENF), and isoflurane (ISO) produce equivalent depression of SA node automaticity, ENF (not HAL or ISO) enhances automaticity of Purkinje fibers exposed to epinephrine (EPI) compared to HAL or ISO. Thus, it would not be suprising if SAP, with electrophysiological characteristics somewhat intermediate between SA node and Purkinje fibers (10,27), might differ with respect to anesthetic effects on automaticity. Additionally, as a model for anesthetic action in patients with sick sinus syndrome, it is required to examine anesthetic and catecholamine effects on the stability of SAP in the absence of a functioning SA node [i.e., following SA node excision (30)].

Acknowledgments

Research was supported in part by the U.S. Public Health Service, National Institutes of Health Grant GM25064.

References

1. Atlee, J. L., and Bosnjak, Z. J. (1990). Mechanisms for cardiac dysrhythmias during anesthesia. *Anesthesiology* **72,** 347–374.
2. Atlee, J. L. (1990). "Perioperative Cardiac Dysrhythmias: Mehanisms, Recognition, Management," 2nd Ed. Year Book Medical Publ., Chicago.
3. Atlee, J. L., and Yeager, T. S. (1989). Electrophysiologic assessment of the effects of

enflurane, halothane, and isoflurane on properties affecting supraventricular reentry in chronically instrumented dogs. *Anesthesiology* **71,** 941–952.
4. Atlee, J. L., Bosnjak, Z. J., and Yeager, T. S. (1990). Effects of diltiazem, verapamil, and inhalation anesthetics on electrophysiologic properties affecting reentrant supraventricular tachycardia in chronically instrumented dogs. *Anesthesiology* **72,** 889–901.
5. Atlee, J. L., and Malkinson, C. E (1982). Potentiation by thiopental of halothane–epinephrine induced arrhythmias in dogs. *Anesthesiology* **57,** 285–288.
6. Atlee, J. L., and Roberts, F. L. (1986). Thiopental and epinephrine-induced arrhythmias in dogs anesthetized with enflurane or isoflurane. *Anesth. Analg. (N.Y.)* **65,** 437–443.
7. Steinbeck, G., Bonke, F. I. M., Allessie, M. A., and Lammers, W. J. E. P. (1980). The effect of ouabain on the isolated sinus node preparation of the rabbit studied with microelectrodes. *Circ. Res.* **46,** 406–414.
8. Randall, W. C., Rinkema, L. E., Jones, S. B., Moran, J. F., and Brynjolfsson, G. (1982). Functional characterization of atrial pacemaker activity. *Am. J. Physiol.* **248,** H98–H106.
9. Lipsius, S. L. (1989). Mechanisms of atrial subsidiary pacemaker function. *In* "Focus on Cellular Pathophysiology" (S. Boca, ed.), pp. 1–40. CRC Press, Boca Raton, Florida.
10. Rozanski, G. J., Lipsius, S. L., and Randall, W. C., (1983). Functional characteristics of sinoatrial and subsidiary pacemaker activity in the canine right atrium. *Circulation* **67,** 1378–1387.
11. Woods, W. T., Urthaler, F., and James, T. N. (1976). Spontaneous action potentials of cells in the canine sinus node. *Circ. Res.* **39,** 76–82.
12. Tuna, I. C., Barragry, T. P., Walker, M., Lillehei, T., Blatchford, J. W., Gornick, C., Ring, W. S., Bolman III, R. M. and Benditt, D. G. (1987). Effects of transplantation on atrioventricular nodal accommodation and hysteresis. *Am. J. Physiol.* **253,** H1514–H1522.
13. Atlee, J. L., Dayer, A. M., and Houge, J. C. (1984). Chronic recording from the His bundle of the awake dog. *Basic Res. Cardiol.* **79,** 627–638.
14. Polic, S., Atlee III, J. L., Laszlo, A., Kampine, J. P., and Bosnjak, Z. J. (1991). Anesthetics and automaticity in latent pacemaker fibers. II. Effects of halothane and epinephrine or norepinephrine on automaticity of the dominant and subsidiary atrial pacemakers in the canine heart. *Anesthesiology* **75,** 298–304.
15. Polic, S., Atlee III, J. L., Laszlo, A., Kampine, J. P., and Bosnjak, Z. J. (1991). Anesthetics and automaticity in latent pacemaker fibers. III. Effects of halothane and ouabain on automaticity of the SA node and subsidiary atrial pacemakers in the canine heart. *Anesthesiology* **75,** 305–312.
16. Boban, M., Atlee III, J. L., Vicenzi, M., Kampine, J. P., and Bosnjak, Z. J. (1993). Anesthetics and automaticity in latent pacemaker fibers. IV. Effects of isoflurane and epinephrine or norepinephrine on automaticity of dominant and subsidiary atrial pacemakers in the canine heart. *Anesthesiology* **79,** 555–562.
17. Woehlck, H. J., Vicenzi, M. N., Bosnjak, Z. J., and Atlee III, J. L. (1993). Anesthetics and automaticity of dominant and latent pacemakers in chronically instrumented dogs. I. Methodology, conscious state and halothane anesthesia: Comparison with and without muscarinic blockade during exposure to epinephrine. *Anesthesiology* **79,** 1304–1315.
18. Vicenzi, M. N., Woehlck, H. J., Bosnjak, Z. J., and Atlee III, J. L. (1993). Anesthetics and automaticity of dominant and latent pacemakers in chronically instrumented dogs. II. Effects of enflurane and isoflurane during exposure to epinephrine with and without muscarinic blockade. *Anesthesiology* **79,** 1316–1323.
19. Jones, S. B., Euler, D. E., Hardie, E., Randall, W. C., and Brynjolfsson, G. (1978). Comparison of SA nodal and subsidiary atrial pacemaker function and location in the dog. *Am. J. Physiol.* **234,** H471–H476.

20. Jones, S. B., Euler, D. E., Randall, W. C., Brynjolfsson, G., and Hardie, E. (1980). Atrial ectopic foci in the canine heart: Hierarchy of pacemaker automaticity. *Am. J. Physiol.* **238,** H788–H793.
21. Hardie, E. L., Jones, S. B., Euler, D. E., Fishman, D. L., and Randall, W. C. (1981). Sinoatrial node artery distribution and its relation to hierarchy of cardiac automaticity. *Am. J. Physiol.* **241,** H45–H53.
22. Edwards, R., Winnie, A. P., and Ramamurphy, S. (1977). Acute hypocapneic hypokalemia: An iatrogenic anesthetic complication. *Anesth. Analg. (N.Y.)* **56,** 786–792.
23. Randall, W. C., Wehrmacher, W. H., and Jones, S. B. (1981). Heirarchy of supraventricular pacemakers. *J. Thorac. Cardiovasc. Surg.* **82,** 1797–1800 (editorial).
24. Baig, M. W., and Perrins, E. J. (1991). The hemodynamics of cardiac pacing: Clinical and phsyiological aspects. *Prog. Cardiovasc. Dis.* **33,** 283–298.
25. Jalife, J. (1984). Mutual entrainment and electrical coupling as mechanisms for synchronous firing of rabbit sino-atrial pace-maker cells. *J. Physiol. (London)* **356,** 221–243.
26. Rubenstein, D. S., Fox, L. M., McNulty, J. A., and Lipsius, S. L. (1987). Electrophysiology and ultrastructure of Eustachian ridge from cat right atrium: A comparison with SA node. *J. Mol. Cell. Cardiol.* **19,** 965–976.
27. James, T. N., Scherf, L., Fine, G., and Morales, A. R. (1966). Comparative ultrastructure of the sinus node in man and dog. *Circulation* **34,** 139–163.
28. Boineau, J. P., Schuessler, R. B., Mooney, C. R., Wylds, A. C., Miller, C. B., Hudson, R. D., Borremans, J. M., and Brockus, C. W. (1978). Multicentric origin of the atrial depolarization wave: The pacemaker complex. Relation to dynamics of atrial conduction, P-wave changes and heart rate control. *Circulation* **58,** 1036–1048.
29. Boineau, J. P., Schuessler, R. B., Hackel, D. B., Miller, C. B., Brockus, C. W., and Wylds, A. C. (1980). Widespread distribution and rate differentiation of the atrial pacemaker complex. *Am. J. Physiol.* **239,** H406–H415.
30. Randall, W. C., Talano, J., Kaye, M. P., Euler, D. E., Jones, S. B., and Brynjolfsson, G. (1970). Cardiac pacemakers in absence of the SA node: Responses to exercise and autonomic blockade. *Am. J. Physiol.* **234,** H465–H470.
31. Kang, P. S., Gomes, J., Kelen, G., and El-Sherif, N. (1981). Role of autonomic regulatory mechanisms in sinoatrial conduction and sinus automaticity in sick sinus syndrome. *Circulation* **64,** 832–838.
32. Rigel, D. F., Lipson, D., and Katona, P. G. (1984). Excess tachycardia: Heart rate after antimuscarinic agents in conscious dogs. *Am. J. Physiol.* **246,** H168–H173.
33. Rigel, D. F., and Katona, P. G. (1986). Effects of antihistaminics and local anesthetics on excess tachycardia in conscious dogs. *J. Pharmacol. Exp. Ther.* **238,** 367–371.
34. Bosnjak, Z. J., and Kampine, J. P. (1983). Effects of halothane, enflurane, and isoflurane on the SA node. *Anesthesiology* **58,** 314–321.
35. Atlee III, J. L., Brownlee, S. W., and Burstrom, R. E. (1986). Conscious-state comparisons of the effects of inhalation anesthetics on specialized atrioventricular conduction times in dogs. *Anesthesiology* **64,** 703–710.
36. Atlee III, J. L., and Yeager, T. S. (1989). Electrophysiologic assessment of the effects of enflurane, halothane, and isoflurane on properties affecting supraventricular re-entry in chronically instrumented dogs. *Anesthesiology* **71,** 941–952.
37. Schmeling, W. T., Bosnjak, Z. J., and Kampine, J. P. (1990). Anesthesia and the autonomic nervous system. *Semin. Anesth.* **9,** 223–231.
38. Laszlo, A., Polic, S., Atlee III, J. L., Kampine, J. P., and Bosnjak, Z. J. (1991). Anesthetics and automaticity in latent pacemaker fibers. I. Effects of halothane and epinephrine or norepinephrine on automaticity and recovery of automaticity from overdrive suppression in Purkinje fibers derived from canine hearts. *Anesthesiology* **75,** 98–105.

The Role of L-Type Voltage-Dependent Calcium Channels in Anesthetic Depression of Contractility

Thomas J. J. Blanck, D. L. Lee, S. Yasukochi, C. Hollmann, and J. Zhang

Department of Anesthesiology
Cornell University Medical College
New York, New York 10021

I. Introduction

The volatile anesthetics alter the Ca^{2+} homeostasis of the myocardial cell (1). Lynch *et al.* demonstrated that the slow action potential was inhibited in a dose-dependent manner by halothane (2). They initially treated guinea pig papillary muscles with isoproterenol to augment the slow calcium current, exposed them to 26 m*M* KCl to partially depolarize the muscle to prevent activation of sodium channels, and showed that the remaining current, presumably due to Ca^{2+}, was reversibly inhibited by halothane. Subsequently, Bosnjak *et al.* demonstrated the reversible inhibition of the slow action potential in SA nodal cells by halothane, enflurane, and isoflurane. (3) The data also suggested that the calcium current through the L-type calcium channel was inhibited by the volatile anesthetics.

With these studies as a background we hypothesized that the L-type calcium channel might be a specific target of the volatile anesthetics. To test this hypothesis we examined whether the volatile anesthetics interfered with the specific binding of L-type calcium channel antagonists to cardiac sarcolemmal membranes. In 1988 we reported that halothane inhibited the binding of the calcium antagonist [^{3}H]nitrendipine to rat

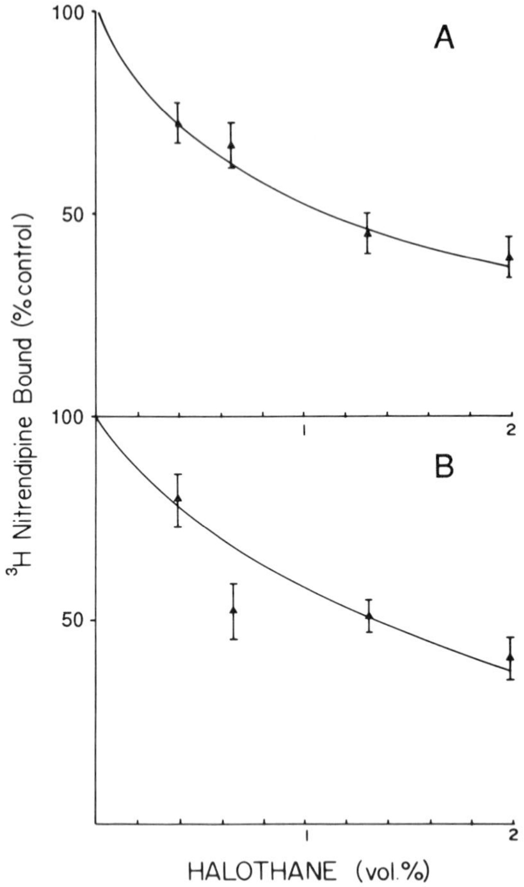

Fig. 1 Depression by halothane of [^{3}H]nitrendipine specifically bound to cardiac membranes. The incubation medium consisted of 5 n*M* [^{3}H]nitrendipine in 50 m*M* Tris-HCl at pH 7.7. (A) Binding in rat cardiac membranes, each data point being the mean of 19 determinations ± SEM (B) Binding in rabbit cardiac membranes, each data point being the mean of 9 observations ± SEM. [From Blanck and Casella (5).]

and rabbit cardiac membranes in a clinically relevant, dose-dependent, reversible manner (Fig. 1) (4,5). We subsequently showed that halothane inhibited the binding of [^{3}H]nitrendipine to purified bovine cardiac sarcolemma by decreasing the density of binding sites (B_{max}) without altering the binding affinity (K_d) (6). We then demonstrated that nifedipine could mimic the effect of halothane on dynamic stiffness and suggested that this was most likely due to the inhibition of entry of barium through L-type calcium channels (7).

Our next step was to examine the effect of halothane on D600 binding. D600, gallopamil, is a member of the phenylalkylamine class of calcium antagonists which blocks Ca^{2+} channels by binding to the α_1 subunit of the voltage-dependent Ca^{2+} channel (VDCC). Hoehner *et al.* showed the reversible, dose-dependent inhibition by halothane of D600 binding to canine cardiac sarcolemma (8). He demonstrated that the density of channels was decreased without any effect on the affinity of D600 binding (Fig. 2). A report by Schmidt *et al.* has demonstrated that this phenomenon also occurs in human myocardial membranes (9).

In this article we present data which show that not only halothane but also enflurane and isoflurane decrease the density of binding sites for the calcium antagonist [^{3}H]isradipine in adult and newborn rabbit myocardial membranes. In addition, we have been concerned about whether this phenomenon occurs *in vivo,* that it is not simply an *in vitro* finding unrelated to the mechanism of anesthetic depression of contractility in intact organisms. To test this, we have designed experiments in the intact, beating, Langendorff perfused rat heart, which is described below.

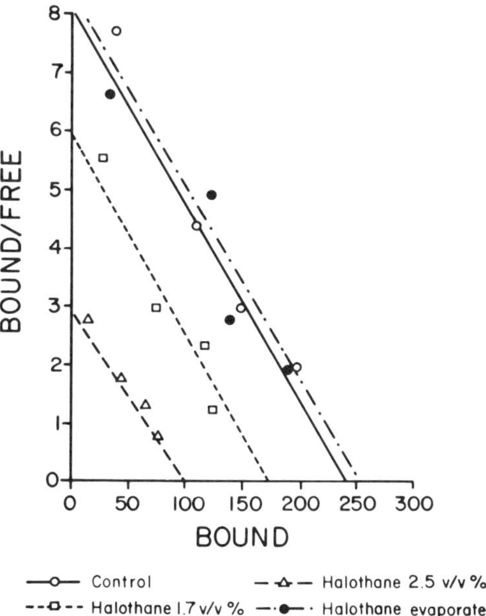

Fig. 2 Scatchard analysis of [^{3}H]D600 specific binding to bovine cardiac sarcolemma. The halothane concentrations are indicated at bottom. The parallel lines indicate a change in $B_{\max}$ with no change in K_d. The effect of halothane was completely reversible on its evaporation. [From Hoehner *et al.* (8).]

The basis on which the perfusion experiments would allow us to examine the interaction of Ca^{2+} channel antgonist binding sites and the volatile anesthetic, halothane, in an intact organ is as follows: (1) halothane, being volatile, would vaporize; and (2) the rate of unbinding of [^{3}H]isradipine at 4 and 25°C was very slow. We postulated that if halothane masked or protected L-type channel antagonist binding sites, perfusion with 100 n*M* isradipine, after equilibration of the heart with halothane, would result in the binding of isradipine to all VDCC except those protected by halothane. If the heart was homogenized and the halothane vaporized during the binding assay, those sites previously protected by halothane would now be available for binding [^{3}H]isradipine. This concept predicts that [^{3}H]isradipine binding in membranes derived from hearts exposed to halothane followed by 100 n*M* isradipine would be greater than those exposed to 100 n*M* isradipine alone. It also predicts that exposure to increasing halothane concentration would result in greater radioligand binding.

II. Isolated Heart Preparation

Experiments were carried out with rat and rabbit hearts following review and approval of the experimental protocol by the Cornell University Medical College Animal Care and Use Committee. Binding studies with [^{3}H]isradipine were performed as previously described with the following modifications. Membranes were prepared from adult and newborn rabbit hearts by the method of Endoh *et al.* (10). Membranes were incubated at 250°C with concentrations of [^{3}H]isradipine of 0.105, 0.263, 0.525, and 1.05 n*M*. Nonspecific binding was determined by incubation of membranes with labeled isradipine and 500 n*M* nitrendipine for 60 min. Total binding was determined as above but in the absence of nitrendipine. Specific binding was defined as the difference between total binding and nonspecific binding.

Langendorff perfusions were performed at 37°C at constant pressure (85 cm H_2O) with hearts excised from rats. The hearts were equilibrated with modified Krebs–Ringer solution for 30 min followed by perfusion with buffer alone, or buffer plus 1.5 or 2.5% (v/v) halothane for 20 min, and 100 n*M* isradipine for an aditional 30 min. The hearts were immediately removed from the perfusion apparatus, homogenized in ice-cold buffer, and membranes prepared as above. Binding studies were performed at 25°C as described above.

III. Effects of Anesthetics on Isradipine Binding

Figure 3 demonstrates the dose-dependent decrease in maximal binding of [^{3}H]isradipine to rabbit adult and newborn cardiac membranes in the presence of halothane, enflurane, and isoflurane. Each of the anesthetics appears equally potent in decreasing the density of calcium channel antagonist (CCA) binding sites but demonstrated no effect on the affinity of binding. (We have assumed that the density of CCA binding site reflects the density of functional VDCC.)

A similar effect has been observed in a skeletal muscle T-tubule preparation. As shown in Fig. 4, halothane and isoflurane each decrease the binding of [^{3}H]isradipine to CCA sites; once again, no effect was observed on the affinity of binding.

The binding results from the Langendorff perfusion experiments are shown in Table I. Isradipine (100 n*M*) and halothane (1.5 and 2.5%) individually depress contractility by 61% and 68 and 82%, respectively, and together by over 90%, which suggests multiple sites of action. Halothane depresses contractility in a dose-dependent fashion. As suggested in the introduction, if halothane actually protects CCA binding sites from the 100 n*M* isradipine perfusion, the binding of [^{3}H]isradipine in the halothane-exposed heart should be greater than that in hearts perfused with isradipine alone. This is shown to be the case in Table I. Furthermore,

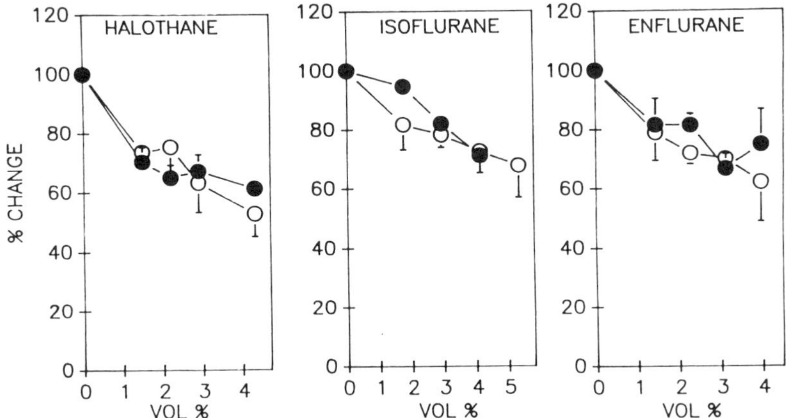

Fig. 3 Effect of halothane, enflurane, and isoflurane on the specific binding of [^{3}H]isradipine to newborn (●) and adult (○) cardiac membranes. Data are presented as percentages of control (no anesthetic) binding. Each data point is the mean of 4 determinations ± SEM.

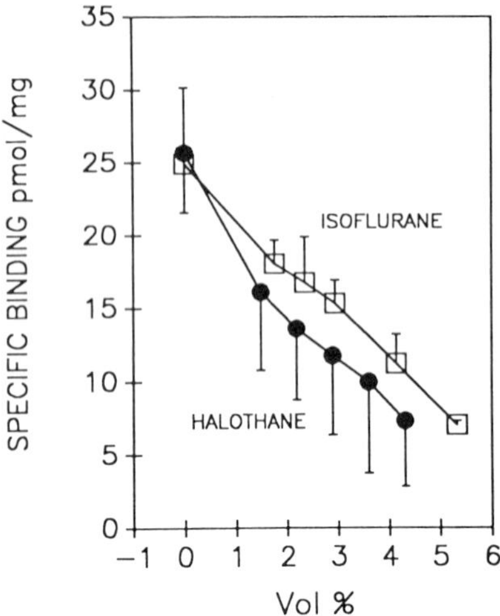

Fig. 4 Effect of increasing concentrations of halothane and isoflurane on the specific binding of [^{3}H]isradipine to skeletal muscle T-tubular membranes. Each data point is the mean of 3 determinations ± SEM.

if halothane is protecting CCA sites, then increasing concentrations should lead to greater binding, as is also shown in Table I.

The data in Table I also demonstrate a 7-fold increase in the K_d (representing a decrease in affinity) for isradipine binding to the sites protected by halothane. This is a distinctly different result from previous *in vitro* data and from the K_d values obtained from control hearts and hearts perfused with isradipine alone.

Table I

Effect of Halothane on Binding of [^{3}H]Isradipine to Voltage-Dependent Calcium Channels

Conditions	B_{max} (fmol/mg)	K_d (nM)
Control	475 ± 26	0.15 ± 0.02
Isradipine	44 ± 8[a]	0.34 ± 0.11[a]
1.5% Halothane	163 ± 7[a]	1.05 ± 0.07[a]
2.5% Halothane	264 ± 51[a]	1.07 ± 0.31[a]

[a] Significant difference ($p < 0.05$) compared to control.

IV. Discussion

This article extends our observation that halothane can interfere with the binding of CCA to cardiac L-type VDCC. We have convincingly demonstrated that the other volatile anesthetics, enflurane and isoflurane, are both able to interfere with the binding of a radiolabeled dihydropyridine, [^{3}H]isradipine, to both cardiac and skeletal muscle membranes.

We have shown that this phenomenon is relevant to anesthetic depression of contractility in an intact organ. Halothane can actually protect VDCC from blockade by isradipine. The experiments took advantage of the volatility of halothane to reveal the CCA binding sites that had been protected by halothane prior to its evaporation. However, the surprising finding is that the sites which were protected by halothane appear to have a markedly lower affinity for [^{3}H]isradipine than the sites which had been untouched by halothane.

The data presented suggest that the volatile anesthetics alter the VDCC; in our assay system this is manifest as a decreased binding of calcium channel antagonists. This phenomenon is relevant to the intact organ as well as isolated membranes and suggests that it is an important mechanism of anesthetic action. VDCC are ubiquitous channels that are found throughout the cardiovascular, endocrine, and peripheral and central nervous systems. It is probable that there is some variability in the sensitivity of VDCC from different organs to the volatile anesthetics, but it is likely that the VDCC are modified in all systems.

The decreased affinity that we have observed in the VDCC that were "protected" by halothane suggests that halothane is either modifying the VDCC or selecting an independent group of VDCC. Isradipine is known to bind to inactive channels with high affinity, and it is possible that halothane "protects" channels that are in the "unavailable" state or that have been modified by phosphorylation or other mechanisms resulting in a change of the K_d for isradipine. Other papers and presentations have suggested a gating mode for VDCC which includes a population of channels that are not immediately available for opening. This population might result from a dephosphorylated state which is augmented by exposure to halothane at 37°C. Modifications of the channel by glycosylation or myristylation might also be invoked, but no evidence currently exists for these possibilities.

In summary we have demonstrated in an *in vitro* system that halothane, enflurane, and isoflurane decrease VDCC density in heart and skeletal muscle membranes. We have also shown in the intact heart that this phenomenon results in a decrease in VDCC density during anesthetic exposure in the beating heart. However, our data also suggest that halothane modifies the VDCC as exemplified by a decrease in affinity of the VDCC for [^{3}H]isradipine.

Acknowledgments

This work was supported by grants from NIGMS (#30799) to T. J. J. Blanck, FAER/AUA to D. L. Lee, and the Harriman Foundation to the Department of Anesthesiology, Cornell University Medical College.

References

1. Rusy, B. F., Komai, H. (1987). Anesthetic depression of myocardial contractility: A review of possible mechanisms. *Anesthesiology* **67,** 745–766.
2. Lynch, C., Vogel, S., and Sperelakis, N. (1981). Halothane depression of myocardial slow action potentials. *Anesthesiology* **55,** 360–368.
3. Bosnjak, Z. J., and Kampine, J. P. (1983). Effects of halothane, enflurane, and isoflurane on calcium current in the SA node. *Anesthesiology* **58,** 314–321.
4. Blanck, T. J. J., Runge, S., and Stevenson, R. L. (1988). Halothane decreases calcium channel antagonist binding to cardiac membranes. *Anesth. Analg. (N.Y.)* **67,** 1032–1035.
5. Blanck, T. J. J., and Casella, E. S. (1989). Interaction of volatile anesthetics with calcium-sensitive sites in the myocardium. *In* "Cell Calcium Metabolism" (G. Fiskum, ed.), pp. 581–591. Plenum, New York.
6. Drenger, B., Quigg, M., and Blanck, T. J. J. (1991). Volatile anesthetics depress calcium channel blocker binding to bovine cardiac sarcolemma. *Anesthesiology* **74,** 155–165.
7. Chung, O. Y., Blanck, T. J. J., and Berman, M. R. (1989). Depression of myocardial force and stiffness without change in crossbridge kinetics: Effects of volatile anesthetics reproduced by nifedipine. *Anesthesiology* **71,** 444–448.
8. Hoehner, P., Quigg, M., and Blanck, T. J. J. 1991). Halothane depresses D600 binding to bovine heart sarcolemma. *Anesthesiology* **75,** 1019–1024.
9. Schmidt, U., Schwinger, R. H. G., Bohm, S., Uberfuhr, P., Kreuzer, E., Reichart, B., Meyer, L. V., Erdman, E., and Bohm, M. (1993). Evidence for an interaction of halothane with the L-type Ca^{2+} channel in human myocardium. *Anesthesiology* **79,** 332–339.
10. Endoh, M., Hiramoto, T., Ishihata, A., Takanashi, M., and Inui, J. (1991). Myocardial α_1-adrenoceptors mediate positive inotropic effect and changes in phosphatidylinositol metabolism. *Circ. Res.* **68,** 1179–1190.

Effects of Inhibition of Transsarcolemmal Calcium Influx on Content and Releasability of Calcium Stored in Sarcoplasmic Reticulum of Intact Myocardium

Hirochika Komai and Ben F. Rusy
Department of Anesthesiology
University of Wisconsin—Madison
Madison, Wisconsin 53792

I. Introduction

Under certain experimental conditions, notably when the muscle is stimulated at a low frequency and the sarcoplasmic reticular (SR) Ca^{2+} content is low, inhibition of the transsarcolemmal Ca^{2+} influx results in marked decrease in the force of contraction of isolated cardiac muscle preparations. It should be noted, however, that the hearts of small- to medium-sized mammals under the physiological condition beat at frequencies of 1–5 Hz or 60 to 300 beats/min. Because anesthetics generally inhibit the transsarcolemmal Ca^{2+} influx, it is important to assess such an effect on their overall negative inotropic effect under conditions close to physiological. We used Ni^{2+} as a model inhibitor of the Ca^{2+} influx, since we wanted an agent which selectively inhibits the Ca^{2+} influx with little use dependence. Note that lowering the extracellular Ca^{2+} concentration has an added effect of altering the transsarcolemmal Ca^{2+} gradient, and Ca^{2+} channel blockers such as verapamil have a marked use dependence. We used the force of contraction measured in the presence of ryanodine as an index of the transsarcolemmal Ca^{2+} influx. We measured postrest contractions and contractures induced by rapid cooling to assess the releasability and content of sarcoplasmic reticular Ca^{2+}. Postrest contractions are largely activated by Ca^{2+} released from the SR in response to electrical stimulation (1), whereas rapid cooling-induced contractures are elicited by Ca^{2+} released from the SR by a mechanism that does not involve membrane depolarization in species like rabbit and guinea pig, (2,3).

II. Isolated Papillary Muscle Preparation

All experiments were carried out with papillary muscles isolated from the right ventricles of rabbits. The medium used was a Krebs–Henseleit bicarbonate buffer (pH 7.4) of the following composition: NaCl, 115 m*M;* KCl, 5.9 m*M;* $CaCl_2$, 2.5 m*M;* $MgCl_2$, 1.2 m*M;* NaH_2PO_4, 1.2 m*M;* Na_2SO_4, 1.2 m*M;* $NaHCO_3$, 25 m*M;* glucose, 5.6 m*M;* EDTA, 50 μM. The medium was equilibrated with a gas mixture of 95% O_2 and 5% CO_2. The muscles were stimulated with a pair of field electrodes, and the isometric force of contraction was measured at 30°C (except for measurements of the rapid cooling-induced contracture which was induced by cooling the muscle to 3°C). Postrest contractions were elicited by stimulating the muscles after 2 sec of rest following interruption of stimulation at 2.0 Hz.

III. Calcium Influx and Release

Figure 1 shows that Ni^{2+} had minimal effect on the postrest contraction measured in the absence of ryanodine but had a marked inhibitory effect in the presence of 1 μM ryanodine (4). We used the force measured in the presence of ryanodine as an index of the transsarcolemmal Ca^{2+} influx, as the extent of inhibition of such a contraction by Ni^{2+} was similar to that of Ca^{2+} current measured by voltage clamp (5,6). Figure 2 shows that Ni^{2+} markedly inhibited the force of steady-state contraction at 2.0 Hz, moderately inhibited the force of postrest contraction elicited after 2 sec of rest, but did not inhibit the force of rapid cooling-induced contracture (4). To account for the observed effect of Ni^{2+} on the force of the steady-state contraction and postrest contraction, we propose an allosteric model of Ca^{2+}-induced Ca^{2+} release in which Ca^{2+} release is

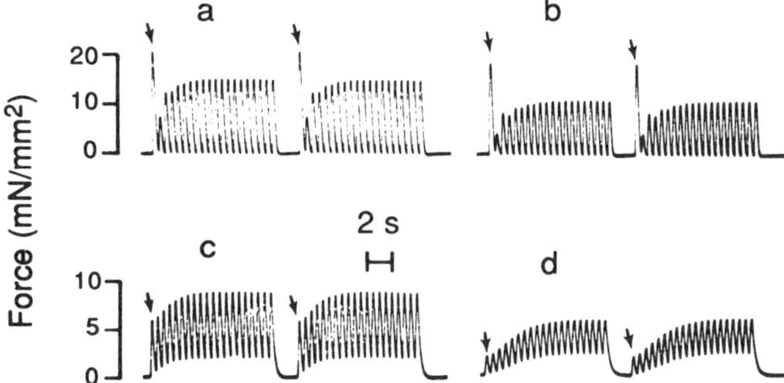

Fig. 1 Effects of 1.0 m*M* Ni^{2+} on the force of the postrest contraction in the absence and presence of ryanodine. The postrest contractions (arrows) are the first contractions elicited after 2 sec of rest following the last train of stimulation at 2.0 Hz. Segments of traces after stabilization of the force of contraction are shown. (a) Control; (b) with Ni^{2+} (1.0 m*M*); (c) with ryanodine (1 μ*M*); (d) with ryanodine (1 μ*M*) plus Ni^{2+} (1.0 m*M*). Note the different scales for the force in the absence (top traces) and presence of ryanodine (bottom traces). [From Komai and Rusy (4), reprinted with permission.]

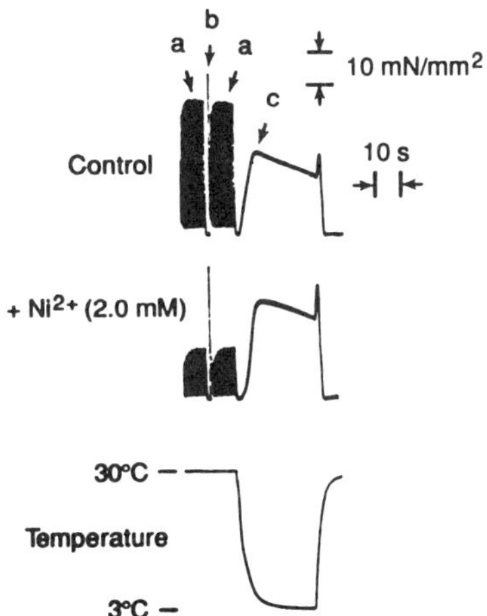

Fig. 2 Effects of Ni^{2+} on the force of the steady-state contraction at 2.0 Hz (a), the postrest contraction (b), and the rapid cooling-induced contracture (c). [From Komai and Rusy (4), reprinted with permission.]

influenced by the releasable SR Ca^{2+} as well as by the transsarcolemmal Ca^{2+} influx, as described by the equation

$$F = 1/\{1 + (A/I) \times 1/[1 - \exp(-kt)]^2\}$$

where F is the force of contraction expressed as a fraction of the maximal value, I is the transsarcolemmal Ca^{2+} influx expressed as the fraction of the maximal value, and $[1 - \exp(-kt)]$ is the releasable fraction of SR Ca^{2+}, where k is the rate constant for the time-dependent process by which SR Ca^{2+} becomes available for release and t is the rest interval. The constant A is empirically determined.

The equation is similar to that of enzyme kinetics as well as to that of pharmacodynamics of drug–receptor interaction. The model predicts that the higher the amount of releasable SR Ca^{2+}, the smaller is the magnitude of the transsarcolemmal Ca^{2+} influx required to generate the same force of contraction. This is consistent with the experimental results of Bass (7) and Lewartowski *et al* (8). It may also account for the higher affinity for extracellular Ca^{2+} of rat muscles as compared to rabbit muscle reported by Capogrossi *et al.* (9).

The negative inotropic effect of many general anesthetics, with the notable exception of halothane, is less pronounced in postrest contraction than in steady-state contraction. In this regard these anesthetics resemble Ni^{2+}. As reported previously (10), thiopental has no effect on the force of the rapid cooling-induced contracture, and it is very likely that thiopental does not decrease the Ca^{2+} content of the SR nor decrease the Ca^{2+} sensitivity of the myofibrils. Thus, we may assume that the negative inotropic effect of thiopental is due to the known inhibitory effect of the

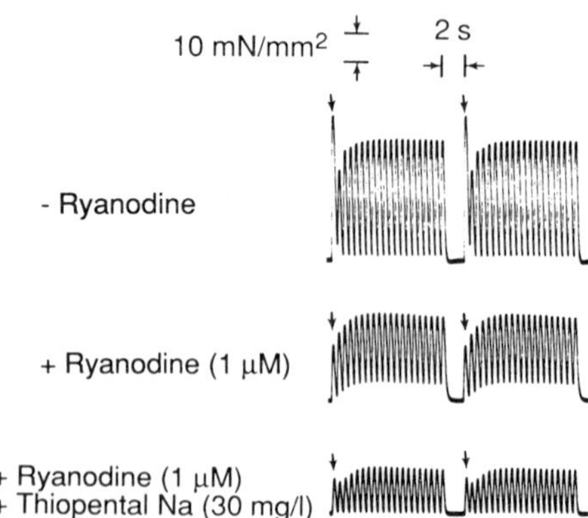

Fig. 3 Effect of thiopental on postrest contraction (arrows) and subsequent contraction at 2.0 Hz measured in the presence of 1 μM ryanodine.

anesthetic on the transsarcolemmal Ca^{2+} influx (11–13) and, possibly, to its effect on SR Ca^{2+} release. We found that, unlike Ni^{2+}, thiopental had a rather small effect on the force of postrest contraction measured in the presence of ryanodine (Fig. 3). In fact, thiopental partially reversed the effect of ryanodine on the postrest contraction relative to subsequent contractions at a relatively high frequency (2.0 Hz).

To test the possibility that ryanodine-induced depletion of SR Ca^{2+} during the rest following high-frequency stimulation is incomplete in the presence of thiopental, we measured the effect of thiopental on the force of steady-state contraction at 0.1 Hz (beat interval 10 sec) in the presence of ryanodine. As can be seen in Fig. 4, thiopental had a smaller inhibitory effect on the postrest contraction as compared to its effect on the steady-state contraction at 0.1 Hz. In contrast, Ni^{2+} had similar effects on the postrest contraction and on the contraction at 0.1 Hz, both measured in the presence of ryanodine. The steady-state contraction at 0.1 Hz in the presence of ryanodine was inhibited to a similar extent by 20 mg/liter thiopental and by 0.5 m*M* Ni^{2+}, whereas the postrest contraction elicited after 2 sec of rest following 2.0 Hz stimulation in the absence of ryanodine was inhibited by 20 mg/liter thiopental but not by 0.5 m*M* Ni^{2+}.

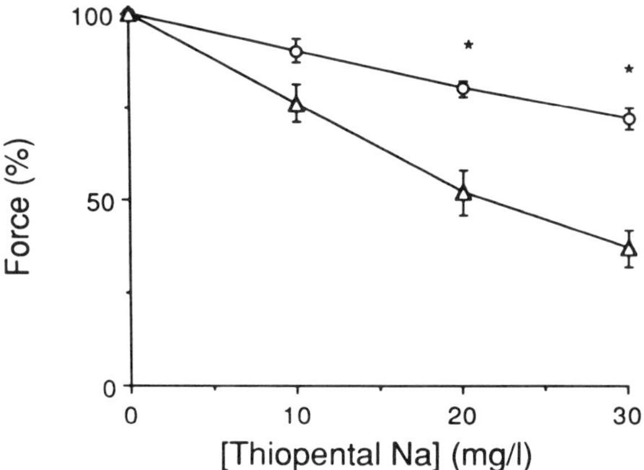

Fig. 4 Effect of thiopental on the force of postrest contraction (○) and steady-state contraction at 0.1 Hz (△) measured in the presence of 1 μM ryanodine. Data are means ± SEM ($n = 6$) of the force expressed as a percentage of the corresponding value in the absence of thiopental (ryanodine present). *Significant difference ($p < 0.01$, paired *t*-test) between the corresponding values of the force (%) of the postrest contraction and the steady-state contraction at 0.1 Hz. The force in the absence of thiopental (ryanodine present) was as follows: postrest, 17.6 ± 2.6 mN/mm^2; steady state at 0.1 Hz, 9.4 ± 1.5 mN/mm^2.

IV. Discussion

The results of the experiments using Ni^{2+} suggest that, in heart muscle beating at close to physiological rates, the inhibition of transsarcolemmal Ca^{2+} influx inhibits Ca^{2+}-induced release of SR Ca^{2+} without decreasing the amount of Ca^{2+} stored in the SR. Furthermore, the negative inotropic effect arising solely from the inhibition of the transsarcolemmal Ca^{2+} influx is rather small for contractions which are highly dependent on SR Ca^{2+}. Note that Terrar and Victory (14) reported that the extent of inhibition of the force of contraction by verapamil was only about half of the extent of inhibition of Ca^{2+} current by the drug. A model in which SR Ca^{2+} release is regulated by the product of the transsarcolemmal Ca^{2+} influx and the amount of releasable SR Ca^{2+} may account for the difference in the effects of an inhibitor of the Ca^{2+} influx on the force of the steady-state contraction and that of the postrest contraction. Even if the total SR Ca^{2+} content is the same in the two types of contractions, the amount of releasable SR Ca^{2+} is most likely smaller in the steady-state contraction at a high frequency (short interval) than in the postrest contraction owing to the time-dependent process by which SR Ca^{2+} becomes available for release (1). The proposed model predicts that the larger the amount of releasable SR Ca^{2+}, the smaller is the transsarcolemmal Ca^{2+} influx required for the same amount of SR Ca^{2+} release.

Although the effects of thiopental and Ni^{2+} were similar, as their negative inotropic effect did not involve a decrease in the SR Ca^{2+} content as determined by the measurements of rapid cooling-induced contracture, the anesthetic differed from Ni^{2+} with respect to relative effect on the postrest contraction compared at the same level of inhibition of the transsarcolemmal Ca^{2+} influx (determined by measurements of the force of contraction at 0.1 Hz in the presence of ryanodine) and with respect to its effect on the postrest contraction measured in the presence of ryanodine. Thus, thiopental appears to have a direct effect on the SR Ca^{2+} release mechanism in such a way that SR Ca^{2+} release is inhibited whether the release is triggered by the transsarcolemmal Ca^{2+} influx or by the action of ryanodine.

Acknowledgments

Research was supported in part by National Institutes of Health Grant GM29527 and Department of Anesthesiology Research and Development Fund, University of Wisconsin.

References

1. Edman, K. A. P., and Jóhannsson, M. (1976). The contractile state of rabbit papillary muscle in relation to stimulation frequency. *J. Physiol. (London)* **254,** 565–581.

2. Kurihara, S., and Sakai, T. (1985). Effects of rapid cooling on mechanical and electrical responses in ventricular muscle of guinea pig. *J. Physiol. (London)* **361,** 361–378.
3. Bridge, J. H. B. Relationship between the sarcoplasmic reticulum and sarcolemmal calcium transport revealed by rapidly cooling rabbit ventricular muscle. *J. Gen. Physiol.* **88,** 437–473.
4. Komai, H., and Rusy, B.F. (1993). Effects of inhibition of transsarcolemmal calcium influx by nickel on force of postrest contraction and on contracture induced by rapid cooling. *Cardiovasc. Res.* **27,** 801–806.
5. Kohlhardt, M., Bauer, B., Krause, H., and Fleckenstein, A. (1973). Selective inhibition of the transmembrane Ca Conductivity of mammalian myocardial fibres by Ni, Co and Mn ions. *Pfluegers Arch.* **338,** 115–123.
6. Kohlhardt, M., Mnich, Z., and Haap, K. (1979). Analysis of the inhibitory effect of Ni ions on slow inward current in mammalian ventricular myocardium. *J. Mol. Cell. Cardiol.* **11,** 1227–1243.
7. Bass, O. (1976). The decay of the potentiated state in sheep and calf ventricular myocardial fibers. Influence of agents acting on transmembrane Ca^{2+} flux. *Circ. Res.* **39,** 396–399.
8. Lewartowski, B., Prokopezuk, A., and Pytkowski, B. (1978). Effect of inhibitors of slow calcium current on rested state contraction of papillary muscles and post rest contractions of atrial muscle of the cat and rabbit hearts. *Pfluegers Arch.* **377,** 167–175.
9. Capogrossi, M. C., Kort, A. A., Spurgeon, H. A., and Lakatta, E. G. (1986). Single adult rabbit and rat cardiac myocytes retain the Ca^{2+}- and species-dependent systolic and diastolic contractile properties of intact muscle. *J. Gen. Physiol.* **88,** 589–613.
10. Komai, H., and Rusy, B. F. (1991). Contribution of the known subcellular effects of anesthetics to their negative inotropic effect in intact myocardium. *In* "Mechanisms of Anesthetic Action in Skeletal, Cardiac, and Smooth Muscle" (T. J. J. Blanck and D. M. Wheeler eds.) pp. 115–123. Plenum, New York.
11. Ikemoto, Y. (1977). Reduction by thiopental of the slow-channel-mediated action potential of canine papillary muscle. *Pfluegers Arch* **372,** 285–286.
12. Frankl, W. S., and Poole-Wilson, P. A. (1981). Effects of thiopental on tension development, action potential, and exchange of calcium and potassium in rabbit ventricular myocardium. *J. Cardiovasc. Pharmacol.* **3,** 554–565.
13. Pak, W. K., and Lynch III, C. (1992). Propofol and thiopental depression of myocardial contractility: A comparative study of mechanical and electrophysiologic effects in isolated guinea pig ventricular muscle. *Anesth. Analg. (N.Y.)* **74,** 395–405.
14. Terrar, D. A., and Victory, J. G. G. (1988). Effects of halothane on membrane currents associated with contraction in single myocytes isolated from guinea pig ventricle. *Br. J. Pharmacol.* **94,** 500–508.

Arrhythmogenic Effect of Inhalation Anesthetics: Biochemical Heterogeneity between Conduction and Contractile Systems and Protein Unfolding

Issaku Ueda and Jang-Shing Chiou
Department of Anesthesia,
University of Utah School of Medicine and
Veterans Affairs Medical Center,
Salt Lake City, Utah 84148

I. Introduction

The conduction system was discovered by brown staining when the excised heart was preserved in iodine solution, because the conduction system contains large amounts of glycogen. The heart, however, is not an ideal place to store glycogen. Cardiac metabolism depends on cytoplasmic glycogenolysis. This contrasts with the contractile system, where fatty acids are the main substrate providing energy via mitochondrial oxidative enzymes. In this regard, the conduction system resembles the central nervous system (CNS), where glucose is the sole source of fuel owing to the discrimination of fatty acids by the blood–brain barrier.

The metabolism of the conduction system depends on glycogenolysis and subsequent hydrolysis of sugars, whereas that of the contractile system depends predominantly on oxidation of lipids. Lipid metabolism is aerobic, while sugar metabolism is anaerobic. Anoxia *inhibits* lipid metabolism, but it *accelerates* sugar metabolism. This acceleration is known as the Pasteur effect.

For this reason, the conduction system is highly resistant to hypoxic stress. Compared to the conduction system, the CNS is vulnerable to

Advances in Pharmacology, Volume 31

hypoxia because it lacks glycogen stores. Glycogenolysis is initiated by the activation of glycogen phosphorylase by cyclic AMP. Cyclic AMP is synthesized by adenylate cyclase and hydrolyzed by cyclic-nucleotide phosphodiesterase. Adenyl cyclase is stimulated by epinephrine, and phosphodiesterase is inhibited by methylxanthines. It is natural to expect that methylxanthines "sensitize" hearts to the effects of epinephrine. The increased susceptibility to epinephrine during administration of the so-called myocardial sensitizing agents, such as cyclopropane, chloroform, and halothane, may share the same mechanism.

Pretreatment of dogs with methylxanthines induced ventricular fibrillation when challenged with 5μg/kg epinephrine. In contrast, when the glycolytic pathway was interrupted at the level of phosphoglucomutase, the methylxanthine–epinephrine combination failed to induce ventricular dysrhythmia. Hyperactive glycolytic activity may be essential for the genesis of ventricular dysrhythmia. Chloroform, which sensitizes the myocardium, inhibited cyclic-nucleotide phosphodiesterase 50% at 30 mbar. Diethyl ether required 425 mbar. Methylxanthines inhibited the enzyme in competition with cyclic AMP, whereas anesthetics inhibited it noncompetitively. Transection of the spinal cord at the level of the foramen magnum did not change the methylxanthine effect. This indicates a peripheral origin of the myocardial sensitization.

The finding of noncompetitive anesthetic action on phosphodiesterase contradicts the claim that anesthetics inhibit firefly luciferase by competition with luciferin. To elucidate the basic mode of anesthetic action on proteins, firefly luciferase was used as a model. We found that anesthetics inhibited firefly luciferase noncompetitively and that the anesthetic binding is endothermic. Proteins undergo order–disorder phase transitions between the N-state (folded) and D-state (unfolded). Infrared studies revealed that anesthetics changed protein conformations from the N-state to the D-state, whereas competitive inhibitors stabilized the N-state.

II. Cyclic AMP and the Conduction System

This section reviews earlier studies, in which pretreatment of dogs with inhibitors of cyclic-nucleotide phosphodiesterase by methylxanthines (theophylline and caffeine) induced ventricular fibrillation when challenged with subarrhythmogenic doses of epinephrine (1). To differentiate between the peripheral effect and the central effect, spinal cord-transsected dogs were used as controls. The effects of cyclic AMP and the second messenger Ca^{2+} in controlling myocardial contractility are now well popularized. To differentiate between the effect of the second messenger Ca^{2+} and the hyperactivity of the glycolytic pathway in the genesis of epinephrine dysrhythmia, control studies were performed by using sodium fluoride to block the glycolytic pathway. This compound is an inhibitor of phosphoglucomutase that catalyzes the conversion of glucose 1-phosphate to glucose 6-phosphate.

The protocol for preparation of dogs for study is detailed in our previous publication (1). Dogs were anesthetized by pentobarbital and intubated, and respiration was mechanically controlled at a P_{CO_2} of 35 torr (the elevation of University of Utah is 1500 m) by monitoring with a Beckman LB1 CO_2 analyzer (Fullerton, CA). Spinal dogs were prepared by bilateral ligation of carotid arteries and vagotomy, followed by transection of the spinal cord at the level of the foramen magnum.

A bolus dose of 5 μg/kg epinephrine induced sinus tachycardia or a short run of bigeminy without showing multifocal ventricular dysrhythmia or ventricular fibrillation in pentobarbital-anesthetized dogs. Dogs pretreated with caffeine or theophylline (50 mg/kg) developed ventricular fibrillation when challenged with 5 μg/kg epinephrine. When pentobarbital was supplemented before the epinephrine challenge, ventricular fibrillation was replaced by multifocal ventricular rhythm. Barbiturates have antiarrhythmic activity. All spinal dogs (which had only the induction dose of pentobarbital) developed ventricular fibrillation after administration of methylxanthines followed by epinephrine. This probably excludes central mechanisms from inducing myocardial sensitization.

The question of whether ventricular dysrhythmia involves cyclic AMP and the sbusequent increase in the secondary messenger Ca^{2+} or anaerobic glycogen metabolism was addressed by administration of sodium fluoride, which is an inhibitor of phosphoglucomutase. Pretreatment of dogs with sodium fluoride (50 mg/kg) suppressed the arrhythmogenic action of epinephrine. These dogs developed hypertension with sinus tachycardia but without a ventricular beat when challenged with 10 μg/kg bolus epinephrine. We contend that ventricular extrasystoles originate only from the conduction cells: contractile cells may not achieve automaticity.

III. Myocardial Sensitizing Agents and Phosphodiesterase

The possible action of a myocardial sensitizing agent, chloroform, on cyclic AMP metabolism was analyzed with cyclic-nucleotide phosphodiesterase (so named because the substrate specificity is broad and both cyclic AMP and cyclic GMP are hydrolyzed by the enzyme) purified from dog hearts. The inhibition kinetics were compared with those of nonsensitizing diethyl ether. Methods for preparation of phosphodiesterase from dog hearts and assay of enzyme activity are detailed in our previous publication (2).

Chloroform inhibited the enzyme. The partial pressure that inhibited the enzyme activity 50% was 30 mbar. The Lineweaver–Burk plot produced noncompetitive inhibition. Contrary to our expectation, diethyl ether also inhibited the enzyme, but the 50% inhibition concentration was 425 mbar.

The present trend in anesthesia research is to identify interacting receptors. This trend is fortified by the availability of a variety of agonists and antagonists. When Ahlquist classified adrenergic receptors as α and β, receptors were clearly defined: excitatory response was α and inhibitory β. Now there are many subtypes, and clearcut classification is impossible. We believe each agonist and antagonist has its own binding characteristics, and there may be as many receptors as the number of ligands.

Hayashi *et al.* (3) proposed a receptor-oriented central origin for halothane–epinephrine arrhythmia. They showed that imidazole-containing α_2-agonists antagonized halothane–epinephrine arrhythmia. The analeptic action site of imidazole was established at the medulopontine area in decerebrated cats at the mid-collicular level (4).

It is interesting to note, however, that Kukovetz and Pöch (5) reported that imidazole inhibited the action of methylxanthines on myocardial glycogen phosphorylase activity. The result of Hayashi *et al.* (3) may be interpreted as follows: the imidazole-containing α_2-agonists antagonized the halothane effect by alleviating the inhibition imposed on phosphodiesterase by halothane. The problem of central versus peripheral effects of myocardial-sensitizing agents, however, is unsettled because Hayashi *et al.* (6) later reported that imidazole-containing antagonists enhanced halothane–epinephrine arrhythmias.

IV. Mode of Anesthetic–Protein Interaction

Anesthetic actions on myocardium involve protein binding. The multifaceted actions of anesthetics on varous functions of the heart favor nonspecific interaction. However, the concept that anesthetics interact with specific hydrophobic pockets on proteins appears to be well accepted among anesthesia researchers. The proposal of specific hydrophobic pockets was based on a study with firefly luciferase by Franks and Lieb (7). They reported that anesthetics competed with luciferin (substrate) at the binding site as revealed by analysis with Lineweaver–Burk plots. We reported that anesthetic inhibition of firefly luciferase was allosteric, inducing conformational changes of the enzyme protein as revealed by analysis with absolute reaction rate theroy (8).

The dispute concerns the basic mechanisms of anesthetic action on any system, including myocardium. The competition theory was derived from the kinetic analysis of anesthetic action, not by binding studies (7). Problems and limitations of application of Lineweaver-Burk plots have been discussed by many (see, for instance, Refs. 9 and 10). The controversy on the mode of anesthetic binding was addressed by directly measuring association and dissociation of luciferin with firefly luciferase to avoid ambiguity in the kinetic procedure. A fluorescence dye, anilinonaphthalenesulfonate (ANS), is known to be a competitive inhibitor of firefly luciferase (11). We found that long-chain fatty acids are also competitive inhibitors.

Anilinonaphthalenesulfonate is a probe to identify hydrophobic patches on protein molecules. The ANS molecule increases its fluorescence intensity more than 150-fold when bound to proteins. When ANS molecules are released from the binding site, the fluorescence intensity practically disappears. The ANS fluorescence has been shown to decrease by a luciferin analog dehydroluciferyl adenylate (11). In contrast, anesthetics did not decrease the ANS fluorescence except at extremely high concentrations.

When another competitive inhibitor, myristate, was added to the system, the fluorescence intensity of luciferase-bound ANS was decreased dose-dependently (Fig. 1). The result agrees with the report by Eckenhoff

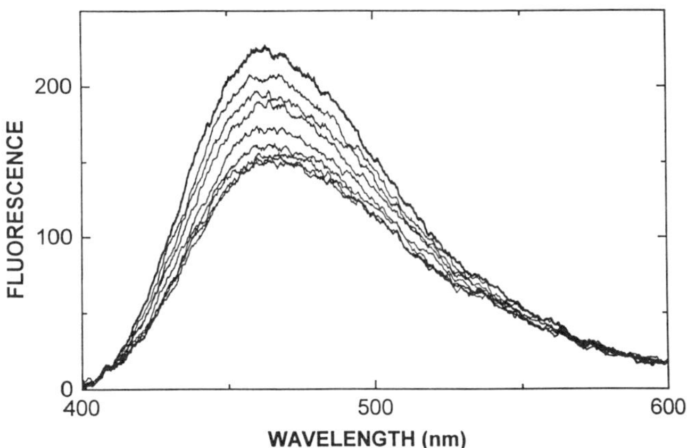

Fig. 1 Release of the competitive inhibitor ANS from firefly luciferase by a competitive inhibitor, myristate. The compound ANS decreases fluorescence at 528 nm when released from firefly luciferase. Consecutive increases of myristate released ANS from the binding site. Volatile anesthetics did not change the ANS fluorescence. The top curve is with ANS at 1.1 μM, and myristate was added to the solution in incremental doses, with each addition increasing the myristate concentration by 0.13 μM.

and Shuman (12), who used photoaffinity labeling to probe anesthetic binding to proteins. They reported that volatile anesthetics did not compete with *n*-decyl aldehyde on bacterial luciferase. For bacterial luciferase, *n*-decyl aldehyde is a counterpart of luciferin for firefly luciferase. This result also supports ruling out the possibility of anesthetic–luciferin competition.

The competition model might have originated from the choice of the steady-state conditions. Because of the double-reciprocal structure and restrictions on the system parameters, the Lineweaver–Burk plot has been criticized by many, from the classic King–Altman approximation (9) to the more recent Roussel–Fraser global analysis (10). There are many conditions to be satisfied in order to apply the Lineweaver–Burk plot.

V. Unfolding of Proteins

Proteins are known to change their conformation according to the temperature (thermotropic phase transition) between the N-state (folded low-entropy state) at low temperatures and the D-state (unfolded high-entropy state) at high temperatures. Differential scanning calorimetry (DSC) showed that firefly luciferase underwent a thermotropic phase transition from the N-state to the D-state at about 42°C. The thermogram is similar to the phase transition of lipid membranes from the solid–gel to the liquid–crystalline states. Anesthetics decreased the phase-transition temperature, whereas luciferin increased it (Fig. 2). Long-chain fatty acids and ANS increased the transition temperature: no transition was observed even when the temperature was increased up to 70°C.

Fourier-transform infrared spectroscopy (FTIR) of luciferase showed that the enzyme started to unfold when the temperature exceeded about 20°C and became stationary at 48°C (Fig. 3A). The phase transition shown by DSC is the all-or-none conformational change, and the FTIR change indicates the mutli-stage conformational change at about 20°C. Figure 3B shows the effect of ANS on the FTIR spectra. The increase in IR intensity was no longer present, and ANS prevented the conformational change. Volatile anesthetics enhanced the thermotropic phase transition (not shown).

These results show that volatile anesthetics preferentially bind to the D-state protein, whereas competitive inhibitors bind to the N-state. To clarify the effect of solutes (anesthetics) on the phase transition of solvents (proteins or lipids), let us consider the effect of salt on the freezing of water. Sodium chloride decreases the freezing temperature of water because NaCl (solute) dissolves into water (high-temperature state) but not

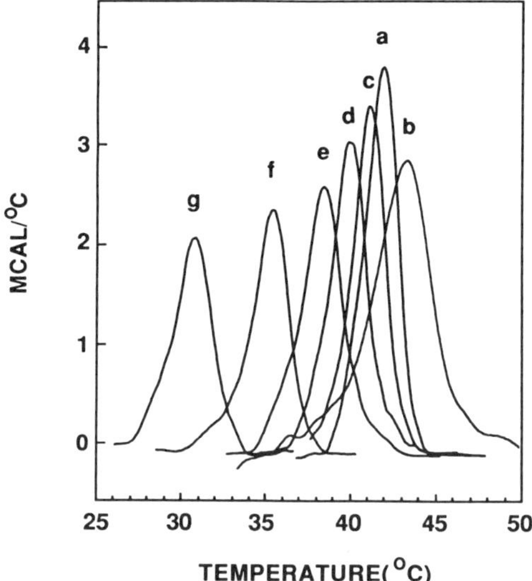

Fig. 2 Differential scanning calorimetry of firefly luciferase. The N–D phase transition is observed at about 42°C. Ethanol decreased the phase-transition temperature dose-dependently. Luciferin slightly increased the transition temperature. Symbols: a, control; b, luciferin at 1 m*M*; c–g, ethanol at 0.15, 0.3, 0.6, 1.2, and 2.3 *M*.

in ice (low-temperature state). When a solute dissolves equally into the high-temperature state (water) and the low-temperature state (ice), then the freezing (transition) temperature does not change. If NaCl dissolves preferentially into ice, the freezing temperature rises.

Anesthetics dissolve (bind) into the high-temperature state (D-state) of the enzyme protein (solvent). Hence, the phase-transition temperature decreases. The competitive inhibitors, namely, ANS and long-chain fatty acids, cannot bind to the unfolded D-state (because of the conformational disparity between the two states). Therefore, the transition temperature increased. When anesthetics change the protein conformation, receptor sites are lost and the binding of ligands may be curtailed. The result that extremely high concentrations of volatile anesthetics displaced ANS molecules from firefly luciferase represents the above situation, where anesthetics damaged the luciferin recognition site. The mode of displacement of ANS by anesthetics is not identical with the competitive inhibitor myristate.

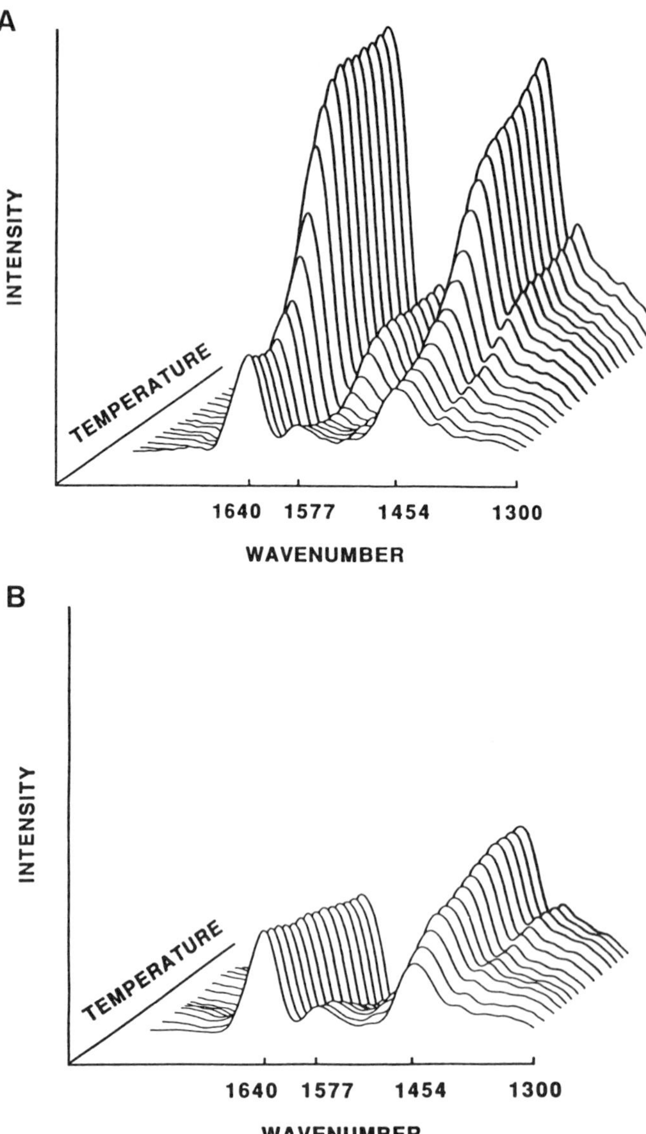

Fig. 3 (A) Fourier-transform infrared spectra of firefly luciferase. The amide-I and amide-II spectra were scanned by temperature between 5 and 70°C. The peak height increased according to the increase in temperature. The protein started to unfold at 20°C. The change in the peak height ceased at about 48°C. Apparently, the enzyme became irreversibly unfolded near 48°C. The peak at 1640 cm^{-1} represents the amide-I band (C═O stretching and N—H bending). The amide-II band (N—H bending coupled with C—N stretching) split into two (1577 and 1454 cm^{-1}). This is caused by the deuteron–proton exchange of amino acid residues located at the surface of the proteins. Those located at the protein core were not affected by the deuteron exchange and appeared at 1577 cm^{-1} (amide-II), and those located at the

VI. Specific Binding: Lack of Definition

The debate on the specific receptor binding of anesthetics suffers a lack of clear definition. The agonists and antagonists of sympathetic receptors do not bind to acetylcholine receptors, whereas those of acetylchloine receptors do not bind to sympathetic receptors. Anesthetics, however, affect almost all neuroreceptors and enzymes. The multiplicity of the action site indicates nonspecificity.

There are many studies on anesthetic effects on enzymes, but we are not aware of reports demonstrating competitive inhibition, except for the report by Franks and Lieb (7). Reports on the anesthetic effects on neuroreceptors often show allosteric inhibition (13,14).

The basic force of anesthetic interaction is the hydrophobic effect, that is, the tendency to be excluded from the extensively hydrogen-bonded aqueous phase. Therefore, anesthetics adsorb to any macromolecular surface: they interact nonselectively with proteins at hydrophobic regions. To designate these hydrophobic areas as receptors is a matter of definition.

Random hydrophobic binding is often expressed by the Langmuir adsorption isotherms, which follows the saturation kinetics. In physical chemistry terms, the Langmuir adsorption follows the Fermi–Dirac statistics. Without specific binding sites, anesthetic binding to lipid membranes still saturates (15). Anesthetics also bind α-helix polypeptides consisting of a single species of amino acid, poly(L-lysine), and partially transform the α-helix structure to β sheet (16). Because the peptide consists of identical amino acids, there is no specific hydrophobic area. Conversely, when the binding follows the Bose–Einstein condensation kinetics (multiple stack adsorption), the binding does not saturate even when interacting with a limited area (17). Thus, saturability of binding is a complex problem as shown by the Bunauer-Emmett-Teller (BET) model (17). Thus, saturability of binding is a complex problem as shown by the BET model (17), where simple hydrophobic adsorption can be saturable, linear to the ligand concentration, or even cooperative (binding increases with the increase in the ligand concentration). The conclusion of specific receptor binding from saturability must be reached with caution. A number of studies have addressed the anesthetic effects on myocardium activities on the basis

---

surface appeared at 1454 cm^{-1} (amide-II′). (B) Effect of the competitive inhibitor ANS on the FTIR spectra of firefly luciferase. The thermotropic change in the secondary conformation of the enzyme shown in (A) disappeared in the presence of ANS. Anesthetics increased the unfolding (not shown). The ANS preferentially binds to the low-entropy N-state of the enzyme and stabilizes the conformation. Anestehtics preferentially bind to a high-entropy D-state and enhance the transition from the N- to the D-state.

of interaction with various enzymes. Presumably, the mode of volatile anesthetic effects on these enzymes is nonselective hydrophobic interaction.

Acknowledgments

Research was supported by Department of Veterans Affairs Medical Research Funds and National Institutes of Health Grant GM25716.

References

1. Ueda, I., Loehning, R. W., and Ueyama, H. (1961). Relationship between sympathomimetic amines and methylxanthines inducing cardiac arrhythmias. *Anesthesiology* **22,** 926–932.
2. Ueda, I., and Okumura, F. (1971). Effects of chloroform, diethyl ether and a propiophenone derivative, 3-dimethylamino-2-methyl-2-phenoxypropiophenone hydrochloride, upon cyclic 3′,5′-nucleotide phosphodiesterase. *Biochem. Pharmacol.* **20,** 1967–1971.
3. Hayashi, Y., Sumikawa, K., Maze, M., Yamatodani, A., Kamibayashi, T., Kuro, M., and Yoshiya, I. (1991). Dexmedetomidine prevents epinephrine-induced arrhythmias through stimulation of central α_2-adrenoceptors in halothane-anesthetized dogs. *Anesthesiology* **75,** 113–117.
4. Fórez, J. (1974). The site of the respiratory stimulant action of imidazole in cats. *Pharmacology* **11,** 308–315.
5. Kukovetz, W. R., and Pöch, G. (1967). The action of imidazole on the effects of methylxanthines and catecholamines on cardiac contraction and phosphorylase activity. *J. Pharmcol. Exp. Ther.* **156,** 514–521.
6. Hayashi, Y., Kamibayashi, T., Maze, M., Yamatodani, A., Sumikawa, K., Kuro, M., and Yoshiya, I. (1993). Role of imidazoline-preferring receptors in the genesis of epinephrine-induced arrhythmias in halothane-anesthetized dogs. *Anesthesiology* **78,** 514–530.
7. Franks, N. P., and Lieb, W. R. (1984). Do general anesthetics act by competitive binding to specific receptors? *Nature* (*London*) **310,** 599–601.
8. Ueda, I., and Kamaya, H. (1973). Kinetic and thermodynamic aspects of the mechanism of general anesthesia in a model system of firefly luminescence *in vitro*. *Anesthesiology* **41,** 425–436.
9. King, E. L., and Altman, C. (1956). A schematic method of deriving the rate laws for enzyme-catalyzed reactions. *J. Phys. Chem.* **60,** 1375–1378.
10. Roussel, M. R., and Fraser, S. J. (1993). Global analysis of enzyme inhibition kinetics. *J. Phys. Chem.* **97,** 8316–8327.
11. DeLuca, M. (1969). Hydrophobic nature of the active site of firefly luciferase. *Biochemistry* **8,** 160–166.
12. Eckenhoff, R. G., and Shuman, H. (1993). Halothane binding to soluble proteins determined by photoaffinity labeling. *Anesthesiology* **79,** 96–106.
13. Young, A. P., and Sigman, D. S. (1981). Allosteric effects of volatile anesthetics on the membrane-bound acetylcholine receptor protein. Stabilization of the high-affinity state. *Mol. Pharmacol.* **20,** 498–505.
14. Kukita, F., and Mitaku, S. (1993). Kinetic analysis of the denaturation process by alcohols of sodium channels in squid giant axon. *J. Physiol.* (*London*) **463,** 523–543.

15. Yoshida, T., Okabayashi, H., Kamaya, H., and Ueda, I. (1989). Saturable and unsaturable binding of a volatile anesthetic enflurane with model lipid vesicle membranes. *Biochim. Biophys. Acta* **979,** 287–293.
16. Shibata, A., Morita, K., Yamashita, T., Kamaya, H., and Ueda, I. (1991). Anesthetic–protein interaction: Effects of volatile anesthetics on the secondary structure of poly(L-lysine). *J. Pharm. Sci.* **80,** 1037–1041.
17. Adamson, A. W. (1982). "Physical Chemistry of Surfaces," pp. 417–600. Wiley, New York.

Potassium Channel Current and Coronary Vasodilatation by Volatile Anesthetics

Nediljka Buljubasic, Jure Marijic, and Zeljko J. Bosnjak

Department of Anesthesiology, The Medical College of Wisconsin Milwaukee, Wisconsin 53226

I. Introduction

Although anesthetics are potent vasodilators, the mechanism of their action is not well understood. Volatile anesthetics cause vasodilation in specific vascular beds, either by direct depressant action on the vessel or by an indirect attenuation of vasoconstrictor activity. In the *in vivo* setting, the mechanism for the altered blood flow to a specific organ during anesthesia is likely to involve interaction among endothelium, vascular smooth muscle, arterial pressure, metabolic requirement of the organ, and the autonomic nervous system. For instance, in the intact coronary circulation, the direct effects of volatile anesthetics on arterial muscle are superimposed on their dominant depressant action on the myocardium, namely, reduced cardiac work and oxygen demand. It appears that isoflurane is a more potent coronary vasodilator than halothane in the isolated perfused heart (1), and a less potent dilator of isolated coronary arterial segments (2). In studies on isolated coronary artery rings (3) and isolated tetrodotoxin-arrested rat heart, halothane caused direct dilation of coronary vessels without affecting oxygen extraction or consumption (4). Other results have shown that isoflurane produces potent dose-dependent relaxation of canine middle cerebral arteries *in vitro*, and that this relaxation is endothelium-independent (5).

The membrane potential of the vascular smooth muscle cell is mainly regulated by the flow of Ca^{2+} and K^+ through specialized channels. Although it is known that K^+ channels act as an endogenous dilating mecha-

Advances in Pharmacology, Volume 31

nism to regulate vascular muscle tone (6), the relative importance of this effect on K^+ channels for the vasorelaxant effects of volatile anesthetics is unclear. In isolated cerebral arteries, halothane-induced relaxation is associated with concentration-dependent membrane depolarization, suggesting electromechanical uncoupling (7). The mechanism by which halothane induces this apparent electromechanical uncoupling, that is, causes membrane depolarization concurrent with vasodilation of vascular muscle cells, remains unclear. One possibility is that halothane decreases K^+ current by causing cell membrane depolarization while at the same time affecting to an even greater degree the tension generation mechanisms, causing relaxation. Indeed, we have reported that volatile anesthetics reduce macroscopic Ca^{2+} and K^+ currents in vascular arterial smooth muscle cells (8,9). However, reduction of Ca^{2+} current was greater than reduction of K^+ current by these agents. These anesthetic-induced effects on the vascular muscle membrane may help to explain the electromechanical uncoupling of vascular smooth muscle by inhalational anesthetic agents.

The role of K^+ current in vascular relaxation induced by volatile anesthetics is not clear. To eliminate potentially confounding indirect effects of anesthetics, an isolated vessel and isolated coronary smooth muscle cell techniques were used. Our hypothesis was that the blockade of K^+ current by antagonists would lead to cell membrane depolarization and enhanced vasoconstriction, but also may potentiate the vasodilatory effect by anesthetics because of the unopposed effect of anesthetics on Ca^{2+} current.

Direct effects of the volatile anesthetics halothane and isoflurane are best determined and compared by using whole-cell and single-channel patch clamp techniques in canine coronary arterial cells. Volatile anesthetics may act on K^+channels by altering mean open time, mean closed time, and/or probability of opening. Before investigating the effects of volatile anesthetics on the K^+ current it is important to characterize the K^+ current by using different agents which are known to activate or block certain types of K^+ channels. Tetraethylammonium (TEA, in concentrations less than 1 m*M*) and charybdotoxin effectively block Ca^{2+}-dependent K^+ channels in vascular smooth muscle. [10–12].

II. Isolated Vessel Ring Experiments

All experimental procedures strictly adhered to the standards of American Association for Accreditation of Laboratory Animal Care and all protocols were approved by the Animal Care Committee of the Medical College of Wisconsin.

Adult mongrel dogs of either sex were killed by exsanguination following anesthesia (sodium pentobarbital, 30 mg/kg i.v.) and hearts were removed. Coronary arteries were identified, carefully dissected, and placed in physiological saline solution (PSS) of the following composition (in m*M*): NaCl, 119; KCl, 4.7; $MgSO_4$, 1.17; $CaCl_2$, 1.6; $NaHCO_3$, 27.8; NaH_2PO_4, 1.18; EDTA, 0.026; glucose, 5.5; and HEPES [*N′* [(2-Hydroxyethyl) piperazine-*N′*-(2-ethanesulfonic acid)], 5. The vessels were cleaned of fat and connective tissue using a dissecting microscope and divided into rings 2 mm in length. The vascular rings were mounted on tungsten triangles [the lower triangle being fixed and the other attached to a force transducer (Model FT 103, Grass Instruments, Quincy, MA)] and suspended in jacketed, temperature-controlled (37°C), tissue baths containing 15 ml of PSS aerated with a mixture of 93.5% O_2 and 6.5% CO_2 (by volume), as described previously by Marijic *et al.* (13). The pH, p_{CO_2} and p_{O_2} of the salt solution were monitored every 30 min and maintained constant at pH 7.38 to 7.42 and p_{CO_2} of 34 to 36 mmHg.

The rings were progressively stretched to an optimal tension which was determined by preliminary length–tension studies using a standard concentration (40 m*M*) of KCl. The rings were then allowed to equilibrate for 90 min before conducting any experiments. Contractile responses were recorded continuously on a polygraph (Grass Model 7). The functional integrity of each ring was confirmed by the contractile response to 40 m*M* KCl in the bath medium. Following a further period of equilibration the rings were constricted to a stable plateau tension.

The relaxant responses to the anesthetics were measured in vessels previously exposed to the voltage-mediated constrictor KCl (40 m*M*). In a random selection of vessels, the procedure was carried out in the presence of TEACl (20 m*M*).

Using separate vaporizers (Dragerwerk, Lubeck, Germany), halothane and isoflurane were bubbled in turn into the tissue baths. The order of administration of the anesthetics was randomized, and the concentration of anesthetic vapor in the carrier gas was measured using a mass spectrometer (Marquette Electronics, Milwaukee, WI). Anesthetic concentrations in the tissue baths were determined by gas chromatography. Vessel rings were exposed to 0.45 ± 0.01 and 0.40 ± 0.02 m*M* of isoflurane and halothane, respectively (mean ± SEM). These figures correspond to anesthetic concentrations of 2.18 and 1.35% at 37°C, and in the dog they represent minimal alveolar concentration (MAC) values of 1.66 ± 0.04 and 1.73 ± 0.09 MAC for isoflurane and halothane, respectively.

Responses of the coronary arteries to inhalational anesthetics were expressed as the percent relaxation (mean ± SEM) of the agonist-induced constriction in order to normalize the data. If more than one ring was used from the same animal with the same protocol, the mean response in these rings was used for statistical analysis. The responses of the vessels to each drug were compared by analysis of variance. When the *F*-test showed significance, a *t*-test was performed for comparison of means. A p value no higher than 0.05 was considered statistically significant.

III. Patch Clamp Experiments

The vascular rings were placed in a vial containing the enzyme solution and dissociated as described previously (9). A drop of dispersed single coronary arterial cells was placed in a perfusion chamber (22°C) on the stage of an inverted microscope (Olympus IMT-2, Leeds Instruments, Minneapolis, MN). At 500× magnification, a hydraulic micromanipulator (Narishige, Tokyo, Japan) was used to position heat-polished borosilicate pipettes with a tip resistance of 4–6 MΩ on the membrane of arterial cells. High-resistance seals (3–30 GΩ) were formed, after which the pipette patch was removed by negative pressure to obtain electrical access to the whole cell as previously described (14). Whole-cell currents were elicited by 200 msec depolarizing pulses generated by a computerized system (pClamp software, Axon Instruments, Foster City, CA) every 5 sec. The currents were amplified by a List EPC-7 patch clamp amplifier (Adams & List Assoc., Great Neck, NY), and the amplifier output was low-pass filtered at 500 Hz. All data were digitized (sampling rate 10,000 per second) and stored on a hard disk to permit analysis at a later time. Leak and capacitative currents were subtracted from each record by linearly summating scaled currents obtained during 10-mV hyperpolarizing pulses. The external solution (Tyrode's) contained the following (in m*M*): $CaCl_2$, 2; NaCl, 135; KCl, 4.7; $MgCl_2$, 1, glucose, 10; and HEPES, 5 (pH 7.4).The pipette solution contained the following (in m*M*): potassium glutamate, 130; $MgCl_2$, 1; EGTA, 1; HEPES, 10; and Na_2ATP, 3 (pH 7.1–7.2).

The cell-attached mode was used to record single-channel currents in order to preserve cell integrity as well as integrity of the cytosolic processes. The cells were superfused with the same Tyrode's solution used for whole-cell current recording. Heat-polished borosilicate glass pipettes were filled with a solution containing the following (in m*M*): KCl, 145; $CaCl_2$, 2; Mg Cl_2, 2; and HEPES, 5 (pH 7.4). High-resistance seals (5–30 GΩ) were obtained by applying negative pressure to the inside of the pipette; however, the membrane was left intact. A List EPC-7 patch clamp amplifier was used to clamp the pipette potential at the desired voltage and to record the single-channel current. Amplifier output was filtered at frequency above 1 kHz by passing it through an 8-pole Bessel filter. Commerically available software (pClamp) was also used for single-channel data analysis.

The inside-out mode was used to investigate direct effects of TEA, Ca^{2+}, and charybdotoxin on a large-conductance K^+ channel. Pipettes were filled with the same solution as in the cell-attached mode. The external (bath) solution contained the following (in m*M*): KCl, 145; $MgCl_2$, 1; EGTA, 1; and HEPES, 10 (PH = 7.2), along with appropriate concentrations of $CaCl_2$ (0.1 or 1 μM).

Cells were exposed to 0.81 ± 0.01 m*M* of halothane (equivalent to 1.5% at 22°C or approximately 2 MAC for the dog) or to 1.03 ± 0.02 m*M* of isoflurane (equivalent to 2.6% at 22°C or approximately 2 MAC for the dog) by changing the inflow perfusate to one containing anesthetic. Effects of isoflurane and halothane reached steady state within 2 min and were reversible on wash-out. This was verified by obtaining similar measurements following removal of anesthetic from the perfusate. Solution was sampled from the perfusion chamber and analyzed by gas chromatography to verify the anesthetic concentration at the cell proximity.

Currents are expressed as means ± SEM. Mean open times, mean closed times, and probability of K^+ channel opening were determined. Channel conductance was estimated by linear regression. Transitions between closed and open channel states were defined as a change of 50% below baseline relative to the predominant channel amplitude. Slopes of the current–voltage curves were compared between different conditions by testing for parallelism (15). Analysis of variance was used to compare amplitude, mean open times, mean closed times, and probabilities of K^+ channel opening among different conditions. If the *F*-test showed significance, Fisher's test for least significant difference was performed, with the level of significance set at $p \leq 0.05$.

IV. Effects of Anesthetics on Isolated Coronary Vessels

The vessel diameter of the coronary arteries ranged from 1.5 to 2.0 mm. The optimal resting tension was determined by the initial response to 40 m*M* KCl at various levels of resting tension and ranged from 1.8 to 3.0 g. Only vessels that had a prolonged, stable constriction in response to KCl were studied.

Tetraethylammonium (TEA, 20 m*M*) had no effect on resting tension in coronary arteries. However, KCl (40 m*M*) produced greater constriction in TEA-pretreated than in non-TEA-pretreated arteries (6.35 ± 0.27 versus 7.46 ± 0.34 g for non TEA- and TEA-pretreated arteries. $p \leq 0.05$). Fig. 1A represents a typical tracing of a chart recording demonstrating the vasodilatory effect of halothane and isoflurane on a single coronary artery preconstricted with KCl in the absence (top trace) and presence (bottom trace) of 20 m*M* TEA. Summarized results for the two anesthetics in

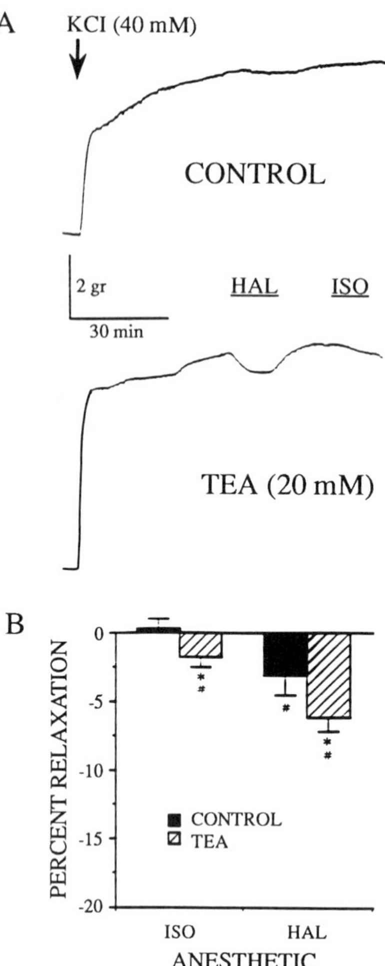

Fig. 1 (A) Actual tracing demonstrating effects of volatile anesthetics on isometric tension in a single canine coronary artery preconstricted with 40 m*M* KCl in the absence (control) and presence of 20 m*M* tetraethylammonium (TEA). The horizontal lines show exposure to halothane (HAL, 1.35%) and isoflurane (ISO, 2.18%). Vessels were exposed to anesthetics for at least 20 min, although the concentration of anesthetics in the bath reached a steady-state level witin 5–10 min. (B) Effect of TEA on changes in isometric tension induced by ISO and HAL (n = 29, 9 dogs) in canine coronary arteries preconstricted with 40 m*M* KCl. Vasodilatation caused by volatile anesthetics was significantly augmented in the presence of TEA.*Significant difference ($p \leq 0.05$) for TEA versus control. # Significant difference ($p \leq 0.05$) versus no anesthetic. [Reprinted with permission from Marijic *et al.* (13).]

coronary vessels are shown in Fig. 1B. Halothane (1.35%) produced significant relaxation of coronary vessels preconstricted with KCl, whereas isoflurane (2.18%) did not produce significant relaxation of coronary arteries preconstricted with KCl. Following treatment with TEA, however, both anesthetics produced significant relaxation.

V. Effects of Anesthetics on Whole-Cell Potassium Current

Cells dialyzed with high-K^+ solution showed a large outward current during 200-msec depolarizing steps from a holding potential (HP) of −60 mV to potentials beyond −30mV.This current was eliminated when Cs^+ was substituted for K^+ in the pipette solution, indicating that K^+ was the mandatory charge carrier for this current.

Concentrations of 1, 10, and 30 m*M* TEA dose-dependently reduced K^+ current (I_K) elicited by depolarization from −60 to +60 mV, by 31 ± 7, 72 ± 2, and 83 ± 4%, respectively, from the control amplitude of 1679 ± 206 pA ($p \leq 0.05$, data not shown). This depression was voltage-independent (no significant difference in percent reduction between different voltages) and completely reversible on washout.

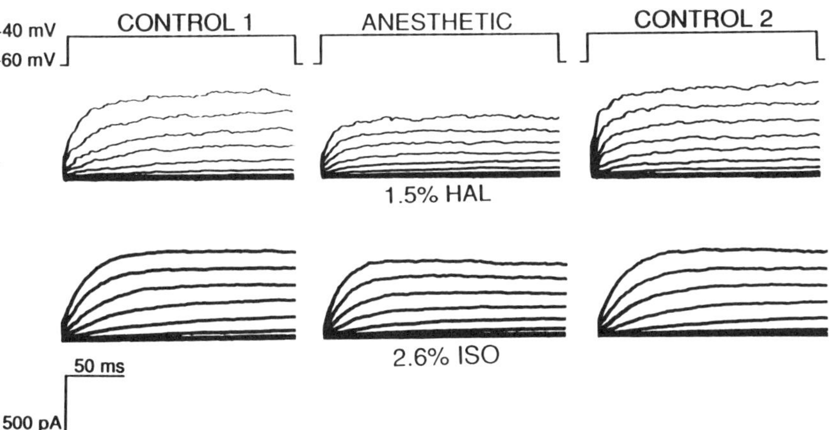

Fig. 2 Actual recordings showing the effects of 1.5% halothane (HAL, top) and 2.6% isoflurane (ISO, bottom) on whole-cell K^+ current (I_K) elicited by progressive 10-mV test pulses from −60 to +40 mV in two canine coronary arterial smooth muscle cells. Recordings were obtained before (control 1), during (anesthetic), and after (control 2) exposure to volatile anesthetics. [Reprinted with permission from Buljubasic *et al.* (9.)]

The effects of halothane and isoflurane on whole-cell I_K were determined by applying a series of stepwise (10 mV) depolarizing test pulses from a holding potential of −60 mV to positive potentials as high as +40 mV. Figure 2 shows actual traces of patch clamp recordings illustrating the reversible depressent effect of 1.5% halothane (HAL) and 2.6% isoflurane (ISO) on I_K amplitude. Peak I_K was plotted as a function of membrane potential to analyze the mean effect of 1.5% halothane (Fig. 3A) and 2.6%

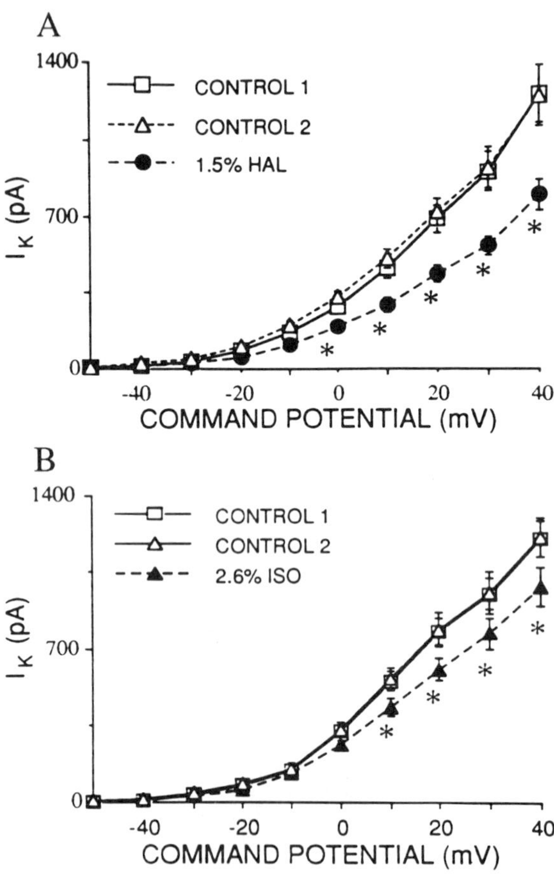

Fig. 3 Current–voltage (I–V) relationship for whole-cell K^+ current (I_K) activation in control solution (control 1), in the presence of 1.5% halothane (HAL) (A, $n = 7$) or 2.6% isoflurane (ISO) (B, $n = 8$), and during wash-out (control 2). Both anesthetics produced significant reduction of I_K which was completely reversible on washout. *Significant difference ($p \leq 0.05$) versus control 1 and control 2. [Reprinted with permission from Buljubasic *et al.* (9).]

isoflurane (Fig. 3B) on I_K amplitude. Halothane at a concentration of 1.5% reduced peak I_K amplitude obtained at +40 mV by 36.4 ± 3% ($p \leq 0.05$, n = 7). Isoflurane (2.6%) reduced I_K amplitude by 15 ± 3% ($p \leq 0.05$, n = 8). At approximately equianesthetic concentrations, I_K reduction by 1.5% halothane was significantly greater than that by 2.6% isoflurane at voltages more positive then 0 mV.

VI. Effects of Anesthetics on Single Potassium Channel Current

Channel conductance in the 37 cell-attached patches was 100 ± 3 pS, and the reversal potential was −60 ± 3 mV. At equianesthetic concentrations, isoflurane and halothane decreased the mean open time at resting membrane potential (pipette potential of 0 mV) from control values of 12 ± 1.4 msec (n = 8 cells) and 11 ± 1.5 msec (n = 10 cells) to 11 ± 1.7 msec (not significant) and 8.0 ± 1.4 msec ($p \leq 0.05$), respectively (data not shown). Isoflurane and halothane increased the mean closed time from a control value of 384 ± 98 to 531 ± 113 msec ($p \leq 0.05$) and from 302 ± 80 to 636 ± 56 msec ($p \leq 0.05$), respectively.

Potassium channel openings were recorded in control solution and compared to those recorded in solution which contained isoflurane or halo-

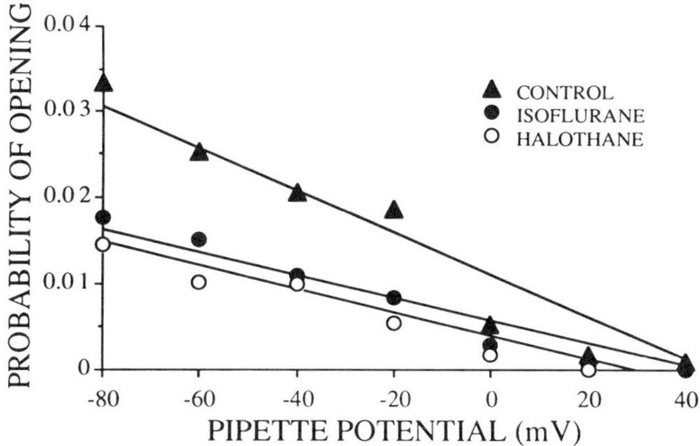

Fig. 4 Relationship between pipette potential and probability of K^+ channel opening in cell-attached patches under control conditions (n = 12 cells) and in the presence of isoflurane (n = 6 cells) or halothane (n = 6 cells). [Reprinted with permission from Buljubasic, N., Marijic, J., Kampine, J. P., and Bonsnjak, Z. J. (1994). The mechanism of isoflurane- and halothane-induced depression of single potassium channel current in isolated coronary smooth muscle cells. *Anesthesiology* in press.

thane. At the resting membrane potential (pipette potential of 0 mV) isoflurane ($n = 6$) and halothane ($n = 6$) decreased the probability of K^+ channel opening from 0.0053 ± 0.0001 to 0.0030 ± 0.0001 ($p \leq 0.05$) and 0.0018 ± 0.0001 ($p \leq 0.05$), respectively. Figure 4 illustrates probability of K^+ channel opening plotted as a function of pipette potential (20-mV increments) for cell-attached patches under control conditions and during

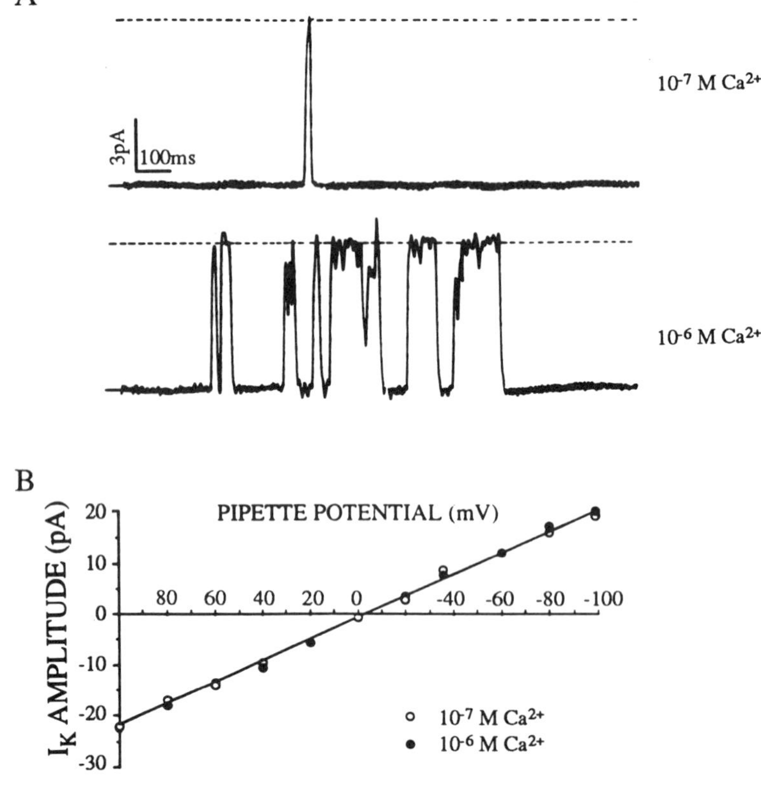

Fig. 5 Effect of internal Ca^{2+} concentration and voltage on K^+ channel activity in inside-out patches. (A) Actual single-channel records were obtained from the same patch held at a pipette potential of −30 mV. The effect of Ca^{2+} concentration on channel kinetics can be observed by comparing the tracings. The solid line at the beginning of the tracings represents a closed channel state, whereas the dashed line represents an open channel state. Upward deflection corresponds to an outward K^+ current. (B) Current–voltage relationships for the K^+ channel in the presence of 0.1 and 1 μM Ca^{2+} are summarized from nine inside-out patches. The slopes of both curves indicate a channel conductance of 230 pS. The pipette solution contained 145 mM KCl, and the bath solution contained 145 mM KCl with Ca^{2+} concentrations as indicated. [Reprinted with permission from Buljubasic, N., Marijic, J., Kampine, J. P., and Bosnjak, Z. J. (1994). Calcium-sensitive potassium current in isolated coronary smooth muscle cells. *Can. J. Physiol. Pharmacol.* **72,** 189–198.]

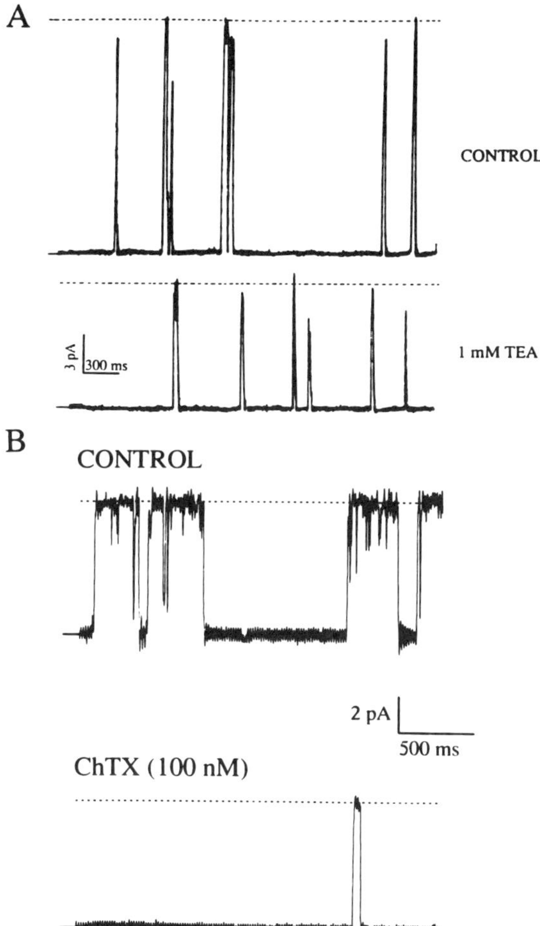

Fig. 6 (A) Effects of 1 mM tetraethylammonium (TEA) on K^+ channel activity. Recordings were obtained from one inside-out patch at a pipette potential of −60 mV. (B) Effects of 100 nM charybdotoxin (ChTX) on K^+ channel activity. An inside-out patch was held at a pipette potential of −60 mV. The solid line at the beginning of the tracings represents a closed channel state, whereas the dashed line represents an open channel state. The pipette solution contained 145 mM KCl, and the bath solution contained 145 mM KCl and 0.1 μM Ca^{2+}. Upward deflection corresponds to an outward K^+ current. [Reprinted with permission from Buljubasic, N., Marijic, J., Kampine, J. P., and Bosnjak, Z. J. (1994). Calcium-sensitive potassium current in isolated coronary smooth muscle cells. *Can. J. Physiol. Pharmacol.* **72,** 189–198.]

exposure to isoflurane or halothane. Because curves for probability of K^+ channel opening in control solution were the same for isoflurane and halothane groups, they were averaged and presented together. In all three cases, the probability of K^+ channel opening increased as patches were progressively depolarized (negative pipette potential), whereas the probability of K^+ channel opening decreased as patches were progresssively hyperpolarized (positive pipette potential). Both anesthetics significantly decreased the probability of K^+ channel opening, and this effect was voltage-independent.

In experiments employing the inside-out patch technique, a large-conductance K^+ channel recorded was sensitive to the Ca^{2+} concentration at the cytoplasmic side of the membrane and blocked by 1 m*M* TEA and 100 n*M* charybdotoxin. The probability of channel opening increased from 0.05 in the presence of 0.1 μM Ca^{2+} to 0.69 ($p \leq 0.05$) in the presence of 1 μM Ca^{2+} in the bath solution at a pipette potential of −30 mV. The mean open time increased from 7.1 ± 0.2 msec in the presence of 0.1 μM Ca^{2+} to 18 ± 1 msec ($p \leq 0.05$) in the presence of 1 μM Ca^{2+} in the bath solution at the same pipette potential (Fig. 5A). Inside-out patches (n = 9) were held at pipette potentials from −100 to +100 mV, resulting in a linear current–voltage curve with calculated slope conductances of 233 ± 4 and 234 ± 6 pS in the presence of 0.1 and 1 μM Ca^{2+}, respectively, and a reversal potential of 0 mV (Fig. 5B). This Ca^{2+}-sensitive, voltage-dependent, large-conductance K^+ channel was blocked by 1 m*M* TEA (Fig. 6A) and 100 n*M* charybdotoxin (ChTX, Fig. 6B) in the bath solution,

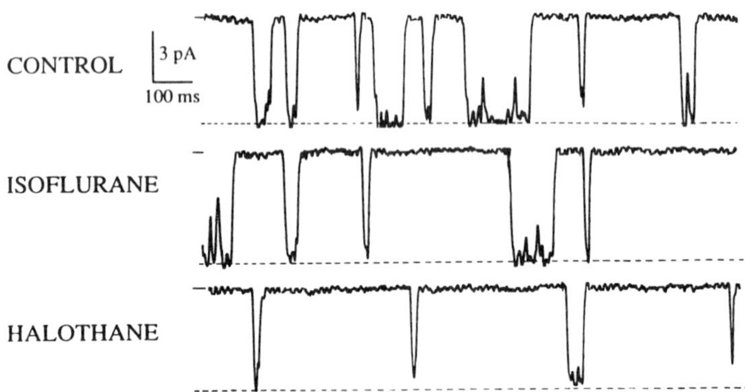

Fig. 7 Orignial tracings recorded at a pipette potential of +20 mV in one inside-out patch under control conditions and after exposure to 2.6% isoflurane or 1.5% halothane. The solid line at the beginning of the tracings represents a closed channel state, whereas the dashed line represents an open channel state. Downward deflection represents an inward K^+ channel current. The pipette solution contained 145 m*M* KCl, and the bath solution contained 145 m*M* KCl and 0.5 μM Ca^{2+}. [Reprinted with permission from Buljubasic, N., Marijic, J., Kampine, J. P., and Bosnjak, Z. J. (1994). The mechanism of isoflurane- and halothane-induced depression of single potassium channel current in isolated coronary smooth muscle cells. *Anesthesiology* in press.]

resulting in decreased open channel current amplitude and decreased frequency of K^+ channel opening, respectively.

A typical inside-out recording of K^+ channels in the presence of isoflurane and halothane at a pipette potential of +20 mV, in the presence of 0.5 μM Ca^{2+} and 145 mM KCl on the cytoplasmic side of the membrane and 145 mM KCl in the pipette solution, is shown in Fig. 7. Under the same experimental circumstances mentioned above, inside-out patches

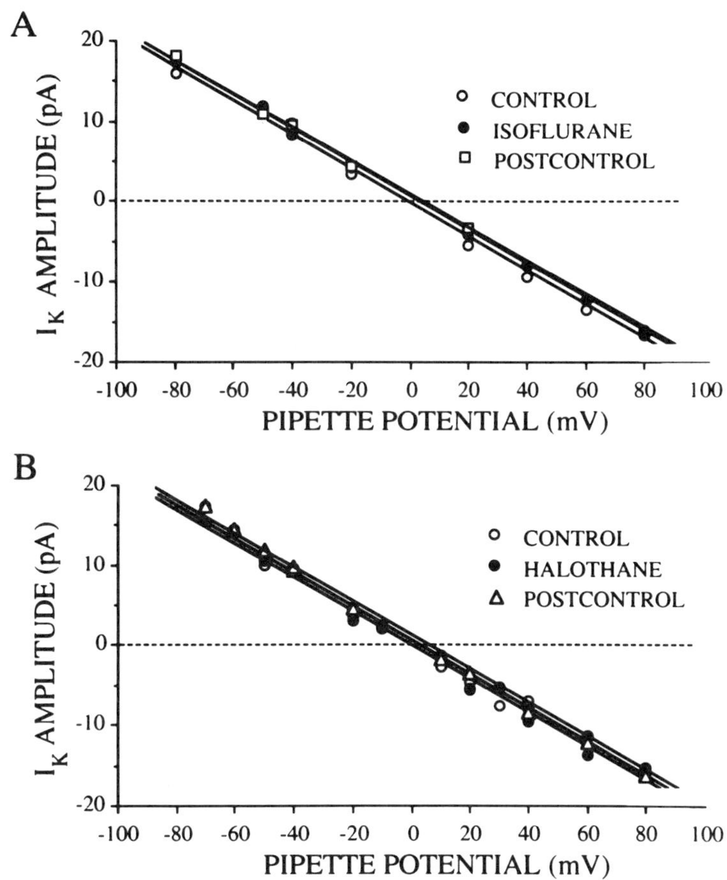

Fig. 8 Comparison of current–voltage relationships for single K^+ channel current (I_K) activation between control inside-out patches, patches exposed to isoflurane (A, $n = 10$) or halothane (B, $n = 10$), and patches after wash-out (postcontrol). The slopes of all curves indicate channel conductances of approximately 230 pS. The pipette solution contained 145 mM KCl, and the bath solution contained 145 mM KCl and 0.5 μM Ca^{2+}. [Reprinted with permission from Buljubasic, N., Marijic, J., Kampine, J. P., and Bosnjak, Z. J. (1994). The mechanism of isoflurane- and halothane-induced depression of single potassium channel current in isolated coronary smooth muscle cells. *Anesthesiology* in press.]

were held at pipette potentials from −80 to +80 mV in the presence of isoflurane ($n = 10$) or halothane ($n = 10$), resulting in linear current–voltage curves with a calculated slope conductances of 233 ± 6 and 235 ± 5 pS, respectively, and a reversal potential of 0 mV (Fig. 8). Control channel conductance was calculated to be 233 ± 7 pS; thus, isoflurane (Fig. 8A) or halothane (Fig. 8B) did not alter the channel conductance. In inside-out patches, the probability of K^+ channel opening decreased from 0.030 ± 0.002 in control solution to 0.015 ± 0.001 ($p \leq 0.05$) and 0.010 ± 0.002 ($p \leq 0.05$) in the presence of isoflurane ($n = 10$) and halothane ($n = 10$), respectively, at a pipette potential of +20 mV (data not shown). In the presence of isoflurane and halothane, the mean open time decreased from 7.0 ± 0.5 to 5.8 ± 0.3 msec (not significant) and from 6.5 ± 0.2 to 4.0 ± 0.1 msec ($p \leq 0.05$), and mean closed time increased from 290 ± 20 to 480 ± 50 msec ($p \leq 0.05$) and from 350 ± 28 to 700 ± 58 msec ($p \leq 0.05$), respectively, at a pipette potential of +20 mV (data not shown).

VII. Discussion

Isoflurane and halothane are both potent coronary vasodilators, although it has been shown that halothane relaxes isolated coronary arterial rings previously constricted with potassium to a greater degree than isoflurane (2). This could suggest that halothane is a more effective suppressor of voltage-dependent Ca^{2+} entry, because voltage-gated Ca^{2+} influx is the mechanism underlying K^+-induced contractions (16) The predominant physiological mechanism for vasodilation in arterial muscle has been demonstrated to be enhanced K^+ efflux, caused by an increased number of open K^+ channels (17). Several distinctive types of K^+ channels have been identified in smooth muscle cells (17). The major outward current measured under the present experimental conditions was a Ca^{2+}-sensitive K^+ current. Activation of membrane K^+ channels by increased $[Ca^{2+}]_i$ may act as a feedback mechanism contributing to the complex regulation of membrane potential in arterial cells and, in addition to Ca^{2+}-induced inhibition of Ca^{2+} influx, may prevent coronary vasospasm.

Tetraethylammonium (TEA) effectively blocks the Ca^{2+}-dependent K^+ channel in vascular smooth muscle, although it may also block other K^+ channel types at higher concentrations (10,11). Our vessel-ring data show that blockade of K^+ current by TEA increases tension of isolated canine coronary arteries; moreover, this increase was significantly augmented when coronary arteries were preconstricted with KCl. This could suggest that the TEA-sensitive K^+ current does not play a major role in regulation of the resting tension. Another possible explanation is that a very small number of open K^+ channels is required for maintenance of resting membrane potential as suggested by Nelson *et al* (17). By contrast, when arteries are exposed to KCl, the TEA-sensitive K^+ channels are activated and may play a more important role, partially antagonizing the effect of the agonists by hyperpolarizing the cell. Therefore, K^+ channel blockade may alter the response of coronary artery rings, resulting in attenuation of vasodilation. We found that both isoflurane and halothane are direct dilators of coronary arteries. The responses to isoflurane and halothane were not equal despite the use of comparable MAC values. Halothane was more effective than isoflurane in producing vessel relaxation. In addition, dramatic augmentation of dilator responses to volatile anesthetics followed a blockade of K^+ current by TEA.

To determine the mechanism of potentiation of the anesthetic-induced coronary dilation in TEA-pretreated coronary vessel rings, studies were performed while measuring the whole-cell K^+ channel current. TEA (1–30 m*M*) dose-dependently decreased the macroscopic K^+ current in isolated coronary arterial smooth muscle cells. Despite the fact that volatile anesthetics decrease the amplitude of both Ca^{2+} and K^+ currents in coronary (9) and cerebral (8) smooth muscle cells, the Ca^{2+} current was considerably more sensitive to anesthetic blockade than the K^+ current. The depression of Ca^{2+} current by halothane and isoflurane may be particularly important in producing coronary dilation, which relies predominantly on influx of external Ca^{2+} for maintenance of contraction (18). On the other hand, the simultaneous block of K^+ current by volatile anesthetics would partially antagonize this effect by causing cell membrane depolarization, as voltage-dependent K^+ current is crucial for membrane repolarization and control of arterial tone. At equianesthetic concentrations, halothane reduced the whole-cell K^+ current more effectively than isoflurane. The depresent effect of volatile anesthetics on the macroscopic K^+ current was voltage-independent, suggesting that halothane and isoflurane are likely to reduce ionic currents at the resting membrane potential in vascular muscle.

To determine the mechanism of anesthetic-induced whole-cell K^+ current depression, the single-channel, patch clamp method was used. This method provides more detail information about the mechanism by which volatile anesthetics alter channel activity at more physiological voltages where the net ionic flux might be too low to be precisely measured by whole-cell techniques. In cell-attached patches a voltage-sensitive, large-conductance K^+ channel detected. At equianesthetic concentrations, isoflurane and halothane decreased the activity of the large-conductance K^+ channel, with halothane being more effective than isoflurane. Decreased freqency of opening and a shorter duration of open time were observed, resulting in a decrease (45–65%) in the probability of K^+ channel opening by both anesthetics. Decreased mean open time, increased mean closed time, and decreased probability of K^+ channel opening explain, at least in part, the anesthetic-induced depression of the macroscopic K^+ current in coronary arterial cells.

The Ca^{2+} sensitivity of the large-conductance K^+ channel was demonstrated in an inside-out recording configuration by exposure of the cytoplasmic side of the membrane patch to increasing free Ca^{2+} concentration. Increase in cytoplasmic free Ca^{2+} resulted in an increase in the probability of K^+ channel opening and mean open time, suggesting calcium dependency. Increased $[Ca^{2+}]_i$ activates the K^+ channel current which produces cell membrane hyperpolarization, which in turn may act as a feedback mechanism leading to a reduction of Ca^{2+} entry, maintenance of the physiological function of the coronary vessel tone, and prevention of the coronary artery spasm. TEA (1 m*M*) and charybdotoxin (100 n*M*) applied at the cytoplasmic side of the membrane decreased the K^+ channel activity by decreasing the open channel current amplitude and probability of K^+ channel opening, respectively. In agreement with our results, it has been reported that application of TEA and charybdotoxin blocks current through Ca^{2+}-sensitive K^+ channels in smooth muscle cells (12,19,20). Strong voltage, TEA, charybdotoxin, and Ca^{2+} sensitivity are characteristics of large-conductance (100–250 pS) K^+ channels (12,19,21). Similar conductances (230 pS) in a 145:145 m*M* KCl system (i.e., 145m*M* KCl in both pipette and bath solutions) under control conditions and in the presence of isoflurane and halothane would suggest that although anesthetics alter channel gating they do not alter ionic flow through the channel once it is open. Because enhanced K^+ efflux induces membrane repolarization or hyperpolarization (17,22), decreased K^+ efflux by volatile anesthetics would favor membrane depolarization, an action that would subsequently lead to vessel constriction.

In summary, K^+ channel blockade by TEA potentiates the vasorelaxing effects of isoflurane and halothane, supporting the hypothesis that blockade of K^+ current enhances vasoconstriction by agonists, but also potentiates the vasodilating effect by anesthetics because of the unopposed effect of anesthetics on the Ca^{2+} current. A significant decrease of the probability of K^+ channel opening, a decrease of the mean open time, and an increase of the mean closed time by volatile anesthetics may be the result of stabilization of the K^+ channel either in the resting state and/or in inactivated state (23). These alterations in ionic membrane fluxes interact with the other *in vivo* actions of anesthetic agents in determining the coronary arterial tone and blood supply to the heart.

Acknowledgments

Research was supported in part by National Institutes of Health Grants HL 01901, HL 34708, and Anesthesiology Research Training Grant GM 08377.

References

1. Stowe, D. E. Marijic, J., Bosnjak, Z. J., and Kampine, J. P. (1991). Direct comparative effects of halothane, enflurane, and isoflurane on oxygen supply and demand in isolated hearts. *Anesthesiology* **74,** 1087–1095.
2. Bollen, B. A., Tinker, J. H., and Hermsmeyer, K. (1987). Halothane relaxes previously constricted isolated porcine coronary artery segments more than isoflurane. *Anesthesiology* **66,** 748–752.
3. Burt, J. M. and Spray, D. C. (1989). Volatile anesthetics block intercellular communication between neonatal rat myocardial cells. *Circ. Res.* **65,** 829–837.
4. Larach, D. R. Schuler, H. G., Skeehan, T. M., and Peterson, C. J. (1990). Direct effects of myocardial depressant drugs on coronary vascular tone: Anesthetic vasodilation by halothane and isoflurane. *J. Pharmacol. Exp. Ther.* **254,** 58–64.
5. Flynn, N. M., Buljubasic, N., Bosnjak, Z. J., and Kampine, J. P. (1992). Isoflurane produces endothelium-independent relaxation in canine middle cerebral arteries. *Anesthesiology* **76,** 461–467.
6. Okabe, K., Kitimura, K., and Kuriyama, H. (1987). Features of 4-aminopyridine sensitive outward current observed in single smooth cells from the rabbit pulmonary artery. *Pfluegers Arch.* **409,** 561–568.
7. Harder, D. R., Graddall, K., Madden, J. A., and Kampine, J. P. (1985). Cellular actions of halothane on cat cerebral arterial muscle. *Stroke* **16,** 680–683.
8. Buljubasic, N., Flynn, N. M. Marijic, J., Rusch, N. J., Kampine, J. P., and Bosnjak, Z. J. (1992). Effects of isoflurane on K^+ and Ca^{2+} conductance in isolated smooth muscle cells of canine cerebral arteries. *Anesth. Analg. (N.Y.)* **75,** 590–596.
9. Buljubasic, N., Rusch, N. J., Marijic, J, Kampine, J. P. and Bosnjak, Z. J. (1992). Effects of halothane and isoflurane on calcium and potassium channel currents in canine coronary arterial cells. *Anesthesiology* **76,** 990–998.
10. Stanfield, P. R. (1983). Tetraethylammonium ions and the potassium permeability of excitable cells. *Rev. Physiol. Biochem. Pharmacol.* **97,** 1–67.
11. Langton, P. D., Nelson, M. T., Huang, Y., and Standen, N. B. (1991). Block of calcium-activated potassium channels in mammalian arterial myocytes by tetraethylammonium ions. *Am. J. Physiol.* **260,** H927–H934.
12. Suggs, E. E., Garcia, M. L., Reuben, J. P., Patchett, A. A., and Kaczorowski, G. J. (1990). Synthesis and structural characterization of charybdotoxin, a potent peptidyl inhibitor of the high conductance Ca^{2+}-activated K^+ channel. *J. Biol. Chem.* **265,** 18745–18748.
13. Marijic, J., Buljubasic, N., Coughlan, M. G., Kampine, J. P., and Bosnjak, Z. J. (1992). Effect of K^+ channel blockade with tetraethylammonium on anesthetic-induced relaxation in canine cerebral and coronary arteries. *Anesthesiology* **77,** 948–955.
14. Hamill, O. P., Marty, A., Neher, E., Sakmann, B., and Sigworth, F. J. (1981). Improved patch-clamp techniques for high-resolution current recording from cells and cell-free membrane patches. *Pfluegers Arch.* **391,** 85–100.

15. Tallarida, R. J., and Murray, R. B. (1981). "Manual of Pharmacologic Calculations with Computer Programs." Springer-Verlag, New York.
16. Haeusler, G. (1983). Contraction, membrane potential, and calcium fluxes in rabbit pulmonary artery muscle. *Fed. Proc.* **42,** 263–268.
17. Nelson, M. T., Patlak, J. B., Worley, J. F., and Standen, N. B. (1990). Calcium channels, potassium channels and voltage dependence of arterial smooth muscle tone. *Am. J. Physiol.* **259,** C3–C18.
18. Mullett, M., Gharaibeh, M., Warltier, D. C., and Gross, G. J. (1983). The effect of diltiazem, a calcium channel blocking agent, on vasoconstrictor responses to norepinephrine, serotonin and potassium depolarization in canine coronary and femoral arteries. *Gen. Pharmacol.* **14,** 259–264.
19. McCann, J. D., and Welsh, M. J. (1986). Calcium-activated potassium channels in canine airway smooth muscle. *J. Physiol. (London)* **372,** 113–127.
20. Brayden, J. E., and Nelson, M. T. (1992). Regulation of arterial tone by activation of calcium-dependent potassium channels. *Science* **256,** 532–535.
21. Pallotta, B. S., Hepler, J. R., Oglesby, S. A., and Harden, T. K. (1987). A comparison of calcium-activated potassium channel currents in cell-attached and excised patches. *J. Gen. Physiol.* **89,** 985–997.
22. Desilets, M., Driska, S. P., and Baumgarten, C. M. (1989). Current fluctuations and oscillations in smooth muscle cells from hog carotid artery. *Circ. Res.* **65,** 708–722.
23. Miller, V. M. (1991). Interactions between neural and endothelial mechanisms in control of vascular tone. *News Physiol. Sci.* **6,** 60–63.
24. Buljubasic, N., Marijic, J., Kampine, J. P., and Bosnjak, Z. J. (1994). The mechanism of isoflurane- and halothane-induced depression of single potassium channel current in isolated coronary smooth muscle cells. *Anesthesiology,* in press.
25. Buljubasic, N., Marijic, J., Kampine, J. P., and Bosnjak, Z. J. (1994). Calcium-sensitive potassium current in isolated canine coronary smooth muscle cells. *Can. J. Physiol. Pharmacol.* **72,** 189–198.

Potassium Channel Opening and Coronary Vasodilation by Halothane

D. R. Larach, H. G. Schuler, K. A. Zangari, and R. L. McCann

Department of Anesthesia,
College of Medicine
The Pennsylvania State University,
Hershey, Pennsylvania 17033

I. Introduction

Volatile anesthetics such as halothane and isoflurane dilate the coronary circulation in part by directly relaxing arterial vascular smooth muscle (VSM) (1–4). The ion-conducting state of membrane K^+ channels constitutes an important contractile regulatory mechanism in VSM; outward K^+ currents hyperpolarize the sarcolemma, reducing Ca^{2+} influx through voltage-sensitive Ca^{2+} channels. The resulting decrease in intracellular $[Ca^{2+}]$ relaxes VSM (5). One K^+ channel subtype in VSM is the ATP-sensitive K^+ channel (K_{ATP} channel); normally it is closed by millimolar intracellular ATP concentrations. Tissue hypoxia, hormones (e.g., endothelium-derived hyperpolarizing factor, vasoactive intestinal peptide), and drugs (e.g., diazoxide, cromakalim) vasodilate primarily by opening K_{ATP} channels (6).

Because halothane hyperpolarizes intact noncerebral VSM cells (7), and volatile anesthetics hyperpolarize neurons by opening membrane K^+ channels (8,9), we postulated that halothane may dilate coronary vessels by opening VSM K^+ channels. We reasoned that if K_{ATP} channel opening is an important mechanism of halothane vasodilation, then channel blockade ought to decrease the effect of halothane. Previously, we reported that K_{ATP} channel blockade with glyburide, but not K_{Ca} channel blockade with tetraethylammonium ion (TEA^+), attenuates halothane vasodilation in rat coronary resistance vessels (10). In contrast, halothane depolarizes cerebral arterial VSM cells (11), and halothane depression of myocardial contractility is not affected by K^+ channel blockade with 4-aminopyridine (4-AP), TEA^+, or glyburide (12). More recently, electrophysiological studies have not revealed evidence of K^+ channel opening by halothane. Thus, Buljubasic *et al.* have shown that halothane reduces whole-cell K^+ currents in isolated coronary VSM cells (13), and Marijic *et al.* reported that K_{Ca} channel blockade potentiates volatile anesthetic vasodilation (14). In addition, in cerebral VSM Eskinder *et al.* report that halothane decreases the activity of a 4-AP-sensitive K^+-channel (15). Thus, there are conflicting data regarding the role of K^+ channels in mediating anesthetic vasodilation.

To establish whether halothane relaxes intact epicardial coronary arteries by opening VSM K^+ channels, we performed studies of halothane vasodilation in the presence and absence of glyburide or 5-hydroxydecanoate (5-HD), pharmacological blockers of K_{ATP} channels, (16,17). In particular, we questioned whether the differences we observed previously between rat and porcine coronary tissues were due to differences in the source of the baseline tone in the blood vessel prior to inducing anesthetic vasodilation. In the arrested perfused heart with intact endothelium, one source of the high coronary vascular resistance may be the endothelial-derived vasoconstrictor peptide endothelin. Because endothelin is associated with the closure of K_{ATP} channels, we examined whether precontracting porcine conducting coronary artery segments *in vitro* with endothelin and other agonists would unmask a K_{ATP} channel-mediated mechanism of halothane vasodilation in this tissue, similar to that observed in the rat coronary resistance vessels *in situ*. Also, we examined the effects of various transmembrane K^+ gradients on halothane responses to further elucidate the importance of K^+ channel opening in anesthetic vasodilation.

This article presents results showing that halothane does vasodilate porcine conducting coronary arteries by opening K_{ATP} channels when tone is provided by endothelin but not methacholine (MCh), demonstrating that halothane vasodilation of coronary VSM by opening K_{ATP} channels is specific for certain contractile stimuli.

II. Isolated Coronary Vessel Preparation

Coronary ring segments were dissected from fresh porcine hearts (87 rings from $n = 11$ hearts), obtained from a slaughterhouse, the endothelium was denuded, and 8 rings from each heart were mounted to monitor isometric force in adjacent tissue baths containing buffer. Rings were conditioned and individually stretched to their optimal length for force development (L_0) as described by us previously (10). The developed force induced by isotonic 50 mM K^+ at L_0 following conditioning, and in the absence of inhibitors or anesthetics, was termed the "initial K^+ force"; this force for each ring (137 ± 65 mN, average ± SD) was used as the reference value (100%) to normalize the subsequent contractile responses of that ring. (Note: A 1-g mass exerts a force of 9.8 mN at sea level.) The rings weighed 19.5 ± 10 mg, and the baseline passive force measured when rings were stretched to L_0 was 22.5 ± 14 mN.

Each of the 8 rings was randomly assigned to a different experimental group (Fig. 1) based on the presence or absence of the following three reagents: indomethacin (10 μM) to block prostanoid generation, glyburide (100 nM) to block K_{ATP} channels, and halothane (0.0175 atm, equivalent to 2.5 MAC, where MAC is the median effective concentration for general anesthesia in the pig). Thus, each heart provided all necessary experimental and control rings for each condition. Glyburide, indomethacin, and anesthetic were allowed to equilibrate with the tissue for 45, 30, and 15 min, respectively before agonist stimulation began. The concentration of halothane in the gas mixture ventilating the baths was monitored by

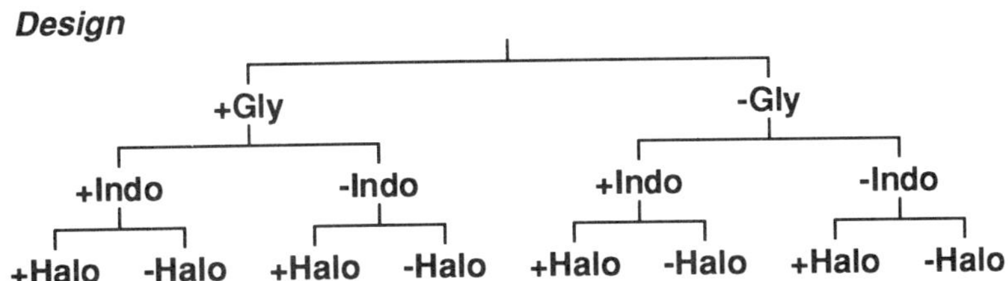

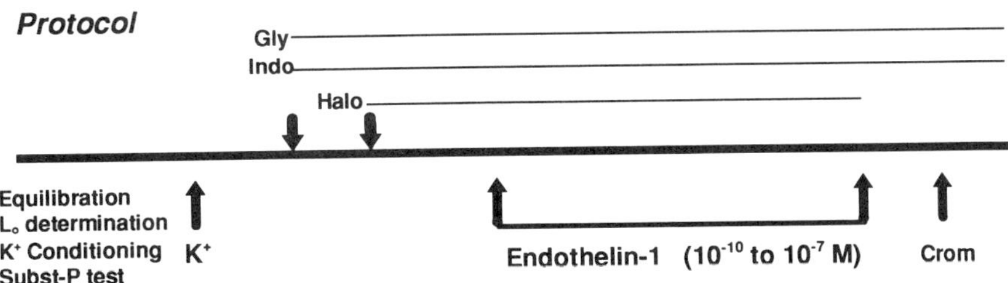

Fig. 1 Experimental design and protocol for the halothane vasodilation experiments in endothelin-contracted epicardial rings of porcine coronary artery. Gly, glyburide; Indo, indomethacin; Halo, halothane; Subst-P, sustance P; Crom, cromakalim.

Raman or mass spectrometry during each experiment; prior experiments show excellent correlation between gas-phase halothane concentrations and buffer-phase values by gas chromatography.

Figure 1 also shows the experimental protocol. After blocker and/or halothane pretreatment, cumulative endothelin-1 (ET) concentration–response curves (0.1 to 100 nM) were generated for each bath. Finally, during a stable contraction at the highest [ET] after halothane washout, the K^+ channel opener cromakalim (1 μM) was used to test for glyburide efficacy.

Endothelial removal was confirmed using substance P (1 μM) or ATP (10 μM) after vessel conditioning during K^+ contraction: the average response was a small contraction (1.0 ± 0.3% change). The irreversible contraction induced by ET prevented us from comparing the force developed by high-K^+ exposure between the beginning and end of an experiment, which we usually use as an index of tissue viability. However, the absolute force generated by high K^+ in the presence of 100 nM ET at the end of an experiment was always at least 111% of that achieved by high K^+ alone at the start of an experiment (mean 138 ± 3% increase); this indicates no gross deterioration of contractile function during these studies.

Thirty-two porcine coronary rings (from n = 4 hearts) were denuded of endothelium and prepared as described above, except three conditioning exposures to 3 μM methacholine (MCh) were required for development of a stable MCh contraction. Figure 2 shows the experimental protocol. After a baseline MCh contraction, two rings from each heart received glyburide (100 nM), two received 5-HD (100 μM), two received both blockers, and two control rings received neither blocker. One of each pair of rings received halothane (1.0, 1.5, and 2.5 MAC, in sequence) for 10 min before being reexposed to MCh, the other serving as a simultaneous control. These rings showed minimal relaxation or contraction to 1 μM substance P as evidence of successful endothelium removal; the passive force at L_0 was 27 ± 4 mN (mean ± SEM), and the initial K^+ force was 124 ± 17 mN. An additional series of 24 rings (from n = 3 hearts) were studied with the identical protocol, except the endothelium was preserved in these rings (1 μM substance P produced an average 23 ± 1% relaxation); the passive force at L_0 was 28 ± 3 mN, and the initial K^+ force was 113 ± 27 mN.

Porcine coronary artery rings from four hearts were prepared as described above, and each ring was exposed sequentially to isotonic buffer containing 10, 20, 30, 50, and 80 mM K^+. Four rings from each heart received halothane, glyburide, and indomethacin prior to the K^+ exposure, using an experimental design and drug concentrations analogous to that of Fig. 1. These rings (weight 21 ± 6 mg, mean ± SD) had a passive force

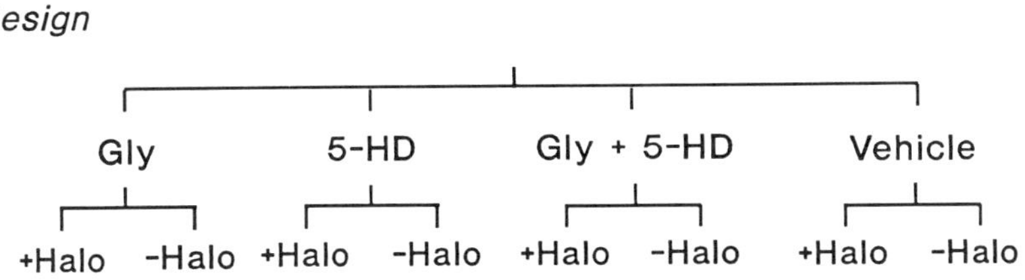

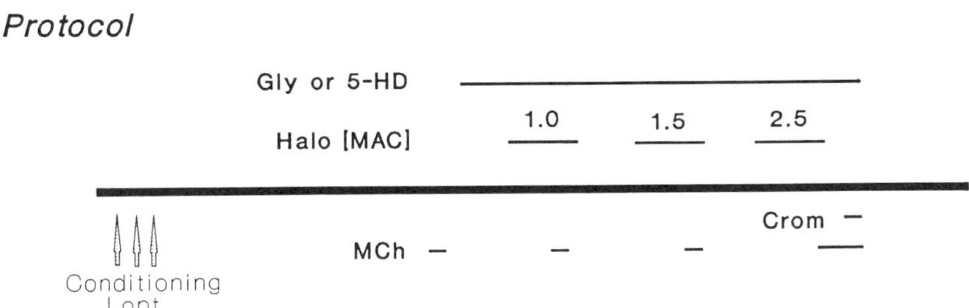

Fig. 2 Experimental design and protocol for the halothane vasodilation experiments with porcine coronary artery rings stimulated by methacholine (MCh).

of 36.0 ± 14.4 mN at L_0, and the initial K^+ force averaged 127 ± 51 mN. The endothelium was denuded, as confirmed by the lack of vasorelaxation in response to substance P.

To facilitate the analysis of the effect of glyburide on the magnitude of halothane-induced relaxation, we calculated the absolute difference in developed tension between the halothane and no-halothane rings (normalized as a percentage of the initial K^+ force) for each experimental condition ("halothane relaxation"). To determine overall effects, data were analyzed by analysis of variance using the area under the force curve for each ring as the response variable. A two-way analysis of variance, random effects model was used for the ET concentration analyses. The MCh and KCl experiments were analyzed using repeated-measures analysis of variance with appropriate correction for multiple inference. Statistical significance was defined at the $\alpha = 0.05$ level. Values are presented as means ± SEM except where stated otherwise.

The composition of modified Krebs–Henseleit buffer was as follows (in m*M*): NaCl, 118; KCl, 4.7, EDTA, 0.5; KH_2PO_4, 1.2; $MgSO_4$, 1.2; $CaCl_2$, 3.0; $NaHCO_3$, 25.0; and glucose, 10.0. The buffer was adjusted to pH 7.4 and bubbled with a mixture of O_2 and CO_2 (95:5, v/v). High-K^+ buffer was isotonic: the $[Na^+]$ was decreased by an equimolar amount to the elevation in $[K^+]$; thus, when KCl was increased to 50 m*M*, NaCl was decreased to 72.7 mM. Halothane (thymol-free) was a gift from Halocarbon Laboratories (Hackensack, NJ), and cromakalim was a gift of Beecham Pharmaceuticals (Surrey, UK). Endothelin-1 was purchased from Sigma (St. Louis, MO), and 5-HD from Research Biochemicals Inc. (Natick, MA). Unless specified, all chemicals were supplied by Fisher, Sigma, or Baker, at the highest available purity. Glyburide was dissolved in 0.1 N NaOH and diluted 1:1000 in buffer before use; all other drugs were dissolved in water and diluted in buffer before use.

III. Effects of Endothelin

Figure 3 shows the ET concentration versus developed force responses in coronary rings without (Fig. 3A) and with indomethacin pretreatment (Fig. 3B). The highest [ET] studied was on the steep portion of the concentration–response curve. Pretreatment with glyburide alone did not signifi-

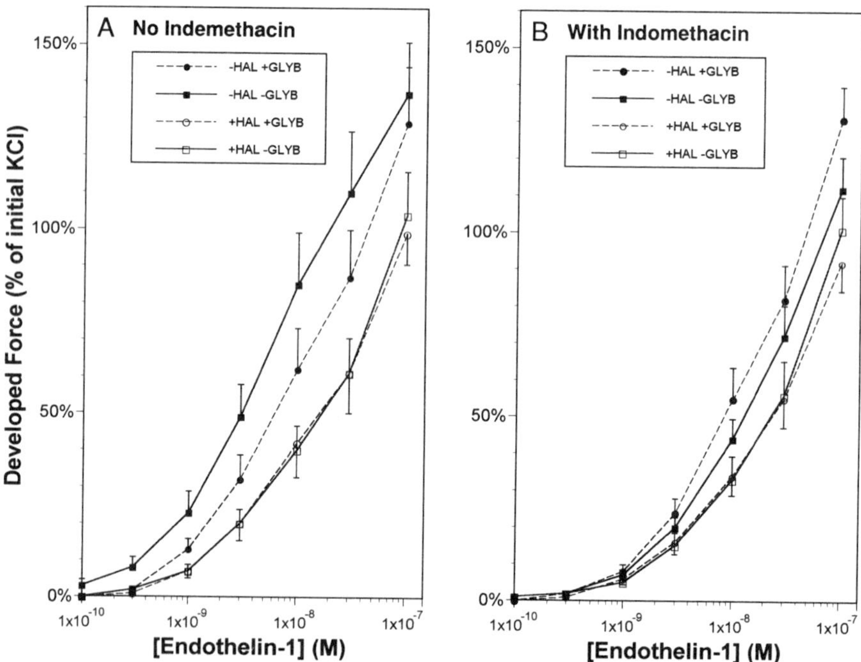

Fig. 3 Endothelin-1 concentration versus developed force responses in epicardial coronary rings (A) without and (B) with indomethacin pretreatment. The effects of halothane treatment (open symbols) and K_{ATP} channel blockade with glyburide (dashed lines) are shown. Data are normalized to the initial K^+ force and are dervied from n = 8 hearts. Symbols indicate means ± SEM.

cantly affect ET contraction, either in the absence or presence of indomethacin. Halothane alone significantly attenuated ET contraction only in rings without indomethacin (see below).

To facilitate the analysis of the effect of glyburide on the magnitude of halothane-induced relaxation, we calculated the absolute difference in developed tension between the halothane and no-halothane rings (normalized as a percentage of the initial K^+ contraction) for each experimental condition. This difference, called halothane relaxation, is plotted in Figs. 4 and 5 as a function of [ET].

In the absence of indomethacin (Fig. 4), the overall effect of glyburide pretreatment was to attenuate halothane relaxation significantly ($p = 0.038$ compared with zero) over the ET range of 0.1 to 30 n*M*. A sigmoid relationship between [ET] and halothane relaxation existed in vehicle-treated (without glyburide) rings, except for 100 n*M;* unless stated otherwise, analyses included only the sigmoid portion of the responses. Inclusion of the 100 n*M* ET data resulted in loss of statistical significance. Analysis of variance revealed that animal, glyburide pretreatment, and ET concentration each significantly affected the halothane relaxation response ($p < 0.001$), but glyburide and [ET] showed no interaction (i.e., the vehicle and glyburide curves were significantly offset vertically but had similar

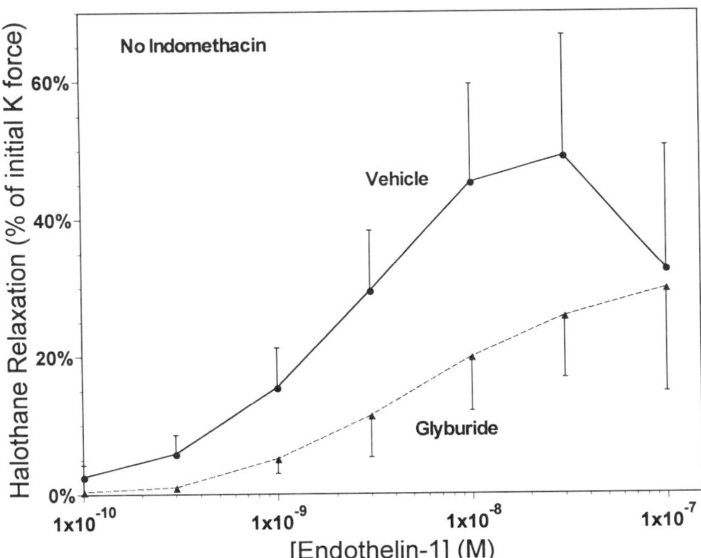

Fig. 4 Data from Fig. 3A replotted as the halothane relaxation response (see text) normalized to the initial K^+ force. Glyburide significantly attenuates the halothane relaxation response ($p = 0.038$). At the [ET] associated with peak halothane relaxation, glyburide pretreatment decreased halothane relaxation by 47% (percent change).

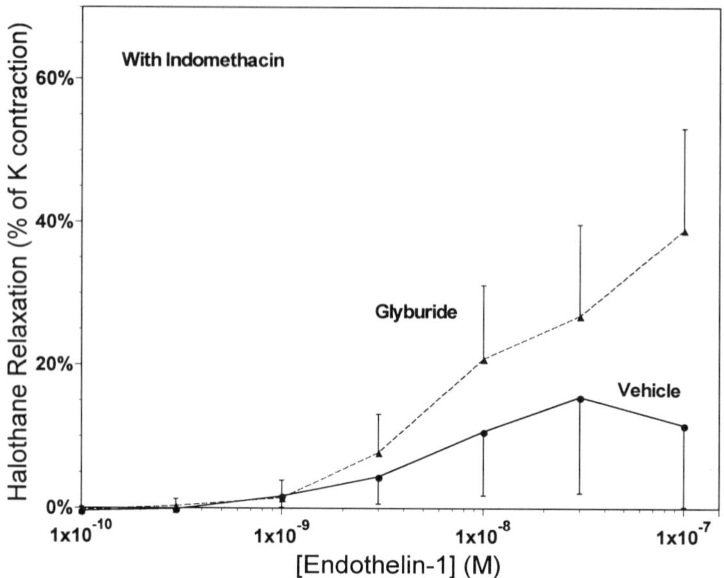

Fig. 5 Data from Fig. 3B replotted as the halothane relaxation response (see Fig. 4 legend). In the presence of indomethacin, halothane had no significant vasodilator action in either the presence or absence of glyburide.

slopes). Thus, glyburide attenuated halothane relaxation by 20.8 ± 4.6% (from 24.6 ± 4.6% with vehicle to 3.8 ± 0.08% with glyburide) of the initial K^+ contraction averaged over all ET concentrations from 0.1 to 30 n*M*. At the [ET] associated with peak halothane relaxation, glyburide pretreatment decreased halothane relaxation by 47% (percent change).

Indomethacin pretreatment (Fig. 5) abolished the halothane relaxation response, both in the absence and presence of glyburide. Cromakalim caused a 5.3 ± 1.1% relaxation in vehicle-treated rings, but glyburide pretreatment abolished cromakalim relaxation (0.2 ± 0.6% contraction). These data indicate that glyburide treatment was effective in blocking the K_{ATP} channels in VSM.

IV. Effects of Methacholine

A. Endothelium-Denuded Rings

In the absence of halothane, following MCh preconditioning, repeated exposures to 3 μM MCh consistently contracted the rings by 39 ± 3% of the initial K^+ response. In control rings without halothane, the magnitude of repeated MCh contractions were unaffected by time or treatment with glyburide. Halothane administration caused a concentration-dependent attenuation of MCh vasoconstriction ($p < 0.005$, Fig. 6). The highest halothane dose decreased the MCh response by 44% (percent change). However, the halothane vasodilation effect was not affected by treatment with either glyburide or 5-HD (Fig. 6). The MCh-induced force was significantly greater with, than without, 5-HD treatment in rings receiving halothane, indicating a 5-HD potentiating effect on MCh vasoconstriction ($p < 0.01$). Cromakalim at 1 μM caused a 43 ± 4% relaxation of the rings, and this response was significantly attenuated by glyburide treatment to 18 ± 4% ($p < 0.005$); however, 5-HD did not affect cromakalim vasodilation. When considering cromakalim vasodilation, there was no interaction between the effect of glyburide and either 5-HD or prior halothane treatment.

B. Endothelium-Intact Rings

Methacholine without halothane contracted the rings by 51 ± 4% of the initial K^+ response in a stable manner that was unaffected by glyburide or 5-HD (mean of all four blocker treatment groups). Halothane caused

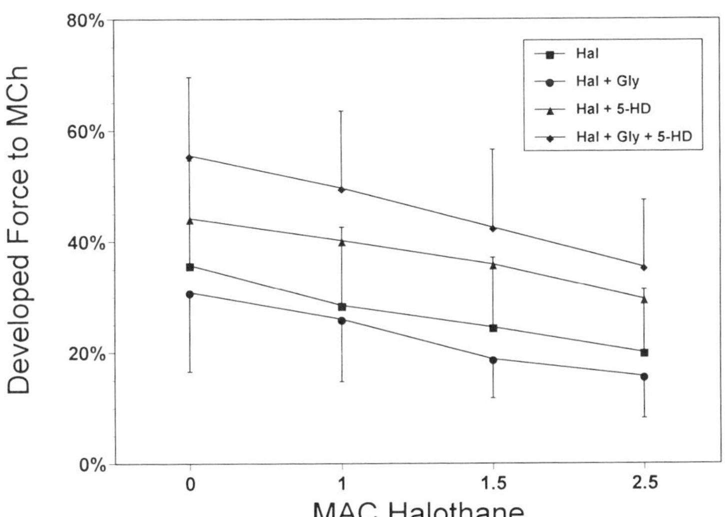

Fig. 6 Halothane concentration–response data in coronary rings repeatedly contracted with methacholine (MCh). The ordinate shows developed force to MCh as a percentage of the initial K^+ force. MCh responses without halothane averaged 39 ± 3% of the initial K^+ force, and halothane (2.5 MAC) attenuated the MCh contraction by 44%. Neither glyburide nor 5-HD pretreatment had any significant effect on halothane attenuation of MCh-induced force or on the slope of [halothane]–force relation. Data are from $n = 4$ hearts.

a significant dose-dependent attenuation of MCh-induced force development (developed force was 55 ± 9, 49 ± 8, 40 ± 7, and 36 ± 6% of the initial K^+ response at 0, 1.0, 1.5, and 2.5 MAC halothane, respectively; $p < 0.005$, data not shown). Similar to the case in endothelium-denuded rings, halothane vasodilation was not significantly altered by the presence of glyburide or 5-HD alone or in combination.

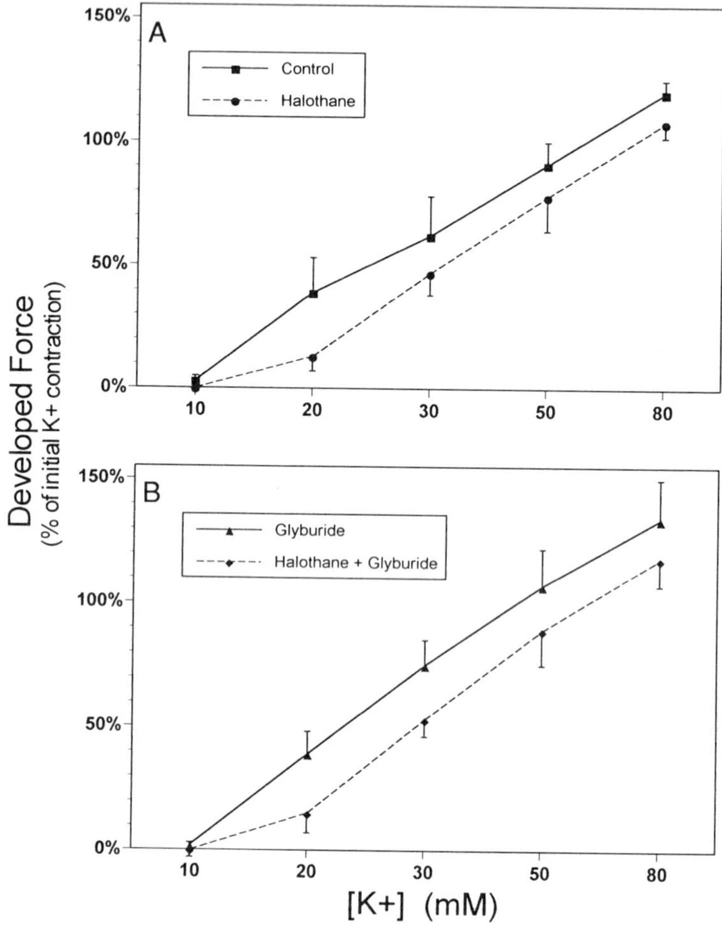

Fig. 7 Potassium vasoconstriction concentration–response curves in epicardial coronary arteries unstimulated by any receptor agonists. Halothane significantly attenuates the K^+-induced force (p = 0.013); however, the slopes of the curves are equivalent in both the presence and absence of halothane (A), which is not consistent with a K^+ channel opening action by the anesthetic. The action of halothane is unaffected by glyburide treatment (B). Data are from n = 4 hearts.

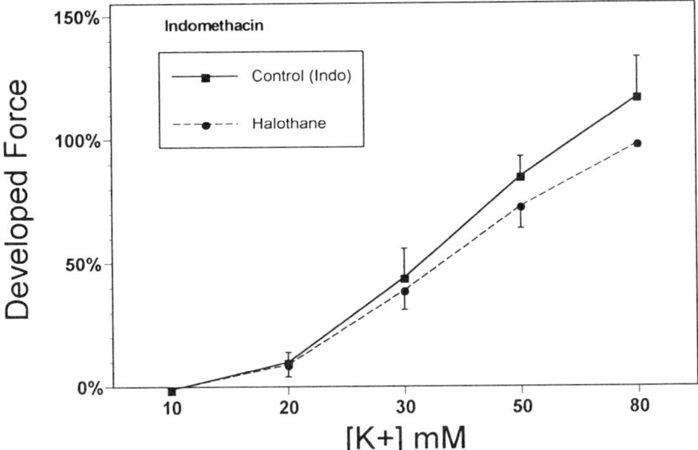

Fig. 8 Potassium concentration–force responses in the presence of indomethacin (see Fig. 7). Halothane relaxation was abolished by the cyclooxygenase blockade. Data are from $n = 4$ hearts.

V. Effects of Potassium

Figure 7 shows that graded K^+ depolarization caused a concentration-dependent vasoconstriction ($p = 0.001$). Halothane-treated rings developed significantly less force than control rings ($p = 0.013$), but the slope of the $[K^+]$–force curve was not affected by halothane (interaction $p > 0.8$). Halothane produced a similar relaxation response in glyburide-treated rings without changing the slope (halothane main effect $p = 0.01$; interaction $p > 0.7$).

Indomethacin treatment abolished the halothane vasodilator response ($p = 0.14$) during K^+-induced contraction (Fig. 8). Likewise, the $[K^+]$–force relationship was not altered by glyburide and/or halothane in the presence of indomethacin ($p > 0.7$, data not shown).

VI. Discussion

In porcine coronary conducting arteries contracted with endothelin-1, we have shown that K_{ATP} channel blockade with glyburide caused up to a 47% reduction in halothane vasorelaxation. This finding confirms and extends our observation of a large glyburide-sensitive component of halothane vasodilation in rat coronary resistance vessels (10). Thus, the data from both rat resistance and ET-contracted porcine conducting coronary preparations are consistent with an effect of halothane to open K_{ATP} channels as a part of its mechanism of action.

This article describes the first demonstration of K_{ATP} channel involvement in anesthetic vasodilation within a macroscopic blood vessel and may be important for understanding volatile anesthetic actions at sites of coronary spasm and eccentric coronary atherosclerotic lesions subject to dynamic stenosis. Furthermore, the availability of a large-vessel model of anesthetic K_{ATP} channel activation will facilitate further studies into the cellular mechanisms of anesthetic action.

Halothane is known to be a less effective vasodilator during high-K^+ coronary contraction than with receptor agonist-induced precontraction (1), and previously we reported that resistance vessel coronary VSM depolarization with high-K^+ buffer markedly attenuates halothane vasodilation in the rat heart (10). These findings are consistent with halothane possessing a K^+ channel opening action, because high external K^+ lowers the transsarcolemmal K^+ gradient, preventing channel-mediated hyperpolarization. Alternatively, halothane could directly decrease Ca^{2+} influx through Ca^{2+} channels. In the current study, we extended our observations by testing the vasodilator action of halothane on the contractions induced by varying concentrations of external K^+ in epicardial arteries. If the vasodilator mechanism of halothane involves K^+ channel opening, then one would expect the magnitude of vasodilation to decrease as the exernal $[K^+]$ increases; that is, the two lines in Fig. 7A would converge toward the right. If the K_{ATP} channel were the responsible K^+ channel, then one would expect pretreatment with glyburide to reverse the halothane vasodilation, particularly at low external $[K^+]$, causing the lines of Fig. 7B to converge at the left-hand side.

Surprisingly, we observed neither of these responses to halothane or glyburide (Fig. 7), suggesting that halothane vasodilation of K^+-contracted epicardial coronary VSM, in the absence of agonists or other stimuli, does not involve processes sensitive to the transsarcolemmal K^+ gradient or membrane potential. This finding is not consistent with halothane possessing either a K^+ channel- or a voltage-sensitive Ca^{2+} channel-dependent mechanism, and it implies that the vasodilator effect of halothane under these conditions may be mediated by other intracellular processes, such as the inositol trisphosphate (IP_3)-regulated sarcoplasmic reticulum (SR) Ca^{2+} release channel or non-Ca^{2+}-dependent processes that regulate myosin light chain phosphorylation and contraction, such as protein kinase C. It would be interesting to see if the coadministration of a K^+ channel-closing agonist such as ET would affect the halothane versus $[K^+]$ response curves and cause the epicardial coronary data to resemble the coronary resistance vessel responses to K^+, in which halothane vasodilation is attenuated at high external K^+.

The glyburide-sensitive nature of halothane vasodilation in the ET-contracted arteries seen in this study stands in marked contrast to our previous results with prostaglandin $F_{2\alpha}$ ($PGF_{2\alpha}$)-contracted coronary rings, in which neither glyburide nor TEA^+ significantly inhibited halothane vasodilation. Similar to our prior report, the current study suggests a lack of K_{ATP} channel involvement in the methacholine (MCh)-contracted arteries and in the K^+ depolarized vessels receiving no agonist. Thus, the K_{ATP} channel-opening action of halothane appears to depend on the nature of the underlying vascular tone. Despite the sensitivity of K_{ATP} channels to closure by normal $[ATP]_i$, there is evidence that K_{ATP} channels in certain VSM preparations may have a relatively high open-state probability (18,19). We hypothesize that the K_{ATP} channel involvement in our perfused arrested rat heart studies was explained by the presence of autologous tonic vasoconstrictor substances that act by closing VSM K_{ATP} channels. Indeed, Miyoshi *et al.* have reported that ET possesses a powerful K_{ATP} channel-closing action (18). It may be tempting to speculate that the presence of a large population of open-state K_{ATP} channels may explain the lack of halothane effect on K_{ATP} channels when epicardial VSM is contracted with K^+, $PGF_{2\alpha}$, or MCh, because the K_{ATP} channel-opening effect of halothane would be masked when these channels are already largely open. However, such a speculation is not supported by the lack of a vasoconstrictor effect of glyburide alone (in concentrations documented to block cromakalim-induced K_{ATP} channel opening) in each of the preparations we have studied. Obviously, patch clamp techniques need to be applied to further extend our pharmacological studies to the channel-population and single-channel level, as well as to understand the nature of the interactions among halothane, glyburide, and ET compared with other agonists.

Of interest is the observation that cyclooxygenase blockade with indomethacin abolishes halothane vasodilation, both in the presence and absence of glyburide. This suggests that generation of a prostanoid intermediate may be a necessary condition for halothane relaxation of ET-contracted coronary VSM. However, studies with indomethacin in rat aorta (20) and rat myocardium (12) do not show cyclooxygenase dependence of halothane action, unlike our finding in coronary arteries. It is tempting to speculate that halothane induces the synthesis of a cyclooxygenase-dependent vasodilating prostanoid such as prostacyclin, which in turn opens K_{ATP} channels in coronary vessels (21). Formal testing of this hypothesis will be required.

Other notable findings from this study include the observation that

100 μM 5-HD, which is reported to completely inhibit K_{ATP} channels of cardiomyocytes (17), did not attenuate cromakalim vasodilation. This suggests that 5-HD is not an effective blocker of K_{ATP} channels in VSM preparations. In addition, 5-HD, but not glyburide, potentiated MCh vasoconstriction, suggesting that K_{ATP} channels are not fully closed during MCh exposure.

In conclusion, data from ET-contracted porcine epicardial coronary arteries and rat resistance coronary vessels suggest that approximately 50% of the halothane vasodilation response is caused by opening of glyburide-sensitive K_{ATP} channels by the anesthetic. In support of this mechanism, we have also reported in rat coronary resistance vessels that high-K^+ depolarization attenuates halothane vasodilation and that halothane treatment nearly obliterates cromakalim vasodilation, which is consistent with halothane and cromakalim both opening the same (K_{ATP}) channel (10). The fact that such a K_{ATP} channel-opening mechanism cannot be detected in arteries precontracted with other agonists, or in K^+-contracted vessels without agonist stimulation, implies that other factors related to the specific cellular mechanisms activated by the various agonists are important in regulating or unmasking the direct anesthetic action. Future studies should address the nature of these agonist-dependent factors. Patch clamp studies of isolated K_{ATP} channels and molecular techniques for identifying K^+ channel subtypes will be useful for defining these cellular actions of halothane. Further research also is needed to define the role of vasodilating prostanoid intermediates in halothane coronary relaxation.

Acknowledgments

The authors thank the Department of Anesthesia, Pennsylvania State University, for supporting this research, Julie Martel for statistical consultation, David M. Fehr for helpful discussions, and Marilyn Green Larach for reviewing the manuscript.

References

1. Bollen, B. A., Tinker, J. H., and Hermsmeyer, K. (1987). Halothane relaxes previously constricted isolated porcine coronary artery segments more than isoflurane. *Anesthesiology* **66,** 748–752.
2. Sill, J. C., Bove, A. A., Nugent, M., Blaise, G. A., Dewey, J. D., and Grabau, C. (1987). Effects of isoflurane on coronary arteries and coronary arterioles in the intact dog. *Anesthesiology* **66,** 273–279.
3. Larach, D. R., Schuler, H. G., Skeehan, T. M., and Peterson, C. J. (1990). Direct effects of myocardial depressant drugs on coronary vascular tone: Anesthetic vasodilation by halothane and isoflurane. *J. Pharmacol. Exp. Ther.* **254,** 58–64.
4. Larach, D. R., and Schuler, H. G. (1991). Direct vasodilation by sevoflurane, isoflurane, and halothane alters coronary flow reserve in the isolated rat heart. *Anesthesiology* **75,** 268–278.

5. Nelson, M. T., Patlak, J. B., Worley, J. F., and Standen, N. B. (1990). Calcium channels, potassium channels, and voltage dependence of arterial smooth muscle tone. *Am. J. Physiol.* **259,** C3–C18.
6. Nichols, C. G., and Lederer, W. J. (1991). Adenosine triphosphate-sensitive potassium channels in the cardiovascular system. *Am. J. Physiol.* **261,** H1675–H1686.
7. Stekiel, T. A., Contney, S. J., Stekiel, W. J., Bosnjak, Z. J., and Kampine, J. P. (1993). Effects of halothane on the *in situ* transmembrane potential (E_m) of vascular smooth muscle. *FASEB J.* **7,** A754 (abstract).
8. Franks, N. P., and W. R. (1988). Volatile general anaesthetics activate a novel neuronal K^+ current. *Nature (London)* **333,** 662–664.
9. Tinklenberg, J. A., Segal, I. S., Tianzhi, G., and Maze, M. (1991). Analysis of anesthetic action on the potassium channels of the *Shaker* mutant of *Drosophila*. *Ann. N.Y. Acad. Sci.* **625,** 532–539.
10. Larach, D. R., and Schuler, H. G. (1993). Potassium channel blockade and halothane vasodilation in conducting and resistance coronary arteries. *J. Pharmacol. Exp. Ther.* **267,** 72–81.
11. Harder, D. R., Gradall, K., Madden, J. A., and Kampine, J. P. (1985). Cellular actions of halothane on cat cerebral arterial muscle. *Stroke* **16,** 680–683.
12. Vulliemoz, Y. (1991). The myocardial depressant effect of volatile anesthetics does not involve arachidonic acid metabolities or pertussis toxin-sensitive G-proteins. *Eur. J. Pharmacol.* **203,** 345–351.
13. Buljubasic, N., Rusch, N. J., Marijic, J., Kampine, J. P., and Bosnjak, Z. (1992). Effects of halothane and isoflurane on calcium and potassium channel currents in canine coronary arterial cells. *Anesthesiology* **76,** 990–998.
14. Marijic, J., Buljubasic, N., Coughlan, M. G., Kampine, J. P., and Bosnjak, Z. (1992). Effect of K^+ channel blockade with tetraethylammonium on anesthetic-induced relaxation in canine cerebral and coronary arteries. *Anesthesiology* **77,** 948–955.
15. Eskinder, H., Kampine, J. P., and Bosnjak, Z. I. (1993). Halothane decreases the opening probability of K^+ channels in dog cerebral arterial muscle cells. *Anesth. Analg. (N.Y.)* **76,** S98 (abstract).
16. Post, J. M., and Jones, A. W. (1991). Stimulation of arterial ^{42}K efflux by ATP depletion and cromakalim is antagonized by glyburide. *Am. J. Physiol.* **260,** H848–H854.
17. Notsu, T., Tanaka, I., Takano, M., and Noma, A. (1992). Blockade of the ATP-sensitive K^+ channel by 5-hydroxydecanoate in guinea pig ventricular myocytes. *J. Pharmacol. Exp. Ther.* **260,** 702–708.
18. Miyoshi, Y., Nakaya, Y., Wakatsuki, T., Nakaya, S., Fujino, K., Saito, K., and Inoue, I. (1992). Endothelin blocks ATP-sensitive K^+ channels and depolarizes smooth muscle cells of porcine coronary artery. *Cir. Res.* **70,** 612–616.
19. Samaha, F. F., Heineman, F. W., Ince, C., Fleming, J., and Balaban, R. S. (1992). ATP-sensitive potassium channel is essential to maintain basal coronary vascular tone *in vivo*. *Am. J. Physiol.* **262,** C1220–C1227.
20. Stone, D. J., and Johns, R. A. (1989). Endothelium-dependent effects of halothane, enflurane, and isoflurane on isolated rat aortic vascular rings. *Anesthesiology* **71,** 126–132.
21. Jackson, W. F., Konig, A., Dambacher, T., and Busse, R. (1993). Prostacyclin-induced vasodilation in rabbit heart is mediated by ATP-sensitive potassium channels. *Am. J. Physiol.* **264,** H238–H243.

Volatile Anesthetics and Coronary Collateral Circulation

Judy R. Kersten,* J. Craig Hartman,† Paul S. Pagel,* and David C. Warltier*,‡,§

*Departments of *Anesthesiology, ‡Pharmacology, and §Medicine*
The Medical College of Wisconsin
Milwaukee, Wisconsin 53226

†The Upjohn Company
Kalamazoo, Michigan 49001

I. Introduction

Volatile anesthetics may alter coronary collateral blood flow to ischemic myocardium by several direct and indirect mechanisms. Decreases in arterial pressure resulting from declines in systemic vascular resistance or cardiac output produced by volatile anesthetics can reduce the driving pressure at the origin of collateral vessels and cause a decrease in collateral blood flow. Reflex increases in heart rate during anesthesia may also decrease coronary collateral flow because of a reduction in the duration of diastole. Direct vasodilation of coronary arterioles induced by volatile anesthetics may result in a redistribution of flow away from collateral-dependent, ischemic zones to normal areas. A decrease in coronary collateral flow during large increases in flow to normal zones via direct vasodilation has been termed "coronary steal."

The purpose of this investigation was to evaluate the effects of a new volatile anesthetic, sevoflurane, on coronary collateral blood flow using chronically instrumented dogs with an experimentally produced "coronary steal prone anatomy" (1). The model included a total occlusion of the left anterior descending coronary artery, a stenosis of the adjacent artery of origin of the collaterals (left circumflex coronary artery), and a

Advances in Pharmacology, Volume 31

well-developed coronary collateral circulation. This anatomical configuration has been previously shown to favor the development of coronary steal induced by potent coronary vasodilators including adenosine and dipyridamole (2,3). The actions of sevoflurane were compared to those produced by isoflurane and the small vessel coronary vasodilator adenosine in separate experimental groups. This multivessel disease model also allowed direct comparison of the conscious and anesthetized states without the potentially confounding influences of acute surgical intervention or the use of baseline anesthetic agents.

II. Chronic Animal Instrumentation

Mongrel dogs of either sex were anesthetized with sodium thiamylal (10 mg/kg i.v.) and halothane (1.0–2.0%) in 100% oxygen. A left thoracotomy was performed under sterile conditions. Catheters were implanted in the left atrium, right atrium, and descending thoracic aorta for administration of radioactive microspheres, drug administration, and withdrawal of reference arterial flow samples used in the calculation of regional myocardial blood flow. A balloon cuff vascular occluder was positioned immediately distal to the aortic catheter for control of arterial pressure via constriction of the thoracic aorta.

The heart was suspended in a pericardial cradle, and 2-cm sections of the proximal left anterior descending and left circumflex coronary arteries were dissected free of surrounding tissue. A Doppler (20 MHz) ultrasonic flow transducer was placed around each vessel for measurement of coronary blood flow velocity. A balloon cuff occluder was positioned immediately distal to the left anterior descending coronary flow transducer. This device was used to produce repetitive brief coronary artery occlusions to induce collateral development. In addition, on the day of the experiment, the balloon cuff was used to produce sustained total coronary artery occlusion. An ameroid constrictor was positioned around the left circumflex coronary artery distal to the flow velocity transducer to form a fixed stenosis of the artery of origin of the coronary collaterals perfusing the left anterior descending bed.

A miniature micromanometer (Model P7, Konigsberg Instruments, Pasadena, CA) was inserted into the left ventricle via a small incision through the apex. A fluid-filled catheter was also inserted into the left ventricle for calibration of the micromanometer *in vivo* and for administration of adenosine in specific experiments. Pairs of ultrasonic segment length transducers were implanted in the subendocardium in the perfusion territories supplied by the left anterior descending and left circumflex coronary arteries for measurement of regional contractile function. Electrodes were sutured to the right atrium for atrial pacing. The instrumentation was exteriorized between the scapulae, and the chest wall was closed in layers. All animals were treated with analgesics as required and antibiotics to prevent infection. Each dog was fitted with a jacket to prevent damage to the implanted instruments, which were enclosed in an aluminum box. Each dog was allowed to recover for 2 days after surgery prior to daily hemodynamic monitoring.

III. Regional Myocardial Function and Perfusion

Regional contractile function was evaluated by ultrasonic segment length transducers. All signals were simultaneously monitored via ultrasonic amplifiers (Crystal Biotech, Hopkinton, MA). Using left ventricular dP/dt, end diastolic segment length (EDL) was measured immediately prior to the onset of left ventricular isovolumic contraction, and end systolic length (ESL) was determined at peak $-dP/dt$. Percent systolic shortening (%SS) was calculated from the following equation:

$$\%SS = [(EDL - ESL)/EDL] \times 100$$

Regional myocardial blood flow in normal, stenotic, and totally occluded (coronary collateral blood flow) regions was measured by the radioactive microsphere technique (4). At specific intervals, radioactive microspheres were injected into the left atrium as a bolus. A withdrawal of arterial blood from the aortic catheter was started at a flow rate of 7 ml/min immediately prior to the injection of microspheres and continued for 3 min. At the conclusion of each experiment, India ink and Monastral blue suspension dye were injected into the coronary circulation immediately distal to the ameroid constrictor and hydraulic occluder, respectively, at pressures of 100 mmHg to delineate the perfusion territories of the totally occluded, stenotic, and normal zones. Tissue samples from these areas were subdivided into subepicardial, midmyocardial, and subendocardial specimens of approximately equal weight. Tissue blood flow (Q_m; ml/min/g) was calculated from the equation

$$Qm = CmQ_r/C_r$$

where Qr is the rate of withdrawal (ml/min) of the reference blood flow sample, Cr is the activity [counts per minute (cpm)] of the reference blood flow sample, and Cm is the activity (cpm/g) of the myocardial tissue sample.

Starting on the second postoperative day, each dog was monitored for changes in systemic and coronary hemodynamics. Once an hour, eight times each day, repetitive, brief (2 min) left anterior descending coronary artery occlusions were performed by inflation and subsequent deflation of the hydraulic balloon cuff coronary occluder. This process has been previously used to induce coronary collateral growth (1,5,6). Coronary collateral development was assessed daily by measurement of the reactive hyperemic response following brief coronary artery occlusion and by assessment of segment shortening in the left anterior descending perfusion territory during the period of coronary artery occlusion. In addition, the severity of left circumflex coronary artery stenosis was also assessed by measurement of coronary reserve on each experimental day (7). The increase in left circumflex flow velocity after a bolus injection of adenosine (25, 50, and 100 μg) into the left atrium was measured.

Repetitive occlusions and stenosis development continued for a variable amount of time (10 to 20 days). On the day that the left circumflex coronary artery stenosis was considered critical (minimal changes in resting coronary blood flow but near abolition of coronary vascular reserve as assessed by the ability to dilate following administration of adenosine), a coronary steal prone anatomy was established. In the presence of a well-developed collateral circulation and stenosis of the artery of origin of the coronary collaterals (left circumflex coronary artery), the left anterior descending coronary artery was totally occluded by inflation of the hydraulic balloon cuff occluder for the duration of the experiment. Systemic and coronary hemodynamics and regional myocardial perfusion and function in collateral-dependent, stenotic, and normal zones were measured in the conscious state 30 min after left anterior descending coronary artery occlusion.

After control measurements were completed, dogs received either isoflurane, sevoflurane, or adenosine in three separate groups of experiments. Anesthesia was induced by inhalation using the volatile anesthetic at high flow rates (8 liters/min oxygen). After tracheal intubation, anesthesia was maintained with isoflurane or sevoflurane in room air enriched with oxygen during positive pressure ventilation using a semiclosed anesthesia circuit. The oxygen flow rate, respiratory rate, and tidal volume were adjusted to maintain arterial blood gas tensions within conscious levels. Two concentrations of each anesthetic (1.0 and 1.5 MAC) producing decreases of approximately 25 and 35 mmHg of arterial pressure were studied. End-tidal gas and anesthetic concentrations were continuously monitored with an infrared gas analyzer (Datex Capnomac, Helsinki, Finland). The first concentration of isoflurane or sevoflurane was allowed to equilibrate for a period of 30 min to provide a steady state, and hemodynamics, myocardial blood flow, and regional contractile function were recorded. Anesthetic depth was then adjusted to produce a greater decrease in arterial pressure, maintained at steady state for an additional 30 min, and measurements repeated. Finally, at the highest concentration of anesthetic agent, hemodynamics were restored to levels present in the conscious control state. Arterial pressure was increased by inflation of the aortic balloon cuff occluder. Baroreflex reduction in heart rate was avoided by atrial pacing to rates present in the conscious state. In this fashion, the direct effects of isoflurane and sevoflurane on the regional distribution of coronary blood flow were assessed independently of changes in systemic hemodynamics.

In a separate group of experiments, the effects of adenosine on regional myocardial blood flow were studied. Systemic and coronary hemodynamics, myocardial perfusion, and contractile function were measured in the conscious control state and during low (range 0.25 to 1 mg/min) and high (range 1.0 to 2.5 mg/min) doses of adenosine. The doses of adenosine were administered through the left ventricular catheter by a continuous infusion. The dose ranges were selected to provide declines of approximately 10 and 20 mmHg in mean arterial pressure. Measurements were also completed during the high dose of adenosine with control of arterial pressure and heart rate as described above. At the conclusion of all experiments, each dog was euthanized by intravenous administration of an overdose of sodium pentobarbital. The heart was excised, stained, and fixed for 24 to 48 h in 10% formaldehyde prior to obtaining tissue specimens for analysis of myocardial blood flow.

A total of 31 dogs were used in the investigation to produce 21 successful experiments for data analysis (group size of 6 to 8 animals). Dogs were excluded because of ventricular fibrillation during coronary artery occlusion or failure of instrumentation. Systemic and coronary hemodynamics, myocardial blood flow, and regional contractile function data were analyzed by analysis of variance with repeated measures (ANOVA) followed by Duncan's multiple range test and were considered significant when p values were below 0.05.

IV. Coronary Steal Prone Anatomy Model

Repetitive left anterior descending coronary artery occlusions resulted in a significant ($p < 0.05$) diminution of peak postocclusive reactive hyperemic flow. The duration of the flow increase above baseline levels and reactive hyperemic debt repayment were also significantly decreased despite a constant oxygen debt accrued during each 2-min occlusion (Fig. 1). In

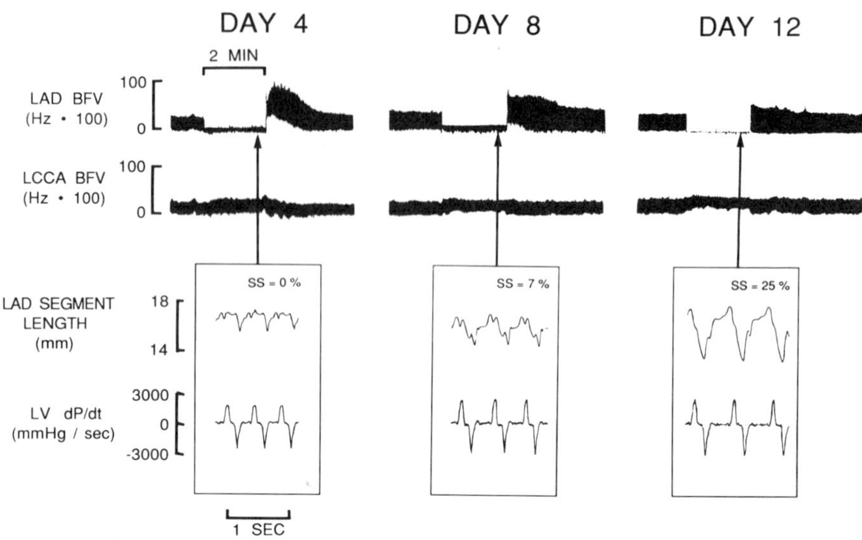

Fig. 1 Left anterior descending (LAD) and left circumflex coronary artery (LCCA) blood flow velocity (BFV) during the reactive hyperemic response after a 2-min coronary artery occlusion. Repetitive coronary artery occlusions were performed once each hour, 8 times each day. Sample data from days 4, 8, and 12 are shown. During repetitive coronary artery occlusions, the reactive hyperemic response is slowly reduced as collateral growth is enhanced. Similarly, segment shortening (SS) during the period of coronary artery occlusion is initially severely depressed; however, with enhancement of collateral growth, contraction is sustained in the ischemic zone during the period of coronary artery occlusion. LV, Left ventricle. [Reprinted with permission from Hartman *et al.* (1).]

some dogs, a total loss of reactive hyperemia following brief coronary artery occlusion occurred. A reduction in reactive hyperemia was indicative of a decrease in the intensity of ischemia during the period of coronary artery occlusion secondary to enhanced collateral development. Examination of regional contractile function during occlusion initially revealed paradoxical systolic aneurysmal bulging during repetitive coronary artery occlusions. Over the period of collateral development, however, aneurysmal bulging was replaced by akinesis and, eventually, by effective systolic shortening in the left anterior descending perfusion territory despite total arterial occlusion (Fig. 1). The progressive improvement in regional contractile function in the ischemic zone also indicated growth and development of coronary collaterals, leading to a reduction in the intensity of ischemia.

As coronary collateral development continued, the ameroid constrictor on the left circumflex coronary artery formed a stenosis that became

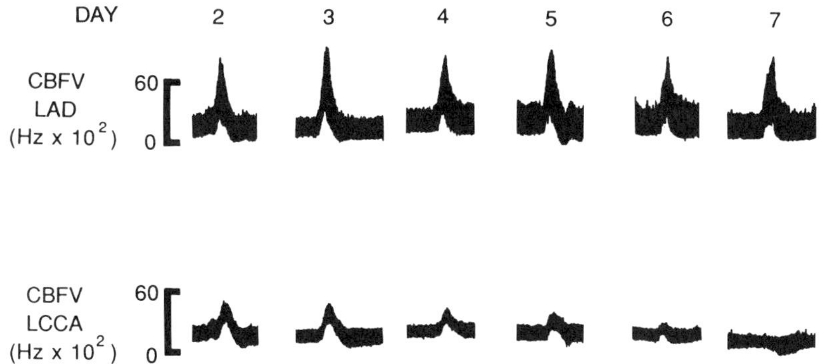

Fig. 2 Coronary blood flow velocity (CBFV) in the left anterior descending (LAD) and left circumflex arteries (LCCA) during left atrial bolus administration of adenosine (100 μg) over 7 days. Note the diminution in flow response to adenosine caused by progressive stenosis of the left circumflex coronary artery via an ameroid constrictor. The flow response in the LAD remains unchanged. [Reprinted with permission from Hartman *et al.* (7).]

continuously more severe over several days. There was a progressive decline in coronary vascular reserve in the left circumflex artery although no change in resting coronary blood flow occurred. This resulted in a diminution of the coronary flow response to bolus injections of adenosine over several days (Fig. 2). In contrast, the flow response to adenosine in the left anterior descending coronary artery remained unchanged over the course of left circumflex stenosis development.

V. Hemodynamic Effects of Isoflurane, Sevoflurane, and Adenosine

The systemic and coronary hemodynamic actions of isoflurane, sevoflurane, and adenosine are summarized in Tables I–III. Isoflurane produced a dose-dependent decline in mean arterial pressure, left ventricular systolic pressure, and left ventricular dP/dt_{50}. Despite a large reduction in mean arterial pressure, left circumflex coronary blood flow velocity remained unchanged, reflecting a reduction in coronary vascular resistance. With the exception of left ventricular dP/dt_{50}, hemodynamics remained unchanged from the conscious state during control of arterial pressure and heart rate. Sevoflurane caused systemic and coronary hemodynamic effects which were similar to those of isoflurane. Sevoflurane produced no change in heart rate and dose-dependent declines in mean arterial pressure, left

Table I

Effects of Isoflurane on Hemodynamics and Transmural Myocardial Perfusion[a]

	Preocclusion	Postocclusion	Isoflurane (MAC)		
			1.0	1.5	1.5 (BP ↑)
HR (bpm)	106 ± 9	113 ± 9	119 ± 6	108 ± 6	116 ± 8
MAP (mmHg)	96 ± 2[b]	102 ± 4	75 ± 4[b]	66 ± 4[b]	100 ± 3
LVSP (mmHg)	123 ± 3	128 ± 5	99 ± 5[b]	89 ± 5[b]	117 ± 4
dP/dt_{50} (mmHg/sec)	1960 ± 130	1890 ± 170	1520 ± 140[b]	1170 ± 140[b]	1170 ± 120[b]
LCCA DBFV (Hz × 10^2)	40 ± 9	46 ± 10	43 ± 10	38 ± 10	44 ± 11
Transmural perfusion (ml/min/g)					
Normal region	—	1.21 ± 0.11	1.07 ± 0.10	0.85 ± 0.15[b]	1.18 ± 0.10
Stenotic region	—	1.05 ± 0.10	1.01 ± 0.11	0.76 ± 0.11[b]	1.06 ± 0.11
Occluded region	—	0.64 ± 0.10	0.50 ± 0.13	0.41 ± 0.11[b]	0.69 ± 0.12

[a] Values are means ± SEM. Abbreviations: HR, heart rate; MAP, mean arterial pressure; LVSP, left ventricular systolic pressure; LCCA DBFV, left circumflex coronary artery diastolic blood flow velocity; (BP ↑), arterial pressure and heart rate adjusted to postocclusion levels.

[b] Significantly different ($p < 0.05$) from postocclusion.

ventricular systolic pressure, and left ventricular dP/dt_{50}. Left circumflex coronary blood flow velocity was unchanged despite sevoflurane-induced decreases in perfusion pressure. Left ventricular dP/dt_{50} remained significantly depressed, and coronary blood flow velocity remained unchanged from the conscious state during restoration of arterial pressure and heart rate to conscious levels.

Adenosine infusions caused decreases in left ventricular systolic pressure and mean arterial pressure which were of lesser magnitude than those observed during isoflurane and sevoflurane anesthesia (Table III). Heart rate and left ventricular dP/dt_{50} were unchanged by the administration of adenosine. Blood flow through the stenotic left circumflex coronary artery was unaffected by vasodilation with adenosine. There was no significant increase in flow through the stenotic left circumflex coronary artery when arterial pressure was controlled to levels obtained prior to the infusion of adenosine.

Table II

Effects of Sevoflurane on Hemodynamics and Transmural Myocardial Perfusion[a]

	Preocclusion	Postocclusion	Sevoflurane (MAC)		
			1.0	1.5	1.5 (BP↑)
HR (bpm)	106 ± 11[b]	142 ± 7	152 ± 11	132 ±9	145 ± 15
MAP (mmHg)	97 ± 5[b]	115 ± 6	73 ± 6[b]	61 ± 4[b]	113 ± 6
LVSP (mmHg)	116 ± 5[b]	131 ± 7	84 ± 6[b]	71 ± 6[b]	126 ± 8
dP/dt_{50} (mmHg/sec)	2140 ± 100	2160 ± 120	1340 ± 140[b]	1050 ± 140[b]	1160 ± 140[b]
LCCA DBFV (Hz × 10^2)	41 ± 9[b]	67 ± 8	60 ± 14	58 ± 9	81 ± 16
Transmural perfusion (ml/min/g)					
Normal region	—	1.54 ± 0.24	1.17 ± 0.28	1.05 ± 0.20	1.69 ± 0.26
Stenotic region	—	1.74 ± 0.27	1.15 ± 0.25	1.02 ± 0.16[b]	1.82 ± 0.25
Occluded region	—	0.29 ± 0.06	0.22 ± 0.07	0.22 ± 0.05	0.51 ± 0.12[b]

[a] Values are means ± SEM. See Table I for explanation of abbreviations.
[b] Significantly different ($p < 0.05$) from postocclusion.

Table III

Effects of Adenosine on Hemodynamics and Transmural Myocardial Perfusion[a]

	Preocclusion	Postocclusion	Adenosine (mg/min)		
			0.68	1.4	1.4 (BP↑)
HR (bpm)	97 ± 7[b]	115 ± 6	118 ± 9	121 ± 7	115 ± 6
MAP (mmHg)	97 ± 3	95 ± 5	92 ± 8	83 ± 9	98 ± 7
LVSP (mmHg)	124 ± 5	117 ± 5	110 ± 6	100 ± 8[b]	117 ± 4
dP/dt_{50} (mmHg/sec)	2210 ± 130	1990 ± 130	2020 ± 160	1700 ± 290	1370 ± 360
LCCA DBFV (Hz × 10^2)	33 ± 6	37 ± 7	41 ± 10	41 ± 11	48 ± 13
Transmural perfusion (ml/min/g)					
Normal	—	1.11 ± 0.10	1.59 ± 0.19	2.36 ± 0.38[b]	2.53 ± 0.61[b]
Stenotic	—	0.86 ± 0.12	1.19 ± 0.18[b]	1.13 ± 0.17[b]	1.26 ± 0.20[b]
Occluded region	—	0.25 ± 0.04	0.23 ± 0.04	0.17 ± 0.06	0.16 ± 0.04[b]

[a] Values are means ± SEM. See Table I for explanation of abbreviations.
[b] Significantly different ($p < 0.05$) from postocclusion.

VI. Regional Myocardial Perfusion: Isoflurane, Sevoflurane, and Adenosine

The effects of isoflurane, sevoflurane, and adenosine on the weighted average of subepicardial, midmyocardial, and subendocardial perfusion (expressed as transmural myocardial blood flow) are summarized in Tables I, II, and III, respectively. Isoflurane decreased transmural perfusion in normal, stenotic, and occluded zones at 1.5 but not 1.0 MAC. When arterial pressure was increased to levels present in the conscious control state, no significant alteration in transmural myocardial perfusion was observed during isoflurane anesthesia (Table I). Sevoflurane produced somewhat different results than isoflurane (Table II). Sevoflurane at 1 and 1.5 MAC had no effect on myocardial perfusion in normal regions. In contrast, blood flow distal to the left circumflex coronary artery stenosis was significantly reduced at 1.5 MAC, but no significant change in coronary collateral blood flow occurred. During control of arterial pressure and heart rate, myocardial perfusion in the normal and stenotic zones was not different than levels observed in the conscious state, but coronary collateral blood flow was increased significantly.

Adenosine caused a dose-related increase in myocardial perfusion of the normal zone (Table III). Small but significant increases in flow were also observed in the area distal to the coronary artery stenosis. The increases in stenosis zone flow were unaffected by control of arterial pressure and heart rate. Adenosine produced a decrease in coronary collateral

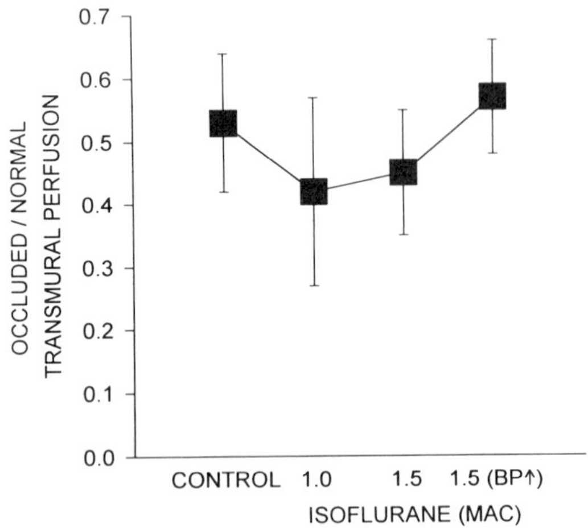

Fig. 3 Ratio of flow between occluded and normal zones as determined by the radioactive microsphere technique in the conscious state and during low and high doses of isoflurane in the absence or presence (BP ↑) of control of arterial pressure.

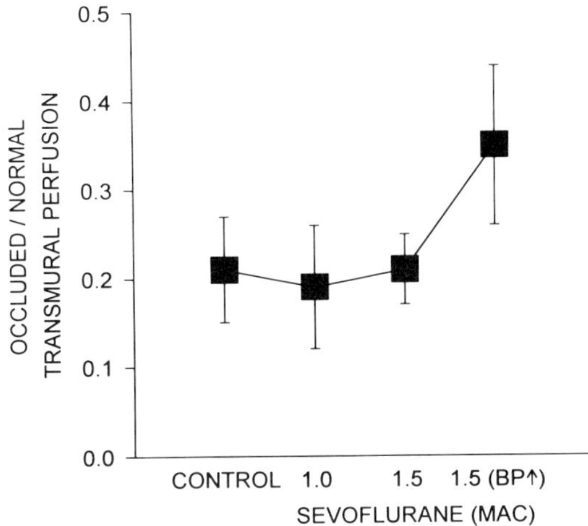

Fig. 4 Ratio of flow between occluded and normal zones as determined by the radioactive microsphere technique in the conscious state and during low and high doses of sevoflurane in the absence or presence (BP ↑) of control of arterial pressure.

blood flow in the occluded zone despite maintenance of arterial pressure and heart rate. Neither isoflurane nor sevoflurane (Fig. 3 and 4) caused redistribution of blood flow away from the collateral-dependent region (decrease in occluded to normal zone flow ratio). In sharp contrast, adenosine caused a dose-related decline in the occluded to normal zone flow

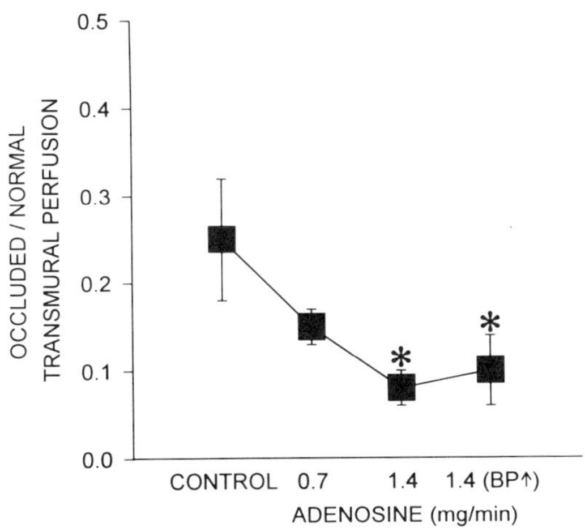

Fig. 5 Ratio of flow between occluded and normal zones as determined by the radioactive microsphere technique in the conscious control state and during low and high doses of adenosine in the absence or presence (BP ↑) of control of arterial pressure. *Significantly different ($p < 0.05$) from control.

ratio (Fig. 5). This maldistribution of perfusion between occluded and normal zones remained present during infusion of the high dose of adenosine despite control of arterial pressure and heart rate, consistent with coronary steal.

VII. Discussion

The results of this investigation demonstrate that isoflurane and sevoflurane cause similar systemic hemodynamic effects during acute coronary artery occlusion in chronically instrumented dogs. Neither agent produces marked coronary vasodilation, and neither agent causes an abnormal redistribution of coronary blood flow away from collateral-dependent zones. Interestingly, coronary collateral blood flow was significantly increased in the presence of sevoflurane during the maintenance of driving pressure for collateral perfusion, suggesting that sevoflurane may provide beneficial effects in myocardium distal to a total coronary artery occlusion. The results obtained with the volatile anesthetics were in direct contrast to those obtained with the small vessel coronary vasodilator adenosine. Adenosine markedly decreased collateral blood flow, resulting in a redistribution of flow away from the collateral-dependent zone during vasodilation of normal regions. This maldistribution of flow was independent of decreases in arterial pressure and occurred despite maintenance of arterial pressure at predrug levels.

The present results were obtained in an experimental model of multivessel coronary artery disease which simulated steal prone coronary anatomy by using a total occlusion of one major coronary artery with adequate distal collateralization arising from an adjacent stenotic coronary artery (1). This model has been shown to correlate well with the abnormal anatomic configuration of the coronary vasculature in approximately 25% of patients presenting for coronary artery bypass graft surgery (8). The tendency for coronary steal to occur with vasodilator drugs is readily apparent in this model, and it occurs independently of alterations in heart rate and diastolic aortic pressure (2). Coronary steal is directly dependent on the degree of vasodilation in the area distal to the coronary artery stenosis and on the severity of the stenotic lesion. Vasodilation in the region supplied by the stenotic artey causes a decrease in coronary artery pressure distal to the stenosis. The decrease in distal coronary pressure results in a reduction in driving pressure for collateral flow into the totally occluded zone. The greater the severity of the stenosis, the larger the drop in perfusion pressure at the origin of the collaterals with any given degree of vasodilation (9). Even minor degrees of vasodilation distal to a severe stenosis can decrease collateral blood flow. Steal of perfusion away from a collateral-dependent zone can also occur in the absence of a stenosis of the artery of origin of coronary collaterals (2,10). During maximal coronary vasodilation in the absence of a stenosis, there is a decrease in distal coronary artery pressure because the length of the epicardial coronary artery becomes a more significant factor in the overall calculation of coronary resistance.

Agents that produce coronary vasodilation can act at one or more regions in the coronary vasculature. Arterioles are the primary vessels controlling resistance in the coronary circulation, and, thus, arteriolar dilators such as adenosine, dipyridamole, and chromonar are capable of producing maximal increases in flow and a redistribution of flow away from collateral-dependent zones (1,2,10–12). Chromonar causes intense and selective coronary vasodilation and may increase myocardial infarct size by a coronary steal mechanism even in the absence of stenosis of the artery of origin of collaterals (10). Dipyridamole may exacerbate recovery of function of stunned myocardium by causing a steal of collateral perfusion during a period of coronary occlusion (13). In contrast, nitroglycerin, a large coronary artery, conduit, and collateral vessel dilator, does not produce coronary steal and may even increase coronary collateral perfusion (14).

Volatile anesthetics, including isoflurane, enflurane, and halothane, have been shown to cause coronary vasodilation, primarily of small resistance vessels (15). This property makes these agents potential candidates for the production of coronary steal. Isoflurane has been particularly scrutinized and has been previously shown to cause a relatively small redistribution of flow away from collateral-dependent regions in certain experimental models (16). Halothane has also been shown to cause direct coronary vasodilation *in vitro* (17,18); however, the possibility that this volatile anesthetic may produce coronary steal has never been seriously entertained because halothane causes several indirect hemodynamic actions which directly oppose the actions of this agent as a coronary vasodilator (19). Potent inhalational anesthetics are direct negative inotropes and decrease aortic blood pressure by the combined effects of reductions in cardiac output and peripheral vascular resistance. Volatile anesthetics also depress sinoatrial (SA) node and cardiac conduction tissue function *in vitro,* and they may cause an overall reduction in heart rate unless baroreflex-mediated tachycardia occurs simultaneously with decreases in arterial pressure. Volatile anesthetic-induced reductions in myocardial contractility, left ventricular afterload, and heart rate contribute to a reduction in myocardial oxygen consumption and lead to a concomitant increase in coronary vascular resistance via metabolic autoregulation. Thus, the sum of indirect and direct effects *in vivo* determines the observed alterations in coronary blood flow and vascular resistance (20).

Several experimental studies indicate that the profound indirect effect of volatile anesthetics on the coronary vasculature offset the direct action of these agents as direct coronary vasodilators. This phenomenon is well established for halothane, has been described for desflurane (21) and, in this investigation, for sevoflurane as well. In contrast, isoflurane produces a small, transient degree of coronary vasodilation in the intact heart; however, the influence of the systemic hemodynamic actions of this agent on myocardial oxygen consumption and coronary autoregulation counterbalance the vasodilator effects of isoflurane so that little overall change in coronary blood flow occurs (19). In addition, minimal alteration in the distribution between coronary blood flow to collateral-dependent, stenotic, and normal zones would be expected with such a small degree of vasodilation during isoflurane administration.

In contrast to the present findings, however, other investigators have shown that volatile anesthetics may cause prominent increases in coronary blood flow when suddenly administered in large doses via an intracoronary route (22). This observation occurs with not only isoflurane but also halothane through an unknown mechanism. Volatile anesthetics may have differential effects on the release of nitric oxide from vascular endothelium (23,24), but it is doubtful whether this property can be solely responsible for the large increases in coronary flow observed by Crystal and coworkers (22). Direct nitric oxide donors, including SIN-1, produce only minimal increases in coronary blood flow. Furthermore, these agents tend to affect large vessel dilation significantly more than resistance vessels.

In the present investigation, sevoflurane was shown to increase coronary collateral blood flow during maintenance of heart rate and arterial pressure at levels observed in the conscious state. In contrast, coronary collateral blood flow remains constant with correction of heart rate and blood pressure during isoflurane or desflurane anesthesia. The basis for the increase in collateral perfusion produced by sevoflurane is unknown; however, the relative vasodilation of collateral vessels as compared to small resistance vessels may represent a potential explanation for the present results. The hypothesis that sevoflurane-induced increases in collateral flow are beneficial for ischemic myocardium has yet to be tested, but the increase in oxygen delivery may contribute to a reduction in the overall severity of myocardial ischemia.

In summary, the present results have demonstrated that isoflurane and a new volatile anesthetic, sevoflurane, do not abnormally redistribute collateral blood flow away from an area distal to a total coronary artery occlusion. In fact, sevoflurane produces an increase in flow to an ischemic zone in a canine model of multivessel coronary artery obstruction. These findings contrast sharply with those obtained with adenosine, which caused marked redistribution of coronary collateral flow away from acutely ischemic myocardium.

References

1. Hartman, J. C., Kampine, J. P., Schmeling, W. T., and Warltier, D. C. (1991). Steal-prone coronary circulation in chronically instrumented dogs: Isoflurane versus adenosine. *Anesthesiology* **74,** 744–756.
2. Gross, G. J., and Warltier, D. C. (1981). Coronary steal in four models of single or multiple vessel obstruction in dogs. *Am. J. Cardiol.* **48,** 84–92.
3. Becker, L. C. (1978). Conditions for vasodilator induced coronary steal in experimental myocardial ischemia. *Circulation* **57,** 1103–1110.
4. Domenech, R. J., Hoffman, J. J., Nobel, M. I., Saunders, K. B., Henson, J. R., and Subijanto, S. (1969). Total and regional coronary blood flow measured by radioactive microspheres in conscious and anesthetized dogs. *Circ. Res.* **25,** 581–596.
5. Fujita, M., McKown, D. P., McKown, M. D., Hartley, J. W., and Franklin, D. (1987). Evaluation of coronary collateral development by regional myocardial function and reactive hyperaemia. *Cardiovasc. Res.* **21,** 377–384.
6. Mohri, M., Tomoike, H., Noma, M., Inove, T., Hisano, K., and Nakamura, M. (1989). Duration of ischemia is vital for collateral development: Repeated brief coronary artery occlusions in conscious dogs. *Circ. Res.* **64,** 287–296.
7. Hartman, J. C., Kampine, J. P., Schmeling, W. T., and Warltier, D. C. (1991). Alterations in collateral blood flow produced by isoflurane in a chronically instrumented canine model of multivessel coronary artery disease. *Anesthesiology* **74,** 120–133.
8. Buffington, C. W., Davis, K. B., Gillispie, S., and Pettinger, M. (1988). The prevalence of steal-prone coronary anatomy in patients with coronary artery disease: An analysis of the Coronary Artery Surgery Study Registry. *Anesthesiology* **69,** 721–727.
9. Warltier, D. C., Hardman, H. F., and Gross, G. J. (1979). Transmural perfusion gradients distal to various degrees of coronary artery stenosis during resting flow or at maximal vasodilation. *Basic Res. Cardiol.* **74,** 494–508.
10. Warltier, D. C., Gross, G. J., and Brooks, H. L. (1980). Coronary steal-induced increase in myocardial infarct size after pharmacologic coronary vasodilation. *Am. J. Cardiol.* **46,** 83–90.
11. Warltier, D. C., Gross, G. J., and Brooks, H. L. (1981). Pharmacologic- vs. ischemia-induced coronary artery vasodilation. *Am. J. Physiol.* **240,** H767–H774.
12. Gross, G. J., Buck, J. D., Warltier, D. C., and Hardman, H. F. (1982). Separation of overlap and collateral perfusion of ischemic canine myocardium: Important considerations in the analysis of vasodilator induced coronary steal. *J. Cardiovasc. Pharmacol.* **4,** 254–263.
13. Kenny, D., Wynsen, J. C., Brooks, H. L., and Warltier, D. C. (1991). Dipyridamole-induced decrement of functional recovery of positischemic reperfused myocardium in conscious dogs with well-developed coronary collateral circulation. *Am. Heart J.* **121,** 1339–1347.
14. Gross, G. J., Warltier, D. C., and Lamping, K. A. (1993). Comparative effects of the potassium channel opener-nitrate, nicorandil, nifedipine and nitroglycerin on coronary collateral blood flow in dogs with well-developed collateral vessels. *Pharmacol. Commun.* **2,** 353–360.
15. Conzen, P. F., Habazettl, H., Vollmar, B., Christ, M., Baier, H., and Peter, K. (1992). Coronary microcirculation during halothane, enflurane, isoflurane, and adenosine in dogs. *Anesthesiology* **76,** 261–270.

16. Buffington, C. W., Romson, J. L., Levine, A., Duttlinger, N. C., and Huang, A. H. (1987). Isoflurane induces coronary steal in a canine model of chronic coronary occlusion. *Anesthesiology* **66,** 280–292.
17. Bollen, B. A., McKlveen, R. E., and Stevenson, J. A. (1992). Halothane relaxes preconstricted small and medium isolated porcine coronary artery segments more than isoflurane. *Anesth. Analg.* **75,** 9–17.
18. Bollen, B. A., McKlveen, R. E., and Stevenson, J. A. (1992). Halothane relaxes previously constricted human epicardial coronary artery segments more than isoflurane. *Anesth. Analg.* **75,** 4–8.
19. Kenny, D., Proctor, L. T., Schmeling, W. T., Kampine, J. P., and Warltier, D. C. (1991). Isoflurane causes only minimal increases in coronary blood flow independent of oxygen demand. *Anesthesiology* **75,** 640–649.
20. Larach, D. R., Schuler, H. G., Skeehan, T. M., and Peterson, C. J. (1990). Direct effects of myocardial depressant drugs on coronary vascular tone: Anesthetic vasodilation by halothane and isoflurane. *J. Pharmacol. Exp. Ther.* **254,** 58–64.
21. Pagel, P. S., Kampine, J. P., Schmeling, W. T., and Warltier, D. C. (1991). Comparison of the systemic and coronary hemodynamic actions of desflurane, isoflurane, halothane, and enflurane in the chronically instrumented dog. *Anesthesiology* **74,** 539–551.
22. Crystal, G. J., Kim, S.-J., Czinn, E. A., Salem, M. R., Mason, W. L., and Abdel-Latif, M. (1991). Intracoronary isoflurane causes marked vasodilation in canine hearts. *Anesthesiology* **74,** 757–765.
23. Blaise, G., Sill, J. C., Nugent, M., Van Dyke, R. A., and Vanhoutte, P. M. (1987). Isoflurane causes endothelium-dependent inhibition of contractile responses of canine coronary arteries. *Anesthesiology* **67,** 513–517.
24. Witzeling, T. M., Sill, J. C., Hughes, J. M., Blaise, G. A., Nugent, M., and Rorie, D. K. (1990). Isoflurane and halothane attenuate coronary artery constriction evoked by serotonin in isolated porcine vessels and in intact pigs. *Anesthesiology* **73,** 100–108.

Myocardial Oxygen Supply–Demand Relations during Isovolemic Hemodilution

George J. Crystal
Departments of Anesthesiology and Physiology and Biophysics
University of Illinois College of Medicine
Illinois Masonic Medical Center
Chicago, Illinois 60680

I. Introduction

The increasing frequency of complex surgical procedures with extensive blood loss, the risk of transmission of disease during donor blood transfusions [e.g., acquired immune deficiency syndrome (AIDS)], the high cost of transfusion therapy, and shortages in blood banks have combined to increase the use of cell-free plasma expanders in anesthesia practice. An increase in cardiac output tends to compensate for the hemodilution-induced reduction in the oxygen-carrying capacity of the blood, and to maintain systemic oxygen supply (1). Thus the tolerance of the body to hemodilution is very much related to how low the hematocrit can be decreased without jeopardizing myocardial oxygen supply, as well as the ability of the heart to sustain an augmented pumping requirement.

Myocardial oxygen supply is the product of coronary blood flow and arterial oxygen content. Coronary blood flow is dependent on perfusion pressure and coronary vascular resistance, which, in turn, is dependent on vasomotor tone and blood viscosity. Hemodilution causes a reduction in arterial oxygen content by decreasing hemoglobin concentration, while it is also reducing blood viscosity by a reduction in hematocrit. Accordingly, the maintenance of myocardial oxygen supply during hemodilution depends on whether the induced decrease in arterial oxygen content is

Advances in Pharmacology, Volume 31

balanced by an increase in coronary blood flow owing to the combined effects of reduced blood viscosity and local metabolic adjustments in vasomotor tone.

This article summarizes several studies in anesthetized dogs to assess myocardial oxygen supply–demand relations during isovolemic hemodilution. The specific objectives of these studies were (1) to determine limits to cardiac compensation during isovolemic hemodilution in normal hearts compared to that in hearts with an acute coronary stenosis, and to define underlying mechanisms; (2) to compare response to isovolemic hemodilution in the left and right ventricles; and (3) to evaluate cardiac effects of controlled hypotension in the presence of isovolemic hemodilution. Detailed descriptions of these studies have been published. (2–7).

II. Experimental Studies

A. Limit to Cardiac Compensation during Isovolemic Hemodilution: Influence of Coronary Stenosis

Although numerous investigators have evaluated cardiac responses during hemodilution (4,8,9), there have been no well-controlled and systematic attempts to define the limits and underlying determinants of cardiac compensation under this condition. The present study was performed to determine the limit of cardiac compensation during isovolemic hemodilution in anesthetized dogs with normal hearts and to describe the hemodynamic and metabolic mechanisms underlying this limit. Because previous studies have suggested that coronary insufficiency may impair the cardiac response to hemodilution because of inadequate coronary flow reserve (2,5), findings in normal hearts were compared to those in hearts in which an acute critical stenosis was created in the left anterior descending coronary artery (LAD).

The studies were conducted in anesthetized (fentanyl–midazolam), open-chest dogs. The LAD was dissected free, just distal to the first major diagonal branch, and fitted with an electromagnetic flow transducer to measure coronary blood flow. A stainless steel, adjustable screw clamp was placed distal to the flow transducer to permit creation of a coronary stenosis. A silk ligature was placed loosely around the LAD and was used to perform intermittent coronary occlusions for evaluation of coronary vasodilator reserve by analysis of reactive hyperemic responses.

Regional and mean transmural myocardial blood flow in the LAD and circumflex (CIRC) regions were measured with 15 μm radioactive microspheres. Venous effluent from the LAD region was obtained from the anterior interventricular vein, whereas that from the CIRC region was obtained from the coronary sinus. The coronary arteriovenous oxygen difference was determined on a regional basis and used to calculate regional oxygen extraction. Myocardial oxygen consumption in the LAD and CIRC regions was calculated using the Fick equation. Convective oxygen supply to the LAD and CIRC regions was calculated by multiplying the values for arterial oxygen content and the respective regional myocardial blood flow.

Paired 1-ml blood samples were obtained from the aorta and the coronary veins and analyzed for plasma lactate concentration using an enzymatic method. Percent lactate extraction was calculated by dividing the coronary arteriovenous lactate difference by the arterial lactate concentration and multiplying by 100. Measurements were obtained by systemic hemodynamic parameters, including aorta pressure, left atrial pressure, heart rate, and cardiac index, using standard methods. The systemic vascular resistance index was computed by dividing mean aortic pressure by cardiac index.

The dogs were randomly divided into two equal groups, the control group and the stenosis group. In the control group, the LAD screw clamp remained open. In the stenosis group, a critical stenosis was created by closing the LAD screw clamp sufficiently to abolish reactive hyperemia without decreasing baseline coronary blood flow.

Hemodynamic measurements were first obtained under control conditions (hematocrit 40%). Then hemodilution was produced progressively by removing blood from the carotid artery at a rate of 20 ml min while replacing it with 6% hetastarch in isotonic saline (Hespan, Du Pont, Wilmington, DE) pumped into the left femoral vein at same rate. Hemodynamic and metabolic measurements were obtained under steady-state conditions as the hematocrit was reduced progressively in 10% decrements until cardiac failure. Cardiac failure was defined as a 20% reduction in mean aortic pressure accompanied by an elevated left atrial pressure. In two dogs of the control group, a small volume of autologous red blood cells (~80–90 ml) was injected in an attempt to reverse the signs of cardiac failure caused by severe hemodilution.

With an intact LAD (control group), mean aortic pressure and mean left atrial pressure were constant until the hematocrit was reduced to 9 ± 1%, when the criteria for cardiac failure were satisfied (Fig. 1A). The cardiac index increased progressively as the hematocrit was reduced (Fig. 2). The increase in cardiac index was attributable to an increase in stroke volume index until cardiac failure, when an increase in heart rate also contributed.

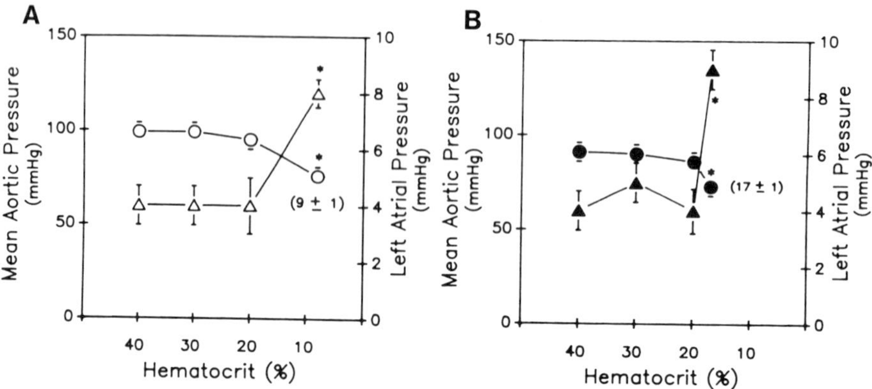

Fig. 1 Comparison of changes in mean aortic pressure (circles) and mean left atrial pressure (triangles) duing graded hemodilution in the control (A) and stenosis (B) groups. Note that the criteria for cardiac failure were satisfied at a hematocrit of 9 ± 1% in the control group and 17 ± 1% in the stenosis group. *Significant difference ($p < 0.05$) from 40% hematocrit (HCT_{40}). [From Levy *et al.* (7) with permission from the American Physiological Society.]

The normal LAD and CIRC regions in control hearts responded similarly with respect to blood flow and oxygen consumption during hemodilution (Table I: Fig. 3). Graded hemodilution caused progressive increases in mean myocardial blood flow within the left ventricular wall. These increases in myocardial blood flow were sufficient to offset the reductions in arterial oxygen content and the arteriovenous oxygen difference; thus, myocardial oxygen supply and myocardial oxygen consumption remained unchanged. The decreases in the coronary arteriovenous oxygen difference paralleled the induced reductions in arterial oxygen content across the entire range of hematocrits studied, with the result that the myocardial oxygen extraction ratio and coronary venous O_2 saturation and pO_2 did not vary. Values for the endocardium/epicardium (endo/epi) flow ratio indicated that the hemodilution-related increases in myocardial blood flow were transmurally uniform in the left ventricular wall until cardiac failure, when the endo/epi flow ratio decreased significantly below 1.0, indicating relative underperfusion of the subendocardium (Fig. 4A). This subendocardial hypoperfusion was coincident with reversal of the coronary arteriovenous lactate gradient (indicating myocardial lactate production) and with a decrease in coronary venous pH (Fig. 4A; Table I). The vasodilator reserve ratio in the LAD bed decreased progressively as the hematocrit was reduced (Fig. 5 and 6). A ratio of 0.9 ± 0.1 at cardiac failure indicated that coronary vasodilation was maximal.

Reinfusion of autologous red blood cells immediately reversed the adverse changes in global cardiac function, endo/epi flow ratio, and lactate extraction ratio caused by extreme hemodilution (Table II). In the presence of LAD stenosis, cardiac failure occurred at a hematocrit of 17 ± 1%, which was significantly higher than that in the control group (Fig. 1A). In contrast to findings in the control group, neither stroke volume index nor cardiac index increased during hemodilution (Fig. 2).

In myocardium supplied by a stenotic LAD, blood flow did not change during hemodilution, which resulted in reductions in both regional myocardial oxygen supply and oxygen consumption (Fig.7). In contrast, in adjacent myocardium supplied by a normal CIRC, blood flow increased pro-

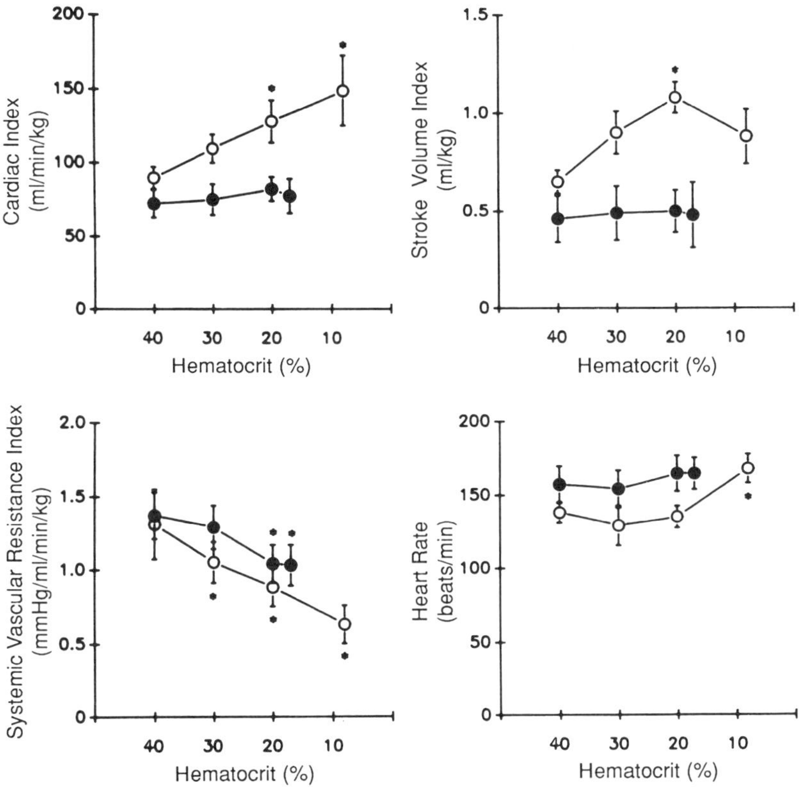

Fig. 2 Comparison of changes in systemic hemodynamic parameters during graded hemodilution in the control (empty circles) and stenosis (filled circles) groups. *Significant difference ($p < 0.05$) from HCT_{40}. [From Levy *et al.* (7) with permission from the American Physiological Society.]

Table I

Blood Gas Values in Anterior Interventricular Vein and Coronary Sinus during Graded Hemodilution until Cardiac Failure[a]

Group and measure	Area	Baseline	Hematocrit 30%	Hematocrit 20%	Cardiac failure
Control					
P_{CO_2}, mmHg	AIV	23 ± 1	23 ± 1	23 ± 1	22 ± 2
	CS	24 ± 1	24 ± 1	23 ± 1	22 ± 2
P_{CO_2}, mmHg	AIV	52 ± 4	49 ± 3	47 ± 2	56 ± 4
	CS	55 ± 2	52 ± 3	49 ± 2	53 ± 2
pH	AIV	7.32 ± 0.01	7.33 ± 0.02	7.34 ± 0.01	7.18 ± 0.03[b]
	CS	7.31 ± 0.01	7.32 ± 0.02	7.32 ± 0.01	7.21 ± 0.03[b]
Stenosis					
P_{CO_2}, mmHg	AIV	23 ± 1	22 ± 1	21 ± 2	17 ± 1[b]
	CS	22 ± 1	21 ± 1	21 ± 1	18 ± 2[b]
P_{CO_2}, mmHg	AIV	49 ± 1	48 ± 4	52 ± 2	55 ± 4
	CS	50 ± 1	47 ± 2	47 ± 2	52 ± 2
pH	AIV	7.34 ± 0.01	7.32 ± 0.03	7.26 ± 0.02[b]	7.21 ± 0.01[b]
	CS	7.33 ± 0.02	7.33 ± 0.01	7.30 ± 0.01	7.26 ± 0.03[b]

[a] Values are means ± SE. From Levy *et al.* (7) with permission. AIV, Anterior interventricular vein; CS, coronary sinus.

[b] Significantly different (p <0.05) from baseline.

gressively during graded hemodilution. These increases in blood flow, which essentially mirrored those in the control hearts, were sufficient to maintain regional myocardial oxygen supply and oxygen consumption constant. Hemodilution of a moderate degree decreased the endo/epi ratio in the stenotic LAD region (Fig. 4). However, hemodilution had no effect on the endo/epi ratio in the normal CIRC bed until cardiac failure, when the ratio decreased. The reductions in the endo/epi ratio were associated with either a reversal of the arteriovenous lactate gradient (LAD region) or a decrease in lactate extraction (CIRC region). In both the LAD and CIRC regions, severe hemodilution caused a decrease in the pH of the local coronary venous effluent (Table I).

The magnitude of the increases in myocardial blood flow in the normal hearts was directly related to the degree of hemodilution and disproportionate to the increases in cardiac index. Inasmuch as coronary driving pressure (aortic pressure) did not increase, the increases in myocardial blood flow resulted from a decreased vascular resistance. Although a reduction in blood viscosity has been shown to contribute to the decreased

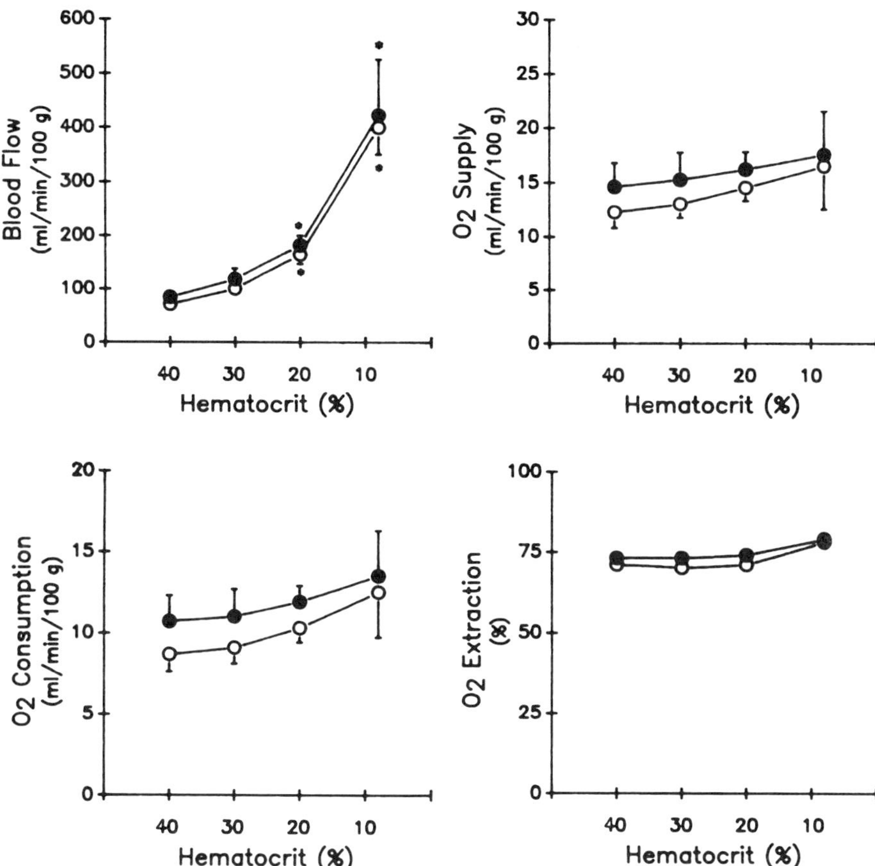

Fig. 3 Similarity of hemodilution-induced changes in oxygen consumption and its determinants in perfusion fields of the intact left anterior descending (empty circles) and circumflex (filled circles) coronary arteries in the control group. *Significantly different ($p < 0.05$) from HCT_{40}. [From Levy *et al.* (7) with permission from the American Physiological Society.]

coronary vascular resistance during hemodilution (2), the rapidly diminishing reactive hyperemic responses suggested that coronary vasodilation via a metabolic mechanism (presumably in response to reduced arterial oxygen content) also played a prominent role. The dependence on coronary vasodilation during hemodilution is heightened by the accentuated inertial pressure losses in the coronary circulation under this condition (10). The present findings are in accord with previous investigations demonstrating that the normal left ventricle depends on increases in blood flow alone to maintain oxygen consumption during isovolemic hemodilution (9).

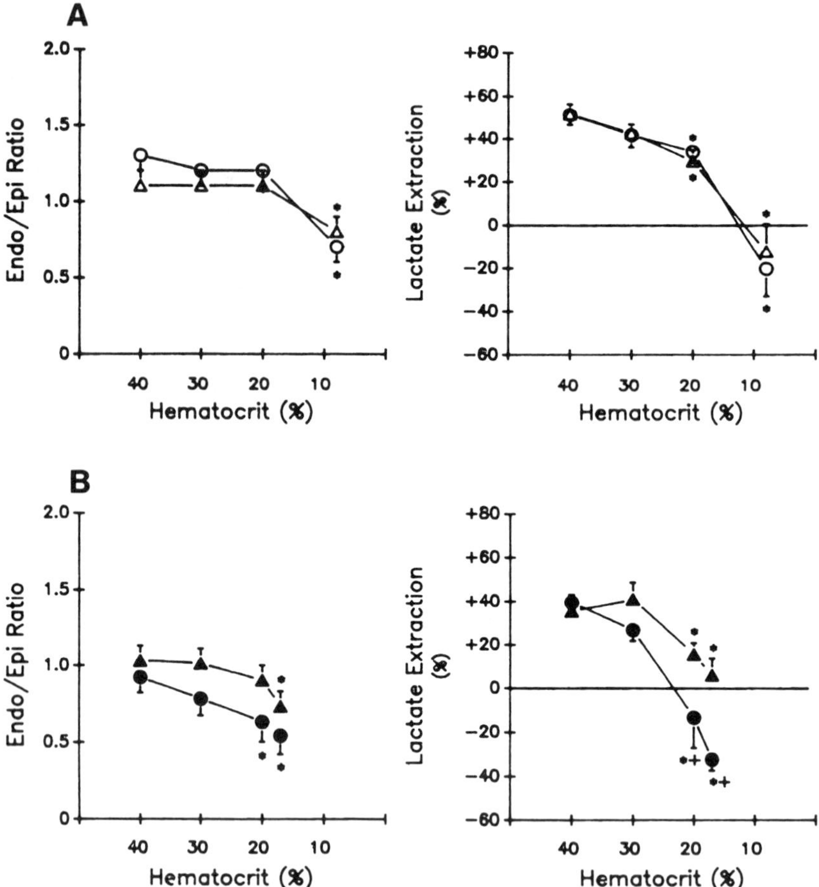

Fig. 4 Comparison of hemodilution-induced changes in endocardial/epicardial flow ratio and lactate extraction in perfusion fields of left anterior descending (circles) and circumflex (triangles) coronary arteries in the control (A) and stenosis (B) groups. *Significantly different ($p < 0.05$) from HCT_{40}. $^{+}$Significantly different ($p < 0.05$) LAD versus CIRC. [From Levy *et al.* (7) with permission from the American Physiological Society.]

The unchanged oxygen extraction in the left ventricle during hemodilution is consistent with its high baseline oxygen extraction (and low value for coronary venous p_{O_2}) and with the strong parallel relationship between oxygen consumption and myocardial oxygen delivery (i.e., blood flow) that is characteristic of the left ventricle under nonhemodiluted conditions (11).

The well-maintained myocardial oxygen supply and oxygen consumption in control hearts during moderate hemodilution did not necessarily

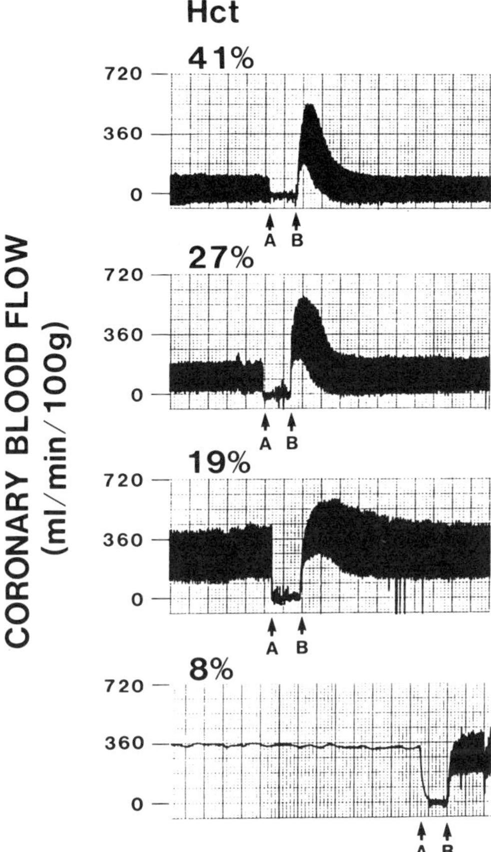

Fig. 5 Representative tracing of reactive hyperemias from one dog in the control group during graded isovolemic hemodilution. Note that the ratio of the peak reactive hyperemic flow to the preocclusion flow (i.e., the vasodilator reserve ratio) decreases as the hematocrit is reduced. The lack of reactive hyperemia at a hematocrit of 8% is evidence for maximal coronary vasodilation. [From Levy *et al.* (7) with permission from the American Physiological Society.]

indicate that the prevailing cardiac work requirements were satisfied. However, several lines of evidence suggest that this was the case. First, indices of global cardiac performance (e.g., aortic pressure and left atrial pressure) were not altered, and the cardiac index increased. Second, the transmurally uniform increases in myocardial blood flow and well-maintained myocardial lactate extraction suggest that myocardial ischemia was not evident. Third, our previous studies demonstrated well-

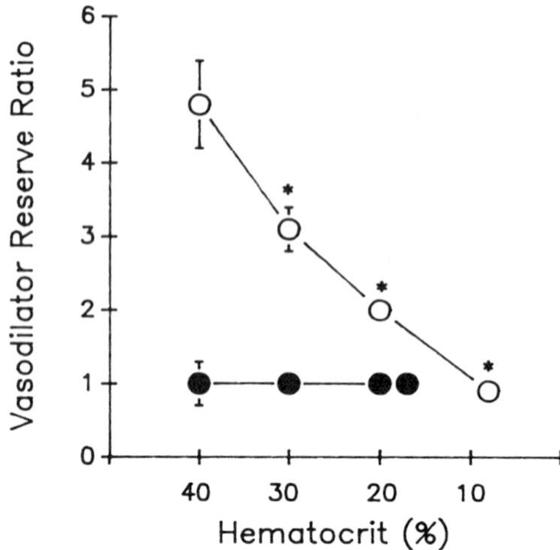

Fig. 6 Comparison of hemodilution-induced changes in coronary vasodilator ratio in intact (control group empty circles) and constricted (stenosis group filled circles) left anterior descending coronary artery. *Significantly different ($p < 0.05$) from HCT_{40}. [From Levy *et al.* (7) with permission from the American Physiological Society.]

maintained local segmental shortening and oxygen consumption when the hematocrit in the LAD was reduced selectively over a similar range using an extracorporeal, controlled-pressure system (Figs. 8–10) (5).

Evidence of cardiac failure was observed at a hematocrit of approximately 9% in hearts with normal coronary circulations. At cardiac failure, mean myocardial blood flow was four times baseline (and was sufficient to maintain myocardial oxygen supply and myocardial oxygen consump-

Table II

Ability of Reinfused Red Blood Cells to Reverse Global Cardiac Dysfunction Caused by Extreme Hemodilution in Control Hearts[a]

Measure	Baseline	Cardiac failure	Reinfusion of cells
HCT, %	38 ± 5	10 ± 0	16 ± 2
MAP, mmHg	119 ± 16	80 ± 7	87 ± 5
MLAP, mmHg	4 ± 1	7 ± 0	4 ± 1
LAC_{ext},%	49 ± 5	−19 ± 8	25 ± 9
Endo/epi	1.2 ± 0.2	0.5 ± 0.1	0.9 ± 0.1

[a] Values are means ± SE in two dogs. From Levy *et al.* (7) with permission. HCT, Hematocrit; MAP, mean aortic pressure; MLAP, mean left atrial pressure; LAC_{ext}, myocardial lactate extraction; Endo/epi, endocardium/epicardium flow ratio.

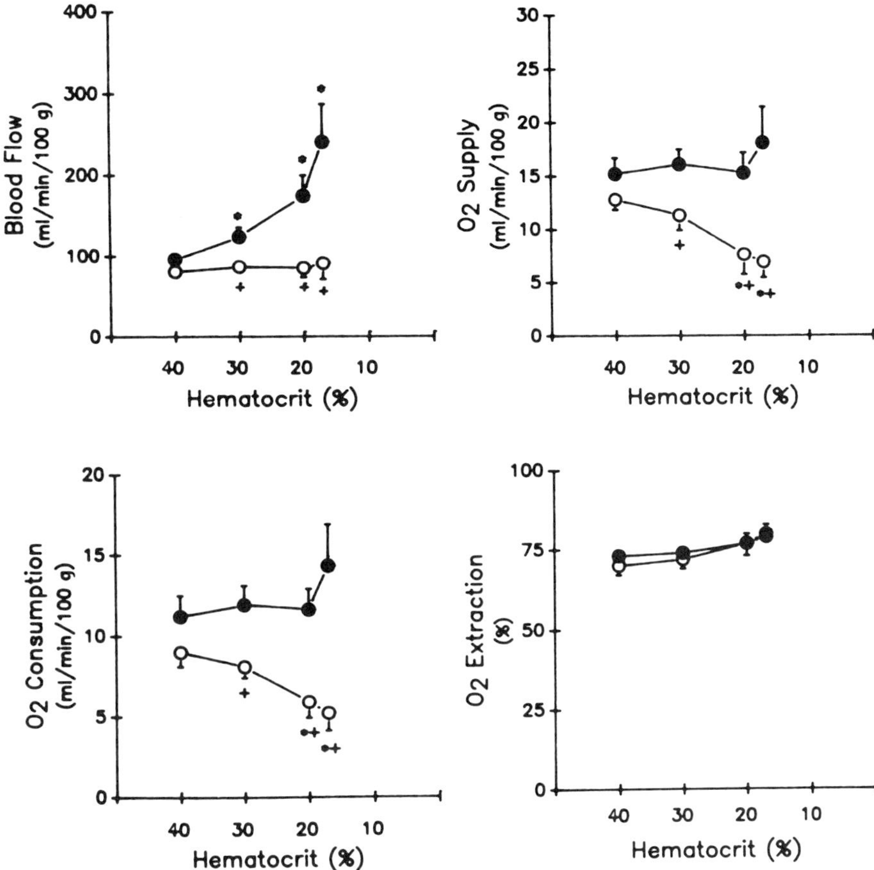

Fig. 7 Difference in hemodilution-induced changes in myocardial oxygen consumption and its determinants in perfusion fields of the stenotic left anterior descending (empty circles) and intact circumflex (filled circle) arteries during graded hemodilution. [From Levy *et al.* (7) with permission from the American Physiological Society.]

tion at the respective baseline values), but this flow was unequally distributed across the left ventricular wall, resulting in relative subendocardial hypoperfusion. The associated findings of a reversed myocardial arteriovenous lactate gradient and a reduction in coronary venous pH support the notion that subendocardial ischemia was evident and that this factor was limiting global cardiac function.

The possibility that cardiac failure during severe hemodilution was due to time-dependent deterioration of the preparation was ruled out by the

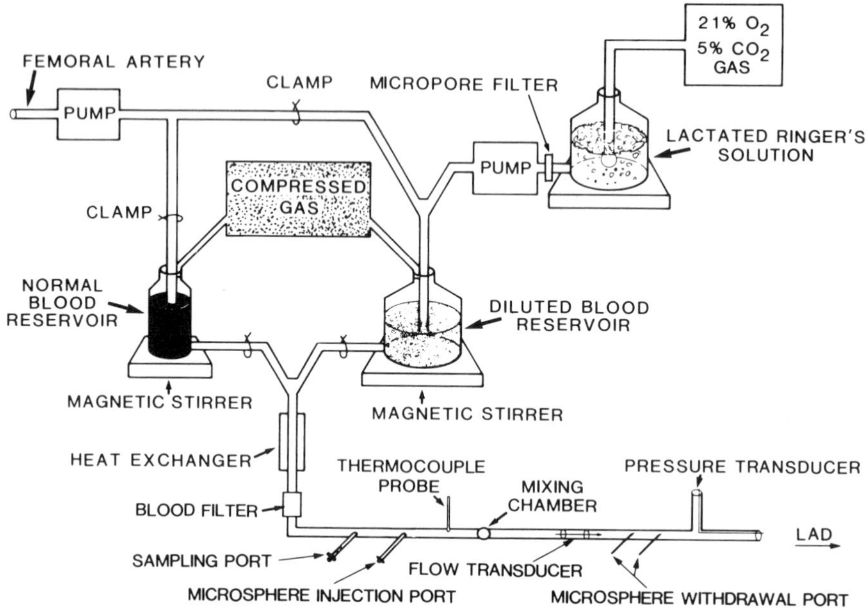

Fig. 8 Extracorporeal system permitting selective perfusion of the LAD with diluted arterial blood at controlled perfusion pressure. [From Crystal (2) with permission from the American Physiological Society.]

ability of small volumes of red blood cells to rapidly reverse the global cardiac dysfunction and the associated hemodynamic and metabolic abnormalities in the myocardium.

The present findings agree with previous studies demonstrating a lower blood flow in the subendocardium than in subepicardium during severe hemodilution (8,12), and they are consistent with reports that hemodilution causes exhaustion of vasodilator reserve in the subendocardium before it does so in the subepicardium (12). Hoffman and Spaan (13) proposed a systolic–diastolic interaction hypothesis to explain the greater vulnerability of the subendocardium to hypoperfusion. According to this hypothesis, the higher systolic tissue pressures (and greater extravascular compressive forces) in the subendocardium both impede forward coronary arterial blood flow and squeeze out blood from microvessels so that refilling of the vasculature is required before diastolic reflow can occur. Because of these factors, the subendocardium must compensate for increases in heart rate (reduced duration of diastole) and left ventricular end-diastolic pressure, and for reductions in perfusion pressure, by arteriolar dilation. Because the subendocardium was maximally dilated during severe hemodilu-

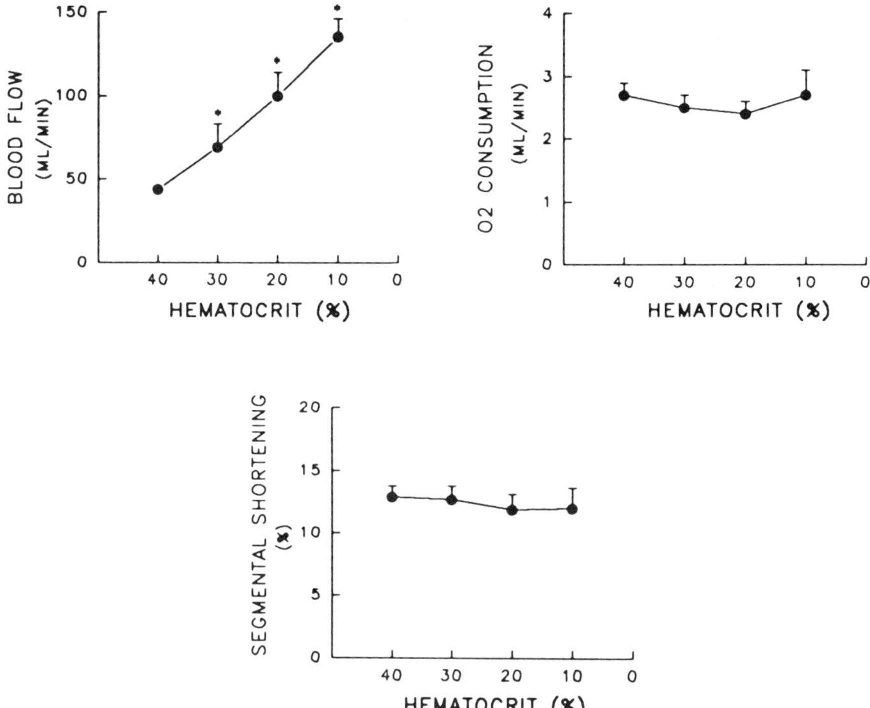

Fig. 9 Effects of graded, selective coronary hemodilution on local blood flow, oxygen consumption, and segmental shortening. *Significantly different (p <0.05) from HCT_{40}. [From Crystal and Salem (5) with permission. © International Anesthesia Research Society.]

tion (loss of hyperemic response), this compensatory vasodilation was not possible, and relative subendocardial hypoperfusion ensued. It is noteworthy that when severe hemodilution was confined to the coronary circulation and hemodynamic conditions were tightly controlled (Fig. 8), the increases in regional myocardial blood flow were transmurall, uniform (Fig. 11) (2).

In light of the facilitating influence of reduced viscosity on blood flow, the failure of hemodilution to cause any increase in blood flow through the stenotic LAD was surprising. We speculate that this finding may reflect a reduction in distal coronary pressure that was sufficient to negate the effect of reduced blood viscosity on coronary blood flow. The tendency for hemodilution to accentuate inertial pressure losses in the coronary circulation would provide a mechanism for such a reduction in distal coronary pressure. Although the coronary stenosis did not itself alter

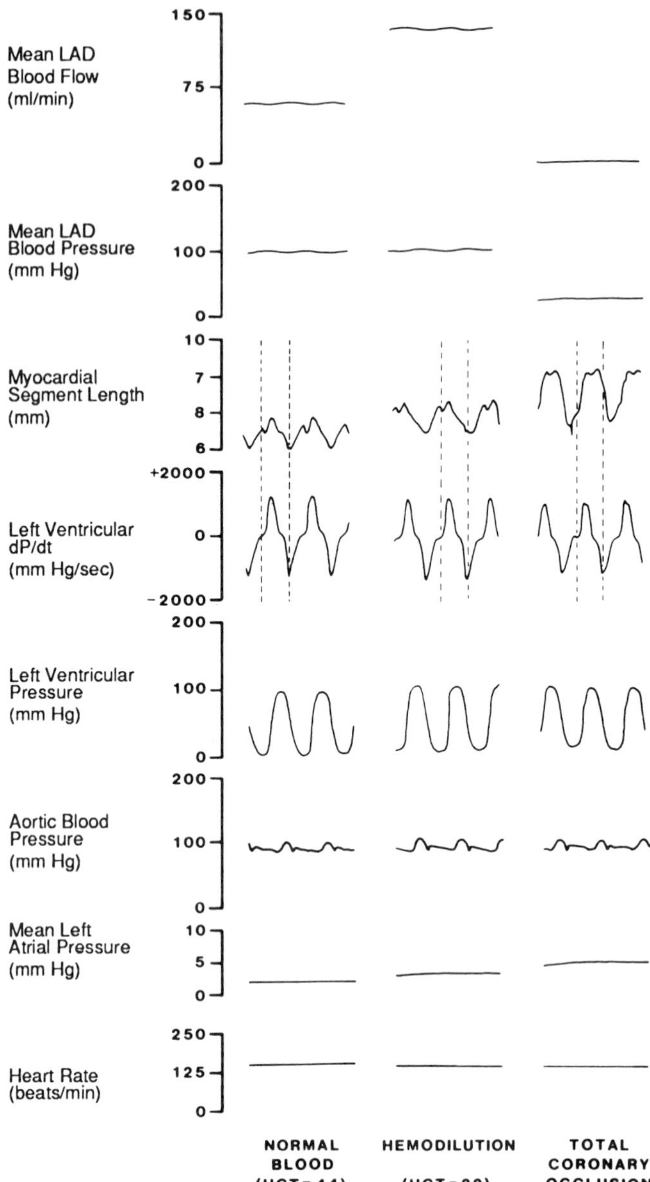

Fig. 10 Actual tracings demonstrating typical changes in monitored hemodynamic parameters during selective coronary hemodilution. With perfusion pressure constant, hemodilution to a hematocrit of 22% caused a doubling of coronary blood flow. End-diastolic and end-systolic lengths increased slightly, but segmental shortening was not affected. Systolic lengthening during occlusion of the inflow line verified that crystals were implanted in the LAD-perfused myocardium. [From Crystal and Salem (5) with permission. © International Anesthesia Research Society.]

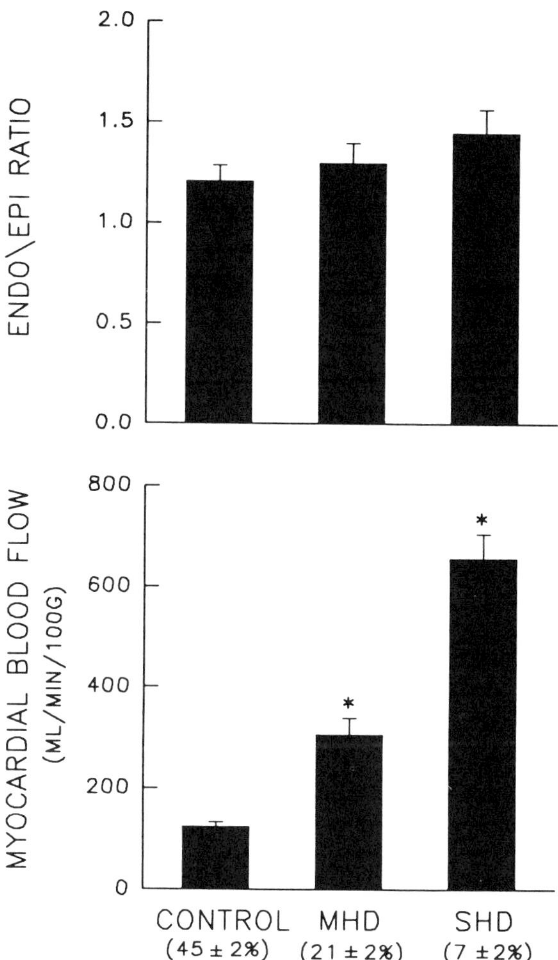

Fig. 11 Moderate (MHD) and severe (SHD) selective coronary hemodilution caused progressive increaes in myocardial blood flow, while leaving unaltered the ratio of endocardial to epicardial blood flow. Numbers in parentheses are hematocrit values. *Significantly different ($p < 0.05$) from control. [From Crystal (2) with permission from the American Physiological Society.]

the transmural distribution of myocardial blood flow (implying that distal coronary pressure remained within the autoregulatory range in both the subendocardium and subepicardium), the subsequent decrease in hematocrit was associated with a greater blood flow in the subepicardium compared to the subendocardium. Such a heterogeneous flow in left ventricular myocardium supplied by a constricted vessel has been demonstrated during increases in cardiac work (14) and during infusion of vasodilators (15), and it has been termed "transmural coronary steal." This phenomenon reflects the lesser ability of the left ventricular subendocardium compared to the left ventricular subepicardium to tolerate reductions in distal coronary pressure.

Although no direct measurements of contractility in myocardium supplied by the stenotic LAD were available, the regional hemodynamic and metabolic responses at a hematocrit of approximately 17% were consistent with myocardial ischemia and an impairment to local contractility. These responses included lack of increased blood flow resulting in reduced transmural O_2 supply and consumption, lactate production, reduced values for pH, p_{O_2}, and hemoglobin saturation in regional coronary venous samples, and relative subendocardial hypoperfusion. This segmental cardiac dysfunction was apparently responsible for preventing the compensatory increase in stroke volume and cardiac index during hemodilution, and it caused the systemic hemodynamic responses (i.e., reduced aortic pressure, increased heart rate, and increased left atrial pressure) which ultimately lead to ischemic changes in the normally perfused CIRC region and global cardiac failure.

In summary, the normal left ventricle tolerated isovolemic hemodilution until the hematocrit was reduced to as low as 10% because increases in myocardial blood flow, sufficient to maintain myocardial oxygen supply, occurred uniformly across the ventricular wall. Coronary vasodilation via metabolic factors was a primary mechanism in this flow increase. Cardiac failure during extreme hemodilution was due to inadequate subendocardial oxygen supply in the left ventricular wall resulting from a maldistribution of myocardial perfusion. A critical stenosis of the LAD increased the hematocrit at which cardiac failure occurred to approximately 17%. This was apparently because the stenosis prevented the normal increase in coronary blood flow during hemodilution, while it was also redistributing blood flow away from the subendocardium. The present findings demonstrate the significant extent to which a single, acute coronary stenosis can reduce the tolerance of the left ventricle to hemodilution. They also underscore the importance of recruitment of coronary vasodilator reserve in the preservation of both total and regional myocardial oxygenation during hemodilution.

B. Comparison of Effects of Isovolemic Hemodilution in Right and Left Ventricles

The right and left ventricles have major functional, hemodynamic, and metbolic differences (16,17). The right ventricle has a thinner wall and generates lower systolic pressure, and thus it normally has a smaller workload, oxygen consumption, blood flow, and oxygen extraction than the left ventricle. The lower baseline oxygen extraction in the right ventricle endows it with an oxygen extraction reserve, which is utilized along with an increased blood flow to meet the increased myocardial oxygen consumption observed under a variety of conditions, including β-adrenergic stimulation with isoproterenol (16). This is in contrast to the left ventricle, which because of its high baseline oxygen extraction is critically dependent on increases in blood flow to maintain myocardial oxygen consumption (11,16). The relative importance of augmented oxygen extraction in the heart ventricles is consistent with their respective capillary reserves. The right ventricle normally has a smaller number of open capillaries and a greater total capillary density; thus, capillary reserve is approximately twice as large in the right ventricle compared to the left ventricle (18). We hypothesized that the reserves for oxygen extraction and capillary recruitment in the right ventricle would render this chamber less dependent than the left ventricle on blood flow increases to maintain myocardial oxygen consumption during isovolemic hemodilution.

In series 1 of the study, changes in myocardial oxygen supply and consumption in the right and left ventricles of the same *in situ* canine hearts were evaluated during progressive isovolemic hemodilution. Particular attention was given to the relative roles of increases in blood flow and oxygen extraction in the maintenance of oxygen consumption in the two heart ventricles. Series 1 demonstrated no augmentation in myocardial oxygen extraction in the right ventricle during hemodilution, which suggested that hemodilution per se may have impaired release of oxygen from red blood cells. To evaluate this possibility, additional studies (series 2) were performed in which isoproterenol was infused (0.1 μg/kg/min i.v.) following hemodilution to a hematocrit 50% of baseline, in an attempt to recruit the oxygen extraction reserve of the right ventricle. In both series 1 and 2, measurements of cardiac output and mixed venous oxygen content were obtained so that changes in oxygen supply–demand relations in the heart ventricles could be compared to those in the systemic circulation.

The animal preparation, methods, and protocol used in this study were essentially similar to those described under Section II,A with two noteworthy exceptions: (1) coronary blood flow was measured electromagnetically in the right coronary artery and (2) venous samples were obtained from a prominent anterior coronary vein draining the right ventricular wall, as well as from the coronary sinus and pulmonary artery.

Baseline values for myocardial oxygen consumption, oxygen supply, blood flow, arteriovenous difference, and percent oxygen extraction were lower in the right ventricle than in the left ventricle (Fig. 12). Graded reductions in hematocrit caused progressive increases in myocardial blood flow in both ventricles until a hematocrit of 10 ± 1%. These increases in

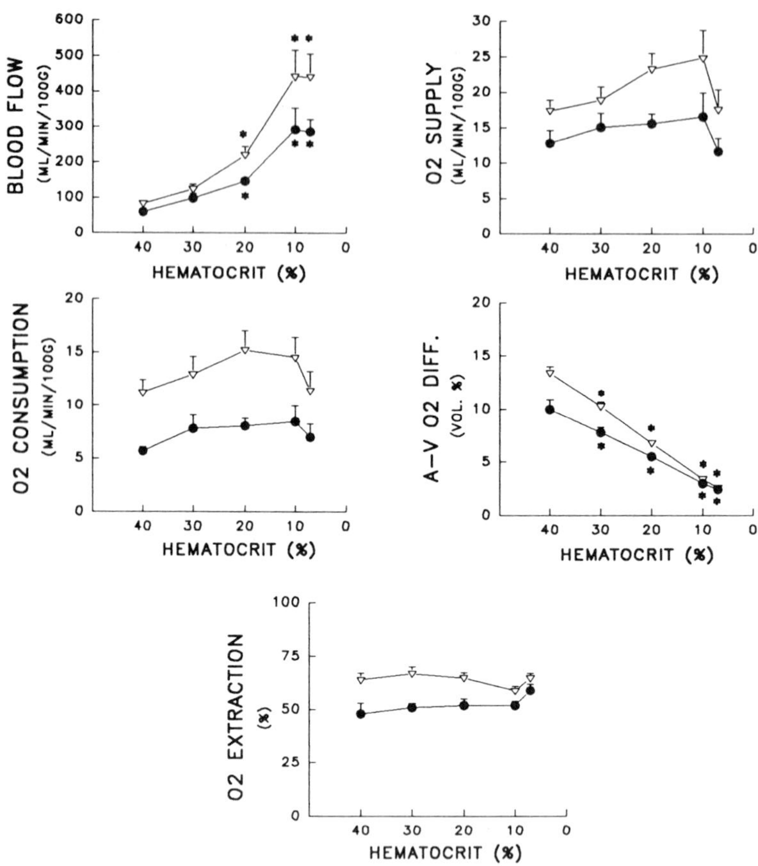

Fig. 12 Comparison of effects of graded isovolemic hemodilution on parameters of oxygen supply and consumption in right (circles) and left (triangles) ventricles. *Significantly different ($p < 0.05$) from HCT_{40}. [From Crystal *et al.* (3) with permission from the American Physiological Society.]

myocardial blood flow were transmurally uniform (data not presented). A further reduction in the hematocrit to a value of 7 ± 1% caused no additional increase in myocardial blood flow in either ventricle.

Hemodilution caused parallel decreases in arterial oxygen content and in the arteriovenous oxygen difference within the ventricles, with the result that local values for oxygen extraction, p_{vO_2}, and SvO_2 did not vary (Fig. 12; Table III). The increases in myocardial blood flow in the ventricles were proportional to these reductions in arterial oxygen content and the arteriovenous oxygen difference, which maintained myocardial oxygen

Table III

Regional Venous Blood Gases during Graded Hemodilution until Cardiac Failure

Area and measure	Baseline	Hematocrit 30%	Hematocrit 20%	Hematocrit 10%	Cardiac failure
Right ventricle					
p_{O_2}, mmHg	34 ± 3	33 ± 1	33 ± 1	33 ± 1	33 ± 1
p_{CO_2}, mmHg	42 ± 1	40 ± 1	40 ± 1	40 ± 1	43 ± 3
pH	7.39 ± 0.01	7.37 ± 0.02	7.36 ± 0.01	7.35 ± 0.01	7.28 ± 0.04[b]
S_{O_2}, %	54 ± 5	51 ± 2	51 ± 2	54 ± 2	53 ± 4
Left ventricle					
p_{O_2}, mmHg	27 ± 2	27 ± 1	28 ± 2	32 ± 2	30 ± 1
p_{CO_2}, mmHg	48 ± 2	45 ± 200	45 ± 2	43 ± 1	45 ± 2
pH	7.36 ± 0.01	7.34 ± 0.02	7.34 ± 0.01	7.33 ± 0.02	7.26 ± 0.02[b]
S_{O_2}, %	37 ± 3	33 ± 3	37 ± 2	46 ± 2	45 ± 3
Systemic					
p_{O_2}, mmHg	56 ± 2	47 ± 2[b]	42 ± 3[b]	35 ± 1[b]	30 ± 2[b]
p_{CO_2}, mmHg	42 ± 1	44 ± 2	44 ± 2	48 ± 1[b]	50 ± 1[b]
pH	7.37 ± 0.01	7.33 ± 0.01	7.32 ± 0.01	7.27 ± 0.01[b]	7.22 ± 0.02[b]
S_{O_2}, %	81 ± 2	68 ± 4[b]	61 ± 3[b]	52 ± 2[b]	45 ± 4[b]

[a] Values are means ± S.E. From Crystal *et al.* (3) with permission.
[b] Significantly different ($p < 0.05$) from baseline.

supply and oxygen consumption at baseline values. Although percent lactate extraction decreased in both ventricles during progressive hemodilution (probably secondary to the elevated blood flow rate), regional myocardial lactate uptake was well maintained as long as the hematocrit remained equal to or greater than 10% (Fig. 13).

Graded hemodilution caused progressive decrease in vasodilator reserve ratio in the right coronary circulation. At a hematocrit of 10% the vasodilator reserve was nearly exhausted (Fig. 14).

The effects of hemodilution on systemic oxygen transport are presented in Fig. 15. Because the increases in cardiac index were modest and much less than proportional to the decreases in arterial oxygen content, hemodilution caused a rapid decline in systemic oxygen supply. Systemic oxygen extraction increased progressively during graded hemodilution, which tended to maintain the systemic arteriovenous oxygen difference. This factor along with the modest increases in cardiac index held systemic oxygen consumption constant. The increased oxygen extraction in the systemic circulation caused progressive decreases in mixed venous p_{O_2} and

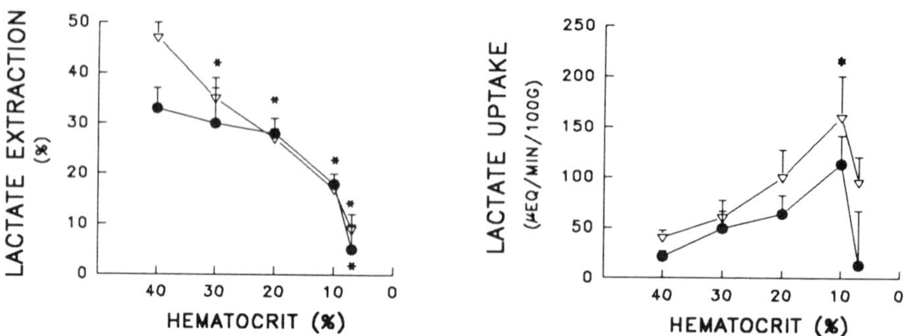

Fig. 13 Comparison of effects of graded isovolemic hemodilution on lactate extraction and uptake in right and left ventricles. *Significantly different (p <0.05) from HCT_{40}. [From Crystal *et al.* (3) with permission from the American Physiological Society.]

S_{O_2} during graded hemodilution (Table III). However, the arterial lactate concentration remained equal to the baseline level of 1.9 ± 0.1 mEq/liter until the hematocrit was reduced to 7 ± 1%, when the lactate concentration increased to 4.0 ± 0.1 mEq/liter.

Myocardial and systemic effects of isoproterenol following hemodilution (series 2) are presented in Fig. 16. Isoproterenol infusion following hemodilution increased myocardial oxygen consumption in both heart ventricles. The isoproterenol-induced increase in myocardial oxygen consumption was satisifed by the combined effect of increased blood flow and oxygen extraction in the right ventricle, and by increased blood flow alone in the left ventricle. The increased oxygen extraction in the right ventricle reduced local venous p_{O_2} and S_{O_2} (data not shown). Because the increase

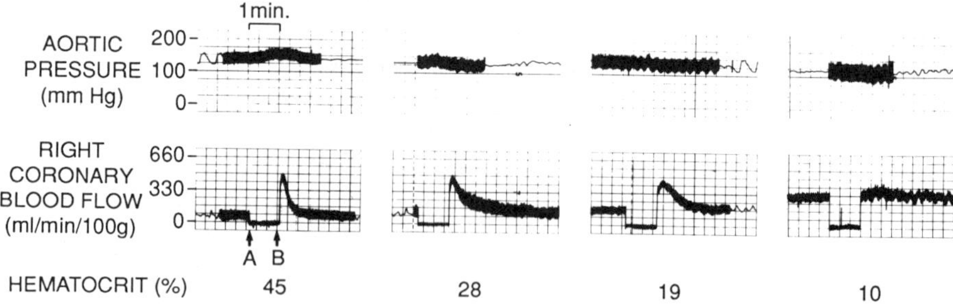

Fig. 14 Original tracings demonstrating effect of graded isovolemic hemodilution on reactive hyperemia in the right coronary artery of one dog. Findings are comparable to those for the left coronary artery presented in Fig. 5. The vasodilator reserve ratio decreased with decreasing hematocrit and was essentially lost at a hematocrit of 10%. [From Crystal *et al.* (3) with permission from the American Physiological Society.]

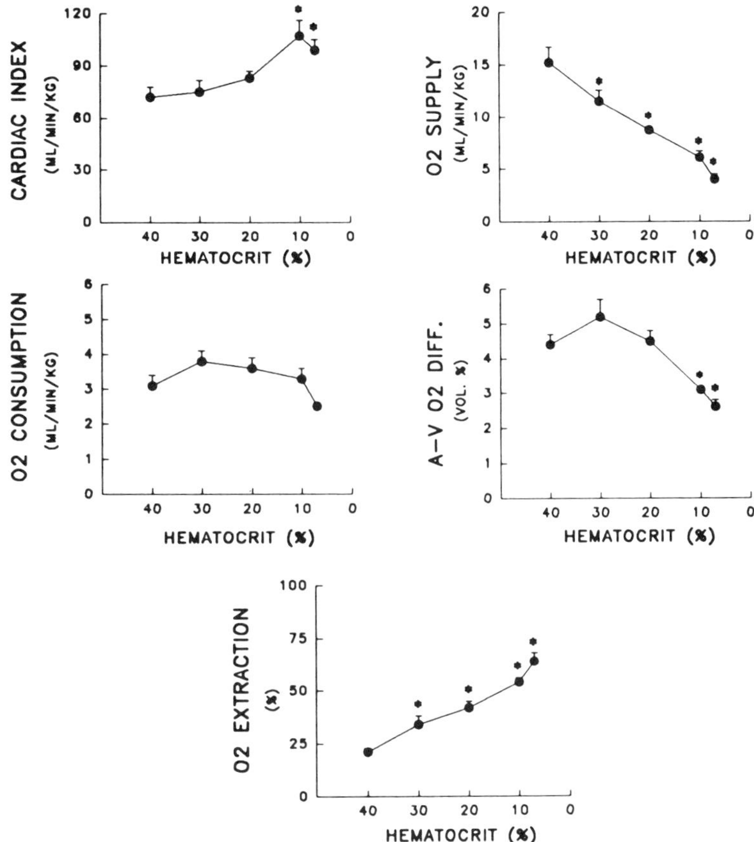

Fig. 15 Effects of graded isovolemic hemodilution on parameters of oxygen supply and demand in systemic circulation. *Significantly different ($p < 0.05$) from HCT_{40}. [From Crystal *et al.* (3) with permission from the American Physiological Society.]

in cardiac index by isoproterenol following hemodilution occurred in the presence of a constant systemic oxygen uptake, there were decreases in systemic arteriovenous oxygen difference and oxygen extraction, as reflected in increased values for mixed venous p_{O_2} and S_{O_2}.

Isovolemic hemodilution caused increases in myocardial blood flow in both ventricular walls. The progressive diminution and virtual exhaustion of vasodilator reserve during graded hemodilution in the right coronary circulation provides evidence that the right coronary circulation, in similarity to the left coronary circulation (see Section II,A), depends on coronary vasodilation for maintenance of oxygen supply during hemodilution.

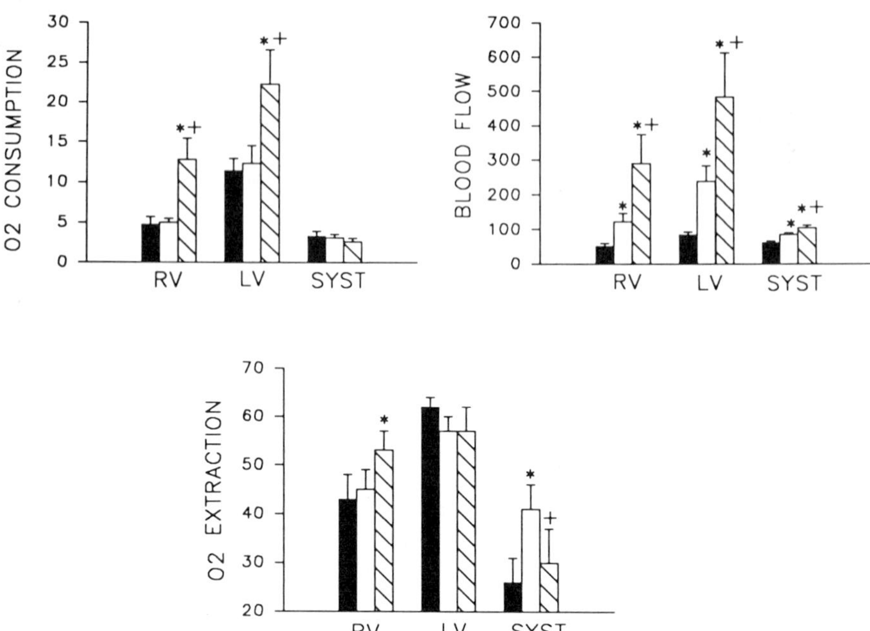

Fig. 16 Effects of hemodilution alone (open bars) and combined with isoproterenol (hatched bars) on oxygen consumption, blood flow, and oxygen extraction in right ventricle (RV), left ventricle (LV), and systemic circulation (SYST) compared to control (solid bars). Units are ml/min/100 g for O_2 consumption and blood flow in RV and LV, and ml/min/kg for O_2 consumption and blood flow in SYST. Units for O_2 extraction are %. *Significantly different ($p < 0.05$) from control. + Significantly different ($p < 0.05$) from hemodilution. [From Crystal *et al.* (3) with permission from the American Physiological Society.]

Several lines of evidence suggest that oxygen supply and consumption were adequate in both heart ventricles as long as the hematocrit exceeded 10%. First, indices of global cardiac performance (e.g., aortic pressure and pulmonary capillary wedge pressure) were not altered, and cardiac index increased. Second, the increases in myocardial blood flow in both ventricular walls were transmurally uniform, implying a lack of selective subendocardial hypoperfusion. Third, myocardial lactate extraction and lactate uptake in both ventricles were maintained, suggesting the absence of anaerobic metabolism and of myocardial ischemia. In both ventricles, oxygen extraction remained constant, and increased blood flow constituted the sole mechanism for the maintenance of oxygen consumption during progressive isovolemic hemodilution. Such flow dependence was expected in the left ventricle, but not in the right ventricle, which, because of a smaller baseline oxygen extraction, has a significant oxygen extraction reserve.

Why the oxygen extraction reserve was not utilized when right ventricular oxygenation was jeopardized by hemodilution is uncertain. One possible explanation was that hemodilution per se impaired release of oxygen from red blood cells by altering plasma protein and buffer content and by diminishing the intracellular convection of oxygen (19). However, the ability of isoproterenol to increase right ventricular O_2 extraction during hemodilution would seem to rule out this explanation. A more likely explanation was that hemodilution did not provide an adequate metabolic stimulus for recruitment of capillaries (18). With blood flow rate increased, this resulted in shortened transit times within capillaries, thus limiting exchange of oxygen across the capillary wall. This mechanism is supported by two lines of evidence. First, Crystal and Salem reported that small vessel blood volume, an index of open capillary density, did not change in the right ventricle during hemodilution (20). Furthermore, Martini and Honig, using stop-motion cinematography in rat hearts beating *in situ,* demonstrated that capillary recruitment in the right ventricle correlated with the decrease in coronary venous p_{O_2} (presumed to be a reflection of tissue p_{O_2}) and was unrelated to changes in coronary venous oxygen content (21). Because hemodilution-related increases in right coronary blood flow were sufficient to maintain coronary venous p_{O_2} constant, no increase in open capillary density would be expected.

In contrast to findings obtained in the heart, the increases in blood flow in the systemic circulation (i.e., cardiac index) during hemodilution were relatively small and not sufficient in magnitude to avert a decline in oxygen supply. Nevertheless, systemic oxygen consumption was maintained because oxygen extraction increased, resulting in falls in both mixed venous oxygen saturation and p_{O_2}. The increase in cardiac index (accompanied by a proportional increase in systemic oxygen consumption) elicited by the isoproterenol infusion during hemodilution indicated that the size of the increase in cardiac index during hemodilution were limited not by the available cardiac reserve but by the level of adrenergic stimulation. The reduced values for mixed venous p_{O_2} during progressive hemodilution reflected decreases in average p_{O_2} in the body tissues. The arterial blood lactate findings imply that tissue p_{O_2} was not reduced sufficiently to stimulate anaerobic metabolism until mixed venous p_{O_2} declined to approximately 30 mmHg (corresponding to a hematocrit of $7 \pm 1\%$).

In summary, the right ventricle, in similarity to the left ventricle, can tolerate decreases in hematocrit to as low as 10% during progressive isovolemic hemodilution. Both heart ventricles depended on increased blood flow alone to maintain oxygen consumption during hemodilution;

oxygen extraction was not augmented. The mobilization of the oxygen extraction reserve in the right ventricle by isoproterenol during hemodilution suggested that the lack of augmented oxygen extraction there during hemodilution alone was apparently not due to impaired unloading of oxygen by diluted red blood cells. We speculate that it was rather due to inadequate capillary recruitment which shortened transit times, thus limiting exchange of oxygen across the capillary wall.

C. Cardiac Effects of Combined Isovolemic Hemodilution and Controlled Hypotension

Combined isovolemic hemodilution and drug-induced controlled hypotension has been proposed as a method of blood conservation. Because the coexistence of depressed driving pressure for coronary blood flow and reduction in oxygen carrying capacity of the arterial blood may risk development of myocardial hypoxia, studies of myocardial oxygen supply–demand relations are necessary before a technique of combined hemodilution and hypotension could be considered for clinical use.

In the initial series described below, cardiac effects of hypotension with sodium nitroprusside (SNP) were evaluated in hemodiluted dogs. Adenosine (ADEN), a metabolic breakdown product of adenosine triphosphate, is an endogenous vasodilator that has been implicated in local regulation of coronary blood flow. Adenosine has favorable characteristics that have stimulated interest in its use as a clinical hypotensive agent. Intravenous infusion of exogenous adenosine causes arterial hypotension that is rapidly achieved, easily controlled, short-lived, and, furthermore, is not associated with any apparent hematologic or biochemical evidence of toxicity. A second series of studies was performed using ADEN to produce hypotension in hemodiluted dogs, and these findings were compared to those in which SNP was used.

The basic methods were essentially similar to those described above. The effects of hemodilution alone (hematocrit reduced to 50% of baseline) and controlled hypotension with either SNP or ADEN (mean aortic pressure reduced to 50% of baseline) were assessed in the left ventricular walls of anesthetized, open-chest dogs.

Findings in the SNP group are presented in Figs. 17 and 18. Administration of SNP during hemodilution caused parallel reductions in myocardial blood flow, oxygen supply, and oxygen consumption, with the result that local values for arteriovenous oxygen difference and oxygen extraction remained constant. Both lactate extraction and lactate uptake were well maintained during combined SNP-induced hypotension and hemodilution. Figure 19 compares findings in the ADEN group to those in the SNP

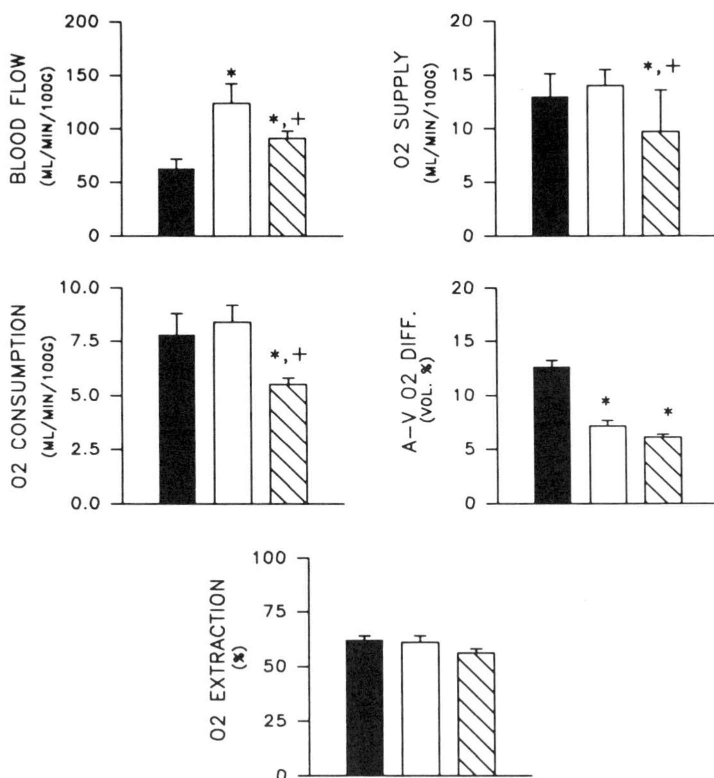

Fig. 17 Effects of hemodilution alone (open bars) and combined with controlled hypotension using sodium nitroprusside (hatched bars) on parameters of oxygen supply and demand in the left ventricle compared to control (solid bars). *Significantly different ($p < 0.05$) from control. $^{+}$ Significantly different ($p < 0.05$) from hemodilution. [From Crystal and Salem (6) with permission. © International Anesthesia Research Society.]

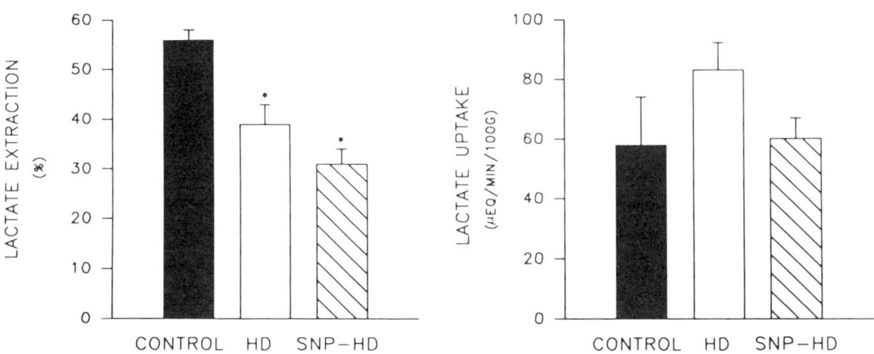

Fig. 18 Effects of hemodilution (HD) alone and combined with controlled hypotension using sodium nitroprusside (SNP) on myocardial lactate extraction and lactate uptake. *Significantly different ($p < 0.05$) from control. [From Crystal and Salem (6) with permission. © International Anesthesia Research Society.]

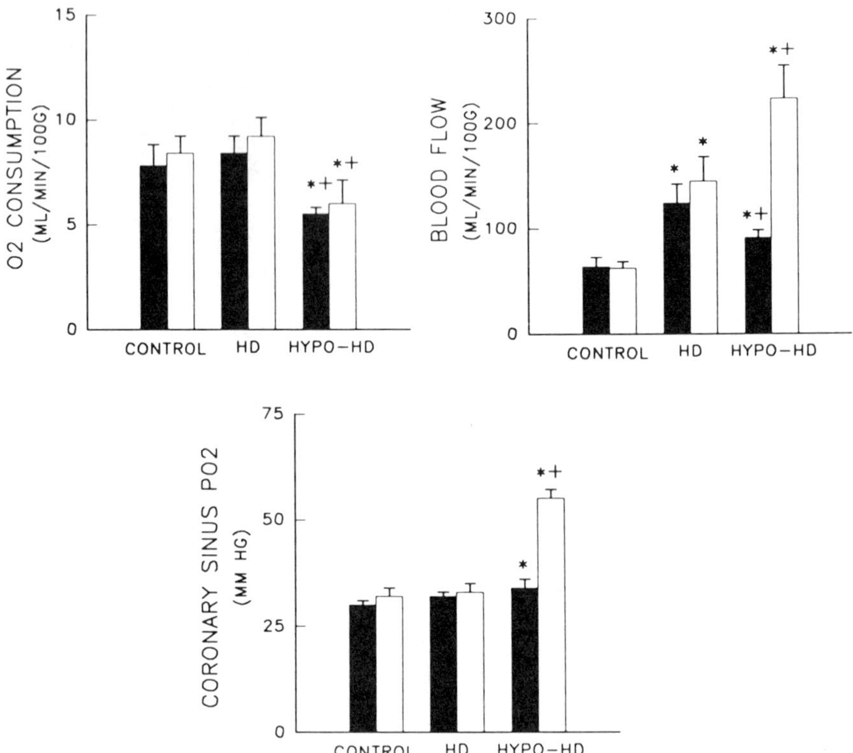

Fig. 19 Effects of hemodilution (HD) alone and combined with controlled hypotension (HYPO) using sodium nitroprusside (solid bars) or adenosine (open bars) on myocardial O_2 consumption, myocardial blood flow, and coronary sinus p_{O_2}. *Significantly different ($p < 0.05$) from control. $^+$ Significantly different ($p < 0.05$) from hemodilution. [From Crystal *et al.* (4) and Crystal and Salem (6) with permission. © International Anesthesia Research Society.

group. In contrast to the results with SNP, the decrease in myocardial oxygen consumption with ADEN was accompanied by an increase in myocardial blood flow and a decrease in myocardial oxygen extraction (reflected in a pronounced rise in coronary sinus p_{O_2}).

The SNP-induced hypotension during hemodilution was accompanied by proportional decreases in myocardial blood flow (oxygen supply) and myocardial oxygen consumption, suggesting intact metabolic control of coronary blood flow. The observed values for coronary sinus p_{O_2} and for lactate extraction and uptake were consistent with well-maintained myocardial oxygenation. On the other hand, ADEN-induced hypotension during hemodilution increased myocardial oxygen supply (blood flow) while it was decreasing myocardial oxygen consumption. This resulted in a reduction in myocardial oxygen extraction and attendant increase in coronary sinus p_{O_2}, which are signs of luxuriant myocardial perfusion owing to a direct vasorelaxing action and an improved balance between myocardial oxygen supply and demand.

These findings suggest that the protocol for combined isovolemic hemodilution and controlled hypotension evaluated in the present study was tolerated well by normal canine hearts, whether SNP or ADEN was used to induce hypotension. However, the pronounced coronary vasodilation caused by ADEN raises concerns about the potential for "coronary steal" if this drug were used for controlled hypotension in the patient with coronary artery disease.

III. Summary

The findings presented in this article pertain strictly to the specific conditions of the studies and should be extrapolated to human patients with caution for several reasons. First, the experimental model of a single acute coronary stenosis is a rare clinical occurrence, and it ignores the influence of the well-developed collateral circulation and impaired endothelial function which may accompany long-standing coronary artery disease (22). Second, the absolute hematocrit at which cardiac dysfunction arises would likely vary if a different anesthetic technique or diluent were used, or if the study were carried out in the absence of thoracotomy and positive-pressure ventilation or at a reduced body temperature. Nevertheless, the present findings provide broad guidelines for assessing when hemodilution may be used safely as a method of blood conservation. They suggest (1) that relatively severe hemodilution alone or moderate hemodilution combined with controlled hypotension with either SNP or ADEN may be used safely if cardiac function is normal and the coronary circulation is not obstructed, but (2) that even moderate hemodilution alone may be unsafe in patients in whom the coronary vasodilator reserve is exhausted or severely depleted by a proximal stenosis.

Acknowledgments

Research was supported in part by a grant from the National Heart, Lung, and Blood Institute (HL-47629).

References

1. Crystal, G. J., Rooney, M. W., and Salem, M. R. (1988). Regional hemodynamics and oxygen supply during isovolemic hemodilution alone and combined with adenosine-induced controlled hypotension. *Anesth. Analg.* **67,** 211–218.
2. Crystal, G. J. (1988). Coronary hemodynamics during local hemodilution in canine hearts. *Am. J. Physiol.* **254,** H525–H531.

3. Crystal, G. J., Kim, S.-J., and Salem, M. R. (1994). Right and left ventricular O_2 uptake during hemodilution and β-adrenergic stimulation. *Am. J. Physiol.* **265:** H1769–1777, 1992.
4. Crystal, G. J., Rooney, M. W., and Salem, M. R. (1988). Myocardial blood flow and oxygen consumption during isovolemic hemodilution alone and in combination with adenosine-induced controlled hypotension. *Anesth. Analg.* **67,** 539–547.
5. Crystal, G. J., and Salem, M. R. (1988). Myocardial oxygen consumption and segmental shortening during selective coronary hemodilution in dogs. *Anesth. Analg.* **67,** 500–508.
6. Crystal, G. J., and Salem, M. R. (1991). Myocardial and systemic hemodynamics during isovolemic hemodilution alone and combined with nitroprusside-induced controlled hypotension. *Anesth. Analg.* **72,** 227–237.
7. Levy, P. S., Kim, S.-J., Eckel, P. K., Chavez, R., Ismail, E. F., Gould, S. A., Salem, M. R., and Crystal, G. J. 1993). Limit to cardiac compensation during acute isovolemic hemodilution: Influence of coronary stenosis. *Am. J. Physiol.* **265,** H340–H349.
8. Brazier, J., Cooper, N., Maloney, J. V., and Buckberg, G. (1974). The adequacy of myocardial oxygen delivery in acute normovolemic anemia. *Surgery* **75,** 508–516.
9. Jan, K.-M., and Chien, S. (1977). Effect of hematocrit variations on coronary hemodynamics and oxygen utilization. *Am. J. Physiol.* **233,** H106–H113.
10. Hofling, B., Restorff, W. V., Holtz, J., and Bassenge, E. (1975). Viscous and inertial fractions of total perfusion energy dissipation in the coronary circulation of the *in situ* perfused dog heart. *Pfluegers Arch.* **358,** 1–10.
11. Feigl, E. M. (1983). Coronary physiology. *Physiol. Rev.* **63,** 1–203.
12. Bagger, H. (1978). Distribution of maximum coronary blood flow in the left ventricular wall of anesthetized dogs. *Acta Physiol. Scand.* **104,** 48–60.
13. Hoffman, J. I. E., and Spann, J. A. E. (1990). Pressure–flow relations in coronary circulation. *Physiol. Rev.* **20,** 331–389.
14. Ball, R. M., and Bache, R. J. (1973). Distribution of myocardial blood flow in the exercising dog with restricted coronary inflow. *Circ. Res.* **38,** 60–66.
15. Gallagher, K. M., Folts, J. D., Shebuski, R. J., Rankin, J. H. G., and Rowe, G. G. (1980). Subepicardial vasodilator reserve in the presence of critical coronary stenosis in dogs. *Am. J. Cardiol.* **46,** 67–73.
16. Kusachi, S., Nishiyama, O., Yasuhara, K., Saito, D., Haraoka, S., and Nagashima, H. (1982). Right and left ventricular oxygen metabolism in open-chest dogs. *Am. J. Physiol.* **243,** H761–H766.
17. Marcus, M. L. (1983). Differences in regulation of coronary perfusion to the right and left ventricles. *In* "The Coronary Circulation in Health and Disease," pp. 337–347. McGraw-Hill, New York.
18. Henquell, L., and Honig, C. R. (1976). Intercapillary distances and capillary reserve in right and left ventricles: Significance for control of tissue p_{O_2}. *Microvasc. Res.* **12,** 35–41.
19. Zander, R., and Schmid-Schonbein, H. (1972). Influence of intracellular convection on the oxygen release by human erythrocytes. *Pfluegers Arch.* **335,** 58–73.
20. Crystal, G. J., and Salem, M. R. (1989). Blood volume and hematocrit in regional circulations during isovolemic hemodilution in dogs. *Microvasc. Res.* **37,** 237–240.
21. Martini, J., and Honig, C. R. (1969). Direct measurements of intercapillary distance in beating rat heart *in situ* under various conditions of O_2 supply. *Microvasc. Res.* **1,** 244–256.
22. Freiman, P. C., Mitchell, G. G., Heistad, D. D., Armstrong, M. I., and Harrison, D. G. (1986). Atherosclerosis impairs endothelium-dependent vascular relaxation to acetylcholine and thrombin in primates. *Circ. Res.* **58,** 783–789.

Plasma Membrane Ca^{2+}-ATPase as a Target for Volatile Anesthetics

Danuta Kosk-Kosicka
The Johns Hopkins University School of Medicine
Department of Anesthesiology and Critical Care Medicine
Baltimore, Maryland 21287

I. Introduction

The molecular mechanisms by which anesthetics exert their effects and the molecular site(s) of action have not been identified. An array of published data suggests that general anesthetics may act on specific hydrophobic sites on protein molecules (1–5). It appears that the most obvious candidate target proteins for general anesthetic effect should be located in brain and heart. Accumulating evidence suggests that the volatile anesthetics at clinically useful concentrations act in heart in a number of specific ways to decrease the magnitude of the intracellular calcium transient leading to depression of contractile force (6). The action of anesthetics appears to include (1) change in Ca^{2+} influx through sarcolemmal Ca^{2+} channels, (2) decrease of Ca^{2+} in sarcoplasmic reticulum (SR), and (3) modification of the responsiveness of the contractile proteins to activation by calcium. In addition to the reported functional alteration of the SR Ca^{2+} pump by anesthetics, their effects on the sarcolemmal Ca^{2+} pump may also contribute to the decline of the Ca^{2+} transient.

Our aim in this study was to determine whether the function of the Ca^{2+} pump in the plasma membrane is indeed altered by anesthetics. We

Advances in Pharmacology, Volume 31

have chosen to start the investigation with the human red cell Ca^{2+} pump, which is the most experimentally accessible plasma membrane Ca^{2+}-ATPase. Simplicity of the system is its strength. In the absence of voltage-regulated Ca^{2+} channels, Na^+/Ca^{2+} exchanger, or endoplasmic reticulum (SR) type Ca^{2+}-ATPase in human red cells, the interference of the other Ca^{2+}-transporting systems could be eliminated. Being the sole Ca^{2+}-transporting system, the Ca^{2+}-ATPase is instrumental in Ca^{2+} homeostasis in the red blood cell. Thus any functional alteration of the Ca^{2+}-ATPase by volatile anesthetics may lead to serious perturbations in Ca^{2+}-regulated processes in the cell. We have evaluated the enzyme as a model for studies of the mechanism of action of volatile anesthetics on membrane proteins. The long-term goal is to establish whether the plasma membrane Ca^{2+}-ATPase in tissues such as brain or heart could be a target for volatile anesthetics *in vivo*.

Over years we have developed conditions providing adequate quantities of purified, functional, well-studied Ca^{2+}-ATPase (7–15). We as well as others have demonstrated that the properties of the purified enzyme are very similar to the properties of the unperturbed enzyme in the red blood cell membrane with respect to the dependency of a multitude of regulating factors (13). Accordingly the purified detergent-soluble Ca^{2+}-ATPase appears well suited for kinetic characterization of the plasma membrane Ca^{2+} pump. At least five modes of activation have been reported for the enzyme. We have characterized extensively two of them, namely, activation of the enzyme by binding of a regulatory protein, calmodulin, to enzyme monomers and activation by a novel calmodulin-independent mode we discovered in which the enzyme is activated by self-association. We have also established a straightforward assay of the effect of volatile anesthetics on Ca^{2+}-ATPase activation by the two modes (16).

To determine whether the Ca^{2+}-ATPase is a suitable model for studies of the mechanism of action of volatile anesthetics on plasma membrane proteins, we have investigated the effects of four volatile anesthetics on enzyme activity, characterized the inhibition quantitatively, and compared the inhibitory potency of the anesthetics to their clinical potency. Furthermore, we have compared the effects of volatile anesthetics with the effects of short-chain alcohols, general anesthetics which have been thoroughly investigated in various biological and model systems in search for the mechanism of anesthetic action. We have established that the observed inhibition of the Ca^{2+}-ATPase meets the following criteria for a good model protein: (1) it is dose dependent, (2) it is reversible, (3) it occurs at clinically relevant concentrations, and (4) the correlation between clinical and inhibitory potency is very good.

II. Isolation and Activity Assay of Ca^{2+}-ATPase

The methods used for preparation of red blood cell membrane, enzyme purification, determination of protein and Ca^{2+} concentrations, and Ca^{2+}-ATPase activity assays have been described in detail elsewhere (7,8,12). Briefly, the Ca^{2+}-ATPase was isolated from human packed red blood cells purchased from the local Red Cross. Red cell membrane proteins were solubilized in the presence of the nonionic detergent $C_{12}E_8$, and the Ca^{2+}-ATPase was purified by calmondulin affinity chromatography. Total calcium was measured by atomic absorption spectrometry, and free Ca^{2+} concentrations were calculated from total calcium and EGTA.

The Ca^{2+}-ATPase activity was determined by colorimetric measurement of inorganic phosphate (P_i) production. In this study the Ca^{2+} -ATPase activity was assayed in the presence of calmodulin. The activity assay was performed in sealed 1.7-ml polypropylene tubes in a total reaction volume of 100 μl. After addition of all reagents and immediately after starting the reaction with 3 m*M* ATP, the volatile anesthetics or alcohols were delivered to the reaction tube in an air-tight Hamilton syringe, following which the tube was sealed and vortexed. The reaction was carried out for 15–30 min at 37 or 25°C. Reactions were terminated by treating the samples with ammonium molybdate/metavanadate at individual times so the anesthetic was not depleted. Steady-state velocities were obtained from plots of inorganic phosphate production, which were linear with time. The aliquots of volatile anesthetics delivered to the assay tube were taken from solutions of saturated volatile anesthetics in reaction mixture and were prepared daily from the stock of pure volatile anesthetics under nitrogen gas. Except for desflurane solutions (prepared at 4°C), all other anesthetics solutions were prepared in room temperature. Tubes run in parallel were used for measurements of the effective volatile anesthetic concentrations in the reaction mixture during the activity assay.

The anesthetic was extracted from the reaction mixture with heptane and measured by gas chromatography. The concentrations of volatile anesthetics in % (v/v) in gas were calculated from measured coefficients as described previously (16). The partition coefficients (between reaction mixture and gas phase) were measured for each volatile anesthetic at 37°C (and additionally at 25°C for halothane), after 2 hr of incubation with 30 sec of vortexing at 15-min intervals; they were similar to the values reported in literature for water/gas partition coefficients. Data are expressed as the mean ± standard error of the mean of independent experiments.

III. Effects of Anesthetics

This article reports on the effects of general anesthetics on the activation of the purified Ca^{2+}-ATPase by the calmodulin-dependent pathway. All four volatile anesthetics studied at 37°C inhibited the enzyme in a dose-dependent manner (Fig. 1). The order of inhibitory potency decreased from halothane to isoflurane, enflurane, and desflurane. In all cases the inhibitory effect was exerted in the range of concentrations routinely used in clinical anesthesia at 37°C. The inhibitory potencies of the volatile anesthetics when plotted against the corresponding clinical potencies (Fig. 2) showed that the higher the minimum alveolar concentration (MAC) value of the anesthetic, the higher was the concentration required to inhibit the Ca^{2+}-ATPase activity. The half-maximal inhibition of the Ca^{2+}-ATPase activity, as established on several enzyme preparations, occurred at 0.34 ± 0.030 % (v/v) halothane, 0.42 ± 0.050 % (v/v) isoflurane, (0.47 ± 0.050% (v/v) enflurane, and 0.56 ± 0.030 % v/v) desflurane. Thus the sensitivity of the enzyme to volatile anesthetics appears to be great: the Ca^{2+}-ATPase activity is half-inhibited at anesthetic concentrations equal to approximately 0.3–0.4 MAC for enflurane, halothane, and isoflurane, and around 0.13 MAC for desflurane.

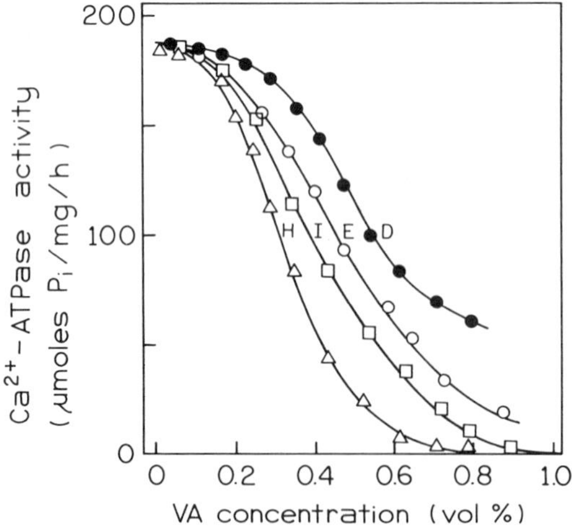

Fig. 1 Inhibition of Ca^{2+}-ATPase activity by volatile anesthetics: halothane (△), isoflurane (□), enflurane, (○), and desflurane (●). The calmodulin-dependent Ca^{2+}-ATPase activity was determined by measurement of P_i production after the reaction was run for 15–30 min at 37°C. The reaction mixture contained 50 m*M* Tris–maleate, pH 7.4, 20 m*M* KCl, 8 m*M* $MgCl_2$, 150 μM $C_{12}E_8$, 1 m*M* EGTA, 3 m*M* ATP, 100 n*M* free Ca^{2+}, 45 n*M* enzyme, 0.6 μg in 100 μl), and 40 n*M* calmodulin. [Data from Kosk-Kosicka and Roszczynska (16).]

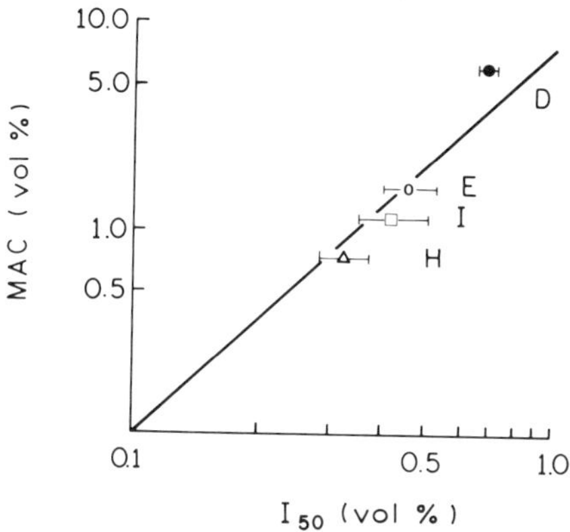

Fig. 2 Correlation between clinical potency of halothane (H), isoflurane (I), enflurane (E), and desflurane (D) and the ability of the anesthetic to inhibit the Ca^{2+}-ATPase activity. The clinical potency is expressed in MAC values (%, v/v); I_{50} is the concentration of the particular anesthetic (%, v/v) that causes 50% inhibition of enzyme activity. Values are means ± SEM from 3 to 6 experiments. [Data from Kosk-Kosicka and Roszczynska (16).]

Inhibition of the Ca^{2+}-ATPase activity caused by volatile anesthetics was reversible. For example, an N_2 purge resulted in 40–50% reversibility of inactivation on estimated 50% removal of halothane.

The effects of short-chain alkanols on the Ca^{2+}-ATPase activity are shown in Fig. 3. Five alkanols, from methanol through *n*-pentanol, were used.The effect was strongly dependent on the length of the carbon chain (Fig. 3A, inset). Whereas methanol did not affect enzyme activity up to a concentration of 2% (v/v), ethanol began to inhibit at 1%, and inhibition by isopropanol, and *n*-pentanol was almost complete at concentrations of 0.5–1.0%. Thus, alkanol concentrations causing half-maximal inhibition of the Ca^{2+}-ATPase activity range from 10% (3.1 *M*) methanol to 0.2% (~23 m*M*) *n*-pentanol (Fig. 3B).

To determine the nature of the interactions between the enzyme and anesthetic that are critical for the observed inhibition, activity patterns were compared at 37 and 25°C in the presence of halothane and ethanol (Fig. 4). At 25°C both agents caused less inhibition than at 37°C. The ratio between anesthetic concentrations at which the enzyme was half-maximally inhibited at 37 versus 25°C was 1.3 for the volatile anesthetic and 2 for the alkanol. These observations are in agreement with the predominance of direct interactions between both kinds of anesthetics and nonpolar surfaces of the Ca^{2+}-ATPase. This diagnostic temperature dependence has been observed in various systems containing proteins or DNA (17,18). At the higher temperatures up to around 37°C, nonpolar

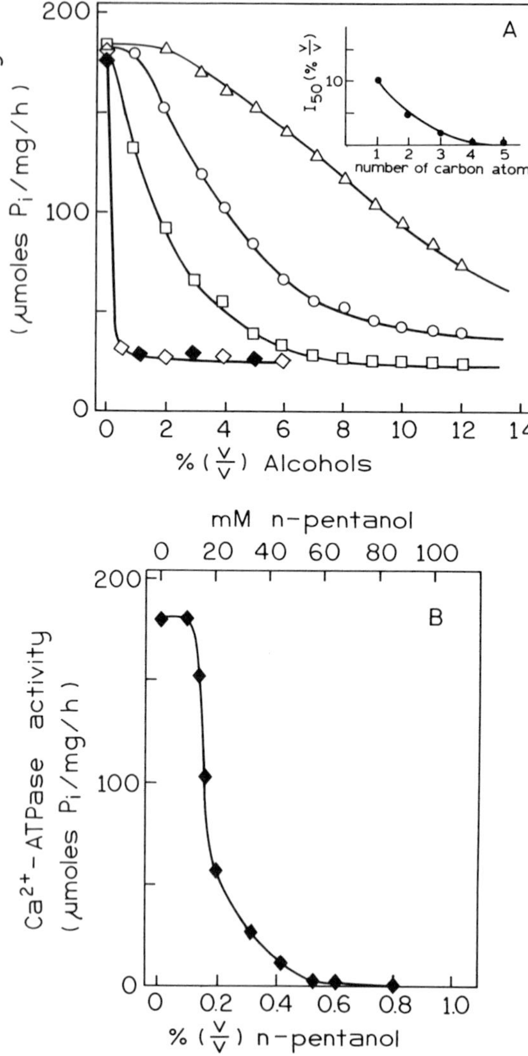

Fig. 3 (A) Inhibition of the Ca^{2+}-ATPase activity by alcohols: methanol (△), ethanol (○), 2-propanol (□), *n*-butanol (◇), and *n*-pentanol (◆). *Inset:* Inhibition of the Ca^{2+}-ATPase activity by alcohols correlates well with the length of carbon chain of alcohols, where I_{50} is the half-maximal inhibition of the Ca^{2+}-ATPase activity. (B) Inhibition of enzyme activity by *n*-pentanol.

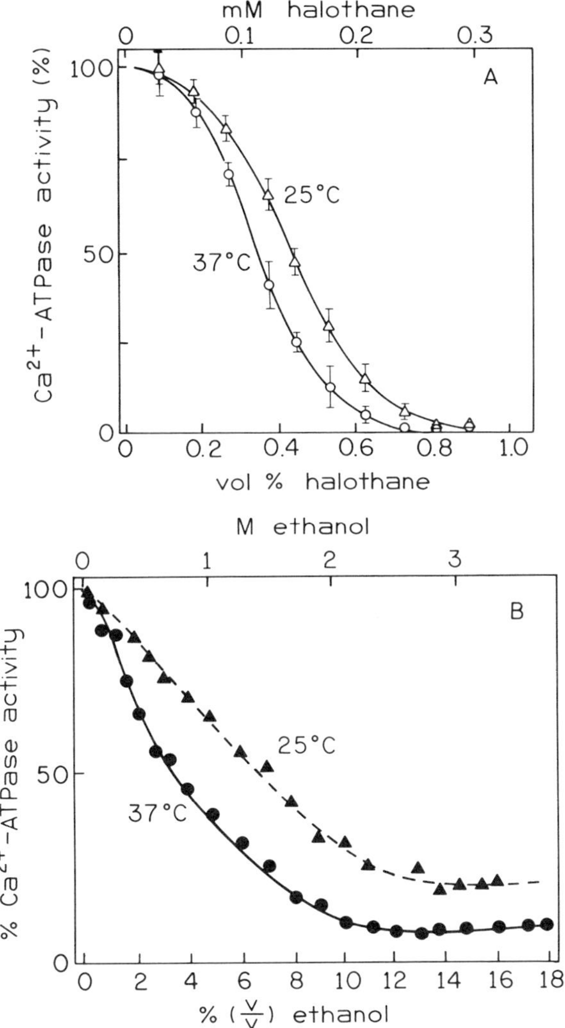

Fig. 4 Comparison of inhibition of the Ca^{2+}-ATPase activity by halothane (A) and ethanol (B) at 37 and 25°C. The data points collected in the presence of either halothane or ethanol are expressed as a percentage of the Ca^{2+}-ATPase activity detected in the absence of the anesthetic at the respective temperature. The specific activity at 37°C was 180 μmol P_i/mg/hr, and that at 25°C was 75 μmol P_i/mg/hr. The measured partition coefficient for halothane was 0.845 ± 0.048 at 37°C and 1.04 ± 0.044 at 25°C. The error bars indicate the standard error of the mean of four replicates and are shown when their dimensions exceeded those of the symbols. [Data from Kosk-Kosicka and co-workers (15,16).]

destabilizing interactions increase. At lower temperatures indirect, water-mediated stabilizing interactions are favored. The direct nonpolar destabilizing interactions between the Ca^{2+}-ATPase and solute apparently lead to increased inactivation of the enzyme. The order of inhibitory potency of volatile anesthetics and alcohols agrees well with their lipid solubility. These findings are in agreement with the postulated action of anesthetics on hydrophobic sites on protein molecules. The involvement of a few specific phospholipid molecules important for enzyme activity is being evaluated.

IV. Discussion

We have demonstrated that the function of red cell Ca^{2+}-ATPase is altered by volatile anesthetics. The observed inhibition of enzyme activity occurs at clinically relevant concentrations of the studied volatile anesthetics, and the correlation between clinical and inhibitory potency is very good. Furthermore, the inhibition is dose dependent and reversible. Thus the enzyme meets the crucial criteria for a target for anesthetic action (4). Several proteins have been reported to be altered by anesthetics in a way suggesting that they could be targets. It has to be pointed out, however, that most proteins are not affected by anesthetics at concentrations which induce general anesthesia. Some proteins are inhibited by certain agents but not by others, indicating that general anesthetics selectively affect certain proteins (2). We have also provided an example supporting the specificity of the anesthetic action on the plasma membrane Ca^{2+}-ATPase by showing that the Ca^{2+}-independent Mg^{2+}-ATPase present in the same red cell membrane is not inhibited by the anesthetics at clinical concentrations (16). Additionally we have demonstrated that the enzyme in the red blood cell membrane is inhibited in a similar manner as the purified enzyme (16). The Ca^{2+}-ATPase is inhibited by all four volatile anesthetics as well as by alcohols.

It is clear that we have established a highly suitable model of an integral membrane protein on which different aspects of anesthetic action can be investigated. An additional advantage of the model is the fact that the enzyme is activated by several modes that are quite frequent in biological systems. These include activation by calmodulin binding (investigated in this article), by self-association of enzyme molecules, as well as activation by lipids, proteolysis, and kinase-dependent phosphorylation. For example, calmodulin is a ubiquitous regulatory protein that modulates a wide range of physiological processes and proteins. Many proteins are active only in oligomeric form, including ion channels. Here and elsewhere (16) we have shown that anesthetics affect the two activation pathways, although to a different extent. It has been reported that kinase C from rat brain is inhibited by halothane, enflurane, and ethanol (19). We are currently investigating whether activation of the Ca^{2+}-ATPase by kinase C-dependent phosphorylation is also altered by anesthetics. Finally, the lipid dependence of anesthetic action on the protein can be investigated, as the lipids could be freely exchanged in the purified enzyme preparation.

Could the plasma membrane Ca^{2+}-ATPase, in addition to being a good model, be an *in vivo* target for volatile anesthetics? I remember meeting Lorin J. Mullins, a biophysicist with great insight, to discuss the findings described here. He was very interested and quite impressed by the data. He gave some valuable suggestions that I plan to incorporate in future studies. Then he asked the inevitable question, the question people working in this field have been trying to answer for years: Is this the general mechanism of anesthesia? Is the principal effect of general anesthetics on ion channels, receptors, or ion pumps? In a paper at the Conference on Molecular and Cellular Mechanisms of Alcohol and Anesthetics, "Viewing Anesthesia Research 1954–1990" (20), Dr. Mullins recounted the answers he got from the participants of the meeting to his questionnaire on "some theories relating to anesthesia." At that time I (a newcomer to the field) was slightly surprised to read how much the opinions of specialists varied and that no concensus could be reached.

In the meantime we have generated some preliminary data showing that the function of plasma membrane Ca^{2+}-ATPase in rat cerebellum synaptosomal membranes is inhibited in a manner similar to that described for the Ca^{2+}-ATPase in red blood cells. We also plan to test the sarcolemmal enzyme. Taking into account the vital role of the Ca^{2+} pump in cellular Ca^{2+} homeostasis in diverse tissues, we are evaluating the prospect of the plasma membrane Ca^{2+}-ATPase being a possible target for anesthetic action *in vivo*.

Acknowledgments

This work was supported by Grant GM 447130 from the National Institutes of Health. The author thanks Grazyna Roszczynska for outstanding technical support. This article is dedicated to the memory of Dr. Lorin J. Mullins, who passed away on April 14, 1993, with appreciation for his sharing of knowledge and enthusiasm, and for encouragement in my quest for the mechanism of anesthetic action.

References

1. Morgan, P. G., Sedensky, M., and Meneely, P. M. (1990). Multiple sites of action of volatile anesthetics in *Caenorhabditis elegans*. *Proc. Natl. Acad. Sci. U.S.A.* **87**, 2965–2969.

2. Franks, N. P., and Lieb, W. R. (1987). What is the molecular nature of general anesthetic target sites? *Trends Pharmacol. Sci.* **8**, 169–174.
3. Firestone, L. L. (1988). General anesthetics. *Int. Anesthesiol. Clin.* **26,** 284–253.
4. Koblin, D. D. (1990). Mechanisms of action. *In* "Anesthesia" (R. D. Miller, ed.), Vol. 1, pp. 51–83. Churchill Livingstone, New York.
5. Dubois, B. W., and Evers, A. S. (1992). F-NMR spin–spin relaxation (T^2) method for characterizing volatile anesthetic binding to protein. Analysis of isoflurane binding to serum albumin. *Biochemistry* **31**, 7069–7076.
6. Bosnjak, Z. J. (1991). Cardiac effects of anesthetics. *Adv. Exp. Med. Biol.* **301**, 91–96.
7. Kosk-Kosicka, D., and Inesi, G. (1985). Cooperative calcium and calmodulin regulation in the calcium–dependent adenosine triphosphatase purified from the erythrocyte membranes. *FEBS Lett.* **189**, 67–71.
8. Kosk-Kosicka, D., Scaillet, S., and Inesi, G. (1986). The partial reactions in the catalytic cycle of the calcium-dependent adenosine triphosphatase purified from erythrocyte membranes. *J. Biol. Chem.* **261**, 3333–3338.
9. Kosk-Kosicka, D., and Bzdega, T. (1988). Activation of the erythrocyte Ca^{2+}-ATPase by either self-association or interaction with calmodulin. *J. Biol. Chem.* **263**, 18184–18189.
10. Kosk-Kosicka, D., Bzdega, T., and Wawrzynow, A. (1989). Fluorescence energy transfer studies of purified erythrocyte Ca^{2+}-ATPase. *J. Biol. Chem.* **264**, 19495–19499.
11. Kosk-Kosicka, D., Bzdega, T., Wawrzynow, A., Scaillet, S., Nemcek, K.,and Johnson, J. D. (1990). Erythrocyte Ca^{2+}-ATPase: Activation by enzyme oligomerization versus by calmodulin. *Adv. Exp. Med. Biol.* **269**, 169–171.
12. Kosk-Kosicka, D., Bzdega, T., and Johnson, J. D. (1990). Fluorescence studies on calmodulin binding to erythrocyte Ca^{2+}-ATPase in different oligomerization states. *Biochemistry* **29**, 1875–1879.
13. Kosk-Kosicka, D. (1990). Comparison of the red cell Ca^{2+}-ATPase in ghost membranes and after purification. *Mol. Cell. Biochem.* **99**, 75–81.
14. Kosk-Kosicka, D., and Bzdega, T. (1991). Regulation of the erythrocyte Ca^{2+}-ATPase by mutated calmodulins with positively charged amino acid substitutions. *Biochemistry* **30**, 65–70.
15. Kosk-Kosicka, D., Wawrzynow, A., and Roszczynska, G. (1992). Different solute sensitivity of the RBC plasma membrane Ca^{2+}-ATPase activation in the calmodulin-dependent and calmodulin-independent pathways. *Ann. N. Y. Acad. Sci.* **671**, 424–427.
16. Kosk-Kosicka, D., and Roszczynska, G. (1993). Inhibition of plasma membrane Ca^{2+}-ATPase activity by volatile anesthetics. *Anesthesiology* **79**, 774–780.
17. Collins, K. D., and Washabaugh, M. W. (1985). The Hofmeister effect and the behaviour of water at interfaces. *Q. Rev. Biophys.* **18**, 1–323.
18. Ha, J.-H. Spolar, R. S., and Record, M. T. (1989). Role of hydrophobic effect in stability of site-specific protein–DNA complexes. *J. Mol. Biol.* **209**, 801–816.
19. Slater, S. J., Cox, K. J. A., Lombardi, J. V., Ho, C., Kelly, M. B., Rubin, E., and Stubbs, C. D. (1993). Inhibition of protein kinase C by alcohols and anesthetics. *Nature (London)* **364**, 82–85.
20. Mullins L. J. (1991). Viewing anesthesia research 1954–1990. *Ann. N.Y. Acad. Sci.* **625**, 841–844.

Enhancement of Halothane Action at the Ryanodine Receptor by Unsaturated Fatty Acids

Jeffrey E. Fletcher and Vincent E. Welter
Department of Anesthesiology
Hahnemann University
Philadelphia, Pennsylvania 19102

I. Introduction

Free unsaturated fatty acids cause a dramatic reduction (~30-fold) in the concentration of halothane required for the rate of Ca^{2+} release through the calcium release channel (ryanodine receptor) to exceed the capacity for uptake in human, porcine, and equine skeletal muscle (1,2). In addition, unsaturated fatty acids reduce the threshold of calcium-induced calcium release in skeletal muscle (3) and directly induce calcium release in cardiac and skeletal muscle (4). These latter actions of unsaturated fatty acids (in the absence of halothane) are not mediated through the ryanodine receptor (4). In contrast to unsaturated fatty acids, saturated fatty acids have no effect on the threshold of calcium-induced calcium release (3) or ryanodine receptor function (5) in the absence of halothane and very little effect on halothane-induced calcium release in skeletal muscle (1). However, saturated fatty acids, existing either as acyl-CoA or acylcarnitine derivatives, open the ryanodine receptor (5).

Intracellular fatty acids are normally associated with fatty acid-binding proteins (molecular weight ~15,000) that bind one fatty acid molecule per molecule of protein (6). Serum albumin (molecular weight ~60,000) binds fatty acids in a manner similar to fatty acid-binding proteins, except that

Advances in Pharmacology, Volume 31

six molecules are bound per molecule of albumin (7). Volatile anesthetics compete with fatty acids for binding to serum albumin (8).

The present study was undertaken to extend published findings in skeletal muscle regarding the actions of fatty acids. First, we examined whether the fatty acid and halothane interaction occurs in cardiac muscle and, if so, whether it is blocked by ruthenium red, an antagonist of the ryanodine receptor. Second, we examined whether halothane could displace fatty acids from fatty acid-binding proteins (the predominant state of fatty acids) in skeletal and cardiac muscle.

II. Calcium Efflux

A heavy sarcoplasmic reticulum fraction (HSRF) was isolated from equine (semimembranosus) or human (vastus lateralis) skeletal muscle, as previously described (1–3). A cardiac microsomal fraction (CMF) was isolated from whole rat hearts (Furth, male, 200–300 g) based on a procedure (9) modified by Dettbarn and Palade (4). The calcium concentration in the extravesicular medium was monitored with antipyrylazo III (250 m*M*; wavelengths 790 and 710 nm), as previously described (2,10). Pulses of Ca^{2+} (30 nmol) were added to the HSRF (40–80 μg) or CMF (~300 μg) suspended in 1.5 ml MOPS/KCl (pH 7.0) buffer containing ATP, an ATP-

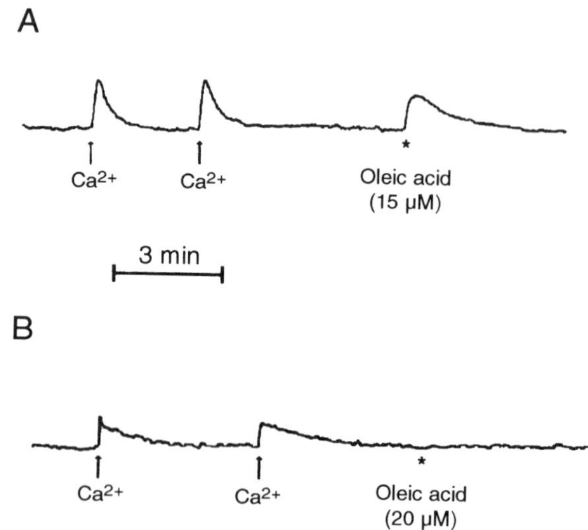

Fig. 1 Effects of oleic acid on Ca^{2+} release from (A) equine skeletal muscle HSRFs and (B) rat cardiac muscle CMFs. Preparations were prepulsed with Ca^{2+} (30 nmol/pulse), and then oleic acid was added. The height of the Ca^{2+} pulses is equivalent to 20 μM Ca^{2+}.

regenerating system, and pyrophosphate to increase the sensitivity of the assay (11); the temperature was maintained at 37°C. Oleic acid, stearic acid, or ruthenium red were added after ATP-stimulated Ca^{2+} uptake had reached equilibrium and after at least two pulses of Ca^{2+} had been added. Ruthenium red, when used, was added 2–3 min before oleic acid. Fatty acids were complexed with albumin (6 : 1 molar ratio) prior to addition to the suspensions. The threshold concentration of halothane at which net Ca^{2+} efflux occurred was determined, as previously described in detail (1,2). Following addition of all other agents, halothane in dimethyl sulfoxide (DMSO) was added at 2-fold increasing increments until a sustained release of Ca^{2+} was observed (1,2). A Teflon-capped cuvette was employed, and the concentration of halothane in the buffer after 5 min was about 40% of the total added (1). The total concentration of halothane added is indicated in the figures. Some tracings have been edited (time between additions reduced, etc.) for greater clarity of presentation.

III. Calcium Release from Heavy Sarcoplasmic Reticulum Fraction

A. Effects of Fatty Acids in Absence of Halothane in Rat Cardiac Muscle

Oleic acid at a concentration as low as 10 μM causes a transient Ca^{2+} release from skeletal muscle HSRFs [see Fletcher *et al.* (1)], and this response is even greater at a concentration of 15 μM (Fig. 1A). At a fatty acid concentration of 20 μM (and less frequently at 15 μm) a sustained Ca^{2+} release was sometimes observed. This action of unsaturated fatty acids is dependent on the fatty acid to membrane ratio and is not blocked by ruthrenium red (4). This action of fatty acids often forced us to use an amount of HSRF at the high end (60–80 μg) of our normal range (20–80 μg) for the studies with halothane, which did not appear to be dependent on the amount of HSRF. In contrast, there was no indication of Ca^{2+} release from CMFs with a concentration of oleic acid as high as 20 μM (Fig. 1B). In addition, stearic acid (20 μM), a saturated fatty acid, had no effect on Ca^{2+} release from CMFs (data not shown).

B. Interaction between Fatty Acids and Halothane in Rat Cardiac Muscle

Under the conditions employed in our assay system, halothane-induced Ca^{2+} release does not exceed the capacity of the Ca^{2+} pump at clinically relevant concentrations of anesthetic in HSRFs from skeletal muscle (1,2)

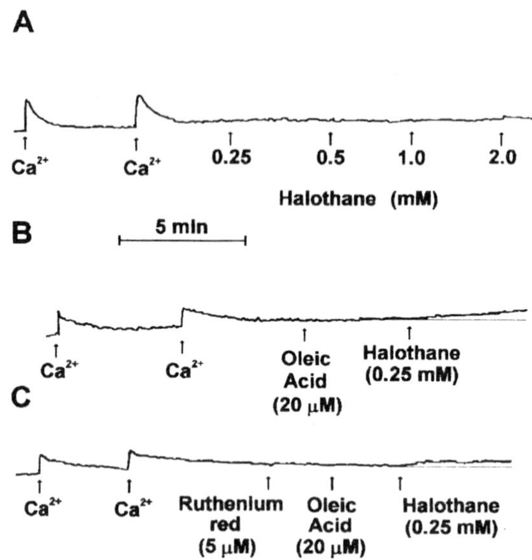

Fig. 2 Effects of oleic acid on halothane-induced net Ca^{2+} efflux from HSRFs. (A) In the absence of oleic acid there is no net Ca^{2+} efflux at up to 2 m*M* halothane. In this preparation net Ca^{2+} efflux was not observed until 16 m*M* halothane. (B) Oleic acid addition reduced the concentration of halothane required for net Ca^{2+} efflux to 0.25 m*M*. (C) Ruthenium red failed to antagonize the oleic acid–halothane interaction. Two pulses of Ca^{2+} (30 nmol each) were added prior to the addition of other agents.

or CMFs from rat cardiac muscle (Fig. 2A). Indeed concentrations of halothane of 8–16 m*M* are required for Ca^{2+} release to exceed uptake in HSRFs (1,2) or CMFs (Fig. 2A). Oleic acid at a concentration of (20 μM), alone having no effect on net Ca^{2+} efflux, lowers the threshold at which the rate of halothane-induced Ca^{2+} release exceeds uptake in CMFs to 0.25m*M* (Fig. 2B). Whereas the interaction between oleic acid and halothane in HSRFs from skeletal muscle is antagonized by ruthenium red (2), this antagonist of the ryanodine receptor has no effect in rat CMFs (Fig. 2C), over a range of 1–10 μM (additional data not shown).

C. Displacement of Fatty Acids from Binding Sites by Halothane

Oleic acid was complexed to fatty acid-free bovine serum albumin at a ratio of six fatty acid molecules per molecule of albumin. Using human skeletal muscle HSRFs, the fatty acid–albumin (6 : 1) complex did not

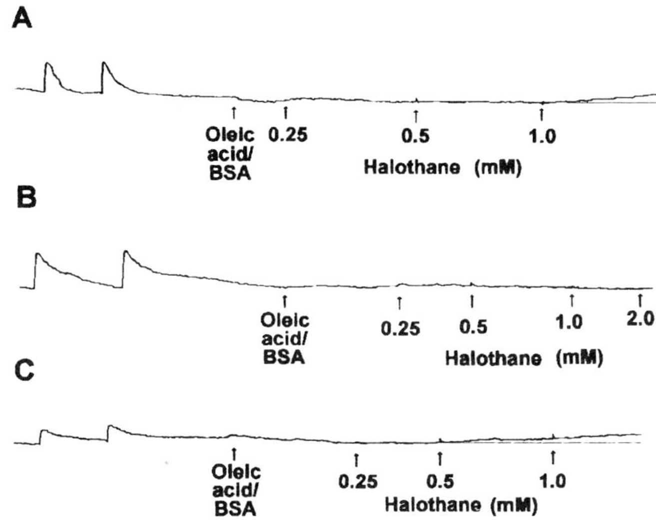

Fig. 3 Displacement by halothane of fatty acids complexed to albumin. (A) Oleic acid–albumin (133 : 22 μM) was added to human skeletal muscle HSRFs prior to halothane. (B) Oleic acid–albumin (264 : 44 μM) was added to rat CMFs prior to halothane. (C) Oleic acid–albumin (396 : 66 μM) was added to rat CMFs prior to halothane. Two pulses of Ca^{2+} (30 nmol each) were added prior to the addition of other agents.

release Ca^{2+} when the concentration of albumin was 22 μM (Fig. 3A), suggesting that the amount of fatty acid dissociating from the albumin and free in solution should be insufficient to interact with halothane. Had a significant amount of fatty acid dissociated, a transient net efflux of Ca^{2+} similar to that in Fig. 1A would have been observed. Also, the threshold of Ca^{2+}-induced Ca^{2+} release was unaffected by this concentration of fatty acid–albumin complex in additional studies (data not shown), and this would have been reduced with free fatty acid levels as low as 10 μM (3). However, on halothane addition the presence of the fatty acid–albumin complex reduced the concentration of halothane required for inducing net Ca^{2+} efflux to 1.0 mM (Fig. 3A). In contrast, even doubling the concentration of fatty acid–albumin complex did not significantly affect the action of halothane in CMFs (Fig. 3B). Even higher concentrations of the complex (66 μM albumin) were required to reduce the concentration of halothane that caused a net Ca^{2+} efflux in CMFs (Fig. 3C). At such high concentrations it is highly likely that sufficient numbers of fatty acids had dissociated from albumin to complicate the interpretation of results.

IV. Discussion

Although there are similarities between skeletal and rat cardiac muscle in the interaction between fatty acids and halothane that induces a net Ca^{2+} efflux, there are also some striking differences. Before attributing this specifically to tissue-related differences, it is important to emphasize that a species-related effect may also play an important role. In other words, in rabbit cardiac muscle unsaturated fatty acids (i.e., arachidonic acid) induce Ca^{2+} release in the absence of anesthetics (4). Such an action was not observed in CMFs from rat hearts in the present study. However, pyrophosphate was not used in the previous studies of cardiac muscle. Indeed, when oxalate, a less effective Ca^{2+} precipitating agent than pyrophosphate, was included in the Dettbarn and Palade studies, the fatty acid-induced Ca^{2+} release was not observed. Additionally, there was a 4- to 8-fold difference in vesicles used for HSRFs and CMFs in the present study. Therefore, fatty acid-induced Ca^{2+} release could exist in rat CMFs.

In the absence of anesthetics, the fatty acids can cause Ca^{2+} release in skeletal muscle and rabbit cardiac muscle by mechanisms independent of the Ca^{2+} release channel (4). Such effects of the fatty acids would most likely involve interactions with hydrophobic regions of a yet to be identified protein, possibly at fatty acid binding domains analogous to those on albumin (7), or fatty acid-binding proteins (6). Similar interactions could occur with other proteins, such as the Na^+ channel, which also has its function altered by halothane (12) and fatty acids (13). However, significant interactions of free fatty acids with the Ca^{2+} release channel and Na^+ channel are somewhat unlikely under normal circumstances in an intact cell, as the fatty acids are bound to fatty acid-binding proteins in the cytoplasm (Fig. 4), keeping the free levels very low. Other interactions could be through covalent bonds via the enzyme-mediated process of acylation (palmitoylation) (14). The Na^+ channel is a palmitoylated protein (15); however, the status of the Ca^{2+} release channel as an acylated protein is unknown. The process of palmitoylation is primarily mediated by attachment of predominantly saturated fatty acids to cysteine residues. A change in the amount or type of fatty acid (i.e., increasingly unsaturated) may alter protein function.

Although fatty acids probably have very little impact on Ca^{2+} regulation in the absence of anesthetics, their augmentation of net halothane-induced Ca^{2+} release is dramatic. Whereas the concentration of fatty acid required for this effect exceeds that of the normally unbound form, halothane could likely displace fatty acids from fatty acid-binding proteins (8). Estimates of the concentration of total fatty acids in cardiac muscle range between 10 and 50 μM, with most being bound to fatty acid-binding proteins, which

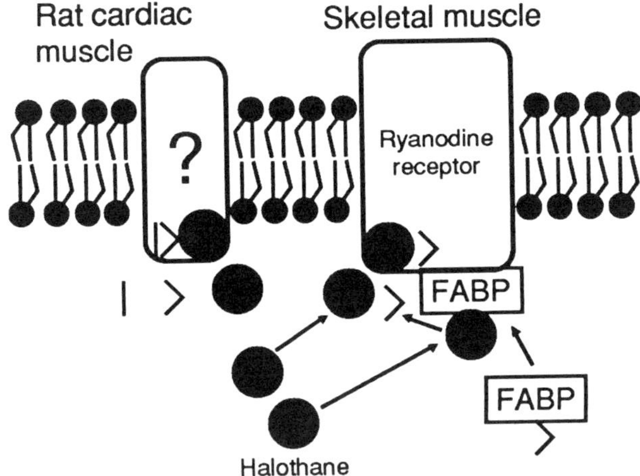

Fig. 4 Model of fatty acid and halothane interaction on Ca^{2+} regulating proteins in skeletal and rat cardiac muscle. The binding sites for halothane (large circles) and fatty acids (saturated fatty acids are straight lines, unsaturated are V-shaped) are shown on two proteins. The function of the skeletal muscle ryanodine receptor is only altered in response to halothane by unsaturated fatty acids. In contrast, the action of halothane on the Ca^{2+} regulating protein in rat cardiac muscle (denoted by ?) is enhanced by saturated or unsaturated fatty acids. Halothane can displace fatty acids from fatty acid-binding protein (FABP). Because this occurs at much lower concentrations of fatty acid and albumin in skeletal muscle, there may be a close association of FABPs with the ryanodine receptor.

themselves are present at about 200–400 μM (16). Considering the low concentration of fatty acids present and the resistance of the cardiac preparation to fatty acid–albumin complexes, it is unlikely that displacement of fatty acids by halothane plays a significant role in anesthetic action in rat cardiac muscle. Human or porcine skeletal muscle contains about 5 to 10 times as much free fatty acid as cardiac muscle (17,18). Therefore, it is highly likely that free concentrations of fatty acids greatly exceeding 20 μM could be achieved in skeletal muscle at the site of halothane action (Fig. 4). The greater potency of fatty acid–albumin complexes in their interaction with halothane in skeletal muscle compared to rat cardiac muscle suggests some actual association of fatty acid-binding proteins with the ryanodine receptor (Fig. 4).

Although the ryanodine receptor appears to be the site for the interaction between fatty acids and halothane in skeletal muscle (2), we were unable to identify the protein involved in rat cardiac muscle, as ruthenium red was ineffective as an antagonist. A role for fatty acids in halothane action may be less likely in cardiac than in skeletal muscle. However, the presence of a site altering the potency of halothane by 30-fold is of interest for its potential involvement in the action of endogenous and exogenous agents. Also, a variety of proteins may possess the potential for similar interactions.

Acknowledgments

The authors are grateful to Dr. Jill Beech for supplying the equine muscle. Research was supported in part by the American Quarter Horse Association (J.B.) and the Hahnemann Anesthesia Research Foundation (J.E.F.).

References

1. Fletcher, J. E., Mayerberger, S., Tripolitis, L., Yudkowsky, M., and Rosenberg, H. (1991). Fatty acids markedly lower the threshold for halothane-induced calcium release from the terminal cisternae in human and porcine normal and malignant hyperthermia susceptible skeletal muscle. *Life Sci.* **49,** 1651–1657.
2. Fletcher, J. E., Tripolitis, L., Rosenberg, H., and Beech, J. (1993). Malignant hyperthermia: Halothane- and calcium-induced calcium release in skeletal muscle. *Biochem. Mol. Biol. Int.* **29,** 763–772.
3. Fletcher, J. E., Tripolitis, L., Erwin, K., Hanson, S., Rosenberg, H., Conti, P. A., and Beech, J. (1990). Fatty acids modulate calcium-induced calcium release from skeletal muscle heavy sarcoplasmic reticulum fractions: Implications for malignant hyperthermia. *Biochem. Cell Biol.* **68,** 1195–1201.
4. Dettbarn, C., and Palade, P. (1993). Arachidonic acid-induced Ca^{2+} release from isolated sarcoplasmic reticulum. *Biochem. Pharmacol.* **45,** 1301–1309.
5. El-Hayek, R., Valdivia, C., Valdivia, H., Hogan, K., and Coronado, R. (1993). Activation of the Ca^{2+} release channel of skeletal muscle sarcoplasmic reticulum by palmitoyl carnitine. *Biophys. J.* **65,** 779–789.
6. Veerkamp, J. H., Peeters, R. A., and Maatman, R. G. H. J. (1991). Structural and functional features of different types of cytoplasmic fatty acid-binding proteins. *Biochim. Biophys. Acta* **1081,** 1–24.
7. Peters, T., Jr. (1985). Serum albumin. *Adv. Protein Chem.* **37,** 161–245.
8. Dubois, B. W., and Evers, A. S. (1992). ^{19}F-NMR spin–spin relaxation (T_2) method for characterizing volatile anesthetic binding to proteins. Analysis of isoflurane binding to serum albumin. *Biochemistry* **31,** 7069–7076.
9. Harigaya, S., and Schwartz, A. (1969). Rate of calcium binding and uptake in normal animal and failing human cardiac muscle. Membrane vesicles (relaxing system) and mitochondria. *Circ. Res.* **25,** 781–794.
10. Fletcher, J. E., Tripolitis, L., and Beech, J. (1992). Bee venom melittin is a potent toxin for reducing the threshold for calcium-induced calcium release in human and equine skeletal muscle. *Life Sci.* **51,** 1731–1738.
11. Palade, P. (1987). Drug-induced Ca^{2+} release from isolated sarcoplasmic reticulum. I. Use of pyrophosphate to study caffeine-induced Ca^{2+} release. *J. Biol. Chem.* **262,** 6135–6141.
12. Wieland, S. J., Fletcher, J. E., Gong, Q.-H., and Rosenberg, H. (1991). Effects of lipid-soluble agents on sodium channel function in normal and MH-susceptible skeletal muscle cultures. *In* "Mechanisms of Anesthetic Action in Muscle" (T. J. J. Blanck and D. M. Wheeler, eds.), pp. 9–19. Plenum, New York.

13. Wieland, S. J., Fletcher, J. E., and Gong, Q.-H. (1992). Differential modulation of a sodium conductance in skeletal muscle by intracellular and extracellular fatty acids. *Am. J. Physiol.* **263,** C308–C312.
14. Grand, R. J. A. (1989). Acylation of viral and eukaryotic proteins. *Biochem. J.* **258,** 625–638.
15. Catterall, W. A. (1992). Cellular and molecular biology of voltage-gated sodium channels. *Physiol. Rev.* **72,** S15–S48.
16. van der Vusse, G. J., Glatz, J. F. C., Stam, H. C. G., and Reneman, R. S. (1992). Fatty acid homeostasis in the normoxic and ischemic heart. *Physiol. Rev.* **72,** 881–940.
17. Fletcher, J. E., Rosenberg, H., Michaux, K., Tripolitis, L., and Lizzo, F. H. (1989). Triglycerides, not phospholipids, are the source of elevated free fatty acids in muscle from patients susceptible to malignant hyperthermia. *Eur. J. Anaesth.* **6,** 355–362.
18. Fletcher, J. E., Rosenberg, H. Michaux, K., Cheah, K. S., and Cheah, A. M. (1988). Lipid analysis of skeletal muscle from pigs susceptible to malignant hyperthermia. *Biochem. Cell Biol.* **66,** 917–921.

Adrenergic Receptors: Unique Localization in Human Tissues

Debra A. Schwinn
Department of Anesthesiology
Duke University Medical Center
Durham, North Carolina 27710

I. Introduction

Adrenergic receptors (ARs) are members of the much larger family of guanine nucleotide-binding proteins (G proteins) characterized by seven transmembrane domains (Fig. 1) (1). Stimulation of adrenergic receptors by the endogenous catecholamines epinephrine and norepinephrine results in increased myocardial chronotropy, inotropy, modulations in vascular tone, bronchodilation, and many endocrine processes. Adrenergic receptors have been classically subdivided into four major groups (α_1, α_2, β_1, β_2) using pharmacological and physiological techniques. However, the advent of molecular biology has led to cloning of genes encoding nine distinct adrenergic receptor subtypes, namely, α_{1A1D}, α_{1B}, α_{1C}, α_{2A}, α_{2B}, α_{2C}, β_1, β_2, and β_3 (Fig. 2) (2,3). Expression of adrenergic receptor subtypes individually in cell lines has facilitated definition of each subtype with regard to pharmacology, G protein coupling, and second messenger production. Mutagenesis of amino acids in specific locations in each receptor protein (by modifying nucleic acid sequences at the DNA level), has provided critical information regarding exact receptor sites important in ligand binding, G protein coupling, and receptor regulation. This process is called structure–function analysis (3).

In addition to pharmacology and second messenger coupling, another method of defining adrenergic receptor subtypes involves localization in

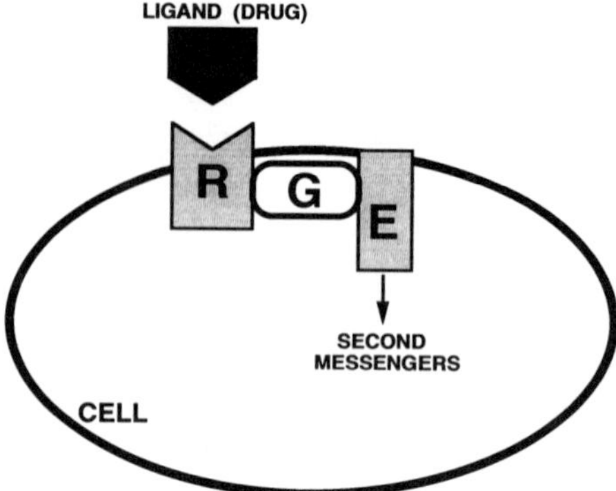

Fig. 1 Schematic of guanine nucleotide (G protein)-coupled receptor in a cell. A hydrophilic ligand or drug circulating in the extracellular fluid binds to the transmembrane receptor (R). This initiates a change in conformation of the receptor enabling it to interact with an intermediary G protein (G). The α subunit of the heterotrimeric G protein then dissociates and interacts with the effector system (E). The effector system is usually an enzyme which catalyzes the formation of a second messenger, but it could also be a channel protein.

mammalian tissues. After all, a receptor-mediated physiological function cannot be assigned to a given subtype unless that receptor subtype is present in the tissue (or cell grouping) of interest. Hence, mammalian tissue distribution (usually in rat tissues) is an integral part of the difinition of a receptor. In this context, an interesting phenomenon was noted when cloned α_1AR subtypes were initially defined. Specifically, the distribution of α_1AR subtypes appeared to be different in rat and rabbit tissues (4). Although minor species variations have been noted in adrenergic receptor physiology and pharmacology experiments over the last several decades,

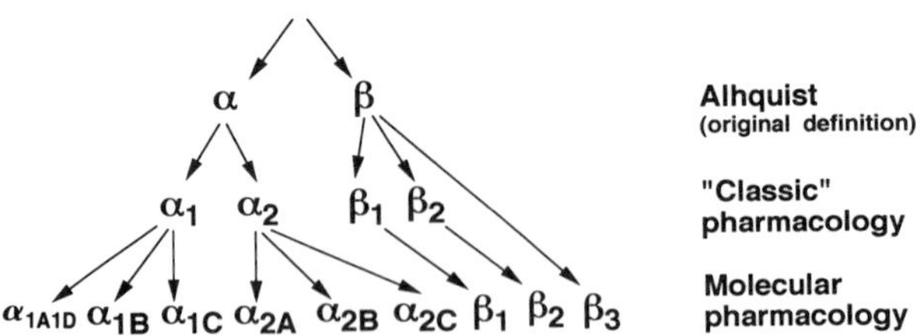

Fig. 2 History of adrenergic receptor (AR) subtypes. Ahlquist first defined adrenergic receptors in 1948 and divided them into two subtypes, α and β. In the 1960s and 1970s, further subdivision into the four classic adrenergic receptors took place. With the advent of molecular biology tools, since 1986 nine genes or cDNAs encoding distinct adrenergic receptors subtypes have been discovered.

these differences were usually thought to be related to the level of receptor expression and not to the complete presence or absence of given subtypes. However, at the level of adrenergic receptor subsubtypes (i.e., α_{1A}AR versus α_{1B}AR), striking species variations apparently occur (4). This observation begs the question, What is the distribution of adrenergic receptor subtypes in human tissues? Because the answer to this question was first described in detail with α_1ARs, it is fitting to examine the distribution of α_1AR subtypes in human tissues.

α_1-Adrenergic receptors are classically present in brain, heart, liver, and spleen, as well as in many other tissues (5). Pharmacological properties of α_1ARs include an agonist potency series of epinephrin $\gg$ norepinephrine $>$ phenylephrine $\gg$ isoproterenol; the specific antagonist is prazosin (Table I). When activated, α_1ARs couple to the newly described and cloned G protein, Gq, a process which results in activation of phospholipase C (PLC) (6). PLC hydrolyzes phosphoinositide bisphosphate (PIP_2) into two predominant products, inositol trisphosphate (IP_3) and diacylglycerol (DAG). IP_3 binds to IP_3 receptors on the sarcoplasmic reticulum, causing release of calcium from intracellular stores. DAG activates protein kinase C, a process which leads to activation of various other proteins and potentially to regulation of receptor function. The primary physiological response to activation of α_1AR subtypes is smooth muscle contraction.

Three distinct α_1AR subtypes are known to exist ($\alpha_{1A/D,}$ $\alpha_{1B,}$ α_{1C}), and genes encoding all three subtypes have been cloned (7–9). Classically the α_{1A1D}AR subtype has high affinity for the antagonists WB4101 and phentolamine as well as the agonist methoxamine, is partially sensitive to inactivation by the akylating agent chloroethylclonidine (CEC), and is present in several rat tissues including vas deferens, cerebral cortex, aorta, and spleen (Table II) (5,10). The α_{1B}AR subtype has relative low affinity

Table I

General Properties of α_1-Adrenergic Receptor

Location	Brain, heart, liver, spleen
Pharmacological properties	
Agonist potency series	EPI $\gg$ NE $>$ PHE $\gg$ ISO
Selective antagonist	Prazosin
Second messengers	$IP_2 \rightarrow IP_3$ and DAG
Physiology	Smooth muscle contraction

[a] EPI, Epinephrine; NE, norepinephrine; PHE, phenylephrine; ISO, isoproterenol; IP_2, inositol bisphosphate; IP_3, inositol trisphosphate; DAG, diacylglycerol.

Table II

Pharmacological Properties of Cloned α_1-Adrenergic Receptor Subtypes[a]

Property	$\alpha_{1A/D}$AR	α_{1B}AR	α_{1C}AR
Antagonist affinity			
WB4101	High	Low	High
Phentolamine	High	Low	High
Prazosin	High	High	High
Agonist affinity			
Methoxamine	High	Low	High
Epinephrine	High	Low	Low
Norepinephrine	High	Low	Low
Inactivation by CEC	Partial	Yes	Partial

[a] Relative affinities are listed. CEC, chloroethylclonidine, an alkylating agent.

to WB4101, phentolamine, and methoxamine, is sensitive to inactivation by CEC, and is present in rat liver, cerebral cortex, and heart (10). The α_{1C}AR subtype was not originally described pharmacologically and was discovered using molecular cloning techniques (8). It has properties quite similar to those of pharmacologically defined α_{1A}AR, having affinity to WB4101, phentolamine, and methoxamine. However, the α_{1C}AR is partially sensitive to the alkylating properties of CEC (60–70% inactivated) and was not present in initial studies in rat tissues using Northern blot analysis (RNA level). [This receptor subtype has been subsequently found to be present in restricted locations in rat brain and heart using more sensitive techniques (11).] The name α_{1C} was given to this α_1AR subtype (8). Follow-up studies of the three cloned α_1ARs using more recently described compounds such as (+)niguldipine, 5-methylurapidil, benoxathian, and spiperone have suggested that the cloned α_{1A}AR might actually be a distinct subtype called the α_{1D}AR based on subtle discrepancies between the cloned receptor and the previously pharmacologically defined α_{1A} subtype from rat tissues (12). Until the debate is resolved, we shall continue to refer to this receptor as the $\alpha_{1A/D}$AR.

As described above, one of the classic methods to define a receptor subtype, in addition to ligand affinities, second messenger assays, and physiological effects, is determining the receptor distribution in rat tissues. As shown in Table III, the cloned $\alpha_{1A/D}$AR is present in all of the rat tissues expected for the pharmacologically defined α_{1A}AR subtype; the α_{1B} receptor is also present in the expected rat tissues, whereas the α_{1C}AR subtype was not originally present in rat tissues by Northern analysis

Table III

Distribution of α_1-Adrenergic Receptor RNA in Rat Tissues by Northern Analysis Using Subtype-Selective Probes[a]

Tissue	$\alpha_{1A/D}$AR	α_{1B}AR	α_{1C}AR
Aorta	+++	+	−
Brain			
Hippocampus	+++	+	−
Brain stem	++	++	−
Cerebellum	+	++	−
Cerebral cortex	+++	+++	−
Heart	+	+++	−
Kidney	−	++	−
Liver	−	++++	−
Lung	+	++	−
Spleen	++	−	−
Vas deferens	++++	−	−

[a] Data from Schwinn *et al.* (4). α_{1C}AR RNA has been described in more sensitive *in situ* hybridization assays in selected rat brain regions and is the only α_1AR subtype present in rabbit liver by Northern analysis.

(4). Because some of the original experiments suggesting linkage of the pharmacologically defined α_{1A}AR to a calcium channel were performed in rabbit aorta, we next investigated the distribution in rat tissues using Northern analysis. To our surprise, the α_{1B}AR subtype was present in most rabbit tissues studied including kidney, spleen, heart, lung, and vascular tissues such as aorta and ear artery, but was conspicuously absent in rabbit liver (4). This is in contrast to the rat where the α_{1B}AR is the only subtype present in liver. In the rabbit, however, the α_{1C}AR is the only α_1AR subtype present in liver. Further, the α_{1A1D}AR subtype (which was present in many rat tissues) was virtually absent in the rabbit tissues studied. These data led us to then examine the distribution of α_1ARs in human tissues.

II. Methods Used to Study Receptor Distribution

To study the tissue distribution of receptors in human tissues, several methods can be used. Ligand binding (interaction of ligands or drugs with receptor protein) and autoradiography are classically used but require subtype-selective ligands which are not currently available for α_1ARs. Labeled antibodies (immunohistochemistry techniques) are also frequently used but require receptor-selective antibodies which are not yet available for α_1ARs. This leaves molecular techniques since genes encoding each α_1AR subtype are available. Although the polymerase chain reaction is frequently used to amplify DNA, this technique may be too sensitive for examining receptor distribution in human tissues since the presence of a single copy (or very few copies) of a gene may not correlate with sufficient receptor protein concentrations to influence physiological function. Hence the predominant subtype present in a given tissue is most important for initially screening human tissues. Ribonuclease (RNase) protection assays (also called solution hybridization when performed in a quantitative fashion) are RNA–RNA hybridization assays which are sensitive but also very specific in defining the presence of RNA encoding a receptor subtype in tissues. Moreover, *in situ* hybridization techniques can be used to further identify the presence of RNA in specific cells in tissue sections. Although determining the presence of RNA in a tissue does not guarantee that receptor protein will be present in identical concentrations, the absence of RNA does suggest that a given receptor subtype is not present in the tissue of interest. However, in general, with a few notable exceptions, RNA expression tends to correlate with receptor protein expression levels.

Because RNase protection assays require a species match of probe, it was necessary to obtain portions of human genes encoding each α_1AR subtype. We have cloned portions of α_1AR cDNAs over the last several years and so were in a unique position to be able to perform these experiments. Human α_1AR probes ranged in length from 0.3 to 0.6 kilobases (kb). In general, RNase protection assays are performed by incubating RNA made from human tissues with a radiolabeled RNA probe generated from a human receptor cDNA fragment. Once hybridization occurs, then RNase is added to the assay; digestion of all single-stranded (nonhybridized) RNA then occurs. The resulting double-stranded protected RNA fragment is electrophoretically separated on an agarose gel and exposed to X-ray film or phosphorimager screens. Bands detected using laser densitometry (for autoradiograms) or directly from the phosphorimager are then normalized for size and incorporation of radiolabeled probe.

III. Localization of Receptors in Human Tissue

Preliminary results of RNase protection experiments with all three α_1AR subtype and human tissues are shown in Table IV. The predominant subtype of α_1AR RNA in human tissues studied is the α_{1C}AR (13). This is in striking contrast to the expression of this receptor subtype in tissues

Table IV

Distribution of α_1-Adrenergic Receptor RNA in Human Tissues: Preliminary Results Using Ribonuclease Protection Assays

Tissue	$\alpha_{1A/D}$AR	α_{1B}AR	α_{1C}AR
Aorta	++	+	+/−
Cerebellum	+/−	++	+++
Cerebral cortex	++	+	+++
Heart	+	+	+++
Kidney	+/−	++	+/−
Liver	+/−	+	++++
Spleen	+	++	+
Vena cava	+/−	+/−	+

from other mammals such as rat and rabbit. In general, α_{1c}AR RNA predominates in human liver, heart, vena cava, cerebellum, and cerebral cortex; α_{1B}AR RNA predominates in kidney and spleen, whereas the α_{1A1D}AR predominates in aorta. This distribution suggests that some selectivity may be obtained using α_1AR subtype-selective agonists and antagonists to treat various human diseases.

In addition to general localization of α_1AR subtype RNA in human tissues, we have also studied individual human tissues where α_1ARs play an important role as therapeutic agents. For example, benign prostatic hypertrophy (BPH) is a condition where enlargement of the prostate gland in males can lead to urinary retention. Classically, therapy for BPH requires surgical resection of portions of the prostate gland. However, recent therapy with α_1AR antagonists has provided symptomatic relief for many patients. The main problem with α_1AR antagonist therapy relates to nonspecific α_1AR blockade, resulting in hypotension and dizziness. If the exact subtype of α_1AR associated with BPH could be identified, then α_1AR subtype-selective agents could be developed for targeted therapy with minimal side effects. BPH is a stromal disease (involved in increased bulk of prostate smooth muscle), as opposed to an epithelial disease (such as prostate cancer). We used RNase protection assays to determine the predominate α_1AR RNA in human prostate and then used *in situ* hybridization to localize the subtype present in stromal tissue. Clearly the predominate α_1AR RNA present in human prostate is the α_{1C} (70–75% of the total), and on tissue sections this subtype is localized to the stroma (14). Smaller amounts of α_{1A1D} RNA are also present in human prostate, followed by very minimal amounts of α_{1B}AR RNA. Hence, although α_1AR therapy for BPH can initially be aimed at the α_{1C}AR subtype, it important to remember that there are two components to BPH, static and active; these components may involve more than one α_1AR subtype. In addition, it will be important to investigate using ligand binding and selective antibodies whether this relationship between α_1AR subtypes holds up at the protein level.

The prostate is not the only human tissue where α_1ARs have important clinical significance. In the cardiovascular system, α_1ARs play an important role in vascular tone, mediate myocardial inotropy (albeit far less than βARs), and may be involved in anesthetic-induced myocardial arrhythmias. Acute therapy for hypotension in the operating room or intensive care unit with an α_1AR subtype-selective agonist could potentially provide restoration of blood pressure without renal vascular compromise if the subtype of α_1AR differs in small resistance vessels and renal artery. Indeed, a mapping of adrenergic receptor subtypes in many human tissues should provide a more rational use of receptor subtype-selective drugs currently being developed and may enhance understanding of the mechanism of human diseases such as hypertension, diabetes, and congestive heart failure.

IV. Summary

The final point to be made is that RNA studies are only the first step in localizing the distribution of adrenergic receptors in human tissues. Although RNA levels tend to correlate well with receptor protein expression in many tissues, this must be confirmed with studies aimed at receptor protein. Selective antibodies are being developed currently by various researchers and selective ligands by several pharmaceutical companies. In the next few years, not only should it be possible to confirm or modify results of adrenergic receptor subtype distribution studies, it also should be possible to design and test specific hypotheses related to adrenergic receptor diseases in whole animal models with newly developed subtype-selective ligands. Because species heterogeneity in adrenergic receptor tissue distribution exists, final testing of adrenergic receptor subtype-selective drugs will have to occur in humans. This is a potentially exciting possibility for anesthesiologists, for what better clinical laboratory is there than the operating room? Hence, anesthesiologists are in a key position to help redefine human adrenergic physiology once new adrenergic receptor subtype-selective agents become available.

References

1. Berkowitz, D. E., and Schwinn, D. A. (1991). New advances in receptor pharmacology. *Curr. Opin. Anesthesiol.* **4**, 486–496.
2. Schwinn, D. A. Adrenoceptors as models for G protein-coupled receptors: Structure/function/regulation. *Br. J. Anaesth.* **71**, 77–85.
3. Schwinn, D. A., Caron, M. G., and Lefkowitz, R. J. (1991). The β-adrenergic receptor as a model for molecular structure–function relationship in G protein-coupled receptors. *In* "The Heart and Cardiovascular System: Scientific Foundations" (H. A. Fozzard, E. Haber, R. B. Jennings, A. M. Katz, and H. E. Morgan, eds.), 2nd Ed., pp. 1657–1684. Raven, New York.
4. Schwinn, D. A., Page, S. O., Middleton, J., Lorenz, W., Liggett, S. B., Yamamoto, K., Caron, M. G., Lefkowitz, R. J., and Cotecchia, S. (1991). The α_{1C}-adrenergic receptor: Characterization of signal transduction pathways and mammalian tissue heterogeneity. *Mol. Pharmacol.* **40**, 619–626.
5. McGrath, J. C., Brown, C. M., and Wilson, V. G. (1989). Alpha-adrenoceptors: A critical review. *Med. Res. Rev.* **9**, 407–533.
6. Wu, D., Katz, A., Lee, C., and Simon, M. I. (1992). Activation of phospholipase C by α_1-adrenergic receptors is mediated by the α subunits of Gq family. *J. Biol. Chem.* **267**, 25798–25802.
7. Cotecchia, S., Schwinn, D. A., Randall, R. R., Lefkowitz, R. J., Caron, M. G., and Kobilka, B. K. (1988). Molecular cloning and expression of the cDNA for the hamster alpha-1-adrenergic receptor. *Proc. Natl. Acad. Sci. U.S.A.* **85**, 7159–7163.
8. Schwinn, D. A., Lomasney, J. W., Lorenz, W., Szklut, P. J., Yang-Feng, T. L., Caron, M. G., Lefkowitz, R. J., and Cotecchia, S. (1990). Molecular cloning and expression of the cDNA for a novel α_1-adrenergic receptor subtype. *J. Biol. Chem.* **265**, 8183–8189.
9. Lomasney, J. W., Cotecchia, S., Lorenz, W., Leung, W.-Y., Schwinn, D. A., Yang-Feng, T. L., Brownstein, M., Lefkowitz, R. J., and Caron, M. G. (1991). Molecular cloning and expression of the cDNA for the α_{1A}-adrenergic receptor, the gene for which is located on human chromosome 5. J. Biol. Chem. **266**, 6365–6369.
10. Ruffolo, R. R., Jr., Nichols, A. J., Stadel, J. M., and Hieble, J. P. (1991). Structure and function of α-adrenoceptors. *Pharmacol. Rev.* **43**, 475–504.
11. McCune, S. K., Voigt, M. M., and Hill, J. M. (1992). Developmental expression of the alpha-1A, alpha-1B and alpha-1C adrenergic receptor subtype mRNAs in the rat brain. *Abstr. Soc. Neurosci.* **18**, 196.2.
12. Schwinn, D. A., and Lomasney, J. W. (1992). Pharmacological characterization of cloned α_1-adrenergic receptor subtypes: Selective antagonists suggest the existence of a fourth subtype. *Eur. J. Pharmacol.* **227**, 433–436.
13. Price, D. T., Lefkowitz, R. J., Caron, M. G., and Schwinn, D. A. (1993). Alpha $_1$-adrenergic receptor mRNA expression in human tissues. *FASEB J.* **7**, A141.
14. Price, D. T., Schwinn, D. A., Lomasney, J. W., Allen, L. E., Caron, M. G., and Lefkowitz, R. J. (1993). Identification, quantification, and localization of mRNA for three distinct alpha$_1$-adrenergic receptors subtypes in human prostate. *J. Urol.* **150**, 546–551.

Volatile Anesthetic Effects on Inositol Trisphosphate-Gated Intracellular Calcium Stores in GH_3 Cells

Alex S. Evers[*,†] and M. Delawar Hossain[*]
Departments of [*]Anesthesiology and [†]Pharmacology
Washington University School Medicine
St. Louis, Missouri 63110

I. Introduction

A wide variety of cellular processes are regulated by changes in intracellular calcium ($[Ca^{2+}]_i$). Included in the calcium-regulated processes are excitation–contraction coupling, smooth muscle constriction, hormonal exocytosis, and neurosecretion. Resting cytoplasmic $[Ca^{2+}]_i$ is usually maintained at approximately 100 nM, four orders of magnitude less than the mM calcium concentrations found in extracellular fluid. In response to specific external stimuli (e.g., hormones, neurotransmitters, electrical stimulation), $[Ca^{2+}]_i$ can abruptly be increased to 1–10 μM. This is accomplished either by opening channels that allow calcium to enter the cytoplasm from the extracellular space or by releasing calcium into the cytoplasm from calcium-sequestering organelles. Interference with stimulus-induced increases in $[Ca^{2+}]_i$ is a potential mechanism for some of the important physiological effects of volatile anesthetics. Indeed, volatile anesthetic-induced vasodilatation results, at least in part, from inhibition of voltage-gated calcium channels. Similarly, volatile anesthetic-induced cardiac depression results in part from anesthetic-induced leak of calcium from sarcoplasmic reticulum (1).

Advances in Pharmacology, Volume 31

Many hormones and neurotransmitters have their actions mediated via cell surface receptors that stimulate production of the chemical second messenger inositol trisphosphate (IP_3). IP_3 binds to a ligand-gated calcium channel (the IP_3 receptor) located in specialized endoplasmic reticulum. Binding of IP_3 opens the channel, releasing sequestered calcium and increasing $[Ca^{2+}]_i$. The effects of anesthetics on IP_3-mediated calcium release have not been defined. This may be an area of some importance, since depression of IP_3-mediated responses could contribute to impaired neurotransmission and to impaired hormonal responsiveness.

We have been studying the effects of volatile anesthetics on various signal transduction pathways in GH_3 cells, a rat clonal pituitary cell line. In these cells, thyrotropin-releasing hormone (TRH) binds to a cell surface receptor which, acting through a G protein, stimulates the generation of IP_3. IP_3 releases sequestered calcium, increasing $[Ca^{2+}]_i$ and ultimately leading to the secretion of stored prolactin. In this article we examine the effects of halothane on the IP_3 signaling pathway in GH_3 cells.

II. Intracellular Calcium Measurements

GH_3 cells were obtained from the American Type Tissue Collection (Rockville, MD). Cells were grown in monolayer culture in 37°C in a humidified, 5% (v/v) CO_2 atmosphere. Growth medium consisted of Ham's F-10, 15% (v/v) horse serum, and 2.5% (v/v) fetal bovine serum. For experiments in which $[Ca^{2+}]_i$ was measured, cells were grown in suspension for 24 hr prior to experimentation. Methods for measurement of inositol phosphate synthesis are as previously described (2,3).

The GH_3 cells were loaded with Fura-2AM using a previously described method (3). Cells loaded with Fura-2 were washed and resuspended in physiological saline. The cells ($1-2 \times 10^6$ cells) were placed in a stirred, temperature-controlled (37°C) 2.5-ml cuvette in a Photon Technology International spectrofluorimeter (South Brunswick, NJ). Fluorescence measurements were made by observing the fluorescence emission at 500 nm generated by exposing the cells to rapidly alternating (100 Hz) excitations of 340 and 390 nm light. To obtain maximum and minimum fluorescence the cells were first permeabilized with Triton X-100 (0.1%), and subsequently EGTA (20 m*M*) was added to give the fluorescence value for zero calcium. The $[Ca^{2+}]_i$ was calculated from the fluorescence measurements using the equations described by Grynkiewicz *et al.* (4). All values of $[Ca^{2+}]_i$ are reported as intracellular concentrations (nanomolar).

III. Effects of Anesthetics

A. TRH-Stimulated Inositol Phosphate Accumulation

To examine the effects of halothane on the proximal portions of the IP_3 signaling pathway, GH_3 cells were labeled to equilibrium with [^{3}H] inositol and stimulated with TRH in the presence of 10 m*M* LiCl. Accumulated inositol phosphates were separated by anion-exchange chromatography and quantified by scintillation counting. Halothane (0.5 m*M*) did not affect the time course of TRH-stimulated inositol phosphate accumulation, and it had minimal effects on the accumulation of inositol phosphates at all TRH concentrations tested (3). These data indicate that clinically relevant concentrations of halothane do not alter the function of TRH receptors, hormone-stimulated phospholipase C (PLC), or the G proteins linking the receptor to PLC.

B. Effects of Halothane on Resting Intracellular Calcium

Clinically relevant concentrations of halothane (<0.75 m*M*) had minimal effects on resting $[Ca^{2+}]_i$ in GH_3 cells. At halothane concentrations of 1.0 m*M* or greater, halothane caused a rapid and concentration-dependent increase in resting $[Ca^{2+}]_i$ (Fig. 1). This increase in resting $[Ca^{2+}]_i$ was unaffected by organic (nimodipine) or inorganic ($LaCl_3$) calcium channel blockers, and it was eliminated by treatments (thapsigargin, see below) which deplete intracellular calcium stores. These data indicate that high

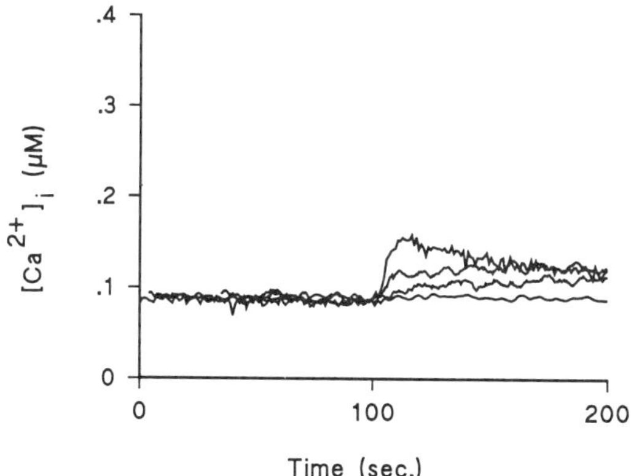

Fig. 1 Effects of halothane on resting $[Ca^{2+}]_i$. GH_3 cells were loaded with the calcium indicator Fura-2 and challenged with various concentrations (0.5, 1.0, 1.5, and 3.0 m*M*) of halothane. Halothane causes a concentration-dependent increase in $[Ca^{2+}]_i$. Reprinted with permission (5).

concentrations of halothane increase resting $[Ca^{2+}]_i$ by releasing calcium from intracellular stores.

C. Effects of Halothane on TRH-Stimulated Increases in Intracellular Calcium

TRH increases $[Ca^{2+}]_i$ both by releasing sequestered calcium from intracellular stores and by stimulating influx through calcium channels. To isolate the component of TRH-stimulated increases in $[Ca^{2+}]_i$ attributable to release from intracellular stores, experiments were conducted in the presence of the inorganic calcium channel blocker La^{3+} (5 μM). La^{3+} eliminated the "plateau" phase of increased $[Ca^{2+}]_i$ normally observed following TRH administration (Fig. 2A).

When cells were stimulated with a saturating concentration of TRH (100 nM) immediately after administration of halothane, there was virtually no effect on the peak $[Ca^{2+}]_i$ response to TRH. As the interval between halothane administration and TRH stimulation was increased, there was

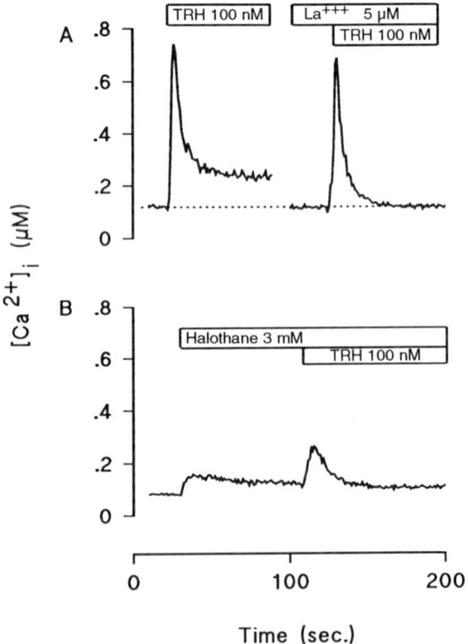

Fig. 2 (A) Continuous application of TRH (100 nM) results in a rapid, transient increase in $[Ca^{2+}]_i$ followed by a sustained increase in $[Ca^{2+}]_i$ (left). The sustained increase in $[Ca^{2+}]_i$ is prevented by pretreatment with 5 μM $LaCl_3$ (right). (B) Halothane increases resting $[Ca^{2+}]_i$ and inhibits the subsequent TRH-induced increase in $[Ca^{2+}]_i$. Fig. 2B is reprinted with permission (5).

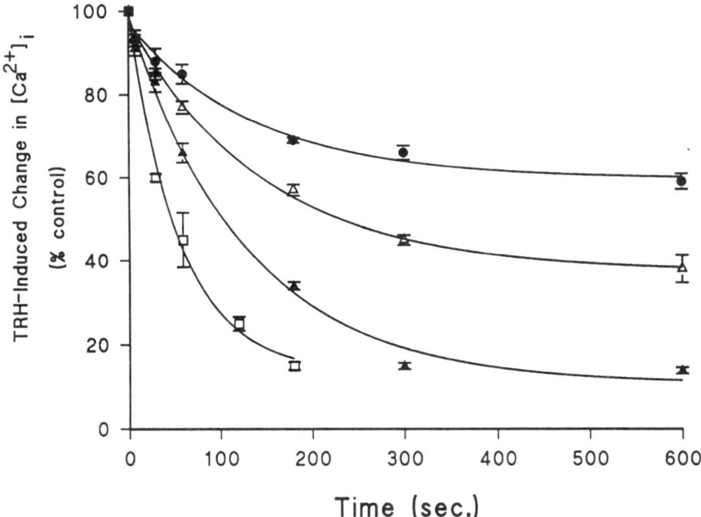

Fig. 3 GH_3 cells were challenged with TRH (100 nM) at various times following the application of halothane [0.5 (○), 1.0 (△), 1.5 (▲), and 3.0 mM (□)]. Peak $[Ca^{2+}]_i$ responses to TRH are expressed as a percentage of the TRH responses elicited in the absence of halothane and are plotted as a function of time following halothane administration. Reprinted with permission (5).

an exponential decline in the peak $[Ca^{2+}]_i$ response to TRH (Figs. 2B and 3). The maximum extent of inhibition of the $[Ca^{2+}]_i$ response to TRH was observed within 10 min of halothane administration and was dependent on halothane concentration (Fig. 3).

D. Mechanism of Halothane Inhibition of Peak Intracellular Calcium Responses to TRH

Inhibition of the peak $[Ca^{2+}]_i$ response to TRH could result from one of several mechanisms. Halothane could act by blocking IP_3 binding or by inhibiting ion flux through the IP_3 receptor/channel. These explanations are unlikely for the following reasons. First, halothane causes an increase in resting $[Ca^{2+}]_i$ by releasing calcium from an intracellular store (presumably but not necessarily the IP_3-gated stores). This response occurs rapidly, indicating that halothane equilibrates rapidly with sites within the cell. If halothane were acting to block the IP_3 receptor/channel, inhibition of the peak $[Ca^{2+}]_i$ response to TRH should have a very rapid time course. In fact, the peak $[Ca^{2+}]_i$ response to TRH decays over tens of seconds following halothane administration, eliminating the possibility that halothane acts to block the IP_3 receptor/channel.

The slow time course of the halothane-induced decay of the TRH response is most consistent with a reduction in the amount of calcium available to be released. The amount of calcium in the IP_3-gated stores could be reduced either by increasing the leak of calcium from the stores, or by reducing the rate of calcium reuptake by the stores. Calcium is normally pumped into the IP_3-gated stores by a Ca^{2+}-ATPase. The Ca^{2+}-ATPase in endoplasmic reticulum is selectivity inhibited by thapsigargin. We have used thapsigargin as a pharmacological tool to determine whether halothane increases the rate of calcium leak from IP_3-sensitive stores, or, alternatively, whether it inhibits calcium uptake by the intracellular stores.

Thapsigargin, like halothane, inhibited the peak $[Ca^{2+}]_i$ response to TRH. When cells were stimulated with a saturating concentration of TRH (100 n*M*) immediately after administration of thapsigargin, there was minimal effect on the peak $[Ca^{2+}]_i$ response to TRH. As the interval between thapsigargin administration and TRH stimulation was increased, there was an exponential decline in the peak $[Ca^{2+}]_i$ response to TRH. The time constant describing the exponential decay of the peak $[Ca^{2+}]_i$ response to TRH decreased as a function of increasing thapsigargin concentration. The time constant reached a limiting value of approximately 100 sec at thapsigargin concentrations above 3 n*M* (Fig. 4). At saturating concentra-

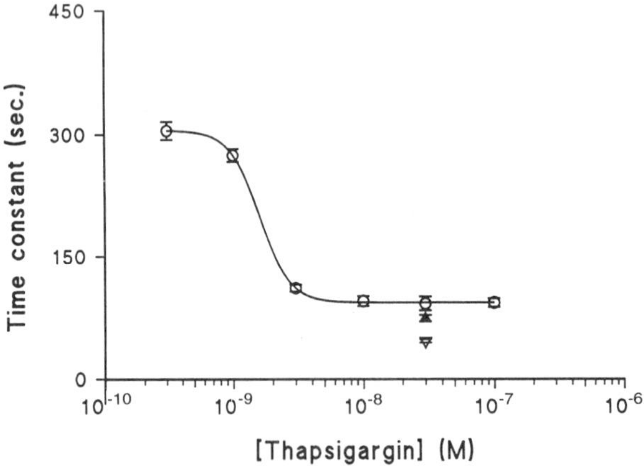

Fig. 4 Time constants of the monoexponential decays of the peak $[Ca^{2+}]_i$ response to TRH plotted as a function of thapsigargin concentration. At thapsigargin concentrations above 3 n*M*, the time constants reach a limiting value, corresponding to the time constant of spontaneous calcium leak from IP_3-gated stores. In the presence of 30 n*M* thapsigargin, halothane [0.5 (▲) and 1.0 m*M* (▽)] causes a concentration-dependent increase in the rate of spontaneous leak of Ca^{2+} from IP_3-gated stores. Time constants are reported with ±95% confidence limits. Reprinted with permission (5).

tions of thapsigargin the Ca^{2+}-ATPase is completely inhibited; thus, the time constant of 100 sec describes the rate of spontaneous leak of calcium from IP_3-gated calcium stores.

If halothane were primarily acting to inhibit the Ca^{2+}-ATPase, the combination of halothane and a saturating concentration of thapsigargin should produce the same exponential decay of the peak $[Ca^{2+}]_i$ response to TRH as thapsigargin alone. Experimentally, halothane in combination with 30 n*M* thapsigargin produced a significantly faster decay of the TRH response than did thapsigargin alone (Fig. 4). At concentrations of 0.5 and 1.0 m*M*, halothane decreased the time constant of calcium leak from a control value of 98 ± 9 sec to 74 ± 12 and 46 ± 6 sec, respectively. These data indicate that clinically relevant concentrations of halothane increase the rate of calicum leak from IP_3-gated stores.

The ability of halothane to increase the rate of leak calcium from intracellular stores provides an explanation for all the observed data. First, the increased rate of leak is likely to be directly responsible for halothane-induced increases in resting $[Ca^{2+}]_i$. The increased rate of calcium leak also causes an imbalance between the rate of leak from the intracellular stores and the rate of reuptake by the Ca^{2+}-ATPase. This imbalance slowly reduces the amount of calcium in the stores. This is manifest by the time-dependent inhibition of the peak $[Ca^{2+}]_i$ response to TRH.

E. Effects of Isoflurane and Octanol on Peak Intracellular Calcium Response to TRH

To determine whether other anesthetics produced effects similar to those observed with halothane, GH_3 cells were challenged with TRH 10 min after administration of various concentrations of isoflurane and octanol. Both octanol and isoflurane produced concentration-dependent inhibition of the peak $[Ca^{2+}]_i$ response to TRH, with 50% inhibition of the response occurring at 0.2 m*M* octanol and 0.5 m*M* isoflurane. These data indicate that anesthetic effects on IP_3-releasable stores of calcium are not limited to halothane.

IV. Summary

The experimental results reviewed in this article indicate that clinically relevant concentrations of halothane (and isoflurane and octanol) cause calcium to leak from IP_3-gated intracellular calcium stores, resulting in depletion of the stores. Depletion of intracellular Ca^{2+} stores should attenuate the actions of a variety of hormones and neurotransmitters that use the IP_3 pathway to produce their effects, potentially contributing to many volatile anesthetic side effects including bronchodilatation, vasodilatation, and unresponsiveness to vasoconstrictive agents.

Acknowledgments

Supported by National Institutes of Health Grant RO1-GM37846 (A.S.E.) and by an Established Investigator Award from the American Heart Association (A.S.E.).

References

1. Herland, J. S., Julian, F. J., and Stephenson, D. G. (1990). Halothane increases Ca^{2+} efflux via Ca^{2+} channels of sarcoplasmic reticulum in chemically skinned rat myocardium. *J. Physiol.* (*London*) **426,** 1–18.
2. Heathers, G. P., Jeuhne, T., Rubin, L. J., Corr, P. B., and Evers, A. S. (1989). Anion exchange chromatographic separation of inositol phosphates and their quantification by gas chromatography. *Anal. Biochem.* **176,** 109–116.
3. Stern, R. C., Herrington, J., Lingle, C. J., and Evers, A. S. (1991). The action of halothane on stimulus-secretion coupling in clonal (GH_3) pituitary cells. *J. Neurosci.* **11,** 2217–2225.
4. Grynkiewicz, G., Poenie, M., and Tsein, R. Y. (1985). A new generation of Ca^{2+} indicators with greatly improved fluorescent properties. *J. Biol.. Chem.* **260,** 3440–3450.
5. Hossain, M. D., and Evers, A. S. (1994). Volatile anesthetic-induced efflux of calcium from IT_3-gated stores in clonal (GH_3) pituitary cells. *Anesthesiology* **80,** 1379–1389.

Differential Control of Blood Pressure by Two Subtypes of Carotid Baroreceptors

Jeanne L. Seagard
Department of Anesthesiology and Zablocki Department of Veterans Affairs Medical Center
The Medical College of Wisconsin
Milwaukee, Wisconsin 53226

I. Introduction

Afferent activity from carotid baroreceptors has been recorded since 1932, when Bronk and Stella first described the pulse-related activity present in the carotid sinus nerve (1). In 1952, Landgren first described the presence of large and small potentials present in the baroreceptor activity recorded from the carotid sinus nerve, and he suggested that there may be different populations of baroreceptors with different afferent fiber types (2). The presence of A- and C-fiber baroreceptors has since been reported by many investigators, and the firing properties of these two populations of carotid baroreceptors have been examined in many studies (3–6). The classification of baroreceptor subtype, based on afferent fiber type, was the only categorization of baroreceptors until 1990, when work from our laboratory described two types of baroreceptors, based on firing patterns obtained from the baroreceptors in response to slow increases in carotid sinus pressure (7). This study identified two types of carotid baroreceptors: type I marked by discontinuous hyperbolic firing patterns, with high firing rates and sensitivities, and type II marked by continuous sigmoidal firing patterns, with spontaneous discharge below threshold, lower sensitivities, and wider operating ranges. The presence of the two firing patterns suggested that each baroreceptor subtype may play a differential role in the

Advances in Pharmacology, Volume 31

control of blood pressure. Thus, the more sensitive type I baroreceptors, with sudden onsets of discharge, may be more important in buffering dynamic changes in pressure, whereas the type II baroreceptors, with lower sensitivity but wider operating ranges, may contribute more to regulation of tonic, resting control of baseline pressure. A series of investigations were initiated to investigate this possible differential regulation of blood pressure.

If the two types of baroreceptors contribute differently to the tonic versus dynamic changes in blood pressure, differences in baroreceptor control of pressure should be seen when afferent input from each of the two subtypes of receptors is blocked. To test this possibility, afferent blocking techniques were utilized that took advantage of the finding that type I baroreceptors generally have larger myelinated afferent A fibers, whereas type II baroreceptors have smaller A and unmyelinated C afferent fibers (7). To examine the effects of blocking larger type I afferent fibers first, an anodal blocking technique was used on an intact carotid sinus nerve from a vascularly isolated carotid sinus preparation that was used to initiate reflex changes in arterial pressure. All other baroreceptors and cardiopulmonary receptors were eliminated to restrict blood pressure control to the one carotid sinus. Anodal blocking consists of application of a small amount of polarizing current to the outside of the nerve, which results in a greater difference between intra- versus extracellular potentials, thereby blocking conduction of the action potential. Careful regulation of blocking conditions can result in a very controllable sequential blockade of nerve fibers, and the nodal conduction of myelinated fibers makes this type of axon more susceptible to the blocking current. The order of blockade is therefore from large A to smaller A to unmyelinated C fibers. To examine the effects of blocking smaller type II afferent fibers first, a local anesthetic block of the carotid nerve, using bupivacaine, was performed in a similar isolated sinus preparation. Local anesthetic blockade of the smaller unmyelinated fibers occurs first, since the anesthetic can more quickly penetrate the thinner myelin layer of those axons. The order of blockade using this technique is from unmyelinated C fibers to small A fibers to large A fibers.

Although both techniques were not completely selective in blocking conduction in one size of fiber, their careful use could be used to minimize input from one of the two subtypes of baroreceptors. When either the blocking current or anesthetic concentration was increased, both blocking techniques ultimately blocked all fibers in the preparations. The effects of the initial selective elimination of afferent input from each baroreceptor type were determined by monitoring arterial blood pressure during constant pressure perfusion of the vascularly isolated carotid sinus (tonic control) and during ramp pressure stimulations of the carotid sinus baroreceptors (dynamic control). Portions of this article have been previously published (8). Details and results of the two studies are presented below.

II. Carotid Baroreceptors

All studies were performed in mongrel dogs anesthetized with thiopental sodium (35 mg/kg induction and 10–12 mg/kg/hr infusion). This technique has been found to produce a stable plane of anesthesia while maintaining baroreflex control of blood pressure. The animal was intubated and ventilated with a Bird Mark 7 respirator with 100% O_2. A femoral artery and vein were cannulated to permit measurement of arterial pressure and infusion of anesthetics and fluids, respectively. The femoral artery pressure was measured via a Statham pressure transducer to a Grass Model 7 polygraph (Quincy, MA). All pressures were recorded on the polygraph and a Vetter Model D FM tape recorder (Rebersburg, PA) for later analysis. Arterial blood gases were measured using a ABL 30 Radiometer Blood Gas Analyzer (Copenhagen, Denmark) and maintained within physiological ranges by adjustment of ventilation and/or infusion of bicarbonate.

To examine carotid baroreflex control of pressure, a vascularly isolated sinus preparation was utilized as previously described (7,9). Briefly, the left carotid sinus and associated vessels were isolated from the rest of the circulation by ligation of the common, external, occipital, lingual, and superior thyroid arteries, as well as any other smaller or additional vessels in the region. The lingual artery was cannulated with polyethylene tubing (PE 90) to permit measurement of carotid sinus pressure (CSP) via a Statham pressure transducer. The common and external carotid arteries were cannulated with metal cannulas connected to Tygon tubing in order to permit a flow-through perfusion of the carotid sinus vascular segment. The sinus was perfused via the common carotid cannula using a Sarns roller pump (Ann Arbor, MI), which drew the perfusate from an oxygenator reservoir. The outflow, via the external carotid cannula, was return to the oxygenator, where the perfusate was warmed and reoxygenated before being reperfused. The perfusate was buffered lactated Ringers solution, oxygenated with 100% O_2 to eliminate any chemoreceptor activity not physically eliminated by the isolation technique. The roller pump was controlled by a servo-controller designed and constructed in the laboratory, so that constant mean pressure perfusion of the sinus was used to condition the sinus and the maintain the desired pressure between pressure ramps used to activate the baroreceptors in a repeatable controlled manner. Pressure ramps were made by infusion of perfusate through the inflow line via a Harvard syringe pump (South Natick, MA) in-line with the roller pump. The roller pump was stopped and the outflow line was clamped, temporarily creating a "blind sac" preparation that permitted slow increases in CSP (1–3 mmHg/sec) from 0 to 300 mmHg. After each pressure ramp, the sinus was always reperfused at the original level of mean pulsatile pressure using the roller pump to prevent acute resetting from occurring.

The afferent blocking techniques were applied to the intact carotid sinus nerve innervating the vascularly isolated sinus. To better expose the nerve fibers to the blocking techniques, the nerve was left intact and only a small length of the outer sheath (2mm) was removed. The contralateral carotid sinus nerve and vagosympathetic trunks, which contain the aortic depressor nerves, were sectioned to eliminate afferent input from other baroreceptors and efferent sympathetic innervation of the isolated carotid sinus. The left carotid sinus nerve was identified by recording of pulse-related activity and isolated from surrounding tissue up to its junction with the glossopharyngeal nerve. Nerve activity was recorded using bipolar tungsten carbide electrodes connected to a high-impedance differential preamplifier (gain 1000; 0.1–10 kHz passband), followed by a filter/amplifier, which provided additional gain (up to 400) and high- and low-pass filtering (fourth order Butterworth, 10 Hz–3 kHz).

A. Anodal Block

To perform anodal blocking of baroreceptor afferent fibers, a modified wick-type electrode was placed on the desheathed segment of the intact left carotid sinus nerve (10). The monopolar electrode consisted of a solid felt tip epoxied into a hollow plastic tube with the bare end of an insulated silver wire threaded down the tube and into the felt wick to serve as the electrode lead. The electrode was soaked in saline for several hours prior to use to ensure complete conduction of the blocking current. The cathodal electrode consisted of a simple alligator clip which was placed in muscle tissue lateral to the blocking site in order to provide multiple current paths from the anode, bidirectionally along the nerve, to the cathode. Current density at the nerve–tissue interface was thus reduced by shunting current through multiple pathways, thereby reducing the excitatory effects of depolarization at the cathode.

Different levels of anodal current were used for each experiment, with each normalized as a percent to 100% maximum blocking current for that animal. Maximum blocking current was defined as level of current needed to produce maximal elevation of baseline blood pressure, which represented complete blocking of all baroreceptor afferent activity. Maximum current ranged from 80 to 150 μA, and, in general, C fibers were not blocked below 60 μA.

B. Anesthetic Block

Anesthetic application was limited to the 2-mm desheathed segment of the sinus nerve, since differential blocking of small versus large fibers using local anesthetics has been reported to be more selective if limited to this length of nerve (11). As reported by Franz and Perry (11), by limiting anesthetic exposure to 2 mm of nerve or less, at most only three nodes of Ranvier will be exposed for the smaller A fibers and two or less for the larger A-fiber neurons. Because three is the minimum number of nodes that must be blocked to eliminate conduction in the myelinated axons, (12), by limiting the exposed length of nerve, a more selective block of small A-fiber and C-fiber afferents can be achieved. The carotid nerve was isolated from surrounding tissue and, if possible, placed in a small slotted plastic chamber that could be sealed around the nerve, minimizing anesthetic exposure to a 2-mm segment of nerve. If there was insufficient room, a wick-type electrode was placed around the nerve and used for anesthetic application. If both methods were impossible, the anesthetic was then directly applied to a 2-mm segment of the nerve that had been desheathed. The remaining sheath on the rest of the nerve served as a diffusion barrier for the anesthetic. All methods appeared to result in differential block, based on the results obtained, but successful block using the last method may have also been dependent on the relatively low doses of anesthetic employed, as well as the short exposed segment. Concentrations of bupivacaine (BUP) used in the study ranged from 5 to 20 mg%, which are less than those used by other investigators in earlier studies.

C. Baroreceptor Activation

The effects of selective blocking of baroreceptor activity on tonic control of baseline blood pressure (BP) were determined by applications of block during constant mean pressure perfusion of the carotid sinus at a conditioning pressure equal to the resting level of the animal prior to sinus isolation. The effects of selective blocking on dynamic control of changes in BP were determined by application of anodal or anesthetic block during the slow ramp increases in sinus pressure. The protocol was as follows. Following a 25-min perfusion of the carotid sinus at the selected conditioning pressure, 30 sec of resting baseline levels of BP were recorded. Immediately after measuring baseline values, a slow ramp increase in sinus pressure was performed. The response to the pressure ramp was used to construct baroreceptor stimulus–response curves by plotting CSP versus BP. Following the ramp, the carotid sinus was again perfused at the original conditioning pressure for 5 min to prevent any acute resetting effects, and either anodal or anesthetic blocking of the baroreceptor afferents was initiated.

For anodal block, baseline and dynamic baroreflex responses were obtained during application of anodal blocking currents of control, 25, 50, 75, and 100% of maximum current to the left sinus nerve, allowing sufficient time for baseline effects of the blocking to stabilize prior to initiation of the pressure ramp. After the tonic and dynamic responses to each current were determined, the current was turned off and the sinus was again perfused at constant pressure for 5 min. New control baseline and reflex responses were determined to ensure that the effects of the previous block were over. The repeated testing was done until control responses returned to normal. After control was reestablished, a new blocking current was tested. This protocol was repeated until all levels of current were tested.

For anesthetic block, baseline and dynamic baroreflex responses were obtained during application of bupivacaine to the left sinus nerve, with responses measured at 7 min after anesthetic application. Bupivacaine was applied in increasing concentrations, starting with 5 mg% and increasing to 7, 10, and 20 mg% in the following manner. After the tonic and dynamic responses at 5 mg% were obtained, the sinus was again perfused at constant pressure for 5 min. Following the reconditioning period, the next level of anesthetic was applied and the response at 7 min was measured. This protocol did not allow recovery between anesthetic exposures, but it did allow sufficient time for the major blocking effects of each level of the anesthetic to occur within the experimental period. This procedure was repeated up to 20 mg% BUP, after which the anesthetic was removed and recovery was allowed to occur. Repeated ramps in CSP were done at varying intervals to ensure that at least 75% of the initial response was obtained at the end of the experiment to ensure that loss of the reflex was not due to time or degradation of the preparation.

For data analysis, analog-to-digital conversion of recorded parameters was performed using a Hewlett-Packard 310 computer. Arterial pressure and carotid sinus pressure were sampled at a frequency of 10 Hz for each control and blocking procedure and stored on disk files for quantitation and statistical analysis. Baseline values of BP were obtained using 30-sec averages of these parameters sampled during the period immediately prior to ramps for control and each blocking condition. To determine baroreflex sensitivity, mean BP was plotted versus ramp changes in CSP to obtain baroreflex response curves. Nonlinear regression was used to curve-fit the sigmoidal response curves (7) and determine pressure threshold (Pth) and maximum slope of the linear portion of the curves. Pressure thresholds, slopes (sensitivities), and baseline BP values for control and each level of anodal or anesthetic block were compared using an analysis of variance. Significantly different means were located using Duncan's Multiple Range Test. All levels of significance were set at $p < 0.05$ *a priori.*

III. Carotid Sinus Nerve Activity

A. Blocking of Normal Activity by Anodal Current

Increasing levels of anodal blocking current applied to the carotid sinus nerve produced changes in tonic control of baseline levels of BP and dynamic baroreflex control of changes in pressure (slope). As shown in the representative example from one animal in Fig. 1, increasing levels of anodal block from 25 to 100% maximum current (100 μA absolute current for this animal) produced stepwise attenuation of the baroreflex-induced inhibition of arterial pressure during slow ramp increases in CSP. The slope of the reflex response was attenuated at all levels of blocking current starting at 25% maximum current. As seen in Fig. 2, attenuation of dynamic baroreflex control of pressure was accompanied by attenuation

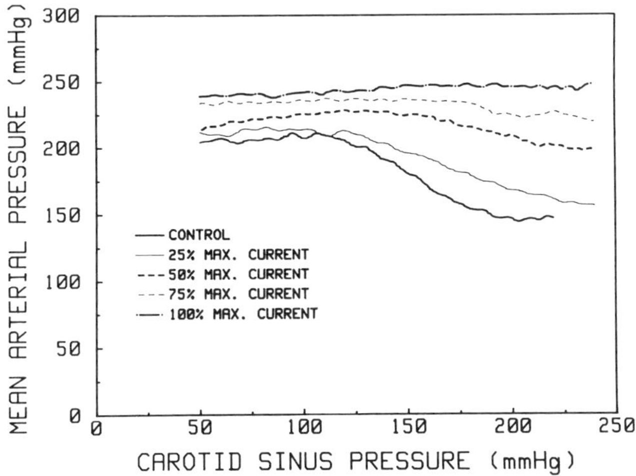

Fig. 1 Effects of anodal block on dynamic baroreflex changes in mean arterial blood pressure in one animal. The typical reflex decrease in arterial pressure was obtained during slow ramp increases in pressure in a vascularly isolated carotid sinus (control). This sensitivity of baroreflex-induced hypotension was attenuated as anodal blocking currents from 25 to 100% maximum blocking current were applied to the carotid sinus nerve from the isolated sinus. The effects of anodal block on control of baseline blood pressure in the same animal are shown in Fig. 2 [From Seagard *et al.* (8).]

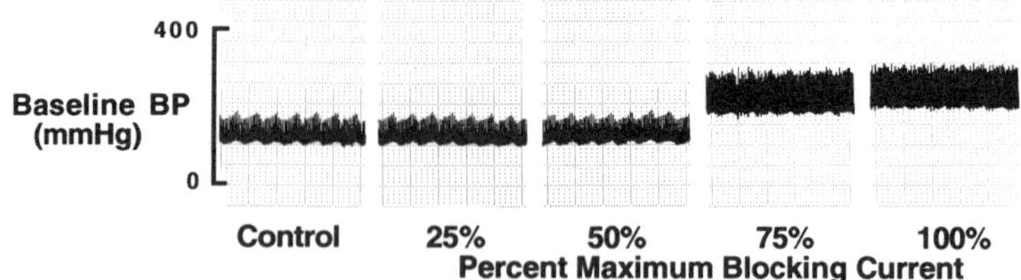

Fig. 2 Effects of anodal block in one animal on tonic control of baseline levels of arterial blood pressure (BP) recorded at constant mean pulsatile perfusion of 135 mmHg of the innervated, isolated carotid sinus. These results were obtained from the same animal whose baroreflex responses are shown in Fig. 1. No changes in baseline blood pressure from control were seen when 25 and 50% of maximum blocking current were applied to the sinus nerve. However, increasing the current to 75 and 100% maximum current produced sequential increases in baseline arterial pressure during the constant pressure stimulation of the sinus baroreceptors. [From Seagard *et al.* (8).]

of tonic control of arterial baseline BP, but at only the higher levels of blocking currents. The level of baseline BP was the same at control and during anodal block of the carotid sinus nerve at 25 and 50% maximum current. Thus, the loss of tonic inhibition of arterial pressure, shown by the increases in baseline BP, occurred at blocking currents higher than those needed to produce attenuation of dynamic baroreflex control of pressure. Attenuation of the control of dynamic baroreflex changes in arterial pressure occurred at blocking current levels that would affect primarily larger A-fiber baroreceptors, whereas attenuation of tonic control of baseline BP occurred at higher blocking currents that would include smaller A-fiber and C-fiber baroreceptors.

The results of anodal blocking of baroreceptors from all animals are shown in Table I. As suggested by the single representative example in Fig. 1 and 2, significant attenuation of baroreflex sensitivity (slope) was obtained at lower levels of blocking current (25% of maximum current) than was significant attenuation of control of baseline BP (75% of maximum). In addition to attenuation of both tonic and dynamic baroreflex control of BP, anodal blocking also produced a significant elevation in Pth. As shown in Table I, a significant increase in Pth from control and 25% maximum current occurred at blocking levels of 50 and 75% of maximum current. This effect on Pth preceded any significant changes in control of tonic baseline BP but occurred at a higher blocking current than that which attenuated dynamic baroreflex sensitivity. Because the baroreflex response was blocked at 100% maximum current, Pth could not be determined for this level of anodal block.

Table I

Effects of Anodal Block on Tonic and Dynamic Baroreflex Control of Blood Pressure, with Carotid Sinus Perfusion Pressure Set to Preisolation Presures[a]

	Blocking current as % maximum				
Measure	Control	25%	50%	75%	100%
BP (mmHg)	158.41 ± 9.5	160.7 ± 9.5	166.4 ± 10.9	181.8 ± 11.1[b]	189.4 ± 13.2[c]
Slope (mmHg MAP/ mmHg CSP)	−0.84 ± 0.11	−0.63 ± 0.10[d]	−0.36 ± 0.09[b]	−0.11 ± 0.05[c]	−0.01 ± 0.01[c]
Pth	129.0 ± 9.4	138.4 ± 10.3	160.3 ± 13.7[b]	174.5 ± 7.6[b]	—

[a] Values are reported as means ± SE (n = 7 dogs). BP, Arterial pressure during constant pressure perfusion of the carotid sinus, Pth, pressure threshold of the baroreflex, MAP, mean arterial blood pressure.
[b] Significantly different from control and 25% of maximum current.
[c] Significantly different from control and 25 and 50% of maximum current.
[d] Significantly different from control.

B. Block of Nerve Activity by Anesthetic

The effects of anesthetic blockade of carotid baroreceptor afferent fibers on dynamic baroreflex control of blood pressure are shown in Fig. 3 and summarized in Table II. In the example from one animal shown in Fig. 3, at 5.0 mg% BUP there was no effect on the slope of the baroreflex-induced decrease in arterial pressure, although there was a slight increase in arterial pressures at all levels of CSP, relative to control. This suggests that there was some inhibition of baroreceptor activity, although the afferents involved in buffering changes in CSP were not significantly affected. Increasing the concentration of BUP to 7.0 mg% attenuated baroreflex slope, and higher levels of BUP produced increasing degrees of reflex attenuation until the entire reflex was blocked at 20.0 mg% BUP. This pattern is reflected in the summed data for all animals in Table II. There was no significant attenuation of baroreflex sensitivity until 7.0 mg% BUP, after which there was an additional decrease in baroreflex slope at 10 and 20 mg%. In most animals (4 of 6) the reflex was almost completely eliminated at 20 mg% BUP.

The effects of BUP on tonic control of pressure in a representative animal are shown in Fig. 4, and data from all animals are summarized in Table II. Figure 4 shows the effects of increasing blocking concentrations

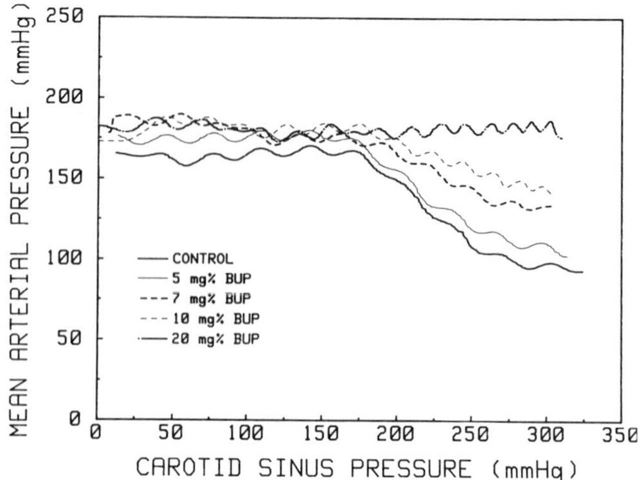

Fig. 3 Effects of anesthetic block on dynamic baroreflex changes in mean arterial blood pressure in one animal. The typical reflex decrease in arterial pressure was obtained during slow ramp increases in pressure in a vascularly isolated carotid sinus (control). The sensitivity of baroreflex-induced hypotension was attenuated as the level of the local anesthetic bupivacaine (BUP) applied to the carotid sinus nerve was increased from 7 to 10 mg%. There was no effect on the slope of baroreflex inhibition of arterial blood pressure at 5 mg% BUP. Effects of anesthetic block on control of baseline blood pressure in the same animal are shown in Fig. 4. [From Seagard *et al.* (8).]

Table II

Effects of Anesthetic Block on Tonic and Dynamic Baroreflex Control of Blood Pressure, with Carotid Sinus Perfusion Pressure Set to Preisolation Presures[a]

	Bupivacaine concentration (mg%)				
Measure	Control	5%	7%	10%	20%
BP (mmHg)	158.8 ± 6.4	169.0 ± 6.5[b]	174.3 ± 5.4[b]	175.7 ± 4.7[c]	175.2 ± 5.9[c]
Slope (mmHg MAP/ mmHg CSP)	−0.85 ± 0.08	−0.73 ± 0.08	−0.64 ± 0.09[b]	−0.25 ± 0.08[d]	−0.12 ± 0.04[d]
Pth	142.7 ± 8.6	137.4 ± 7.8	145.0 ± 9.1	149.0 ± 7.9	—

[a] Values are reported as means ± SE (n = 6 dogs). BP, Arterial pressure during constant pressure perfusion of the carotid sinus; Pth, pressure threshold of the baroreflex; MAP, mean arterial blood pressure.

[b] Significantly different from control.

[c] Significantly different from control and 5 mg% bupivacaine.

[d] Significantly different from control and 5 and 7 mg% bupivacaine.

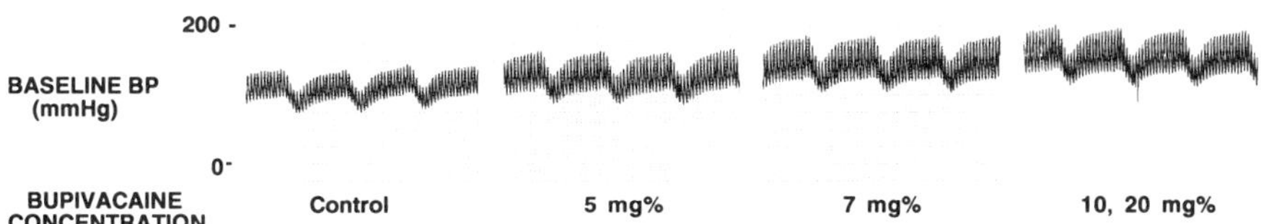

Fig. 4 Effects of anesthetic block in one animal on tonic control of baseline levels of arterial blood pressure (BP) recorded at constant mean pulsatile perfusion of 132 mmHg of the innervated, isolated carotid sinus. The results were obtained from the same animal whose baroreflex responses are shown in Fig. 3. As the level of the local anesthetic bupivacaine (BUP) applied to the carotid sinus nerve was increased from 5 to 10 mg%, there were sequential increases in baseline arterial BP during the constant pressure stimulation of the sinus baroreceptors. The level of baseline BP during blockade with 20 mg% BUP was the same as that during blockade with 10 mg%. [From Seagard *et al.* (8).]

of BUP on the baseline level of BP recorded during constant pressure perfusion of the carotid sinus at 132 mmHg from the same dog whose reflex responses are shown in Fig. 3. As seen in Fig. 4, as BUP concentration was increased, there was an elevation in baseline BP at 5, 7, and 10 mg% BUP, indicating inhibition of baroreceptor control of tonic BP. However, there was no additional increase in baseline BP at 20 mg% BUP. These results are similar to those for all animals (Table II). As seen in Table II, there were no significant changes in Pth in response to any level of BUP exposure. Because the baroreflex response was blocked at 20 mg% BUP in most animals, Pth was not determined for this level of anesthetic exposure.

IV. Discussion

Based on the earlier finding that type I baroreceptors generally have larger afferent fibers than type II baroreceptors (7), the present study utilized two techniques which block fibers based on size to try to establish roles for two different types of functionally characterized baroreceptors. Results from the present study indicate that selective blocking of baroreceptors unmasks differential roles for two types of baroreceptors in the control of blood pressure. Large A-fiber baroreceptors appear to contribute more to control of dynamic baroreflex regulation of changes in arterial pressure, whereas small A- and C-fiber baroreceptors appear to be the primary regulators of tonic baseline levels of BP. This differential contribution to control of two aspects of blood pressure regulation is not absolute, for baroreceptors with smaller afferent fibers were found to contribute to some degree to dynamic pressure control. However, each baroreceptor type appears to contribute preferentially to either dynamic or tonic control of blood pressure.

Results from the present study agree with results from previous studies. Activation of arterial baroreceptors, either through pressure stimulation of the receptor directly or by electrical stimulation of afferent fibers, is known to initiate depressor responses (13–16), but most studies have focused on the role of larger A-fiber, or type I, baroreceptors. However, the role of C-fiber baroreceptors, which are primarily type II baroreceptors when functionally classified, has been examined in a few studies by other investigators (14,16–18). These studies suggest that although both A- and C-fiber baroreceptors mediate depressor responses, prolonged control of blood pressure requires C-fiber baroreceptor participation, a finding that agrees with the preferential control of tonic pressure by baroreceptors with smaller afferents observed in the present study.

One of the earliest studies by Douglas *et al.* (17) showed that the greatest hypotension produced by electrical stimulation of aortic baroreceptor afferent fibers in rabbits required activation of C-fiber baroreceptors. C-fiber baroreceptor activation was also required to maintain the hypotension for the duration of the stimulation, leading the investigators to propose that A-fiber baroreceptors may be involved in rapid adjustment of systemic blood pressure, but the C-fiber baroreceptors may play a longer lasting role. In a study by Aars *et al.*(3) that utilized anodal block to elucidate the role of aortic C-fiber baroreceptors in the rabbit, the investigators found that A-fiber baroreceptors produced a greater reflex decrease in sympathetic activity, especially at low pressures, than C-fiber baroreceptors in response to a change in aortic pressure. In the study, tracings indicate no change in baseline levels of pressure or sympathetic activity in response to anodal block of A-fiber baroreceptors at resting aortic pressures, although the reflex effects to increases in aortic pressure were eliminated or greatly attenuated. Finally, an earlier study by Wiemer and Kiwull (19), which used cold block to attempt a selective block of carotid sinus afferents, suggested that there were different types of baroreceptors, one characterized by larger spikes and a greater sensitivity toward blood pressure changes and a second, less-sensitive type with smaller spikes.

In the present study, to determine the differential blocking capabilities of the two techniques, a length of vagus nerve approximately the diameter of the carotid sinus nerve was used to test the selectively of each type of block (8). Application of anodal block found that conduction of A fibers with velocities of 9.8 to 98.9 m/sec could be blocked before significant blocking of smaller A fibers and C fibers. However, the selectivity of the anesthetic blocking technique was not so discrete. Anesthetic application to the vagus nerve blocked C-fiber conduction before altering conduction in fast A fibers. However, many A fibers with conduction velocities that ranged from 4.1 to 8.5 m/sec were also blocked at the same concentration of anesthetic that blocked C fibers. Therefore, although both techniques allowed fairly selective blockade of large A fibers versus small C fibers, the ability to block selectively medium A fibers versus C fibers with local anesthetic blockade was found to be less distinct.

Gissen *et al.* (20) found that 250 mg% BUP produced an 80% blockade of C fibers but also a 100% blockade of A fibers in the rabbit vagus nerve in 3 min. However, lowering the level of BUP to a range of 1.3 to 9.3 mg% (0.03–0.40 m*M*) allowed more selective blockade of C fibers. At these lower concentrations, 79.2 min was required before A-fiber blockade exceeded C-fiber block, although the extent of blockade of both fibers was only approximately 20% at that time. This study indicated that two factors were involved in initiating fiber block by BUP, namely, both the concentration of the anesthetic used and the time allowed for block to occur. A later study by Palmer *et al.* (21) also examined the blocking ability of BUP on the dog vagus nerve. In that study, 20 mg% BUP produced a 78% blockade of C fibers in 3.2 min following exposure, whereas 5.4 min was required to block 86% of A fibers. Owing to the lower lipid solubility characteristics and higher pK_a of BUP, the selectivity of blocking by BUP was found to be superior to that for lidocaine (21) or etidocaine (20), two other local anesthetics. Finally, as stated earlier, differential anesthetic block was found to be enhanced when 2 mm of nerve or less was exposed to the anesthetic (21). Thus, four factors must be considered when attempting differential blockade using local anesthetics: (1) anesthetic used, (2) concentration of the anesthetic, (3) time of exposure, and (4) length of nerve exposed. However, in spite of the difficulty in separating the block of A and C fibers, careful application of BUP did permit a relatively greater block of C fibers at the lower anesthetic concentrations. Moreover, because the two nerve blocking techniques reversed the order of blocking of afferent fiber types, their use allowed the different roles of the baroreceptor types to be discerned.

The techniques of anodal block and local anesthetic block have been used to examine the reflex effects of receptors with different afferent fiber types. Anodal block has been successfully used to differentially block A versus C fibers in the aortic (3,15) and vagal nerves (10,22). Local anesthetic blockade was performed in a study to separate the contributions of afferent activity from larger myelinated (groups I and II) versus small myelinated (group III) and unmyelinated (group IV) fibers from a muscle to the reflex cardiovascular and respiratory responses initiated by exercise of the muscle (23). Careful application and monitoring of evoked potentials indicated that although some conduction block of large A fibers occurred during the more complete blockade of the type III and type IV fibers, reflex attenuation correlated well with loss of conduction in the smaller fibers. Thus, although anesthetic blockade was not completely selective, careful use of the technique permitted discrimination of the reflex pathway to the smaller myelinated and unmyelinated fibers.

V. Anesthetic Implications

Inhalational anesthetics including halothane (9,24,25) and isoflurane (26) have been found to increase discharge of arterial baroreceptors. However, the effects of anesthetics on each subtype of baroreceptor has not been examined. Some evidence based on reflex studies suggests that there may be a differential effect of the anesthetics on the two baroreceptor subtypes, for some studies have found that resting baseline blood pressure may be decreased while baroreflex sensitivity is not significantly changed (26,27). This suggests that there may be a greater anesthetic effect on type II baroreceptors, leading to changes in tonic blood pressure control, while sensitivity of the type I baroreceptors, and therefore dynamic pressure control, is not altered. However, other studies have found that anesthesia decreases reflex sensitivity at levels that also depress baseline levels of blood pressure (25). Some differences in the levels or combinations of anesthetics may explain the differences between these studies in regard to tonic and dynamic control of blood pressure. In addition, in studies where both tonic and dynamic baroreflex control was attenuated, increased baroreceptor input due to exposure to inhalational anesthetics may have tonically lowered the overal level of outflow of sympathetic activity, resulting in these cases in an attenuated ability of the sympathetic system to respond to produce reflex changes in blood pressure.

A preliminary study was initiated to determine the effects of halothane on the firing of type I versus type II baroreceptors. The left carotid sinus was vascularly isolated as described above and perfused at a constant mean pressure of 125 mmHg. The left carotid sinus nerve was isolated, sectioned at its junction with the glossopharyngeal nerve, and placed in a mineral oil pool constructed of surrounding tissue. The nerve was then desheathed and dissected into consecutively smaller bundles until a single-fiber preparation was obtained. Nerve activity was recorded using tungsten carbide electrodes connected to a high-impedance differential preamplifier (gain 1000; passband, 0.1–10 kHz), followed by a filter/amplifier, which provided additional gain (up to 400) and high- and low-pass filtering (fourth order Butterworth, 10 Hz–3 kHz). Amplifier output was recorded on the FM tape recorder and averaged and displayed on the Grass recorder.

The response of the single baroreceptor to a slow ramp increase in CSP was obtained in order to characterize the baroreceptor as either type I or type II. After the control response was obtained, constant pressure perfusion was reinstated, and 1.5% halothane was administered to only the perfusate via an oxygenator reservoir system using a vaporizer. The response of the baroreceptor to a ramp increase in CSP was again obtained after 7 min of halothane exposure. The baroreceptor response curves for control and halothane exposure were plotted as shown in Fig. 5. As can be seen, halothane had no effect on the discharge pattern of the type I baroreceptor, but increased the spontaneous discharge and sensitivity of the type II baroreceptor. Similar results were obtained for one other baroreceptor of each type. These preliminary results suggest that halothane may have a differential effect on the two subtypes of baroreceptors and this effect may contribute to the altered control of blood pressure during anesthesia.

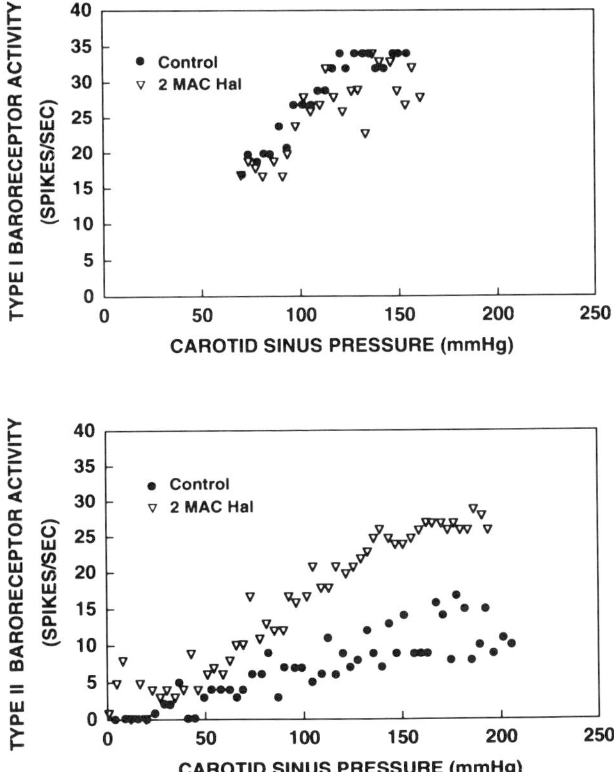

Fig. 5 Effects of 2 MAC halothane (Hal) on the firing patterns of a type I and type II carotid sinus baroreceptors. (Top) Exposure to halothane had no effect on the response curve of a type I baroreceptor. (Bottom) Exposure to halothane increased the saturation firing rate and sensitivity of the type II baroreceptor. MAC, minimum alveolar concentration.

References

1. Bronk, D. W., and Stella, G. (1932). Afferent impulses in the carotid sinus nerve. *J. Cell. Comp. Physiol.* **1,** 113–130.
2. Landgren, S. (1952). On the excitation mechanism of the carotid baroreceptors. *Acta Physiol. Scand.* **26,** 1–34.
3. Aars, H., Myhre, L., and Haswell, B. A. (1978). The function of baroreceptor C fibres in the rabbit's aortic nerve. *Acta Physiol. Scand.* **102,** 84–93.
4. Coleridge, H. M., Coleridge, J. C. G., and Schultz, H. D. (1987). Characteristics of C fibre baroreceptors in the carotid sinus of dogs. *J. Physiol. (London)* **394,** 291–313.
5. Thoren, P., Saum, W. R., and Brown, A. M. (1977). Characteristics of rat aortic baroreceptors with nonmedullated afferent nerve fibers. *Circ. Res.* **40,** 231–237.
6. Yao, T., and Thoren, P. (1983). Characteristics of brachiocephalic and carotid sinus baroreceptors with non-medullated afferents in the rabbit. *Acta Physiol. Scand.* **117,** 1–8.
7. Seagard, J. L., Van Brederode, J. F. M., Dean, C., Hopp, F. A., Gallenberg, L. A., and Kampine, J. P. (1990). Firing characteristics of single-fiber carotid sinus baroreceptors. *Circ. Res.* **66,** 1499–1509.
8. Seagard, J. L., Hopp, F. A., Drummond, H. A., and Van Wynsberghe, D. M. (1993). Selective contribution of two types of carotid sinus baroreceptors to the control of blood pressure. *Circ. Res.* **72,** 1011–1022.
9. Seagard, J. L., Hopp, F. A., Bosnjak, Z. J., Elegebe, E. O., and Kampine, J. P. (1983). Extent and mechanism [illegible]ane sensitization of the carotid sinus baroreceptors. *Anesthesiology* **58,** 432–4[illegible].
10. Hopp, F. A., Zuperku, F. A., Coon, R. L., and Kampine, J. P. (1980). Effect of anodal blockade of myelinated fibers on vagal C-fiber afferents. *Am. J. Physiol.* **239,** R454–R462.
11. Franz, D. E., and Perry, R. S. (1974). Mechanisms for differential block among single myelinated and nonmyelinated axons by procaine. *J. Physiol. (London)* **236,** 193–210.
12. Tasaki, I. (1959). "Handbook of Physiology, Section 1: Neurophysiology," Vol. 1, pp.75–122. American Physiological Society, Washington, D.C.
13. Douglas, W. W., and Ritchie, J. W. (1956). Cardiovascular reflexes produced by electrical excitation of non-medullated afferents in the vagus, carotid sinus and aortic nerves. *J. Physiol. (London)* **134,** 167–178.
14. Kardon, M. B., Peterson, D. F., and Bishop, V. S. (1973). Reflex bradycardia due to aortic nerve stimulation in the rabbit *Am. J. Physiol.* **225,** 7–11.
15. Kardon, M. B., Peterson, D. F., and Bishop V. S. (1975). Reflex heart rate control via specific aortic nerve afferents in the rabbit. *Circ. Res.* **37,** 41–47.
16. Kendrick, J. E., and Matson, G. L. (1971). Depressor and pressor afferent fibers in the carotid sinus nerve of the dog. *Proc. Soc. Exp. Biol. Med.* **138,** 175–177.
17. Douglas, W. W., Ritchie, J. M., and Schaumann, W. (1956). Depressor reflexes from medullated and nonmedullated fibres in the rabbit's aortic nerve. *J. Physiol. (London)* **132,** 187–198.
18. Heistad, D. D., Abboud, F. M., Mark, A. L., and Schmid, P. G. (1975). Effect of baroreceptor activity on ventilatory response to chemoreceptor stimulation. *J. Appl. Physiol.* **39,** 411–416.
19. Wiemer, W., and Kiwull, P. (1967). Blocking temperature and spike heights of baroreceptor afferents in the carotid sinus nerve. *In* "Baroreceptors and Hypertension" (P. Kezdi, ed.), pp. 51–67. Pergamon, London.
20. Gissen, A. J., Covino, B. G., and Gregus, J. (1982). Differential sensitivity of fast and

slow fibers in mammalian nerve III. Effect of etidocaine and bupivacaine on fast/slow fibers. *Anesth. Analg. (N.Y.)* **61,** 570–575.

21. Palmer, S. K., Bosnjak, Z. J., Hopp, F. A., Von Colditz, J. H., and Kampine, J. P. (1983). Lidocaine and bupivacaine differential blockade of isolated canine nerves. *Anesth. Analg. (N.Y.)* **62,** 754–757.
22. Thoren, P. N., Donald, D. E., and Shepherd, J. T. (1976). Role of heart and lung receptors with nonmeddullated vagal afferents in circulatory control. *Circ. Res.* **38,** (Suppl. 2), II-2–II-9.
23. McCloskey, D. I., and Mitchell, J. H. (1972). Reflex cardiovascular and respiratory responses originating in exercising muscle. *J. Physiol. (London)* **224,** 173–186.
24. Biscoe, T. J., and Millar, R. A. (1964). The effect of halothane on carotid sinus baroreceptor activity. *J. Physiol. (London)* **173,** 24–37.
25. Seagard, J. L., Hopp, F. A., Donegan, J. H., Kalbfleisch, J. H., and Kampine, J. P. (1982). Halothane and the carotid sinus reflex: Evidence for multiple sites of action. *Anesthesiology* **57,** 191–202.
26. Seagard, J. L., Elegbe, E. O., Hopp, F. A., Bosnjak, Z. J., Von Colditz, J. H., Kalbfleisch, J. H., and Kampine, J. P. (1983). Effects of isoflurane on the baroreceptor reflex. *Anesthesiology* **59,** 511–520.
27. Seagard, J. L., Hopp, F. A., Bosnjak, Z. J., Osborn, J. L., and Kampine, J. P. (1984). Sympathetic efferent nerve activity in conscious and isolfurane-anesthetized dogs. *Anesthesiology* **61,** 266–270.

Sympathetic Activation with Desflurane in Humans

Thomas J. Ebert and Michael Muzi
Department of Anesthesiology
Veterans Affairs Medical Center and
The Medical College of Wisconsin
Milwaukee, Wisconsin 53295

I. Introduction

Desflurane is the first new volatile anesthetic introduced into clinical practice in the United States in two decades. Published data from U.S. Food and Drug Administration Phase II and III trials with desflurane have suggested that this agent is remarkably similar to the parent compound, isoflurane. For example, at steady state, desflurane reduces blood pressure primarily by reducing systemic vascular resistance while maintaining cardiac output (1). However, several intriguing preliminary reports appeared in late 1992 suggesting that during the initial administration of desflurane to the inspired gas or when deepening the level of anesthesia with desflurane, undesirable increases in heart rate and blood pressure might occur (2,3). If substantiated, this effect might account for the unusually high incidence of myocardial ischemia noted in coronary artery disease patients receiving desflurane anesthesia for elective coronary artery bypass graft surgery (4). Although the parent compound isoflurane has not been associated with sympathetic activation or hyperdynamic circulation, one report indicated that when intubated patients were briefly exposed to 5% isoflurane, signs of sympathetic excitation occurred (5). These results raise the possibility that higher inspired concentrations of the two anesthetics trigger sympathetic activation. Thus, the present study employed sympathetic microneurography in humans to record directly efferent sympathetic vasoconstrictor activity during the incremental administration of isoflurane or desflurane.

II. Sympathetic Microneurography

Sixteen normotensive male volunteers, aged 20–32, provided informed consent and were studied while supine. Monitoring included heart rate (HR) from the electrocardiogram, mean arterial pressure (MAP) from a radial artery catheter, and end-tidal carbon dioxide and anesthetic concentrations by infrared spectrometry. Neuromuscular function was evaluated by train-of-four responses from the ulnar nerve. Sympathetic microneurography was performed in the peroneal nerve of the right leg immediately inferior to the bony prominence at the proximal head of the fibula on the lateral aspect of the leg. The location of the nerve was identified by applying brief electrical impulses to a 32-gauge, epoxy-coated, tungsten needle with an exposed tip diameter of approximately 5 μm. Sympathetic neural activity (SNA) directed to blood vessels within skeletal muscle were identified as previously described (6). Briefly, spontaneous bursts of neural activity were evident on the amplified signal, and these could be increased by prolonged breath holding, during Phase II and III of the Valsalva maneuver, and by a brief hypotensive stimulus induced with sodium nitroprusside. These signals were not altered by startle maneuvers or embarrassing comments, indicating that the activity was not from sympathetic efferent fibers directed to the skin vasculature or sweat glands.

Subjects received either desflurane ($n=7$) or isoflurane ($n=9$). Once an acceptable and stable sympathetic recording was obtained, a 10-min rest period was observed followed by 5 min of hemodynamic (HR, MAP) and neural (SNA) measurements. After preoxygenation, subjects were induced with sodium thiopental (5 mg/kg) and vecuronium (0.2 mg/kg). Ventilation was controlled through the face mask, without an oral airway, for 12 min. Exactly 2 min after injection of the thiopental, the vaporizer was activated to 0.5 minimum alveolar concentration (MAC) (3.6% desflurane or 0.6% isoflurane). In the two subsequent 1-min periods (4 and 5 min after thiopental), the vaporizer setting was increased to 1.0 MAC and then to 1.5 MAC while mask ventilation continued and end-tidal CO_2 concentrations were maintained at conscious baseline. On completion of the 12-min period following anesthetic induction, the trachea was intubated and ventilation was controlled to maintain end-tidal carbon dioxide at awake levels. The vaporizer setting was reduced to 0.5 MAC for an additional 20 min of anesthesia, providing a 32-min, postinduction interval before additional hemodynamic and neural data were collected. This reduced the likelihood that the initial administration of thiopental influenced subsequent measurements.

Neurocirculatory responses during steady-state periods of 0.5, 1.0, and 1.5 MAC, as well as the transition between vaporizer settings, were determined. Because we employed a high fresh gas flow rate (6–8 liters/min), the transitions were carried out by simply advancing the vaporizer setting to the next desired percent inspired concentration. At least 15 min was allowed between steady-state measurements to allow stabilization of end-tidal concentrations. In five subjects, the response to an increase in isoflurane concentration from 1.2% (1.0 MAC) to 5% was determined.

Consecutive neurocirculatory measurements were compared with analysis of variance for repeated measures, and post hoc *t*-test analyses were performed with Bonferroni correction. Probability values less than 0.05 were considered sufficient to reject the null hypothesis.

III. Anesthetics and Sympathetic Activation

Resting neurocirculatory parameters in conscious subjects are displayed in Fig. 1. There were no differences at rest between the two treatment groups. The administration of induction doses of thiopental resulted in significant increases in HR but no changes in MAP and SNA (Fig. 1). As desflurane was carefully titrated into the inspired gas beginning 2 min after thiopental administration, significant increases in SNA were noted (Fig. 1). This peaked, on average, 4–5 min after initiating the desflurane, at a time when end-tidal concentrations were 6–8%. This sympathoexcitation led to significant increases in HR and MAP, and these hemodynamic changes did not return to baseline for 8–10 min. End-tidal CO_2 was not significantly different from conscious baseline at any time during anesthesia.

Steady-state effects of desflurane and isoflurane are depicted in Figs. 2 and 3. Compared to conscious baseline, 0.5, 1.0, and 1.5 MAC anesthesia reduced MAP similarly in both groups. The HR and MAP measures did not differ between groups at any steady-state period of observation. With increasing MAC of desflurane, there was a significant trend for higher levels of SNA during the steady state. This contrasted to the relatively stable levels of SNA recorded with increasing isoflurane MAC.

The neurocirculatory responses during the first 12 min after initiating a vaporizer change to the next desired concentration of desflurane and isoflurane also are depicted in Figs. 2 and 3. The transition from 0.5 to 1.0 MAC was relatively unremarkable and basically resulted in a gradual

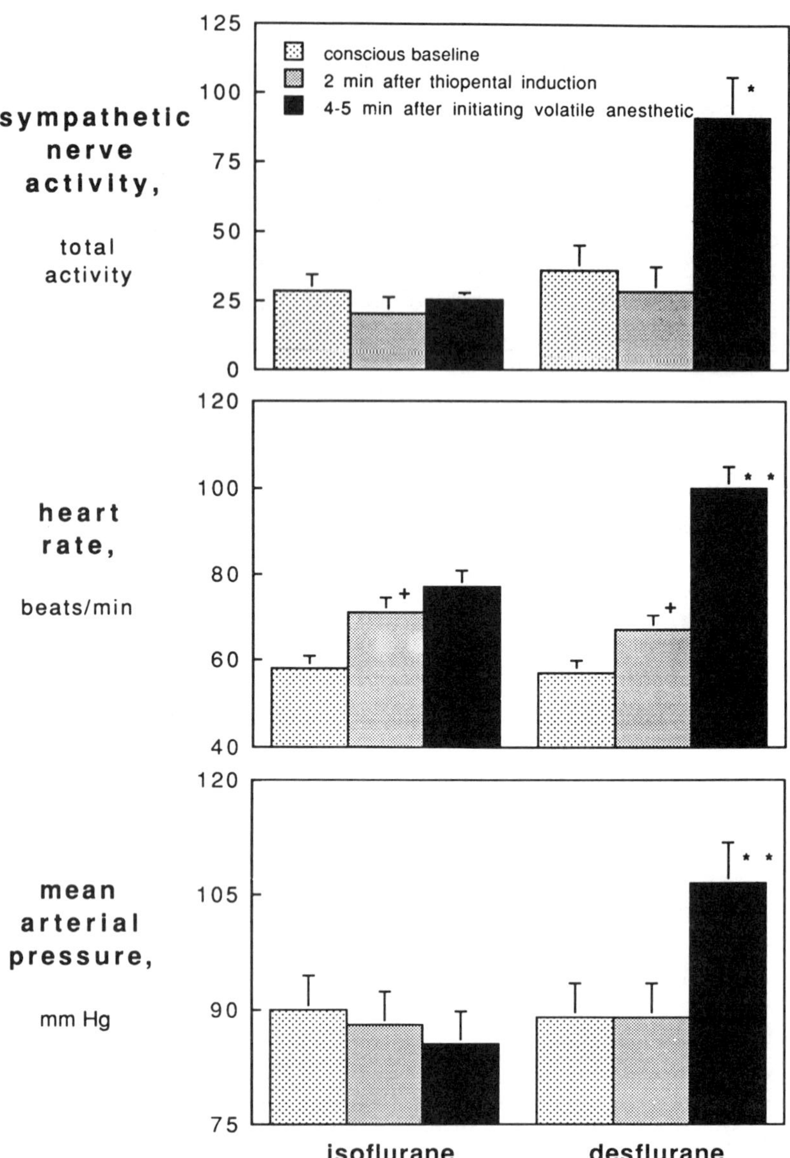

Fig. 1 Neurocirculatory responses to induction of anesthesia. Sodium thiopental increased heart rate. Peak sympathetic and hemodynamic responses triggered by desflurane occurred 4–5 min after adding it to the inspired gas at a time when the inspired concentration was 10.9% and the expired concentration was 6–8%. Similar responses did not occur in subjects receiving isoflurane. * Significant difference ($p < 0.01$) versus conscious and thiopental baseline. + Significant difference ($p < 0.05$) versus conscious baseline.

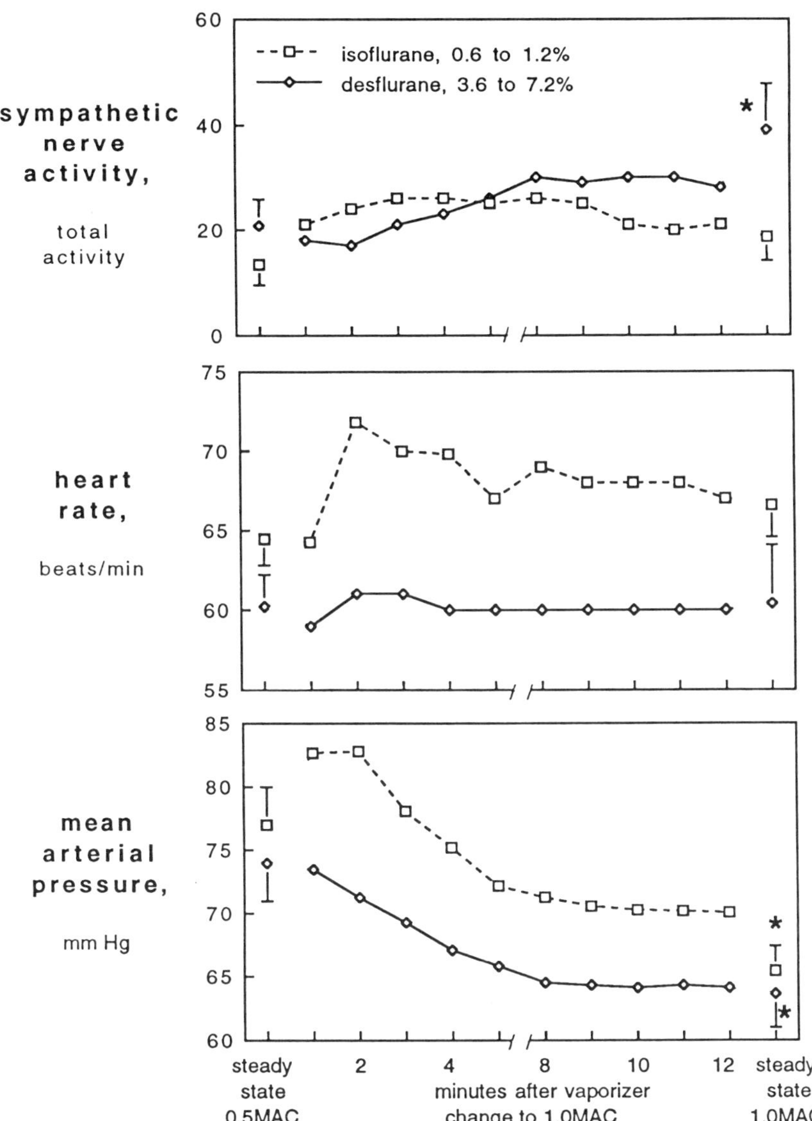

Fig. 2 Steady-state neurocirculatory responses to desflurane and isoflurane at 0.5 and 1.0 MAC. In addition, the moment-to-moment sympathetic and hemodynamic responses that occurrred during the first 12 min after advancing the vaporizer from 3.6 to 7.2% desflurane or from 0.6 to 1.2% isoflurane (0.5 to 1.0 MAC) are displayed. Mean arterial pressure was significantly lower at 1.0 MAC compared to 0.5 MAC (*, $p < 0.05$). Sympathetic nerve activity was increased at 1.0 MAC desflurane, but the absolute level of SNA did not differ from isoflurane at 1.0 MAC.

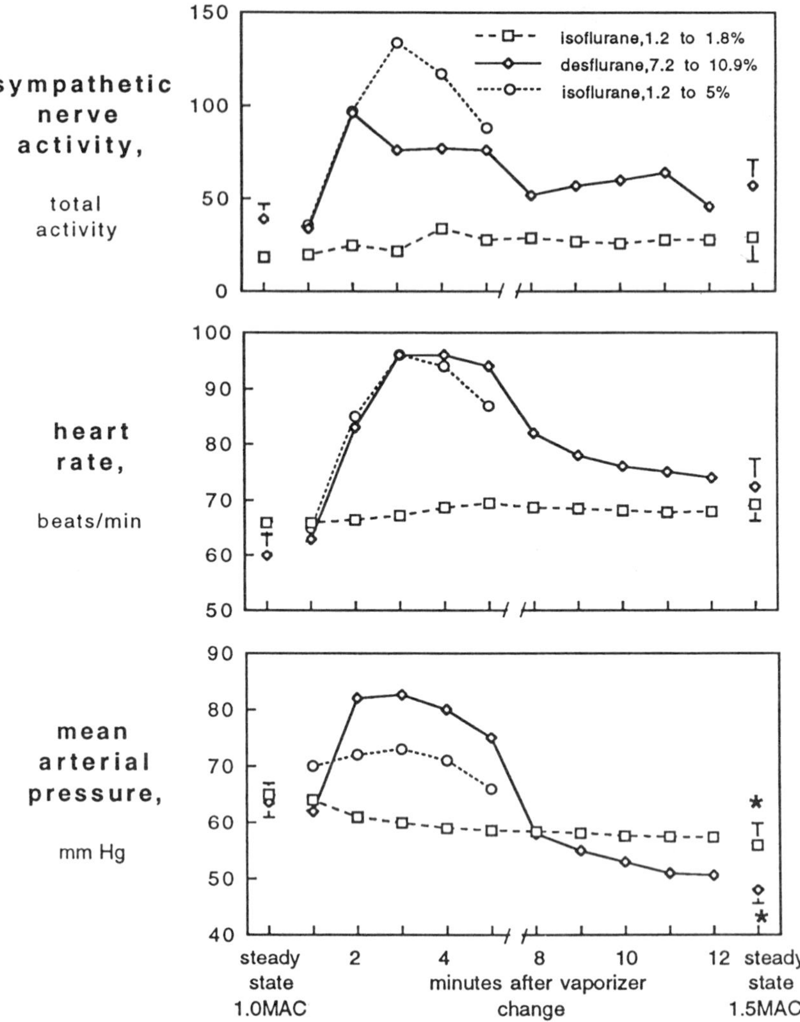

Fig. 3 Steady-state neurocirculatory responses to desflurane and isoflurane at 1.0 and 1.5 MAC. In addition, the moment-to-moment sympathetic and hemodynamic responses that occurred during the first 12 min after advancing the vaporizer from 7.2 to 10.9% desflurane or from 1.2 to 1.8% isoflurane (1.0 to 1.5 MAC), or briefly advancing the vaporizer from 1.2 to 5% isoflurane, are displayed. Mean arterial pressure was significantly lower at 1.5 MAC compared to 1.0 MAC (*, $p < 0.05$). Sympathetic nerve activity at 1.5 MAC did not differ from SNA at 1.0 MAC in either group. The transition to 1.5 MAC disflurane resulted in a 3 to 5-min period of tachycardia and hypertension owing to an increase in sympathetic outflow. The transition from 1.0 to 1.5 MAC isoflurane was uneventful, but when the vaporizer was advanced to 5%, a brief but profound period of sympathetic activation and tachycardia occurred.

decline in MAP in both treatment groups. However, the transition from 1.0 to 1.5 MAC desflurane led to significant increases in SNA, HR, and MAP (Fig. 3). This did not occur in the isoflurane group during a similar change in anesthetic concentration. However, in a small subset of patients briefly exposed to a 5% inspired concentration of isoflurane, substantial increases in SNA and HR occurred (Fig. 3). Interestingly, MAP did not increase, probably because of the powerful direct effect of isoflurane on vascular smooth muscle.

IV. Discussion

The present research demonstrates profound sympatho-excitation during the initial mask administration of desflurane to unpremedicated volunteers anesthetized with thiopental. This was not apparent when equi-MAC levels of isoflurane were administered following anesthetic induction. A second remarkable period of sympathetic activation also was noted when the inspired desflurane concentration was acutely increased from 7.25 to 10.9% (1.0 to 1.5 MAC) in intubated subjects. In both situations, the sympathetic activation led to large and potentially undesirable increases in heart rate and mean arterial pressure. Thus, during non-steady-state situations, the neurocirculatory responses to desflurane were not identical to isoflurane, and caution is advised when administering desflurane to patients whose cardiac or neurological risk might be worsened by these responses.

A. Induction Responses

Despite a generous induction dose of thiopental, which we have previously shown to be sympatho-inhibitory (7), the titration of desflurane into the inspired gas during mask ventilation resulted in marked sympathetic activation. This peaked 4–5 min after initiating desflurane at a time when end-tidal concentrations were 7–9%. The tachycardia and hypertension resulting from this sympathetic activation during the initial administration of desflurane has been observed by others. In one retrospective study, two groups of surgical patients pretreated with sedative doses of midazolam and induced with thiopental were evaluated (2). Patients were mask ventilated while either isoflurane or desflurane was gradually administered to the inspired gas. Five to 6 min after induction, heart rate had increased 30–40 beats/min (bpm) and mean arterial pressure had increased 20–26 mmHg in the desflurane group, whereas only minimal changes were noted in patients receiving isoflurane. In another study, significant increases in heart rate, blood pressure, systemic vascular resistance, pulmonary artery pressures, and pulmonary wedge pressures were noted in a population of coronary artery disease patients who had been premedicated with intramuscular morphine and midazolam, anesthetized with midazolam and thiopental, and given 1.0–2.0 MAC desflurane by mask for 15 min prior to intubation (4). These changes were associated with a 9–13% incidence of Holter and transesophageal echocardiographic evidence of ischemia. In a control group of patients receiving sufentanil instead of desflurane, signs of ischemia did not occur.

Aside from the obvious undesirable hemodynamic effects produced during the initial administration of desflurane that were reported in these studies, a second observation is of interest. It appears that neither midazolam nor morphine sulfate pretreatment abolishes these hyperdynamic responses. Whether responses would have been greater without these adjuvants cannot be ascertained from the data.

B. Steady-State Responses

At equi-MAC settings, the steady-state reductions in mean arterial pressure were similar between desflurane and isoflurane. Isoflurane tended to produce an earlier tachycardia (at 0.5 and 1.0 MAC) compared with desflurane, whereas 1.5 MAC desflurane triggered increases in heart rate similar to isoflurane. Subtle differences between isoflurane and desflurane were noted in the SNA responses. Although isoflurane produced only minor and insignificant changes in basal sympathetic outflow during progressively increasing MAC of anesthesia, desflurane was associated with a gradual increase in the sympathetic outflow with increasing MAC. This could be due to several unexplored mechanisms. First, desflurane might augment tonic central sympathetic outflow by an, as yet, undefined mechanism. Second, desflurane may better maintain baroreflex function; that is, the reduction in mean arterial pressure caused by direct vascular effects of desflurane might trigger increases in sympathetic outflow. Such an effect might also explain the tachycardia noted at 1.5 MAC desflurane.

C. Transition Responses

When the anesthetic depth was increased from 0.5 to 1.0 MAC, the neurocirculatory responses were relatively unremarkable in both the isoflurane and desflurane groups. There was a gradual and significant decrease in blood pressure which was undoubtedly due to the direct vascular effects of the potent volatile anesthetics. In contrast, increasing the anesthetic depth from 1.0 to 1.5 MAC desflurane resulted in profound sympathetic activation, tachycardia, and increases in mean arterial pressure. This was not evident in the isoflurane group during the transition to 1.5 MAC.

Although increasing the isoflurane vaporizer to 1.8% (1.5 MAC) was unremarkable aside from a progressive reduction in MAP, when the isoflurane vaporizer was advanced from 1.2 to 5% for a brief period of time, increases in SNA and HR were noted. This observation confirms another publication in which plasma catecholamine concentrations, HR, and MAP were significantly elevated during a rapid increase in the inspired isoflurane concentration to 5% (5).

Because desflurane is a relatively weak anesthetic, higher concentrations are necessary to achieve stable surgical planes of anesthesia. Changes in the inspired desflurane concentration above 7% triggered sympathetic responses. It appears that slightly lower concentrations of isoflurane (5%) also trigger responses. If sympathetic activation is due to irritation of airway receptors, then the irritant properties of isoflurane may be equal to or greater than those of desflurane. Interestingly, both isoflurane and desflurane triggered sympathetic responses in intubated subjects, implying that stimulation of the upper airway (pharynx and larynx) is not necessary to trigger responses. It therefore seems likely that either stimulation of lower airway receptors or rapid uptake of the volatile anesthetics into the central nervous system may be implicated in inciting the sympathetic discharge.

V. Summary

Although the blood pressure lowering effects of desflurane and isoflurane were similar at equi-MAC, we noted a different pattern of response during intervals of rapidly increasing the inspired concentration of desflurane, when substantial increases in SNA, HR, and MAP occurred. Because of the lower potency of desflurane compared to isoflurane, higher concentrations of desflurane are necessary to establish an adequate surgical plane of anesthesia. Although clinically relevant concentrations of isoflurane did not trigger sympathetic activation, isoflurane triggered responses at an inspired concentration (~5%) nearly equal to that of desflurane. The present research demonstrates that the initial exposure to desflurane in clinically relevant concentrations following anesthetic induction and the deepening of anesthesia with higher concentrations of desflurane can be profoundly sympatho-excitatory. Considerable caution should be taken when administering desflurane to patients who may be placed at risk by these responses.

Acknowledgment

Research was supported in part by a National Institutes of Health training grant (GM 08377).

References

1. Weiskopf, R. B., Cahalan, M. K., Eger II, E. I., Yasuda, N., Rampil, I. J., Ionescu, P., Lockhart, S. H., Johnson, B. H., Freire, B., and Kelley, S. (1991). Cardiovascular actions of desflurane in normocarbic volunteers. *Anesth. Analg. (N. Y.)* **73**, 143–156.
2. Fleischer, L. H., Young, W. L., Ornstein, E., and Smiley, R. S. (1992). Systemic hemodynamic changes of desflurane vs isoflurane during anesthetic induction. *Anesthesiology* **77**, A334 (abstract).
3. Ostapkovich, N., Ornstein, E., Jackson, L., and Young, W. L. (1992). Hemodynamic changes with rapid increases in desflurane or isoflurane dose. *Anesthesiology* **77**, A333 (abstract).
4. Helman, J. D., Leung, J. M., Bellows, W. H., Pineda, N., Roach, G. W., Reeves III, J. D., Howse, J., McEnany, M. T., Mangano, D. T., and SPI Research Group (1992). The risk of myocardial ischemia in patients receiving desflurane versus sufentanil anesthesia for coronary artery bypass graft surgery. *Anesthesiology* **77**, 47–62.
5. Jones, R. M., Cashman, J. N., and Mant, T. G. K. (1990). Clinical impressions and cardiorespiratory effects of a new fluorinated inhalation anaesthetic, desflurane (I-653), in volunteers. *Br. J. Anaesth.* **64**, 11–15.
6. Ebert, T. J., and Muzi, M. (1993). Sympathetic hyperactivity during desflurane anesthesia in healthy volunteers. *Anesthesiology* **79**, 444–453.
7. Ebert, T. J., Kanitz, D. D., and Kampine, J. P. (1990). Inhibition of sympathetic neural outflow during thiopental anesthesia in humans. *Anesth. Analg. (N. Y.)* **71**, 319–326.

Randomized, Prospective Comparison of Halothane, Isoflurane, and Enflurane on Baroreflex Control of Heart Rate in Humans

Michael Muzi and Thomas J. Ebert

Department of Anesthesiology
Veterans Affairs Medical Center and
The Medical College of Wisconsin
Milwaukee, Wisconsin, 53295

I. Introduction

The volatile anesthetics are known to impair the baroreflex regulation of heart rate in humans (1–4). However, because, at steady state, equipotent concentrations of isoflurane, halothane, and enflurane provoke different effects on basal heart rate, it has been suggested that baroreflex mechanisms regulating heart rate may not be diminished equally by the anesthetics. For example, halothane induces a dose-dependent hypotension without a compensatory increase in heart rate, suggesting an impairment of baroreflex function (5). In contrast, isoflurane and enflurane produce similar dose-dependent decreases in blood pressure but, compared to halothane, trigger increases in heart rate, suggesting a preserved or minimal impairment of baroreflex function (6,7). In support of this possibility, previous work from our laboratory has provided indirect evidence that isoflurane may, in fact, be the least depressant volatile anesthetic on baroreflex control of heart rate in humans (2). This conclusion was based on a retrospective comparison of isoflurane to previously published reports on halothane and enflurane (2). To our knowledge, all three anesthetics

Advances in Pharmacology, Volume 31

have never been compared in a single, randomized study in humans. Therefore, the objective of this study was to compare the effects of equipotent concentrations of halothane, isoflurane, and enflurane on the arterial baroreflex regulation of heart rate in a single population of healthy humans in order to substantiate the possibility that there are subtle differences between these agents.

II. Baroreflex Function in Humans

After human studies review board approval and informed consent, 22 healthy young male volunteers were instrumented with intravenous and radial artery catheters as well as American Society of Anesthesiology (ASA) standard monitors. The subjects refrained from oral intake for 8 hr prior to the study and did not receive any preanesthetic medication. Each underwent a trial exposure to the baroreceptor test to familiarize them with the flushed sensation associated with nitroprusside and the transient headache associated with phenylephrine.

After a 20-min accommodation period, a 5-min sampling epoch was obtained to derive the average conscious resting heart rate and mean arterial pressure. Baroreflex function was then assessed by a combined depressor and pressor test. This was accomplished by the intravenous injection of 100 μg of nitroprusside followed 60 sec later by 150 μg of phenylephrine. These doses were chosen to lower blood pressure 20 to 25 torr below baseline and subsequently elevate pressure 10 to 15 torr above baseline. After a 10-min recovery period, patients were preoxygenated for 5 min and induced with 0.3 mg/kg of etomidate followed by 0.2 mg/kg of vecuronium for neuromuscular inhibition. Patients were then endotracheally intubated and mechanically ventilated. Both end-tidal CO_2 and confirmatory arterial blood gases were employed to maintain p_{CO_2} at physiological levels (conscious baseline) throughout the study.

Nine patients received isoflurane, seven patients received halothane, and six received enflurane anesthesia. The anesthetic agent and order of anesthetic concentration (either 0.5 or 1.0 minimum alveolar concentration (MAC) were randomly assigned. Subjects were maintained at each anesthetic level for approximately 20 min or until 5 min after the end-tidal gas concentrations had achieved target values. Steady-state hemodynamic measures were determined by averaging beat-to-beat heart rates and mean arterial blood pressures over a 5-min sampling period. Baroreflex testing was then repeated in an identical fashion to conscious baseline. Following this test, the anesthetic concentration was changed and, after equilibration, hemodynamic data collection and baroreceptor testing repeated.

Each 3-min data file containing baroreflex responses was analyzed by averaging data in 2 mmHg increments of mean blood pressure. This averaging was applied to all data that encompassed the fall and rise in blood pressure. Because data collection began precisely at the time of nitroprusside injection, the initial 15–20 sec of each data file included baseline data immediately prior to the onset of the hypotension. Data were plotted as average R–R interval (ordinate) versus average mean arterial pressure (abscissa) for each 2 mmHg increment in pressure. The data from the initial period of falling pressure were analyzed separately from the data during the subsequent rise in pressure. The linear portion of each response relationship was then identified, and the line of best fit was calculated. Because of the well-known sigmoid relationship between mean arterial pressure (MAP) and R–R interval, the falling pressure response (slope) was typically less steep than the rising pressure slope as it encompassed the lower, flat portion of this sigmoid curve (Fig. 1).

Baseline hemodynamics and the slope of the regression line corresponding to the depressor and pressor baroreflex response curves were averaged within each anesthetic group, and an unpaired *t*-test was used to compare differences between each anesthetic group at each anesthetic level. A p value below 0.05 was considered sufficient to reject the null hypothesis.

III. Effects of Anesthetics

Each anesthetic produced a dose-dependent, progressive fall in mean arterial blood pressure with the exception of enflurane at 0.5 MAC (Table I). The blood pressure decline at 1.0 MAC of isoflurane was significantly larger than the decrease produced by halothane at 1.0 MAC, but there were no blood pressure differences between these two agents at 0.5 MAC (Table I). Both enflurance and isoflurane triggered significant decreases in R–R interval (increases in heart rate) at 0.5 and 1.0 MAC, whereas no significant changes were noted in the halothane group.

Each anesthetic depressed both the falling and rising limb of the cardiac baroreceptor reflex (Table II and Figs. 2 and 3). There were no significant differences between groups with regard to this depressant effect. At both anesthetic concentrations, the magnitude of decrease in the baroslope derived from the falling pressure period was significantly greater when compared to the decrement in the baroslope derived during the rising pressure response. In fact, the rising pressure baroslopes at 0.5 MAC were not significantly changed from control for any of the three anesthetics but the falling pressure slopes were markedly diminished at 0.5 MAC (Table II and Figs. 2 and 3).

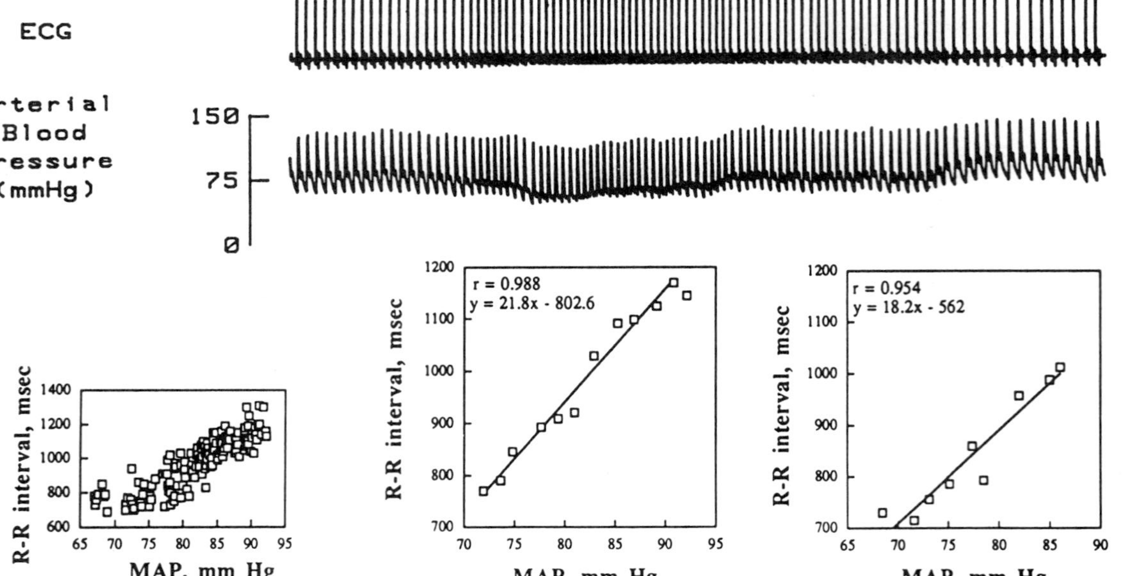

Fig. 1 Representative 2-min segment of data that displays the electrocardiograph (ECG) and radial artery blood pressure responses to the sequential administration of nitroprusside (100 μg) and phenylephrine (150 μg). Nitroprusside was given through a peripheral vein at a time corresponding to the initial (left) portion of the tracing, whereas phenylephrine was given at a time corresponding to the end of the first one-third of the tracing (during peak hypotension). Each cardiac interval and the corresponding MAP during the recording period are displayed at bottom left. The response to nitroprusside is displayed in the bottom right-hand graph. The data for R–R interval and MAP were averaged for each 2 mmHg increment of MAP. The regression line equation is displayed (slope 18.2 msec/mmHg). The response to phenylephrine is averaged and displayed in the middle graph (slope 21.8 msec—mmHg).

Table I

Hemodynamic Variables[a]

Variable	Anesthetic agent	Conscious baseline	0.5 MAC	1.0 MAC
R–R interval, msec	Halothane	1028 ± 49.6	1038 ± 65.3	985 ± 45.7
	Isoflurane	1076 ± 39.4	918 ± 27.9[b,c]	920 ± 42.7[b]
	Enflurane	972 ± 75.4	783 ± 54.8[b,c]	761 ± 41.4[b–d]
Mean arterial pressure, mmHg	Halothane	86 ± 2.6	76 ± 3.6[b]	71 ± 3.3[b]
	Isoflurane	90 ± 1.5	79 ± 2.9[b]	65 ± 2.0[b,c]
	Enflurane	84 ± 2.2	78 ± 4.4	64 ± 4.4[b]

[a] Data are means ± SEM. MAC, Minimum alveolar concentration.
[b] Significantly changed from conscious baseline ($p < 0.05$).
[c] Change from conscious baseline significantly different compared to halothane response ($p < 0.05$).
[d] Significantly changed from 0.5 MAC ($p < 0.05$).

IV. Discussion

This study is unique in that all three potent volatile anesthetic agents were studied in a randomized population of age-matched, healthy humans receiving no preanesthetic sedation and anesthetized at approximately the same hour of the day. As a result, resting heart rate, blood pressure, and rising and falling baroreflex slopes varied only slightly and nonsignificantly

Table II

Cardiac Baroreflex Slopes[a]

Baroslope (msec/mmHg)	Anesthetic agent	Conscious baseline	O.5 MAC	1.0 MAC
Rising pressure	Halothane	26.3 ± 4.5	23.2 ± 4.8	20.7 ± 3.9[b]
	Isoflurane	26.0 ± 4.3	20.5 ± 3.6	15.5 ± 2.8[b]
	Enflurane	18.4 ± 2.1	14.7 ± 3.0	11.5 ± 1.9[b]
Falling pressure	Halothane	20.2 ± 3.4	13.9 ± 2.7[b]	7.3 ± 1.7[b,c]
	Isoflurane	19.1 ± 3.1	7.8 ± 0.8[b]	5.5 ± 1.1[b]
	Enflurane	16.8 ± 2.8	6.7 ± 0.4[b]	4.7 ± 1.0[b,c]

[a] Data are means ± SEM. MAC, Minimum alveolar concentration.
[b] Significantly changed from conscious baseline ($p < 0.05$).
[c] Significantly changed from 0.5 MAC ($p < 0.05$).

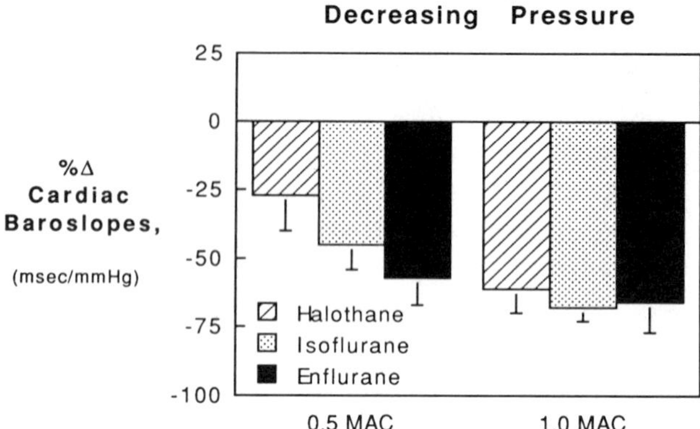

Fig. 2 Percentage change (from conscious baseline) in baroreflex slopes. The decreasing pressure baroreflex slopes were determined during sodium nitroprusside administration. The slope was calculated as the linear relationship between mean arterial pressure and R–R interval. There were no significant differences between the three anesthetic agents.

between groups. The progressive decline in blood pressure with increasing MAC of anesthesia was, in general, similar between anesthetics. In response to the hypotension, there were significant increases in heart rate (decreases in R–R interval) in the isoflurane and enflurane groups but no change in heart rate in the halothane group. The increase in heart rate in the isoflurane group was not incremental and therefore not dose-related to either the anesthetic concentration or the blood pressure decline. Addi-

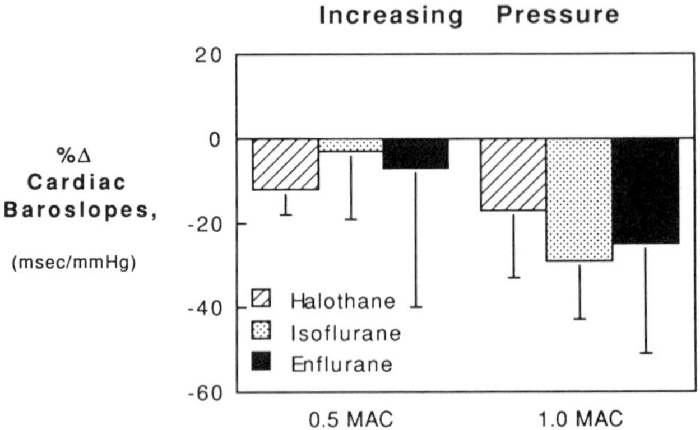

Fig. 3 Percentage change (from conscious baseline) in baroreflex slopes. The rising pressure baroreflex responses were obtained during phenylephrine administration. The slope was calculated as the linear relationship between MAP and R–R interval. There were no significant differences between the three anesthetic agents.

tionally, in the enflurane group, the increase in heart rate appeared to be anesthetic dose-related, but it was clearly not reciprocally related to the nonlinear decline in blood pressure. This was particularly evident at 0.5 MAC when blood pressure was not changed from baseline but heart rate had increased significantly. These observations call into question whether steady-state heart rates during anesthesia bear any relationship to blood pressure via baroreflex mechanisms.

In previous work performed in our laboraory, we have demonstrated that both isoflurane and halothane attenuate the cardiac baroreflex in humans (2,8). This is in agreement with numerous animal studies delineating the same effect (9–11). The findings of the present study differ from our previous report (2) suggesting a preservation of baroreflex function during anesthesia with isoflurane. The falling pressure baroslopes were progressively decreased with increasing MAC, and this attenuation did not differ from the attenuation noted with equipotent doses of halothane or enflurane. In contrast, our data demonstrate a better preservation of baroreflex responses to increasing pressure ramps, especially at 0.5 MAC, and this agrees with the work of Seagard *et al.* in dogs (9,10). Because there were no discernible differences between anesthetics, resting heart rate differences at equipotent concentrations of the three anesthetics cannot be attributed to a differential effect on the baroreflex.

The apparent greater resistance of the rising pressure baroslopes compared with the falling pressure baroslopes to anesthetics has not been discussed previously. Because in healthy humans, the baroreflex regulation of heart rate is primarily under efferent vagal control mechanisms (12,13), the reflex inhibition of cardiac–vagal activity by unloading baroreceptors may be more sensitive to anesthetics than the reflex augmentation in this activity during baroreceptor loading. However, this might also be related to a relatively low level of basal cardiac–vagal activity that exists under anesthesia, thereby making further withdrawal of this activity during a hypotensive stimulus rather limited.

A limitation of this study is that the technique of baroreceptor testing, namely, an acute lowering and raising of blood pressure, does not test the various components of the reflex arc. The work of Seagard and others have documented that anesthetics work at multiple sites along the reflex chain, including carotid baroreceptors, afferent nerve pathways within the central nervous system, ganglia, efferent nerve pathways, and neuroeffector junctions (9,10,14). In defense of the pharmacological technique for the evaluation of baroreflex function employed in the present research, the stress provokes the simultaneous loading and unloading of carotid, aortic, and low-pressure baroreceptors, much like the true physiological situation.

The neural signals from baroreceptors involved with maintaining the absolute level of blood pressure may differ from those involved in responding to a rapid change in blood pressure. Seagard *et al.* have documented the existence of two cell types within the carotid baroreceptors which act to control blood pressure (15). Type I barorecptor cells contribute to the regulation of dynamic pressure changes, whereas type II baroreceptor cells have a continuous firing pattern and may regulate baseline, resting levels of blood pressure. It would be reasonable to assume that this cell architecture would be consistently repeated throughout the baroreceptor reflex pathways. Our technique of loading and unloading baroreceptors may provoke only type I cell responses, whereas the differences in steady-state heart rate responses to the volatile anesthetics may be more importantly regulated by the type II cell response to the prevailing blood pressure.

In summary, in contrast to our previous work suggesting that isoflurane might maintain the integrity of baroreflex regulation of heart rate based on a retrospective analysis of work in other laboratories (2), the present randomized prospective study demonstrates that reflex heart rate responses to both decreasing and increasing blood pressure do not differ between the three potent inhalational anesthetic agents studied, halothane, enflurane, and isoflurane.

Acknowledgments

The authors acknowledge the technical assistance of Jill Barney, Toni Uhrich, and Linda Messana.

References

1. Duke, P. C., Fownes, D., and Wade, J. G. (1977). Halothane depresses baroreflex control of heart rate in man. *Anesthesiology* **46,** 184–187.
2. Kotrly, K. J., Ebert, T. J., Vucins, E., Igler, F. O., Barney, J. A., and Kampine, J. P. (1984). Baroreceptor reflex control of heart rate during isoflurane anesthesia in humans. *Anesthesiology* **60,** 173–179.
3. Morton, M., Duke, P. C., and Ong, B. (1980). Baroreflex control of heart rate in man awake and during enflurane and enflurane–nitrous oxide anesthesia. *Anesthesiology* **52,** 221–223.
4. Takeshima, R., and Dohi, S. (1989). Comparison of arterial baroreflex function in humans anesthetized with enflurane or isoflurane. *Anesth. Analg.* (*N. Y.*) **69,** 284–290.
5. Heiberg, J. K., Wiberg-Jorgensen, F., and Skovsted, P. (1978). Heart rate changes caused by enflurane and halothane anesthesia in man. *Acta Anaesthesiol. Scand.* **67,** 59–62.
6. Shimosato, S., Iwatsuki, N., and Carter, J. G. (1979). Cardio-circulatory effects of enflurane anesthesia in health and disease. *Acta Anaesthesiol. Scand.* **71,** 69–70.
7. Wade, J. G., and Stevens, W. C. (1981). Isoflurane: An anesthetic for the eighties? *Anesth. Analg.* (*N. Y.*) **60,** 666–682.

8. Ebert, T. J., Kotrly, K. J., Vucins, E. J., Pattison, C. Z., and Kampine, J. P. (1985). Halothane anesthesia attenuates cardiopulmonary baroreflex control of peripheral resistance in humans. *Anesthesiology* **63,** 668–674.
9. Seagard, J. L., Elegbe, E. O., Hopp, F. A., Bosnjak, Z. J., von Colditz, J. H., Kalbfleisch, J. H., and Kampine, J. P. (1983). Effects of isoflurane on the baroreceptor reflex. *Anesthesiology* **59,** 511–520.
10. Seagard, J. L., Hopp, F. A., Donegan, J. H., Kalbfleisch, J. H., and Kampine, J. P. (1982). Halothane and the carotid sinus reflex: Evidence for multiple sites of action. *Anesthesiology* **57,** 191–202.
11. Sellgren, J., Biber, B., Henriksson, B.-A., Martner, J., and Ponten, J. (1992). The effects of propofol, methohexitone and isoflurane on the baroreceptor reflex in the cat. *Acta Anaesthesiol. Scand.* **36,** 784–790.
12. Greene, N. M., and Bachand, R. G. (1971). Vagal component of the chronotropic response to baroreceptor stimulation in man. *Am. Heart J.* **82,** 22–27.
13. Leon, D. F., Shaver, J. A., and Leonard, J. J. (1970). Reflex heart rate control in man. *Am Heart J.* **80,** 729–739.
14. Seagard, J. L., Hopp, F. A., Bosnjak, Z. J., Osborn, J. L., and Kampine, J. P. (1984). Sympathetic efferent nerve activity in conscious and isoflurane-anesthetized dogs. *Anesthesiology* **61,** 266–270.
15. Seagard, J. L., Hopp, F. A., Drummond, H. A., and Van Wynsberghe, D. M. (1993). Selective contribution of two types of carotid sinus baroreceptors to the control of blood pressure. *Circ. Res.* **72,** 1011–1022.

Baroreflex Modulation by Isoflurane Anesthesia in Normotensive and Chronically Hypertensive Rabbits

Leonard B. Bell
Departments of Anesthesiology and Physiology
The Medical College of Wisconsin
and Zablocki Veterans Affairs Medical Center
Milwaukee, Wisconsin 53226

I. Introduction

The arterial baroreflex is important for moment-to-moment regulation of arterial pressure (AP). In response to transient changes in AP, the baroreflex produces changes in both heart rate (HR) and peripheral vascular tone (mediated through efferent sympathetic nerves) in an effort to return AP to the original level. Maintenance of AP during surgery is of primary concern for the anesthesiologist. However, it is well known that volatile anesthetics attenuate baroreflex control of AP. There is clear evidence that baroreflex control of HR is depressed by isoflurane (ISO) (1–4), halothane (5–12), and enflurane (4–13), and baroreflex control of efferent sympathetic nerve activity is attenuated by ISO (2,14–16), halothane (5,17) and enflurane (18).

However, although baroreflex control of HR has been compared between awake and anesthetized patients and animals (1,2,4,8,10,13), there have been no reports comparing baroreflex control of efferent sympathetic nerve activity in the conscious and anesthetized patient or animal. Therefore, the first purpose of this study was to determine simultaneously baroreflex control of HR and efferent sympathetic nerve activity in the conscious and ISO-anesthetized rabbit.

Advances in Pharmacology, Volume 31

It has also been demonstrated that chronic hypertension attenuates baroreflex control of HR (19–28). However, there have been conflicting reports concerning baroreflex control of efferent sympathetic nerve activity in hypertension. Following the development of chronic hypertension, baroreflex control of efferent sympathetic nerve activity has been shown to be either preserved (22,23,27,29,30) or attenuated (29).

Although several studies have investigated baroreflex conrol of HR in awake hypertensive (HT) patients and animals (19–22,24,26,28), there is little information concerning baroreflex control of efferent sympathetic nerve activity in the conscious HT model (27,30). The effect of chronic hypertension on baroreflex control of visceral efferent sympathetic nerve activity over a wide range of AP has not been studied. Therefore, the second purpose of this study was to determine simultaneously baroreflex control of HR and renal efferent sympathetic nerve activity (RSNA) in conscious normotensive (NT) and chronically HT animals over the full range of physiologically relevant AP.

Even less understood is baroreflex control of HR and efferent sympathetic nerve activity during anesthesia in the HT individual. Prys-Roberts *et al.* (31) demonstrated that in NT and untreated HT subjects, halothane attenuated the baroreflex control of HR equally. However, the effect of anesthesia on baroreflex control of efferent sympathetic nerve activity in NT and HT individuals has not been studied. Therefore, the third purpose of this study was to determine if chronic hypertension modified the effects of ISO anesthesia on baroreflex control of HR and efferent sympathetic nerve activity.

II. Chronically Hypertensive Rabbit Model and Experimental Protocol

Twelve New Zealand White rabbits were divided into two groups. Group I rabbits ($n = 6$) were NT and used as controls. Group II rabbits ($n = 6$) were made HT and used as the experimental group. Studies were conducted on each rabbit while conscious and following four levels of ISO anesthesia. Prior to experimentation, each rabbit underwent four separate surgical procedures, each separated by at least 2 weeks. For each surgical procedure, rabbits were induced with Telazol (titelamine–zolazepam, 30 mg/kg, i.m.), intubated with a cuffed endotracheal tube (3 mm I D), and ventilated with a small animal respirator using halothane or ISO (1–2%) in room air.

The first two procedures involved implanting perivascular occlusion cuffs around the inferior vena cava and descending aorta. Slow inflation of the cuffs produced ramped decreases and increases in AP, respectively. By recording HR and RSNA during changes in AP, AP–HR and AP–RSNA baroreflex function curves could be determined.

During the third surgical procedure, 19–38 days prior to experimentation, each kidney in group II rabbits (HT) was exposed through an abdominal incision and encapsulated with an inert latex capsule (Molding Compound #3361-6, Dural Company Inc., Milwaukee, WI) fabricated to fit snugly around the kidney leaving the hilar region intact so as not to disturb renal innervation. A sham surgery was performed in group I (NT) rabbits. Arterial pressure was recorded in both groups of rabbits while conscious before and 3–5 weeks after renal wrap/sham surgery.

Last, a renal nerve recording electrode was implanted in all rabbits 3–5 days prior to the experiment to measure whole RSNA as previously described (32). In brief, two or three renal nerves were placed in the exposed, spiral wound ends of two single-standed (0.003 inch) stainless steel wires. A third wire was embedded in a fat pad, acting as a ground. The entire nerve–electrode–ground complex was then embedded in silicone gel. The electrode leads and ends of the balloon cuffs were secured in subcutaneous pockets. Whole RSNA was amplified, rectified, and time averaged. Potentials were monitored through a loudspeaker, viewed on a storage oscilloscope, and recorded on an Hewlett-Packard computer and videocassette recorder (VCR)-based data recorder for later analysis. Signals were also displayed on a Grass polygraph.

In rabbits, activation of the nasopharyngeal reflex with cigarette smoke produces a dramatic increase in RSNA (33) and complete cessation of renal blood flow (34). The RSNA response to nasopharyngeal stimulation is considered to be the "maximum" RSNA that can be elicited by physiological means in a conscious rabbit. Maximal RSNA was determined by averaging the 10-sec interval with the highest RSNA following activation of the nasopharyngeal reflex.

Minimum RSNA, or "baseline noise," was obtained at the end of the experiment by ganglionic blockade with trimethaphan camphorsulfonate (10 mg/kg) or hexamethonium hydrochloride (10 mg/kg). The value obtained following ganglionic blockade was subtracted from the activity obtained during nasopharyngeal stimulation, and the resultant activity was considered to be 100%. All responses described are a percentage of this maximum activity.

Figure 1 represents typical recordings of AP, HR, and RSNA in a NT (Fig. 1, top) and HT (Fig. 1, bottom) rabbit during slow ramped inflation (1–2 mmHg/sec) of the vena caval (decreases in AP) and aortic (increases in AP) balloons. The left-hand side of Fig. 1 represents recordings in the conscious rabbit (control), and the right-hand side of Fig. 1 represents

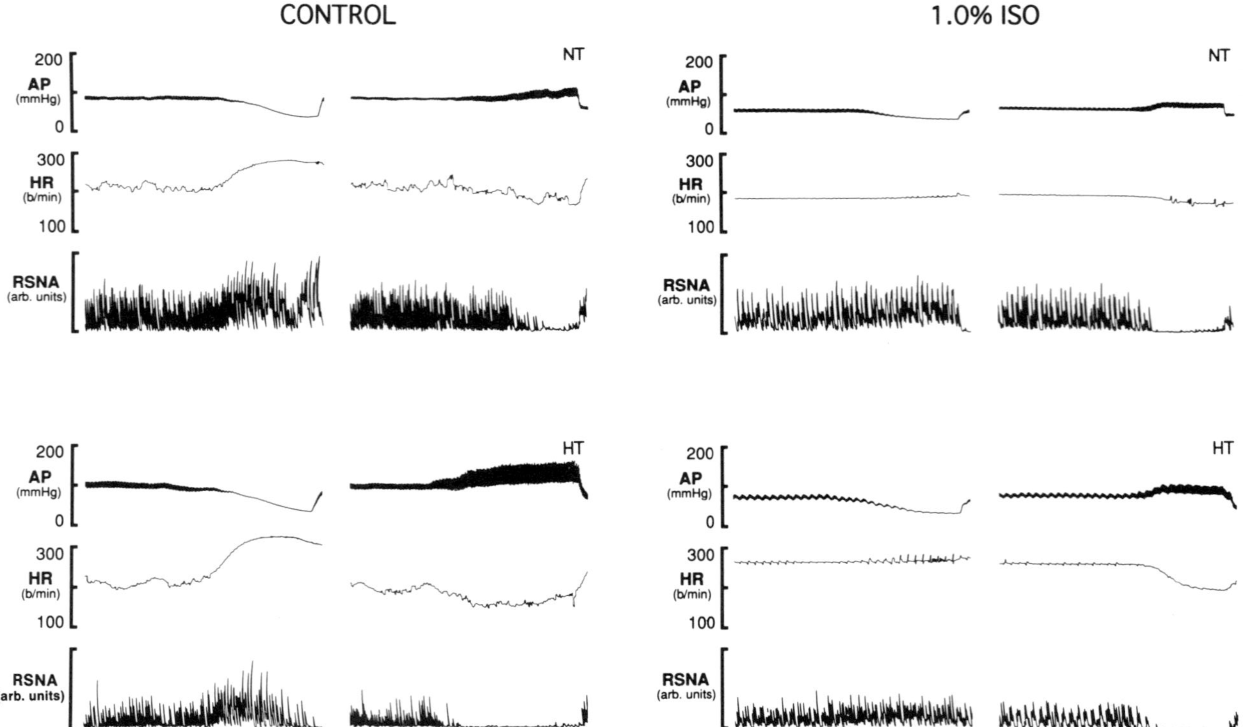

Fig. 1 Original recordings of AP, HR, and RSNA in a NT (top) and HT (bottom) rabbit during slow ramped inflation of the vena caval (decreases in AP) and aortic (increases in AP) balloons. The left-hand side represents recordings in conscious animals, and the right-hand side represents recordings during 1.0% ISO anesthesia in the same animals.

recordings during 1.0% ISO anesthesia in the same animals. As expected, caval balloon inflation produced a reflex rise in HR and RSNA which reached a plateau at moderate decreases in AP. Aortic balloon inflation produced reflex bradycardia and inhibition of RSNA. Note that following ISO anesthesia reflex responses to caval balloon were attenuated in both NT and HT rabbits.

Each pair of caval and aortic balloon inflations were used to construct one HR and RSNA baroreflex curve. All MAP–RSNA barocurves were fit with the following sigmoid logistic function curve:

$$\text{RSNA} = P_1/\{1 + \exp[P_2(\text{MAP} - P_3)]\} + P_4$$

as shown in Fig. 2, where P_4 is the lower plateau; P_1 is the range between the upper and lower plateaus; P_2 is a range-independent measure of slope; and P_3 is the mean arterial pressure (MAP) at one-half the range of RSNA (BP_{50} in Fig. 2). The upper plateau is equal to $P_1 + P_2$. The best fit was determined by iterative least squares regression. The MAP–HR barocurves in conscious rabbits were also fit with this sigmoid function curve. Following ISO anesthesia, however, MAP–HR barocurves were better described by an exponential than sigmoid function since the HR did not reach a plateau level with increases in AP (see Figs. 6 and 7).

Three pairs of baroreflex curves were recorded while conscious, and three more pairs were recorded at each of the four levels of ISO anesthesia in each rabbit. The curve parameters of each of the three curves under each condition were averaged to obtain a single barocurve for each experimental condition for each animal. Comparisons between the averaged baroreflex curves were made within and between groups at each level of ISO concentration using repeated measures two-way analysis of variance. Post hoc

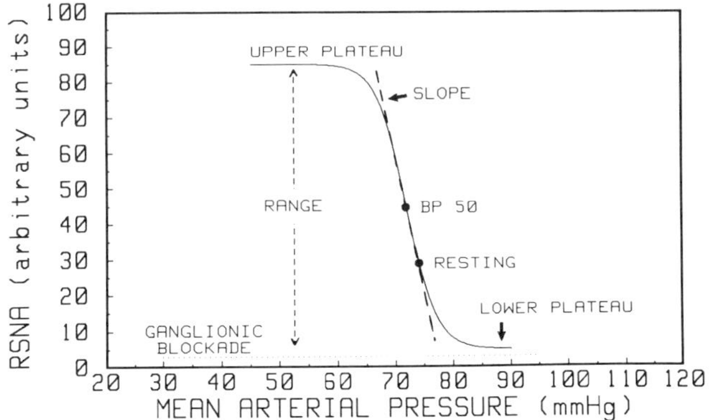

Fig. 2 Representative sigmoid function curve derived from combining baroreflex responses elicited by caval and aortic balloon inflations.

comparisons were made with Fisher's Protected Least Significant Difference test. Comparisons between RSNA obtained during nasopharyngeal stimulation with smoke and ISO induction were compared using a paired *t*-test. The location of baroreflex curves in NT and HT rabbits along the pressure axis (i.e., BP_{50}) were made using an unpaired *t*-test. Differences were considered significant if *P* values were below 0.05. Values presented represent means plus or minus the standard error of the mean (SEM).

On the experimental day, a peripheral ear vein and central ear artery were cannulated percutaneously (Angiocath 24 gauge) for infusion of drugs and recording AP, respectively. Heart rate was determined from the pulsatile AP. A 5-min sample of resting, steady-state AP, HR, and RSNA values was recorded while the animal sat quietly in a Plexiglas box. Maximum RSNA was then determined by nasopharyngeal stimulation with smoke. Triplicate barocurves were then obtained in each rabbit while conscious and used as control curves.

The Plexiglas box was then made airtight and a gas mixture of oxygen and room air was forced through the box (2–3 liters/min). Exhaust gases were continually monitored to ensure adequate oxygen concentrations inside the box. Anesthetic induction was accomplished by spontaneous ventilation of ISO vapors by slowly increasing the inflow concentration of ISO until a steady-state level of 4% was obtained in the exhaust gas for 15 min. During induction, AP, RSNA, ISO concentration, and HR were continuously recorded. Following induction, the animal was removed from the box, intubated, and allowed to breathe controlled concentrations of anesthetic gases spontaneously. The end-tidal ISO concentration was measured by mass spectrometry.

Drummond (35) has determined that the minimum alveolar concentration (MAC) for ISO is 2.05 ± 0.18% in the rabbit. To test the effects of ISO on the baroreflex at concentrations above and below the MAC, four levels of end-tidal ISO concentration, namely, 1.0, 1.5, 2.0, and 2.5%, were achieved in each animal in random order. Following 20 min of steady-state end-tidal ISO concentration at each of the four levels, triplicate baroreflex curves were obtained. Finally, minimum RSNA was determined by infusion of a sympathetic ganglionic blocker.

III. Effect of Isoflurane and Hypertension on Arterial Pressure–Renal Sympathetic Nerve Activity and Arterial Pressure–Heart Rate Barocurves

Renal encapsulation resulted in a 34.8 ± 8.7 mmHg increase in mean AP (range 17–72 mmHg) in group II (HT) rabbits. Figure 3 shows the resting group values for MAP, HR, and RSNA in NT (Fig. 3, hatched bars) and

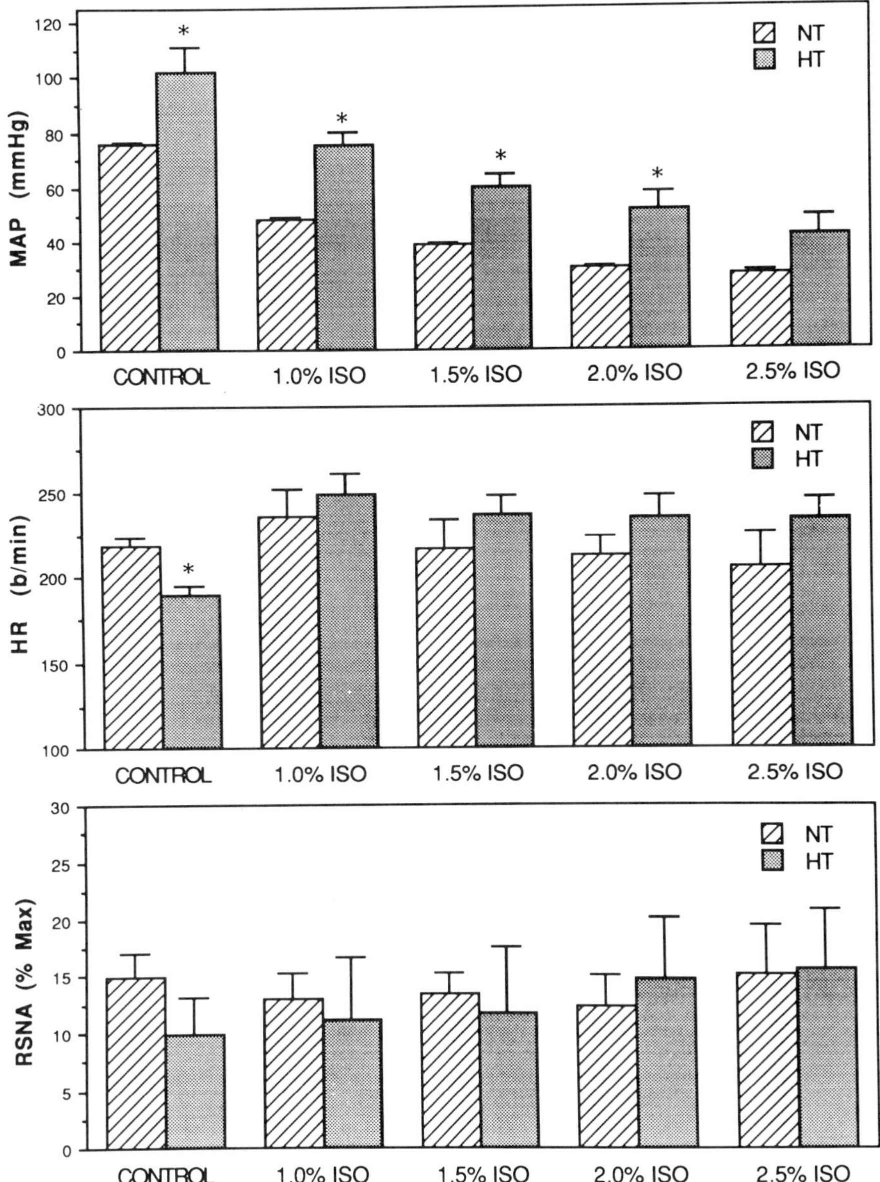

Fig. 3 Group averages (±SE) for steady-state values of MAP, HR, and RSNA in NT (hatched bars) and HT (shaded bars) rabbits. Control represents values obtained while conscious. An asterisk (*) indicates that the mean is significantly different from NT.

HT (Fig. 3, shaded bars) rabbits. The MAP in HT rabbits was significantly elevated compared to NT rabbits at rest and following 1.0, 1.5, and 2.0% ISO. Administration of ISO significantly reduced MAP from control in both NT and HT rabbits at all levels of ISO. In general, the higher the concentration of ISO, the lower the MAP. Resting HR in conscious HT rabbits was significantly lower than that in NT rabbits. However, steady-state HR in HT rabbits was not different from NT rabbits at any of the four concentrations of ISO anesthesia studied. Resting RSNA, on the other hand, was not different between conscious NT and HT rabbits. Furthermore, steady-state RSNA was not affected by ISO anesthesia at any of the four concentrations used in either NT or HT animals.

A. Mean Arterial Pressure–Renal Sympathetic Nerve Activity Baroreflexes

Figure 4 shows the average AP–RSNA baroreflex curves for NT (Fig. 4, top) and HT (Fig. 4, bottom) conscious animals (control) and following 20 min of ISO anesthesia at four different concentrations. Each curve in Fig. 4–7 represents an average of 18 curves (i.e., triplicate baroreflex determinations in each of 6 rabbits). Resting AP, RSNA, and HR are indicated by the filled circles.

When compared to control curves, ISO curves (all concentrations) were shifted to the left as indicated by not only a decrease in BP_{50}, but also a decrease in the minimum and maximum blood pressures that could be elicited by caval and aortic balloon inflations, respectively, in both NT and HT animals. In addition, the upper plateau, slope, and the range of the baroreflex (upper plateau minus lower plateau) were significantly attenuated by ISO anesthesia in both NT and HT animals. However, the resting level of RSNA was not significantly different between control and ISO baroreflex curves. Because the increase in RSNA that could be elicited by caval balloon inflation (upper plateau minus resting RSNA) was attenuated by ISO anesthesia, resting RSNA was located on or near the upper plateau of the baroreflex in both NT and HT rabbits when the ISO concentration exceeded 1.0%. The inhibition of RSNA associated with increases in AP was preserved at all levels of ISO.

When comparing ISO baroreflex curves, ISO anesthesia produced a concentration-dependent decrease in resting AP and BP_{50} in both NT and HT animals with no significant change in resting RSNA. In addition, there was a concentration-dependent decrease in the blood pressure range over which the baroreflex functioned in both NT and HT animals. This was due primarily to a concentration-dependent decrease in the maximum AP that could be produced by aortic balloon inflation. There was also a

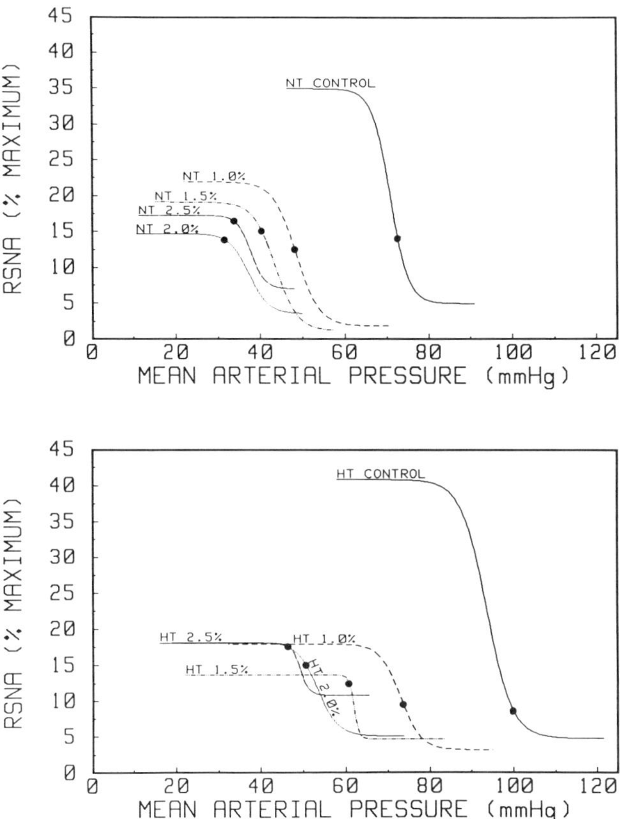

Fig. 4 Mean AP–RSNA baroreflex curves for NT (top) and HT (bottom) rabbits while conscious (control) and during four levels of ISO anesthesia (1.0, 1.5, 2.0, and 2.5%). Each curve represents an average of 18 curves (i.e., triplicate determinations in each of 6 rabbits). Filled circles represent resting levels of AP and RSNA for each curve.

significant concentration-dependent decrease in the RSNA range of the baroreflex in NT rabbits, but not in HT rabbits.

Figure 5 shows the average MAP–RSNA baroreflex curves for NT and HT animals while conscious (control) and during 1.0 and 2.0% ISO (Fig. 5, top), and during 1.5 and 2.5% ISO (Fig. 5, bottom). These are the same curves shown in Fig. 4, but regrouped to more readily compare NT and HT animals. When comparing NT and HT baroreflex curves, there was a significant rightward shift in HT animals along the pressure axis while conscious and during 1.0, 1.5, and 2.0% ISO anesthesia as indicated by a significant difference in BP_{50}. Mean AP was also significantly higher in HT conscious and ISO-anesthetized rabbits. In addition, the blood pres-

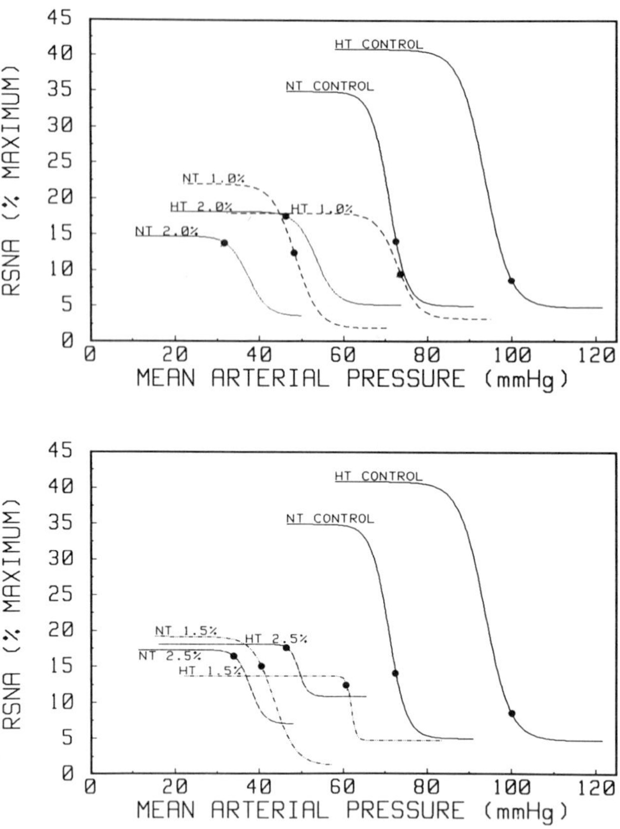

Fig. 5 Mean AP–RSNA baroreflex curves redrawn from Fig. 4 to allow comparison of NT and HT animals at each level of ISO anesthesia. The MAP–RSNA baroreflex curves are shown for NT and HT animals while conscious (control) and during 1.0 and 2.0% ISO (top) and 1.5 and 2.5% ISO (bottom).

sure range over which the AP–RSNA baroreflex functions in HT rabbits was significantly larger than in NT rabbits in the conscious state as well as at each level of ISO anesthesia. This was due primarily to a significantly higher maximum AP produced by aortic balloon inflation in HT rabbits, since the minimum AP produced by caval balloon inflation was only different in conscious rabbits.

B. Mean Arterial Pressure–Heart Rate Baroreflexes

Figure 6 shows the average MAP–HR baroreflex curves for NT (Fig. 6, top), and HT (Fig. 6, bottom) in conscious (control) animals and following 20 min of ISO anesthesia at four different concentrations. When compared

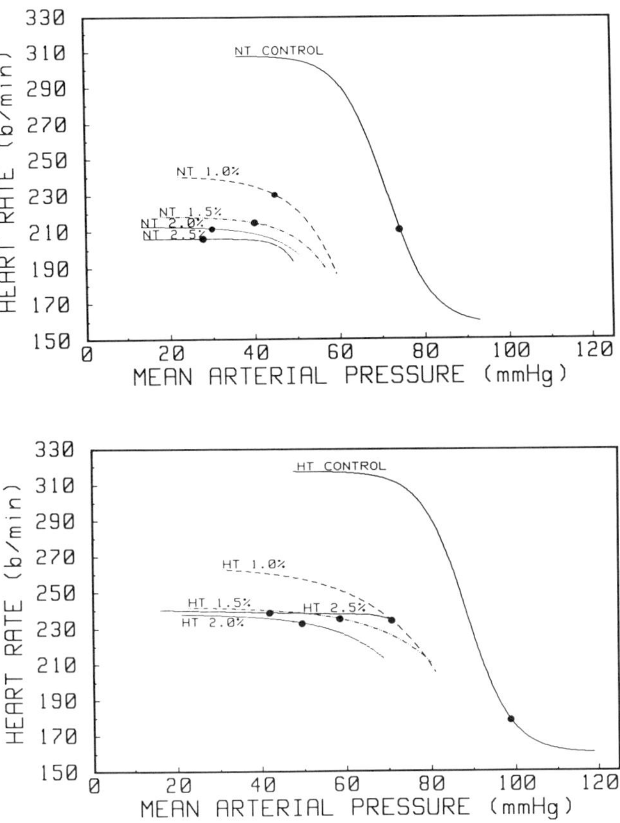

Fig. 6 Mean AP–HR baroreflex curves for NT (top) and HT (bottom) rabbits while conscious (control) and during four levels of ISO anesthesia (1.0, 1.5, 2.0, and 2.5%). Each curve represents an average of 18 curves (i.e., triplicate determinations in each of 6 rabbits). Filled circles represent resting levels of AP and HR for each curve.

to control curves, ISO curves (all concentrations) were shifted to the left as indicated by a significant decrease in BP_{50} and the minimum and maximum blood pressures that could be elicited by caval and aortic balloon inflations, respectively, in both NT and HT animals. In addition, the upper plateau and the range of the baroreflex were significantly attenuated by ISO anesthesia in both NT and HT animals. The operating point of the AP–HR baroreflex was located on or near the upper plateau at all levels of ISO anesthesia in both NT and HT animals. Therefore, with caval occlusion, there was little or no increase in HR at any level of ISO.

When comparing ISO curves, the maximum HR achieved with caval occlusion was higher following 1.0% ISO than any other level of ISO in both NT and HT animals. However, with the exception of the HR range following 1.0% ISO in NT rabbits, there was no effect of ISO concentration on the AP or HR range over which the baroreflex functioned in either NT or HT animals.

Figure 7 shows the average MAP–HR baroreflex curves for NT and HT animals while conscious (control) and during 1.0 and 2.0% ISO (Fig. 7, top) and 1.5 and 2.5% ISO (Fig. 7, bottom). These are the same curves shown in Fig. 6, but regrouped to more readily compare NT and HT animals. When comparing NT and HT baroreflex curves, there was a significant rightward shift in HT animals along the pressure axis while conscious as indicated by a significant increase in BP_{50}. Resting HR was

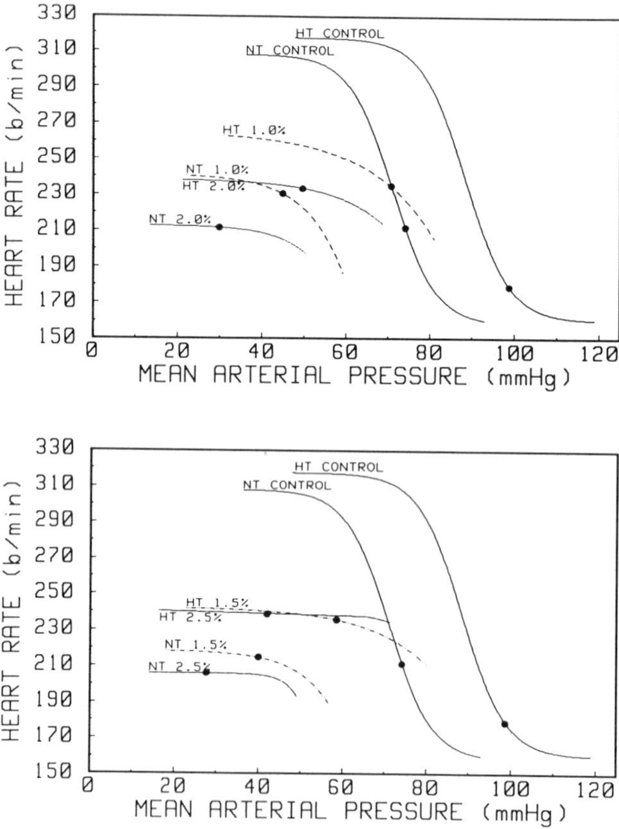

Fig. 7 Mean AP–HR baroreflex curves redrawn from Fig. 6 to allow comparison of NT and HT animals at each level of ISO anesthesia. The MAP–HR baroreflex curves are shown for NT and HT animals while conscious (control) and during 1.0 and 2.0% ISO (top) and 1.5 and 2.5% ISO (bottom).

significantly reduced in HT rabbits compared to NT rabbits, and, therefore, the increase in HR elicited by caval occlusion was significantly greater in conscious HT rabbits.

IV. Discussion

Patients with untreated chronic hypertension present unique problems to the anesthesiologist. Untreated HT patients are more likely to encounter postintubational hypertension (36), perioperative complications including enhanced hypotension (31), dysrythmias (31), myocardial ischemia (31,37), and postoperative hypertension (38). Hemodynamic instability has been well documented in both humans and animals in the awake hypertensive and is presumed to be at least partly due to impaired baroreflex control of efferent sympathetic activity. Control of AP is further complicated during anesthesia (39), since anesthetics attenuate arterial baroreflexes. This study was designed to investigate the individual effects of ISO anesthesia and chronic HT on baroreflex control of HR and efferent sympathetic nerve activity. In addition, the integrative effects of HT and anesthesia are discussed.

A. Effect of Isoflurane Anesthesia on Arterial Pressure–Heart Rate Baroreflex

Until relatively recently, the effect of anesthesia on baroreflex function was limited to the study of the HR baroreflex in awake and anesthetized subjects using the technique of Smyth *et al.* (40). Changes in heart period were recorded during graded intravenous infusions of vasoactive agents used to produce transient changes in AP. The slope of the baroreflex was then determined using linear regression. With these limitations in mind, it has previously been shown that the slope of the AP–HR baroreflex was decreased with ISO (1,2,4), halothane (10), and enflurane (4). In most of these studies, however, only increases in AP are induced by infusing pressor agents. This can lead to results that are difficult to interpret. For example, the results of this study indicate that the increase in HR that could be elicited with decreases in AP was reduced, such that the set point of the baroreflex was located on or near the upper plateau at most concentrations of ISO. In addition, as the ISO concentration increased, the lower plateau of the AP–HR baroreflex progressively disappeared, such that at 2.0 and 2.5% ISO the baroreflex was essentially flat. Therefore, to fully appreciate the effect of anesthetics on baroreflex function, it is essential to record the baroreflex over the full range of physiologically relevant arterial pressures.

B. Effect of Isoflurane Anesthesia on Arterial Pressure–Renal Sympathetic Nerve Activity Baroreflex

Two studies have demonstrated that increases in the level of ISO in already anesthetized animals produce progressive depression of the sympathetic baroreflex (2,16). The AP–sympathetic nerve activity baroreflex has not been compared between awake and anesthetized animals. The results reported here demonstrate that, compared to conscious animals, the slope of the AP–RSNA baroreflex was not altered by low levels of ISO (1.0%) but was significantly depressed at higher levels of ISO anesthesia. In addition, the upper plateau and range of the AP–RSNA baroreflex was significantly reduced at all levels of ISO. The AP range over which the baroreflex operated was also significantly lower than that in conscious animals. This result is similar to that seen by Skovsted *et al.* (16), who demonstrated that the baroreflex operated over a lower pressure range at higher levels of ISO (1.8%) but not at low levels of ISO (0.69%).

The results of the present study also demonstrate that the AP–RSNA baroreflex remains functional during ISO anesthesia, although the ability to respond to transient falls in AP is decreased with increasing levels of ISO. These results are similar to those of Skovsted *et al.* (16,41) and Seagard *et al.* (2) in halothane- and ISO-anesthetized animals, respectively.

C. Effect of Isoflurane on Steady-State Resting Levels of Renal Sympathetic Nerve Activity and Heart Rate

Of interest in the present study is the finding that resting steady-state levels of RSNA were not different between conscious and ISO-anesthetized animals (at any level of ISO). These results are in contrast to those reported by others. Seagard *et al.* (15) demonstrated that resting sympathetic nerve activity was not altered by low levels of ISO (1.5%) but was reduced at 2.5% ISO. Similarly, Skovsted *et al.* (41) demonstrated that cervical sympathetic nerve activity was not reduced at low levels of halothane anesthesia, but was slightly reduced at high levels of halothane.

Resting steady-state heart rates were also unaffected by ISO anesthesia in NT animals (when compared to awake) in the present study. This is similar to results observed by Seagard *et al.* (15) in the dog. In contrast, Skovsted *et al.* (16) showed an increase in HR at low levels of ISO and a decrease in HR at higher levels.

D. Effect of Chronic Hypertension on Mean Arterial Pressure–Heart Rate Baroreflex

Several studies have shown that the bradycardic response to increases in AP is blunted in the conscious chronically HT rat (42), rabbit (19,20), and human (43,44). In each of these studies, however, only the HR response to increases in pressure were studied. In studies recording the HR responses to both increases and decreases in AP (22–24,28,45), it is the bradycardic response to increases in AP that is most attenuated. Guo and Abboud (45) have shown that the tachycardic response to transient falls in AP was not altered by chronic hypertension in the rabbit. Taken together these results suggest that is is the vagal arm of the baroreflex that is modified in chronic hypertension (24).

Previous studies have also shown that the range (21,24,28) and slope (21,24,42) of the HR baroreflex is attenuated in chronic HT. These results differ from the present study which demonstrates that the range and slope of the HR baroreflex do not differ between NT and HT rabbits (Fig. 6, control curves). It is likely that the reason the present results differ from previous reports is that the experiments reported here were conducted before (i.e., 3–5 weeks after renal encapsulation) all of the cardiac and peripheral modifications to hypertension were manifested. Angell-James *et al.* (20) recorded baroreflex function every 4 weeks after the development of renal hypertension and found that the HR baroreflex was not attenuated after 4 weeks of hypertension, but was attenuated 8 weeks after renal encapsulation. This result was supported by Guo and Abboud (45), who demonstrated that although some attenuation of the HR baroreflex was present after 6 weeks, the attenuation after 4 months of renal encapsulation was much greater. Furthermore, Angell-James and George (46) determined that there was a closer relationship between the length of time the rabbit had been HT and the reduction in baroreflex sensitivity than the level to which the AP rose. Therefore, the results from the present study may better describe a transitional stage in the development of chronic hypertension than that which is present after established hypertension.

It has also been observed that, following the development of hypertension, pressor agents produce larger changes in AP in HT than in NT animals (22,23,43). These results are consistent with the present study in which the increase in AP owing to aortic balloon inflation was larger in the HT rabbits than in the NT rabbits. Presumably this is due to modifications that occur in peripheral vessels (i.e., decreased distensibility).

E. Effect of Chronic Hypertension on Mean Arterial Pressure–Renal Sympathetic Nerve Activity Baroreflex

Although acute hypertension has been shown to attenuate baroreflex control of RSNA (47), this is the first study to record baroreflex-mediated changes in RSNA in the chronic HT conscious animal following both increases and decreases in AP. The results of the present study indicate that the range and slope of the AP–RSNA baroreflex is not altered by hypertension in conscious rabbits. Previous studies have reported conflicting results. Several studies in conscious (27,48) and anesthetized (22,23) animals have reported that the visceral sympathetic nerve activity response to increases in AP is not different between NT and HT. Rea and Hamdam (30) have also shown that the AP–muscle sympathetic nerve activity baroreflex was similar in NT and HT humans. On the other hand, the AP–sympathetic nerve activity baroreflex has been reported to be attenuated following hypertension in the rabbit (49), pig (29), and human (43,44). It is unclear why there are conflicting results. One possible explantation is that in each of the studies in which an attenuated baroreflex has been observed, only increases in AP are used to determine the slope of the baroreflex. Thames *et al.* (49) suggest that the reason they saw an attenuated baroreflex response may have been that they did not determine the entire baroreflex curve and that the resting sympathetic nerve activity was located on a different portion of the barocurve. This interpretation would be consistent with the results of the present study.

F. Effect of Chronic Hypertension on Steady-State Resting Levels of Renal Sympathetic Nerve Activity and Heart Rate

In the present study the resting RSNA was not different between conscious NT and HT rabbits. These results are similar to those of Thames and colleagues (22,49), who observed no difference in renal or lumbar sympathetic nerve activity in renal HT rabbits when compared to NT rabbits. Notvest and Zambraski (29), on the other hand, recorded an increased RSNA in the anesthetized HT miniature swine.

Resting HR in the present study was significantly lower in HT rabbits than in NT rabbits. This is in contrast to the observations made in other studies in conscious (20,22,50) and anesthetized (23) rabbits in which there was no difference in resting HR between NT and HT animals, as well as the observations in conscious spontaneously HT rats in which hypertension was associated with an increase in resting HR. It is possible that the reduced resting HR in HT rabbits in this study represents the HR of animals in transition to established hypertension where the mechanisms controlling the set point and gain of the HR baroreflex are not fully adapted to the increase in AP. As the AP–HR baroreflex becomes attenuated with a longer period of hypertension, there would be less baroreflex suppression of HR, and resting HR in the HT would move toward that seen in the NT. The effect of time on changes in the AP–HR and AP–RSNA baroreflexes needs further investigation.

G. Interaction of Isoflurane Anesthesia and Chronic Hypertension on Baroreflex Function

The results of the present study indicate that ISO anesthesia depressed both baroreflex curves (i.e., AP–HR and AP–RSNA) to the same degree in NT and HT rabbits. The only difference was that baroreflex curves in HT rabbits were displaced to the right along the pressure axis. Both NT and HT animals were able to respond to increases in AP by decreasing both HR and RSNA. Responses to decreases in AP were limited to small increases in RSNA at low levels of ISO (1.0%) in both NT and HT animals.

Only one other study has compared the effect of anesthesia on baroreflex function in NT and HT subjects. Prys-Roberts *et al.* (31) have shown that halothane anesthesia attenuated the baroreflex to the same degree in HT subjects as in NT subjects. However, they only determined the HR responses to increases in AP. They concluded that the baroreflex in conscious HT subjects was reduced. However, as discussed above, recording only one half of the baroreflex curve an lead to results that are difficult to interpret. Therefore, it is difficult to compare their results to the results of the present study. There have been no previous reports of the AP–sympathetic nerve activity baroreflex in the conscious and anesthetized HT animal.

In summary, the findings reported in this article demonstrate that both chronic hypertension and ISO anesthesia modified baroreflex control of HR and RSNA. Chronic hypertension shifts the baroreflex to the right along the pressure axis and operates around a higher AP. Although resting HR is decreased in the HT rabbit, the baroreflex is still able to compensate for alterations in AP by changing both HR and RSNA. In contrast, ISO anesthesia shifts the baroreflex to the left along the pressure axis and operates around a lower AP in both NT and HT animals. At low levels of ISO, the baroreflex can compensate somewhat for decreases in AP through small increases in HR and sympathetic nerve activity. At higher levels of ISO, the ability to compensate for decreases in AP is abolished. The RSNA responses to increases in AP is preserved following ISO anesthesia, whereas HR responses to increases in AP are progressively attenuated as the ISO concentration increases.

Acknowledgements

This work was supported by National Institutes of Health FIRST Award (1-R29-HL42455), NIH Anesthesia Research Training Grant (GM08377), and the Department of Veterans Affairs, Medical Research Service.

References

1. Kotrly, K. J., Ebert, T. J., Vucins, E., Igler F. O., Barney J. A., and Kampine, J. P. (1984). Baroreceptor reflex control of heart rate during isoflurane anesthesia in humans. *Anesthesiology* **60**, 173–179.
2. Seagard J. L., Elegbe E. O., Hopp, F. A., Bosnjak, Z. J., von Colditz, J. H., Kalbfleisch, J. H., and Kampine J. P. (1983). Effects of isoflurane on the baroreceptor reflex. *Anesthesiology* **59**, 511–520.
3. Sellgren, J., Biber, B., Henriksson, B.-A., Martner, J., Ponten, J. (1992). The effects of propofol, methohexitone and isoflurane on the baroreceptor reflex in the cat. *Acta Anaesthesiol. Scand.* **36**, 784–790.
4. Takeshima, R., and Dohi, S. (1989). Comparison of arterial baroreflex function in humans anesthetized with enflurane or isoflurane. *Anesth. Analg. (N. Y.)* **69**, 284–290.
5. Biscoe, T. J., and Millar, R. A. (1966). The effects of cyclopropane, halothane and ether on central baroreceptor pathways. *J. Physiol. London* **184**, 535–559.
6. Bristow, J. D., Honour, A. J., Pickering, G. W., Sleight, P., Smyth, H. S. (1969). Diminished baroreflex sensitivity in high blood pressure. *Circulation* **39**, 48–54.
7. Duke, P. C., Fownes, D., and Wade, J. G. (1977) Halothane depresses baroreflex control of heart rate in man. *Anesthesiology* **46**, 184–187.
8. Ngai, S. H., and Bolme, P. (1966). Effects of anesthetics on circulatory regulatory mechanisms in the dog. *J. Pharmacol. Exp. Ther.* **153**, 495–504.
9. Price, H. L., Linde, H. W., and Morse, H. T. (1963). Central nervous actions of halothane affecting the systemic circulation. *Anesthesiology* **24**, 770–778.
10. Seagard, J. L., Hopp, F. A., Donegan, J. H., Kalbfleisch, J. H., and Kampine, J. P. (1982). Halothane and the carotid sinus reflex. *Anesthesiology* **57**, 191–202.
11. Van Leeuwen, A. F., Evans, R. G., and Ludbrook, J. (1990). Effects of halothane, ketamine, propofol, and alfentanil anaesthesia on circulatory control in rabbits. *Clin. Exp. Pharmacol. Physiol.* **17**, 781–798.
12. Wilkinson, P. L., Stowe, D. F., Glantz, S. A., and Tyberg, J. V. (1980). Heart rate–systemic blood pressure relationship during halothane anesthesia. *Acta Anaesthesiol. Scand.* **24**, 181–186.
13. Morton, M., Duke, P. C., and Ong, B. (1980). Baroreflex control of heart rate in man awake and during enflurane and enflurane–nitrous oxide anesthesia. *Anesthesiology* **52**, 221–223.
14. McCallum, J. B., Stekiel, T. A., Bosnjak, Z. J., and Kampine, J. P. (1993). Does isoflurane alter mesenteric venous capacitance in the intact rabbit? *Anesth. Analg. (N. Y.)* **76**, 1095–1105.
15. Seagard, J. L., Hopp, F. A., Bosnjak, Z. J., and Kampine, J. P. (1984). Sympathetic efferent nerve activity in conscious and isoflurane-anesthetized dogs. *Anesthesiology* **61**, 266–270.
16. Skovsted, P., and Sapthavichaikul, S. (1977). The effects of isoflurane on arterial pressure, pulse rate, autonomic nervous activity, and barostatic reflexes. *Can. Anaesth. Soc. J.* **24**, 304–314.
17. Stekiel, T. A., Ozono, K., McCallum, J. B., and Kampine, J. P. (1990). The inhibitory action of halothane on reflex constriction in mesenteric capacitance veins. *Anesthesiology* **73**, 1169–1178.
18. Stadnicka, A., Stekiel, T. A., Bosnjak, Z. J., and Kampine, J. P. (1993). Inhibition by enflurane of baroreflex mediated mesenteric venoconstriction in the rabbit ileum. *Anesthesiology* **78**, 928–936.
19. Alexander, N., and DeCuir, M. (1966). Loss of baroreflex bradycardia in renal hypertensive rabbits. *Circ. Res.* **19**, 18–25.

20. Angell-James, J. E., George, M. J., Peters, C. J. (1980). Baroreflex sensitivity in rabbits during the development of experimental renal hypertension and medial sclerosis. *Clin. Exp. Hypertens.* **2**, 321–340.
21. Eckberg, D. L. (1979). Carotid baroreflex function in young men with borderline blood pressure elevation. *Circulation* **59**, 632–636.
22. Guo, G. B., and Thames, M. D. (1983). Abnormal baroreflex control in renal hypertension is due to abnormal baroreceptors. *Am. J. Physiol.* **245**, H420–H428.
23. Guo, G. B., Thames, M. D., and Abboud, F. M. (1983). Arterial baroreflexes in renal hypertensive rabbits: Selectivity and redundancy of baroreceptor influence on heart rate, vascular resistance, and lumbar sympathetic nerve activity. *Circ. Res.* **53,** 223–234.
24. Korner, P. I., West, M. J., Shaw, J., and Uther, J. B. (1974). 'Steady state' properties of the baroreceptor–heart rate reflex in essential hypertension in man. *Clin. Exp. Pharmacol. Physiol.* **1**, 65–76.
25. Moreira, E. D., Ida, F., Oliveira, V. L. L., and Krieger, E. M. (1992). Early depression of the baroreceptor sensitivity during onset of hypertension. *Hypertension* **19**(Suppl. 2), II-17–II-21.
26. Oliveria, V. L. L., Irigoyen, M. C., Moreira, E. D., Strunz, C., and Krieger, E. M. (1992). Renal denervation normalizes pressure and baroreceptor reflex in high renin hypertension in conscious rats. *Hypertension* **19**(Suppl. 2), II17–II21.
27. Ricksten, S. E., and Thoren, P. (1981). Reflex control of sympathetic nerve activity and heart rate from arterial baroreceptors in conscious spontaneously hypertensive rats. *Clin. Sci.* **61**, 169s–172s.
28. Widdop, R. E., Verberne, A. J. M., Jarrott, B., and Louis, W. J. (1990). Impaired arterial baroreceptor reflex and cardiopulmonary vagal reflex in conscious spontaneously hypertensive rats. *J. Hypertens.* **8**, 269–275.
29. Notvest, R. R., and Zambraski, E. J. (1985). Baroreflex control of renal sympathetic nerve activity in hypertensive miniature swine. *Hypertension* **7**, 879–885.
30. Rea, R. F., and Hamdam, M. (1990). Baroreflex control of muscle sympathetic nerve activity in borderline hypertension. *Circulation* **82**, 856–862.
31. Prys-Roberts, C., Meloche, R., and Foex, P. (1971). Studies of anaesthesia in relation to hypertension. I: Cardiovascular responses of treated and untreated patients. *Br. J. Anaesth.* **43**, 122–137.
32. Bell, L. B., O'Hagan, K. P., and Clifford, P. S. (1993). Cardiac but not pulmonary receptors mediate depressor response to IV phenyl biguanide in conscious rabbits. *Am. J. Physiol.* **264**, R1050–R1057.
33. Dorward, P. K., Riedel, W., Burke, S. L., Gipps, J., Korner, P. I. (1985). The renal sympathetic baroreflex in the rabbit: Arterial and cardiac baroreceptor influences, resetting, and the effect of anesthesia. *Circ. Res.* **57**, 618–633.
34. White, S. W., and McRitchie, R. J. (1973). Nasopharyngeal reflexes: Integrative analysis of evoked respiratory and cardiovascular effects. *Aust. J. Exp. Biol. Med. Sci.* **51**(1), 17–31.
35. Drummond, J. C. (1985). MAC for halothane, enflurane and isoflurane in the New Zealand White rabbit; and a test for the validity of MAC determinations. *Anesthesiology* **62**, 336–338.
36. Bedford, R. F., and Feinstein, B. (1980). Hospital admission blood pressure: A predictor for hypertension following endotracheal intubation. *Anesth. Analg. (N. Y.)* **59**, 367–370.
37. Stone, J. G., Foex, P., Sear, J. W., Johnson, L. L., Khambatta, J. H., and Triner, L. (1988). Myocardial ischemia in untreated hypertensive patients: Effect of a single small oral dose of a beta-adrenergic blocking agent. *Anesthesiology* **68**, 495–500.
38. Asiddao, C. B., Donegan, J. H., Whitesell, R. C., and Kalbfleisch, J. H. (1982). Factors associated with perioperative complications during carotid endarterectomy. *Anesth. Analg. (N. Y.)* **61**, 631–637.

39. Reves, J. G., and Knopes, K. D. (1988). Adrenergic component of the stress response. *Anesth. Rep.* **1**, 175–197.
40. Smyth, H. S., Sleight, P., and Pickering, G. W. (1969). Reflex regulation of arterial pressure during sleep in man. *Circ. Res.* **24**, 109–121.
41. Skovsted, P., Price, M. L., and Price, H. L. (1969). The effects of halothane on arterial pressure, preganglionic sympathetic activity and barostatic reflexes. *Anesthesiology* **31**, 507–514.
42. Edmunds, M. E., Russell, G. I., Burton, P. R., and Swales, J. D. (1990). Baroreceptor–heart rate reflex function before and after surgical reversal of two-kidney, one clip hypertension in the rat. *Circ. Res.* **66**, 1673–1680.
43. Matsukawa, T., Gotoh, E., Miyajima, E., Yamada, Y., Shionoiri, H., Tochikubo, O., and Ishii, M. (1988). Angiotensin II inhibits baroreflex control of muscle sympathetic nerve activity and heart rate in patients with essential hypertension. *J. Hypertens.* **6**(Suppl. 4) S501–S504.
44. Yamada, Y., Miyajima, E., Tochikubo, O., Matsukawa, T., Shionoiri, H., Ishii, M., and Kaneko, Y. (1988). Impaired baroreflex changes in muscle sympathetic nerve activity in adolescents who have a family history of essential hypertension. *J. Hypertens.* **6**(Suppl. 4), S525–S528.
45. Guo, G. B., and Abboud, F. M. (1984). Impaired central mediation of the arterial baroreflex in chronic renal hyprtension. *Am. J. Physiol.* **246**, H720–H727.
46. Angell-James, J. E., and George, M. J. (1976). Time-course of the reduction of baroreceptor sensitivity in experimental hypertensive rabbits. *Clin. Sci. Mol. Med.* **51**, 369S–372S.
47. Dorward, P. K., Bell, L. B., and Rudd, C. D. (1990). Cardiac afferents attenuate renal sympathetic baroreceptor reflexes during acute hypertension. *Hypertension* **16**, 131–139.
48. Ricksten, S. E., and Thoren, P. (1980). Reflex inhibition of sympathetic activity during volume load in awake normotensive and spontaneously hypertensive rats. *Acta Physiol. Scand.* **110**, 77–82.
49. Thames, M. D., Gupta, B. N., and Ballon, B. J. (1984). Central abnormality in baroreflex control of renal nerves in hypertension. *Am. J. Physiol.* **245**, H843–H850.
50. Flasher, J., and Drury, D. R. (1949). Effects of removal of the 'ischemic' kidney in rabbits with unilateral renal hyertension, as compared to unilateral nephrectomy in normal rabbits. *Am. J. Physiol.* **158**, 438–443.

Effects of Isoflurane on Regulation of Capacitance Vessels under Normotensive and Chronically Hypertensive Conditions

Thomas A. Stekiel, Leonard B. Bell, Zeljko J. Bosnjak, and John P. Kampine

Departments of Anesthesiology and Physiology
The Medical College of Wisconsin
and Zablocki Veterans Affairs Medical Center
Milwaukee, Wisconsin 53295

I. Introduction

Chronic hypertension is a condition frequently encountered in anesthetic practice. Perioperative management of the hypertensive patient continues to be an important clinical problem because the anesthetic course of chronically hypertensive patients is characterized by a poorly understood circulatory instability and an increased perioperative morbidity (1–3). Although data linking hypertension alone to adverse perioperative outcome have not been conclusive (4) when groups of patients undergoing particular anesthetic procedures are studied, there is clearly an increased incidence of perioperative morbidity (2,3). Most of these procedures are associated with disease processes common in hypertension such as coronary artery disease and peripheral vascular disease (5). In addition, wide or sustained fluctuations in hemodynamic control may occur during the administration of general anesthesia, and the incidence of such circulatory instability is itself increased in the presence of chronic hypertension (6).

Essential hypertension, that is, where no specific etiology has been established, makes up approximately 85% of all cases of hypertension (2). Several distinct factors have been associated with the development of

Advances in Pharmacology, Volume 31

hypertension in experimental models, and prominent among these is an elevation in the level of sympathetic regulation of the vasculature (7). Evidence for such an elevation includes an increased catecholamine uptake and release by neuronal tissue innervating arterial vascular smooth muscle (VSM) and an increased contractile response to endogenously released neurotransmitter (8). More recently, direct measurements of vascular smooth muscle transmembrane potentials demonstrated enhanced sympathetic nerve input in two different experimental models of essential hypertension (9,10). Of particular importance is that in both studies the increase in neural control was present in venous vessels regulating capacitance as well as the small arteries regulating vascular resistance. Likewise, elevated sympathetic efferent regulation of venous capacitance as well as arterial resistance beds has been demonstrated in a coarctation rabbit model of hypertension (8).

Although the majority of studies investigating the regulation of the circulation and the hemodynamic effects of anesthetics have focused on myocardial and arterial resistance mechanisms, increasing emphasis is now being placed on the importance of factors controlling venous capacitance which, in turn, also regulates overall hemodynamic stability. The importance of active (neurally mediated) as well as passive constriction of mesenteric capacitance veins in circulatory control has been described by Rothe (11). The splanchnic venous circulation contains approximately 25% of the total blood volume and is known to be a major regulator of the total capacitance system; thus it contributes significantly to the hemodynamic regulation of venous return, cardiac output, and, hence, arterial blood pressure. For example, maximal constriction of capacitance veins may increase cardiac output up to 50% (12). Abdominal vascular capacitance is known to be under sympathetic control (13,14).

We have demonstrated active sympathetic reflex-mediated constriction of mesenteric capacitance vessels by direct *in situ* measurement of vein diameter and intravenous pressure (15). In subsequent studies we demonstrated that volatile anesthetics can significantly inhibit mesenteric venoconstriction and related circulatory responses to both carotid sinus and chemoreceptor-mediated reflex stimuli (16–18). Also, studies in other laboratories have demonstrated differences in the hemodynamic effects of volatile anesthetics in normotensive and chronically hypertensive animals (19). For example, in genetically spontaneous hypertensive rats, systemic vascular resistance was preserved but cardiac output was reduced in the presence of inhaled isoflurane, whereas the reverse was true in normotensive controls (20).

Thus, clinical studies and laboratory studies have both demonstrated that the hormonal and hemodynamic responses to anethetics are altered in hypertension. However, the pharmacological and physiological mechanisms underlying such altered responses have not been clarified. Furthermore, studies to date that have investigated the effects of anesthetics in the chronically hypertensive condition have focused primarily on changes in myocardial function and arterial resistance. Given the significant role of capacitance vessels in the regulation of hemodynamic stability, the objective of the present study was to advance our overall understanding of the interactions between anesthetics and the hypertensive condition in two ways: (1) to quantify any differences in the level of resting or reflex sympathetic regulation of vascular tone in mesenteric capacitance veins under normotensive versus chronically hypertensive conditions; (2) to quantify the effect of inhaled isoflurane on the level of sympathetic control of mesenteric capacitance veins under normotensive versus chronically hypertensive conditions.

II. Hypertensive Rabbit Preparation

We studied a total of 16 New Zealand White rabbits (1–2 kg body weight). Of these, 7 animals were made hypertensive using a latex renal wrap procedure as described previously (21). Each rabbit was anesthetized with 30 mg/kg of intramuscular Telazol and was intubated for maintenance of general endotracheal halothane anesthesia (1–2 MAC) (22). After surgical exposure via lateral retroperitoneal incisions under sterile surgical conditions, each kidney was encapsulated with inert latex wraps and replaced into the retroperitoneal space. Following surgical closure of the incisions the animals were allowed to emerge from anesthesia; postoperative analgesia was maintained with Buprenorphine (0.01–0.02 mg/kg i.m. every 12 hr as needed).

The onset of hypertension was monitored at 3–4 weeks by means of percutaneous cannulation of the ear artery. The criterion for hypertension was defined arbitrarily as a 20% or greater increase in resting mean arterial blood pressure over that time period. Animals not achieving that level of blood pressure were rejected from the study. A total of nine sham-wrapped rabbits served as normotensive controls. These animals underwent a surgical procedure identical to the hypertensive animals except that no latex kidney encapsulation was performed.

III. Mesenteric Reflex Measurements

Within six weeks after the surgical renal wrap procedure, each animal was subjected to an acute reflex response protocol utilizing the mesenteric vessel preparation that we have used extensively in our previous studies and which has been described in detail (15–18). Briefly, surgical preparation was initiated following anesthetic induction with thiamylal (10–20 mg/kg) via the ear vein and maintenance with α-chloralose (12.5–37.5 mg/hr). Surgical sites were infiltrated with a total of 3–5 ml of 1% (w/v) lidocaine. After tracheotomy, one femoral artery and vein were cannulated for arterial blood pressure measurement and blood sampling, and continuous intravenous infusion, respectively. The carotid arteries were dissected bilaterally and were isolated *in situ* for carotid artery occlusion. Aortic depressor nerves were also dissected and isolated *in situ* for subsequent stimulation.

For all rabbit preparations a midline laparotomy was performed and a 13-cm loop of terminal ileum was externalized and mounted in a 37–38°C temperature-regulated polystyrene tissue chamber mounted on a movable microscope stage. The ileum and associated mesentery were superfused continuously with a physiological salt solution formulated by Bohlen to simulate the peritoneal environment (23). The mesentery was pinned to a layer of clear Silastic rubber coating the chamber floor, and short *in situ* segments of 500–1000 μm OD mesenteric veins were used for diameter measurements.

Ventilation was controlled immediately after tracheotomy, and systemic muscle relaxation was produced with a vecuronium infusion (0.1 mg/kg/hr) to suppress spontaneous ventilatory efforts. Arterial blood gases were sampled periodically during each experiment, and normocarbia and normal pH were maintained via ventilator adjustments and 1–2 mg bolus doses of $NaHCO_3$. In addition, a baseline infusion of 0.5 to 1.5 mg/hr of $NaHCO_3$ was maintained to correct for the metabolic acidosis that typically occurs under α-chloralose anesthesia (24). End-tidal CO_2 was also monitored with a medical gas analyzer mass spectrometer and maintained between 30 and 40 mmHg. This is the approximate range observed for spontaneously breathing rabbits immediately following tracheotomy. Rectal temperature was maintained between 36.5–37.5°C with a heating blanket.

Blood pressure was monitored continuously via a transducer connected to the femoral arterial cannula and heart rate via conversion of arterial pressure pulse frequency to voltage. Mesenteric vein diameter was measured continuously by means of an on-line videomicrometer system whose design, accuracy, and precision have been described previously (25). All data were recorded on videocassette tapes and printed on an eight-channel strip-chart recorder. Blood concentrations of isoflurane were measured by gas chromatography. Differences in the level of sympathetic efferent neural control of small mesenteric veins and related reflex responses between hypertensive and normotensive rabbits as well as any differential effects of inhaled isoflurane were assessed by measuring the above-described cardiovascular variables in response to (1) carotid sinus activation (40-sec periods of bilateral carotid artery occlusion); (2) aortic depressor nerve stimulation (10-sec trains of pulses, 20 Hz, 1 msec duration, 0.05–5 mA intensity); and (3) chemoreflex activation [induced by sequential 40-sec periods of administration of 10, 5, 2.5, and 0% inspired oxygen (FiO_2), each separated by a 6-min reequilibration period of 21% oxygen). Carotid artery occlusion, aortic nerve stimulation, and acute graded hypoxia were performed on the same animal before, during, and after administration of 1.5% isoflurane (i.e., 0.75 MAC) (22). The anesthetic was delivered for 45 min through an isoflurane vaporizer using a 21% O_2–79% N_2 mixture (% v/v) as a carrier gas at a flow rate of 5 liters/min.

Changes in heart rate and mean arterial blood pressure and percent changes in mesenteric vein diameter in response to baroreflex and chemoreflex activation were calculated before, during, and after inhaled isoflurane. The statisical significance of the differences for each variable between hypertensive versus the normotensive animals as well as the effects of isoflurane was determined by multiple analysis of variance for repeated measures at $p \leq 0.05$.

IV. Circulatory Responses to Baroreflex and Chemoreflex Responses in Normotensive and Hypertensive Animals

In normotensive sham animals, 40 sec of bilateral carotid occlusion produced the classic reflex increase in blood pressure and heart rate coupled with a simultaneously measured reflex mesenteric venoconstriction. Conversely, stimulation of the aortic depressor nerve produced reflex

Table I

Baseline (Prestimulation) Measurements[a]

	Normotensive shams			Hypertensives		
Measure	Pre-Iso	1.5% Iso	Δ	Pre-Iso	1.5% Iso	Δ
Mesenteric vein diameter (μm)	786 ± 13	843 ± 15[b]	56.4	756 ± 17	778 ± 19[b]	25.6[c]
Heart rate (bpm)	238 ± 5	195 ± 7[b]	−42.9	271 ± 5	220 ± 5[b]	−53.2
Mean arterial pressure (mmHg)	79 ± 2	31 ± 2[b]	−47.6	110 ± 2	44 ± 2[b]	−66.1[c]

[a] Values represent mean ± SEM baseline (prestimulation) measurements taken before, during, and after inhaled 1.5% isoflurane (Iso).

[b] Significantly different ($p \leq 0.05$) versus preisoflurane.

[c] Significantly different ($p \leq 0.05$) versus normotensive change.

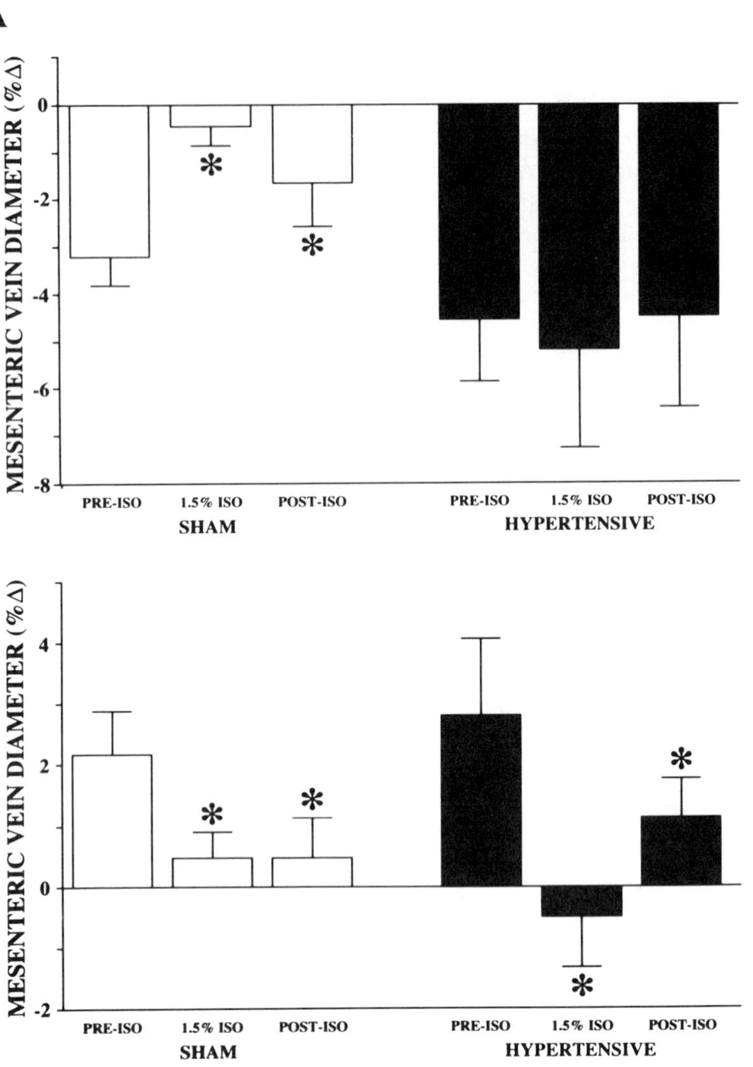

Fig. 1 Effect of bilateral carotid occlusion (BCO) and aortic depressor nerve stimulation (ANS) on reflex responses before, during, and after 1.5% inhaled isoflurane in normotensive (sham) and hypertensive (renal wrap) animals. (A) Differential effect during BCO (top), and similar effect during ANS (bottom), of isoflurane on reflex change in mesenteric vein diameter in normotensive and hypertensive animals. Each column represents the mean ± SEM percent change. *Significant difference $p \leq 0.05$ versus preceding preisoflurane control.

bradycardia and hypotension coupled with a simultaneously measured mesenteric venodilation. Also, in these animals, the administration of acute graded hypoxia produced the classic chemoreflex response of hyper-

B

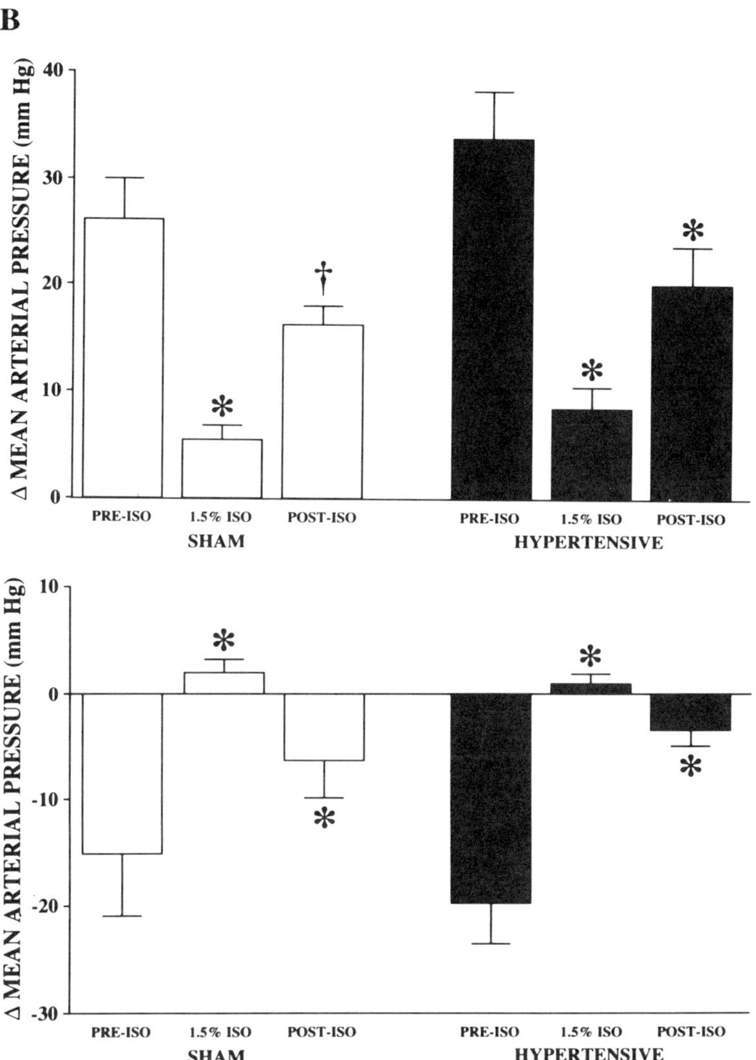

(B) Inhibitory effect of isoflurane on reflex change in mean arterial blood pressure in response to BCO (top) and ANS (bottom) in both hypertensive and normotensive animals. Each column represents the mean ± SEM change. *Significant difference ($p \leq 0.05$) versus preceding preisoflurane control. †Significant difference ($p \leq 0.05$) versus 1.5% isoflurane.

tension and bradycardia coupled with a simultaneously measured reflex mesenteric venoconstriction. Prior to isoflurane administration, the hypoxia-mediated responses in the animals tended to be proportional to the intensity of the hypoxic stimulus. Except for the smallest hypoxia

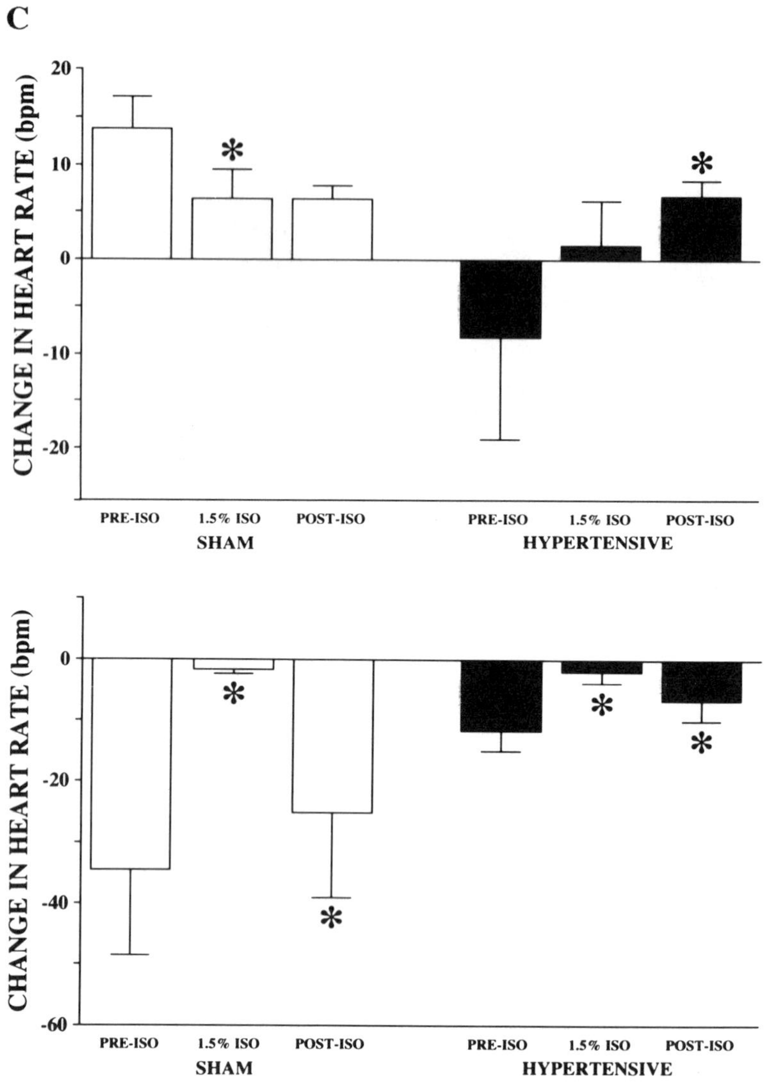

(C) Inhibitory effect of isoflurane on reflex change in heart rate response to ANS (bottom) in both hypertensive and normotensive animals and on change in heart rate response to BCO (top) in normotensives. The initial preisoflurane control response of hypertensive was inconsistent. Columns and asterisks have same meaning as in (B).

stimulus (10% FiO_2), all reflex blood pressure responses to graded steps of hypoxia were significant (Fig. 3).

The average initial mean arterial blood pressure and heart rate for both the normotensive sham animals and the hypertensive renal wrap animals are illustrated in Table I. In addition to a significantly greater mean arterial pressure, the hypertensive animals also had a significantly greater average resting heart rate than the normotensive controls. Despite these differences, the baroreflex and chemoreflex-mediated vein diameter and arterial

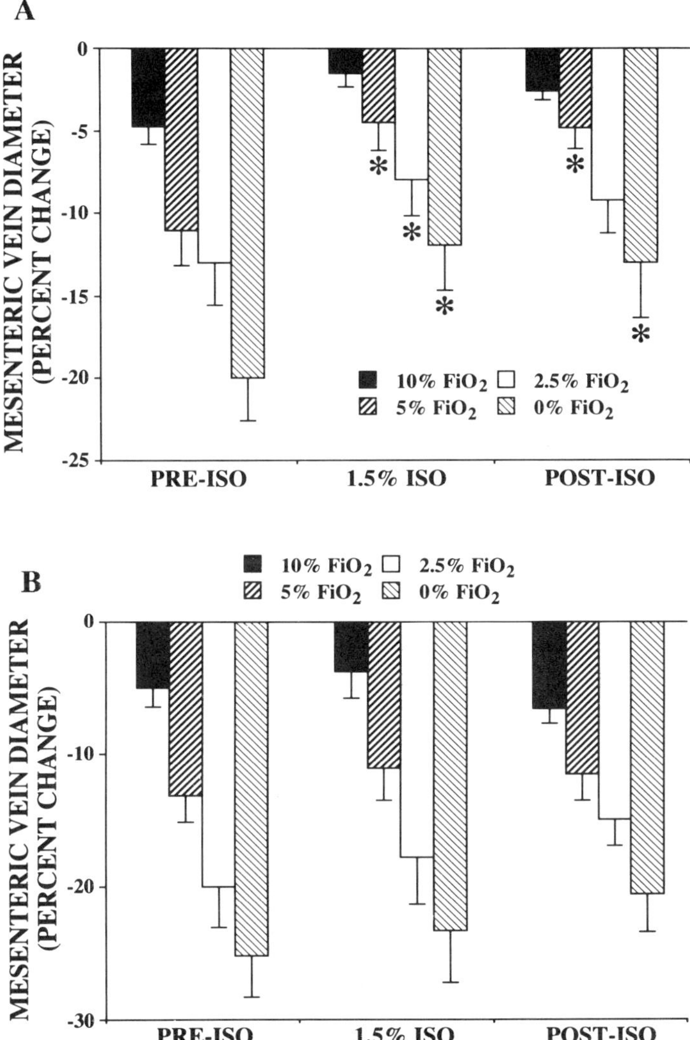

Fig. 2 Differential effect of 1.5% inhaled isoflurane on reflex mesenteric venoconstriction in normotensive sham (A) and hypertensive renal wrap (B) animals. Each column represents the mean ± SEM percent change. *Significant difference ($p \leq 0.05$) versus corresponding preisoflurane control.

pressure responses prior to isoflurane administration in the hypertensive animals were virtually identical to the preisoflurane responses observed in the normotensive controls (Figs. 1–3). Prior to isoflurane administration, heart rate response to aortic nerve stimulation in hypertensive ani-

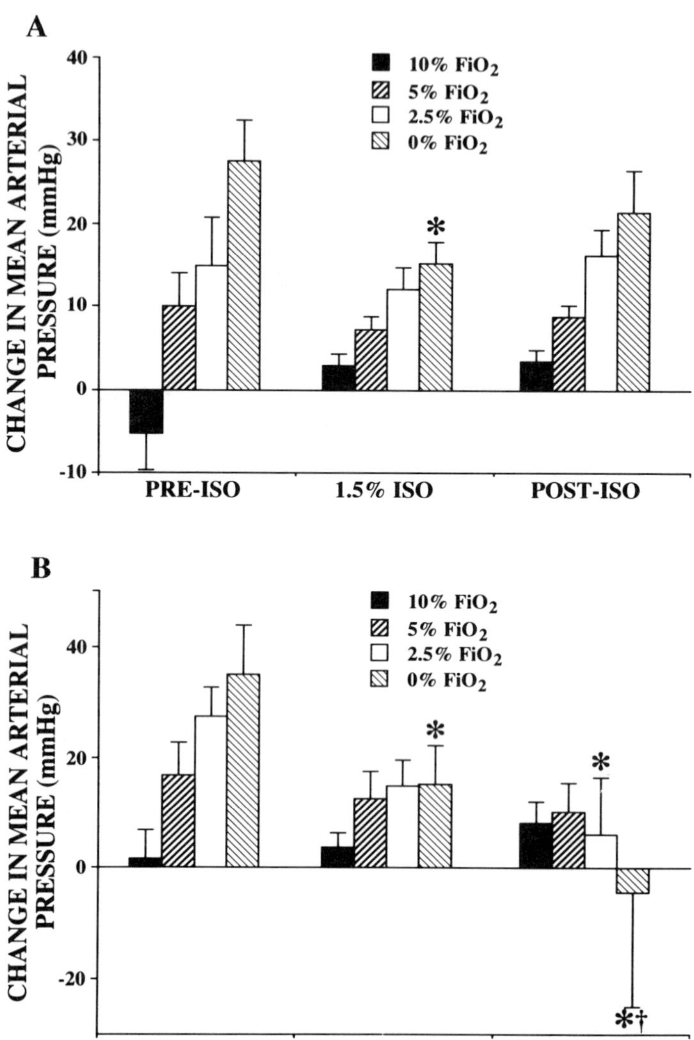

Fig. 3 Similar (minimal) effect of 1.5% inhaled isoflurane on reflex mean arterial blood pressure increase in normotensive sham (A) and hypertensive renal wrap (B) animals. Each column represents the mean ± SEM change. *Significant difference ($p \leq 0.05$) versus corresponding preisoflurane control. †Significant difference ($p \leq 0.05$) versus corresponding 1.5% iso column.

Table II
Arterial Blood Gas Analysis

FiO_2	pH	pO_2	pCO_2	HCO_3^-	Hemoglobin O_2 saturation
Control (room air, %)	7.42 ± 0.01	86.7 ± 2.9	35.2 ± 1.5	22.7 ± 1.1	95.5 ± 0.5
10	7.43 ± 0.01	40.5 ± 1.2	35.0 ± 1.6	22.8 ± 1.0	76.8 ± 1.8
5	7.43 ± 0.01	29.6 ± 0.9	31.6 ± 1.4	21.0 ± 0.6	59.0 ± 1.8
2.5	7.44 ± 0.01	23.8 ± 1.0	31.2 ± 1.5	21.1 ± 0.9	45.4 ± 2.8
0	7.45 ± 0.01	18.3 ± 1.1	29.1 ± 1.5	20.1 ± 0.8	29.6 ± 3.1

Note. Values are means ± SEM.

mals was also identical to that observed in normotensive animals. However, no significant preisoflurane heart rate response to carotid occlusion was observed in hypertensive animals in contrast to the tachycardic response in normotensive animals (Figs. 1C and 4). Table II illustrates the average arterial blood gas values for both hypertensive and normotensive animals during each level of graded hypoxia administration.

V. Effects of Isoflurane on Hypoxia and Baroreflex-Mediated Responses

In the normotensive control animals, the administration of 1.5% inhaled isoflurane significantly attenuated the reflex change in heart rate, blood pressure, and mesenteric vein diameter in response to bilateral carotid occlusion and aortic depressor nerve stimulation (Fig. 1). Likewise, in the normotensive sham animals, the mesenteric vein diameter and heart rate changes in response to all levels of acute graded hypoxia were significantly inhibited by 1.5% isoflurane (Figs. 2A and 4A). As illustrated in Fig. 3A, in normotensive control animals only the blood pressure increase in response to the most severe hypoxic step (0% FiO_2) was attenuated by 1.5% isoflurane administration.

In the hypertensive renal wrap animals, 1.5% inhaled isoflurane produced an attenuation of the heart rate and blood pressure responses to aortic nerve stimulation and an attenuation of the blood pressure responses to carotid occlusion similar to that observed in the normotensive control animals (Figs. 1B,C). Likewise, the level of attenuation of heart rate and blood pressure response to acute graded hypoxia by 1.5% isoflurane in the hypertensive group was identical to that observed in the normotensive group (Figs. 3B and 4B). Isoflurane also similarly attenuated the reflex

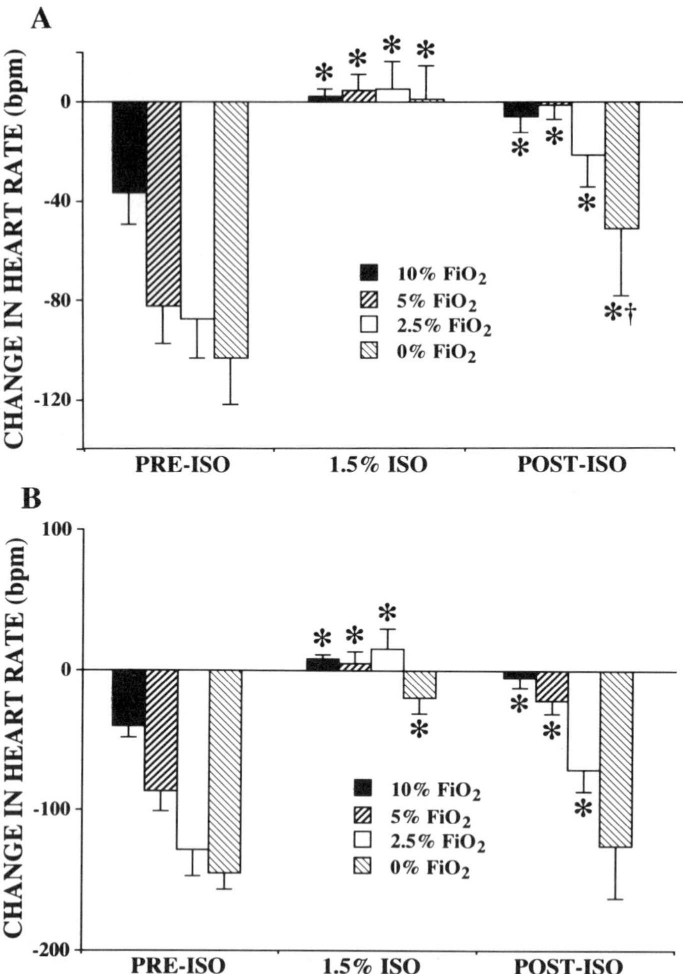

Fig. 4 Virtual elimination of reflex bradycardia during 1.5% inhaled isoflurane in normotensive sham (A) and hypertensive renal wrap (B) animals. Columns and symbols have same meaning as in Fig. 3.

mesenteric venodilation response to aortic depressor nerve stimulation in both the hypertensive and normotensive animal groups. The most important observation of the study was that, in direct contrast to the attenuation by 1.5% isoflurane of blood pressure and heart rate responses in both normotensive and hypertensive animals, the reflex mesenteric venoconstriction in response to bilateral carotid occlusion (Fig. 1A) and acute graded hypoxia (Fig. 2B) in the hypertensives (but not the normotensives) was completely resistant to the inhibitory effects of 1.5% inhaled isoflurane. Also, as illustrated in Figs. 1–4, most of the responses that were inhibited by 1.5% inhaled isoflurane in both normotensive and hypertensive animals tended not to recover to preisoflurane baseline levels following isoflurane removal.

VI. Effects of Isoflurane on Prestimulation Baseline Measurements

The effects of inhaled isoflurane on baseline mesenteric vein diameter, heart rate, and mean arterial pressure (i.e., measurement before each episode of baroreflex or hypoxic stimulus) are reported in Table I. Resting heart rate was equally reduced by 1.5% inhaled isoflurane in both hypertensive and normotensive control animals. Likewise, baseline mean arterial pressure was also significantly reduced in both groups of animals. However, the decrease for the hypertensive group was greater in magnitude than that observed for the normotensive group. Finally, 1.5% inhaled isoflurane also significantly dilated mesenteric capacitance veins in both hypertensive and normotensive control animals. Again, there was a difference between the two groups of animals, with the amount of dilation being greater in the normotensive than in the hypertensive animals.

The isoflurane concentration in blood resulting from 1.5% inhaled isoflurane was 0.75 ± 0.04 m*M*. The corresponding postisoflurane washout concentration was not significantly different from zero (0.01 ± 0.00 m*M*. These concentrations represent means of data pooled from all animals studied ($n = 16$).

VII. Discussion

These studies demonstrate that significant differences exist in the reflex regulation of mesenteric capacitance veins in chronic renal wrap hypertensive versus normotensive rabbits. Prior to the administration of isoflurane, the baroreflex and hypoxia-mediated reflex blood pressure increase and venoconstriction were similar to both groups of animals. During 1.5% inhaled isoflurane, the baroreflex and the hypoxia-mediated reflex mesenteric venoconstrictions were preserved in hypertensive animals, whereas these responses were significantly attenuated by isoflurane in normotensive animals. In contrast, reflex mesenteric venodilation in response to aortic depressor nerve stimulation was attenuated in both hypertensive and normotensive animals during 1.5% inhaled isoflurane. Likewise, the systemic (i.e., blood pressure) response to each of the three reflex stimuli (carotid occlusion, aortic depressor nerve stimulation, and graded hypoxia) was significantly attenuated by the administration of 1.5% inhaled isoflurane in both hypertensive and normotensive animals. The heart rate response to aortic depressor nerve stimulation and to graded hypoxia was also equally attenuated by isoflurane in both groups of animals. The two groups of animals also differed significantly in the baseline (i.e., prestimulation) responses to isoflurane administration (i.e., differences in heart rate, blood pressure, and mesenteric vein diameter measurements before and during 1.5% inhaled isoflurane but immediately prior to carotid occlusion, aortic nerve stimulation, or graded hypoxia). In both groups of animals, 1.5% inhaled isoflurane significantly reduced heart rate and mean arterial pressure and dilated mesenteric capacitance vein diameter. However, in the hypertensive animals, the isoflurane-mediated reduction in arterial pressure was significantly greater and the isoflurane-mediated increase in mesenteric vein diameter was significantly less than that observed in the normotensive animals. As discussed below, these differences between hypertensive and normotensive animals during isoflurane administration in both reflex responses and prestimulation (i.e., baseline) measurements support the hypothesis for a higher level of sympathetic neural regulation in chronic hypertension.

A substantial amount of literature (including reports from our laboratory) exists to demonstrate conclusively that changes in the diameter of mesenteric capacitance veins, as well as changes in systemic variables (i.e., heart rate and blood pressure during decreased carotid sinus pressure or aortic depressor nerve stimulation), are neurally mediated reflex responses (13,15,26,27). Baroreflex-mediated changes in mesenteric arterial tone will alter mesenteric venous inflow. As such, these alterations can and do produce passive changes in mesenteric capacitance vein diameter which are of hemodynamic significance (12,28). Despite this fact, the predominance of active venoconstriction in the mesenteric bed has been demonstrated by several observations in previous studies utilizing a preparation identical to that employed in the current study. First, during carotid sinus-mediated mesenteric venoconstriction, mesenteric venous pressure (measured in the same vessel segment) increased, whereas during reflex mesenteric venodilation this pressure decreased (15–17). Such pressure responses are indicative of active (i.e., sympathetic neurally mediated) reflex diameter changes as opposed to passive changes attributable to changes in transmural distending pressure. Second, these carotid sinus-initiated mesenteric vein diameter changes are completely blocked by local application of tetrodotoxin to the mesenteric vein preparation. This agent is a selective blocker of sodium channels in nerve fiber membranes and thus effectively abolishes neuronal conduction (29). The inhibition of venous reflexes following topical administration of tetrodotoxin strongly suggests a sympathetic neurally mediated regulation. Third and most importantly we have observed in previous studies that sympathetic efferent nerve activity to the mesentery increases simultaneously with carotid sinus-mediated reflex mesenteric venoconstriction and conversely decreases simultaneously with reflex mesenteric venodilation (16,17).

Just as carotid sinus-initiated mesenteric vein diameter changes reflect an algebraic summation between active and passive forces, the hypoxia-initiated changes in heart rate, blood pressure, and mesenteric vein diameter observed in the present group of studies also reflect the net result of an interaction between hypoxia-initiated neural activation of these variables and the direct effects of hypoxia on cardiovascular tissues (30). During controlled ventilation, hypoxia produces a reflex bradycardia which is the result of a predominant chemoreflex vagally mediated mechanism (31) coupled to a direct myocardial depression resulting from the lowered p_{O_2} (32). However, when hypoxia produces hyperventilation, increased activation of pulmonary stretch receptors will inhibit vagally mediated bradycardia resulting in a hypoxia-induced tachycardia (31). Likewise, under other experimental conditions, chemoreflex-mediated sympathetic tachycardia, possibly coupled with a vagally mediated tachycardia, can be made to predominate (33,34). The hypoxia-mediated changes in blood pressure and mesenteric vein diameter observed in the present group of studies are also the net result of opposing forces acting on the peripheral vasculature. Systemic hypoxia produces a sympathetic neurally mediated increase in peripheral vascular resistance as well as a reduction in abdominal vascular capacitance (35). This produces the reflex mesenteric venoconstriction as well as the systemic hypertension that we observed. However, this hypoxia-mediated sympathetic activation is apparently unevenly distributed (36). In coronary, cerebral, and cutaneous beds, direct vasodilatory effects of hypoxia predominate over sympathetic neurally mediated vasoconstriction (30). Thus, during hypoxia, the sympathetic neurally mediated increase in peripheral vascular resistance and decrease in abdominal vascular capacitance [the major changeable component of the capacitance systems (12)] produces a redistribution of the unstressed blood volume to the relatively vasodilated cardiac and cerebral beds. Presumably this represents a mechanism for optimizing and sustaining oxygenation of the relatively metabolically active vital tissues during conditions of hypoxic stress (30).

Reflex sympathetic responses such as those described above are clearly recognized as being enhanced in the clinical situation in hypertensive patients undergoing anesthesia (1–3). Experimentally, the most impressive aspect of hypertension seems to be the exceptional size of the literature and the number of different models that have emerged over the past 60 years to explain the pathophysiology of this condition, particularly essential hypertension. These models can be broadly categorized into different groups such as renal, vascular, genetic, endocrine, and neurogenic (37). Each of these groups emphasizes a particular physiological aberration which is felt to be of critical importance in the onset and/or maintenance of the hypertensive condition. There is evidence to support each of the different types of models in explaining particular aspects of hypertension in the clinical setting. However, no single model or group of models comprehensively explains the pathophysiology of essential hypertension.

The complexity of the problem is further enhanced by the high degree of overlap between different models, such that abnormalities characteristic of one form of experimental hypertension are frequently found to varying degrees in other forms. The existence of such convincing and yet often contradictory data has produced numerous controversies, primarily concerning the cause and effect relationships in hypertension. Given the current level of understanding and the state of the knowledge in this field, perhaps the best recommendation has come from Page, who suggested that essential hypertension should be regarded as a condition with multiple causes each with varying levels of importance in individual cases. (37).

There are even controversial issues within similar types of experimental models of hypertension. The renal wrap model utilized in the present study is a modification of the kidney wrap renal model of hypertension developed by Page in 1939 (38). This model, which was originally designed to produce hypertension in experimental animals over several weeks in order to mimic the more typical long-term development of essential hypertension in humans, was considered to be more appropriate than the renal artery occlusion models which produce hypertension within several hours (37). Plasma renin and angiotensin II levels were not measured in the current study; however, these levels are reported to be elevated in the renal wrap model (39). Despite this elevation, there does not appear to be complete agreement as to the role of renin and angiotensin II in maintaining an elevated blood pressure in renal wrap hypertension. Denton and Anderson concluded that angiotensin II had major effects in producing the blood pressure increase in this form of hypertension (40). In contrast, Fletcher *et al.* reported just the opposite (41).

Perhaps one of the most controversial issues in the study of hypertension is the contribution of the sympathetic nervous system not only to development but also to the maintenance of a sustained increase in blood pressure. A comprehensive review of the causes of hypertension is beyond the scope of this article, particularly since no single pathophysiological aberration can be invoked to explain the initiation and maintenance of essential hypertension. Nevertheless, a large body of evidence in support of the sympathetic component (be it as a cause or an effect) in many if not most forms of hypertension cannot be denied (42–44). Even with the kidney wrap procedure utilized in the present study (which is basically regarded as a renal model of hypertension) there is evidence that augmented sympathetic nervous system function is significantly involved in the maintenance of sustained elevated blood pressure (45). Furthermore, in the clinical setting it has been recognized by anesthesiologists for years that the interoperative course of patients with essential hypertension is associated with more circulatory instability than normal patients and exaggerated hemodynamic responses that are characteristic of enhanced sympathetic neural activity. Thus, the present study of the interaction between a volatile anesthetic and the hypertensive condition focused on changes in reflex responses that are mediated primarily by an elevation of sympathetic neural efferent activity.

We observed an inhibition of baroreflex and chemoreflex changes in heart rate, blood pressure, and mesenteric vein diameter during the administration of 0.75 MAC isoflurane in normotensive control animals. Such inhibition is in agreement with what we reported in previous studies of normal animals (17). The mechanism by which such inhibition occurs is a more complicated issue. Seagard *et al.* examined the effects of halothane and isoflurane on baroreflex responses and concluded that inhibition occurs at multiple sites along the reflex pathway. (46,47). In other studies reported from our laboratory isoflurane has been shown to inhibit directly membrane ion channel activity in vascular smooth muscle cells from coronary arteries (48). Also, using *in vitro* vessel ring preparations taken from mesenteric veins similar to those utilized in the present study, other investigators in our laboratory have observed that isoflurane directly inhibits contractile responses to perivascular nerve stimulation as well as to exogeneously administered norepinephrine (49). In addition, isoflurane has been shown to inhibit the relaxation effect of endothelium-derived nitric oxide on vascular smooth muscle cells in *in vitro* mesenteric vein ring preparations (50). Initially, this might be regarded as a contradiction to the isoflurane-mediated inhibition of venoconstriction that we observed in the present study. However, the possibility still exists that volatile anesthetics might alter the *in situ* release and/or actions of nonadrenergic noncholinergic mediators (such as those released from endothelial cells or from nitroxidergic neurons). Such alterations may account for some of the inhibitory actions of isoflurane that we have observed in the present study.

In our earlier study (17) we concluded that isoflurane inhibited baroreflex-mediated mesenteric venoconstriction at a site proximal to the postganglionic neuron. This was based on the observation that systemically administered (i.e., inhaled) isoflurane inhibited venoconstriction to reflex stimuli but had no effect on the venoconstriction response to direct sympathetic postganglionic stimulation. Likewise, superfused isoflurane (i.e., applied directly to the vein segments being measured) had no effect on the diameter response to either baroreflex or sympathetic postganglionic stimulation, suggesting that, in that preparation, the direct vascular inhibitory effects of the anesthetic were less significant. Finally, Ponte and Sadler reported that isoflurane inhibited reflex responses to hypoxia by direct action at the level of the carotid body receptor organ (51).

In the hypertensive animals, the inhibitory effect of isoflurane on the heart rate responses to chemoreflex activation, on the blood pressure responses to both baroreflex and chemoreflex activation, and on the mesenteric venodilation in response to aortic depressor nerve stimulation was the same as in the normotensive control animals. In sharp contrast, reflex mesenteric venoconstriction in the hypertensive animals in response to bilateral carotid occlusion and hypoxia persisted throughout the course of inhaled isoflurane administration. Because such reflex-mediated constriction as well as the isoflurane-mediated inhibition of this constriction in the current preparation (as discussed in detail above) primarily involves autonomic mechanisms, we conclude that in the hypertensive animals, sympathetic regulation of mesenteric venous tone is enhanced to a level that negates the inhibitory effects of isoflurane that we observed in the normotensive animal. Such a mechanism, if true, would require that in the hypertensive animal a disproportionate enhancement of sympathetic neural input would occur to the mesenteric capacitance vessels. The autonomic regulation of heart rate and mean arterial pressure (the latter predominantly via increased neural tone in resistance vessels) would be similar to that in the normotensive animal. In fact, some evidence exists to support such redistribution of sympathetic neural activity even in renal models of hypertension. Edmunds *et al.* observed that decreased vascular capacitance was associated with the maintenance of elevated blood pressure in two-kidney, one-clip hypertension. Following clip removal a reversal of the capacitance decrease occurred as reflected by an increase in the unstressed vascular volume and an associated fall in blood pressure (52). Additional support for an enhanced (possibly neurogenic) tone of the capacitance veins, at least in volume-expanded models of hypertension, can be based on the following argument. Volume expansion without an increased venous tone to decrease (or at least maintain) vascular capacity, and hence respectively increase (or at least maintain) venous return and cardiac output, would not maintain an increase in arterial blood pressure. The unstressed vascular volume would simply increase (53).

In the present group of studies, the effects of isoflurane on baseline (i.e. prestimulation) mesenteric vein diameter, heart rate, and mean arterial blood pressure were consistent with its inhibitory effects on reflex responses discussed above. Heart rate and mean arterial blood pressure decreased in both hypertensive and normotensive animals during isoflurane administration. However, the fall in blood pressure was greater in the hypertensive group. As discussed above, this is certainly consistent with the circulatory instability that characterizes anesthetic administration in hypertensive patients in the clinical setting (5,6). Also, baseline mesenteric vein diameter was increased in both hypertensive and normotensive animals, but the dilation was less in the hypertensive group. This is certainly in agreement with the relative increase in mesenteric venous tone and its proposed resistance to an isoflurane-mediated inhibition in the hpertensive animal. In contrast to what was observed in conscious (unanesthetized) hypertensive rabbits (see Chapter 27) resting heart rates in the present group of studies was greater in hypertensive than in normotensive animals. This suggests there was an anesthetic effect, probably vagolytic. The fact that resting heart rate was greater in hypertensive than in normotensive animals in the present study could also be a reflection of greater resistance to sympathetic inhibition under anesthesia. This may also account for the lack of reflex tachycardia in response to carotid occlusion in the hypertensive animals. If the heart rate was already close to maximial, a further sympathetic stimulation may be unable to further increase heart rate and may actually tend to cause some compensatory slowing because of the simultaneous increase in arterial pressure.

Finally, it should be noted that the action of different anesthetic agents on sympathetically mediated reflex responses, particularly reflex regulation of venous tone, may vary. In preliminary data (not presented in this article) we have observed that venous reflex responses do not appear to be resistant to inhibition by halothane as they are to isoflurane. Despite the fact that volatile anesthetics tend to be grouped in one pharmacological classification, there are differences in cardiovascular effects among the different agents, and it is entirely possible if not likely that halothane could cause greater inhibition of sympathetic control of the vasculature than isoflurane (54).

We should also note that there tended to be a lack of recovery in reflex responses following the administration of 1.5% isoflurane in both the normotensive and the hypertensive animals. Although we have not discussed this in detail, such persistent effects of anesthetics on vascular responses are consistent with what has been observed previously in our laboratory (18,47), as well as others (55).

In summary, we have presented evidence to indicate that the responses measured in the present group of studies were largely the result of sympathetically mediated reflexes. Also, we used a nephrogenic experimental model of hypertension to demonstrate that, compared with the normotensive control condition, sympathetic reflex responses on the arterial side of the circulation (i.e., heart rate and especially blood pressure changes) tended to be similar, but reflex control of mesenteric capacitance veins was greater and more resistant to isoflurane-mediated inhibition. It is possible that similar changes may be present in the clinical setting in patients with essential hypertension (as compared with normotensive patients), and if so this could, in part, account for some of the hemodynamic abnormalities that are associated with the administration of anesthesia in these individuals.

Acknowledgments

The authors thank Anita Tredeau for assistance with preparation of this article. This work was supported by Veterans Affairs Medical Research Funds, U.S. Public Health Service Grant HL 01901, and Anesthesiology Research Training Grant GM 08377.

References

1. Longnecker, D. E. (1987). Alpine anesthesia: Can pretreatment with clonidine decrease the peaks and valleys? *Anesthesiology* **67,** 1–2.
2. Estafanous, F. G. (1989). Hypertension in the surgical patient: Management of blood pressure and anesthesia. *Cleveland Clin. J. Med.* **56,** 385–393.
3. Stone, J. G., Foëx, P., Sear, J. W., Johnson, L. L., Khambatta, H. J., and Triner, L. (1988). Risk of myocardial ischaemia during anaesthesia in treated and untreated hypertensive patients. *Br. J. Anaesth.* **61,** 675–679.
4. Ross, A. F., and Tinker, J. H. (1992). "Cardiovascular Disease, Risk and Outcome in Anesthesia" (D. L. Brown, ed.), 2nd ed, pp. 39–76. Lippincott, Philadelphia, Pennsylvania.
5. Stoelting, R. K., and Dierdorf, S. F. (1993). Hypertension. *In* "Anesthesia and Coexisting Disease," 3rd ed, pp. 79–86. Churchill Livingstone, New York.
6. Goldman, L., and Caldera, D. L. (1979). Risks of general anesthesia and elective operation in the hypertensive patient. *Anesthesiology* **50,** 285–292.
7. Nilsson, H., and Folkow, B. (1982). Vasoconstriction nerve influence on isolated mesenteric vessels from normotensive and spontaneously hypertensive rats. *Acta Physiol. Scand.* **116,** 205–208.
8. Bevan, R. D., Purdy, R. E., Su, C., and Bevan, J. A. (1975). Evidence for an increase in adrenergic nerve function in blood vessels from experimental hypertensive rabbits. *Circ. Res.* **37,** 503–508.
9. Stekiel, W. J., Contney, S. J., and Lombard, J. H. (1991). Sympathetic neural control of vascular muscle in reduced renal mass hypertension. *Hypertension* **17,** 1185–1191.
10. Stekiel, W. J., Contney, S. J., and Lombard, J. H. (1986). Small vessel membrane potential, sympathetic input, and electrogenic pump rate in SHR. *Am. J. Physiol.* **250,** C547–C556.
11. Rothe, C. F. (1983). Venous system: Physiology of the capacitance vessels. *In* "Handbook of Physiology, Section II: The Cardiovascular System" (J. T. Shepherd and F. M. Abboud, vol. eds.), Vol. 3 (Peripheral Circulation, Part I), pp. 397–452. American Physiological Society, Bethesda, Maryland.
12. Hainsworth, R. (1990). The importance of vascular capacitance in cardiovascular control. *News Physiol. Sci.* **5,** 251–254.
13. Hainsworth, R., and Karim, F. (1976). Responses of abdominal vascular capacitance in the anesthetized dog to changes in carotid sinus pressure. *J. Physiol. (London)* **262,** 659–677.

14. Ford, R., Hainsworth, R., Rankin, A. J., and Soladoye, A. O. (1985). Abdominal vascular responses to changes in carbon dioxide tension in the cephalic circulation of anaesthetized dogs. *J. Physiol (London)* **358,** 417–431.
15. Ozono, K., Bosnjak, Z. J., and Kampine, J. P. (1989). Reflex control of mesenteric vein diameter and pressure *in situ* in rabbits. *Am. J. Physiol.* **256,** H1066–H1072.
16. Stekiel, T. A., Ozono, K., McCallum, J. B., Bosnjak, Z. J., Stekiel, W. J., and Kampine, J. P. (1990). The inhibitory action of halothane on reflex constriction in mesenteric capacitance veins. *Anesthesiology* **73,** 1169–1178.
17. McCallum, J. B., Stekiel, T. A., Bosnjak, Z. J., and Kampine, J. P. (1993). Does isoflurane alter mesenteric venous capacitance in the intact rabbit? *Anesth. Analg. (N.Y.)* **76,** 1095–1105.
18. Stekiel, T. A., Tominaga, M., Bosnjak, Z. J., and Kampine, J. P. (1992). The inhibitory effect of halothane on mesenteric venoconstriction and related reflex responses during acute graded hypoxia in rabbits. *Anesthesiology* **77,** 709–720.
19. Miller, E. D. Jr., Beckman, J. J., and Althaus, J. S. (1985). Hormonal and hemodynamic responses to halothane and enflurane in spontaneously hypertensive rats. *Anesth. Analg. (N. Y.)* **64,** 136–142.
20. Seyde, W. C., Durieux, M. E., and Longnecker, D. E. (1987). The hemodynamic response to isoflurane is altered in genetically hypertensive (SHR), as compared with normotensive (WKY), rats. *Anesthesiology* **66,** 798–804.
21. Alexander, N., and DeCuir, M. (1966). Loss of baroreflex bradycardia in renal hypertensive rabbits. *Circ. Res.* **19,** 18–25.
22. Drummond, J. C. (1985). MAC for halothane, enfluran, and isoflurane in the New Zealand White rabbit; and a test for the validity of MAC determinations. *Anesthesiology* **62,** 336–338.
23. Bohlen, H. G. (1980). Intestinal tissue P_{O_2} and microvascular responses during glucose exposure. *Am. J. Physiol.* **238,** H164–H171.
24. Holzgrefe, H. H., Everitt, J. M., and Wright, E. M. (1987). Alpha-chloralose as a canine anesthetic. *Lab. Anim. Sci.* **37,** 587–595.
25. Bell, L. B., Hopp, F. A., Seagard, J. L., von Brederode, H. F. M., and Kampine, J. P. (1988). A continuous noncontact method for measuring *in situ* vascular diameter with a video camera. *J. Appl. Physiol..* **64,** 1279–1284.
26. Shoukas, A. A., and Sagawa, K. (1973). Control of total systemic vascular capacity by the carotid sinus baroreceptor reflex. *Circ. Res.* **33,** 22–33.
27. Hainsworth, R. (1986). Vascular capacitance: Its control and importance. *Rev. Physiol. Biochem. Pharmacol.* **105,** 102–173.
28. Greenway, C. V. (1983). Role of splanchnic venous system in overall cardiovascular homeostasis. *Fed. Proc.* **42,** 1678–1684.
29. Gershon, M. D. (1967). Effects of tetrodotoxin on innervated smooth muscle preparations. *Br. J. Pharmacol. Chemother.* **29,** 259–279.
30. Heistad, D. D., and Abboud, F. M. (1980). Circulatory adjustments to hypoxia. *Circulation* **61,** 463–470.
31. Daly, M. de B. (1985). Chemoreceptor reflexes and cardiovascular control. *Acta Physiol. Pol.* **36,** 4–20.
32. Marijic, J., Stowe, D. F., Turner, L. A., Kampine, J. P., and Bosnjak, Z. J. (1990). Differential protection effects of halothane and isoflurane against hypoxic and reoxygenation injury in the isolated guinea pig heart. *Anesthesiology* **73,** 976–983.
33. Marshall, J. M. (1987). Analysis of cardiovascular responses evoked following changes in peripheral chemoreceptor activity in the rat. *J. Physiol. (London)* **394,** 393–414.
34. Katzin, D. B., and Rubinstein, E. H. (1976). Reversal of hypoxic bradycardia by halothane or midcollicular decerebration. *Am. J. Physiol.* **231,** 179–184.

35. Hainsworth, R., Karim, F., McGregor, K. H., and Wood, L. M. (1983). Responses of abdominal vascular resistance and capacitance to stimulation of carotid chemoreceptors in anaesthetized dogs. *J. Physiol. (London)* **334,** 409–419.
36. Pelletier, C. L. (1972). Circulatory responses to graded stimulation of the carotid chemoreceptors in the dog. *Circ. Res.* **31,** 431–443.
37. Page, I. H. (1987). Overview. *In* "Hypertensive Mechanisms," Chap. 18, pp. 95–103. Grune & Stratton, Orlando, Florida.
38. Page, I. H. (1939). The production of persistent arterial hypertension by cellophane perinephritis. *J. Am. Med. Assoc.* **113,** 2046–2048.
39. Anderson, W. P., Woods, R. J., Denton, K. M., and Alcorn, D. (1987). Renal actions of angiotensin II in renovascular hypertension. *Can. J. Physiol. Pharmacol.* **65,** 1559–1565.
40. Denton, K. M., and Anderson, W. P. (1985). Role of angiotensin II in renal wrap hypertension. *Hypertension* **7,** 893–898.
41. Fletcher, P. J., Korner, P. I., and Angus, J. A. (1976). Changes in cardiac output and total peripheral resistance during development of renal hypertension in the rabbit. *Circ. Res.* **39,** 633–639.
42. Abboud, F. M. (1982). The sympathetic system in hypertension. State-of-the-art review. *Hypertension* **4**(*Suppl. 2), II-208–II-225.*
43. Mathias, C. J. (1991). Role of sympathetic efferent nerves in blood pressure regulation and in hypertension. *Hypertension* **18**(Suppl. 3), III-22–III-30.
44. Dzau, V. J. (1990). Sympathetic nervous system: The possible pathophysiologic link among the cardiovascular risk factors. *J. Vasc. Med. Biol.* **2,** 219–224.
45. Hinojosa-Laborde, C., Guerra, P., and Haywood, J. R. (1992). Sympathetic nervous system in high sodium one-kidney, figure-8 renal hypertension. *Hypertension* **20,** 96–102.
46. Seagard, J. L., Hopp, F. A., Donegan, J. H., Kalbfleisch, J. H., and Kampine, J. P. (1982). Halothane and the carotid sinus reflex: Evidence for multiple sites of action. *Anesthesiology* **57,** 191–202.
47. Seagard, J. L., Elegbe, E. O., Hopp, F. A., Bosnjak, Z. J., von Colditz, J. H., Kalbfleisch, J. H., and Kampine, J. P. (1983). Effects of isoflurane on the baroreceptor reflex. *Anesthesiology,* **59,** 511–520.
48. Buljubasic, N., Rusch, N., Marijic, J., Kampine, J. P., and Bosnjak, Z. J. (1992). Effects of halothane and isoflurane on calcium and potassium channel currents in canine coronary arterial cells. *Anesthesiology* **76,** 990–998.
49. Stadnicka, A., Flynn, N. M., Bosnjak, A. J., and Kampine, J. P. (1993). Enflurane, halothane, and isoflurane attenuate contractile responses to exogenous and endogenous norepinephrine in isolated small mesenteric veins of the rabbit. *Anesthesiology* **78,** 326–334.
50. Stone, D. J., and Johns, R. A. (1989). Endothelium-dependent effects of halothane, enflurane and isoflurane on isolated rat aortic vascular rings. *Anesthesiology* **71,** 126–132.
51. Ponte, J., and Sadler, C. L. (1989). Effect of halothane, enflurane and isoflurane on carotid body chemoreceptor activities in the rabbit and the cat. *Br. J. Anaesth.* **62,** 33–40.
52. Edmunds, M. E., Russell, G. L., and Swales, J. D. (1989). Vascular capacitance and reversal of 2-kidney, 1-clip hypertension in rats. *Am. J. Physiol.* **256,** H502–H507.
53. Ferrario, C. M., and Carretero, O. A. (1984). Hemoydnamics of experimental renal hypertension. *In* "Handbook of Hypertension, Volume 4: Experimental and Genetic Models of Hypertension." (W. de Jong, ed., pp. 54–80. Elsevier, New York.
54. Eger II, E. I. (1981). Isoflurane: A review. *Anesthesiology* **55,** 559–576.
55. Spiss, C. K., Smith, C. M., Tsujimoto, G., Hoffman, B. B., and Maye, M. (1985). Prolonged hyporesponsiveness of vascular smooth muscle contraction after halothane anesthesia in rabbits. *Anesth. Analg. (N. Y.)* **64,** 1–6.

Effect of Volatile Anesthetics on Baroreflex Control of Mesenteric Venous Capacitance

J. Bruce McCallum, Thomas A. Stekiel, Anna Stadnicka, Zeljko J. Bosnjak, and John P. Kampine

Department of Anesthesiology
The Medical College of Wisconsin
Milwaukee, Wisconsin 53226

I. Introduction

The capacity of mesenteric veins to actively constrict and dilate in response to baroreflex stimulation is an important mechanism for the control of hemodynamic changes (1–7). Ozono *et al.* used videomicroscopy and direct intravenous pressure measurements to demonstrate the neurally mediated reflex control of mesenteric venous capacitance for discrete pressure–diameter relationships (8). Subsequently, this technique was used to demonstrate neural control of mesenteric venous capacitance and compliance over the entire physiological range of distending pressures (9). Our overall hypothesis in these studies was that volatile anesthetics increase venous capacitance by reducing neurally mediated venous tone (10–12). To test this hypothesis, we used the method of Ozono *et al.* (8) to determine the influence of volatile anesthetics on the reflex capability of mesenteric capacitance veins to actively constrict and dilate in response to changes in mesenteric sympathetic efferent nerve activity produced by barostatic reflex activation or direct electric stimulation.

Advances in Pharmacology, Volume 31

II. Measurement of Venous Capacitance

Mesenteric venous capacitance was determined by measuring the changes in vein diameter and intravenous pressure in correlation with the changes in sympathetic efferent nerve activity according to the methods described previously (8,13,14). In brief, after institutional Animal Care Committee approval, 93 male New Zealand White rabbits, aged 6–8 weeks, with an average weight of 1–2 kg, were studied after a 24 hr water-only fast. After induction of anesthesia, an arterial catheter placed in the femoral artery was connected to a pressure transducer for aortic pressure measurements, and heart rate was electronically derived from the time intervals between peak systoles. A steady plane of anesthesia was achieved by continuous infusion of α-chloralose (15–20 mg/kg/hr) through the venous catheter. Ventilation was controlled using a forced air ventilator supplemented with 0.25 liter/min O_2. Normal pH was maintained by continuous infusion of $NaHCO_3$. Core temperature was kept at 36.5–37.5°C. In one group of 43 rabbits, the carotid arteries were dissected free for occlusion, and the aortic depressor nerves of 21 rabbits from this group were exposed bilaterally for electric stimulation. In 18 rabbits from this group, a postganglionic mesenteric nerve derived from the celiac ganglion was detached from the surrounding tissue, and bipolar recording electrodes, composed of two stainless steel wires in silastic tubes, were fixed to the intact nerve with Wacker SilGel 604A (Wacker, Munich, Germany) in order to obtain simultaneous measurements of efferent nerve activity and mesenteric vein responses. In a second group of 38 rabbits, bipolar silver stimulating electrodes were sewn distal to the celiac ganglion, thereby making it possible to activate directly the sympathetic postganglionic nerve fibers leading to the mesenteric veins.

Our study of the changes in mesenteric vein diameter was conducted on a 13-cm loop of terminal ileum externalized through a laparotomy incision and mounted in a temperature regulated plastic perfusion chamber placed on a movable microscope stage. The ileal loop was continuously superfused with a physiologic salt solution originally formulated by Bohlen (15) to simulate the peritoneal environment, and it was kept at pH 7.4 ± 0.05 by continuous bubbling with a gas mixture of 5% O_2, 5% CO_2, and 90% N_2 (by volume) in correspondence with conditions in the peritoneal cavity. The mesentery was carefully pinned to the chamber floor and transilluminated with a fiber-optic lamp. Vein diameter was continually monitored through a videomicrometer system composed of a Reichert Stereo Star Zoom microscope (Cambridge Instruments, Buffalo, NY) and an on-line digital-to-analog converter which produced an analog signal proportional to the vessel diameter. Vein pressure measurements were obtained with a servo-null bellows which achieved equilibrium at the tip of a glass micropipette penetrating the vein wall. Compensations in resistance were converted to changes in pressure by a strain gauge bridge transducer. Sympathetic efferent nerve activity (SENA) was measured using a high-impedance differential preamplilfier and a voltage-to-frequency converter. Nerve activity was analyzed in terms of the summation of changes in amplitude, frequency, and duration of individual depolarizations within a segment of multifiber nerve tissue. At the end of the study, tetrodotoxin (1.5 μM) was applied topically to block the local vein smooth muscle responses to neural activation. Thereafter, a zero level reference baseline for nerve activity was obtained by blockade of ganglionic transmission with hexamethonium (10 mg/kg i.v.), which abolished all efferent activity.

Aortic pressure, heart rate, vein diameter, vein pressure, and SENA were recorded on a Vetter Model 820 digital videocassette recorder (Vetter, Rebersberg, PA) and subsequently printed on an eight-channel strip-chart recorder. End-tidal gases were monitored with an anesthesia/respiratory monitoring system specifically calibrated for the volatile anesthetics used in this study, and samples of anesthetic concentrations in blood and superfusate were obtained for subsequent analysis by gas chromatography.

Two experimental protocols were used to determine the effect of anesthetics on baroreflex control of mesenteric venous capacitance. The first group, consisting of 43 rabbits, underwent baroreceptor stimulation. Baroreceptor stimulation consisted of bilateral carotid occlusion (BCO) for 30 sec and aortic depressor nerve stimulation (ANS) for 10 sec (0.5 mA, 20 Hz, 1-msec pulse). Thirty-three members of the group received baroreceptor stimulation (either BCO or ANS) during inhaled anesthetic administration. Halothane, isoflurane, and enflurane were delivered to different rabbits in this group in equipotent doses of 0.36 and 0.72 minimal alveolar concentration (MAC) in random order. In the rabbit, these doses are equivalent to vaporizer settings of 0.5 and 1.0% (v/v) for halothane, 0.75 and 1.5% (v/v) for isoflurane, and 1.0 and 2.0% (v/v) for enflurane (16). Anesthetic exposure lasted 30 min for each dose, with a recovery period of 1 hr between different doses. Baroreceptor stimulation was begun approximately 15 min after the onset of anesthetic administration when a steady state was observed in hemodynamic variables, and equilibration of end-tidal gases was observed on the mass spectrometer. Thirteen rabbits from this group underwent baroreceptor stimulation (either BCO or ANS) during administration of anesthetics in the superfusate. Halothane and isoflurane were equilibrated in the superfusate for 30 min at concentrations of 0, 3, and 0% and then delivered directly to the mesenteric vessels of different rabbits in the tissue bath for 15 min before initiating baroreceptor stimulation. A washout period of 1 hr was allowed between measurements.

The concentrations of anesthetics in the superfusate at these vaporizer settings were approximately the same as the amounts absorbed in the blood at the high doses as determined by gas chromatography. Five of the rabbits exposed to isoflurane received both inhaled and superfused isoflurane to eliminate differences between rabbits.

Another group of 38 rabbits was studied during direct electric stimulation of the postganglionic fibers of the celiac ganglion (CGS) under the same conditions of inhaled and superfused isoflurane administration. A frequency-dependent venoconstriction was elicited in response to 60-sec sequential stimulations of 5, 10, and 20 Hz (1-msec pulse width; 3–6 mA current) with a recovery period of 3–4 min between stimulations. These stimulation parameters are within the physiological range (17). Inhaled halothane or isoflurane was administered to 13 separate rabbits in equipotent doses of 0, 0.72, and 0 MAC. Halothane, isoflurane, and enflurane equilibrated in the superfusate was administered in sequential concentrations of 0, 5, and 0% to separate rabbits totalling 33 experiments from this group. The higher concentration of superfused anesthetics was required because electric stimulation of the postganglionic nerve tissue is more direct than physiological stimulation of the reflex arc. The time course was the same as above. The presence of postganglionic stimulation was verified by hexamethonium blockade at the end of the study. Five rabbits from the group exposed to isoflurane received both inhaled and superfused isoflurane to eliminate the possibility of differences between rabbits.

III. Response to Baroreceptor Stimulation

Under control conditions before anesthetic administration, BCO produced reflex increases in arterial blood pressure and heart rate, as well as reflex mesenteric venoconstriction and intravenous pressure increases. For example, mean arterial pressure in rabbits before exposure to isoflurane rose from 90 ± 4 to 115 ± 5 mmHg during BCO, a significant increase of 28% (Fig. 1A), whereas heart rate increased from 273 ± 8 to 228 ± 8 beats/min, a significant increase of 5.5% (Fig. 1B). Simultaneously, mesenteric vein diameter in all the rabbits constricted during BCO between 5 and 7% before anesthetic exposure (Fig. 2A), and intravenous pressure rose between 8 and 13% with decreased carotid sinus pressure during BCO as compared to the resting state before stimulation (Fig. 3A). Inhaled anesthetics equally attenuated the reflex increases in arterial blood pressure and heart rate, as well as the reflex mesenteric venoconstriction and intravenous pressure increases. However, volatile anesthetics equilibrated in the superfusate did not inhibit the reflex responses of mesenteric vein diameter and vein pressure to decreased carotid sinus pressure (Figs. 2B and 3B).

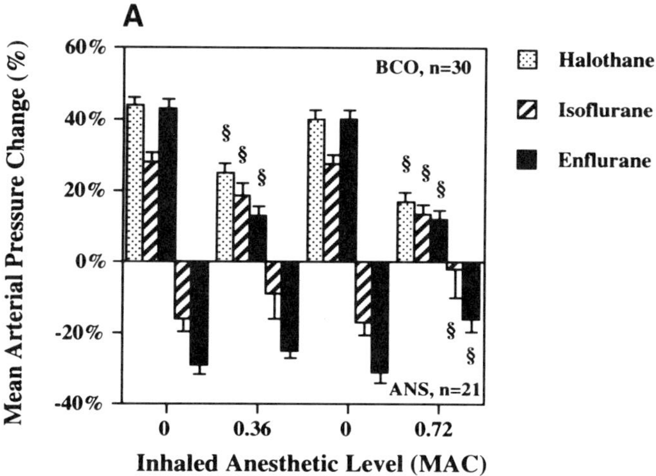

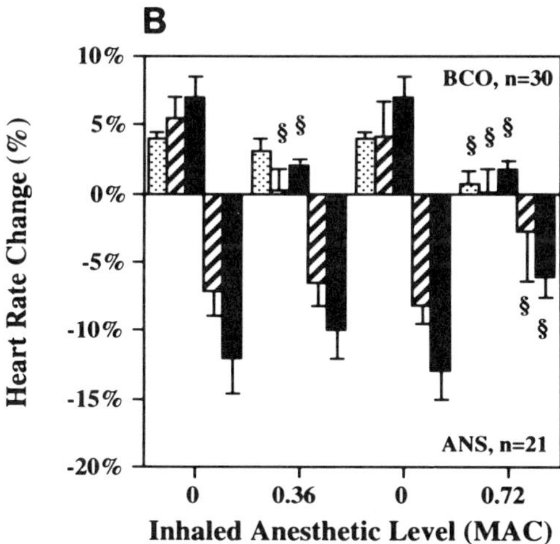

Fig. 1 (A) Significant attenuation of mean arterial pressure increase occurred in response to bilateral carotid occlusion by both doses of anesthetics, whereas only the high dose of anesthetics impaired the response to aortic nerve stimulation. (B) Both doses of isolfurane and enflurane (halothane only at the high dose) suppressed heart rate increase after bilateral carotic occlusion, whereas the high dose of anesthetics was required to suppress the bradycardia in response to aortic nerve stimulation. Each column represents the mean ± SEM. §Significant difference ($p \leq 0.05$) relative to 0 MAC inhaled anesthetics.

Aortic nerve stimulation produced a converse depression in arterial blood pressure and heart rate, whereas the mesenteric veins dilated and intravenous pressure dropped. For example, in rabbits exposed to isoflur-

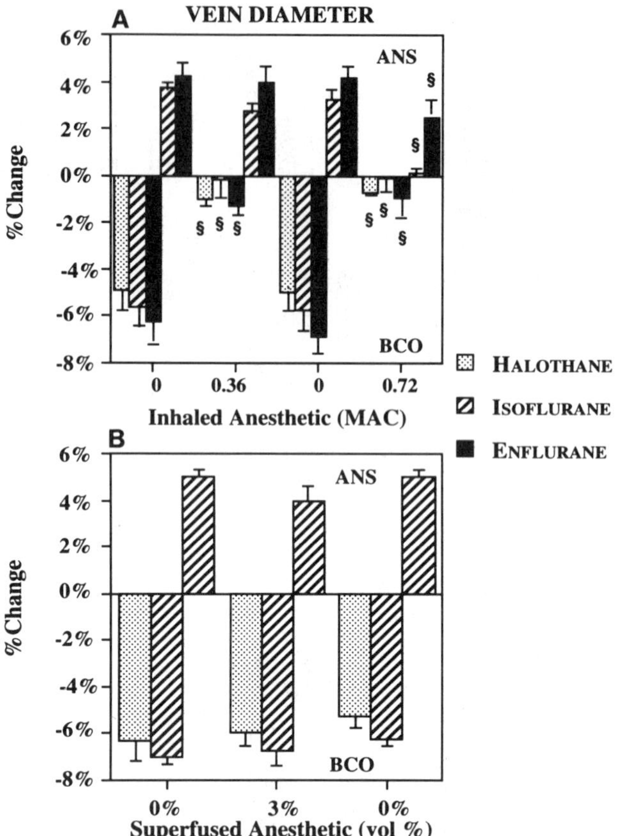

Fig. 2 Equal attenuation of vein diameter decrease in response to bilateral carotid occlusion occurred during both levels of inhaled anesthetics, whereas the vein diameter increase in response to aortic nerve stimulation was abolished by only the high dose of inhaled anesthetics. (B) Vein diameter responses to baroreceptor stimulation were unaffected by a 3% concentration of anesthetics in the superfusate. §Significant difference ($p \leq 0.05$) versus 0 MAC (n = 4–30) rabbits for each group).

ane, mean arterial pressure decreased from 83 ± 4 to 70 ± 4 mmHg, a significant difference of 15.6% (Fig. 1A), and heart rate decreased from 281 ± 6 to 261 ± 6 beats/min, a significant depression of 7.1% (Fig. 1B). Vein diameter increased around 4% in response to ANS (Fig. 2A), and vein pressure decreased between 4 and 8.5% compared to the resting state before stimulation (Fig. 3A). Only the high dose of 0.72 MAC inhaled isoflurane and enflurane attenuated the responses of mean arterial pressure, heart rate, and mesenteric vein diameter to ANS (Figs. 1 and 2A). A significant attenuation of the decrease in vein pressure was observed

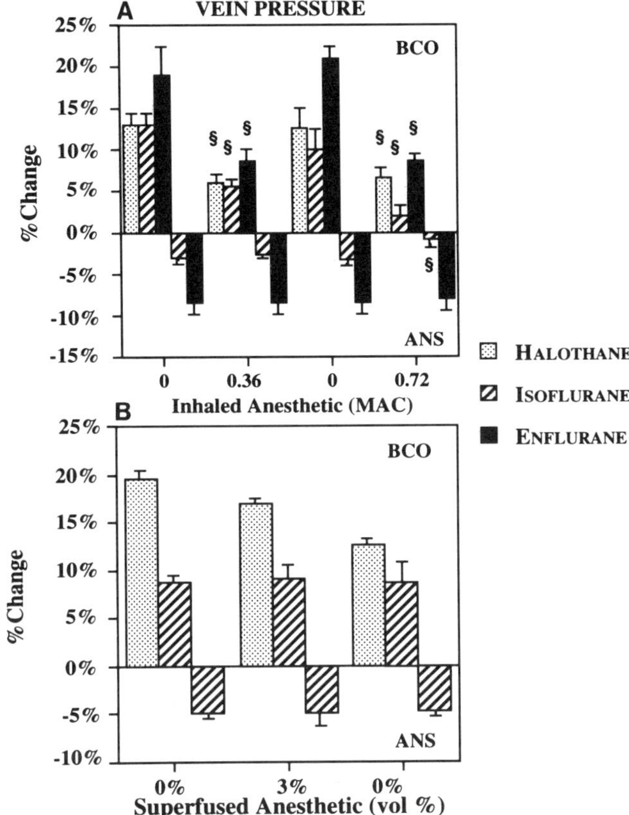

Fig. 3 Intravenous pressure increase in response to bilateral carotid occlusion was reduced by both doses of inhaled anesthetics, whereas the vein pressure decrease in response to aortic nerve stimulation was abolished by the high dose of enflurane and isoflurane alone. (B) Intravenous pressure responses to baroreceptor stimulations were unaffected by 3% superfused anesthetics. §Significant difference ($p \leq 0.05$) versus 0 MAC (n = 4–30 rabbits for each group).

in rabbits exposed to the high dose of inhaled isoflurane but not enflurane (Fig. 3A). However, isoflurane equilibrated in the superfusate did not attenuate the reflex response to stimulation of the aortic depressor nerve (Fig. 2B and 3B).

Bilateral carotid occlusion resulted in reflex increases in sympathetic efferent nerve activity of between 30 and 40% as compared to averaged nerve activity before stimulation. Both the low and high doses of volatile anesthetics significantly reduced the level of averaged nerve activity before stimulation, and the reflex increase in nerve activity in response to baroreceptor stimulation was abolished (Fig. 4). The most remarkable

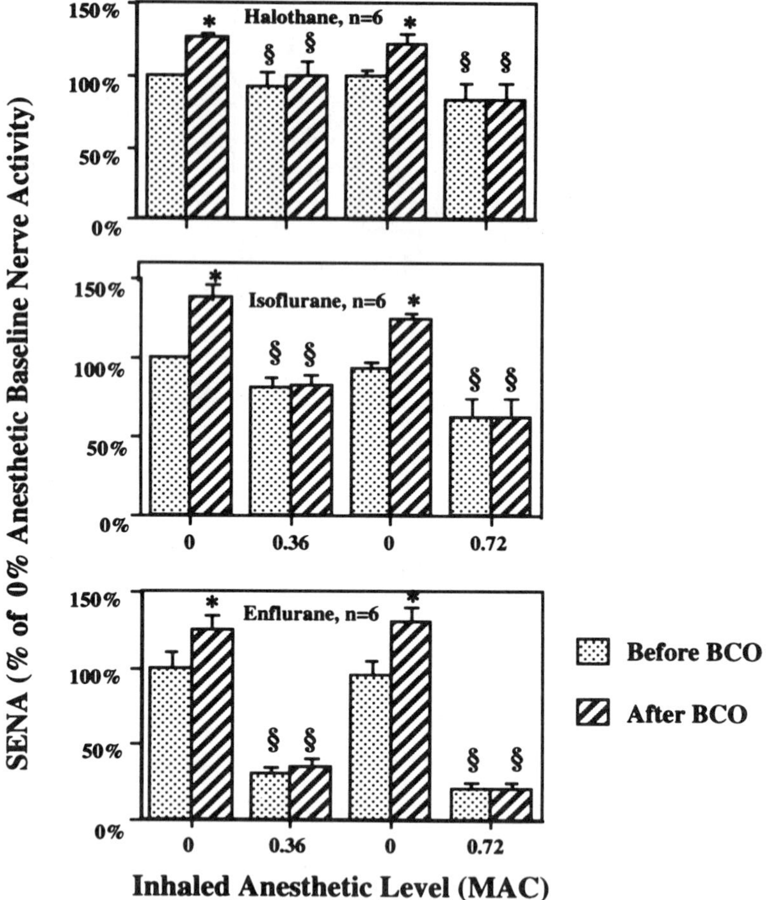

Fig. 4 Equal attenuation by both doses of anesthetics of the prestimulation nerve activity and the reflex increase in nerve activity after bilateral carotid occlusion. *Significant difference ($p \leq 0.05$) in average nerve activity during bilateral carotid occlusion versus prestimulation average activity. §Significant difference ($p \leq 0.05$) versus the preceding 0 MAC dose.

inhibition was observed in the presence of enflurane. Resting splanchnic nerve activity was reduced approximately 70–80% in the presence of both 0.36 and 0.72 MAC inhaled enflurance. In contrast to enflurane, however, halothane and isoflurane demonstrated more dose-dependent differences. Inhaled isoflurane at the low dose of 0.36 MAC inhibited sympathetic efferent nerve activity by 20% and 0.72 MAC reduced SENA by 40%, whereas inhaled halothane at the low and high doses attenuated SENA by 10 and 30%, respectively.

Figure 5 illustrates the typical response of vein diameter and intravenous pressure to BCO during 0% inhaled isoflurane, 1.5% inhaled isoflurane,

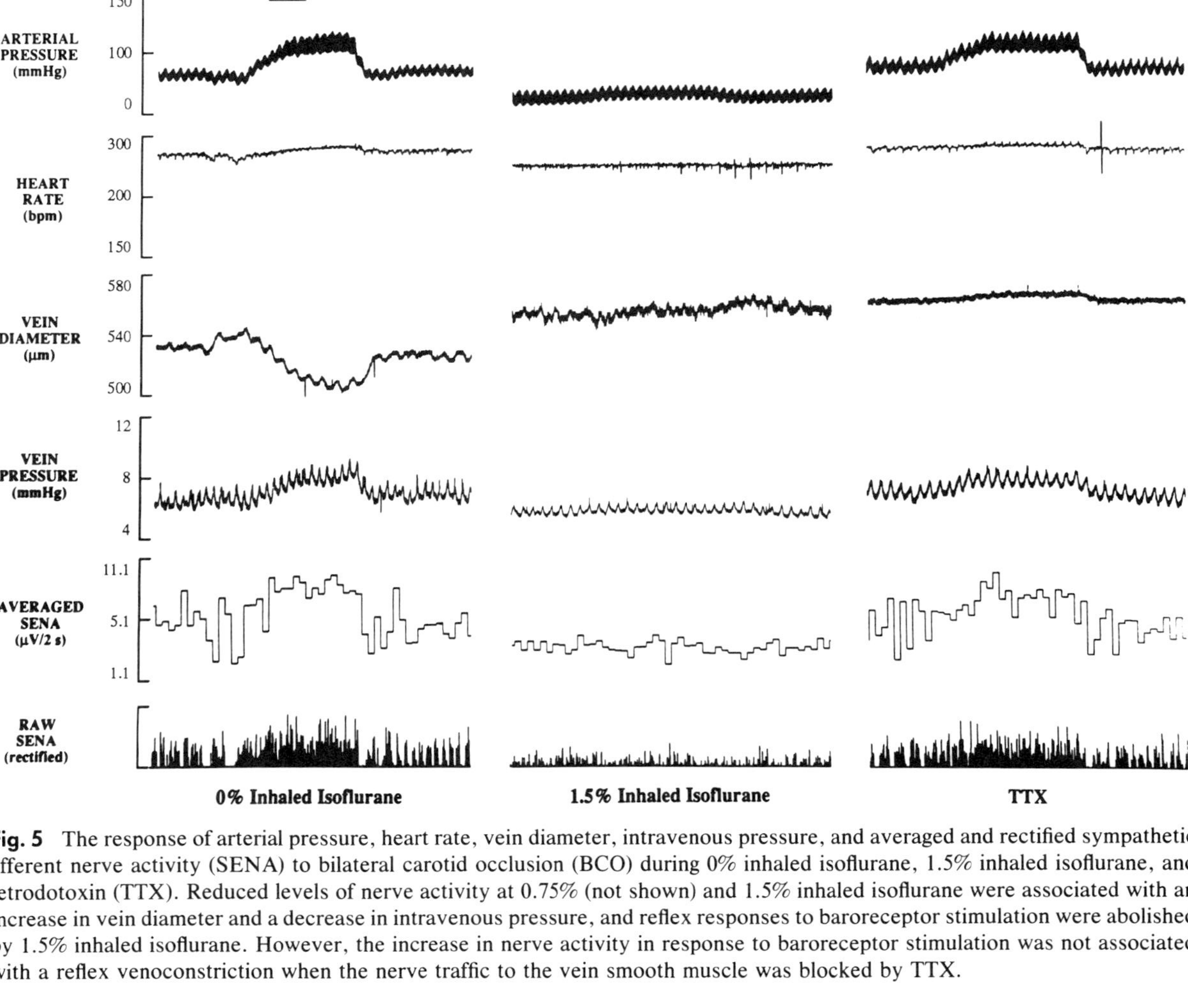

Fig. 5 The response of arterial pressure, heart rate, vein diameter, intravenous pressure, and averaged and rectified sympathetic efferent nerve activity (SENA) to bilateral carotid occlusion (BCO) during 0% inhaled isoflurane, 1.5% inhaled isoflurane, and tetrodotoxin (TTX). Reduced levels of nerve activity at 0.75% (not shown) and 1.5% inhaled isoflurane were associated with an increase in vein diameter and a decrease in intravenous pressure, and reflex responses to baroreceptor stimulation were abolished by 1.5% inhaled isoflurane. However, the increase in nerve activity in response to baroreceptor stimulation was not associated with a reflex venoconstriction when the nerve traffic to the vein smooth muscle was blocked by TTX.

and tetrodotoxin application. During 0% inhaled isoflurane administration, vein diameter decreased and vein pressure increased in response to sympathetic activation caused by BCO. After the inhalation of 1.5% isoflurane, vein diameter increased and vein pressure fell before stimulation owing to reduced sympathetic tone to the vessel, and 1.5% inhaled isoflurane abolished reflex responses to BCO. However, local denervation of the mesenteric vein by tetrodotoxin abolished only the reflex venoconstriction in response to BCO, whereas the other responses remained essentially as they were during 0% inhaled isoflurane; indeed, vein diameter increased slightly during BCO as a result of improved filling pressure.

IV. Response to Electric Stimulation

Direct electric stimulation of the postganglionic axons leading from the celiac ganglion to the mesenteric bed caused a decrease in vein diameter, and the level of this venoconstriction was frequency dependent (Fig. 6). Neither the high dose of inhaled anesthetics nor an extremely high concentration of anesthetics in the superfusate produced any significant changes in the frequency-dependent venoconstriction following celiac ganglion stimulation. Yet systemic administration of anesthetics through the rebreathing circuit produced a significant reduction in mean arterial pressure and heart rate. For example, in the rabbits exposed to inhaled isoflurane, mean arterial pressure dropped from 70 ± 4 mmHg before anesthetics to 37 ± 4 mmHg after anesthetics, whereas heart rates declined from 276 ± 16 beats/min before anesthesia to 247 ± 15 beats/min after anesthesia. Furthermore, pilot studies (data not reported) had shown that intravenous hexamethonium blockade of ganglionic transmission did not attenuate the venoconstriction observed in response to direct electric stimulation of the postganglionic nerve fibers.

Mean anesthetic concentrations are shown in Table I. For inhaled anesthetics, values are expressed as millimolar in blood, whereas concentration of anesthetics equilibrated in the physiological salt solution (PSS) appear as millimolar in the superfusate. Blood concentrations during superfusion of anesthetics was essentially null because no systemic uptake occurred as a result of anesthetics in the superfusate.

V. Discussion

Our study demonstrates that inhaled anesthetics impede the reflex ability of mesenteric veins to actively constrict and dilate in reaction to baroreflex stimulation. Inhaled anesthetics depressed mean arterial pressure, heart rate, and sympathetic efferent outflow. Both low and high doses of inhaled

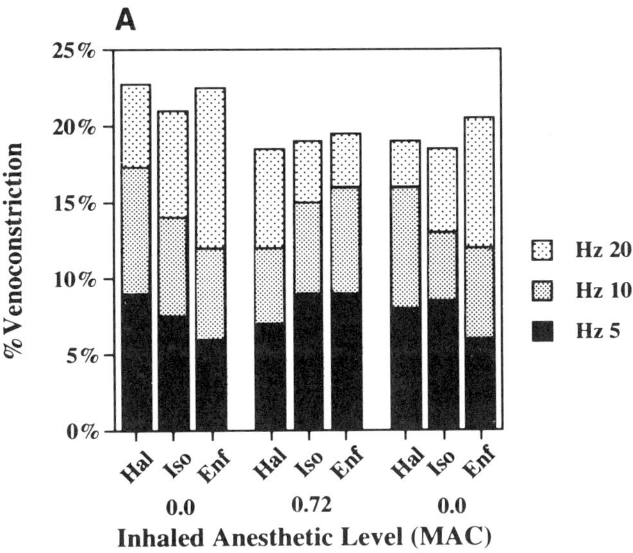

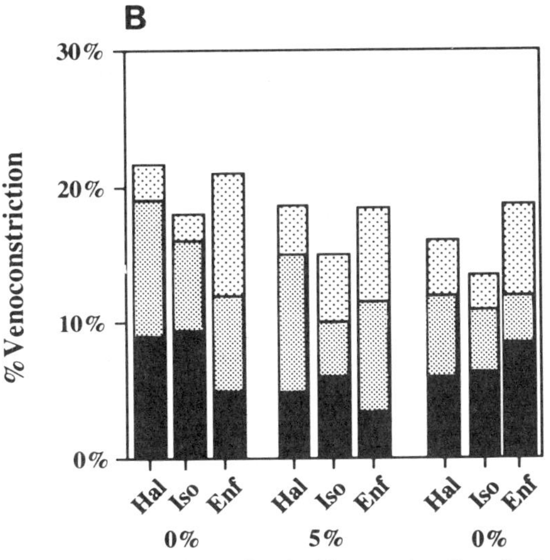

Fig. 6 (A) High-dose inhaled anesthetics did not abolish the frequency-dependent venoconstriction in response to stepwise increases in the frequency of electric stimulation (n = 13). (B) An extremely high concentration of anesthetics in the superfusate did not attenuate the frequency-dependent level of venoconstriction (n = 33).

anesthetics attenuated the reflex constriction of mesenteric veins caused by lowered carotid sinus pressure, whereas reflex dilation in response to aortic depressor nerve stimulation was reduced by the higher dose of inhaled anesthetics. Furthermore, volatile anesthetics dramatically re-

Table I

Concentrations of Anesthetics in Blood and Physiological Saline Solution[a]

Vapor concentration	Concentration in blood (m*M*)	Concentration in PSS (m*M*)
Halothan		
0.5% inhaled	0.51 ± 0.03	0
1% inhaled	0.88 ± 0.03	0
3% dissolved in PSS	0	0.91 ± 0.06
5% dissolved in PSS	0	1.97 ± 0.11
Isoflurane		
0.75% inhaled	0.47 ± 0.04	0
1.5% inhaled	0.80 ± 0.05	0
3% dissolved in PSS	0	0.84 ± 0.06
5% dissolved in PSS	0	1.13 ± 0.06
Enflurane		
1% inhaled	0.74 ± 0.04	0
2% inhaled	1.54 ± 0.11	0
5% dissolved in PSS	0	1.64 ± 0.11

[a] Values are reported as means ± SEM.

duced the reflex increase in sympathetic efferent nerve activity as a result of baroreceptor stimulation. On the other hand, direct celiac sympathetic postganglionic activation of the mesenteric veins was unaffected by even the highest dose of inhaled anesthetics. Likewise, equal or greater concentrations of anesthetics delivered directly to the veins in the superfusate failed to suppress sympathetically induced venomotion. Changes in vein diameter and vein pressure arising from baroreceptor stimulation and direct electric activation of the celiac postganglionic axons were the same before, during, and after direct exposure to equal or greater concentrations of anesthetics in the superfusate. In short, only inhaled isoflurane influenced baroreceptor control of mesenteric venous reflexes in our studies, but neither inhaled nor directly administered anesthetics interfered with the changes produced by postganglionic stimulation of the mesenteric veins.

These findings support the hypothesis that volatile anesthetics may influence mesenteric vascular capacitance through inhibition of the active sympathetic control of veins. An active change in venous capacitance is to be distinguished from a passive reduction of arterial inflow or a redistribution of blood between vascular beds (18,19). Venous capacitance is a term which describes the relationship between a given volume of blood and a specific filling pressure. Passive changes in volume occur when the vessel wall complies to the transmural distending pressure accompanying changes in inflow or outflow. An active change in venous capacitance refers to the ability of vein smooth muscle to mobilize the unstressed blood volume present in the vein (1). It is hard to detect the active contribution of vein smooth muscle to changes in total blood flow in an intact animal because it is necessary to obtain stable volumes of venous outflow and inflow before measurements can be taken. In our studies it was impossible to isolate completely the splanchnic circulation. Therefore, it is necessary to conceive the dynamic relationship between active and passive factors in order to determine whether the changes in vein diameter and intravenous pressure represent active changes in mesenteric venous capacitance.

The passive elements of smooth muscle compliance to hemodynamic changes are either (a) decreased distending pressure as a consequence of greater resistance and reduced arteriolar inflow or (b) increased distending pressure owing to the increase in resistance of the hepatic sphincter and liver system (2). Added together, these mutually exclusive passive effects produce little or no net difference in the distending pressure of the vein at normal flows. An active response would therefore be present if a small increase in vein pressure occurred together with a decrease in vein diameter. For withdrawal of sympathetic tone, the active aspects would be just the reverse: both a small decrease in vein pressure and an increase in vein diameter. The combination of changes in vein pressure and vein diameter observed in our experiments after sympathetic stimulation indicates an active change in venous capacitance (20). These findings are in agreement with anatomic studies which have shown that second-order mesenteric veins similar to those used in these studies are innervated (21). In our studies, postganglionic sympathetic efferent nerve activity measured in the mesenteric bed corresponded to the reflex changes in vein diameter and intravenous pressure, and the decrease in neural activity during inhaled anesthetics was reversible. Moreover, reflex-mediated venoconstriction was abolished by locally applied tetrodotoxin, a specific sodium channel neural blocker, yet the increase in neural activity after baroreflex stimulation was not abolished. The combination of active changes in vein pressure and vein diameter coupled with the changes in sympathetic efferent nerve activity in our studies indicate that the changes in venous capacitance are most likely the result of neurally mediated active smooth muscle contractions. Inhaled anesthetics, but not superfused anesthetics, inhibited these active responses.

Because the primary site of anesthetic attenuation of the active venous response to baroreceptor stimulation is caudal to the postganglionic nerve fibers, the mechanisms involved may originate from central neural control, including inhibition of sympathetic ganglionic transmission, central activation, or alterations in baroreceptor organ activity (22). Abundant evidence exists for the inhibitory action of anesthetics on neural activity at a number of sites. Skovsted and Sapthavichaikul found a suppression of preganglionic sympathetic activity in conjunction with diminished barostatic reflexes, arterial pressure, and heart rate during halothane and isoflurane anesthesia (23). Seagard *et al.* demosntrated multiple sites of sympathetic inhibition in a series of studies of baroreflexes during isoflurane and halothane anesthesia (24–27). Bosnjak *et al.* exhibited the influene of halothane on pre- and postganglionic sympathetic activity (28,29). Price *et al.* demonstrated central inhibition of baroreflexes, heart rate, and blood pressure by confining halothane administration to the cephalic circulation (30, 31). Studies on the effects of anesthetics on the postsynaptic neuromuscular junction also support the finding of central control. In an isolated ring preparation taken from vessels indentical to those used in our studies, vessels exposed to halothane, isolfurane, and enflurane had reduced constrictions in response to endogenous neurotransmitter release as a result of direct electric stimulation (32). *In situ* studies of the differential effects of α_1- and α_2-adrenoceptor activation on the constriction of mesenteric arteries demonstrated the predominance of α_1-receptors in response to endogenous electric stimulation of the celiac ganglion and exogenous application of phenylephrine (33,34).

The predominance of central control is still a valid assumption despite a number of serious misgivings about the experimental protocol used in our studies. The dramatic decrease in mean arterial pressure raises a question about the overall function of experimental animals in light of the possibility of reduced cerebral blood flow and peripheral organ perfusion. Sperry *et al.* addressed the problem of regional blood flow in rats during deliberate hypotension combined with profound isoflurane anesthesia and hypovolemia. The investigators indicated that hypotension during isoflurane anesthesia does not affect cerebral blood flow (35). Brain blood flow decreased during isoflurane anesthesia only when hypovolemia (20%) accompanied hypotension (59% decrease in mean arterial pressure). The maximum depression of mean arterial pressure in our studies of around 43% should not have affected cerebral blood flow nor interfered with the response to baroreceptor stimulation. Indeed, Seagard *et al.* showed that halothane actually sensitized baroreceptor function in the isolated carotid sinus preparation (36).

Another concern about the validity of our findings might be the presence of direct mechanisms of venomotion through the endothelium. Rorie *et al.* suggested that the mechanism for halothane-induced inhibition may involve the activation of prejunctional muscarinic receptors that, once activated, reduce the norepinephrine release from the postganglionic nerve terminal (37). Kobayashi *et al.* proposed that enflurane may directly affect vascular smooth muscle by inhibiting norepinephrine release from sympathetic nerve endings and by impeding the interaction between norepinephrine and postjunctional α_1-adrenergic receptors in vascular smooth muscle (38). Indeed, our investigations found a significant inhibition of vein constrictions in response to exogenously applied norepinephrine in the isolated ring preparation exposed to halothane, isoflurane, and enflurane (32). Despite these findings of direct, noncentral effects of anesthetics on vein smooth muscle motion, our studies agree with the findings of others (39) that the primary site of control lies caudal to the postganglionic nerves. Both baroreceptor and direct electric stimulation were able to overcome whatever direct inhibitory effects of anesthetics were present as a result of administration of anesthetics in the superfusate surrounding the veins.

The contribution of venous capacitance to overall blood flow remains somewhat controversial. Most studies on the cardiovascular effects of anesthetic inhibition to autonomic stimuli emphasize the depression of myocardial contractility, reductions in cardiac output, and arterial blood pressure (40–44). Our studies indicate the importance of anesthetics on intestinal vascular tone and the resultant reduction in splanchnic and hepatic blood flow. These findings are in agreement with earlier reports of regional blood flow in the splanchnic circulation (45–48). Numerous studies of blood flow measured by changes in blood volume in an external reservoir, entrapment of radioactive microspheres, or flow probes have reported that anesthetics reduce blood flow through the mesenteric vascular bed. However, few previous studies have considered the possibility that reduced flow through the mesenteric circulation was a result of increased venous capacitance. The importance of venous capacitance is underscored by the fact that veins contain as much as 75% of the blood volume in the mesenteric circulatory system compared to only 18% in arteries, according to anatomic studies (49). The splanchnic circulation accounts for approximately 27–50% of the total blood volume shift mobilized by sympathetic stimulation (2,39). Greene and Shoukas have concluded that changes in vascular capacitance in response to carotid sinus baroreceptor activation were the primary mechanism responsible for changes in cardiac output (50). The present studies have contributed new evidence from direct observation of the mesenteric veins, indicating that anesthetics inhibit baroreflex regulation of vein diameter and subsequent changes in intravenous pressure, and that the capacitance of the mesenteric bed is thereby increased.

In conclusion, systemically administered halothane, isoflurane, and enflurane increased mesenteric venous capacitance, and this action was primarily through inhibition of baroreflex control of venous tone. The clinical importance of this finding in terms of cardiac disease is difficult to assess, but it appears that patients with autonomic neuropathy undergoing anesthesia may not tolerate the hemodynamic instability associated with the perioperative period because the baroreflex control over venous capacitance is already impaired. Further degradation of the reflex responses during anesthesia could expose these patients to the risk of bradycardia, hypotension, and cardiopulmonary arrest. Reports of perioperative hypotension have been made in studies of patients with diabetes (51,52) and age-related changes in response to circulatory stress (53). Patients with clinical signs of neuropathy may require perioperative management of pressor responses.

References

1. Rothe, C. F. (1983). Venous system: Physiology of the capacitance vessels. *In* "Handbook of Physiology, Section II: The Cardiovascular System" (J. T. Shepherd and F. M. Abboud, vol. eds.), Vol. 3 (Peripheral Circulation, Part I), pp. 397–452. American Physiological Society, Bethesda, Maryland.
2. Greenway, C. V. (1983). Role of splanchnic venous system in overall cardiovascular homeostasis. *Fed. Proc.* **42,** 1678–1684.
3. Hadjiminas, J., and Oberg, B., (1968). Effects of carotid baroreceptor reflexes on venous tone in skeletal muscle and intestine of the cat. *Acta Physiol. Scand.* **72,** 518–532.
4. Granger, D. N., Richardson, P. D., Kvietys, P. R., and Mortillaro, N. A. (1980). Intestinal blood flow *Gastroenterology* **78,** 837–863.
5. Rothe, C. F. (1983). Reflex control of veins and vascular capacitance *Physiol. Rev.* **63,** 1281–1342.
6. Hainsworth, R. (1986). Vascular capacitance: Its control and importance. *Rev. Physiol. Biochem. Pharmacol.* **105,** 101–173.
7. Longnecker, D. E. (1984). Effects of general anesthetics on the microcirculation. *Microcirculation, Endothelium, Lymphatics* **1,** 129–150.
8. Ozono, K., Bosnjak, Z. J., and Kampine, J. P. (1989). Reflex control of mesenteric vein diameter and pressure *in situ* in rabbits. *Am. J. Physiol.* **256,** H1066–H1072.
9. Ozono, K., Bosnjak, Z. J., and Kampine, J. P. (1991). Effect of sympathetic tone on pressure–diameter relation of rabbit mesenteric veins *in situ*. *Circ. Res.* **68,** 888–896.
10. Stekiel, T. A., Ozono, K., McCallum, J. B., Bosnjak, Z. J., Stekiel, W. J., and Kampine, J. P. (1990). The inhibitory action of halothane on reflex constriction in mesenteric capacitance veins. *Anesthesiology* **73,** 1169–1178.
11. McCallum, J. B., Stekiel, T. A., Bosnjak, Z. J., and Kampine, J. P. (1993). Does isoflurane alter mesenteric venous capacitance in the intact rabbit? *Anesth. Analg. (N. Y.)* **76,** 1095–1105.
12. Stadnicka, A., Stekiel, T. A., Bosnjak, Z. J., and Kampine, J. P. (1993). Inhibition by enflurane of baroreflex mediated mesenteric venoconstriciton in the rabbit ileum. *Anesthesiology* **78,** 928–936.
13. Bell, L. B., Hopp, F. A., Seagard, J. L., Van Brederode, H. F., and Kampine, J. P. (1988). A continuous noncontact method for measuring *in situ* vascular diameter with a video camera. *J. Appl. Physiol.* **64,** 1279–1284.
14. Hopp, F. A., Seagard, J. L, and Kampine, J. P. (1986). Comparison of four methods of averaging nerve activity. *Am. J. Physiol.* **251,** R700–R711.
15. Bohlen, H. G., (1980). Intestinal tissue P_{O_2} and microvascular responses during glucose exposure. *Am. J. Physiol.* **238,** H164–H171.
16. Drummond, J. C. (1985). MAC for halothane, enflurane, and isoflurane in the New

Zealand White rabbit; and a test for the validity of MAC determinations. *Anesthesiology* **62,** 336–338.
17. Nishi, S. (1974). Ganglionic transmission. *In* "The Peripheral Nervous System" (J. I. Hubbard, ed.), pp. 225–255. Plenum, New York.
18. Shoukas, A. A., and Sagawa, K. (1973). Control of total systemic vascular capacity by the carotid sinus baroreceptor reflex. *Circ. Res.* **33,** 22–33.
19. Hainsworth, R. and Linden, R. J. (1979). Reflex control of vascular capacitance. *In* "Cardiovascular Physiology" (A. C. Guyton and D. B. Young, eds.), pp. 67–124. Univ. Park Press, Baltimore, Maryland.
20. Karim, F. and Hainsworth, R. (1976). Responses of abdominal vascular capacitance to stimulation of splanchnic nerves. *Am. J. Physiol.* **231,** 434–440.
21. Furness, J. B. (1973). Arrangement of blood vessels and their relation with adrenergic nerves in the rat mesentery. *J. Anat.* **115,** 347–364.
22. Seagard, J. L., Bosnjak, Z. J., Hopp, F. A., Kotrly, K. J., Ebert, T. J., and Kampine, J. P. (1985). Cardiovascular effects of general anesthesia. *In* "Effects of Anesthetics" (B. G. Covina, H. A. Fozzard, K. Rehder, and G. Strichartz, eds.), pp. 149–177. Waverly, Baltimore, Maryland.
23. Skovsted, P., and Sapthavichaikul, S. (1977). The effects of isoflurane on arterial pressure, pulse rate, autonomic nervous activity, and barostatic reflexes. *Can. Anaesth. Soc. J.* **24,** 304–314.
24. Seagard, J. L., Hopp, F. A., Bosnjak, Z. J., Osborn, J. L., and Kampine, J. P. (1984). Sympathetic efferent nerve activity in conscious and isoflurane-anesthetized dogs. *Anesthesiology* **61,** 266–270.
25. Seagard, J. L., Elegbe, E. O., Hopp, F. A., Bosnjak, Z. J., von Colditz, J. H., Kalbfleisch, J. H., and Kampine, J. P. (1983). Effects of isoflurane on the baroreceptor reflex. *Anesthesiology* **59,** 511–520.
26. Seagard, J. L., Hopp, F. A., Donegan, J. H., Kalbfleisch, J. H., and Kampine, J. P. (1982). Halothane and the carotid sinus reflex: Evidence for multiple sites of action. *Anesthesiology* **57,** 191–202.
27. Bachhuber, S. R., Seagard, J. L., Bosnjak, Z. J., and Kampine, J. P. (1981). The effect of halothane on reflexes elicited by acute coronary artery occlusion in the dog. *Anesthesiology* **54,** 481–487.
28. Bosnjak, Z. J., Seagard, J. L., Wu, A., and Kampine, J. P. (1982). The effects of halothane on sympathetic ganglionic transmission. *Anesthesiology* **57,** 473–479.
29. Bosnjak, Z. J., Dujic, Z., Roerig, D. L., and Kampine, (1988). Effects of halothane on acetylcholine release and sympathetic ganglionic transmission. *Anesthesiology* **69,** 500–506.
30. Price, H. L., Linde, H. W., and Morse, H. T. (1963). Central actions of halothane affecting the systemic circulation. *Anesthesiology* **24,** 770–778.
31. Price, H. L., and Price, M. L., (1966). Has halothane a predominant circulatory action? *Anesthesiology* **27,** 764–769.
32. Stadnicka, A., Flynn, N. M., Bosnjak, Z. J., and Kampine, J. P. (1993). Enflurane, halothane, and isolfurane attenuate contractile responses to exogenous and endogenous norepinephrine in isolated small mesenteric veins of the rabbit. *Anesthesiology* **78,** 326–334.
33. Mizuno, K., Stekiel, T. T., Bosnjak, Z. J., and Kampine, J. P. (1993). Differential effects of $\alpha 1$ and $\alpha 2$ adrenoceptor activation on constriction of mesenteric arteries in rabbits. *FASEB J.* **7,** A779.
34. Tominaga, M. Stekiel, T. A., Bosnjak, Z. J., and Kampine, J. P. (1991). The role of the α-adrenoceptor subtype during mesenteric venoconstriction in response to celiac ganglion stimulation. *Anesthesiology* **75,** A570.

35. Sperry, R. J., Monk, C. R., Durieux, M. E., and Longnecker, D. E. (1992). The influence of hemorrhage on organ perfusion during deliberate hypotension in rats. *Anesthesiology* **77,** 1171–1177.
36. Seagard, J. L., Hopp, F. A., Bosnjak, Z. J., Elegbe, E. O., and Kampine, J. P. (1983). Extent and mechanism of halothane sensitization of the carotid sinus baroreceptors. *Anesthesiology* **58,** 432–437.
37. Rorie, D. K., Tyce, G. M., and Mackenzie, R. A. (1984). Evidence that halothane inhibits norepinephrine release from sympathetic nerve endings in dog saphenous vein by stimulation of presynaptic inhibitory muscarinic receptors. *Anesth. Analg. (N. Y.)* **63,** 1059–1064.
38. Kobayashi, Y., Yoshida, K., Noguchi, M., Wakasugi, Y., Ito, H., and Okabe, E. (1990). Effect of enflurane on contractile reactivity in isolated canine mesenteric arteries and veins. *Anesth. Analg. (N. Y.)* **70,** 530–560.
39. Brunner, M. I., Greene, A. S., Frankle, A. E., and Shoukas, A. A. (1988). Carotid sinus baroreceptor control of splanchnic resistance and capacity. *Am. J. Physiol.* **255,** H1305–H1310.
40. Eger, E. (1984). The pharmacology of isoflurane. *Br. J. Anaesth.* **56,** 71S–99S.
41. Black, G. W. (1979). Enflurane. *Br. J. Anaesth.* **51,** 627–640.
42. Pagel, P. S., Kampine, J. P., Schmeling, W. T., and Warltier, D. C. (1991). Comparison of the systemic and coronary hemodynamic actions of desflurane, isoflurane, halothane, and enflurane in the chronically instrumented dog. *Anesthesiology* **74,** 539–551.
43. Henriksson, B. A., Biber, B., Martner, J., Ponten, J., and Werner, O. (1985). Cardiovascular studies during controlled baroreflex activation in the dog: I. Effects of enflurane. *Acta Anaesthesiol. Scand* **29,** 90–94.
44. Morton, M., Duke, P. C., and Ong. B. (1980). Baroreflex control of heart rate in man awake and during enflurane and enflurane–nitrous oxide anesthesia. *Anesthesiology* **52,** 221–223.
45. Bagshaw, R. J., and Cox, R. H. (1988). Baroreceptor control of central and regional hemodynamics with isoflurane in the dog. *Acta Anaesthesiol. Scand.* **32,** 82–92.
46. Gelman, S. Fowler, K. C., and Smith, L. R., (1984). Regional blood flow during isoflurane and halothane anesthesia. *Anesth. Analg. (N. Y.)* **63,** 557–565.
47. Lundeen, G., Manohar, M., and Parks, C. (1983). Systemic distribution of blood flow in swine while awake and during 1.0 and 1.5 MAC isoflurane anesthesia with or without 50% nitrous oxide. *Anesth. Analg. (N. Y.)* **62,** 499–512.
48. Tverskoy, M., Gelman, S., Fowler, K. C., and Bradley, E. L. (1985). Intestinal circulation during inhalation anesthesia. *Anesthesiology* **62,** 462–469.
49. Green, H. D. (1950). Circulatory system. *In* "Medical Physics" (O. Glasser, ed.), p. 231. Yearbook, Chicago.
50. Greene, A. S., and Shoukas, A. A. (1986). Changes in canine cardiac function and venous return curves by the carotid baroreflex. *Am. J. Physiol.* **251,** H288–H296.
51. Page, M. M., and Watkins, P. J. (1978). Cardiorespiratory arrests and diabetic autonomic neuropathy. *Lancet* **1,** 14–16.
52. Burgos, L. G., Ebert, T. J., Asiddao, C., Turner, L. A., Pattison, C. Z., Wang-Cheng, R., and Kampine, J. P. (1989). Increased intraoperative cardiovascular morbidity in diabetics with automonic neuropathy. *Anesthesiology* **70,** 591–597.
53. McDermott, D. J., Tristani, F. E., Ebert, T. J., Porth, D. J., and Smith, J. J. (1980). Age-related changes in cardiovascular response to diverse circulatory stresses. *In* "Arterial Baroreceptors and Hypertension" (P. Sleight, ed.), pp. 361–364.

Effect of General Anesthesia on Modulation of Sympathetic Nervous System Function

Margaret Wood
Department of Anesthesiology
Vanderbilt University Medical Center
Vanderbilt University School of Medicine
Nashville, Tennessee 37232

I. Introduction

The effects of volatile and intravenous anesthetics on sympathetic nervous function have been investigated for many years, and it is generally recognized that halogenated anesthetics exhibit profound cardiovascular depressant effects which may be accompanied by a decrease in plasma catecholamine concentrations (1–5). Plasma norepinephrine (NE) concentrations have been used as a measure of sympathetic activity not only during anesthesia, but also in congestive cardiac failure and ischemia and to define sympathetic responses during vasodilator therapy. It is important to recognize that plasma NE is determined by NE clearance from the circulation in addition to the rate of NE release into the circulation. Thus, change in NE concentration may result from change in NE clearance from the circulation and/or in the rate of NE release into the circulation (systemic NE spillover). NE spillover can be defined as the total rate of entry of NE into the circulation and occurs primarily from sympathetic nerve endings. NE spillover rate and NE clearance *in vivo* can be assessed by an isotope dilution technique using trace infusions of $[^3H]$NE (6).

Advances in Pharmacology, Volume 31

II. Effect of Intravenous and Inhalational Anesthetics on Norepinephrine Kinetics

Norepinephrine spillover and clearance rates were determined in a chronically instrumented dog model by isotope dilution (7). Cannulae were implanted into the right femoral artery and vein during pentobarbital anesthesia 7 days before the study was performed. Studies were performed during halothane, isoflurane, enflurane, and propofol anesthesia. For each determination of NE kinetics, a separate 50-min infusion of a trace amount of [^{3}H]NE was performed. Plasma [^{3}H]NE concentrations reach steady state by 20 min (Fig. 1). Arterial blood samples were taken just before the start of the infusion and 20, 30, 40, and 50 min after the start of the infusion and assayed for [^{3}H]NE and endogenous NE. NE concentrations were measured by high-performance liquid chromatography (HPLC). The HPLC effluent coinciding with the NE peak was collected and counted by liquid scintillation to determine plasma [^{3}H]NE concentration [disintegrations/minute (dpm)/ml] without interference from tritiated metabolites (8).

Figure 1 illustrates plasma concentrations of endogenous and [^{3}H]NE during propofol and halothane anesthesia. During anesthesia with both propofol and halothane, endogenous NE concentrations decreased mark-

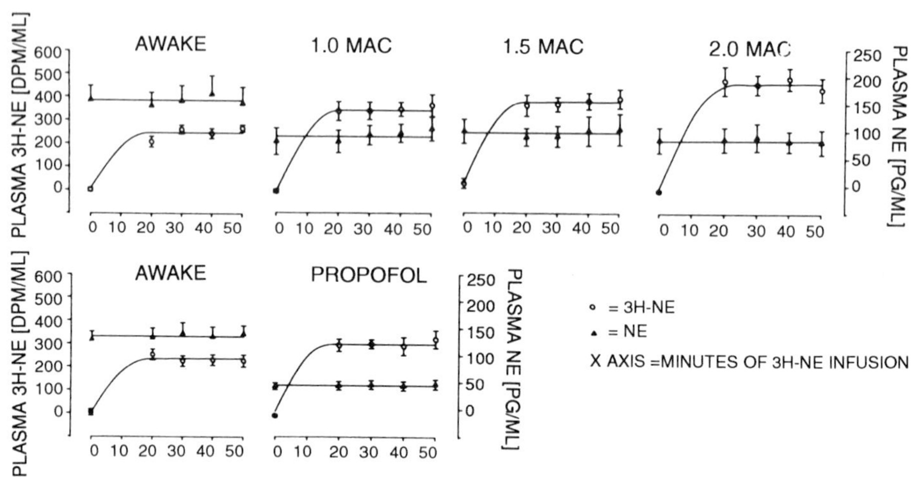

Fig. 1 Plasma concentrations of endogenous NE (pg/ml) (triangles) and [^{3}H]NE (dpm/ml) (circles) ($\pm$ SEM) in awake dogs and in dogs anesthetized with halothane (group A: 1.0, 1.5, and 2.0 MAC) and propofol. The *x* axis represents minutes of [^{3}H]NE infusion. [From Deegan *et al.* (7).]

edly whereas [^{3}H]NE plasma concentrations rose, indicating reduced [^{3}H]NE clearance. Calculated NE spillover and NE clearance are depicted in Fig. 2 for dogs anesthetized with halothane and in Fig. 3 for dogs

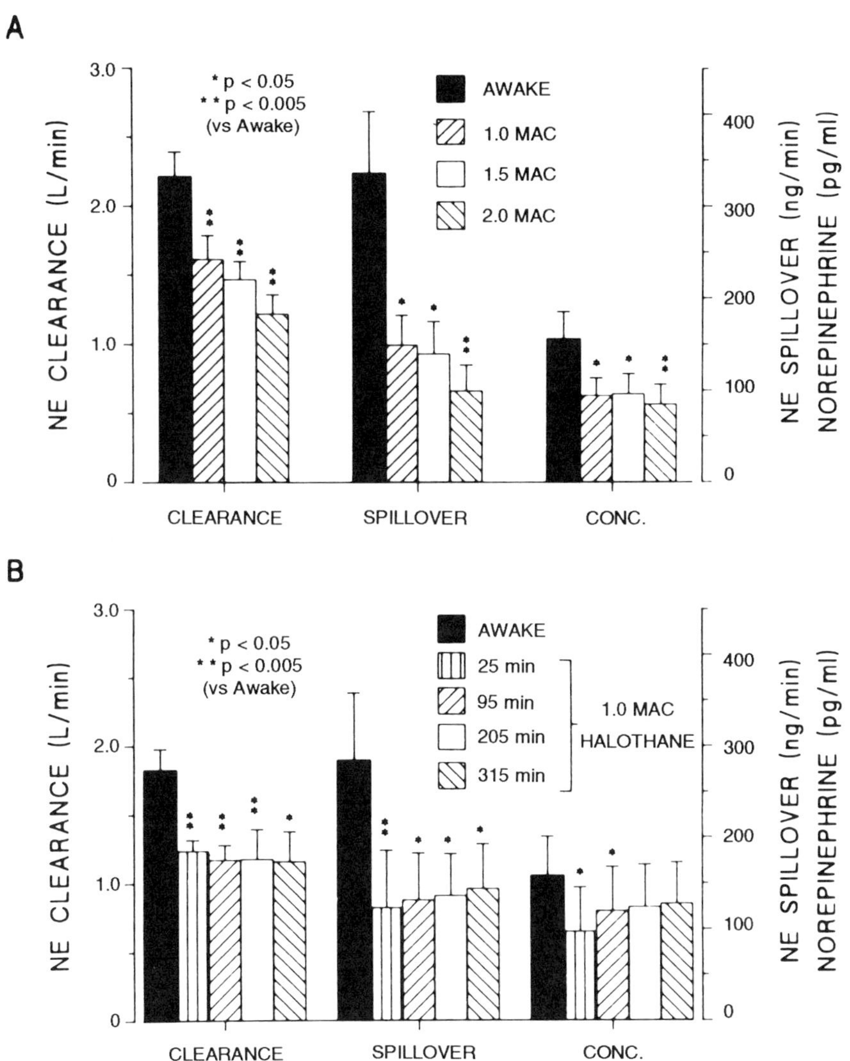

Fig. 2 Effect of halothane anesthesia on NE concentration (pg/ml) [^{3}H]NE clearance (liters/min), and NE spillover (ng/min) in dogs. Data are expressed as mean values ± SEM. (A) Group A: Awake and during 1.0,1.5, and 2.0 MAC halothane. (B) Group B: Awake and during 1.0 MAC halothane anesthesia for up to 5 hr. [From Deegan *et al.* (7).]

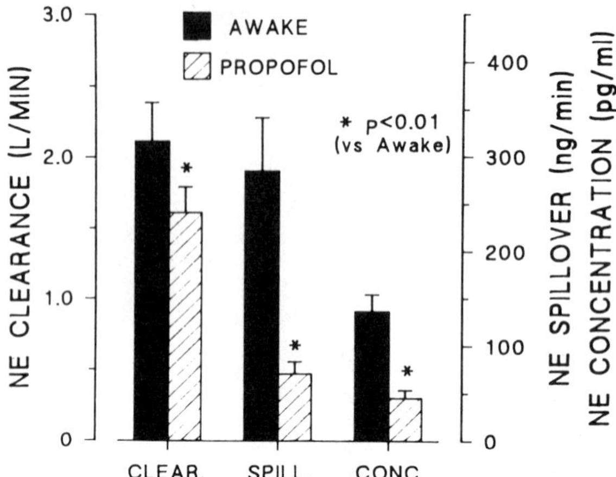

Fig. 3 Effect of propofol anesthesia on NE concentration (pg/ml), [^{3}H]NE clearance (liters/min), and NE spillover (ng/min) in dogs. Data are expressed as mean values ± SEM. [From Deegan *et al.* (7).]

anesthetized with propofol. Plasma NE concentrations reflect both the clearance and spillover of endogenous NE. NE spillover rates decreased during halothane, resulting in decreases of 52, 59, and 72% at 1.0, 1.5, and 2.0 MAC, respectively, from awake values (Fig. 2). Halothane anesthesia also resulted in a dose-dependent decrease in [^{3}H]NE clearance. Thus, halothane anesthesia produced a decrease in NE spillover, but because of the simultaneous reduction in NE clearance the magnitude of the decrease in NE spillover was not fully reflected in changes in endogenous NE concentration. Propofol anesthesia produced a significant (65%) reduction in NE plasma concentration (Fig. 3); in addition, NE spillover (73%) and NE clearance (22%) both exhibited a significant decrease.

We have also investigated the effect of two other inhalational anesthetics (isoflurane and enflurane) on NE kinetics (Fig. 4) (9). During anesthesia with either enflurane or isoflurane, endogenous NE concentrations decreased markedly. In contrast [^{3}H]NE concentrations increased or remained unchanged. There was a significant reduction in NE spillover with both anesthetics at all doses studied. NE spillover decreased by 54 and 68% at 1.0 MAC enflurane and 1.5 MAC enflurane, respectively. Isoflurane decreased NE spillover by 54, 59, and 71% at 1.0, 1.5, and 2.0 MAC isoflurane, respectively. NE clearance was reduced by both anesthetics; enflurane reduced NE clearance at all concentrations studied, whereas NE clearance was reduced by 24% from awake values only at the highest dose of isoflurane (2.0 MAC). For enflurane, NE clearance was decreased

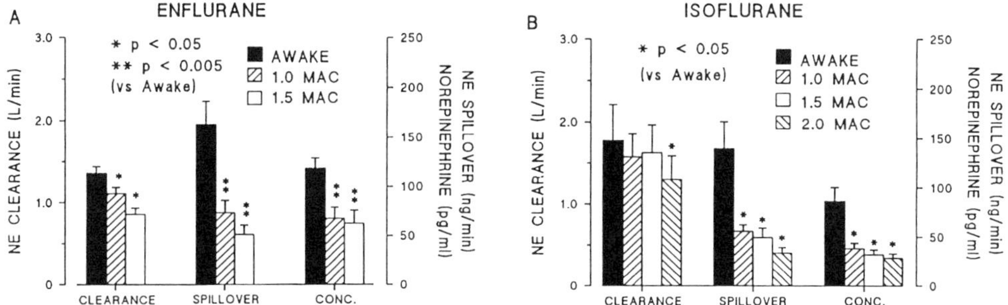

Fig. 4 Norepinephrine (NE) concentration (pg/ml), NE clearance (liters/min), and NE spillover (ng/min) during (A) enflurane and (B) isoflurane anesthesia, compared with the corresponding awake values. Data are expressed as mean values ± SEM. [From Deegan *et al.* (9).]

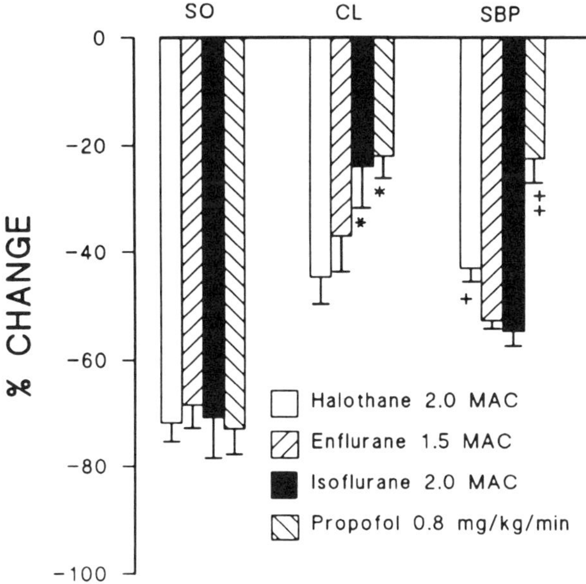

Fig. 5 Comparison of the effects (percent changes from awake) of halothane, enflurane, isoflurane, and propofol on norepinephrine (NE) clearance (CL) and systolic blood pressure (SBP) at doses which produce equal inhibition of NE spillover (SO). *Significant difference ($p < 0.5$) versus halothane. + Significant difference ($p < 0.01$) versus enflurane, isoflurane, and propofol. + + Significant difference ($p < 0.01$) versus halothane, enflurane, and isoflurane. [From Deegan *et al.* (9).]

by 17% at 1.0 MAC and 37% at 1.5 MAC enflurane. Because decreased NE clearance would otherwise have increased NE plasma concentration, our measured decreases in endogenous plasma NE concentrations probably did not reflect the actual extent of sympathetic inhibition produced by enflurane and isoflurane. Thus, all the inhalational anesthetics studied and the intravenous anesthetic propofol produced marked sympathetic inhibition. We have also shown that NE spillover is a more accurate measure of sympathetic activity during anesthesia than is plasma NE concentration alone.

We postulated that measurement of NE kinetics might allow more reasonable comparison of hemodynamic effects of anesthetics because doses producing equal reductions in sympathetic activity could be chosen. Figure 5 shows a comparison of effects of all four anesthetics at similar levels of suppression of NE spillover, expressed as the change from awake values. At similar decreases in NE spillover, the effects on NE clearance were not the same.

III. β-Adrenergic Receptor-Mediated Release of Norepinephrine

The release of NE from peripheral sympathetic nerves is subject to presynaptic modulation by α-adrenergic, dopaminergic, and angiotensin receptors (10). A facilitative effect on peripheral noradrenergic sympathetic transmission has been demonstrated during and after exposure to epinephrine. Presynaptic β_2-adrenergic receptors have been postulated to facilitate release of NE. Thus, direct activation of presynaptic or prejunctional β_2-adrenergic receptors by circulating (humoral) epinephrine facilitates norepinephrine release as a local positive feedback mechanism. There is evidence that epinephrine also undergoes neuronal uptake and acts as cotransmitter, and it is released with norepinephrine at a later time (i.e., autofacilitation), whereby epinephrine augments its own release via autoreceptor stimulation.

The role of presynaptic β_2-adrenergic receptors in facilitating NE release in humans is difficult to define since (1) measurement of plasma NE is an inadequate measure of local NE release and (2) β-agonist administration produces systemic effects with consequent vasodilation and hypotension with resultant secondary reflex stimulation of NE release. To determine whether stimulation of prejunctional β_2-receptors alters local release of NE, Stein and co-workers examined the effects of 60 and 400 ng/min infusion of intraarterial (brachial) isoproterenol on NE kinetics in seven volunteers, using radiotracer methods (11).

Forearm blood flow was measured using mercury-in-silicone elastomer strain gauge plethysmography. Low doses of isoproterenol were administered directly into the brachial artery to stimulate β_2-receptors in the forearm; the doses chosen had minimal systemic effects and produced no change in blood flow in the contralateral forearm.

Isoproterenol administered via the brachial artery resulted in a significant dose-related increase in forearm blood flow (Fig. 6)—a postsynaptic effect—whereas stimulation of presynaptic adrenergic receptors by intraarterial isoproterenol increased local NE spillover into the forearm (Fig. 7). Isoproterenol at 60 ng/min increased forearm NE spillover 6-fold, whereas 400 ng/min isoproterenol produced a 29-fold increase in NE spillover above baseline (Fig. 7). In addition, coinfusion of the β_1- and β_2-antagonist propranolol attenuated the isoproterenol-induced increase in local NE spillover. In keeping with the hypothesis that the increase in forearm NE spillover was due to the local effect of isoproterenol was the lack of effect of the low dose of isoproterenol (60 ng/min) on heart rate or blood pressure. In contrast to the 540% and 2844% increase in forearm NE spillover during the two infusions (60 and 400 ng/min) of isoproterenol, systemic NE spillover increased by only 17 and 50%, respectively, at the two infusion rates.

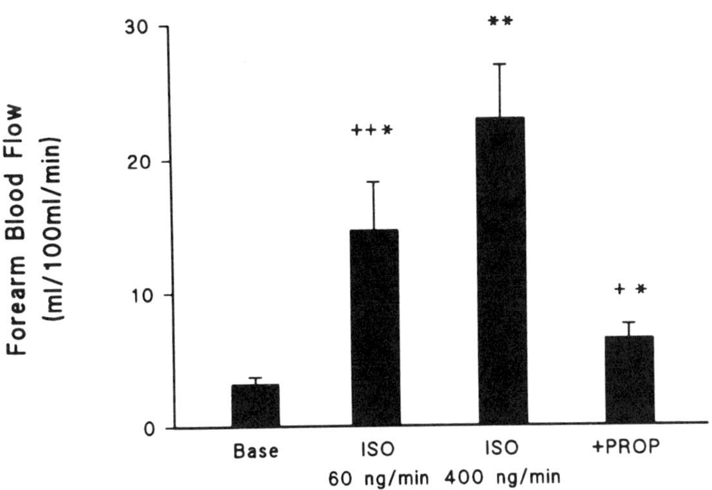

Fig. 6 Forearm blood flow (ml/100 ml/min) at baseline (Base) and after intraarterial infusion of 60 ng/min isoproterenol, 400 ng/min isoproterenol ($n = 7$), and coinfusion of 20 to 40 μ/min propranolol and 400 ng/min isoproterenol (+PROP) ($n = 4$). *Significant difference ($p < 0.05$) versus baseline. **Significant difference ($p < 0.005$) versus baseline. +Significant difference $p < 0.05$) versus ISO 400 ng/min. ++Significant difference ($p < 0.005$) versus ISO 400 ng/min. [From Stein *et al.* (11).]

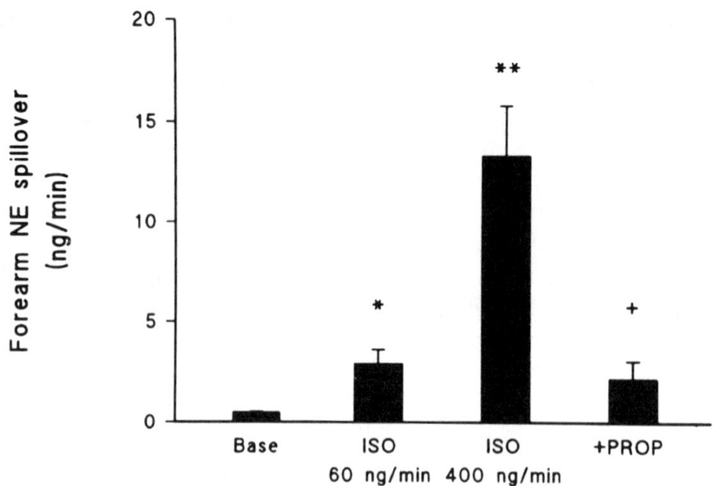

Fig. 7 Forearm norepinephrine spillover (ng/min) baseline (Base) and after intraarterial infusion of 60 ng/min isoproterenol, 400 ng/min isoproterenol ($n = 7$), and coinfusion of 20 to 40 μ/min propranolol and 400 ng/min isoproterenol (+PROP) ($n = 4$). **Significant difference ($p < 0.05$) versus baseline. **Significant difference ($p < 0.005$) versus baseline. + Significant difference ($p < 0.05$) versus ISO 400 ng/min. [From Stein *et al.* (11).]

Increased blood flow may increase local NE spillover, and therefore flow-independent measures of spillover and clearance have also been calculated (versus plasma appearance rate and intrinsic clearance). The forearm NE plasma appearance rate increased from 0.74 ng/min at baseline to 4.61 and 20.0 ng/min after infusion of 60 and 400 ng/min isoproterenol, respectively, indicating that the profound increase in local NE spillover seen after intrabrachial infusion of isoproterenol is not due to changes in forearm blood flow. Thus, isoproterenol mediates local release of NE *in vivo* in humans, presumably secondary to presynaptic β_2-adrenergic receptor stimulation.

IV. Effect of Inhalational Anesthesia on Prejunctional Norepinephrine Release *in Vivo*

The aim of the next series of studies was to determine whether general anesthetics interact with peripheral presynaptic NE release mechanisms, in particular to ask the question whether one of the many sites at which anesthetics affect NE release might be at the presynaptic β_2-receptor. To study the effect of anesthesia on presynaptic regulation of NE spillover we used an experimental dog hind limb model and a similar protocol to the one described in the previous section to determine the effect of isoproterenol on NE spillover in the human forearm. Direct femoral intraarterial infusions of low doses of isoproterenol were administered, and local hind limb NE spillover in awake animals and during halothane anesthesia were determined using [^{3}H]NE infusion methodology (12).

Based on our studies described in humans in the previous section, intraarterial isoproterenol would be expected to increase local NE spillover into the hind limb, whereas postsynaptic β_2-adrenoceptor stimulation by intraarterial isoproterenol would be expected to result in an increase in hind limb blood flow. In awake dogs, isoproterenol increased hind limb blood flow and also increased NE spillover into the hind limb (12). Halothane had no effect on baseline or isoproterenol-stimulated hind limb blood flow (postjunctional β_2-effect) but markedly inhibited hind limb NE spillover rate (presynaptic β_2 effect) (12). Thus, these preliminary studies support our hypothesis that beta-presynaptic β_2-stimulation results in increased NE spillover in both humans and the laboratory animal, and that this increase is impaired by halothane (12).

V. Conclusions

The inhalational anesthetics halothane, enflurane, and isoflurane and the intravenous anesthetic propofol produce marked sympathetic inhibition as measured by the decrease in NE spillover. In addition, they decrease NE clearance from the circulation. The mechanisms by which anesthetics inhibit sympathetic nervous system activity are probably multifactorial (both central and peripheral), but we have demonstrated that halothane inhibits NE release at the prejunctional level.

Acknowledgments

Supported in part by U.S. Public Health Service Grants HL14192 and GM31304 and by a grant-in-aid from the Tennessee Affiliate of the American Heart Association.

References

1. Price, H. L., Linde, H. W., Jones, R. E., Black, G. W., and Price, M. L. (1959). Sympathoadrenal responses to general anesthesia in man and their relation to hemodynamics. *Anesthesiology* **20**, 563–575.
2. Muldoon, S. M., Cress, L., and Freas, W. (1989). Presynaptic adrenergic effects of anesthetics. *Int. Anesthesiol. Clin.* **27**, 248–258.
3. Perry, L. B., Van Dyke, R. A., and Theye, R. A. (1974). Sympathoadrenal and hemodynamic effects of isoflurane, halothane, and cyclopropane in dogs. *Anesthesiology* **40**, 465–470.

4. Roizen, M. F., Moss, J., Henry, D. P., and Kopin, I. J. (1974). Effects of halothane on plasma catecholamines. *Anesthesiology* **41**, 432–439.
5. Derbyshire, D. R., and Smith, G. (1984). Sympathoadrenal responses to anaesthesia and surgery. *Br. J. Anaesth.* **50**, 725–739.
6. Esler, M., Jennings, G., Korner, P., Willett, I., Dudley, F., Hasking, G., Anderson, W., and Lambert, G. (1988). Assessment of human sympathetic nervous system activity from measurements of norepinephrine turnover. *Hypertension* **11**, 3–20.
7. Deegan, R., He, H. B., Pharm, D., Wood, A. J. J., and Wood, M. (1991). Effects of anesthesia on norepinephrine kinetics. *Anesthesiology* **75**, 481–488.
8. He, H. B., Deegan, R. J., Wood, M., and Wood, A. J. J. (1992). Optimization of high-performance liquid chromatographic assay for catecholamines. Determination of optimal mobile phase composition and elimination of species-dependent differences in extraction recovery of 3,4-dihydroxybenzylamine. *J. Chromatogr.* **574**, 213–218.
9. Deegan, R., He, H. B., Wood, A. J. J., and Wood, M. (1993). Effect of enflurane and isoflurane on norepinephrine kinetics: A new approach to assessment of sympathetic function during anesthesia. *Anesth. Analg. (N. Y.)* **77**, 49–54.
10. Floras, J. S. (1992). Epinephrine and the genesis of hypertension. *Hypertension* **19**, 1–18.
11. Stein, M., Deegan, R., He, H. B., and Wood, A. J. J. (1993). β-Adrenergic receptor-mediated release of norepinephrine in the human forearm. *Clin. Pharmacol. Ther.* **54**, 58–64.
12. Deegan, R., Krivoruk, Y., Wood, A. J. J., and Wood, M. (1992). Halothane inhibits presynaptic β_2 norepinephrine release *in vivo*. *Anesthesiology* **77**, A622.

Inhibition of Nitric Oxide-Dependent Vasodilation by Halogenated Anesthetics

Ming Jing,* Jayne L. Hart,† Saiid Bina,* and Sheila M. Muldoon*

**Department of Anesthesiology*
Uniformed Services University of the Health Sciences
F. Edward Hébert School of Medicine
Bethesda, Maryland 20814

†Department of Biology
George Mason University
Fairfax, Virginia 22030

I. Introduction

The vascular actions of halogenated hydrocarbons have been studied extensively since the initial use of these agents as anesthetics, and a variety of actions both direct and indirect have been reported. When the importance of endothelial-derived substances in the control of blood vessels became apparent in the 1980s, we investigated the possibility that the halogenated anesthetics may also influence this aspect of vascular control (1).

Because of the prominent hypotensive effects of the halogenated anesthetics, we originally hypothesized that they may stimulate the release of, or in some way enhance the actions of, endothelial-derived relaxing factor (EDRF) as had been suggested by Blaise *et al.* (2). However, this was not the case. In fact, our original studies (1) reported that endothelial-dependent relaxations of norepinephrine (NE)-contracted isolated dog and rabbit arteries, induced by either acetylcholine (ACh) or bradykinin (BK), were attenuated by halothane. Later studies using isolated rat aortas supported these findings (3). At that time, we suggested that halothane could

Advances in Pharmacology, Volume 31

be interfering with the generation, transit, or action of nitric oxide (NO) within the vascular smooth muscle.

Since then, we have made observations indicating that a major site of action of halothane (and to a lesser extent isoflurane and enflurane) in interfering with NO-dependent relaxation is on the enzyme-soluble guanylate cyclase (sGC) within the vascular smooth muscle. sGC is a ferrous heme-containing enzyme found in most tissues which is the receptor for NO. NO activates sGC by nitrosation of the ferrous atom, which forms a coordinate complex resulting in the changing of the iron–protein–ligand interaction. NO has a high affinity for ferrous heme proteins as is evident from its ease of binding to other heme proteins such as hemoglobin and myoglobin (4). In the target cells, NO reacts with and activates the sGC, thereby elevating intracellular production of cyclic guanosine 3′,5′-monophosphate (cGMP) levels. The latter causes reduction in cytosolic Ca^{2+} and thereby interferes with the contractile process. The following is a summary of the studies which we have done that support this hypothesis.

II. Methods

Vascular rings were isolated from dogs, rabbits, and rats, placed in tissue baths (37°C) containing Krebs solution, and aerated with 95% O_2–5% CO_2 at the optimal passive tensions, and isometric tensions were recorded. Protocols started with a cumulative NE concentration–response curve from which the concentration causing 60–70% of the maximum contraction (EC_{60-70}) for each ring was determined. This EC_{60-70} was used to contract the rings to a stable plateau at which time the appropriate relaxant was added cumulatively: ACh (1×10^{-9}–$1 \times 10^{-5} M$), BK (3×10^{-9}–$2 \times 10^{-8} M$), NO (3×10^{-9}–$1 \times 10^{-7} M$) nitroglycerin (NG) (2×10^{-9}–$3 \times 10^{-7} M$), or carbon monoxide (CO) (26–180 μM). The NE contraction–relaxation procedure was repeated successively three times in paired rings from the same animal. One ring was exposed to the appropriate anesthetic 10 min before and during the second contraction–relaxation response, whereas the other ring which was not exposed to anesthetics served as a time control. Different concentrations of each anesthetic were tested on separate rings. The endothelium was intact for the ACh and bradykinin studies, but removed prior to testing of the other relaxants. Endothelium was removed when necessary by rotating the rings around the tip of a pair of small forceps. The effects of superoxide dismutase (SOD, 120 units/ml), a superoxide scavenger, and the cyclooxygenase inhibitor indomethacin (10 M) on the halothane attenuation of ACh relaxations were also investigated.

Solutions of nitric oxide and carbon monoxide gases were prepared daily prior to use. NO gas was synthesized by addition of sodium nitrite (10 ml, 0.27 g/ml) at a rate 30 drops/min to 20 ml of a solution of $FeCl_2 \cdot 4H_2O$ dissolved in deoxygenated 4 *M* HCl. A saturated solution (1.9 m*M*) of NO was prepared by using a gas-tight syringe to inject 10 ml NO gas into 100 ml deoxygenated water, from which an aliquot of NO was removed and added to the bath to produce appropriate concentrations. A stock solution of CO (1.7 m*M*) was prepared by injecting 10 ml CO gas into a sealed flask of water (0°C). Appropriate concentrations were produced by adding an aliquot of CO stock solution to the bath from the stock flask.

Halothane, enflurane, and isoflurane were delivered from calibrated vaporizers to the O_2/CO_2 mixture aerating the Krebs solution. The concentrations in the resulting gas mixtures were monitored by an infrared analyzer (Datax, Model 254, Helsinki, Finland) which was calibrated using standard calibration gas mixtures. Bath concentrations of the anesthetics in Krebs solution were confirmed by gas chromatography as previously described (1).

For tests of possible interactions of halothane and nitric oxide, various concentrations of NO (1–3 μM) were mixed under anaerobic conditions with different concentrations of halothane (0.38–11.4 m*M*) in the absence of any tissue. Samples were taken repeatedly up to 120 min and analyzed for both NO and halothane concentrations. NO was measured spectrophotometrically at 548 nm (5).

Four pairs of cyclic CMP analyses were done on aortic rings without endothelium. These rings were placed in separate organ baths containing aerated Krebs solution at 37°C for 60 min. NE (EC_{60-70}) was added to all of the baths for 10 min. The rings were treated with NO (50 n*M*–5 μM) alone, or NO in the presence of the anesthetic (15 min pretreatment). Tissues were frozen in liquid nitrogen and the cGMP was extracted with ethanol. The extracts were analyzed for cGMP content using an Amersham cGMP radioimmunoassay (RIA) kit (Arlington Heights, IL). Results were expressed as nano-moles per gram of tissue wet weight.

The following chemicals were used: acetylcholine chloride, bradykinin, indomethacin, norepinephrine hydrochloride, superoxide dismutase (Sigma, St. Louis, MO); carbon monoxide (Aldrich, Milwaukee, WI); nitroglycerin (Sterling Drugs, McPherson, KS); isoflurane, enflurane (Anaquest, Madison, WI); and halothane (Halocarbon, N. Augusta, NC).

Relaxations caused by ACh, NO, NG, or CO were expressed as a percentage of the active tension produced by EC_{60-70} NE, and designated as percent relaxation of NE contraction on the graphs. All data are expressed as means ± SEM. Data were analyzed by analysis of variance (ANOVA) for repeated measures within groups. Where significant effects were determined, multiple comparisons were made using the Student Newman–Keuls test; $p \leq 0.05$ was considered significant.

III. Effects of Anesthetics on Endothelium-Dependent Relaxations of Isolated Blood Vessels

Halothane significantly attenuated the ACh-induced relaxations of NE-contracted canine carotid and femoral arteries and rabbit and rat aortas, as well as the bradykinin-induced relaxation of canine carotid arteries (1,3). The effects of halothane at levels equivalent to one and two minimal alveolar concentrations (MAC) used for anesthesia on ACh-induced relaxation of rat aortas are shown in Fig. 1. At concentrations similar to 1 MAC (0.37 m*M*, rat equivalent) halothane, neither enflurane (0.5 MAC, 0.34 m*M*) nor isoflurane (1 MAC, 0.33 m*M*) significantly inhibited ACh-induced relaxations of rat aorta. At higher concentrations (1 MAC, 0.68 m*M*) enflurane inhibited relaxations at low ACh concentrations, and isoflurane (2 MAC, 0.65 m*M*) attenuated relaxations at most ACh concentrations (Fig. 2).

Indomethacin did not significantly modify the inhibition of ACh relaxation of NE-contracted rat aortic rings by halothane. Pretreatment of the rings with SOD did attenuate (by approximately 25%) the action of 2 MAC halothane on ACh-induced relaxation, but the inhibitory effect of the anesthetic on ACh-induced relaxation was still significant.

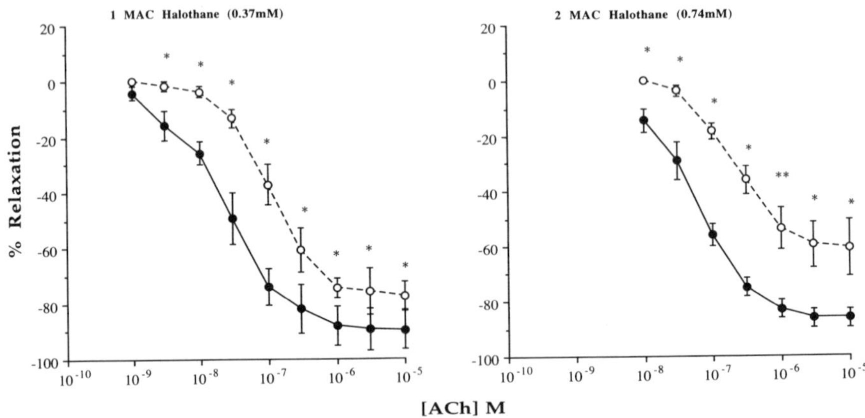

Fig. 1 Effects of 1 and 2 MAC halothane on ACh-induced relaxations of rat aortic rings. Values are means ± SEM ($n = 5$) and are expressed as the percent relaxation of NE (EC_{60-70})-induced tension. ●, ACh-induced relaxation in the absence of halothane; ○, ACh-induced relaxation in the presence of halothane. *Significant difference from preanesthetic response ($p < 0.05$); ** ($p < 0.01$).

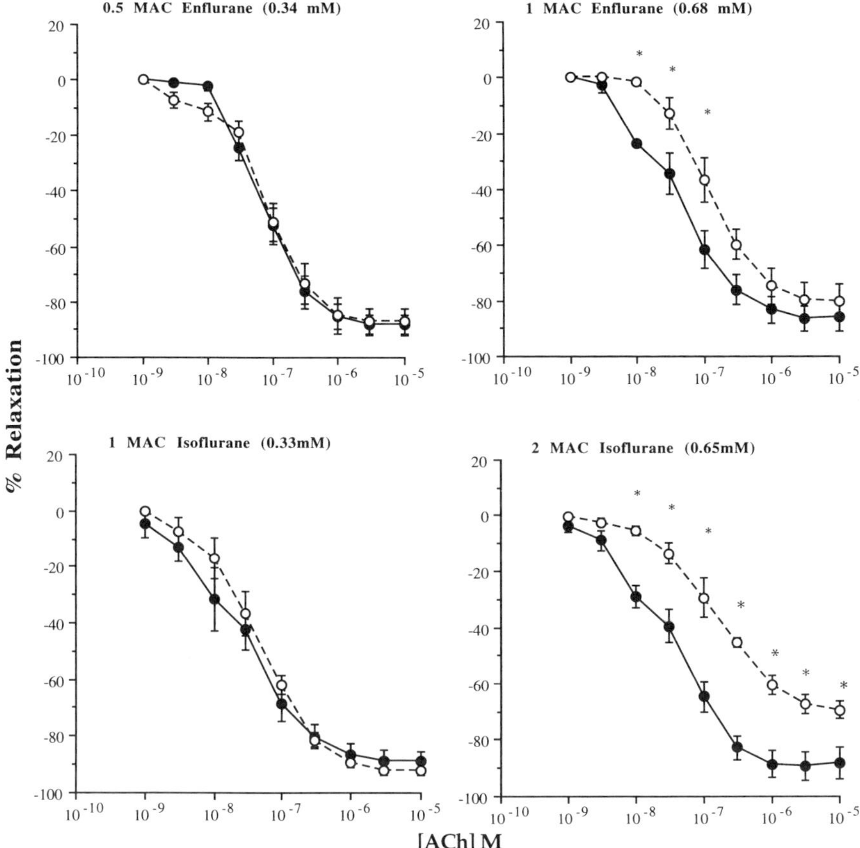

Fig. 2 Effects of 1 and 2 MAC isoflurane and 0.5 and 1 MAC enflurane on ACh-induced relaxations of rat aortic rings. ●, ACh-induced relaxation in the absence of anesthetics; ○, ACh-induced relaxation in the presence of anesthetics. Values are means ± SEM (n = 5) and are expressed as the percent relaxation of NE (EC_{60-70})-induced tension. *Significant difference from preanesthetic response ($p < 0.05$).

IV. Effects of Halothane and Isoflurane on Nitric Oxide-, Nitroglycerin-, and Carbon Monoxide-Induced Relaxations of Rat Aorta

Both 1 and 2 MAC halothane inhibited relaxations induced by NO, whereas isoflurane had significant inhibitory effects only at 2 MAC (Fig. 3). Nitroglycerin- (Fig. 4) and carbon monoxide-induced (Fig. 5) relaxations of rat aorta were significantly reduced at most concentrations by 2 MAC halothane.

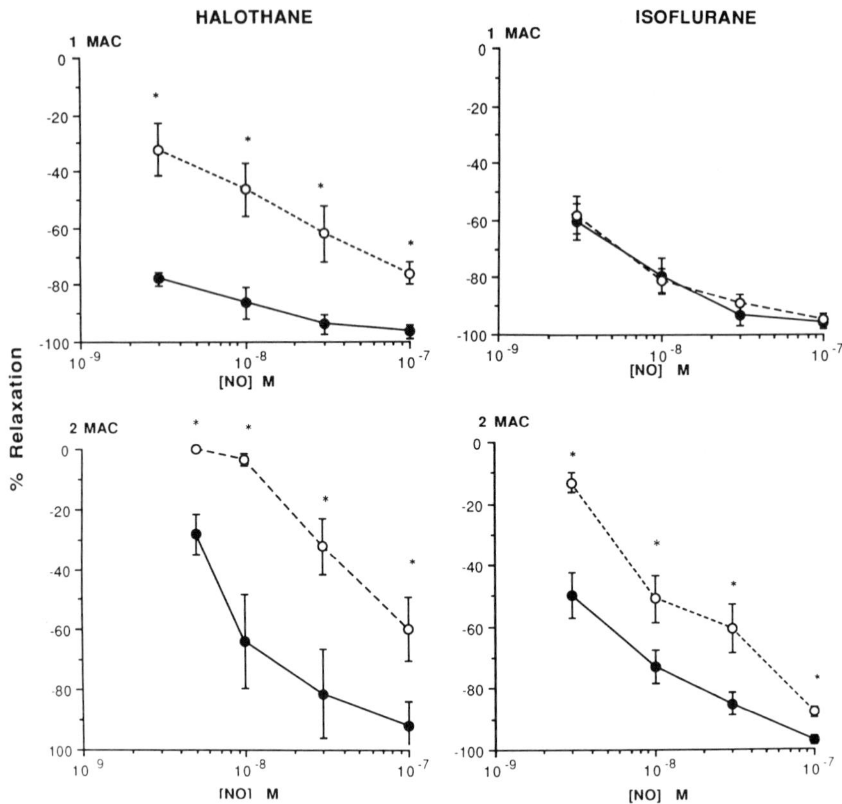

Fig. 3 Effects of 1 and 2 MAC halothane and isoflurane on NO-induced relaxations of endothelium-denuded rat aortic rings. Values are means ± SEM ($n = 5$) and are expressed as the percent relaxation of NE (EC_{60-70})-induced tension. ●, NO-induced relaxation in the absence of anesthetics; ○, NO-induced relaxation in the presence of anesthetics. *Significant difference from preanesthetic response ($p < 0.05$).

V. Effects of Halothane on Nitric Oxide-Stimulated Cyclic GMP

Figure 6 shows the effects of 2 MAC halothane on cGMP content following NO stimulation of isolated rat aortic rings. Halothane did not have significant effects on cGMP content in the absence of NO. Stimulation with NO (5×10^{-8}–$5 \times 10^{-6} M$) increased cGMP content in a concentration-related manner, and 2 MAC halothane significantly inhibited NO stimulation of cGMP at NO concentrations of $10^{-7} M$ and above.

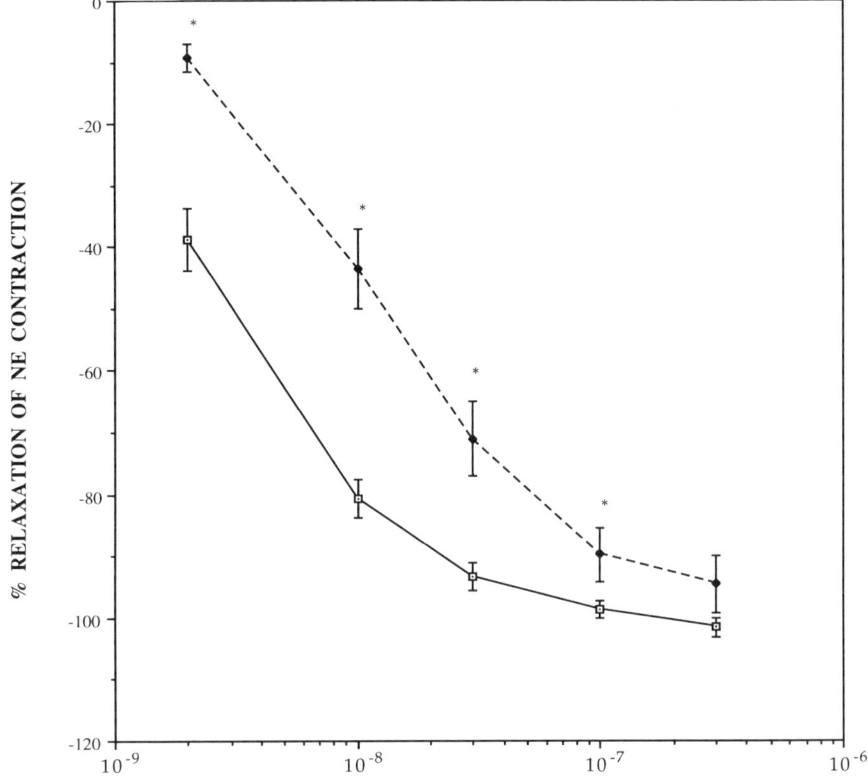

Fig. 4 Comparison of nitroglycerin-induced relaxation of endothelium-denuded rat aortic rings contracted with NE in the presence and absence of 2 MAC halothane. □, NG-induced relaxation in the absence of halothane; ◆, NG-induced relaxation in the presence of halothane. *Significant difference from time control ($p < 0.05$, $n = 5$). [Reprinted from Hart *et al.* (3) with permission.]

VI. Interactions of Halothane and Nitric Oxide in Absence of Tissues

Incubation mixtures of several concentrations of NO (1–3 μM) and halothane (0.38–11.4 mM) were stable for up to 120 min. There were no detectable changes in either the halothane or NO content of these mix-

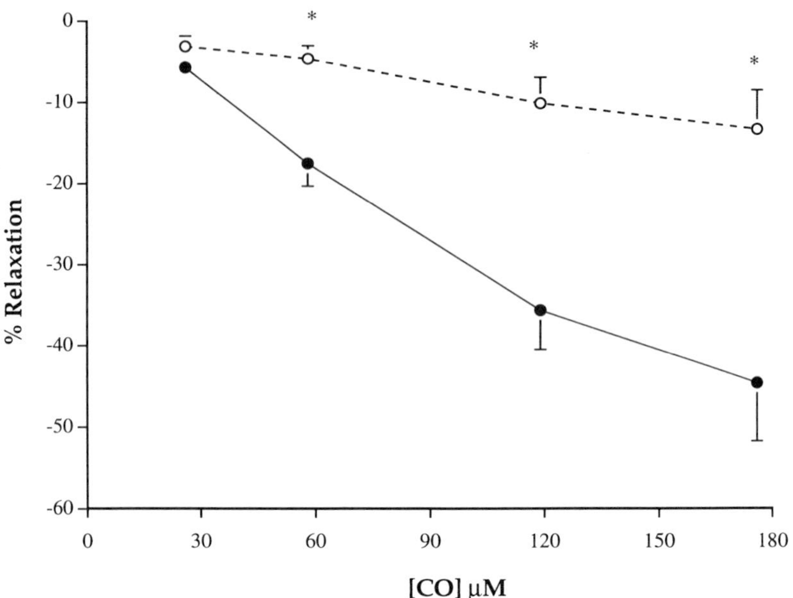

Fig. 5 Comparison of carbon monoxide-induced relaxation of rat aortic rings contracted with NE in the presence and absence of 2 MAC halothane. ●, CO-induced relaxation in the absence of halothane; ○, CO-induced relaxation in the presence of halothane. *Significant difference from preanesthetic treatment ($p < 0.05$, $n = 5$).

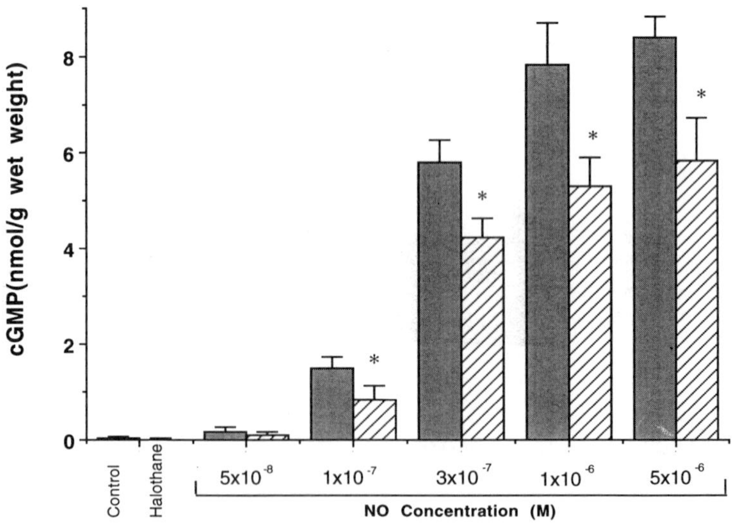

Fig. 6 Effects of halothane, NO (shaded bars), and NO plus 2 MAC halothane (hatched bars) on cGMP content (nmol/g wet weight) of rat aortic rings without endothelium. *Significant difference from NO-treated vessels ($p < 0.05$, $n = 4–5$). [Reprinted from Hart *et al.* (3) with permission.]

tures, demonstrating that there was not a direct chemical interaction between the two compounds.

VII. Discussion

In 1988 we reported that halothane interfered with ACh- and bradykinin-induced vasodilatory mechanisms and proposed that the anesthetic could interfere with EDRF/NO activity at a number of sites, namely, synthesis/release of EDRF, transit of EDRF, and/or interference with its site of action in the vascular smooth muscle (1). EDRF is now known to be NO. The synthesis of NO from L-arginine is catalyzed by an apparently widely distributed enzyme, nitric oxide synthase (NOS). The availability of calcium is the major mode of regulation of this NO synthesizing enzyme (constitutive NOS) in the vascular endothelium. Based on what is known about the action of halothane in endothelial cells (6), neurons (7), and other cell types (8), the activation of the enzyme by calcium is a logical site of action for anesthetics. We have not examined this interaction, but others have reported (9) that in homogenates of rat cerebellum halothane has an inhibitory action on NOS activity. Whether halothane could interfere with other aspects of NOS activation is unknown.

It is also possible that halothane and other anesthetics could inactivate NO after its production either by a direct mechanism or during transit to its receptor site in the smooth muscle. Our investigations do not support such a chemical inactivation of NO by halothane since in aqueous media, in the absence of tissue, NO and halothane did not chemically react. Thus, attenuation of ACh-induced relaxations by halothane cannot be explained by a chemical reaction or bonding between two compounds at the concentrations tested.

Another possibility is that the anesthetics act indirectly to produce or enhance free radical activity within the endothelial cell. Yoshida and Okabe (10) suggested that the formation of superoxide radicals was the mechanism underlying the inactivation by sevoflurane of endothelial-dependent relaxation. However, this does not appear to be the primary mechanism with halothane, since the addition of SOD reduced but did not eliminate the inhibitory action of halothane on ACh-induced relaxation.

In order to bypass the endogenous NO production and release by the endothelium, we examined the actions of anesthetics in vessels without endothelium. Exogenous NO, CO, and NO-generating agents were used to activate soluble guanylate cyclase (sGC) directly. We found that halothane significantly inhibited vascular relaxation induced by exogenous NO and the NO donor drug nitroglycerin in concentrations ranging from (3×10^{-9}–$1 \times 10^{-7} M$). Because the relaxant responses to the dilators result from activation of guanylate cyclase and production of cGMP, it was suggested that halothane and, to a lesser extent, enflurane and isoflurane have an inhibitory effect on sGC in the smooth muscle cell. Blaise *et al.* (11) have also reported that in endothelial-denuded vessels halothane significantly suppressed relaxation induced by exogenous NO.

Like NO, carbon monoxide activates GC by binding to the heme group in the enzyme (12). Relaxation induced by CO was also inhibited by halothane. This provides further support for an interaction between the anesthetic and GC. In addition, studies using hepatic tissue provide direct evidence that halothane and isoflurane inhibit NO activation of this enzyme (13). These studies are of particular interest because they examine the effect of the anesthetics on the enzyme activation and do not depend on muscle activity.

We propose that the mechanism(s) whereby the anesthetics could interfere with activation of sGC is dependent on their chemical structure. As a consequence of their specific halogen content, halogenated anesthetics are attracted to ferrous heme non-uniformly. Because halothane contains bromine, in addition to chlorine and fluorine, it has the greatest affinity for ferrous heme proteins. Halothane is attracted to and reduced by cytochrome *P*-450 as is evidenced by its extensive metabolism (14). In contrast, isoflurane and enflurane, which contain chlorine and fluorine but not bromine, do not show the same affinity for ferrous heme proteins, and this is supported by their more limited metabolism. Halothane, at both 1 and 2 MAC, inhibited NO-induced relaxation and decreased NO stimulation of cGMP. In contrast, 1 MAC isoflurane was without effect, but 2 MAC isoflurane significantly inhibited those responses. These observations suggest that isoflurane has a lower affinity than halothane for the ferrous heme of sGC; thus, a higher concentration of isoflurane is required to achieve the same inhibition.

In conclusion, we report inhibitory effects of halothane, enflurane, and isoflurane on ACh-, NO- and CO-induced relaxations in vascular smooth muscle. Our results support the hypothesis that this inhibition is the result of competition between NO and the anesthetics for the heme moiety on the soluble guanylate cyclase. The higher heme affinity of halothane compared to isoflurane and enflurane may be the basis, therefore, of the greater inhibitory effects on NO-mediated relaxation. How this action may contribute to the overall effects of these anesthetics on vascular control remains to be determined.

Acknowledgments

The valuable advice and assistance of Dr. R. A. Van Dyke are gratefully acknowledged. These studies were supported by Uniformed Services University of the Health Sciences Grants CO-8038 and RO-8006.

References

1. Muldoon, S. M., Hart, J. L., Bowen, K. A., and Freas, W., (1988) Attenuation of endothelium-mediated vasodilation by halothane. *Anesthesiology* **68,** 31–37.
2. Blaise, G. A., Still, J. C., Nugent, M., Van Dyke, R. A., and Vanhoutte, P. M. (1987). Isoflurane causes endothelium-dependent inhibition of contractile response of canine coronary artery rings. *Anesthesiology* **67,** 513–517.
3. Hart, J. L., Jing, M., Bina, S., Van Dyke, V. A., Freas, W., and Muldoon, S. M. (1993). Effects of halothane on EDRF/cGMP-mediated vascular smooth muscle relaxation. *Anesthesiology* **79,** 323–331.
4. Ignarro, L. J., Adams, J. B., Horwitz, P. M., and Wood, K. S. (1986). Activation of soluble guanylate cyclase by NO-hemeproteins involve NO-heme containing and heme-deficient enzyme forms. *J. Biol. Chem.* **261,** 4997–5002.
5. Bina, S., Freas, W., Hart, J. L., Jing, M., Gantt, R. M., and Muldoon, S. M. (1992). Interaction between nitric oxide and halothane. *FASEB J.* **6,** A1865.
6. Loeb, A. L., O'Brien, D. K., and Longnecker, D. E. (1993). Alteration of calcium mobilization in endothelial cells by volatile anesthetics. *Biochem Pharmacol.* **45,** 1173–1142.
7. Puil, E., El-Beheiry, H., and Baimbridge, K. G. (1990). Anesthetic effects on glutamate-stimulated increase in intraneuronal calcium. *J. Pharmacol. Exp. Ther.* **255,** 955–961.
8. Johnson, M. E., Sill, J. C., Uhl, C. B., and Van Dyke, R. A. (1993). Effect of halothane on hypoxic toxicity and glutathione status in cultured rat hepatocytes. *Anesthesiology* **79,** 1061–1071.
9. Tobin, J. R., Martin, L. D., Breslow, M. J., and Traystman, R. J. (1993). Anesthetic inhibition of brain nitric oxide synthase. *FASEB J.* **7,** A257.
10. Yoshida, K. I., and Okabe, E. (1992). Selective impairment of endothelium dependent relaxation by sevoflurane: Oxygen free radicals participation. *Anesthesiology* **76,** 440–447.
11. Blaise, G. A., Quy, T. O., Parent, M., and Asenjo, F. (1992). Does halothane inhibit the release or the action of EDRF? *Anesthesiology* **77,** A692.
12. Verma, A., Hirsch, D. J., Glatt, C. E., Ronnett, G. V., and Snyder, S. H. (1993). Carbon monoxide: A putative neural messenger. *Science* **259,** 381–384.
13. Van Dyke, R. A., Masaki, E., Muldoon, S. M., and Marsh, H. M. (1993). The effects of halothane and isoflurane on nitric oxide stimulated soluble guanylate cyclase. *Anesthesiology* **79,** A397.
14. Van Dyke, R. A. (1982). Hepatic centrilobular necrosis in rats after exposure to halothane, enflurane, or isoflurane. *Anesth. Analg.* **61,** 812–819.

Effects of Epidural Anesthesia on Splanchnic Capacitance

Quinn H. Hogan, Anna Stadnicka, and John P. Kampine
Department of Anesthesiology
The Medical College of Wisconsin
Milwaukee, Wisconsin 53226

I. Introduction

Regional anesthetic techniques, using local anesthetic injection to block the function of major neural trunks, provide an alternative to general anesthesia for most surgical procedures. Study of the physiological effects of neural blockade, however, has been less intense than research on general anesthesia. It is clear that most of the physiological changes following neuraxial (epidural and spinal) blockade are due to decreased sympathetic activity, but a major limitation is the lack of an adequate means to measure sympathetic blockade. Studies have used catecholamine levels, skin blood flow and temperature, sweating, sympathogalvanic response, and the extent of sensory blockade as indirect indicators of sympathetic interruption, producing a confusing picture with little agreement on the extent and completeness of sympathetic blockade.

Hypotension from sympathetic block is a deleterious feature of extensive neuraxial blockade. This can be attributed to three sources. Decreased peripheral vascular resistance owing to arteriolar vasodilatation is a well-documented factor and accounts for the unchanged cardiac output that usually accompanies block-induced hypotension. Myocardial contractility decreases with blocks that extend to spinal segments T1–T5, but this is usually a minor contribution to hypotension after blockade. A third factor,

Advances in Pharmacology, Volume 31

systemic venodilatation, is of particular importance because it produces hypotension through decreased venous return and therefore decreased cardiac output. It is a variable and sometimes abrupt feature of extensive blocks.

The relative contribution of various venous beds to increased capacitance during regional anesthesia is uncertain. The limbs have veins with negligible sympathetic innervation and contribute minimally to reflex changes in capacitance or to sympathetic denervation with neural blockade (1–4). The splanchnic veins, however, have extensive sympathetic innervation which maintains a baseline tone in the vessels and produces the majority of the total systemic reflex capacitance changes (2,5,6). Early authorities speculated that splanchnic venodilatation was a key component causing hypotension during blockade (7), but no evidence has been forthcoming. Our studies therefore have used direct measures of sympathetic efferent nerve activity (SENA) and vein diameter (VD) to examine the venous capacitance effects of epidural anesthesia in rabbits.

II. Epidural Anesthesia

Anesthesia was maintained with α-chloralose, and vecuronium provided neuromuscular blockade. Surgical preparation of the animals (8) included tracheostomy for mechanical ventilation, intraarterial blood pressure measurement, and the placement of a catheter (1 mm outer diameter) into the epidural space through a small dorsal midline incision. A bipolar recording electrode was placed on a postganglionic branch from the celiac ganglion for measuring SENA. A loop of small bowel was bathed in warmed artificial peritoneal fluid in a plexiglass chamber designed to allow transillumination of mesenteric veins with diameters of 500–800 μm. Videomicrography provided continuous measurement of VD. Prior to local anesthetic injections and at the end of the protocol, the adequacy of the VD and SENA preparation was examined by exposing the animal to 40 sec of hypoxia (F_IO_2 of 0). A typical response shows sympathetic hyperactivity and a decrease in VD of about 20%.

Epidural injections of lidocaine were made with a volume (0.4 ml/kg) large enough to reach most of the preganglionic sympathetic fibers (T1–L4) to minimize variability in the extent of blockade. Lidocaine concentrations of 0.5, 1, and 1.5% were used for epidural injection, and results were compared to controls receiving intramuscular lidocaine at doses of 6 or 15 mg/kg which produced a similar range of serum lidocaine concentrations (1–3.5 μg/ml). Groups included between 6 and 8 animals. Observations were made for 120 min. Hexamethonium (50 mg/kg i.v.) is administered at the termination of the protocol to confirm that the nerve recording is from a postganglionic trunk and to establish a baseline for calculation of SENA.

In other experiments, intraluminal pressure in the mesenteric veins was measured with a servo-null micropipette system. Epinephrine and norepinephrine levels were also examined in control and blocked animals by high-performance liquid chromatography. Topical lidocaine (10 and 100 μg/ml for 15 min) or tetrodotoxin (TTX, 1 *M*) was applied to the mesenteric vessels by administration in the superfusate. Epidural anesthesia (1.0% lidocaine) was induced during the continued exposure of the vessel to TTX. The vertebral columns of the animals were dissected at the completion of the protocol to confirm epidural catheter placement and extent of fluid passage.

III. Splanchnic Capacitance

Epidural lidocaine produced a prompt decrease in mean arterial pressure (MAP) of about 60%, compared to 8 and 13% in the control groups (Fig. 1). Hypotension lasted longer with higher epidural lidocaine concentrations, probably owing to the larger mass of drug providing a reservoir for more prolonged block. Despite the marked degree of hypotension, bicarbonate requirements did not differ from the control groups, indicating that adequate perfusion was maintained. With higher epidural doses, heart rate decreased by about 10%, most likely secondary to blockade of preganglionic cardioaccelerator fibers.

The onset of epidural anesthesia in this model was accompanied by a dramatic decrease in SENA (Fig. 2) averaging between 75 and 86%, in agreement with direct microneurographic measurement of sympathetic activity in the leg in humans during epidural anesthesia (9,10). SENA also decreased with intramuscular lidocaine, but only at the higher dose and by only 26%, similar to the changes in SENA to the heart, kidney, and leg in other studies of systemic lidocaine (11–13).

Concurrent with the SENA decrease, mesenteric veins dilated by 7.5 to 10% (Fig. 3), corresponding to a capacitance increase of about 20%. Complete blockade of sympathetic fibers with intravenous hexamethonium produced a similar venodilatation, indicating that the vessels in the epidural animals were almost fully denervated during epidural blockade. The congruent time courses of VD and systemic blood pressure changes suggests that venodilatation was a factor contributing to hypotension.

Only one other study has examined splanchnic capacitance changes with regional anesthesia, using scintigraphy to measure the changes in

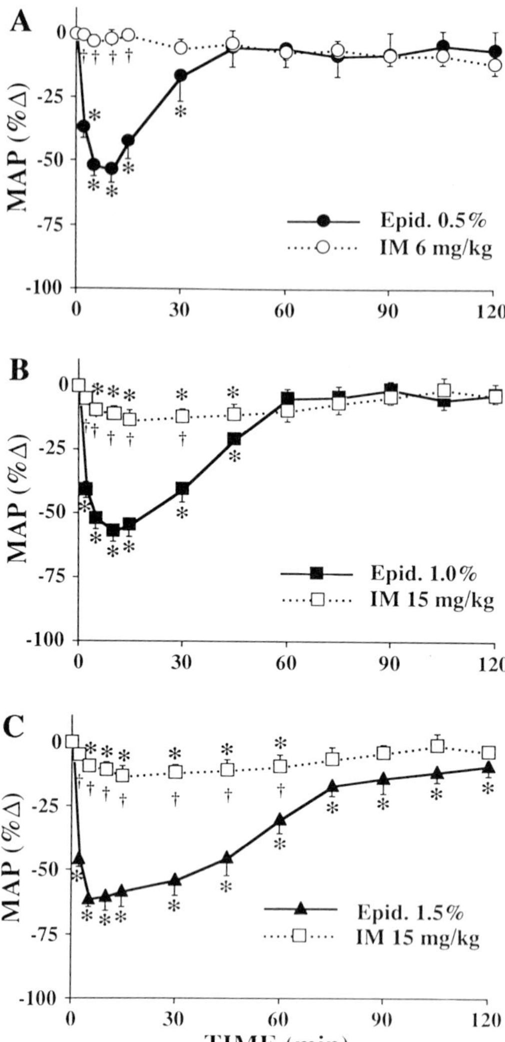

Fig. 1 Mean arterial pressure changes following epidural anesthesia with three concentrations of lidocaine compared to changes after intramuscular (IM) injection of lidocaine at two different doses. Values are means ± SEM. *Significant difference ($p < 0.05$) compared to baseline (time = 0). †Significant difference ($p < 0.05$) in comparing the epidural to intramuscular value at that time point.

distribution of radiolabeled red cells during epidural anesthesia in human volunteers (14). Although the block extended to the fourth thoracic dermatome, splanchnic blood volume decreased in six subjects in whom hemody-

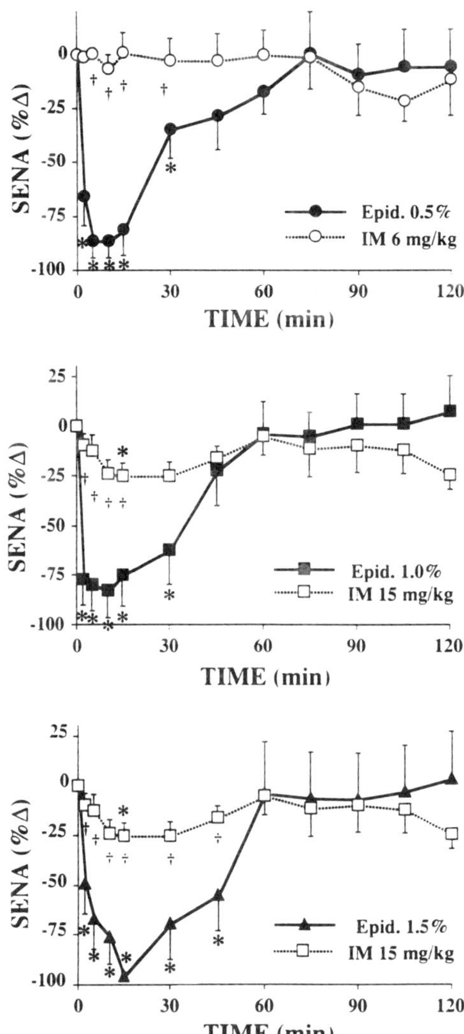

Fig. 2 Sympathetic efferent nerve activity (SENA) changes following epidural anesthesia with three concentrations of lidocaine compared to changes after intramuscular injection of lidocaine at two different doses. Values are means ± SEM. *Significant difference ($p < 0.05$) compared to baseline (time = 0). †Significant difference ($p < 0.05$) in comparing the epidural to intramuscular value at that time point.

namic changes were minor. In two subjects, however, splanchnic volume increased associated with abrupt decreases in systemic blood pressure. The variability in splanchnic capacitance response to segmental denerva-

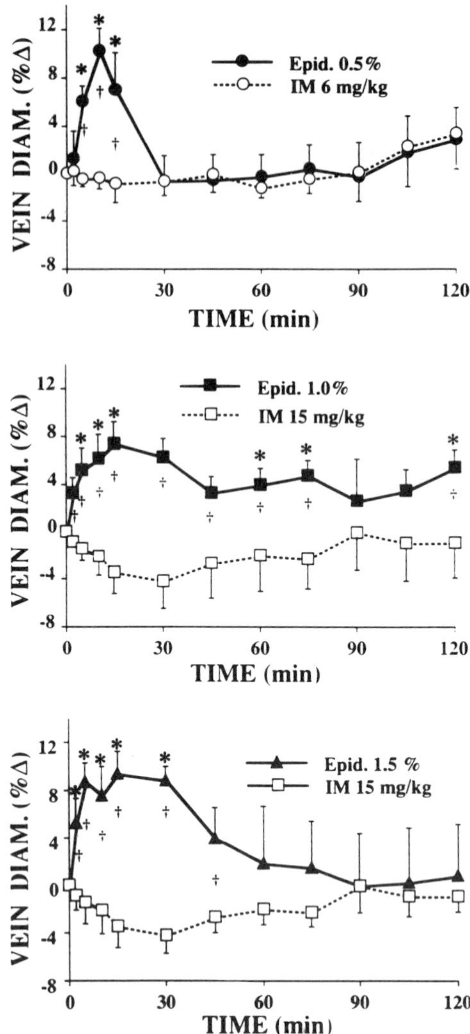

Fig. 3 Vein diameter (VD) changes following epidural anesthesia with three concentrations of lidocaine compared to changes after intramuscular injection of lidocaine at two different doses. Values are means ± SEM. *Significant difference ($p < 0.05$) compared to baseline (time = 0). †Significant difference ($p < 0.05$) in comparing the epidural to intramuscular value at that time point.

tion can be explained by the balance of two opposing influences. Hypotension from the block results in baroreceptor-driven increased central sympathetic activity, but blockade of the preganglionic sympathetic fibers by local anesthetic decreases the SENA reaching the splanchnic veins. In our model, extensive and intense blockade predominates, whereas clinical blocks in humans may often be inadequate to conceal increased sympathetic drive to the veins.

Of 70 epidural injections in rabbits, one resulted in death soon after blockade. The animal was unique in that the preepidural hypoxia test resulted in a VD decrease of 40%, indicating much greater sympathetic reactivity than in other animals. Within 3 min after epidural lidocaine, VD increased by 19%, SENA decreased 100%, heart rate decreased by 25%, and MAP decreased from 80 to 20 mmHg with a flat arterial wave form. Necropsy showed the epidural catheter to be properly placed. This sequence of events may reflect similar changes in occasional patients following spinal and epidural anesthesia. Unexpected cardiovascular collapse with bradycardia and low cardiac output may precipitously arise in the absence of apparent dosing or technical error (15).

Other mechanisms which might contribute to mesenteric VD changes with regional blockade include a passive response to increased transmural pressure, altered circulating catecholamines, or a direct effect of local anesthetics on the vessel. Increased VD with the onset of blockade (Fig. 4) was accompanied by a 42% decrease in vein pressure (from 8.75 to 5.79 mmHg), eliminating the possibility of a passive contribution to venodilatation. There were no changes in vein pressure with intramuscular lidocaine.

Epidural anesthesia has been shown to decrease circulating catecholamines, and *in vitro* study of mesenteric veins has documented their responsiveness to exogenous catecholamines (16). This suggests that altered catecholamine levels could be a contributing factor to VD changes during blockade. Plasma norepinephrine (NE) and epinephrine levels in our rabbit model (Fig. 5) were found to be very high, probably because of the extensive surgical preparation and the use of α-chloralose as the anesthetic, which minimally suppresses catecholamine production. There were no differences in epinephrine levels between the epidural blockade group and intramuscular lidocaine rabbits, but NE levels were significantly lower during epidural blockade compared to intramuscular controls. Because the decrease in NE was only 28% with onset of blockade, and because levels remained very high, this slight withdrawal of endogenous vasoconstrictor does not likely play a large role in splanchnic venodilatation during epidural anesthesia.

The direct effects of local anesthetics on mesenteric veins has been studied *in vitro,* examining the contraction of vein rings in response to exogenously applied NE and to NE released by the sympathetic nerve terminals during field stimulation (17). Lidocaine slightly amplified con-

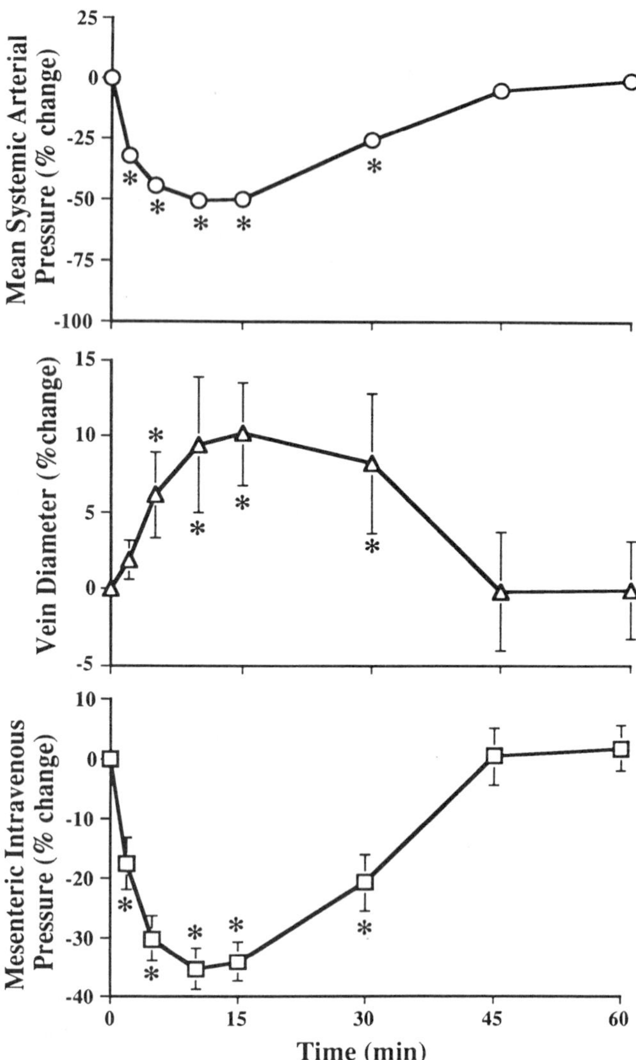

Fig. 4 Mesenteric intravenous pressure, mesenteric vein diameter, and mean systemic arterial pressure changes after epidural block with lidocaine (1.0%). Values are means ± SEM. *Significant difference ($p < 0.05$) compared to baseline (time = 0).

traction from exogenous NE at lidocaine concentrations of 20–100 μg/ml, but otherwise had no effect. Response to field stimulation was decreased 25% by lidocaine concentrations of 5 μg/ml and progressively more at higher concentrations. Although this implies a possible contribu-

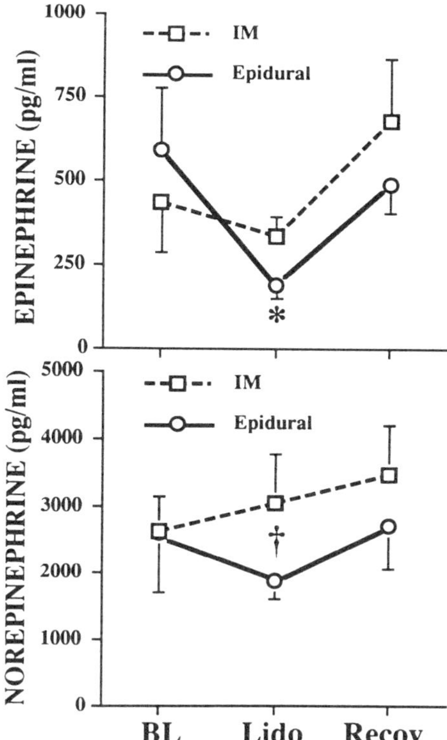

Fig. 5 Catecholamine changes after epidural anesthesia (lidocaine at 1% or 0.4 ml/kg) versus intramuscular injection of lidocaine (15 mg/kg). Values are means ± SEM. *Significant difference ($p < 0.05$) compared to baseline (time = 0). †Significant difference ($p < 0.05$) in comparing the epidural to intramuscular value at that time point.

tion by circulating lidocaine to venodilatation, the absence of increased VD in control animals with plasma lidocaine concentrations similar to the epidural animals indicates at most a minor role. Furthermore, lidocaine in the superfusate bathing mesenteric vessels *in situ* produced no VD change at concentrations of 10 μg/ml and dilated veins by 3.5% only at 100 μg/ml.

We hypothesize that decreased sympathetic activity to the veins during blockade is the principal factor causing mesenteric venodilatation during eppidural anesthesia. This predicts an absence of response in veins exposed to tetrodotoxins (TTX, a voltage-gate Na^+ channel blocker), which blocks NE release from sympathetic neural terminals. In the same rabbit *in situ* mesenteric veins preparation described above, TTX in the superfusate bath produced an increase in VD of 7.6%, comparable to the dilatation

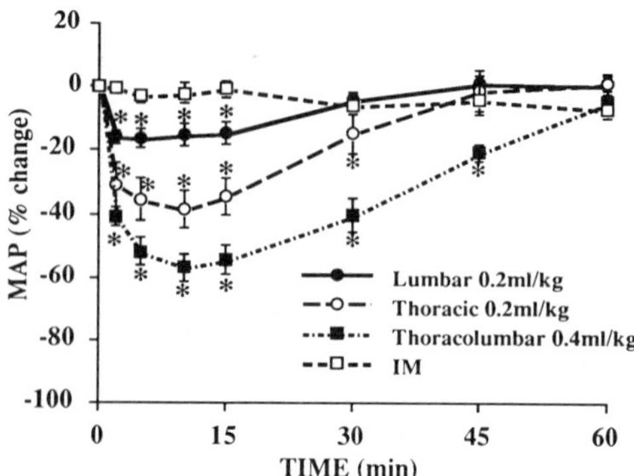

Fig. 6 Changes in mean arterial pressure after lumbar, thoracic, or thoracolumbar epidural block, compared to intramuscular lidocaine (6 mg/kg). Values are means ± SEM. *Significant difference ($p < 0.05$) compared to baseline (time = 0).

produced by epidural anesthesia without TTX. This is further evidence for the near-complete sympathetic denervation produced by epidural lidocaine. Epidural anesthesia during exposure of the mesentery to TTX resulted in a decrease of VD by 1.41%, probably owing to falling transmural pressures with unchanged wall tension. The absence of venodilatation supports our supposition that neural blockade of sympathetic tone is the major mechanism in epidural venodilatation: if circulating NE had contributed to vessel tone prior to blockade and NE withdrawal was important in producing VD changes with blockade, this would by evident by dilatation even after TTX.

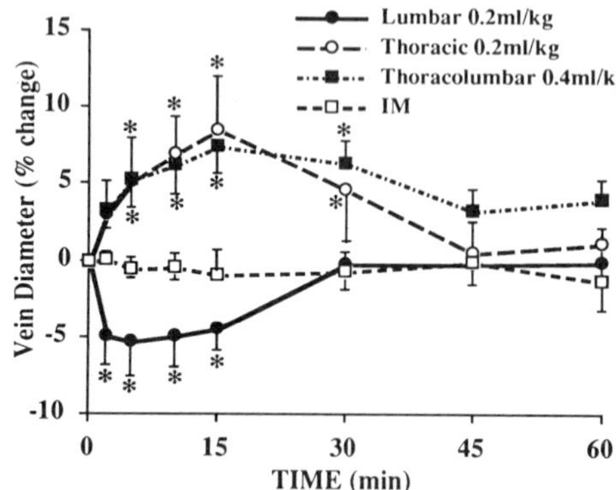

Fig. 7 Changes in vein diameter after lumbar, thoracic, or thoracolumbar epidural block, compared to intramuscular lidocaine (6 mg/kg). Values are means ± SEM. *Significant difference ($p < 0.05$) compared to baseline (time = 0).

In a clinical setting, different hemodynamic responses are seen with blockade of different ranges of spinal nerves. Minimal changes take place with blocks isolated to the lower extremities. Extensive blocks involving the splanchnic innervation often result in hypotension and hypoperfusion, indicating the pivotal nature of the splanchnic vascular bed. Therefore, we compared the effects on VD and SENA of epidural blocks centered on the lumbar and thoracic segments in the rabbit model, using a diminished lidocaine dose of 0.2 ml/kg to limit the extent of anesthetic spread. Maximal decrease in MAP (Fig. 6) was 16.5% in the lumbar group and 38.3% in the thoracic group, compared to 56.9% with extensive thoracolumbar blocks using 0.4 ml/kg epidural lidocaine. The greater fall of MAP in the thoracolumabar group may be expained by the more extensive sympathetic denervation. The importance of splanchnic capacitance is again evident; the maximal VD change in the lumbar group (Fig. 7) was −4.9% (constriction), which would tend to limit MAP fall in this group, as opposed to +8.6% (dilatation) with thoracic blocks and +7.5% with thoracolumbar blocks. Perhaps the venodilatation in the thoracolumbar group is diminished by a vein pressure fall greater than in the thoracic block animals. Reciprocal changes in SENA were seen (Fig. 8), with a maximal change in the lumbar group of +15%, −47% in the thoracic group, and −83% in the thoracolumbar group. SENA most likely increased during lumbar block in response to baroreceptor stimulation unimpeded by splanchnic neural blockade, and may have limited the MAP decrease. Thoracic block produced a decrease in SENA less than the decrease during thoracolumbar block because of less completely blocked fibers with the smaller dose.

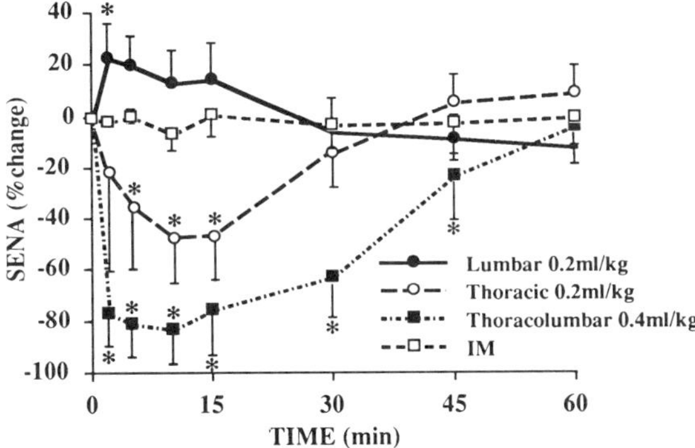

Fig. 8 Changes in sympathetic efferent nerve activity after lumbar, thoracic, or thoracolumbar epidural block, compared to intramuscular lidocaine (6 mg/kg). Values are means ± SEM. *Significant difference ($p < 0.05$) compared to baseline (time = 0).

Our investigations delineate an interaction of reflex activation, neural blockade, and possibly direct vascular effects of catecholamines and local anesthetics contributing to hemodynamic changes during neuraxial anesthesia. Splanchnic vascular capacitance is subject to tonic sympathetic neural tone, and changes in splanchnic capacitance play a role in hemodynamic adjustments to epidural blockade. Future work will examine the modification during epidural anesthesia of responses to stress such as hypoxia, hypercarbia, and baroreceptor stimulation.

IV. Summary

Splanchnic veins play an important role in the active control of total body circulatory capacitance. The effects of epidural anesthesia on splanchnic venous capacitance have not previously been examined. A rabbit model using direct measures of mesenteric vein diameter and sympathetic efferent nerve activity was used to test the response to epidural lidocaine at three different doses and to intramuscular lidocaine at two doses. Epidural anesthesia produced hypotension, mesenteric venodilatation, and interruption of sympathetic activity. Maximal changes of these parameters were comparable in the three epidural dosage groups but were more prolonged with increasing dose. High-dose systemic lidocaine caused smaller changes in arterial pressure and sympathetic acitivity.

Further experiments were done to investigate the mechanism of splanchnic venodilatation. Passive vein distension and effects of circulating lidocaine or catecholamines are not likely contributing factors. Blocks limited to thoracic segments, but including the origin of splanchnic preganglionic fibers, produce comparable mesenteric venodilatation and sympathetic interruption as extensive thoracolumbar blocks. Blocks limited to lumbar segments, however, showed mesenteric venoconstriction and increased splanchnic sympathetic activity. The variable responses in splanchnic capacitance with the onset of epidural anesthesia are the result of the competing influences of increased sympathetic activity from decreasing blood pressure and blockade of sympathetic fibers to the splanchnic veins.

References

1. Fuxe, K., and Sedvall, G. (1965). The distribution of adrenergic nerve fibers to the blood vessels in skeletal muscle. *Acta Physiol. Scand.* **64,** 75–86.
2. Hainsworth, R. (1990). The importance of vascular capacitance in cardiovascular control. *News in Physiological Sciences* **5,** 250–254.
3. Modig, J., Malmberg, P., and Karlstrom, G. (1980). Effect of epidural versus general anaesthesia on calf blood flow. *Acta Anaesthesiol. Scand.* **24,** 305–309.

4. Perhoniemi, V., and Linko, K. (1987). Effect of spinal versus epidural anaesthesia with 0.5% bupivacaine on lower limb blood flow. *Acta Anaesthesiol. Scand.* **31,** 117–121.
5. Bohlen, H. G., and Gore, R. W. (1977). Comparison of microvascular pressures and diameters in the innervated and denervated rat intestine. *Microvasc. Res.* **14,** 2551–2564.
6. Rothe, C. F. (1983). Reflex control of veins and vascular capacitance. *Physiol. Rev.* **63,** 1281–1342.
7. Labat, G. (1927). Circulatory disturbances associated with subarachnoid block. *Long Island Med. J.* **21,** 573.
8. Hogan, Q. H., Stadnicka, A., Stekiel, T. A., Bosnjak, Z. J., and Kampine, J. P. (1994). Effects of epidural and systemic lidocaine on sympathetic activity and mesenteric circulation in rabbits. *Anesthesiology* **79,** 1250–1260.
9. Lundin, S., Wallin, B., and Elam, M. (1989). Intraneural recording of muscle sympathetic activity during epidural anesthesia in humans. *Anesth. Analg.* (*N.Y.*) **69,** 788–793.
10. Lundin, S., Kirno, K., Wallin, B., and Elam, M. (1990). Effects of epidural anesthesia on sympathetic nerve discharge to the skin. *Acta Anaesthesiol. Scand.* **34,** 492–497.
11. Nishikawa, K., Fukada, T., Yukioka, H., and Fujimori, M. (1990). Effects of intravenous administration of local anesthetics on the renal sympathetic nerve activity during nitrous oxide and nitrous oxide–halothane anesthesia in the cat. *Acta Anaesthesiol. Scand.* **34,** 231–236.
12. Miller, B. D., Thames, M. D., and Mark, A. L. (1983). Inhibition of cardiac sympathetic nerve activity during intravenous administration of lidocaine. *J. Clin. Invest.* **71,** 1247–1253.
13. Ebert, T., Mohanty, P., and Kampine, J. P. (1991). Lidocaine attenuates efferent sympathetic responses to stress in humans. *J. Cardiothorac. Vasc. Anesth.* **5,** 437–443.
14. Arndt, J., Hock, A., Stanton-Hicks, M., and Stuhmeier, K.-D. (1985). Peridural anesthesia and the distribution of blood in supine humans. *Anesthesiology* **63,** 616–623.
15. Caplan R., Ward, R., Posner, K., and Cheney, F. (1988). Unexpected cardiac arrest during spinal anesthesia: A closed claims analysis of predisposing factors. *Anesthesiology* **68,** 5–11.
16. Stadnicka, A., Flynn, N., Bosnjak, Z. J., and Kampine, J. P. (1993). Enflurane, halothane, and isoflurane attenuate contractile responses to exogenous and endogenous norepinephrine in isolated small mesenteric veins of the rabbit. *Anesthesiology* **78,** 326–334.
17. Stadnicka, A., Hogan, Q., Stekiel, T., Bosnjak, Z. J., and Kampine, J. P. (1993). Dose-dependent effects of lidocaine in isolated small mesenteric veins of the rabbit. *Reg. Anesth.* **18**(S), 89.

Anesthetic Modulation of Pulmonary Vascular Regulation

Paul A. Murray

Department of Anesthesiology and Critical Care Medicine
The Johns Hopkins University School of Medicine
Baltimore, Maryland 21287

I. Introduction

A variety of vasoactive mechanisms, both intrinsic and extrinsic to the lung, can modulate the pulmonary circulation. Anesthetic agents have the potential to alter pulmonary vasomotor tone directly, or to modify the complex array of vasoactive mechanisms that regulate the pulmonary circulation. Our laboratory is systematically investigating the effects of general anesthesia on neural, humoral, and local mechanisms of pulmonary vasoregulation. We have developed several techniques that allow the generation of multipoint pulmonary vascular pressure–flow ($P/\dot{Q}$) plots in chronically instrumented conscious dogs (1,2). These techniques allow us to measure the pulmonary vascular $P/\dot{Q}$ relationship in the conscious state, and again in the same animal during general anesthesia. This experimental approach avoids the confounding effects of background anesthetics and acute surgical trauma. It also allows us to distinguish between active and passive (flow-dependent) effects of anesthetics on the pulmonary circulation. This article reviews some of the work from our laboratory concerning anesthetic modulation of pulmonary vascular regulation in chronically instrumented dogs.

Advances in Pharmacology, Volume 31

II. Measurement of Pulmonary Vasoregulation

A. Surgical Preparation for Chronic Instrumentation

Conditioned, microfilaria-free mongrel dogs weighing approximately 30 kg are sedated with morphine sulfate (10 mg i.m.) and anesthetized with pentobarbital sodium (20 mg/kg i.v.) and fentanyl citrate (15 μg/kg i.v.). The dogs are intubated, placed on positive pressure ventilation, and prepared for sterile surgery. The heart and great vessels are exposed through the fifth intercostal space. The pericardium is incised ventral to the phrenic nerve. Heparin-filled Tygon catheters (1.02 mm ID, Norton) are implanted in the descending thoracic aorta, left and right atrium, and main pulmonary artery. Hydraulic occluders (18–22 mm ID, Jones) are positioned around the right main pulmonary artery and thoracic inferior vena cava. An electromagnetic flow probe (10–12 mm ID, Zepeda) is positioned around the left main pulmonary artery. The edges of the pericardium are loosely apposed. The catheters, hydraulic occluders, and flow probe are exteriorized through the seventh intercostal space and tunneled subcutaneously to a final position between the scapulae. The thorax is closed and evacuated. A chest tube placed in the left thorax prior to closure is removed on the first postoperative day. Antibiotics (cephazolin, 2 g i.v.) are administered intraoperatively and are continued for 10 days postoperatively (cephalexin, 2 g/day p.o.). Postoperative analgesia is achieved with morphine sulfate (10 mg i.m.) as required. A minimum of 2 weeks is allowed for recovery prior to experimentation.

B. Physiological Measurements

Vascular pressures are measured by attaching the fluid-filled catheters to strain gauge manometers (Gould Statham P23). Vascular pressures are referenced to atmospheric pressure with the transducers positioned at the level of the spine. Heart rate is calculated from the aortic pressure trace. Cardiac output is measured by the thermal dilution technique (American Edwards Laboratories Model 9520A). A 7-Fr Swan-Ganz catheter is inserted percutaneously into the external jugular vein under local anesthesia (xylocaine spray). The catheter is advanced into the main pulmonary artery with the aid of pressure monitoring. The injectate consists of 5 ml of sterile iced 5% dextrose in water. Reported values of cardiac output represent the mean of at least three consecutive measurements after discarding the first measurement and are indexed to body weight (ml/min/kg). Pulmonary arterial wedge pressure (PAWP) is measured by acute inflation of the balloon at the tip of the Swan-Ganz catheter until a stable pressure is achieved over at least three respiratory cycles. Left pulmonary blood flow is measured by connecting the flow probe to an electromagnetic flowmeter (model SWF-4rd, Zepeda). Aortic and pulmonary arterial catheters are used to obtain blood samples for systemic arterial and mixed venous blood gases, which are analyzed by standard methods (Radiometer ABL-3 and Radiometer OSM-3 hemoximeter).

C. Generation of Pulmonary Vascular Pressure–Flow Plots

Multipoint pulmonary vascular $P/\dot{Q}$ plots are obtained using either the inferior vena caval constriction (IVCC) technique (1) to measure the $P/\dot{Q}$ relationship for the entire lung, or the right pulmonary artery occlusion (RPAO) technique (2) to measure selectively the left pulmonary $P/\dot{Q}$ relationship. All measurements are made at end-expiration. With the IVCC technique, multipoint $P/\dot{Q}$ plots are generated over approximately 30 min by gradual stepwise inflation of the hydraulic occluder implanted around the thoracic inferior vena cava to decrease venous return and cardiac output. The pulmonary vascular pressure gradient (PAP − PAWP) and cardiac output are measured approximately 5 min after each incremental inflation of the hydraulic occluder, when pressures and cardiac output have achieved new steady-state values. With the RPAO technique, left pulmonary $P/\dot{Q}$ plots are generated by continuously measuring the pulmonary vascular pressure gradient (PAP − LAP) and left pulmonary blood flow during gradual (~1 min) inflation of the hydraulic occluder around the right pulmonary artery, which directs total cardiac output through the left pulmonary circulation.

D. General Protocols

All experiments are performed with each healthy dog lying on its right side in a quiet laboratory environment. $P/\dot{Q}$ plots are measured in the conscious or anesthetized states without prior drug administration (no drug) and during administration of pharmacological agonists and antagonists. Cumulative dose–effect relationships are obtained in some protocols. Anesthesia is achieved with either pentobarbital sodium (30 mg/kg i.v.), halothane (~1.2%, end-tidal), or isoflurane (-1.6%, end-tidal). Halothane and isoflurane anesthesia are induced by mask and are supplemented with a subanesthetic dose of thiopental sodium (3 mg/kg i.v.) to minimize excitatory behavior. After induction, the dogs are placed on positive pressure ventilation (without positive end-expiratory pressure) to match systemic arterial and mixed venous blood gases to conscious values. Tidal volume is fixed at 15 ml/kg. Supplemental O_2 is administered, and respiratory rate is adjusted to between 10 and 13 breaths/min. End-tidal CO_2 is monitored continuously during the experiment (Beckman LB-2). After induction, halothane and isoflurane are allowed to equilibrate for approximately 1 hr to achieve steady-state conditions and to allow the plasma thiopental sodium concentration to decrease to negligible values.

III. Anesthesia and Pulmonary Vasoregulation

A. Effects of Anesthesia on Baseline Pressure–Flow Relationship

The intravenous anesthetic pentobarbital sodium is seldom used clinically, but it is widely used as a background anesthetic in experimental studies of the pulmonary circulation. With ventilation controlled to match blood gases to conscious values, pentobarbital has no net effect on the baseline $P/\dot{Q}$ relationship compared to conscious dogs (3). Although pentobarbital has no net effect on the baseline $P/\dot{Q}$ relationship, a number of individual mechanisms of pulmonary vasoregulation are altered during pentobarbital anesthesia. These are discussed in subsequent sections.

The inhalational anesthetic halothane is extensively used as the anesthetic agent of choice in a wide range of surgical procedures. In contrast to pentobarbital, halothane significantly modifies the baseline $P/\dot{Q}$ relationship compared to that measured in the conscious state (4). Halothane causes active pulmonary vasoconstriction (Fig. 1), which is not attenuated by sympathetic α_1-adrenoreceptor block, angiotensin-converting enzyme inhibition, combined arginine vasopressin V_1 and V_2 receptor block, or by cyclooxygenase inhibition (Fig. 2). Thus, halothane-induced pulmonary

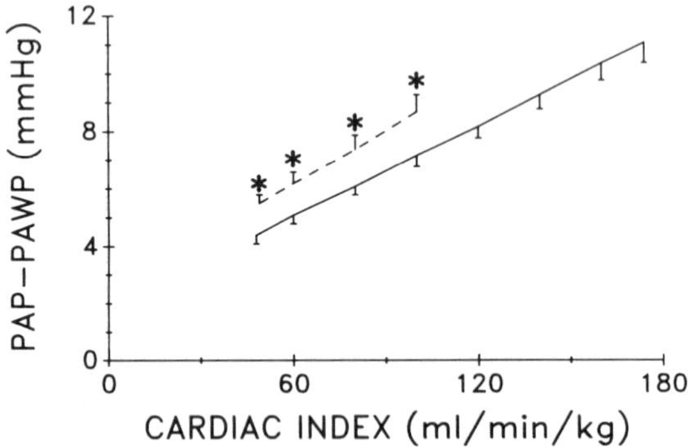

Fig. 1 Composite baseline pulmonary vascular pressure–cardiac index ($P/\dot{Q}$) plots in intact dogs in the conscious state (solid line) and during halothane anesthesia (dashed line). Halothane increases ($*p < 0.01$) the pulmonary vascular pressure gradient (PAP − PAWP) over the entire range of cardiac index compared with the conscious state; that is, halothane results in active pulmonary vasoconstriction. Values are means ± SE in this and all other figures. [Data from Chen *et al.* (4).]

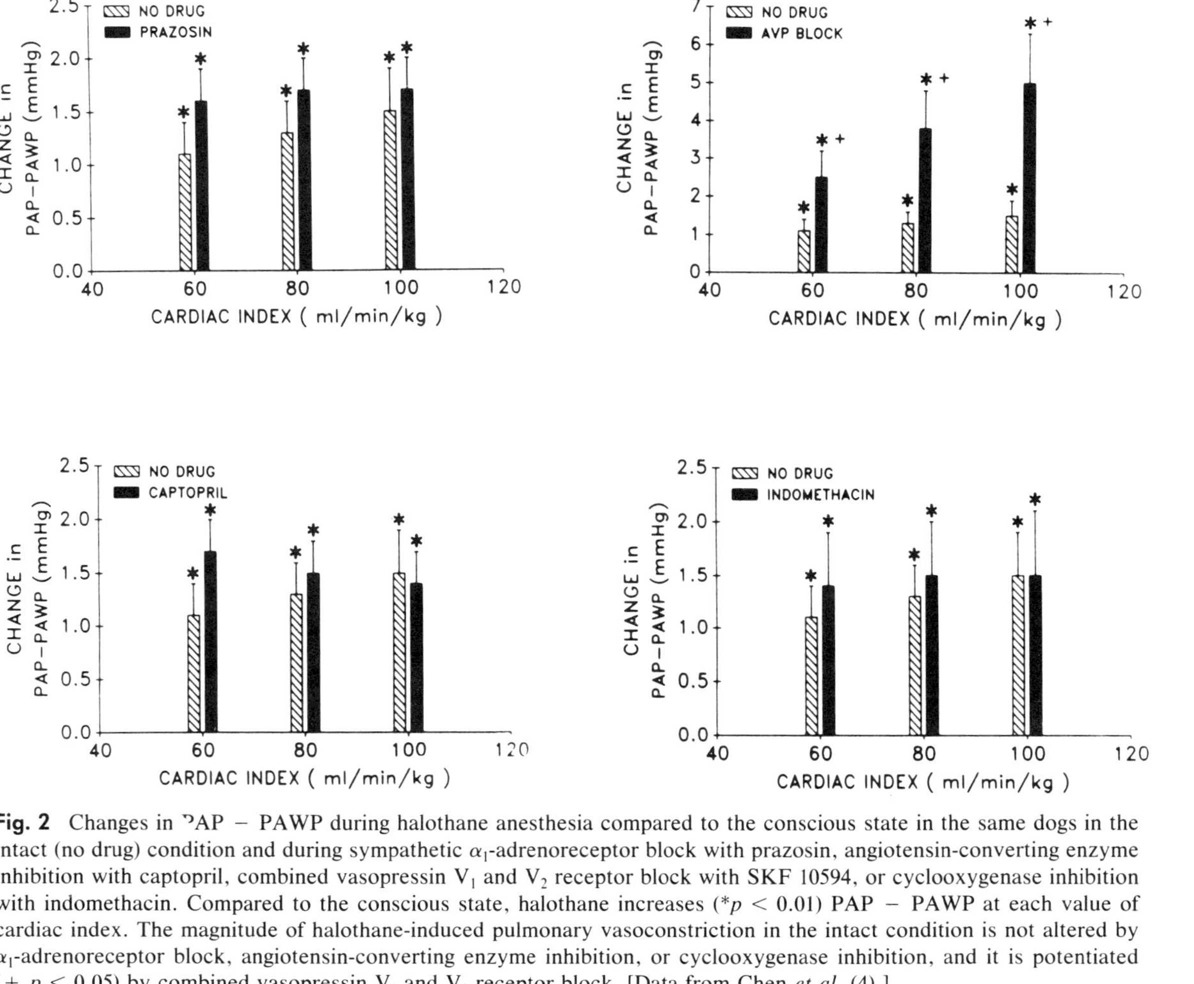

Fig. 2 Changes in PAP – PAWP during halothane anesthesia compared to the conscious state in the same dogs in the intact (no drug) condition and during sympathetic α_1-adrenoreceptor block with prazosin, angiotensin-converting enzyme inhibition with captopril, combined vasopressin V_1 and V_2 receptor block with SKF 10594, or cyclooxygenase inhibition with indomethacin. Compared to the conscious state, halothane increases ($*p < 0.01$) PAP – PAWP at each value of cardiac index. The magnitude of halothane-induced pulmonary vasoconstriction in the intact condition is not altered by α_1-adrenoreceptor block, angiotensin-converting enzyme inhibition, or cyclooxygenase inhibition, and it is potentiated ($+\ p < 0.05$) by combined vasopressin V_1 and V_2 receptor block. [Data from Chen *et al.* (4).]

vasoconstriction is not mediated by reflex neurohumoral activation secondary to systemic hypotension, or by metabolites of the cyclooxygenase pathway.

Isoflurane is a rapid-onset inhalational anesthetic that is frequently used for surgical procedures that require controlled hypotension to minimize bleeding. In contrast to halothane, isoflurane has no net effect on the baseline $P/\dot{Q}$ relationship (5). However, it is possible that isoflurane may have multiple offsetting effects on the pulmonary circulation, including a direct vasoconstrictor effect as well as a secondary vasodilator influence owing to the release of cyclooxygenase metabolites (6).

In summary, halothane, but not pentobarbital or isoflurane, alters the baseline $P/\dot{Q}$ relationship compared to that measured in the conscious state. It is important to emphasize that even when an anesthetic has no net effect on the baseline pulmonary circulation, it can modify individual mechanisms of vasoregulation.

B. Anesthesia and Autonomic Nervous System Regulation of Pressure–Flow Relationship

The autonomic nervous system (ANS) actively modulates the intact baseline $P/\dot{Q}$ relationship in conscious dogs (7). We investigated the extent to which pentobarbital anesthesia alters ANS regulation of the baseline $P/\dot{Q}$ relationship (8), and the results are summarized in Fig. 3. Sympathetic α_1-adrenoreceptor block causes active pulmonary vasodilation in conscious dogs but results in pulmonary vasoconstriction during pentobarbital anesthesia. Sympathetic β-adrenoreceptor block results in pulmonary vasoconstriction in both the conscious and pentobarbital-anesthetized states, but the magnitude of pulmonary vasoconstriction is attenuated during pentobarbital. Finally, cholinergic receptor block causes pulmonary vasodilation in the conscious state, whereas pulmonary vasoconstriction is observed during pentobarbital. Thus, ANS regulation of the intact baseline $P/\dot{Q}$ relationship is markedly altered during pentobarbital anesthesia.

We have also investigated the effects of halothane anesthesia on ANS regulation of the baseline $P/\dot{Q}$ relationship (9). As summarized in Fig. 4, in contrast to the prominent role of the ANS in the conscious state, sympathetic α_1- and β-adrenoreceptor and cholinergic receptor regulation of the baseline $P/\dot{Q}$ relationship is abolished during halothane anesthesia.

In preliminary studies, we have investigated the effects of isoflurane anesthesia on the pulmonary vascular responses to the sympathetic α_1-adrenoreceptor agonist phenylephrine and the sympathetic β-adrenoreceptor agonist isoproterenol (5). Compared to values measured

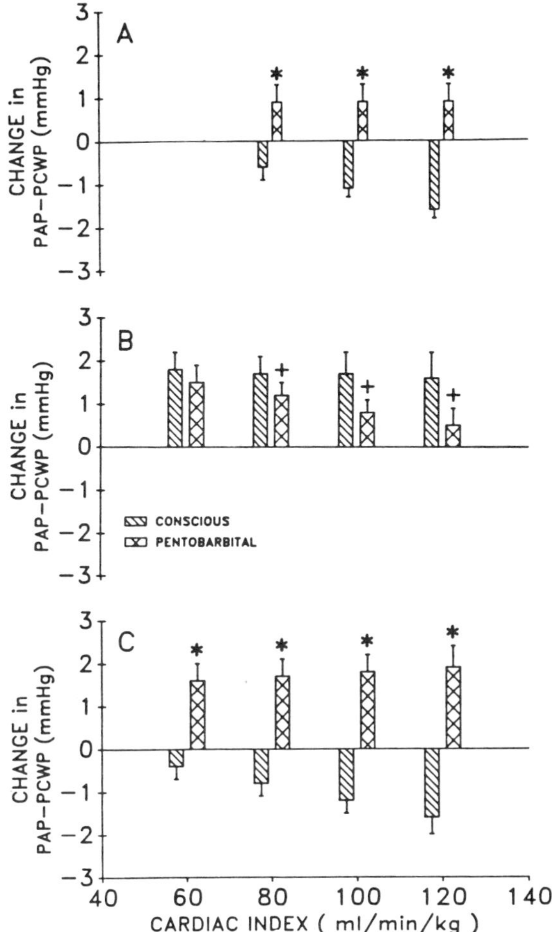

Fig. 3 Effects of pentobarbital anesthesia on pulmonary vascular responses to α_1-adrenoreceptor block with prazosin (A), β-adrenoreceptor block with propranolol (B), and cholinergic receptor block with atropine (C) compared to responses measured in the conscious state. Changes in the pulmonary vascular pressure gradient (PAP − PCWP) from the intact condition in response to the neural antagonists are shown over the range of cardiac index studied. Symbols ($*p < 0.01$; $+ \, p < 0.05$) indicate significant differences in responses to neural antagonists during pentobarbital anesthesia compared with the conscious state. [Reprinted from Nyhan *et al.* (8) with permission.]

in the conscious state, the dose–response relationship to phenylephrine is unchanged during isoflurane anesthesia,wherease the dose–response relationship to isoproterenol is shifted to the left. In other words, isoflurane

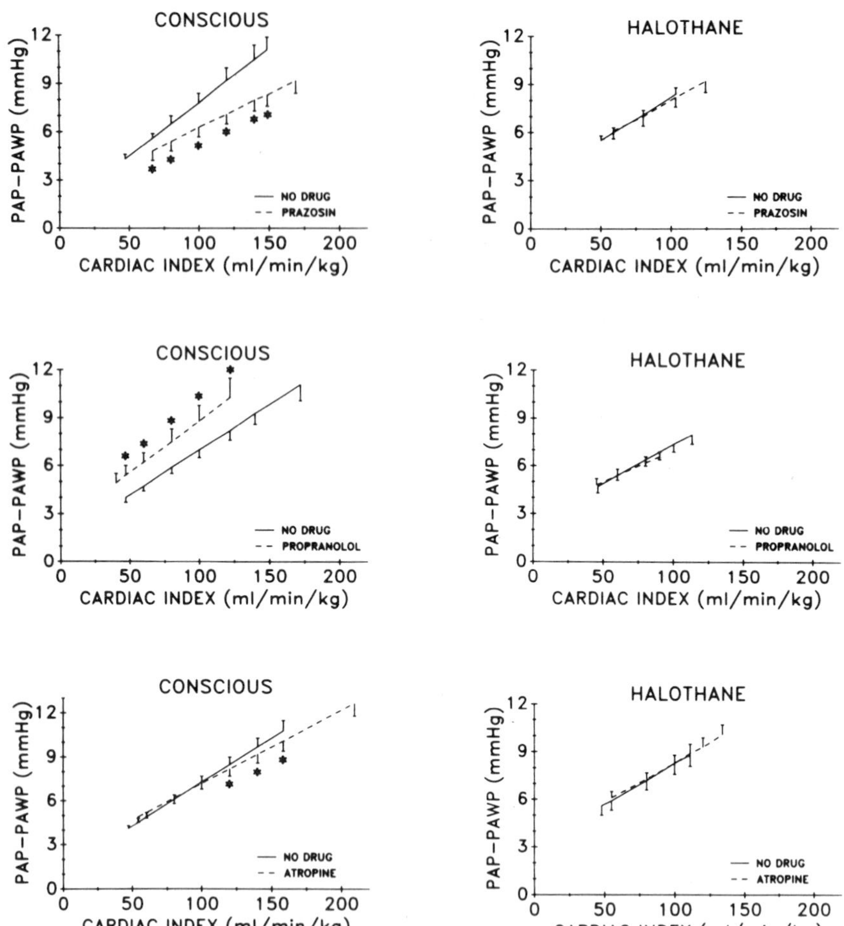

Fig. 4 Composite $P/\dot{Q}$ plots in conscious and halothane-anesthetized dogs measured before (no drug) and after sympathetic α_1-adrenoreceptor block with prazosin, β-adrenoreceptor block with propranolol, or cholinergic receptor block with atropine. In contrast to the significant ($p < 0.05$) effects of the neural antagonists on the baseline $P/\dot{Q}$ relationship in conscious dogs, the responses are abolished during halothane anesthesia. [Data from Chen *et al.* (9).]

has no effect on the pulmonary vasoconstrictor response to the α_1-agonist, but potentiates the pulmonary vasodilator response to the β-agonist.

In summary, pentobarbital and halothane anesthesia markedly alter ANS regulation of the baseline $P/\dot{Q}$ relationship compared to that measured in conscious dogs, and isoflurane anesthesia has differential effects on the pulmonary vascular responses to sympathetic α_1- and β-adrenoreceptor activation.

C. Anesthesia and Humoral Regulation of Pressure–Flow Relationship

In conscious dogs, the exogenous administration of angiotensin II results in active pulmonary vasoconstriction, whereas administration of the angiotensin-converting enzyme inhibitor captopril causes pulmonary vasodilation (10). As summarized in Fig. 5, the magnitude of pulmonary vasoconstriction normally observed in conscious dogs in response to angiotensin II is unaltered during pentobarbital anesthesia but is markedly attenuated during halothane anesthesia (11). Moreover, the pulmonary vasodilator response to captopril observed in conscious dogs is actually reversed to pulmonary vasoconstriction during pentobarbital anesthesia, and is abolished during halothane anesthesia (Fig. 6) (11).

Exogenously adminstered arginine vasopressin (AVP) can result in either pulmonary vasoconstriction or pulmonary vasodilation in conscious dogs depending on the integrity of AVP V_1 receptors (12,13). In preliminary studies, we have assessed the effects of pentobarbital and halothane anes-

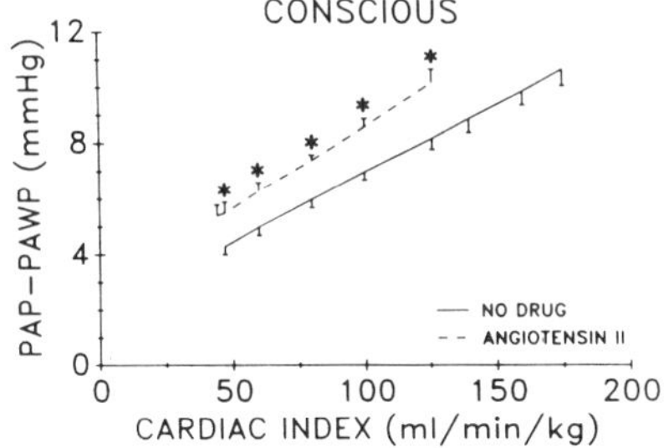

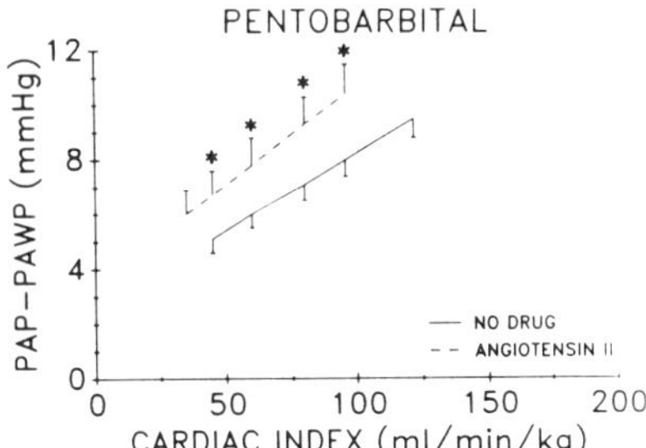

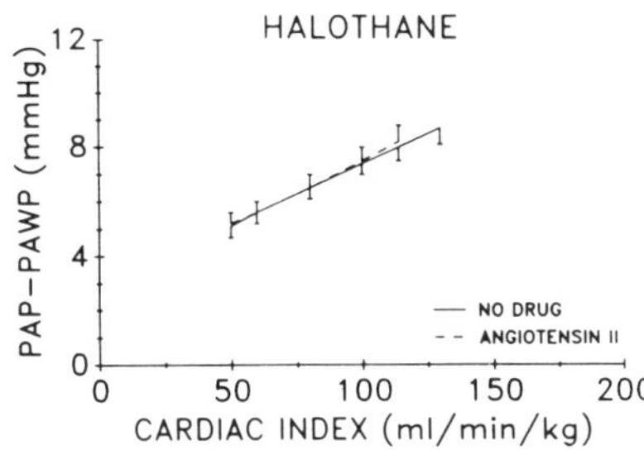

Fig. 5 Composite $P/\dot{Q}$ plots measured in conscious, pentobarbital-anesthetized, and halothane-anesthetized dogs before (no drug) and during the continuous intravenous infusion of angiotensin II. PAP − PAWP is increased ($*p < 0.01$) over the empirically measured range of cardiac index during angiotensin administration in the conscious state and during pentobarbital anesthesia. This vasoconstrictor response is abolished during halothane anesthesia. [Data from Nyhan *et al.* (11).]

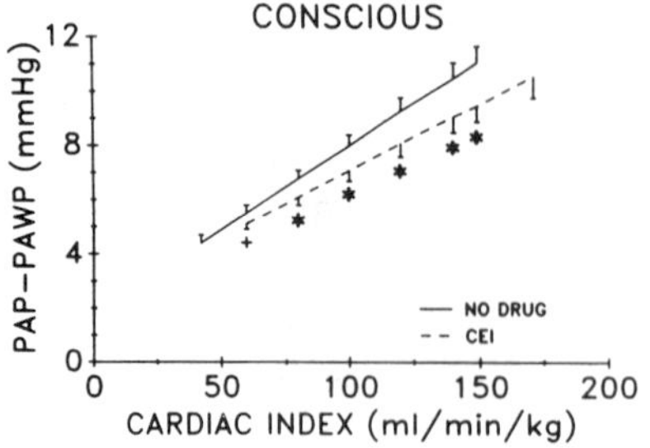

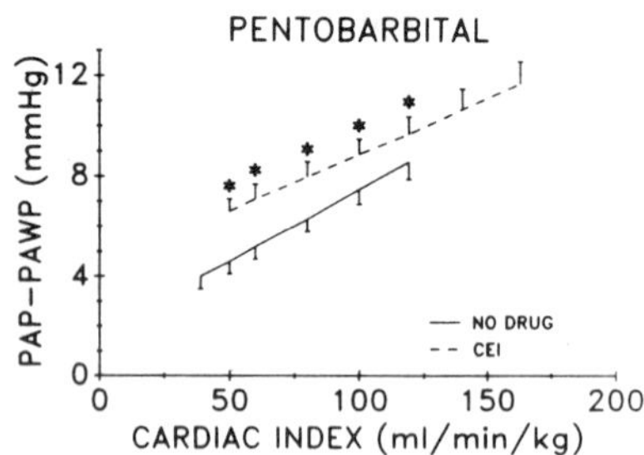

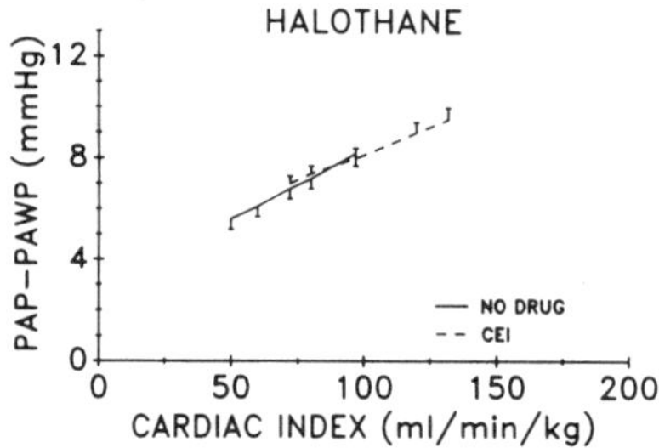

Fig. 6 Composite $P/\dot{Q}$ plots measured in conscious, pentobarbital-anesthetized, and halothane-anesthetized dogs before (no drug) and after angiotensin-converting enzyme inhibition (CEI) with captopril. CEI decreases (*$p < 0.01$) PAP − PAWP in the conscious state. This vasodilator response is reversed to vasoconstriction (*$p < 0.01$) during pentobarbital and is abolished during halothane anesthesia. [Data from Nyhan *et al.* (11).]

thesia on the pulmonary vascular response to AVP following preconstriction of the pulmonary circulation with the thromboxane analog U46619 (14). Following U46619 preconstriction, AVP results in pulmonary vasodilation in conscious dogs. However, AVP-induced pulmonary vasodilation is reversed to pulmonary vasoconstriction during both pentobarbital and halothane anesthesia.

In summary, both pentobarbital and halothane anesthesia alter the pulmonary vascular effects of angiotensin II and AVP compared to that measured in the conscious state.

D. Anesthesia and Regulation of Pressure–Flow Relationship by Cyclooxygenase Metabolites

The continuous release of prostacyclin has been hypothesized to exert a tonic vasodilator influence on the pulmonary circulation. Consistent with this hypothesis, we have observed that cyclooxygenase pathway inhibition results in active pulmonary vasoconstriction in conscious dogs (15). In the presence of pentobarbital anesthesia, the magnitude of the pulmonary

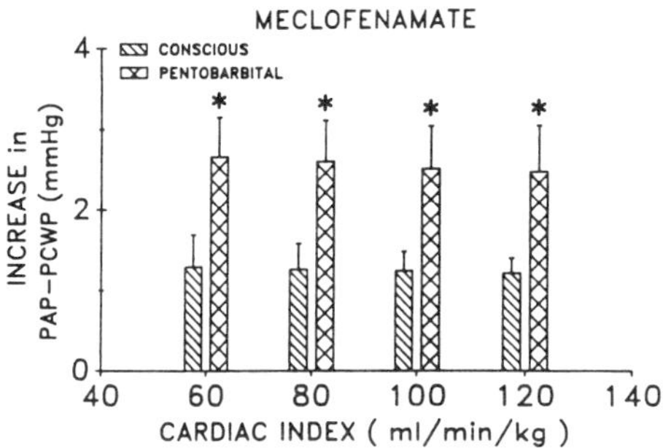

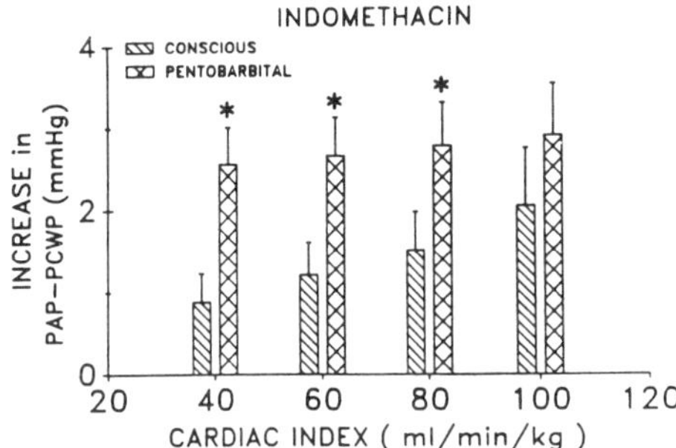

Fig. 7 Effects of pentobarbital anesthesia on pulmonary vasoconstrictor response to cyclooxygenase inhibition with either meclofenamate or indomethacin. Increases in PAP − PCWP induced by cyclooxygenase inhibition are shown over the range of cardiac index studied. Pulmonary vasoconstriction in response to cyclooxygenase inhibition is increased ($^*p < 0.05$) during pentobarbital anesthesia compared to the conscious state. [Data from Nyhan *et al.* (16).]

vasoconstrictor response to cyclooxygenase inhibition with either meclofenamate or indomethacin is significantly augmented (Fig. 7) (16). In contrast, the response to cyclooxygenase inhibition is virtually abolished during halothane anesthesia (17).

In summary, pentobarbital and halothane anesthesia exert diametrically opposite effects on pulmonary vascular regulation by endogenous cyclooxygenase metabolites compared to that measured in conscious dogs.

E. Anesthesia and Pulmonary Vascular Repsonse to Hypoperfusion

We investigated the role of the ANS in the pulmonary vascular response to increasing pulmonary blood flow after a 10 to 15-min period of hypoperfusion in conscious dogs (18). Surprisingly, the $P/\dot{Q}$ relationship is unchanged during the posthypoperfusion period compared to baseline in intact dogs (Fig. 8). However, pulmonary vasoconstriction is unmasked posthypoperfusion following sympathetic β-adrenoreceptor block, and this effect is abolished by subsequent sympathetic α_1-adrenoreceptor block (Fig. 9). Thus, the ANS actively regulates the pulmonary circulation during the posthypoperfusion period in intact conscious dogs, in that β-adrenergic vasodilation offsets α_1-adrenergic vasoconstriction to prevent pulmonary vasoconstriction during the posthypoperfusion period.

In contrast to the conscious state, striking pulmonary vasoconstriction is observed during the posthypoperfusion period in otherwise intact pentobarbital-anesthetized dogs (Fig. 8) (3). The magnitude of posthypoperfusion pulmonary vasoconstriction is not enhanced by sympathetic β-adrenoreceptor block, but it is significantly attenuated by α_1-adrenoreceptor block (Fig. 9). These results indicate that sympathetic β-adrenergic vasodilation fails to offset α_1-adrenergic vasoconstriction during the posthypoperfusion period in pentobarbital-anesthetized dogs.

We have also investigated the effects of halothane anesthesia on ANS regulation of the pulmonary circulation during the posthypoperfusion period (19). Just as in conscious dogs, the $P/\dot{Q}$ relationship is unaltered compared to baseline during the posthypoperfusion period under halothane anesthesia. However, the magnitude of posthypoperfusion pulmonary vasoconstriction unmasked by sympathetic β-adrenoreceptor block is significantly attenuated during halothane anesthesia compared to the conscious state.

As summarized in Fig. 10, in conscious dogs vasodilator metabolites of the cyclooxygenase pathway offset the vasoconstrictor influence of angiotensin II to prevent posthypoperfusion pulmonary vasoconstriction (20). These competing mechanisms of pulmonary vasoregulation are abol-

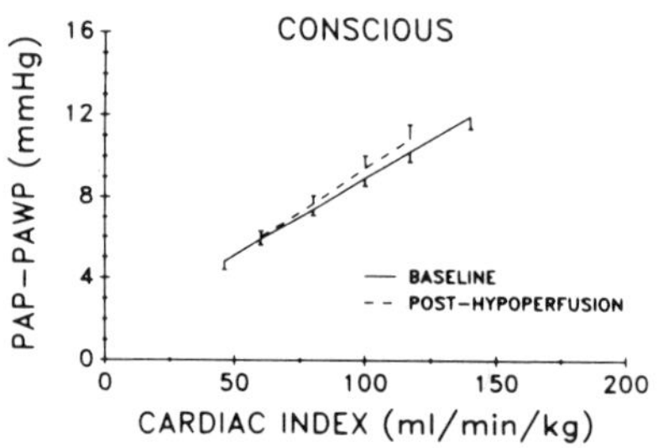

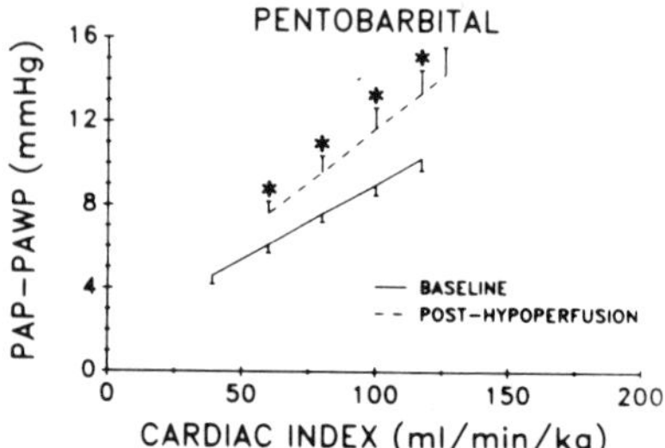

Fig. 8 Composite $P/\dot{Q}$ plots measured at baseline and during the posthypoperfusion period in conscious and pentobarbital-anesthetized dogs. In contrast to the conscious state, pulmonary vasoconstriction ($*p < 0.01$) is observed posthypoperfusion during pentobarbital anesthesia. [Data from Chen *et al.* (3) and Clougherty *et al.* (18).]

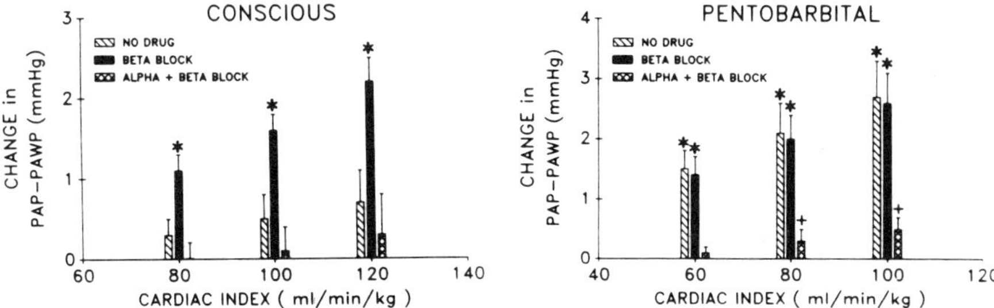

Fig. 9 Changes in PAP − PAWP during posthypoperfusion period compared to baseline in conscious and pentobarbital-anesthetized dogs in the intact (no drug) condition, during β-adrenoreceptor block with propranolol, and during combined β-adrenoreceptor and α_1-adrenoreceptor block with prazosin. In conscious dogs, pulmonary vasoconstriction ($^*p < 0.01$) posthypoperfusion is unmasked by β-block and abolished by α_1-block. During pentobarbital anesthesia, posthypoperfusion pulmonary vasoconstriction ($^*p < 0.01$) in the no drug condition is not potentiated by β-block, but is attenuated ($+\ p < 0.05$) by α_1-block. [Data from Chen *et al.* (3) and Clougherty *et al.* (18).]

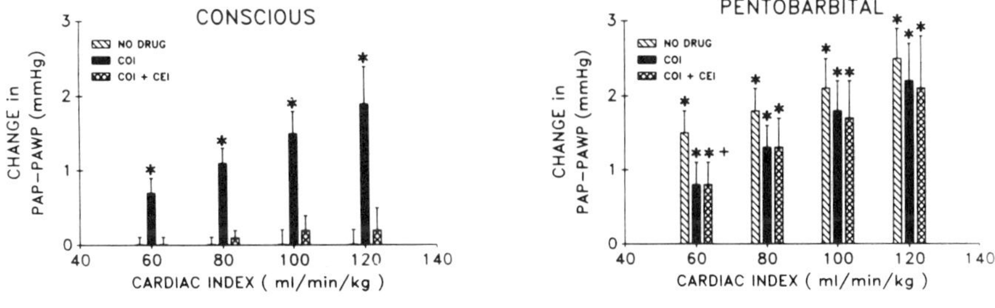

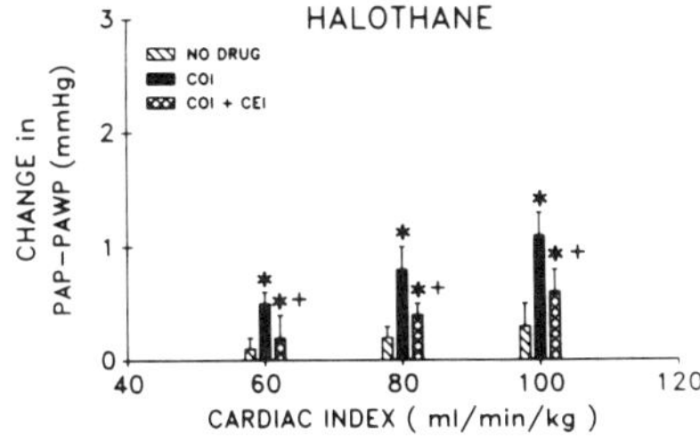

Fig. 10 Changes in PAP − PAWP during posthypoperfusion period compared to baseline in conscious, pentobarbital-anesthetized, and halothane-anesthetized dogs in the intact (no drug) condition, during cyclooxygenase inhibition (COI) with indomethacin, and during combined COI and angiotensin coverting-enzyme inhibition (CEI) with captopril. In conscious dogs, pulmonary vasoconstriction ($^*p < 0.01$) posthypoperfusion is unmasked by COI and inhibited by CEI. These competing mechanisms are abolished during pentobarbital and attenuated ($+\ p < 0.05$) during halothane anesthesia. [Data from Fehr *et al.* (20).]

ished during pentobarbital anesthesia and attenuated during halothane anesthesia (Fig. 10) (20).

In summary, both pentobarbital and halothane anesthesia alter pulmonary vasoregulation by the ANS, cyclooxygenase metabolites, and angiotensin II during the posthypoperfusion period compared to that measured in the conscious state.

F. Anesthesia and Endothelium-Dependent and -Independent Pulmonary Vasodilation

In conscious dogs, bradykinin causes marked pulmonary vasodilation, an effect which is not altered by cyclooxygenase pathway inhibition (21). We investigated the effects of an inhibitor of nitric oxide (NO) synthase, N^{ω}-nitro-L-arginine (NNLA), on the pulmonary vascular responses to the putative endothelium-dependent vasodilator bradykinin and the endothelium-independent vasodilator sodium nitroprusside in conscious dogs (2). NNLA has no effect on the baseline $P/\dot{Q}$ relationship (Fig. 11), which indicates that tonic release of NO is not responsible for low resting vasomotor tone in conscious dogs. NNLA causes a leftward shift in the dose–response relationship to the thromboxane mimetic, U46619, which indicates that the endogenous release of NO modulates the pulmonary vascular response to this vasoconstrictor (Fig. 12). Finally, NNLA abolishes the pulmonary vasodilator response to bradykinin but has no effect on the vasodilator response to sodium nitroprusside (Fig. 13). These results clearly demonstrate that, in conscious dogs, the pulmonary vasodilator response to bradykinin is entirely mediated by the endogenous release of the endothelium-derived relaxing factor, EDRF–NO, whereas the pul-

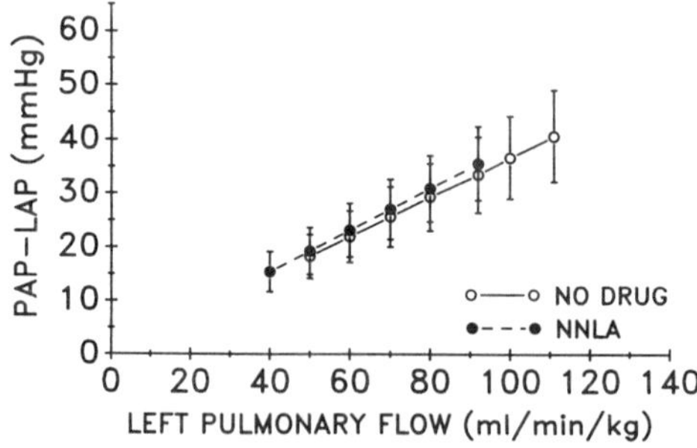

Fig. 11 Composite $P/\dot{Q}$ plots in conscious dogs in the intact (no drug) condition and following the administration of the nitric oxide (NO) synthase inhibitor N^{ω}-nitro-L-arginine (NNLA). Inhibition of NO synthase has no effect on the baseline $P/\dot{Q}$ relationship in conscious dogs. [Reprinted from Nishiwaki *et al.* (2) with permission.]

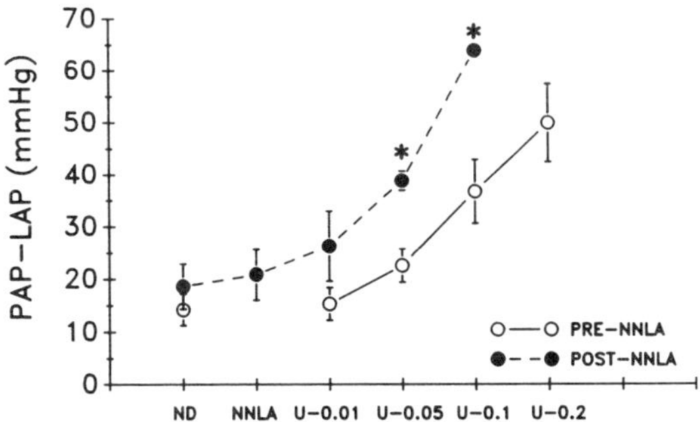

Fig. 12 The pulmonary vascular pressure gradient (PAP − LAP) at a value of left pulmonary blood flow of 60 ml/min/kg in conscious dogs is shown for the no drug (ND) condition, following NO synthase inhibition with NNLA, and during increasing doses (μg/kg/min i.v.) of the thromboxane analog, U46619 (U). The dose–response relationship to U46619 is shifted (*$p < 0.01$) leftward post-NNLA compared to pre-NNLA. [Reprinted from Nishiwaki *et al.* (2) with permission.]

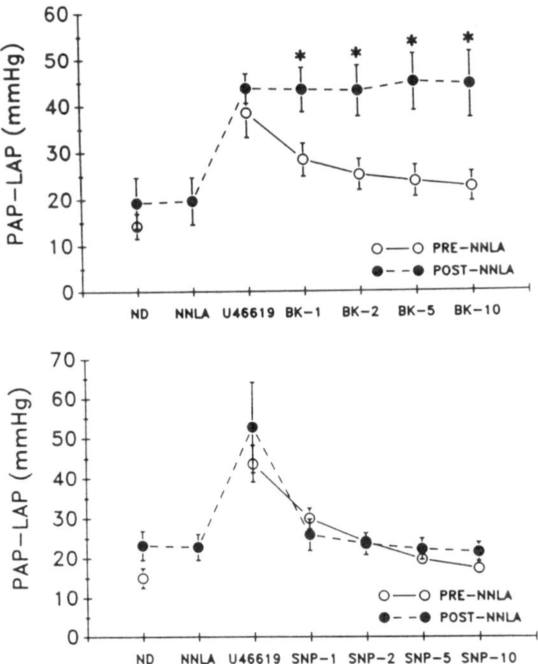

Fig. 13 The PAP − LAP at a value of left pulmonary flow of 60 ml/min/kg in conscious dogs is shown for the no drug (ND) condition, following NNLA, following preconstriction with U46619, and during increasing doses (μg/kg/min i.v.) of either bradykinin (BK) or sodium nitroprusside (SNP). The pulmonary vasodilator response to BK in the presence of U46619 preconstriction is abolished (*$p < 0.01$) post-NNLA, whereas the pulmonary vasodilator response to SNP is unchanged post-NNLA. [Data from Nishiwaki *et al.* (2).]

monary vasodilator response to sodium nitroprusside is independent of EDRF–NO.

We next assessed the effects of pentobarbital and halothane anesthesia on the pulmonary vasodilator responses to bradykinin and sodium nitroprusside (22). Compared to the conscious state, the pulmonary vasodilator responses to bradykinin and sodium nitroprusside are unaltered during pentobarbital anesthesia (Figs. 14 and 15). In contrast, the pulmonary vascular responses to both of these vasodilators are abolished during halothane anesthesia (Figs. 14 and 15).

In preliminary studies, we have assessed the extent to which isoflurane anesthesia alters the pulmonary vascular responses to endothelium-dependent (bradykinin) and endothelium-independent (SIN-1) vasodilators (23). Compared to the conscious state, isoflurane causes a rightward shift in the dose–response relationship to bradykinin, which indicates that isoflurane (like halothane) also attenuates EDRF–NO-mediated pulmonary vasodilation. However, isoflurane is associated with a leftward shift

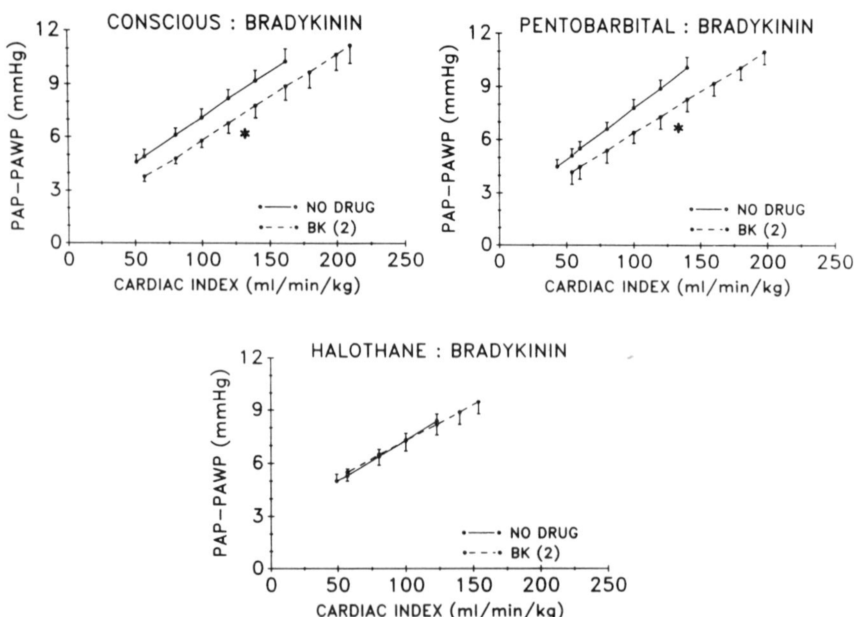

Fig. 14 Composite $P/\dot{Q}$ plots in conscious, pentobarbital-anesthetized, and halothane-anesthetized dogs in the no drug condition and during the continuous administration of bradykinin (BK, 2 μg/kg/min i.v.). Bradykinin-induced pulmonary vasodilation ($^*p < 0.01$) in the conscious state is not altered during pentobarbital but is abolished during halothane anesthesia. [Data from Murray *et al.* (22).]

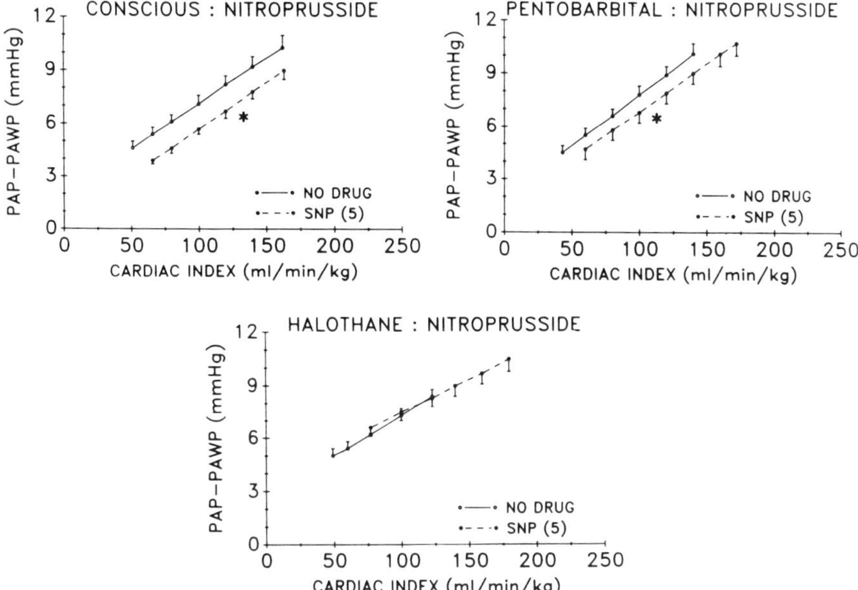

Fig. 15 Composite $P/\dot{Q}$ plots in conscious, pentobarbital-anesthetized, and halothane-anesthetized dogs in the no drug condition and during the continuous administration of sodium nitroprusside (SNP, 5 μg/kg/min i.v.). Nitroprusside-induced pulmonary vasodilation ($^{*}p < 0.01$) in the conscious state is not altered during pentobarbital but is abolished during halothane anesthesia. [Data from Murray *et al.* (22).]

in the dose–response relationship to SIN-1,which indicates that isoflurane potentiates the pulmonary vasodilator response to this endothelium-independent vasodilator. These results are in striking contrast to the attenuated response to sodium nitroprusside observed during halothane anesthesia (22).

Therefore, both halothane and isoflurane anesthesia, but not pentobarbital anesthesia, significantly attenuate endothelium-dependent EDRF–NO-mediated pulmonary vasodilation compared to that measured in the conscious state. However, the anesthetics have differential effects and act to either attenuate (halothane), potentiate (isoflurane), or not alter (pentobarbital) the pulmonary vascular responses to endothelium-independent vasodilators. These results raise the possibility that endothelium-independent pulmonary vasodilation induced by sodium nitroprusside and SIN-1 may not be entirely mediated by the same cellular mechanisms.

IV. Summary

These results provide compelling evidence that general anesthesia can alter neural, hymoral, and local mechanisms of pulmonary vascular regulation. A clear strength of these studies is that we have assessed these various mechanisms of pulmonary vascular regulation in the conscious state, and then again in the same animal during general anesthesia. Our studies in chronically instrumented conscious dogs have documented that regulation of the pulmonary vascular $P/\dot{Q}$ relationship is multifactorial, and that these vascular mechanisms are interactive. These results in conscious animals serve as a database to assess the specific effects of general anesthesia on pulmonary vascular regulation. A striking finding in these studies is that different anesthetics can have differential effects on neural, humoral, and local mechanisms of pulmonary vascular regulation. Without this fundamental information, it is difficult to interpret previous experimental studies that have utilized background anesthetics. Moreover, this information is essential for the adequate clinical management of patients in the perioperative period.

These *in vivo* studies have allowed us to identifly the overall effects of general anesthetics on the integrative response of the intact organism to these diverse pulmonary vasoregulatory pathways. Our next objective is to elucidate the cellular mechanisms that mediate the effects of general anesthetics on these various mechanisms of pulmonary vasoregulation. To that end, we are now performing *in vivo* studies utilizing isolated pulmonary arterial rings to elucidate abnormalities in the various signal transduction pathways that could be responsible for the effects of general anesthetics. We propose to identify the loci of dysfunction in the signal transduction pathways that mediate the pulmonary vascular effects of general anesthetics on cAMP-, cGMP-, and $K^+{}_{ATP}$ channel-mediated pulmonary vasodilation, as well as α_1- and angiotensin-mediated pulmonary vasoconstriction. These combined *in vivo* and *in vitro* approaches are complementary and can be used to delineate the effects and mechanisms of action of general anesthetics on pulmonary vasoregulation.

Acknowledgments

The author is indebted to the numerous investigators in the Department of Anesthesiology and Critical Care Medicine, Johns Hopkins University School of Medicine, who have been directly involved in the conduct of these studies. The author thanks Lisa DeLoriers for outstanding secretarial skills,and Rosie Cousins for tireless technical support. This work is supported by National Heart, Lung, and Blood Institute Grant HL-38291.

References

1. Lodato, R. F., Michael, J. R., and Murray, P. A. (1985). Multipoint pulmonary vascular pressure–cardiac output relationships in conscious dogs. *Am. J. Physiol.* **249**, H351–H357.
2. Nishiwaki, K., Nyhan, D. P., Rock, P., Desai, P. M., Peterson, W. P., Pribble, C. G., and Murray, P. A. (1992). N^{ω}-nitro-L-arginine and the pulmonary vascular pressure–flow relationship in conscious dogs. *Am. J.Physiol.* **262**, H1331–H1337.
3. Chen, B. B., Nyhan, D. P., Goll, H. M., Clougherty, P. W., Fehr, D. M., and Murray, P. A.(1988). Pentobarbital anesthesia modifies pulmonary vasoregulation following hypoperfusion. *Am. J. Physiol.* **255**, H569–H576.
4. Chen, B. B., Nyhan, D. P., Fehr, D. M., Goll, H. M., and Murray, P. A. (1990). Halothane anesthesia causes active, flow-independent pulmonary vasoconstriction. *Am. J. Physiol.* **259**, H74–H83.
5. Lennon, P. F., Fujiwara, Y., and Murray, P. A. (1993). Isoflurane anesthesia has no effect on the baseline pulmonary circulation but potentiates cAMP-mediated vasodilation compared to the conscious state. *Anesthesiology* **79**, A619. [Abstract]
6. Naeije, R., Lejeune, P., Leeman, M., and Deloof, T. (1987). Pulmonary arterial pressure–flow plots in dogs: Effects of isoflurane and nitroprusside. *J. Appl. Physiol.* **63**, 969–977.
7. Murray, P. A., Lodato, R. F., and Michael, J. R. (1986). Neural antagonists modulate pulmonary vascular pressure–flow plots in conscious dogs. *J. Appl. Physiol.* **60**, 1900–1907.
8. Nyhan, D. P., Goll, H. M., Chen, B. B., Fehr, D. M., Clougherty, P. W., and Murray, P. A. (1989). Pentobarbital anesthesia alters pulmonary vascular response to neural antagonists. *Am. J. Physiol.* **256**, H1384–H1392.
9. Chen, B. B., Nyhan, D. P., Fehr, D. M., and Murray, P. A. (1992). Halothane anesthesia abolishes pulmonary vascular responses to neural antagonists. *Am. J. Physiol.* **262**, H117–H122.
10. Goll, H. M., Nyhan, D. P., Geller, H. S., and Murray, P. A. (1986). Pulmonary vascular responses to angiotensin II and captopril in conscious dogs. *J. Appl. Physiol.* **61**, 1552–1559.
11. Nyhan, D. P., Chen, B. B., Fehr, D. M., Rock, P., and Murray, P. A. (1992). Anesthesia alters pulmonary vasoregulation by angiotensin II and captopril. *J. Appl. Physiol.* **72**, 636–642.
12. Nyhan, D. P., Geller, H. S., Goll, H. M., and Murray, P. A. (1986). Pulmonary vasoactive effects of exogenous and endogenous AVP in conscious dogs. *Am. J. Physiol.* **251**, H1009–H1016.
13. Nyhan D. P., Clougherty, P. W., and Murray, P. A. (1987). AVP-induced pulmonary vasodilation during specific V_1 receptor block in conscious dogs. *Am. J. Physiol.* **253**, H493–H499.
14. Trempy, G. A., Lennon, P. E., Peterson, W. P., Nyhan, D. P., and Murray, P. A. (1993). Halothane anesthesia reverses the pulmonary vascular response to arginine vasopressin following preconstriction with U46619. *FASEB* **7**, A542 (Abstract).
15. Geller, H. S., Nyhan, D. P., Goll, H. M., Clougherty, P. W., Chen, B. B., and Murray, P. A. (1988). Combined neurohumoral block modulates pulmonary vascular $P/\dot{Q}$ relationship in conscious dogs. *Am. J. Physiol.* **255**, H1084–H1090.
16. Nyhan, D. P., Chen, B. B., Fehr, D. M., Goll, H. M., and Murray, P. A. (1989). Pentobarbital augments pulmonary vasoconstrictor response to cyclooxygenase inhibition. *Am. J. Physiol.* **257**, H1140–H1146.

17. Nyhan, D. P., Chen, B. B., Fehr, D. M., Rock, P., and Murray, P. A. (1990). Pulmonary vascular response to cyclooxygenase inhibition is abolished during halothane anesthesia. *Anesthesiology* **73**, A627 (Abstract).
18. Clougherty, P. W., Nyhan, D. P., Chen, B. B., Goll, H. M., and Murray, P. A. (1988). Autonomic nervous system pulmonary vasoregulation after hypoperfusion in conscious dogs. *Am. J. Physiol.* **254**, H976–H983.
19. Chen, B. B., Fehr, D. M., Nyhan, D. P., and Murray, P. A. (1989). Sympathetic adrenergic regulation of pulmonary vascular pressure–flow relationship is altered during halothane anesthesia. *Anesthesiology* **71**, A529 (Abstract).
20. Fehr, D. M., Nyhan, D. P., Chen, B. B., and Murray, P. A. (1991). Pulmonary vasoregulation by cyclooxygenase metabolites and angiotensin II after hypoperfusion in conscious, pentobarbital-anesthetized and halothane-anesthetized dogs. *Anesthesiology* **75**, 257–267.
21. Nyhan, D. P., Clougherty, P. W., Goll, H. M., and Murray, P. A. (1987). Bradykinin actively modulates pulmonary vascular pressure–cardiac index relationships in conscious dogs. *J. Appl. Physiol.* **63**, 145–151.
22. Murray, P. A., Fehr, D. M., Chen, B. B., Rock, P., Esther, J. W., Desai, P. M., and Nyhan, D. P. (1992). Differential effects of general anesthesia on cGMP-mediated pulmonary vasodilation. *J. Appl. Physiol.* **73**, 721–727.
23. Gambone, L. M., Fujiwara, Y., and Murray, P. A. (1993). Endothelium-dependent cGMP-mediated pulmonary vasodilation is selectively attenuated during isoflurane anesthesia compared to the conscious state. *Anesthesiology* **79**, A681. [Abstract]

Pulmonary Mechanics Changes Associated with Cardiac Surgery

Ron Dueck

Department of Anesthesiology
Veterans Affairs Medical Center
and University of California at San Diego
La Jolla, California 92161

I. Introduction

Pulmonary mechanics changes associated with cardiac surgery are usually of only modest clinical significance. When lung compliance and gas exchange are seriously deranged, however, it is most often associated with prolonged cardiopulmonary bypass, with difficult weaning from cardiopulmonary bypass, or from a protamine reaction. Fortunately, the introduction of on-line automated spirometry has facilitated a more systematic review of what cardiopulmonary bypass does to lung function. The initial introduction to such spirometry lead to a surprisingly basic question, What happens to lung function if you open the thorax by sternotomy?

These studies were started with the assumption that compliance should decrease with sternotomy, since the elasticity of the chest wall provides a distending force to the lung while the elasticity of the lung provides a retracting or deflationary force. At functional residual capacity (FRC), these opposing forces are perfectly in balance. Removing the chest wall should therefore shift the static respiratory pressure–volume curve to the right, so that more pressure would be required to inflate the lung and thereby reduce lung compliance.

Advances in Pharmacology, Volume 31

II. Dynamic Lung Compliance

The clinical test of this hypothesis addressed changes in total dynamic compliance with opening and closing of the chest after sternotomy in 18 patients during open-heart surgery for coronary artery bypass graft, valvulotomy, or both. The patients were 55 to 75 years of age and had a body mass index [BMI = weight (kg)/height $(m)^2$] of 20–36, that is, ranging from asthenic to moderate clinical obesity. Preoperative forced expired volume in 1 second (FEV_1) values ranged from 36 to 86% of predicted. Anesthesia consisted of high-dose fentanyl, midazolam, and pipecuronium. Ventilation was set at tidal volumes of 10–12 ml/kg with a rate sufficient to produce an end-tidal p_{CO_2} of 30–35 mmHg. Sidestream spirometry with the Datex (Helsinki, Finland) Ultima flowmeter was used to derive dynamic total compliance (C_T), percent of tidal volume exhaled in 1 sec (V1%), and the presence or absence of "auto-PEEP" (onset of the next inspiration before expiratory flow rate has fallen to zero), which implies either stacking or gas trapping (1). Statistical significance of the changes was determined by factorial and repeated measures analysis of variance. The studies were approved by the University of California, San Diego, Committee on Human Subjects Investigation.

III. Effects of Sternotomy

The normal flow–volume curve shows a characteristic plateau during inspiration, determined principally by the flow characteristics of the mechanical ventilator and the fixed resistance of the endotracheal tube. Expiratory flow shows a characteristic convex deflationary curvature produced by the elasticity of the lung with normal airflow resistance. In our typical cardiac surgery patient with chronic obstructive pulmonary disease (COPD), on the other hand, we saw an abnormally flattened expiratory flow–volume curve as well as incomplete exhalation before onset of the next inspiration, as shown in Fig. 1A. This patient required a relatively high tidal volume and frequency to produce a normal p_{CO_2} because of preexisting lung disease, and showed a distinctly abnormal compliance (only 45 ml/cm H_2O compared to a normal anesthetized value of at least 60 ml/cm H_2O). The expiratory flow rate was also abnormal (only 62% exhaled in 1 sec when it should have been 70%). After the chest was opened, there was a noticeable change in the shape of the expiratory flow–volume curve: it was now convex, as shown in Fig. 1B. Exhalation was complete enough to reach zero flow, except for the apparent cardiac oscillations, before onset of the next breath. The V1% value of the patient

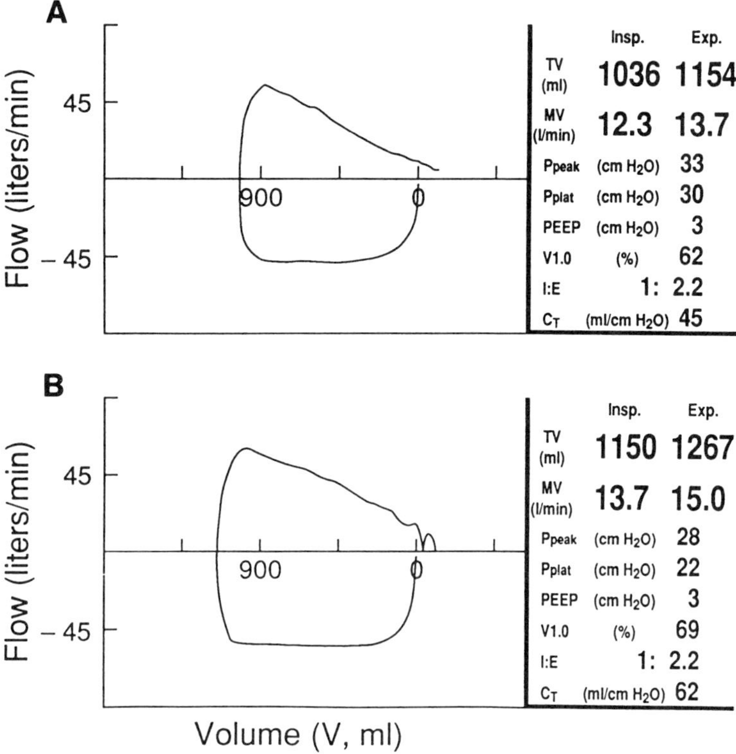

Fig. 1 (A) Inspiratory and expiratory flow–volume curves (clockwise from zero) for a 114-kg (BMI = 31.4) patient before sternotomy (chest closed). Expiratory flow did not fall to zero before onset of the next breath, which implies "stacking" or "gas trapping." This interpretation is supported by the abnormally low compliance (C_T) and the low percent tidal volume (TV) exhaled in 1 sec (V1.0%). (B) Flow–volume curve for the same patient after chest opening showed improved expiratory flow with complete exhalation before onset of the next breath as well as improved compliance and V1.0%, suggesting that the amount of gas trapping improved significantly.

improved after sternotomy, although other patients with gas trapping showed reduced V1%. However, all patients who had gas trapping before sternotomy showed improved compliance with chest opening, as in the above case from 45 to 62 ml/cm H_2O.

The average total compliance value of 67 ml/cm H_2O before sternotomy (for our 18 subjects) was lower than expected for perfectly healthy, young, normal awake subjects; however, it was within the expected range for surgical patients with COPD, heart disease, and varying degrees of relative obesity (see Fig. 2 and Table 1). The majority of patients showed improved

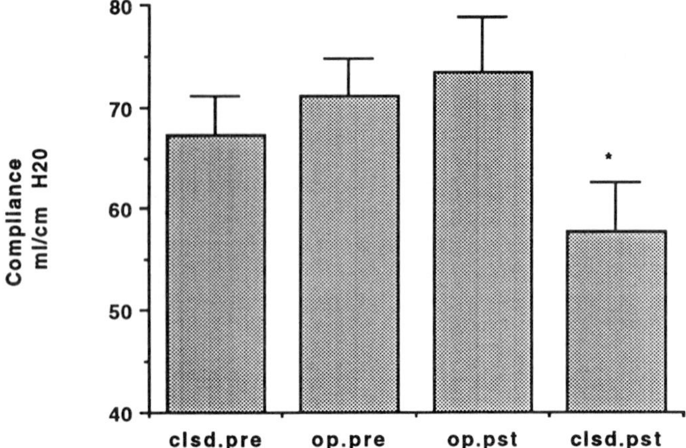

Fig. 2 Mean ± SE of dynamic compliance is shown for chest closed precardiopulmonary bypass (CPB, clsd.pre), chest open pre-CPB (op.pre), chest open post-CPB (op.pst), and chest closed post-CPB (clsd.pst) conditions. The value for closed chest post-CPB was significantly lower than those for the three other conditions ($p < 0.05$).

compliance with sternotomy and sternal retraction, although others showed reduced compliance, so that the change was not statistically significant. This variability in compliance was in large part related to the interacting effects of relative body weight, since changes in compliance were directly related to body mass index, as shown in Fig. 3. Inspiratory plateau airway pressure changes with sternotomy, on the other hand, were statistically significant, as shown in Table I.

Table I

Pulmonary Mechanics during Chest Open/Closed, Pre-/Postcardiopulmonary Bypass[a]

Condition	P_{plat} (cmH_2O)	C_T (ml/cmH_2O)	V1%	Gas trap (n)
Closed/pre	17.4 ± 1.1	67.2 ± 3.9	67 ± 2	3[b]
Open/pre	15.2 ± 0.7[c]	71.2 ± 3.7	70 ± 2	1
Open/post	15.3 ± 1.0[c]	73.5 ± 5.3	66 ± 2	2
Closed/post	19.7 ± 1.3	57.8 ± 20.2[d]	68 ± 2	7[b,e]

[a]The mean ± SE of inspiratory plateau airway pressure (P_{plat}), total compliance (C_T), percent of tidal volume exhaled in 1 sec (V1%), and number of subjects who showed gas trapping (onset of next inspiration before expiratory flow reached zero) is presented for four conditions: closed chest before cardiopulmonary bypass (CPB), open chest before CPB, open chest after CPB, and closed chest after CPB.

[b]Significant effect on V1% ($p < 0.05$).

[c]Significant difference from closed ($p < 0.05$).

[d]Significant difference from closed/pre ($p < 0.05$).

[e]Significant effect on C_T ($p < 0.05$).

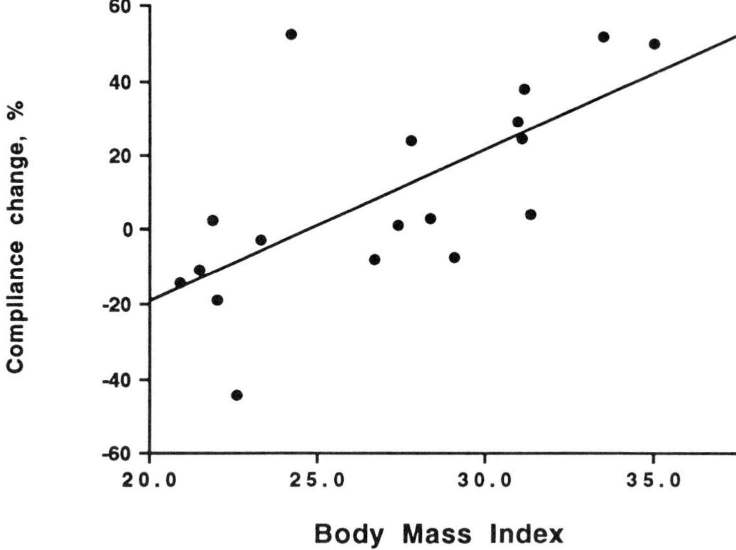

Fig. 3 Regression of change in compliance from chest closed to chest open, precardiopulmonary bypass, showed significant correlation with body mass index ($r = 0.68$, $p < 0.001$).

After cardiopulmonary bypass, compliance again showed no significant change, possibly owing in part to the fact that we often placed patients in a 10°, reverse Trendelenberg position to allow reinfusion of an extra volume of blood from the cardiopulmonary bypass circuit. Then, when the chest was closed, the compliance decreased to significantly lower than the presternotomy value.

Expiratory flow rate, on the other hand, tended to show reciprocal changes with respect to compliance change as shown in Fig. 4. Thus, patients with the greatest improvement in compliance after sternotomy showed reduced expiratory flow rate or V1%, whereas, subjects whose compliance was reduced after sternotomy showed an increase in V1%. More importantly, however, we noted that incomplete exhalation before onset of the next breath (i.e., gas trapping) showed significant correlation with V1%, as shown in Table I.

IV. Discussion

This investigation of flow–volume and pressure–volume changes associated with cardiac surgery led to several important observations. First, the changes in dynamic compliance with sternotomy were not readily

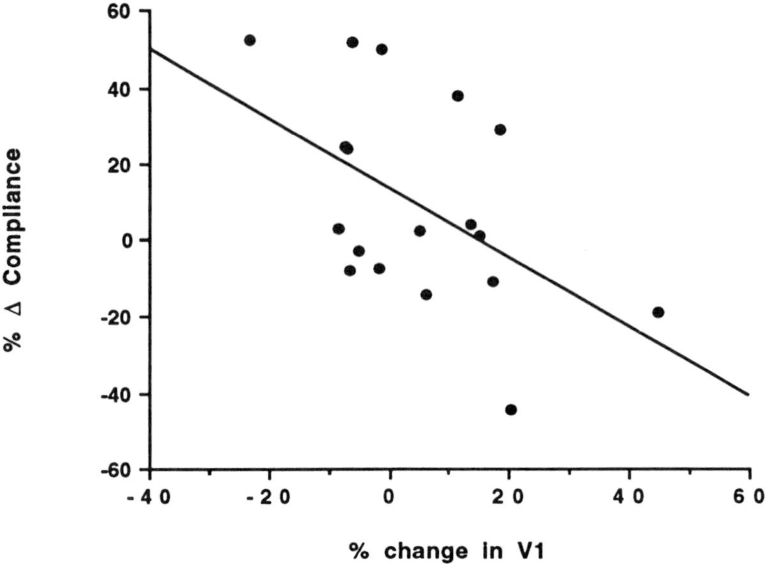

Fig. 4 Change in compliance with chest open before cardiopulmonary bypass showed significant inverse correlation with change in V1.0% ($r = -0.52, p < 0.001$). In other words, subjects with significant improvement in compliance tended to show reduced expiratory flow rate, whereas subjects with reduced compliance tended to show increased expiratory flow rate.

predicted from consideration of "normal" static total respiratory pressure–volume curves. This was due in part to the significant interacting effects of relative body weight and in part to the effects of preexisting chronic obstructive lung disease on dynamic compliance. The effects of increasing body weight on compliance were significantly ameliorated by chest opening. In addition, the presence of expiratory airflow obstruction was compounded by increased body weight, resulting in significant gas trapping. Gas trapping was therefore also improved by sternotomy. However, the expiratory flow rate was often reduced in association with the reduced peak and plateau airway pressures. Finally, the presence of altered pulmonary mechanics associated with cardiopulmonary bypass was effectively masked by these interacting body weight and preexisting lung disease variables. Bypass-induced compliance changes were therefore apparent only after sternotomy closure.

Analysis of these findings immediately raises the question, Why did relative body weight or body mass index have such a significant effect on compliance? In 1960 Naimark and Cherniack published a compliance study of healthy young volunteer subjects whose weight was either within normal limits of ideal or more than 20% above ideal, and therefore considered as obese (2). They found that normal subjects had an average 120 ml/cmH_2O compliance regardless of whether they were seated or supine. The individual component compliances of the lung and chest wall also showed no difference in the seated versus supine position. However, more than 20% above ideal body weight was sufficient to produce a 50% reduction in compliance. This substantial reduction in compliance was almost entirely due to the reduced compliance of the chest wall. In the obese population the supine position was therefore associated with a significant exacerbating effect on chest wall and total compliance. Obviously, one wonders whether removing the chest wall from the lungs, as in sternotomy with sternal retraction, would result in a relatively normal compliance. The answer in our surgical patient population was a qualified "yes."

The qualifying factor was obviously the presence of preexisting obstructive lung disease. The characteristic flow–volume curve from a COPD patient before sternotomy showed significant flow limitation. The failure of exhalation to end before the next breath started also suggested stacking or auto-PEEP (1). When the Narcomed ventilator rate was instantaneously turned to 0.0 breaths/min at end exhalation, it prevented onset of the next inspiratory breath. This enabled sufficient time for an additional 550 ml exhaled tidal volume. However, the very next breath showed an approximately 550 ml lower exhaled tidal volume. Apparently the lung had to reaccumulate volume by stacking, or auto-PEEP, to raise alveolar pressure sufficiently and overcome the abnormal expiratory flow resistance.

What happened with chest opening and sternal spreading, on the other hand, was analagous to the loss of auto-PEEP. In other words, the loss of chest wall-induced lung compression or reduced expiratory force led to reduced alveolar pressure, so that the next several expired breaths were decreased in size until the lung hyperinflated sufficiently to regain an adequate alveolar pressure and overcome the inherent expiratory flow resistance associated with the preexisting lung disease. It was therefore highly probable that the reduced compliance and gas trapping with chest closed conditions were mechanistically related to the low expiratory flow rates.

What produced the expiratory flow rate and auto-PEEP change with sternotomy? The chest wall was no longer limiting the size of the lung. The lungs therefore expanded, allowing airway diameter to increase both because of the change in size and the removal of dynamic airway compression. The evidence for such an increase in lung volume comes from a study by Jonmarker *et al.* in Lund, Sweden (3). They found an average FRC increase of 1.5 liters with sternotomy, along with an increase in arterial p_{O_2}. It is likely, however, that at least some of the FRC increase was in terms of gas communicating space from regions in the lung which previously constituted the volume of trapped gas, rather than simply an increase in overall gas-containing space.

Another interesting feature was the higher expiratory flow rates or V1% along with low compliance for some patients during closed chest conditions. Presumably these higher flow rates derived from a high driving force; that is, high alveolar pressure was produced by the high inflation pressure in the presence of a low chest wall compliance (or high chest wall elastance). When the compliance improved with open-chest conditions, the reduced alveolar pressure led to reduced expiratory flow rate and therefore the lower proportion of breath exhaled in 1 sec (V1%).

In summary, it appears that sternotomy with sternal retraction leads to reduced peak and plateau airway pressures along with improved compliance, and chest closure produced enchanced changes in the opposite direction. The greater effect of chest closure was almost certainly due to pulmonary changes associated with cardiopulmonary bypass. Furthermore, we found evidence of gas trapping or auto-PEEP in patients with COPD that improved with sternotomy and became even worse after chest closure. These findings may be interpreted as evidence that anesthesia may be associated with dynamic airway compression or gas trapping, especially in patients with chronic obstructive pulmonary disease. This phenomenon is significantly enhanced by increased body weight. This type of auto-PEEP should be recognized before the decision is reached regarding early wake up for weaning and extubation. Unfortunately, the earliest indication of significant pulmonary mechanics changes induced by or in association with cardiopulmonary bypass may appear after chest closure, because they are masked by the improved compliance produced by chest open conditions.

References

1. Pepe, P. E., and Marini, J. J. (1982). Occult positive end-expiratory pressure in mechanically ventilated patients with airflow obstruction. *Am. Rev. Respir. Dis.* **126,** 166–170.
2. Naimark, A., and Cherniack, R. M. (1960). Compliance of the respiratory system and its components in health and obesity. *J. Appl. Physiol.* **15,** 377–382.
3. Jonmarker, C., Nordström, L., and Werner, O. (1986). Changes in functional residual capacity during cardiac surgery. *Br. J. Anaesth.* **58,** 428–432.

Inhaled Nitric Oxide in Adult Respiratory Distress Syndrome and Other Lung Diseases

Warren M. Zapol and William E. Hurford

Department of Anesthesia
Massachusetts General Hospital
Boston, Massachusetts 02114

I. Introduction

In 1987, nitric oxide (NO) was reported to be an endothelium-dependent relaxing factor. When inhaled as a gas at low levels, NO selectively dilates the pulmonary circulation. Significant systemic vasodilation does not occur because nitric oxide is inactivated by rapidly binding to hemoglobin. In an injured lung with pulmonary hypertension, inhaled NO produces local vasodilation of well-ventilated lung units and may "steal" blood flow away from unventilated regions. This reduces intrapulmonary shunting and may improve systemic arterial oxygenation. In patients with adult respiratory distress syndrome (ARDS), inhaled NO reduces pulmonary hypertension and improves arterial oxygenation without reducing systemic arterial pressure. Tachyphylaxis to NO inhalation has not been observed. Although additional chronic toxicology studies need to be performed, significant pulmonary toxicity and methemoglobinemia has not been observed at low inhaled concentrations [<80 parts per million by volume (ppm)]. Inhaled nitric oxide is likely to be a valuable therapy in patients with acute lung injury and/or reversible pulmonary hypertension.

Advances in Pharmacology, Volume 31

II. Pulmonary Hypertension in Adult Respiratory Distress Syndrome

Acute pulmonary hypertension consistently occurs in severe ARDS. In survivors, pulmonary vascular resistance progressively decreases over time. Nonsurvivors tend to have a persistently increased pulmonary vascular resistance. The increased pulmonary artery pressure is independent of changes of cardiac output and persists after correction of systemic hypoxemia (1).

The pulmonary vascular changes in ARDS are probably produced by a complex combination of primary lung injury (i.e., aspiration, trauma, infection), consequences of the pulmonary inflammatory response to injury (hypoxia, acidosis, and release of cytokines, components of the complement system and the arachidonic acid pathway, as well as inhibitors of fibrinolysis), and iatrogenic complications of intensive care therapy (oxygen toxicity and barotrauma). In severe ARDS, thromboembolic occlusion of the pulmonary vasculature is common (2). Pulmonary hypertension in patients with severe ARDS is likely to increase edema formation, produce right ventricular dysfunction, and limit cardiac output, and it is associated with a poor prognosis (3–5).

Various vasodilator therapies aimed at reducing pulmonary hypertension have been tested in patients with ARDS. Systemic vasodilation and hypotension occur with all the currently available intravenous vasodilators tested in dosages sufficient to reduce the pulmonary artery pressure. Intravenous infusion of systemic vasodilators such as nitroprusside or prostacyclin (PGI_2) markedly increases the venous admixture and may decrease oxygen delivery to peripheral tissues (6–9).

III. Nitric Oxide

Nitric oxide is a powerful endogenous vasodilator and has been identified as an endothelium-derived relaxing factor (EDRF) (10–13). Nitric oxide is an ideal local transcellular messenger because of its small size, lipophilic nature, and short duration of action (14). In vascular endothelial cells, nitric oxide is synthesized from the terminal guanidino nitrogen of L-arginine and diffuses rapidly into subjacent vascular smooth muscle (15,16). There, NO binds to the heme iron complex of soluble guanylate cyclase. The resulting nitrosyl-heme activates guanylate cyclase, stimulating the production of cyclic guanosine 3′,5′-monophosphate (cGMP) and producing relaxation of vascular smooth muscle (15,17,18). When NO diffuses into the intravascular space, its biological activity is limited by avid binding to hemoglobin (19). Commonly used chemical nitrovasodilators, such as nitroglycerin and nitroprusside, appear to act by releasing NO (20).

IV. Rationale for Use of Inhaled Nitric Oxide in Patients with Adult Respiratory Distress Syndrome

Zapol and co-workers hypothesized that nitric oxide, when inhaled as a gas at low concentrations, should diffuse into the pulmonary vasculature of ventilated lung regions and cause relaxation of pulmonary vascular smooth muscle, thereby decreasing pulmonary hypertension in ARDS (21,22). Because the NO is inhaled, the gas should be distributed predominantly to well-ventilated alveoli and not to collapsed or fluid-filled areas of the lung. In the presence of increased vasomotor tone, local vasodilation of well-ventilated lung regions should cause a "steal" or diversion of pulmonary artery blood flow toward well-ventilated alveoli, improving the matching of ventilation to perfusion and augmenting arterial oxygenation during ARDS. Such an effect would be in marked contrast to the effects of intravenously administered conventional vasodilators (such as nitroprusside, nitroglycerin, or prostacyclin). These intravenous agents also decrease pulmonary artery (PA) pressure, but by nonselectively dilating the pulmonary vasculature they augment blood flow to nonventilated regions, thereby increasing right-to-left shunting (Q_{VA}/Q_T) and reducing arterial P_{O_2}. Also, unlike available intravenous vasodilators, inhaled NO, because it is avidly bound to hemoglobin and rapidly inactivated, should not produce systemic vasodilation.

V. Laboratory Studies of Inhaled Nitric Oxide

A. Acute Pulmonary Hypertension

Inhaled nitric oxide decreases pulmonary hypertension without altering the systemic vascular resistance (21,22). In studies of normal awake lambs without pulmonary hypertension, inhaling 80 ppm NO produced no hemodynamic alterations. The pulmonary artery pressure, cardiac output, systemic arterial pressure, and systemic vascular resistance remained unchanged. When the pulmonary artery pressure was acutely increased, either by hypoxia, infusing the thromboxane endoperoxide analog U46619, or inducing the heparin–protamine reaction, the pulmonary hypertension was reversed by inhalation of 40 to 80 ppm NO (21,22). Pulmonary vasodi-

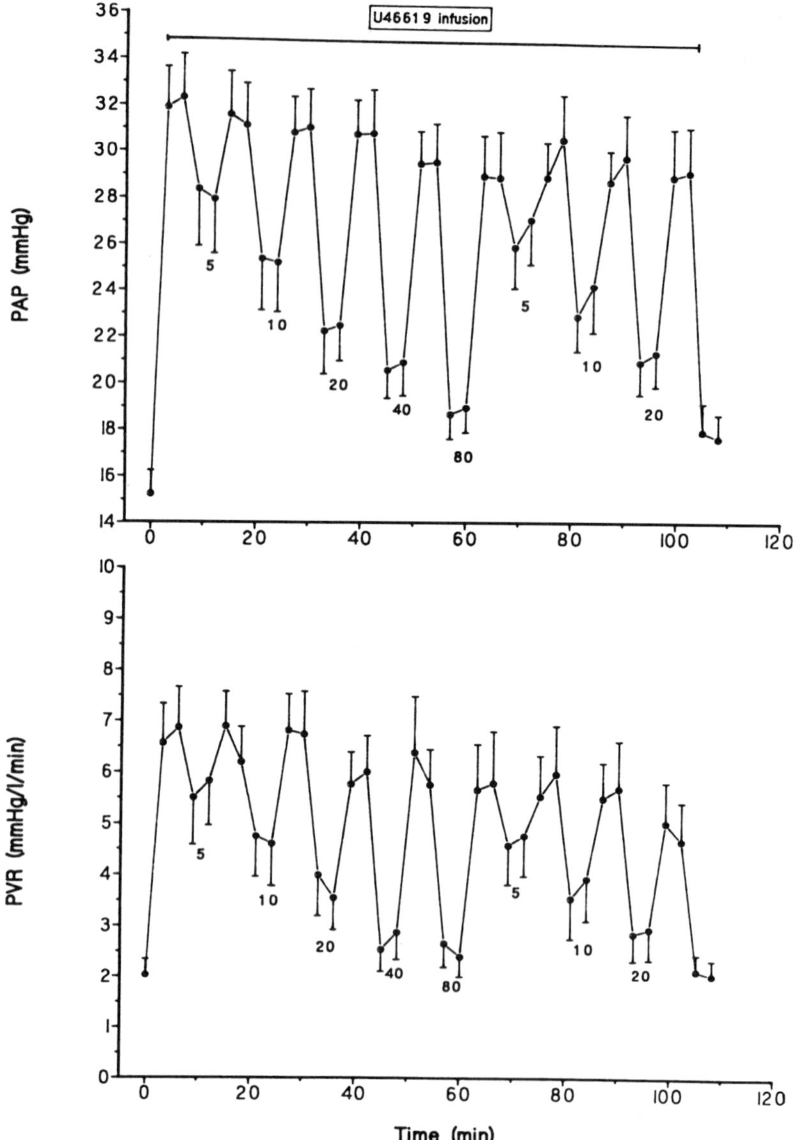

Fig. 1 Plots of mean pulmonary artery pressure (PAP) and pulmonary vascular resistance (PVR) during continuous infusion of U46619. Lambs breathed various levels of nitric oxide (5–80 ppm) at FiO_2 0.6 for 6 min, then breathed a gas mixture without nitric oxide at FiO_2 0.6 for 6 min (n = 8, means ± SEM). [Reprinted from Frostell *et al.* (21).]

lation occurred within 3 min, lasted throughout the duration of NO inhalation, and disappeared within 3 min after the discontinuation of NO (Fig. 1). Importantly, the systemic vascular resistance was unchanged. The pulmonary vasodilator effect occured at low levels of inhaled NO (i.e., 5 ppm). Potent vasodilation (65% of the maximal effect) occurred at 20 ppm inhaled NO. This concentration is less than the National Institute for Occupational Safety and Health (NIOSH) standard for 8-hr working exposures (25 ppm).

During continuous inhalation of 80 ppm NO for 1 hr, no tolerance was observed. That inhaled NO reverses hypoxic pulmonary vasoconstriction has been confirmed by others (23–25). Pison *et al.* studied the matching of ventilation to perfusion using the multiple inert gas technique in mechanically ventilated normal sheep (23). They reported that inhaling 20 ppm NO redistributed blood flow toward well-ventilated alveoli and reversed the pulmonary hypertension caused by breathing a hypoxic gas mixture.

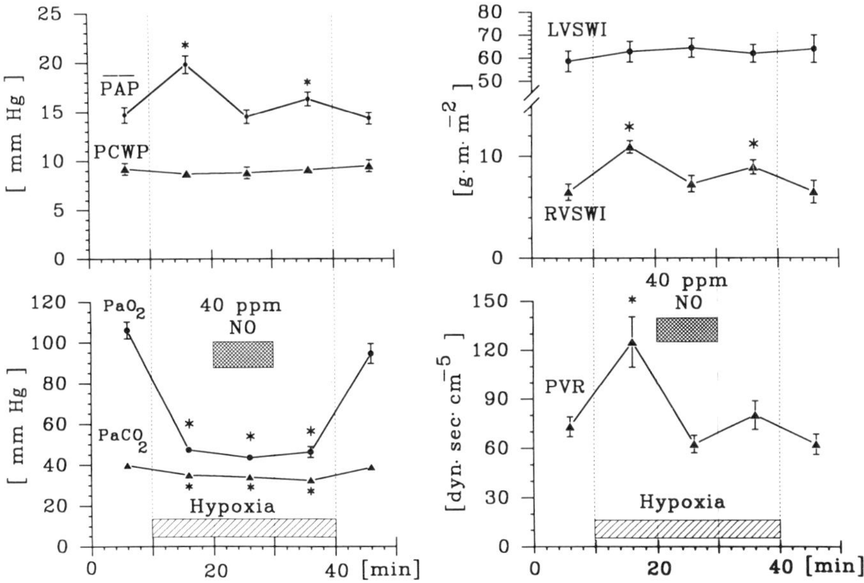

Fig. 2 Nitric oxide (NO) at 40 ppm was inhaled during hypoxia by nine normal volunteers. The effects of NO on mean pulmonary artery (PAP) and wedge (PCWP) pressures, pulmonary vascular resistance (PVR), arterial oxygen (PaO_2) and carbon dioxide ($PaCO_2$) tensions, as well as right (RVSWI) and left (LVSWI) ventricular stroke work indexes are shown. All data are means ± SEM. *Value differs ($p < 0.01$) from first control breathing air. [Reprinted from Frostell *et al.* (26).]

Frostell *et al.* subsequently studied the effects of breathing 40 ppm NO for 10 min in nine awake, healthy human volunteers exposed to hypoxic conditions (26). During air breathing, 40 ppm NO had no pulmonary or systemic vasodilatory effect. When the subjects breathed 12% oxygen to produce mild pulmonary hypertension, 40 ppm NO completely reversed the hypoxic increase of pulmonary artery pressure and vascular resistance (Fig. 2). In two volunteers, the hypoxic pulmonary vasoconstriction was completely reversed by breathing 10 ppm NO. Systemic blood pressure and vascular resistance were unchanged, and methemoglobin levels remained below 1%.

Inhaled NO probably mediates pulmonary vasodilation during lung injury via increasing cGMP levels within smooth muscle. This increase may be reflected by an increased plasma cGMP concentration. Rovira *et al.* studied a model of acute lung injury induced by bilateral lung lavage in anesthetized lambs (27). When endogenous NO production was inhibited by infusing N^G-nitro-L-arginine methyl ester (L-NAME), a consistent increase of aortic over pulmonary artery plasma cGMP concentration was measured within 5 min of breathing 60 ppm NO. Increased aortic plasma cGMP levels were associated with selective pulmonary vasodilation, reduced Q_{VA}/Q_T, and an increased P_aO_2. Plasma levels of plasma cGMP returned to baseline within 10 min of discontinuing NO breathing.

B. Bronchodilation

Because inhaled NO can diffuse through pulmonary tissue to relax upstream pulmonary artery smooth muscle cells, it seems reasonable that inhaled NO might also dilate constricted airways. However, airways have a thick epithelium and are covered by a mucus coating. Diffusion of NO, a highly lipid-soluble substance with a low aqueous solubility (28), might be impeded by the mucus barrier. Dupuy *et al.* examined the effect of inhaled NO on airway mechanics in anesthetized guinea pigs (29). In animals bronchoconstricted with methacholine, inhalation of 5 to 300 ppm NO produced a rapid and dose-related reduction of airway resistance and an increase of dynamic compliance (Fig. 3). An inhaled NO concentration of 15 ppm reduced pulmonary resistance by 50%. The bronchodilator effects of NO were additive to the effects of inhaled terbutaline. Inhaling NO produced stable bronchodilation for over 1 hr. Additional studies demonstrated that inhaled NO in guinea pigs can reverse bronchoconstriction caused by an intravenous infusion of leukotriene D_4, histamine, or neurokinin A (30). Preliminary clinical studies suggest that inhaled NO may have a bronchodilator effect in some patients with asthma (31).

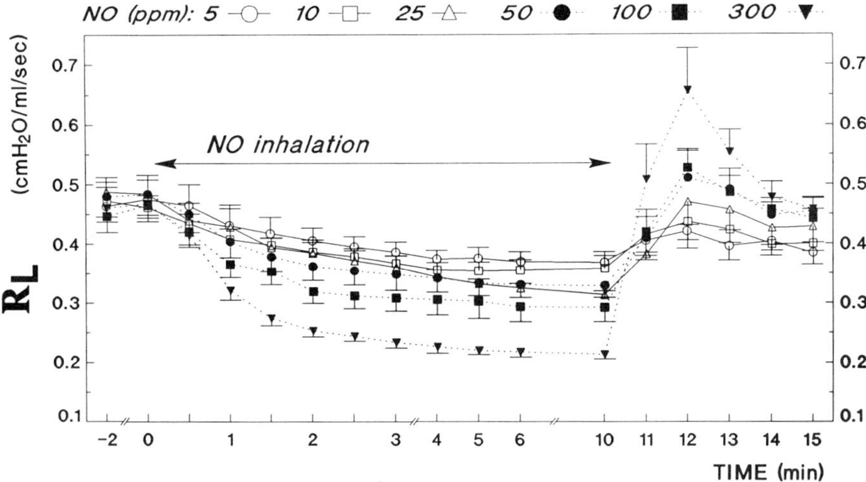

Fig. 3 Time course of changes in pulmonary resistance after ventilation with various concentrations of NO (5–300 ppm) and after ceasing NO gas inflow during a continuous infusion of methacholine ($n = 8$, means ± SE). [Reprinted from Dupuy *et al.* (29).]

VI. Clinical Studies of Nitric Oxide Inhalation in Adult Respiratory Distress Syndrome

Rossaint *et al.* compared the effects of NO inhalation (18 and 36 ppm) to intravenously infused prostacyclin in nine patients with ARDS (32). Nitric oxide selectively reduced mean pulmonary artery pressure from 37 ± 3 to 30 ± 2 mmHg (means ± SE). Oxygenation improved owing to a decreased Q_{VA}/Q_T. During NO breathing, the P_aO_2/FiO_2 ratio increased from 152 ± 15 to 199 ± 23 mmHg (Fig. 4). Although the intravenous infusion of prostacyclin also reduced pulmonary artery pressure, the mean systemic arterial pressure and P_aO_2 decreased as Q_{VA}/Q_T increased. Subsequent reports documented that inhalation of lower concentrations of NO (<20 ppm) effectively reduced pulmonary artery pressure and improved P_aO_2 (33–36). Even very small inhaled NO concentrations [as low as 60–250 parts per billion (ppb) by volume] may effectively raise P_aO_2 in some ARDS patients (37).

A marked variation has been reported for the hemodynamic and respiratory effects of NO inhalation, both among ARDS patients and within the same patient at different times during illness (33,38,39). It is possible that preexisting pulmonary disease as well as the concomitant administration

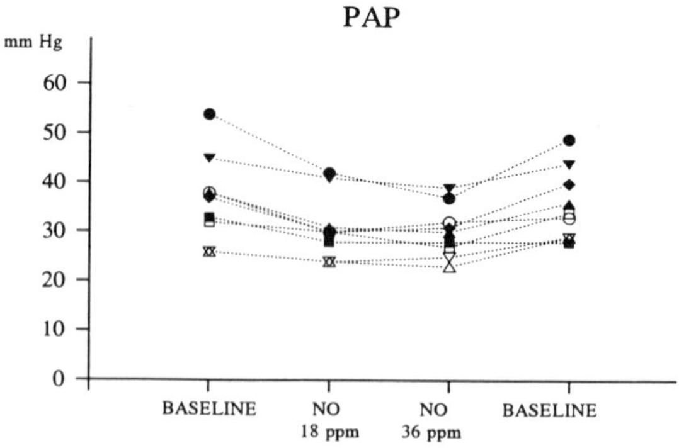

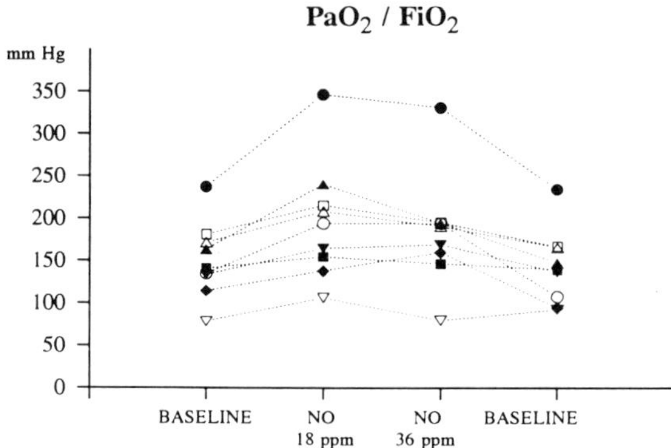

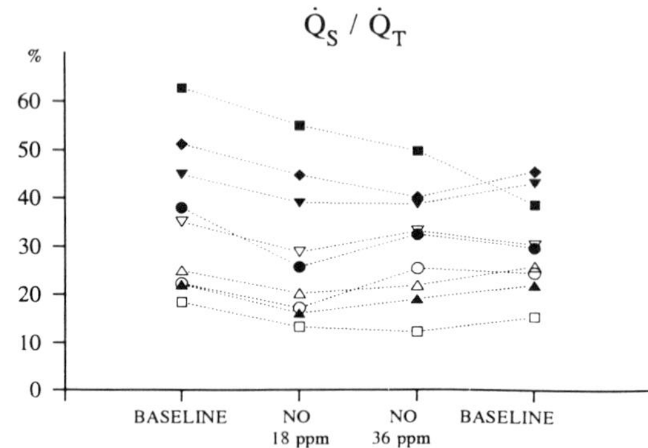

Fig. 4 Mean pulmonary artery pressure (PAP), arterial oxygenation efficiency (PaO_2/FiO_2), and intrapulmonary shunting (Q_S/Q_T) in nine patients with ARDS during inhalation of nitric oxide. Filled symbols represent patients treated with extracorporeal membrane oxygenation. [Reprinted from Rossaint *et al.* (32).]

of other vasoactive drugs may contribute to the observed variability. In general, the baseline level of pulmonary vascular resistance (PVR) appears to predict the degree of pulmonary vasoconstriction reversible by NO inhalation. Those with the greatest degree of pulmonary hypertension appear to respond the most to NO inhalation (33,39). It has been reported that administering the pulmonary vasoconstrictor drug almitrine can further reduce the shunt during NO breathing, possibly by enhancing vasoconstriction in shunting lung regions (39a). Tachyphylaxis has not been observed even when NO inhalation was continued for up to 53 days (32). However, pulmonary artery pressure and P_aO_2 rapidly return to baseline values after discontinuation of the gas (Fig. 5). Occasionally, sudden discontinuation of inhaled NO can produce severe pulmonary vasocon-

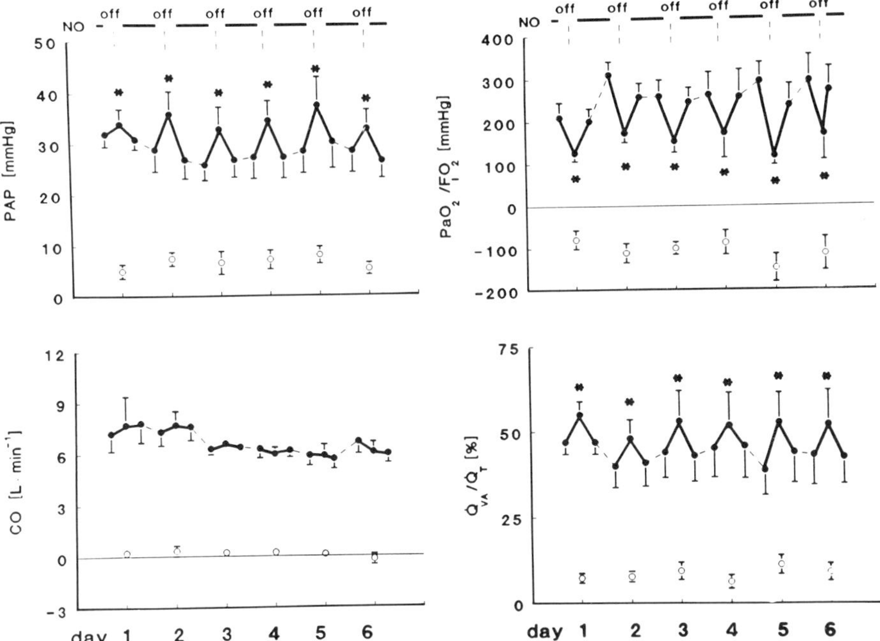

Fig. 5 Hemodynamic function and gas exchange before, during, and after brief interruptions (off) of nitric oxide inhalation (bars) during the first 6 days of treatment in seven patients with ARDS. Values are means ± SE (filled symbols). Also shown (open symbols) are the means ± SE of the individual differences between the values for the effect of treatment and the means of the values determined before and after interrupting NO therapy. The standard errors for the treatment effects were small, indicating that the effects of withdrawal of nitric oxide were clear and precisely estimated. Each asterisk denotes a significant difference from the mean of the values determined before and after interrupting NO therapy. [Reprinted from Rossaint *et al.* (32).]

striction (33,37,40). The mechanism for this is uncertain. Possibly, the addition of exogenous NO may decrease NO synthase activity or increase tissue cGMP phosphodiesterase activity.

VII. Inhaled Nitric Oxide in Other Lung Diseases Associated with Pulmonary Hypertension

A. Neonatal Respiratory Failure

At birth, there is a sustained decrease of pulmonary vascular resistance and an increased pulmonary blood flow, in part due to increasing oxygen tensions. If this does not occur, persistent pulmonary hypertension of the newborn (PPHN) may result. Persistent pulmonary hypertension of the newborn is a syndrome characterized by an increased pulmonary vascular resistance, augmented right-to-left shunting across the ductus arteriosus and foramen ovale, and severe hypoxemia. Extracorporeal membrane oxygenation (ECMO) is often used to support these infants, because conventional vasodilator therapy is limited by severe systemic hypotension and may reduce the P_aO_2 by increasing right-to-left shunting. It has been hypothesized that endogenous production of NO by the pulmonary vasculature might be decreased in PPHN. If so, inhaled NO might provide an effective therapy for these severely ill infants (41–45).

Several investigators have studied the effects of inhaled NO on the neonatal pulmonary circulation. In hypoxic newborn lambs, Roberts *et al.* reported that NO inhalation decreased pulmonary artery pressure and increased pulmonary blood flow without reducing systemic vascular resistance (46). Ventilation with 80 ppm NO at an FiO_2 of 0.21 caused a 3-fold increase of both lung tissue and preductal plasma cGMP concentrations. Severe respiratory acidosis did not attenuate the pulmonary vasodilation caused by inhaled NO. Zayek *et al.* studied the effects of inhaled NO (6 to 100 ppm) in a model of PPHN created by prenatal ligation of the ductus arteriosus in lambs (45). They reported that inhaled nitric oxide caused dose-dependent decreases of pulmonary artery pressure and pulmonary vascular resistance. During 100 ppm NO inhalation there was reduced right-to-left shunting through the foramen ovale and an increased P_aO_2 (from 43 ± 16 to 185 ± 72 mmHg, mean $\pm$ SEM) and arterial oxygen saturation (from 74 ± 8 to $96 \pm 2\%$) with a decreased P_aCO_2. Systemic blood pressure was unaffected by breathing NO (45). In a piglet model of group B streptococcal sepsis, a common cause of neonatal respiratory failure studied by Berger *et al.*, inhalation of 150 ppm NO for 30 min reversed the pulmonary hypertension caused by infusion of group B streptococcus (25). The inhalation of NO did not alter ventilation/perfusion abnormalities caused by group B streptococcal sepsis or alter systemic hemodynamics in this study.

In clinical studies by Roberts *et al.*, critically ill infants with PPHN rapidly increased preductal oxygen saturation in response to NO inhalation at concentrations up to 80 ppm. In five of six infants studied, postductual oxygen saturation also increased. In one newborn, the resulting improvement in P_aO_2 persisted after the discontinuation of NO, eliminating the need for ECMO (46). Kinsella *et al.* reported the effects of inhaling 10–20 ppm NO in nine infants with severe PPHN. All nine infants demonstrated a rapid improvement in oxygenation without systemic hypotension. Clinical improvement continued during treatment with 6 ppm NO for 24 hr in 6 of the infants (41). Subsequently, this group reported that 13 of 15 patients with PPHN who were candidates for support with ECMO were successfully treated with inhaled NO; ECMO therapy was avoided in these patients (44). Nitric oxide inhalation in babies with PPHN is currently being studied in a blinded randomized multicenter trial. Inhaled NO has also been shown to be useful in the management of an infant with a congenital diaphragmatic hernia (46a).

B. Chronic Pulmonary Hypertension

Clinical studies of brief periods of NO inhalation have been performed in patients with chronic pulmonary hypertension. In 18 patients with either chronic pulmonary hypertension or cardiac disease, Pepke-Zaba *et al.* reported that inhalation of 40 ppm NO in air decreased pulmonary vascular resistance by 5–68% without affecting systemic vascular resistance (47). Similar hemodynamic effects have been reported in response to inhaled NO in six patients with pulmonary hypertension following mitral valve replacement for mitral stenosis (48), and in patients with chronic obstructive pulmonary disease (COPD) complicated by pulmonary hypertension (49,50).

C. Congenital and Acquired Heart Disease

Inhaled nitric oxide has also been evaluated as a potential therapy for patients with pulmonary hypertension due to cardiac disease. Roberts *et al.* studied 10 children from 3 months to 6.5 years of age with chronic pulmonary hypertension caused by congenital heart defects. Inhaling 20 to 80 ppm NO at FiO_2 0.9 for 10 min promptly decreased pulmonary artery pressure and pulmonary vascular resistance (51). Compared to breathing 0 ppm NO at FiO_2 0.9, NO decreased the mean pulmonary artery pressure from 48 ± 19 to 37 ± 11 mmHg (mean $\pm$ SD). Under the same conditions, NO decreased the pulmonary vascular resistance from 536 ± 376 to 308 ± 260 $dyne \cdot sec \cdot cm^{-5} \cdot m^{-2}$, whereas systemic arterial pressure and resistance remained unchanged. Inhaling 80 ppm NO at FiO_2 0.9 increased pulmonary blood flow in all six patients with intracardiac shunts.

Rich *et al.* have studied the effects of inhaling 20 ppm NO in 20 adult patients undergoing various types of cardiac surgery requiring cardiopulmonary bypass and 5 patients requiring ventricular assist devices (39). They report that NO inhalation decreased the pulmonary artery pressure from 36 ± 3 (mean ± SEM) to 29 ± 2 mmHg and from 32 ± 2 to 27 ± 1 mmHg, before and after cardiopulmonary bypass, respectively, and from 68 ± 12 to 55 ± 9 mmHg in patients requiring ventricular assist devices. The decrease of pulmonay vascular resistance during NO breathing was proportional to the baseline pulmonary vascular resistance of the patients. Once again, systemic hemodynamics were unaffected by NO inhalation.

VIII. Toxicity of Nitric Oxide

The interest generated by reports of the successful therapeutic use of inhaled nitric oxide is tempered by concerns over its toxicity. Nitric oxide is a common pollutant. It is produced in nature by lightning and the burning of fossil fuels and exists in the atmostphere at levels near 10 ppb. The major atmospheric breakdown pathway at these low levels is via combination with ozone (52). Nitric oxide is present in cigarette smoke and is routinely inhaled for short periods in concentrations of 400 to 1000 ppm by millions of people (53). Although the U.S. Occupational Safety and Health Administration has set an NO exposure limit of 25 ppm when breathed for 8 hr/day in the workplace, few long-term studies of NO toxicology have been reported (54).

Nitric oxide may form several toxic products. In oxygen mixtures, NO is oxidized to NO_2. The rate of oxidation is in proportion to the oxygen concentration and the square of the NO concentration (55). Nitrogen dioxide is cytotoxic (56) and is converted in aqueous solutions to nitric and nitrous acids. Occupational safety and health standards limit the exposure of workers to 5 ppm NO_2 (54). In aqueous solutions, nitric oxide reacts rapidly with $O_2^{\cdot -}$ to form peroxynitrite ($OONO^-$) (57), a molecular species which is also highly cytotoxic. In addition, nitric oxide forms complexes with transition metal complexes, including those in metalloproteins such as hemoglobin. In tissues, nitrosation of iron-containing enzymes and iron–sulfur proteins of target cells may be responsible for the cytotoxic action of NO generated by activated macrophages (58,59). In the circulation, nitric oxide combines extremely rapidly with hemoglobin to form nitrosyl Fe (II)–hemoglobin and then methemoglobin (19).

High inhaled concentrations of nitric oxide have been reported to cause acute pulmonary injury, methemoglobinemia, asphyxia, and death (56,60,61). Greenbaum *et al.* exposed anesthetized dogs to 5000–20,000 ppm NO. Death ensued secondary to methemoglobinemia, acidemia, and alveolar adema (62).

Subsequent controlled studies demonstrated that much of the direct pulmonary toxicity of inhaled NO is due to the NO_2 contained in the gas mixtures. In studies which meticulously controlled for the presence of NO_2 during NO exposure, rats breathing 1000 ppm NO for 30 min or 1500 ppm for 15 min showed no changes of lung wet weight or histological structure (56). Histological evidence of lung injury, however, was noted when the animals were exposed to 25 ppm NO_2 for 30 min, and lung wet weight was increased after 30-min exposures to 50 ppm NO_2 as well as after 5- to 15-min exposures to 100 ppm NO_2 (56). In studies of low-level NO exposure, rabbits breathing 43 ppm NO and 3.6 ppm NO_2 for 6 days showed no pulmonary pathological changes by light or electron microscopy or gravimetric techniques (63).

Methemoglobinemia may occur if its production is increased or its reduction via methemoglobin reductase is diminished (64). Clinically significant methemoglobinemia has not been reported following low-level NO exposure. For example, mice exposed to 10 ppm NO for 6 months had a methemoglobin concentration of 0 to 0.3%, identical to unexposed animals (65). Also, von Nieding *et al.* found that methemoglobin levels did not rise above 0.7% in 48 normal volunteers breathing 40 ppm NO for up to 15 min (66). Furthermore, Frostell *et al.* reported that breathing 80 ppm NO for 3 hr did not increase blood methemoglobin levels in awake spontaneously ventilating lambs (21). It is possible, however, that certain patients with decreased methemoglobin reductase activity may develop methemoglobinemia in the face of an increased rate of hemoglobin oxidation. The activity of methemoglobin reductase may be decreased as a result of a hereditary deficiency and is normally low in newborn infants (64).

Nitric oxide also inhibits platelet adhesion to endothelial cells and reverses platelet aggregation *in vitro* (67). Högman *et al.* reported a 20% increase of the bleeding time when rabbits inhaled 80 ppm NO; this effect was not noted at 10 ppm (67a).

IX. Guidelines for Nitric Oxide Inhalation

Inhalation circuits designed to deliver NO must ensure the accurate delivery of nitric oxide while maintaining low levels of nitrogen dioxide (NO_2) (54). Because conversion of NO to NO_2 is offset by minimizing the residence time of NO in the inhaled gas mixture, mixing of a stock gas of NO in nitrogen with the carrier gases (usually oxygen and air) should take place immediately before inhalation. Several studies have used large gas collection bags to store and deliver NO mixed with oxygen-containing gases. Such NO–O_2 mixtures produce steadily increasing levels of NO_2 with concomitantly decreasing levels of NO (68). This method of administration is unsuitable for clinical application because it does not allow prolonged breathing of constant levels of NO with minimal levels of NO_2. Nitrogen dioxide can be removed from gas mixtures by soda lime (69).

Several safety issues should be addressed in experimental and clinical studies of inhaled NO (37):

1. Calibrated, commercially purchased stock tanks of nitric oxide in nitrogen should be used. Stock NO concentrations should not exceed 1000 ppm.
2. Systems used to blend nitric oxide in breathing circuits should be carefully designed and tested for accurate NO delivery and minimal NO_2 production. Concentrations of NO within a breathing circuit will vary with the NO and oxygen concentrations used and the residence time of NO within the lungs and breathing circuit.
3. To minimize contamination of the environment, exhaust gases from the breathing circuit should be scavenged from the patient environment.
4. Nitric oxide and nitrogen dioxide concentrations should be monitored continuously. Soda lime is useful in removing NO from the breathing circuit. Continuous or intermittent monitoring can be performed with commercially available chemiluminescence (70) or electrochemical analyzers (71).
5. Blood methemoglobin levels should be measured frequently in each patient.
6. The lowest effective concentration of inhaled NO should be used (68,72,73).
7. In some patients the sudden discontinuation of inhaled NO may produce severe arterial desaturation and pulmonary hypertension (40). A supplemental breathing circuit capable of delivering inhaled NO should be available to allow manual ventilation of the patient during tracheal suctioning, transport, etc.

At present, NO inhalation therapy remains experimental. In our opinion, several important areas remain to be studied. For how long and at what levels is it safe for the normal or acutely injured lung to breathe NO? What is the effect of inhaled NO on pulmonary structure and function? What are the best methods to deliver NO to the critically ill patient? Will NO inhalation therapy measurably change outcome for patients with various types of pulmonary hypertension? The study of nitric oxide biology has provided us with many new insights into pulmonary vasomotor control. The use of inhaled NO is the first of several therapies based on these recent basic science studies which may benefit patients with reversible lung disease.

Acknowledgments

Research was supported by a grant from the National Heart, Lung, and Blood Institute (HL-42397).

References

1. Zapol, W. M., and Snider, M. T. (1977). Pulmonary hypertension in severe acute respiratory failure. *N. Engl. J. Med.* **296,** 476–480.
2. Zapol, W. M., and Jones, R. (1987). Vascular components of ARDS: Clinical pulmonary hemodynamics and morphology. *Am. Rev. Respir. Dis.* **136**, 471–474.
3. Gattinoni, L., Pesenti, A., Bombino, M., *et al.* (1988). Relationships between lung computed tomographic density, gas exchange, and PEEP in acute respiratory failure. *Anesthesiology* **69,** 824–832.
4. Vlahakes, G. J., Turley, K., and Hoffman, J. I. E. (1981). The pathophysiology of failure in acute right ventricular hypertension: Hemodynamic and biochemical correlations. *Circulation* **63,** 87–95.
5. Sibbald, W. J., Driedger, A. A., Myers, M. L., *et al.* (1983). Biventricular function in the adult respiratory distress syndrome. *Chest* **84,** 126–134.
6. Zapol, W. M., Snider, M. T., Rie, M. A., *et al.* (1985). Pulmonary circulation during adult respiratory distress syndrome. *In* "Acute Respiratory Failure" (W. M. Zapol and K. J. Falke, eds.), pp. 241–270. Dekker, New York.
7. Radermacher, P., Santak, B., and Falke, K. J. (1989). Comparison of prostaglandin E_1 and nitroglycerin in patients with ARDS. *Prog. Clin. Biol. Res.* **301,** 267–270.
8. Radermacher, P., Santak, B., Wust, H. J., *et al.* (1990). Prostacyclin for the treatment of pulmonary hypertension in the adult respiratory distress syndrome: Effects on pulmonary capillary pressure and ventilation–perfusion distributions. *Anesthesiology* **72,** 238–244.
9. Radermacher, P., Santak, B., Wust, H. J., *et al.* (1990). Prostacyclin and right ventricular function in patients with pulmonary hypertension associated with ARDS. *Intensive Care Med.* **16,** 227–232.
10. Ignarro, L. J., Buga, G. M., Wood, K. S., *et al.* (1987). Endothelium-derived relaxing factor produced and released from artery and vein is nitric oxide. *Proc. Natl. Acad. Sci. U.S.A.* **84,** 9265–9269.
11. Palmer, R. M. J., Ferrige, A. G., and Moncada, S. (1987). Nitric oxide release accounts for the biological activity of endothelium-derived relaxation factor. *Nature (London)* **327,** 524–526.
12. Ignarro, L. J., Buga, G. M., Byrns, R. E., *et al.* (1988). Endothelium-derived relaxing factor and nitric oxide possess identical pharmacologic properties as relaxants of bovine arterial and venous smooth muscle. *J. Pharmacol. Exp. Ther.* **246,** 218–226.
13. Archer, S. L., Rist, K., Nelson, D. P., *et al.* (1990). Comparison of the hemodynamic effects of nitric oxide and endothelium-dependent vasodilators in intact lungs. *J. Appl. Physiol.* **68,** 735–747.

14. Ignarro, L. J. (1991). Signal transduction mechanisms involving nitric oxide. *Biochem. Pharmacol.* **41,** 485–490.
15. Palmer, R. M. J., Ashton, D. S., and Moncada, S. (1988). Vascular endothelial cells synthesize nitric oxide from L-arginine. *Nature (London)* **333,** 664–666.
16. Moncada, S., Palmer, R. M. J., and Higgs, E. A. (1989). Biosynthesis of nitric oxide for L-arginine: A pathway for the regulation of cell function and communication. *Biochem. Pharmacol.* **38,** 1709–1715.
17. Furchgott, R. F., and Vanhoutte, P. M., (1989). Endothelium-derived relaxing and contracting factors. *FASEB J.* **3,** 2007–2018.
18. Ignarro, L. J. (1989). Biological actions and properties of endothelium-derived nitric oxide formed and released from artery and vein. *Circ. Res.* **65,** 1–21.
19. Doyle, M.L., and Hoekstra, J. W. (1981). Oxidation of nitrogen oxides by bound dyoxygen in hemoproteins. *J. Inorg. Biochem.* **14,** 351–358.
20. Gruetter, C. A., Gruetter, D. Y., Lyon, J. E., *et al.* (1981). Relationship between cyclic guanosine 3′:5′-monophosphate formation and relaxation of coronary artery arterial smooth muscle by glyceryl trinitrate, nitroprusside, nitrite and nitric oxide. *J. Pharmacol. Exp. Ther.* **219,** 181–186.
21. Frostell, C., Fratacci, M.-D., Wain, J. C. *et al.* (1991). Inhaled nitric oxide: A selective pulmonary vasodilator reversing hypoxic pulmonary vasoconstriction. *Circulation* **83,** 2038–2047.
22. Fratacci, M. D., Frostell, C., Chen, T.-Y., *et al.* Inhaled nitric oxide: A selective pulmonary vasodilator of heparin–protamine vasoconstriction in sheep. *Anesthesiology* **75,** 990–999.
23. Pison, U., Lopez, F. A., Heidelmeyer, C. F., *et al.* (1993). Inhaled nitric oxide reverses hypoxic pulmonary vasoconstriction without impairing gas exchange. *J. Appl. Physiol.* **74,** 1287–1292.
24. Tönz, M., von-Segesser, L. K., and Turina, M. (1993). Selective pulmonary vasodilation with inhaled nitric oxide (letter). *J. Thorac. Cardiovasc. Surg.* **105,** 760–762.
25. Berger, J. I., Gibson, R. L., Redding, G. J., *et al.* (1993). Effect of inhaled nitric oxide during group B streptococcal sepsis in piglets. *Am. Rev. Respir. Dis.* **147,** 1080–1086.
26. Frostell, C. G., Blomqvist, H., Hedenstierna, G., *et al.* (1993). Inhaled nitric oxide selectively reverses human hypoxic pulmonary vasoconstriction without causing systemic vasodilation. *Anesthesiology* **78,** 427–435.
27. Rovira, I., Chen, T. Y., Winkler, M., *et al.* (1994). Effects of inhaled nitric oxide on pulmonary hemodynamics and gas exchange in an ovine model of ARDS. *J. Appl. Physiol.* 1994; 76:345–355.
28. Shaw, A. W., and Vosper, A. J. (1977). Solubility of nitric oxide in aqueous and nonaqueous solvents. *J. Chem. Soc. Faraday Trans.* **73,** 1239–1244
29. Dupuy, P. M., Shore, S. A., Drazen, J. M., *et al.* (1992). Bronchodilator action of inhaled nitric oxide in guinea pigs. *J. Clin. Invest.* **90,** 421–428.
30. Dupuy, P. M., Shore, S. A., Kim, K., *et al.* (1993). Bronchodilator action of inhaled nitric oxide in histamine, leukotriene D_4, and neurokinin A constricted guinea pigs. *Am. Rev. Respir. Dis.* **147,** A288
31. Frostell, C., Högman, M., Hedenström, H., and Hedenstierna, G. (1993). Is nitric oxide inhalation beneficial for the asthmatic patient? *Am. Rev. Respir. Dis.* **147,** A720.
32. Rossaint, R., Falke, K. F., Lopez, F., *et al.* (1993). Inhaled nitric oxide for the adult respiratory distress syndrome. *N. Engl. J. Med.* **328,** 399–405.
33. Bigatello, L. M., Hurford, W. E., Kacmarek, R. M., *et al.* (1993). The hemodynamic and respiratory response of ARDS patients to prolonged nitric oxide inhalation. *Am. Rev. Respir. Dis.* **147,** A720.
34. Payen, D. M., Gatecel, C., and Guinard, N. (1993). Inhalation of low dose of nitric

oxide (NO) and iv L-arg in ARDS: Effect on pulmonary hemodynamic, gas exchange and NO metabolites. *Am. Rev. Respir. Dis.* **147,** A720.
35. Grover, R., Smithies, M., and Bihari, D., (1993). A dose profile of the physiological effects of inhaled nitric oxide in acute lung injury. *Am. Rev. Respir. Dis.* **147,** A350.
36. Wysocki, M., Vignon, P., Roupie, E., *et al.* (1993). Improvement in right ventricular function with inhaled nitric oxide in patients with the adult respiratory distress syndrome (ARDS) and permissive hypercapnia. *Am. Rev. Respir. Dis.* **147,** A350.
37. Zapol, W. M., Falke, K. J., and Roissant, R. (1993). Inhaled nitric oxide for the adult respiratory distress syndrome (letter). *N. Engl. J. Med.* **329,** 207.
38. Ricou, B., and Suter, P. M. (1993). Variable effects of nitric oxide (NO) in ARDS patients. *Am. Rev. Respir. Dis.* **147,** A350.
39. Rich, G. F., Murphy, G. D., Roos, C. M., and Johns, R. A. (1993). Inhaled nitric oxide: Selective pulmonary vasodilation in cardiac surgical patients. *Anesthesiology* **78,** 1028–1035.
39a. Payen, D. M., Gatecel, C., and Plaisance, P. (1993). Almitrine effect on nitric oxide inhalation in adult respiratory distress syndrome (letter). *Lancet* **341,** 1664
40. Grover, R., Murdoch, I., Smithies, M., *et al.* (1992). Nitric oxide during hand ventilation in patient with acute respiratory failure (letter). *Lancet* **340,** 1038–1039.
41. Kinsella, J. P., Neish, S. R., Shaffer, E., *et al.* (1992). Low-dose inhalational nitric oxide in persistent pulmonary hypertension of the newborn. *Lancet* **340,** 819–820.
42. Kinsella, J. P., McQueston, J. A., Rosenberg, A. A., and Abman, S. H. (1992). Hemodynamic effects of exogenous nitric oxide in ovine transitional pulmonary circulation. *Am. J. Physiol.* **263,** H875–H880.
43. Roberts, J. D., Jr., Polaner, D. M., Lang, P., *et al.* (1992). Inhaled nitric oxide in persistent pulmonary hypertension of the newborn. *Lancet* **340,** 818–819.
44. Kinsella, J. P., and Abman, S. H. (1993). Inhalational nitric oxide therapy for persistent pulmonary hypertension of the newborn. *Pediatrics* **91,** 997–998.
45. Zayek, M., Cleveland, D., and Morin III, F. C. (1993). Treatment of persistent pulmonary hypertension in the newborn lamb by inhaled nitric oxide. *J. Pediatr.* **122,** 743–750.
46. Roberts, J. D., Jr., Chen, T. Y., and Kawai, N. (1993). Inhaled nitric oxide reverses pulmonary vasoconstriction in the hypoxic and acidotic newborn lamb. *Circ. Res.* **72,** 246–254.
46a. Frostell, C. G., Lonnqvist, P. A., Sonesson, S. E., *et al.* (1993). Near fatal pulmonary hypertension after surgical repair of congenital diaphragmatic hernia. *Anaesthesia* **48,** 679–683.
47. Pepke-Zaba, J., Higenbottam, T. W., Dinh-Xuan, A. T., *et al.* (1991). Inhaled nitric oxide as a cause of selective pulmonary vasodilation in pulmonary hypertension. *Lancet* **338,** 1173–1174.
48. Girard, C., Lehot, J. J., Pannetier, J. C., *et al.* (1992). Inhaled nitric oxide after mitral valve replacement in patients with chronic pulmonary artery hypertension. *Anesthesiology* **77,** 880–883.
49. Fratacci, M. D., Defouilloy, C., Andrivet, P., *et al.* (1992). Responses to infusion of acetylcholine and inhalation of nitric oxide in patients with chronic obstructive lung disease and pulmonary hypertension. *Am. Rev. Respir. Dis.* **145,** (4), A722.
50. Adatia, I., Thompson, J., Landzberg, M., and Wessel, D. L. (1993). Inhaled nitric oxide in chronic obstructive lung disease (letter). *Lancet* **341,** 307–308.
51. Roberts, J. D., Jr., Lang, P., Bigatello, L. M., *et al.* (1993). Inhaled nitric oxide in congenital heart disease. *Circulation* **87,** 447–453.
52. Levine, J. S. (1989). Photochemistry of biogenic gases. *In* "Global Ecology" (M. B. Rambler, L. Margulis, and R. Fester, eds.), pp. 51–74. Academic Press, Orlando, Florida.

53. Norman, V., and Keith, C. (1988). Nitrogen oxides in tobacco smoke. *Nature (London)* **205,** 915–916.
54. Centers for Disease Control. (1988). NIOSH recommendations for occupational safety and health standards. *Morbid. Mortal. Wkly. Rep.* **37,** (Suppl. S-7), 21.
55. Glasson, W. A., and Tuesday, C. S. (1963). The atmospheric thermal oxidation of nitric oxide. *J. Am. Chem. Soc.* **85,** 2901–2904.
56. Stavert, D. M., and Lehnert, B. E. (1990). Nitric oxide and nitrogen dioxide as inducers of acute pulmonary injury when inhaled at relatively high concentrations for brief periods. *Inhalation Toxicol.* **2,** 53–67.
57. Beckman, J. S., Beckman, T. W., Chen, J., *et al.* (1990). Apparent hydroxyl radical production by peroxynitrite: Implications for enothelial injury from nitric oxide and superoxide. *Proc. Natl. Acad. Sci. U.S.A.* **87,** 1620–1624.
58. Hibbs, J. B., Taintor, R. R., Vavrin, Z., and Rachlin, E. M. (1988). Nitric oxide: A cytotoxic activated macrophage effector molecule. *Biochem. Biophys. Res. Commun.* **157,** 87–94.
59. Stuehr, D. J., and Nathan, C. F. (1989). Nitric oxide: A macrophage product responsible for cytostasis and respiratory inhibition in tumor target cells. *J. Exp. Med.* **169,** 1543–1555.
60. Clutton-Brock, J. (1967). Two cases of poisoning by contamination of nitrous oxide with higher oxides of nitrogen during anaesthesia. *Br. J. Anaesth.* **39,** 388–392.
61. Austin, A. T. (1967). The chemistry of higher oxides of nitrogen as related to the manufacture, storage and administration of nitrous oxide. *Br. J. Anaesth.* **39,** 345–350.
62. Greenbaum, R., Bay, J., Hargreaves, M. D., *et al.* (1967). Effects of higher oxides of nitrogen on the anesthetized dog. *Br. J. Anaesth.* **39,** 393–404.
63. Hugod, C. (1979). Effect of exposure to 43 ppm nitric oxide and 3.6 ppm nitrogen dioxide on rabbit lung. *Int. Arch. Occup. Environ. Health* **42,** 159–167.
64. Beutler, E. (1977). Methemoglobinemia and sulfhemoglobinemia. *In* "Hematology" (W. J. Williams, E. Beutler, A. J. Erslev, and R. W. Rundles, eds.), 2nd Ed., pp. 491–494. McGraw-Hill, New York.
65. Oda, H., Nogami, H., Kusumoto, S., *et al.* (1976). Long-term exposure to nitric oxide in mice. *J. Jpn. Soc. Air Pollut.* **11,** 150–160.
66. von Nieding, G., Wagner, H., and Kockeler, H. (1975). Investigation of the acute effects of nitrogen monoxide on lung function in man. *Staub–Reinhalt Luft* **35,** 175–178.
67. Radomski, M. W., Palmer, R. M. J., and Moncada, S. (1987). Endogenous nitric oxide inhibits human platelet adhesion to vascular endothelium. *Lancet* **2,** 1057–1058.
67a. Högman, M., Frostell, C., Arnberg, H., *et al.* (1993). Bleeding time prolongation and NO inhalation (letter). *Lancet* **341,** 1664–1665.
68. Bouchet, M., Renaudin, M. H., Raveau, C., *et al.* (1993). Safety requirement for use of inhaled nitric oxide in neonates (letter). *Lancet* **341,** 968–969.
69. Oda, H., Kusumoto, S., and Nakajima, T. (1975). Nitrosyl-hemoglobin formation in the blood of animals exposed to nitric oxide. *Arch. Environ. Health* **30,** 453–456.
70. Fontijin, A., Sabadell, A. J., and Ronco, R. J. (1970). Homogeneous chemiluminescent measurement of nitric oxide and ozone. *Anal. Chem.* **42,** 575–579.
71. Petros, A. J., Cox, P. B., and Bohn, D. (1992). Simple method for monitoring concentration of inhaled nitric oxide (letter). *Lancet* **340,** 1167.
72. Foubert, L., Fleming, B., Latimer, R., *et al.* (1992). Safety guidelines for use of nitric oxide. *Lancet* **339,** 1615–1616.
73. Laguenie, G., Berg, A., Saint-Maurice, J.-P., *et al.* (1993). Measurement of nitrogen dioxide formation from nitric oxide by chemiluminescence in ventilated children. *Lanet* **341,** 969.

First Pass Uptake in the Human Lung of Drugs Used during Anesthesia

David L. Roerig,[*,†] Susan B. Ahlf,[‡] Christopher A. Dawson[§] John H. Linehan,[||] and John P. Kampine[*]

Departments of [*]Anesthesiology and [†]Pharmacology and Toxicology and [§]Physiology
The Medical College of Wisconsin
Milwaukee, Wisconsin 53226

[||] Department of Biomedical Engineering
Marquette University
Milwaukee, Wisconsin 53233

[‡] Department of Veterans Affairs
Zablocki Veterans Affairs Medical Center
Milwaukee, Wisconsin 53295

I. Introduction

It is well established that the pulmonary circulation has important functions other than gas exchange. This includes a pharmacokinetic function in which cells of the pulmonary endothelium have been shown to accumulate a wide variety of biogenic and xenobiotic substances (1–4). Based on animal drug distribution studies, amine drugs exhibit high pulmonary accumulation, with lung tissue to plasma concentration ratios as high as 400 (4). Isolated perfused animal lung studies have indicated that this high uptake appears to be rapid, extensive, saturable, an is a result of simple diffusion of the drug from the vascular space into the lung tissue (5–13). The physicochemical properties of the drug appear to determine in large part the extent of pulmonary accumulation, with drugs that contain a basic

Advances in Pharmacology, Volume 31

amine moiety (p$K > 8.0$) and that have moderate to high lipid solubility accumulating in lung tissue to the greatest extent (5,6,9,10).

The lung is uniquely situated between systemic arterial and venous circulations. It receives the entire cardiac output and has a very large capillary surface area. For these reasons, the lung can play an important role in regulating the arterial blood concentrations of compounds that exhibit high pulmonary accumulation. In the short term, immediately after intravenous bolus injection of a drug, high pulmonary uptake moderates peak concentrations of the drug during the first pass through the systemic circulation and thus other organ systems. At longer times after administration, the lung can act as a significant tissue reservoir and thus participate in the dispositional phase of plasma pharmacokinetics.

The accumulation of certain drugs in pulmonary tissue after intravenous bolus administration can also reveal information about the lungs in dilution studies. Removal of drugs from the circulation on passage through the lungs occurs at the perfused vascular surface. Because 90% of the vascular surface is associated with the pulmonary capillary bed, it is possible that the extent or character of the pulmonary uptake of some compounds can provide information about the permeability of the capillary endothelium and perfused capillary surface area. Changes in the pulmonary extraction of a number of biogenic substances by the human lung have been used to detect lung injury or disease (14–17). Similarly, it has been suggested that uptake of the drug propranolol in the human lung may be a useful indicator of change in capillary endothelial function and detection of early pulmonary microvascular damage (18).

Our studies have focused on understanding the extent and character of uptake of drugs, both in the human lung and in isolated perfused animal lung preparations. For studies on pulmonary drug uptake in the human lung, we have employed multiple indicator dilution methods which involve the intravenous or intrapulmonary arterial injection of a bolus containing a reference indicator that is confined to the vascular space and a test indicator which is the drug of interest that is taken up by the pulmonary endothelium as the blood flows through the pulmonary vascular bed. The amount of drug extracted by the lung is detected as a difference in the fractional recovery between the reference indicator and the test indicator at the systemic arterial sampling site.

II. First Pass Drug Uptake in Human Lung

Details of double indicator dilution methods have been previously reported (19–21). Briefly, all subjects were ASA physical status I–III patients and were studied prior to elective surgery. All studies were performed in accordance with institutional policies on human experimentation with informed consent. Indocyanine green dye (ICG) was used as the reference indicator. The 2.0 ml bolus contained 10 mg ICG, 41 mg human serum albumin, and the test drugs at the doses shown in Table I. The bolus was prepared immediately before use and loaded into a 2-foot length of plasic tubing connected to the central venous catheter as depicted in Fig. 1. The ICG–drug bolus was injected within 2 sec, with a 10-ml saline flush. Blood was pumped from the radial artery (60 ml/min) and collected in 1-sec fractions for 45 sec after injection. The 1-second (1 ml) blood samples were diluted with 4 ml of water, and the ICG concentration was determined spectrophotometrically. Whole blood drug concentrations were determined by radioimmunoassay, high-performance liquid chromatography, or gas–liquid chromatography (19–21). The ICG and drug concentration in each blood sample was expressed as the fraction of the injected dose recovered per milliliter of blood with time after injection. The ICG curve represents the fraction of the injected dose versus time curve wherein no extraction by the lung occurs, and the difference in area under the ICG curve and the drug curve is equal to the total fraction of injected drug extracted from the blood into the lung during the first pass. For comparative purposes the percentage of drug taken up into the lung was calculated at the time when 95% of the injected ICG had passed through the lung as described by Jorfeldt *et al.* (22,23).

Figures 2, 3, and 4 show typical first pass uptake curves for the human lung for fentanyl, meperidine, and morphine (19). Each graph represents the fraction of injected ICG and drug per millilter of blood with time after injection of the dye–drug bolus. In Figs 2–4 the ICG curves are similar, with a peak between 19 and 24 sec after injection and a smaller second peak at 40–42 sec representing the second pass of the ICG through the lung. The extent of first pass pulmonary uptake of these drugs represents a wide range that is consistent with differences in the lipophilic nature of

Table I

Summary Data for First Pass Uptake in Human Lung

Drug	Number of patients	Dose	Age (years)	CO (liters/min)	Lung uptake (% injected)
Fentanyl	10	75 μg	57 ± 6	5.29 ± 0.63	82.6 ± 1.4
Meperidine	7	25 mg	58 ± 3	6.58 ± 0.51	64.5 ± 7.8
Verapamil	6	2.5 mg	64 ± 2	5.31 ± 0.31	48 ± 3.4
Diazepam	9	7.5 mg	57 ± 4	6.11 ± 0.78	33.8 ± 3.7
Thiopental	8	25 mg	49 ± 5	6.23 ± 0.57	15.8 ± 2.4
Morphine	4	10 mg	71 ± 3	5.65 ± 0.57	3.5 ± 7.1

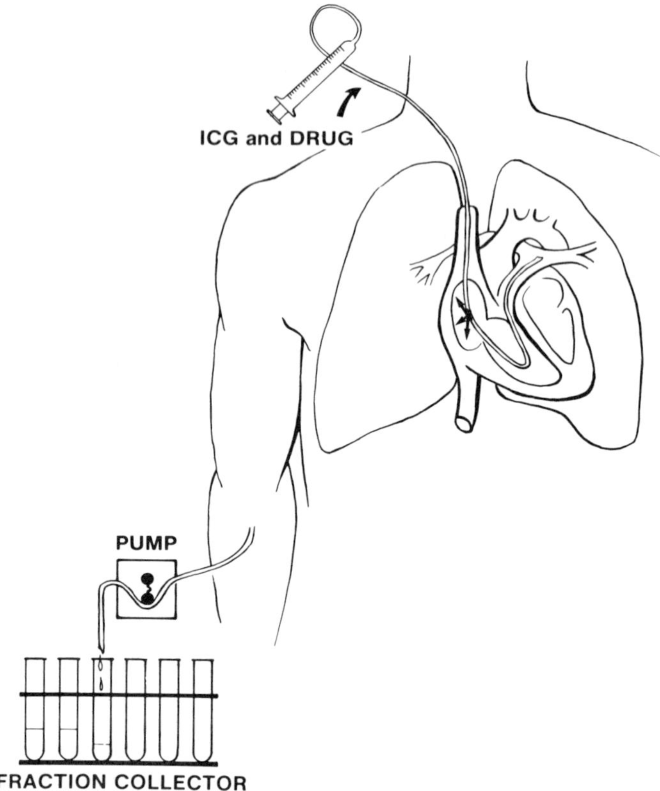

Fig. 1 Schematic representation of drug and dye administration and blood collection for determination of first pass drug uptake in the human lung. All blood samples were collected in heparinized tubes in the fraction collector, and the concentration of each drug was determined per milliliter of whole blood.

the drugs. For example, comparing the area under the fentanyl and ICG curves in Fig. 2, it is apparent that the majority of the fentanyl was extracted by the lung in a single pass through the human lung. In this particular patient, 86.6% of the injection dose was taken up in the first pass, resulting in only about 13% of the injected dose reaching the systemic circulation immediately after intravenous drug administration. The mean fentanyl uptake for all patients we have studied was 82.6 ± 1.4% of the injected dose (see Table I). Meperidine, like fentanyl and morphine, is a basic amine, but it is less lipophilic than fentanyl as judged by octanol–water partition coefficients (see Table II). Figure 3 indicates significant uptake of meperidine in the human lung with a mean uptake of

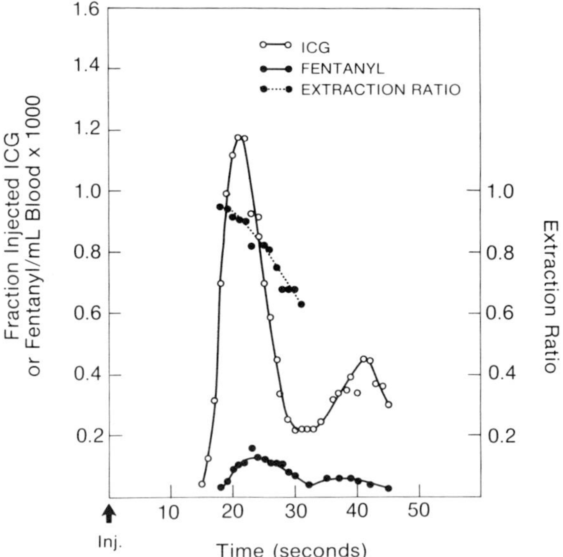

Fig. 2 Fraction of injected dose (solid lines) of ICG and fentanyl per milliliter of arterial blood versus time (seconds) after intravenous injection of the dye–drug bolus. The dashed line represents the extraction ratio for fentanyl with time. The difference in area under the ICG and fentanyl curves at 95% ICG recovery indicate 86.6% uptake of fentanyl during the first pass through the lung. Initial extraction of fentanyl was 94.7%.

64.5 ± 7.8% of the injected dose (Table I). The least lipophilic of these narcotic analgesics is morphine, and Fig. 4 demonstrates almost no pulmonary accumulation, with a mean first pass uptake of 3.5 ± 7.1% of the injected dose (Table I).

For fentanyl and meperidine, Figs. 2 and 3, respectively, show the change in the instantaneous extraction ratio (*ER*) with time after injection. The extraction ratio represents the fraction of drug in the blood taken up into the lung at each time point and was calculated from the formula

$$ER = 1 - \mathrm{F}_{\mathrm{drug}}/F_{\mathrm{ICG}}$$

where *F* is the fraction of injected ICG or drug in arterial blood at each time. The *ER* for both fentanyl and meperidine was very high (>90%) in the initial part of the first pass through the lung, but it decreased during the duration of the first pass. This apparent decrease in *ER* with time is indicative of the returning flux of the accumulated fentanyl and meperidine back out of the lung into the blood, although the net flux of either drug is into the lung during the entire first pass.

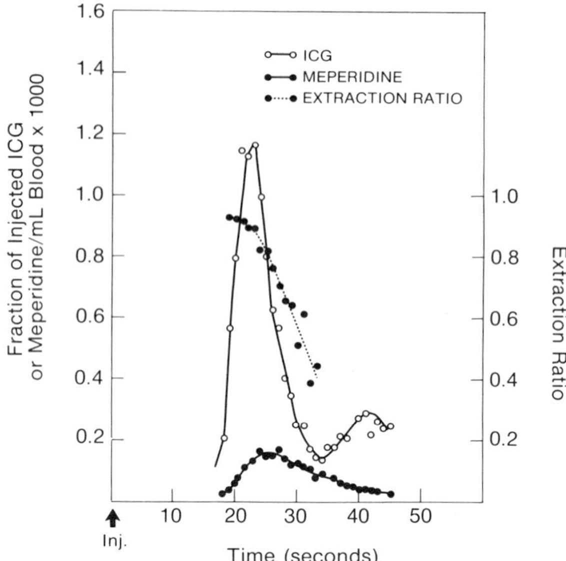

Fig. 3 Fraction of injected dose (solid lines) of ICG and meperidine per milliliter of arterial blood versus time (seconds) after intravenous injection of the dye–drug bolus. The dashed line represents the extraction ratio for meperidine with time. The difference in area under the ICG and meperidine curves at 95% ICG recovery indicated 80.7% uptake of meperidine during the first pass through the lung. Initial extraction of meperidine was 93.1%.

Such diffusion of accumulated drug back out of the lung after the bolus has passed reflects on the contribution of the pulmonary drug reservoir to blood drug concentration with time. However, because of the second and subsequent passes of drug through the lung, the rate of diffusion of this large pulmonary drug reservoir out of the lung is difficult to assess from these studies. Taeger *et al.* measured the difference in arterial and venous blood concentrations of fentanyl for 14 min after the first pass through the human lung and estimated that 60% of the fentanyl accumulated during the first pass diffused back out of the lung during the first 10 min (24). Based on our measurements, 15–20% of the injected fentanyl enters the systemic circulation immediately after injection, and, together with the estimate by Taeger *et al.*, another 45% of the injected dose would enter the systemic circulation over the next 10 min. The early or distributional plasma half-life of fentanyl has been estimated at between 1.2 and 4 min (25). In view of the extensive first pass uptake and the data of Taeger *et al.* (24), this early phase of plasma disappearance of fentanyl reflects both disappearance of drug owing to uptake by other tissues and

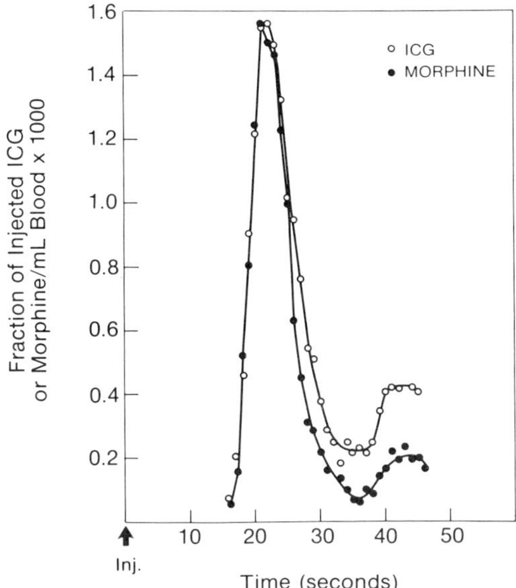

Fig. 4 Fraction of injected dose ICG and morphine per milliliter of arterial blood versus time (seconds) after intravenous injection of the dye–drug bolus. The difference in area under the ICG and morphine curves at 95% ICG recovery indicated that 92.5% of the morphine remained in the blood after the first pass through the lung.

addition of drug to the blood from the large pulmonary pool. Thus, the lungs act as a capacitor modulating the rate of plasma disapperance of the drug.

We have determined the first pass pulmonary uptake of drugs having diverse physicochemical properties to provide insight into how these properties influence pulmonary accumulation. Uptake of verapamil, diazepam,

Table II

Physicochemical Properties of Drugs Studied in Human Lung

Drug	pK_a	Octanol–water partition coefficient	Amount bound to plasma protein (%)
Fentanyl	8.4	816	84.4
Meperidine	8.5	39	70.0
Verapamil	8.4	103	90.0
Diazepam	3.4	261	97.0
Thiopental	7.5[a]	179	85.0
Morphine	7.9	1.4	30.0

[a] Acidic compound.

and thiopental by the human lung is summarized in Table I. Based on octanol–water partition coefficients (Table II), verapamil is a moderately lipophilic basic amine. About 50% of the intravenous bolus of verapamil was taken up by the lungs on the first pass. Diazepam represents an interesting class of compounds in that it has a higher lipid solubility than verapamil but is a nonbasic amine with a pK_a of 3.4 (Table II). Its lower first pass uptake of only about 30% would be consistent with the idea that high pulmonary accumulation is favored for drugs with a higher pK_a. Thiopental, on the other hand, has a high lipid solubilty and high pK_a, but it is anionic at physiological pH. The uptake of this moderately lipophilic compound was only about 14% of the injected dose (Table I).

III. Factors Affecting Pulmonary Drug Uptake

The impact of pulmonary drug uptake during the first pass through the lungs on the plasma pharmacokinetics can be very different for different classes of drugs as demonstrated by the six drugs we have studied in the human lung. For all drugs, the greatest impact of pulmonary uptake is on the early plasma pharmacokinetics; this impact depends on the extent of first pass uptake as already discussed with fentanyl. There are several factors that influence the extent of first pass uptake of a drug in the lung tissue, some of which are discussed below.

A. Saturability

As already mentioned, animal studies have indicated that pulmonary drug uptake is saturable (5–13). Furthermore, displacement of one lipophilic drug from the lung by injection of another has been reported for a number of drugs (6,26). It is possible, therefore, that prior administration of a high dose of a lipophilic amine could decrease the first pass uptake of itself or another lipophilic amine. As an example of this, Geddes *et al.* (18) found that the first pass uptake of propranolol in the human lung was reduced from 80 to 33% of the injected dose in patients receiving chronic propranolol therapy. To determine whether such competition might occur between basic lipophilic amines in general in the human lung, we compared the first pass pulmonary uptake of fentanyl in two groups of patients, one group receiving no prior medication and another group receiving a daily dose of propranolol of 30 to 120 mg/day for at least 1 month prior to surgery (20). Figures 5 and 6 show typical first pass uptake curves for fentanyl in the two patient groups. The smaller difference in area between the dye and fentanyl curves in patients on a daily 120 mg dose of proprano-

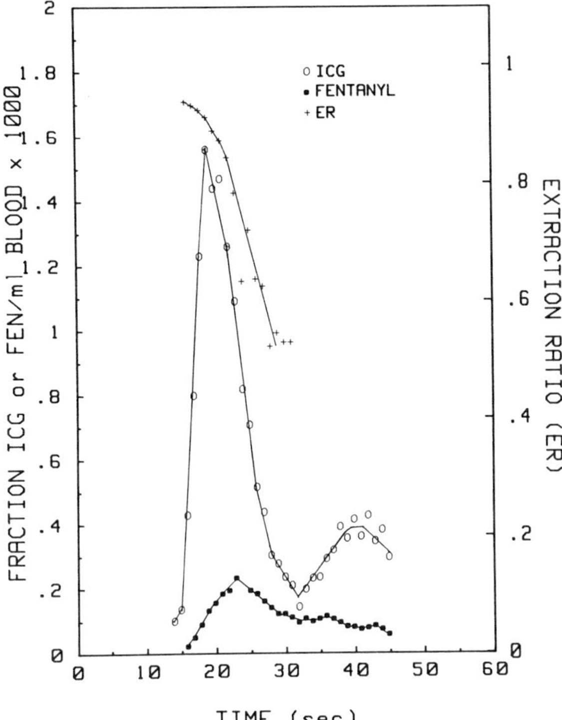

Fig. 5 Fraction of injected dose of ICG and fentanyl per milliliter of arterial blood versus time (seconds) after intravenous injection of the dye–drug bolus. Points denoted by + symbols represent the extraction ratios (*ER*) for fentanyl with time. Data are from a patient in the group that received no propranolol. The differences in area under the ICG and fentanyl curves at 95% ICG recovery indicate 82.3% uptake of fentanyl during the first pass through the lung. The initial extraction ratio of fentanyl was 0.94.

lol reflects a lower first pass uptake of fentanyl (82 versus 20.6% of injected dose). Figure 7 shows the percent first pass uptake of fentanyl in all patients as a function of daily propranolol dose, and there was a significant negative correlation between propranolol dose and first pass fentanyl uptake. This study emphasizes the saturable nature of first pass pulmonary drug uptake and how this can be a pharmacodynamic mechanism for a drug–drug interaction.

The existence of a group of patients receiving chronic daily administration of propranolol in which the pulmonary uptake of another basic lipophilic amine (fentanyl) could be studied was fortuitious in general. It can be difficult to assess the saturability of pulmonary uptake of a drug in

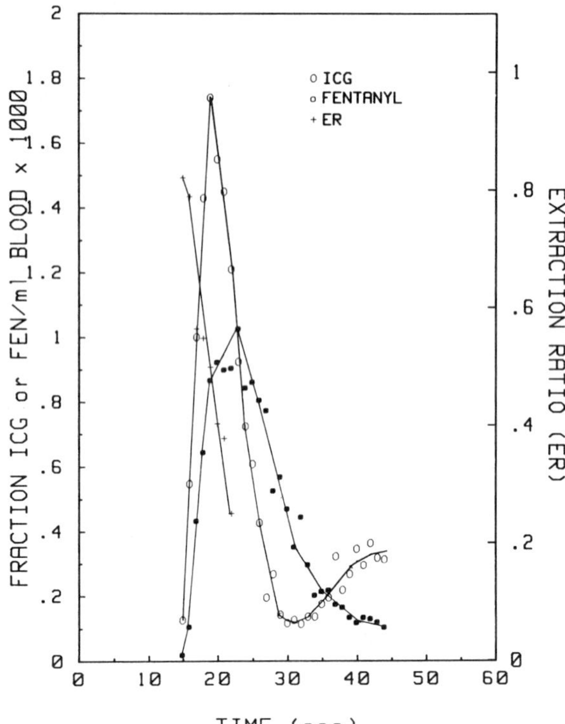

Fig. 6 Fraction of injected dose of ICG and fentanyl per milliliter of arterial blood versus time (seconds) after intravenous injection of the dye–drug bolus. Points denoted by + represent the extraction ratios (*ER*) for fentanyl with time. Data are from a patient that received 120 mg propranolol per day for 1 month. The difference in area under the ICG and fentanyl curves at 95% ICG recovery indicated 20.6% uptake of fentanyl during the first pass through the lung. The initial extraction ratio of fentanyl was 0.82.

patients because of dose constraints. With a highly potent narcotic analgesic such as fentanyl, where the total mass of administered drug is small, it does not appear likely that such saturation could be attained within a safe therapeutic dose range. Jorfeldt *et al.* (22,23) did measure the first pass pulmonary uptake of lidocaine in the human patient for two 50 mg bolus doses given 10 min apart. They found a small decrease in the percent first pass uptake of lidocaine after the second bolus. We found a significant decrease in the first pass pulmonary uptake of a 100 mg lidocaine bolus compared to 30 mg bolus in human patients (27). From these findings we estimate that a 3.3-fold increase in dose resulted in a 5- to 6-fold increase in the amount of drug reaching the systemic circulation immediately after

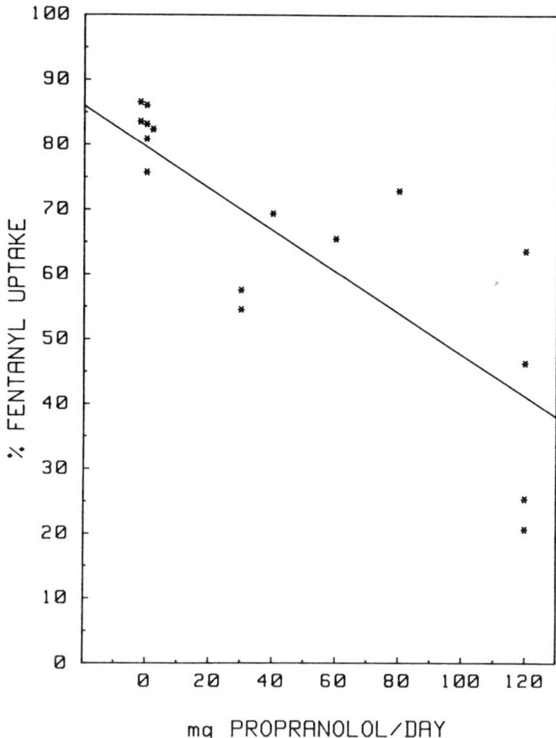

Fig. 7 Percent first pass uptake of fentanyl in the human lung versus daily dose of propranolol. Each point is the percent first pass uptake of fentanyl at 95% ICG recovery for each patient studied. The slope of the regression line was −0.332 ($p < 0.001$, $r = -0.817$).

injection. To what extent such saturation of first pass pulmonary uptake occurs with other drugs is difficult to predict, but it is most likely to occur if the drugs exhibit a high first pass pulmonary uptake and the pharmacological properties are such that the therapeutic doses are of high milligram amounts.

Some quantitative differences may exist in the pulmonary accumulation of diazepam compared to basic amine drugs that are significantly ionized at physiological pH. For example, diazepam uptake in the rat lung appeared to be linearly related to the total amount infused, whereas in the same dose range methadone uptake was curvilinear (Fig. 9). Thus, methadone uptake was saturable at lower concentrations than diazepam. A second difference between methadone and the nonbasic diazepam was observed in the kinetics of efflux from the rat lung. With basic lipophilic amines such as methadone and three acetylmethadol derivatives we found evidence of a "slowly effluxable pool" that had a half-life of efflux from the lung greater than several hours (11,13). A similar slowly effluxable pool has been reported for imipramine and methadone in the isolated perfused rabbit lung (8,9). This slowly effluxable pool contained 10–25% of the total drug accumulated in the lung. With diazepam, all of the drug accumlated in the lung during a 10-min infusion and was recovered within a 30-min drug-free perfusion period so that no slowly efflxable pool was evident (13). It may be that the cationic character of basic lipophilic amines allows access to an additional highly sequestered pool.

B. Plasma Protein Binding

Binding of a drug to plasma proteins may also limit uptake into the lung. For those drugs whose binding to plasma proteins is in a rapid equilibrium (where "rapid" refers to the pulmonary vascular transit time), accumulation in pulmonary tissue can be viewed as partitioning of the drug between the vascular space and the lung tissue. The partition coefficient is determined by the affinity of the drug for structures in the vascular (plasma protein) and extravascular space (tissue protein, lipid, etc.). The drug–protein interaction in the vascular space can be represented by

$$\text{Protein} + \text{drugs} \underset{K_2}{\overset{K_1}{\rightleftarrows}} \text{Protein–drug complex}$$

The rate constants K_1 and K_2 describe the rate of association and dissociation of the protein–drug complex, and $K_1/K_2 = K_{eq}$, which is a measure of the degree to which the drug is bound to plasma protein at equilibrium. In the simplest case wherein K_1 and K_2 are large, then, an equilibrium between the free and bound forms is maintained during passage through the lung. The influence of binding on the amount of drug taken up by the lung on a single pass would then be determined by K_{eq}. Furthermore, if the uptake of the free drug into the tissue were rapidly equilibrating, the fraction of drug taken up would be relatively insensitive to the blood flow rate (cardiac output) through the lung. Conversely if K_1 is small, plasma protein binding would not be a factor during the first pass of the drug. For intermediate values of K_1 and K_2, the fractional uptake by the lung in a single pass would decrease with increasing flow, assuming for the moment that perfused capillary surface area were constant. Measurement of drug–plasma protein binding by equilibrium dialysis or centrifugal ultrafiltration can give a measure of K_{eq} but not of K_1 and K_2. Some insight, however, can be gained from indicator dilution studies of the uptake of the drugs as they pass through the lung (e.g., the effect of changing cardiac output).

Without knowledge of K_1 and K_2, the extent of drug–plasma protein binding does not have much predictive value for the extent of pulmonary drug uptake. This is shown by our findings with fentanyl, meperidine, and morphine (Figs. 2–4). It is the unbound drug that diffuses into lung tissue, and Meuldermans *et al.* (28) determined that the free fraction of fentanyl, meperidine, and morphine in human plasma was 16, 30, and 70%, respectively. This order is exactly opposite the order of extent of first pass uptake in the human lung (Table I). This suggests that the affinity of fentanyl for lung tissue is much greater than for plasma protein. At the other extreme, morphine appears to have a relatively low affinity for either tissue or plasma protein. The very high single pass uptake observed with fentanyl would also suggest that K_1 and K_2 are large (rapid association/dissociation with plasma protein) relative to the pulmonary capillary transit time. For these three drugs the extent of pulmonary uptake appears to be most closely related to the lipid solubility as measured by the octanol–water partition coefficients (Table II).

Plasma protein binding may be a major determinant of pulmonary accumulation of some compounds. The simplest example is the indocyanine green (ICG) dye used as the vascular reference indicator. In the isolated dog lung lobe perfused with a protein-free artificial perfusate, ICG was rapidly taken up by the lung, whereas no uptake is observed when lungs are perfused with blood or plasma (29) wherein virtually 100% of ICG is bound to plasma protein and thus completely confined to the vascular space.

The relatively low first pass uptake in the human lung for diazepam may also be related to its very extensive binding to plasma protein. For example, in rat lung tissue slices (Fig. 8) and isolated perfused rat lungs (Fig. 9), diazepam uptake was very small if 4.5% bovine serum albumin (BSA) were present (13). Under these conditions diazepam is about 93 to 95% bound to BSA. If the BSA were omitted from the perfusate, there was about a 10-fold increase in the pulmonary accumulation of diazepam. In the same studies, the high pulmonary accumulation of the basic lipophilic amine methadone showed only a slight dependence on the presence of BSA in the perfusate. The effect of plasma protein binding on the pulmonary accumulation of diazepam was also evident in the isolated dog lung lobe preparation (30). We found that single pass uptake of diazepam was inversely proportional to the concentration of BSA in the perfusate. In the same studies we observed that the perfusate flow rate through the lung had little effect on the uptake of diazepam, suggesting a rapid equilibrium between diazepaam bound in the plasma and lung tissue spaces.

In contrast to the case for diazepam we found that a rapid equilibrium for binding to plasma protein in relation to single pass transit time

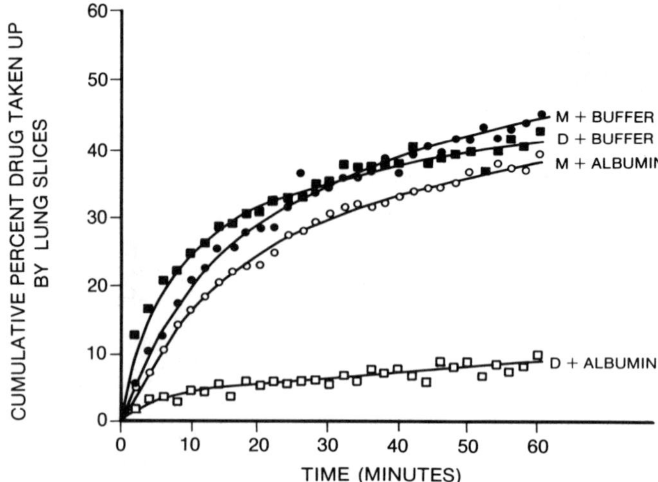

Fig. 8 Uptake of methadone (M) and diazepam (D) in lung slices with and without 4.5% BSA (albumin) in the incubation medium. Lung tissue slices equivalent in total weight to one rat lung were prepared and incubated with [^{3}H]methadone or [^{14}C]diazepam. Each point represents the cumulative percentage of total radioactivity taken up by the lung tissue slices with time.

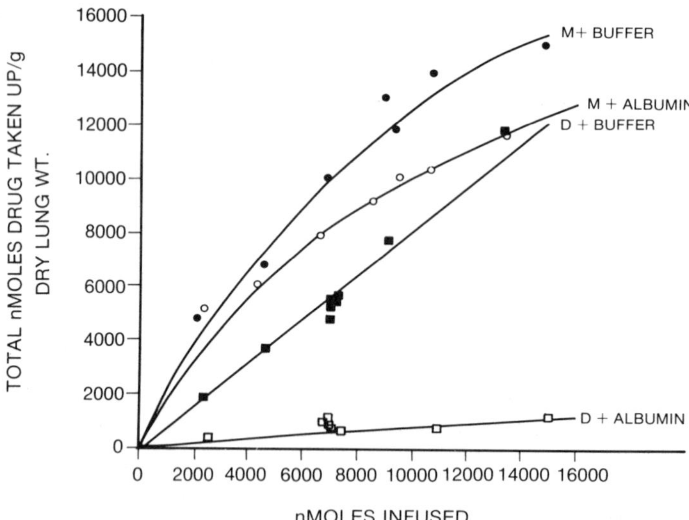

Fig. 9 Total amount (nmol/g dry lung) of methadone (M) and diazepam (D) taken up by the isolated perfused rat lung during the 10-min drug perfusion as a function of the total (nmol) infused. Open symbols represent lungs perfused with 4.5% BSA, and the filled symbols represent data from lungs perfused with buffer only (no BSA).

through isolated perfused rabbit lung was not achieved for the synthetic angiotensin-converting enzyme (ACE) substrate benzoylphenylalanyalanylproline (BPAP) (31). When BPAP was incubated for increasing lengths of time prior to passage through the lung, the extent of BPAP hydrolysis by pulmonary ACE was decreased. From this we concluded that multiple binding sites for BPAP on albumin existed, and that for certain sites (22% of total binding) K_1 was slow with respect to transit time through the lung.

Any factor that alters plasma protein binding of a drug and could alter the partitioning of the drug between the plasma and lung tissue. As an example, we observed a decrese in doxorubicin binding to BSA with increased temperature (32). Of course, changes in drug–plasma protein binding would be expected to influence the uptake of some drugs more than other. For example, with two highly plasma protein-bound drugs, decreased plasma protein binding of fentanyl would have little effect on fentanyl uptake, whereas a large increase in uptake of diazepam is potentially possible.

C. Effect of Cardiac Output

For compounds that pass from the blood into lung tissue via simple diffusion, the effect of blood flow rate through the lung on the rate or extent of pulmonary uptake will be dependent on the rate at which an equilibrium between the drug in the blood and tissue take place relative to the transit time through the pulmonary capillary bed. For example, uptake of the basic lipophilic amines imipramine, methadone, and three acetylmethadols was found to be flow-limited in isolated rabbit or rat lung preparations (8,9,11,12). Flow-limited uptake was indicated by the fact that during constant infusion the initial rate of uptake was approximately equal to the rate of drug infusion. Figure 10 shows the initial velocities of methadone and *l*-α-acetylmethadol (LAAM) uptake as a function of infusate concentration in the rat lung, and comparison with the line of identity reveals that velocity of uptake was equal to the rate at which the drug flowed into the lung (11). As defined by Goresky, such "flow limited distribution describes that situation in which the capillary walls present no barrier to diffusion and in which the tissue is so well perfused that diffusion equilibration occurs virtually instantaneously between continuous vascular and extravascular spaces at each point along the length of a capillary (33). For an idealized bolus injection, however, all of the drug is presented to the pulmonary capillary surface at a single instant. If the rapid equilibrium is achieved (flow-limited), the uptake of the drug after intravenous bolus administration should be independent of the rate of passage through the lung. That is, uptake of such a drug would be independent of cardiac output.

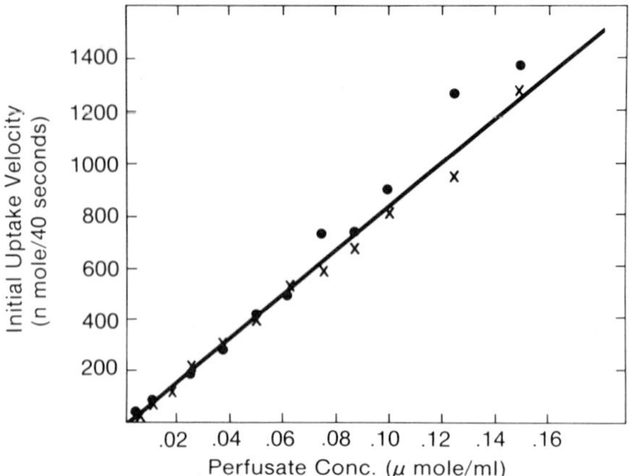

Fig. 10 Initial uptake velocities of LAAM (×) and methadone (●) versus the concentration of each drug in the perfusate. The solid line is the linear regression line for combined LAAM and methadone data, and the slope, normalized to 1, was 0.891 ± 0.062.

The impact of cardiac output is difficult to determine from studies in the human lung. For fentanyl, for which we have the most patient data, we could find no statistical correlation between cardiac output (range 2.5 to 7.6 liters/min) and extent of first pass uptake of fentanyl (19,20). Likewise, in the 9 diazepam patients studied we detected no correlation between cardiac output (range 3.4 to 10.3 liters/min) and first pass uptake of diazepam in the human lung (21). More complete studies of the influrence of flow can be carried out in animal lung preparations. For example, our data from isolated rat and rabbit lungs indicated that diazepam rapidly equilibrates between the vascular and tissue spaces, and that after bolus injection the fractional uptake was independent of flow (13,30). For the other compounds listed in Table I, it will be necessary to determine the effect of flow on pulmonary uptake kinetics to gain some insight into their dependence on cardiac output. For compounds that do not instantaneously equilibrate between the vascular and tissue space on a single passage through the lung, we expect their extraction from the blood would be dependent on cardiac output. Uptake of these drugs is commonly referred to as being barrier-limited. Again, this is difficult to evaluate from studies on patients. This could be important, however, when considering the potential use of therapeutic agents in indicator dilution studies to detect changes in the function or state of the pulmonary endothelium.

IV. Conclusions

The pulmonary circulation is capable of a very rapid and extensive extraction of certain drugs from the blood during a single transit through the lung after intravenous administration of the drug bolus. The extent of uptake depends on the physicochemical characteristics of the drug and its binding affinity to plasma proteins and lung tissue. For some basic lipophilic amines the high first pass uptake in the humna lung controls the initial rate of entry of the major portion of the dose into the systemic circulation and plays a complex role in moderating plasma drug concentrations during the early distributional phase of plasma pharmacokinetics. The extent of first pass uptake in the lung also predicts the relative magnitude of the lung as a tissue reservoir of the drug at later times after administration during the dispositional phase of the pharmacokinetics of the drugs.

Pulmonary uptake kinetics can reflect the effect of rate process such as association and dissociation from plasma protein and diffusional rates. The lungs, especially isolated animal lung preparations, can be used as a reactor to observe nonequilibrium phenomena of drugs and biological structures that have rate constants small in relation to the time of a single transit through the organ.

The fact that the pulmonary capillary bed accumulates so many different drugs presents an opportunity for using these therapeutic agents as indicators in multiple indicator dilution studies of the pulmonary circulation in humans. Technically, such studies are neither difficult nor unnecessarily invasive as shown by our own studies in patients prior to surgery. The use of therapeutic agents as indicators has the advantage that the pharmacology and toxicology are known, and in many cases relatively inexpensive and sensitive assays for determining the blood concentration of the drug are routinely available. Such studies hold promise of allowing one to monitor change associated with injury to the pulmonary circulation. To this end, we need additional insight into the nature of the interaction of drugs with the lung tissue as they pass through the pulmonary capillary bed.

References

1. Vane, J. R. (1968). "The Importance of Fundamental Principles in Drug Evaluation," pp. 217–235. Raven, New York.
2. Gillis, C. N. (1973). Metabolism of vasoactive hormones by the rat lung. *Anesthesiology* **39,** 626–632.
3. Junod, A. F. (1976). Uptake, release and metabolism of drugs in the lungs. *Pharmacol. Ther. Part B* **2,** 511–521.

4. Anderson, M. W., Philopt, R. M., Bend, J. R., Wilson, A. G. E., and Eling, T. E. (1976). Pulmonary uptake and metabolism of chemicals by the lung. *In* "Clinical Toxicology" (W. A. M. Duncan, ed.) pp. 85–105. Leonary, Amsterdam.
5. Orton, T. C., Anderson, M. W., Pickett, R. D., Eling, T. E., and Fouts, J. R. (1973). Xenobiotic accumulation and metabolism by isolated perfused rabbit lungs. *J. Pharmacol. Exp. Ther.* **186,** 482–497.
6. Anderson, M. W., Orton, T. C., Pickett, R. D., and Eling, T. E. (1974). Accumulation of amines in the isolated perfused rabbit lung. *J. Pharmacol. Exp. Ther.* **189,** 456–466.
7. Junod, A. F. (1975). Mechanism of drug accumulation by the lung. *In* "Lung Metabolism" (A. F. Junod and R. Deltalles, eds.), pp. 214–227. Academic Press, New York.
8. Wilson, A. G. E., Law, F. C. P., Eling, T. E., and Anderson, M. W. (1976). Uptake, metabolism and efflux of methadone in "single pass" isolated perfused rabbit lung. *J. Pharmacol. Exp. Ther.* **199,** 360–367.
9. Eling, T. E., Pickett, R. D., Orton, T. C., and Anderson, M. W. (1975). A study of dynamics of imipramine accumulation in the isolated perfused rabbit lung. *Drug Metab. Dispos.* **3,** 389–400.
10. Wilson, A. G. E., Pickett, R. D., Eling, T. E., and Anderson, M. W. (1980). Studies on the persistence of basic amines in the rabbit lung. *Drug Metab. Dispos.* **7,** 402–424.
11. Roerig, D. L., Dawson, C. A., and Wang, R. I. H. (1982). Uptake and efflux of *l*-alpha-acetyl-methadol (LAAM) and methadone in the isolated perfused rat lung. *Drug Metab. Dispos.* **10,** 230–235.
12. Roerig, D. L., Dawson, C. A., and Wang, R. I. H. (1983). Uptake of *l*-alpha-acetyl-methadol (LAAM) and its analgesically active metabolites, nor-LAAM and dinor-LAAM, in the isolated perfused rat lung. *Drug Metab. Dispos.* **11,** 411–416.
13. Roerig, D. L., Dahl, R. R., Dawson, C. A., and Wang, R. I. H. (1984). Effect of plasma protein binding on the uptake and efflux of methadone and diazepam in the isolated perfused rate lung. *Drug Metab. Dispos.* **12,** 536–542.
14. Gillis, C. N., Greene, N. M., Gronau, L. H., and Hammond, G. L. (1972). Pulmonary extraction of 5-hydroxytryptamine and norepinephrine before and after cardiopulmonary bypass in man. *Circ. Res.* **30,** 666–674.
15. Gillis, C. N., Grounau, L. H., Greene, N. M., and Hammond, G. L. (1974). Removal of 5-hydroxytryptamine and norepinephrine from the pulmonary vascular space of man. Influence of cardiopulmonary bypass and pulmonary arterial pressure on the process. *Surgery* **76,** 608–616.
16. Gills, C. N., Gronau, L. H., Mandel, S., and Hammond, G. L. (1979). Indicator dilution measurement of 5-hydroxytryptamine clearance by human lung. *J. Appl. Physiol.* **46,** 1178–1183.
17. Gillis, C. N., and Catravas, J. D. (1982). Altered removal of vasoactive substances in the injured lung: Detection of lung microvascular injury. *Ann. N.Y. Acad. Sci.* **77,** 458–474.
18. Geddes, D. M., Nesbitt, K., Traill, T., and Blackburn, J. P. (1979). First pass uptake of 14C-propranolol by the lung. *Thorax* **34,** 810–813.
19. Roerig, D. L., Kotrly, K. J., Vucins, E. J., Ahlf, S., Dawson, C. A., and Kampine, J. P. (1987). First pass uptake of fentanyl, meperidine and morphine in the human lung. *J. Anesthesiol.* **64,** 466–472.
20. Roerig, D. L., Kotrly, K. J., Ahlf, S. B., Dawson, C. A., and Kampine, J. P.(1989). Effect of propranolol on the first pass uptake of fentanyl in the human lung. *Anesthesiology* **71,** 62–68.
21. Roerig, D. L., Kotrly, K, J., Dawson, C. A., Ahlf, S. B., Gualtieri, J. F., and Kampine, J. P. (1989). First pass uptake of verapamil, diazepam and thiopental in the human lung. *Anesth. Analg. (N.Y)* **69,** 461–466.

22. Jorfeldt, L., Lewis, D. H., Lofstrom, J. B., and Post, C. (1979). Lung uptake of lidocaine in healthy volunteers. *Acta Anaesthesiol. Scand.* **23,** 567–574.
23. Jorfeldt, L., Lewis, D. H., Lofstrom, J. B., and Post, C. (1983). Lung uptake of lidocaine in man as influenced by anaesthesia, mepivacaine infusion or lung insufficiency. *Acta Anaesthesiol. Scand.* **27,** 5–9.
24. Taeger, K., Wenninger, E., Franke, M., and Finsterer, U. (1984). Uptake of fentanyl by the human lung. *Anesthesiology* **61,** A246.
25. Hug, C. C. (1984). Pharmacokinetics and dynamics of narcotic analgesics, "Pharmacokinetics of Anaesthesia" (C. Prys-Roberts and C. C. Hugs, Jr. eds). pp. 187–234. Blackwell, London.
26. Ohmura, S., Yamamoto, K., Kobayashi, T., and Murakami, S. (1993). Displacement of lidocaine from the lung after bolus injection of bupivacaine. *Can. J. Anesth.* **40,** 676–680.
27. Roerig, D. L., Kotrly, K. J., Lennertz, R., Ahlf, S. B., Dawson, C. A., and Kampine, J. P. (1988). Effect of dose on the first pass uptake of lidocaine in the human lung. *FASEB J.* **2,** A951.
28. Meuldermans, W. E. G., Hurkmans, R. M. A., and Heykants, J. J. P. (1982). Plasma protein binding and distribution of fentanyl, sufentanil, alfentanil and lofentanil in blood. *Arch. Int. Pharmacodyn. Ther.* **257,** 4–19.
29. Dawson, C. A., Linehan, J. H., Rickaby, D. A., and Roerig, D. L. (1984). Influence of plasma protein on the inhibitory effects of indocyanine green and bromcresol green on pulmonary prostaglandin extraction. *Br. J. Pharmacol.* **81,** 449–455.
30. Dawson, C. A., Roerig, D. L., Rickaby, D. A., Nelin, L. D., Linehan, J. H., and Krenz, C. S. (1992). The use of diazepam for interpreting changes in extravascular lung water. *J. Appl. Physiol.* **72,** 686–693.
31. Linehan, J. H., Dawson, C. A., Bongard, R. D., Bronikowski, T. A., and Roerig, D. L. (1989). Plasma protein binding in endothelial enzyme interactions in the lung. *J. Appl. Physiol.* **66,** 2617–2628.
32. Bongard, R. D., Roerig, D. L., Johnson, M. R., Linehan, J. H., and Dawson, C. A. (1993). Influence of temperature and plasma protein in doxorubicin uptake by isolated lungs. *Drug Metab. Dispos.* **21,** 428–434.
33. Goresky, C. A. (1963). A linear method for determining liver sinusoidal and extravascular volumes. *Am. J. Physiol.* **204,** 626–640.

Lactic Acidosis and pH on the Cardiovascular System

Yuguang Huang*, James B. Yee†, Wang-Hin Yip‡, and K. C. Wong†

**Peking Union Medical College*
Peking, China

†Department of Anesthesiology
University of Utah School of Medicine
Salt Lake City, Utah 84132

‡Kaohsiung Medical College
Kaohsiung, Taiwan, Republic of China

I. Introduction

It is generally believed that metabolic acidosis is harmful for the cardiovascular system because acidosis may make cardiac defibrillation more difficult, may prevent the catecholamines from working effectively, and may worsen cardiac contractility and conduction, thus making overall resuscitation of the patient under shock less effective (1–3). Unfortunately, because of differences in study designs, the experimental results are conflicting. The results show that metabolic acidosis may decrease the pressure response of epinephrine (EPI) by 30–50%. However, one human study of cardiac arrest victims showed no relationship between the venous or arterial pH and the response of the coronary perfusion pressure to large doses of EPI (4). Other data also suggest that although metabolic acidosis does lower the threshold at which the heart fibrillates, it does not affect the amount of energy required to defibrillate the heart (5). Thus, controversy still exists with regard to whether metabolic acidosis in shock should be treated.

Advances in Pharmacology, Volume 31

While trying to develop a usable animal acidotic model, we observed that dogs succumb to the rapid infusion (over 1 hr) of lactic acid when we attempted to produce a pH of 7.1. On the other hand, dogs were more tolerant toward the infusion of hydrochloric acid with a target of producing a pH of 7.1. Therefore, we postulated that there are physiological differences of systemic acidosis produced by lactic acid and hydrochloric acid. Epinephrine is the time-honored endogenous catecholamine with potent α and β cardiovascular stimulatory effects, whereas dobutamine (DOB) is a synthetic catecholamine with predominant β_1 stimulatory actions on the heart. In the present study we evaluated the cardiovascular effects of acute infusion of lactic acid or hydrochloric acid and compared the cardiovascular effects of DOB and EPI as well as norepinephrine (NE) on the cardiovascular system.

II. Induction of Lactic Acidosis

Fifty-four dogs of either sex (22.6 ± 3.2 kg, mean ± standard deviation) were used for the study. Telazol at 0.5 mg/kg (a combination of 50 mg zolazipam and 50 mg tellatamine) and pancuronium bromide at 0.1 mg/kg were given intravenously to facilitate endotracheal intubation, and the lungs were ventilated with nitrous oxide (70% in oxygen) while maintaining the p_{CO_2} at around 35 mmHg. Anesthesia was maintained with isoflurane (1–2.5%, v/v), sufficient to keep a mean arterial blood pressure at approximately 100 mmHg. Sodium chloride (0.9%) was infused at 10 ml/kg/hr to all subjects during the experiment. Electrocardiographic recordings were taken from standard lead II, and the femoral artery was cannulated for arterial pressure monitoring and blood sampling. A 7.5-French thermodilution Swan-Ganz catheter was inserted through the right external jugular vein and positioned in the pulmonary artery for mixed venous blood sampling and measurement. A 4.5-French Royal Flush II Pigtail catheter was inserted via the carotid artery and positioned in the left ventricle for measurement of left ventricular end diastolic pressure. The temperature of the animal was maintained at approximately 35°C with an electric heating pad and lamp.

Electrocardiogram and heart rate (HR) were recorded continuously, as well as the following blood pressures: systolic blood pressure, diastolic blood pressure, mean arterial pressure (MAP), mean pulmonary arterial pressure (MPAP), pulmonary capillary wedge pressure, central venous pressure, and left ventricular end diastolic pressure. The cardiac output was measured by thermodilution using 5 ml of 5% dextrose at 0°C with an Edwards 9520 cardiac output computer. The cardiac index (CI), stroke volume index (SVI), left ventricular stroke work index (LVSWI) (coronary perfusion pressure), and total systemic vascular resistances (SVR) were calculated from the measured variables by standard hemodynamic formulas (6). The p_{O_2}, p_{CO_2}, pH, and percent oxygen saturation were measured, and base excess, hemoglobin, and bicarbonate were calculated (Radiometer ABL-2, Copenhagen, Denmark).

Fifty-four dogs were randomly allocated to one of three different experimental groups: control (n = 18), lactic acidosis (n = 18), and hydrochloric acidosis (n = 18). Within each group, six dogs received one of three different catecholamines: EPI, NE, and DOB. The instrumentation and procedure were identical for all animals.

Following instrumentation, the animal was stabilized for 30 min. The experimental models were set as follows. In the lactic acidosis (LAC) group a 2 *M* lactic acid solution was infused intravenously at a rate of approximately 4 mg/kg/hr until the arterial pH was reduced to about 7.0. In the hydrochloric acid (HCl) group a 2 *M* HCl solution was infused intravenously at a rate of approximately 1 ml/kg/hr until the arterial pH was reduced to about 7.0. Infusion rates were titrated individually to avoid cardiovascular collapse during the administration of acid, and the arterial pH of 7.0 was then maintained by adjusting the lactic or hydrochloric infusion rate. The arterial blood gases were measured every 15–20 min to determine the degree of acidosis. In the control group normal saline was infused for 3 hr in lieu of the acid solution to simulate the effects of increased volume and the duration of anesthesia.

Each animal was then given one of three catecholamines at six different infusion rates. Epinephrine and NE infusion rates were 0.1, 0.2, 0.4, 0.8, 1.6, and 3.2 μg/kg/min. Dobutamine infusion rates were 5, 10, 20, 40, 80, and 160 μg/kg/min.

Eight sets of measurements were collected during each experiment. The hemodynamic and metabolic variables were measured at the following occasions: (1) 30 min after instrumentation and before the induction of the acidotic models; (2) when the target arterial pH values were reached, or after 3 hr in the control group; (3) 10 min after initial infusion of the catecholamine; and (4) 10 min after each incremental infusion rate.

The data were expressed as means ± standard deviation. Analysis of variance was used to distinguish between the group means for animals receiving each of the three catecholamines. The Boneferroni test was used for individual differences within groups. The differences were considered statistically significant for p values of 0.05 or less. The results were analyzed using the 1990 BMDP Statistical Software for repeated measurements.

III. Cardiovascular Effects of Acute Acidosis

There were no significant differences in animal weight (22.6 ± 3.2 kg) or premodel pH (7.36 ± 0.05) between any of the groups ($p > 0.05$). The acidosis model was created by the infusion of either LAC or HCl intravenously over a 3-hr period. The arterial pH at the end of the infusion was between pH 7.01 and 7.05 for all study groups. It was observed in preliminary studies that rapid infusion (over 1 hr) of LAC produced cardiovascular collapse in some dogs. The animals were more tolerant toward the infusion of HCl.

The production of acidosis by either LAC or HCl resulted in significant increases in MPAP (Fig. 1). Infusion of EPI, NE, or DOB led to further elevation of MPAP as compared to the control group. There was also a significant reduction in the arterial oxygen saturation with infusion of all the drugs in the dogs who were acidotic (Table I). The oxygen saturation did not change in the animals with normal pH.

The production of acidosis by LAC or HCl did not significantly change MAP, SVR, CI, LVSWI, or HR compared to control animals with normal pH ($p > 0.05$). There in general appeared to be a trend toward stimulation of the monitored cardiovasular parameters with lower doses of all the study drugs (Figs. 2–4). However, as the infused concentrations of the three drugs were increased, differences appeared between the effects of the natural catecholamines and DOB. In the presence of severe lactic acidosis infusion of EPI resulted in a dose-dependent decrease in CI and LVSWI (Figs. 3 and 4). This depressant effect of EPI or CI and LVSWI was less if the acidosis was the result of HCl infusion. Epinephrine treatment resulted in a dose-dependent increase in MAP (Fig. 2) and a variable effect on HR (Fig. 5).

The effects of NE on CI and LVSWI are quite different from those of EPI. Norepinephrine had no effect on these parameters in the animals treated with HCl (Figs. 3 and 4). However, in animals treated with LAC, in high doses of NE depressed CI and LVSWI (Figs. 3 and 4). Both EPI and NE produced dose-related increases in MAP, but variable effects on HR. Systemic vascular resistance increased significantly following infusions of EPI and NE in both the control and acidotic groups (Table II).

Interestingly, DOB produced a dose related increase in CI and HR in both the control and acidotic groups (Figs. 3 and 5). Unlike the case in EPI- and NE-treated dogs, SVR tended to decrease in the animals treated with DOB (Table II). Mean arterial pressure was increased at lower doses, whereas at higher DOB doses MAP returned toward baseline pressure in all groups (Fig. 2).

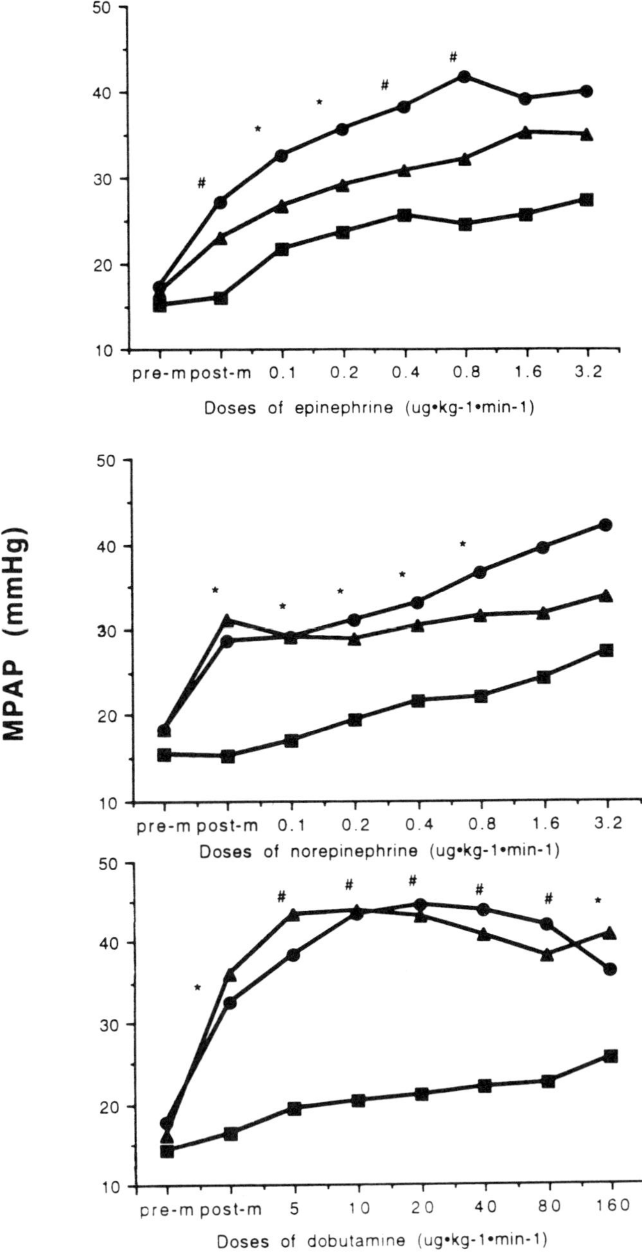

Fig. 1 Mean pulmonary artery pressures during the infusion of increasing doses of EPI, NE, or DOB under conditions of normal pH (■) and LAC (●) or HCl (▲) acidosis. *Significant difference ($p < 0.05$) compared to control. #Significant difference ($p < 0.01$) compared to control.

Table I

Changes in Arterial Oxygenation during Normal Acid–Base and Acute Lactic or Hydrochloric Acidotic Conditions[a]

Drug	Control			Lactic acidosis			HCl acidosis		
	Pre-m	Post-m	Post-d	Pre-m	Post-m	Post-d	Pre-m	Post-m	Post-d
Epinephrine	98.2 ± 0.4	98.1 ± 0.4	98.1 ± 0.3	98.3 ± 0.3	96.8 ± 1.2	92.9 ± 4.2[b]	98.0 ± 1.2	97.6 ± 1.2	94.3 ± 1.9[b]
Norepinephrine	97.1 ± 2.9	97.3 ± 3.1	97.4 ± 1.0	98.2 ± 1.1	96.4 ± 1.3	94.5 ± 1.7	97.6 ± 1.1	90.5 ± 8.5	84.7 ± 8.5[b]
Dobutamine	99.1 ± 0.4	98.6 ± 0.4	97.6 ± 0.5	98.3 ± 0.7	96.4 ± 1.3	91.3 ± 4.0[b]	98.5 ± 0.7	95.3 ± 3.2[b]	86.3 ± 16.6[b]

[a] Values are means ± SD. Pre-m, Premodel; Post-m, postmodel; Post-d, postdrug.
[b] Significant difference compared to premodel ($p < 0.05$).

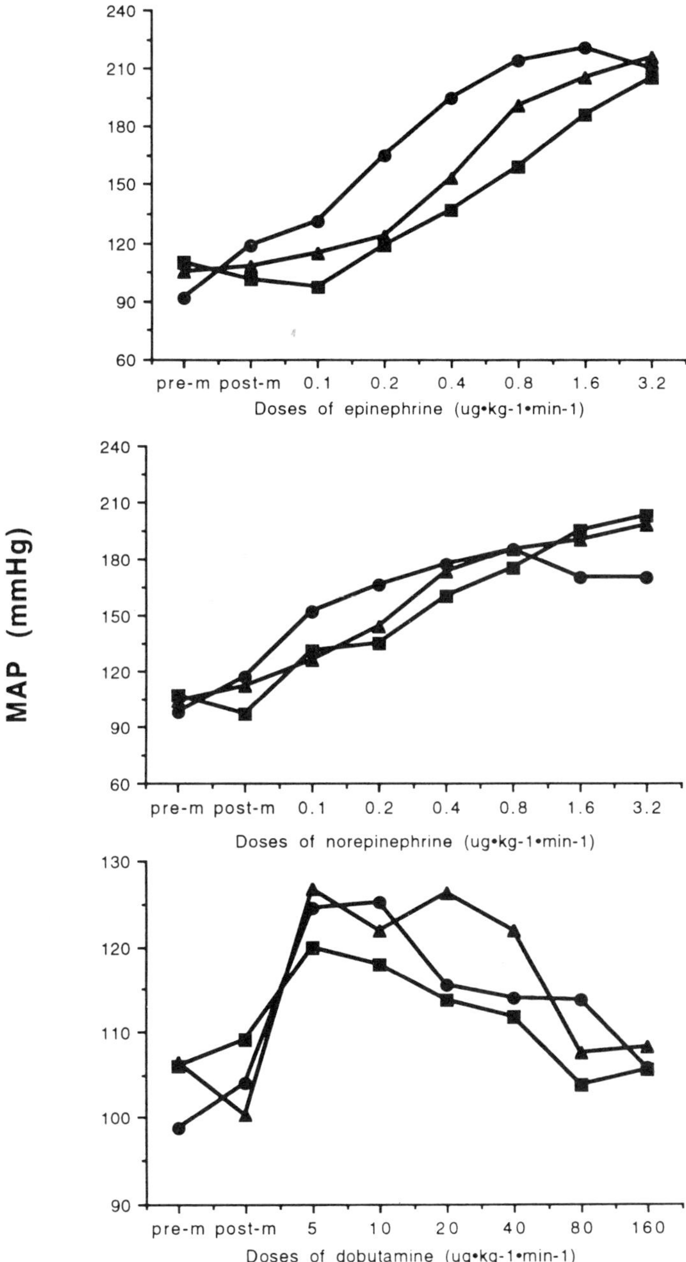

Fig. 2 Mean arterial pressures during the infusion of increasing doses of EPI, NE, or DOB under conditions of normal pH (■) and LAC (●) or HCl (▲) acidosis.

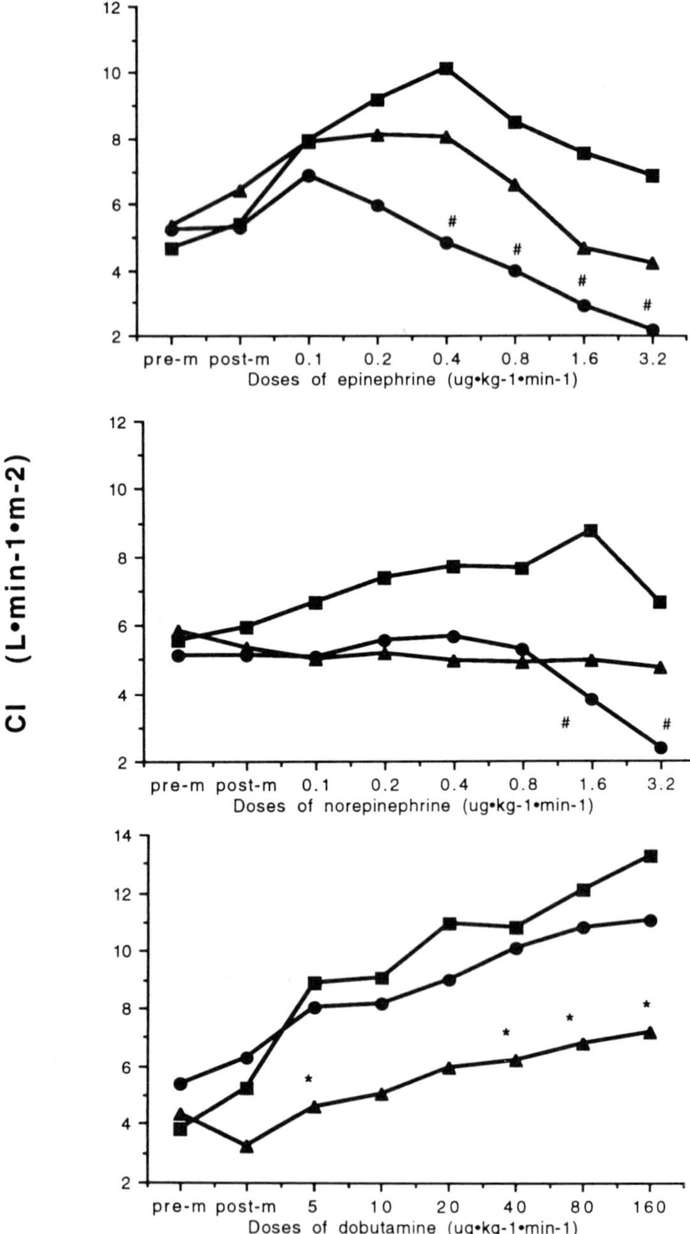

Fig. 3 Cardiac indexes during the infusion of increasing doses of EPI, NE, or DOB under conditions of normal pH (■) and LAC (●) or HCl (▲) acidosis. *Significant difference ($p < 0.05$) versus control. #Significant difference ($p < 0.01$) versus control.

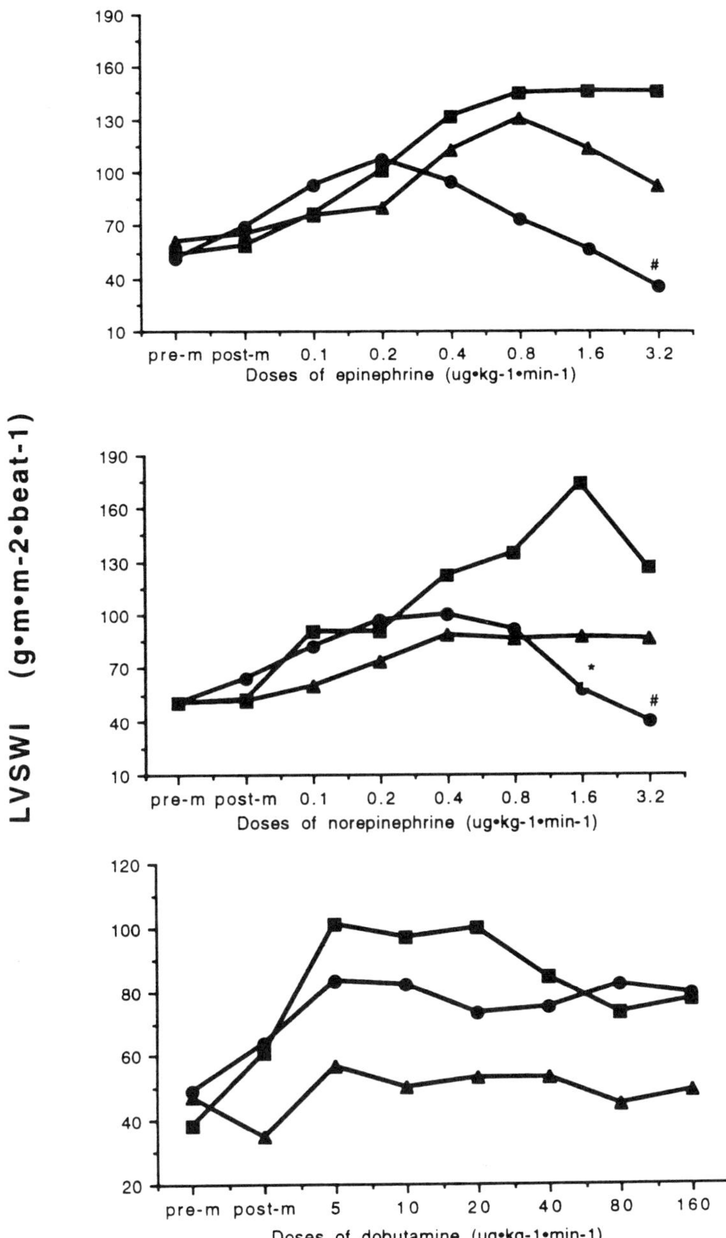

Fig. 4 Left ventricular stroke work indexes during the infusion of increasing doses of EPI, NE, or DOB, under conditions of normal pH (■) and LAC (●) or HCl (▲) acidosis. *Significant difference ($p < 0.05$) versus control. #Significant difference ($p < 0.01$) versus control.

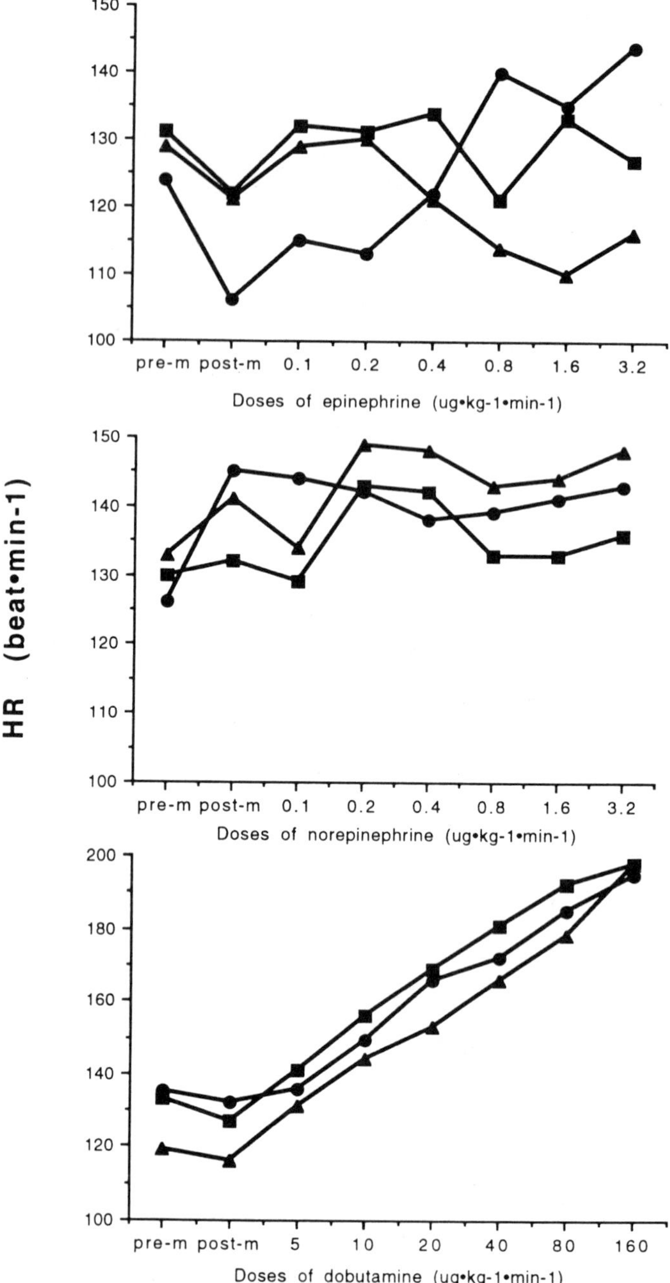

Fig. 5 Heart rates during the infusion of increasing doses of EPI, NE, or DOB under conditions of normal pH (■) and LAC (●) or HCl (▲) acidosis.

Table II

Systemic Vascular Resistance Changes with Epinephrine, Norepinephrine, and Dubutamine during Normal Acid–Base Status and during Lactic or Hydochloric Acidosis[a]

Experimental group	Pre-m	Post-m	Epinephrine (μg/kg/min)					
			0.1	0.2	0.4	0.8	1.6	3.2
Control	2254 ± 129.6	1775 ± 98.4	1150 ± 58.6	1256 ± 89.8	1252 ± 64.8	1889 ± 166.8	2399 ± 129.4	2768 ± 82.6
LAC	1683 ± 108.8	2044 ± 112.2	1702 ± 90.2	2456 ± 101	3547 ± 100.4[b]	4860 ± 199.6[b]	7253 ± 412.8[b]	8941 ± 474[b]
HCl	1847 ± 168.8	1730 ± 201.1	1481 ± 186.2	1628 ± 236	1942 ± 229.8[c]	3100 ± 326	4315 ± 309.6[d]	5329 ± 504.6[d]
			Norepinephrine (μg/kg/min)					
			0.1	0.2	0.4	0.8	1.6	3.2
Control	1826 ± 122.2	1591 ± 116.6	1853 ± 54.6	1695 ± 68.2	1921 ± 68.8	2373 ± 318.4	2137 ± 146.6	2955 ± 244.4
LAC	1748 ± 81	2574 ± 358.2	3039 ± 359.2	2782 ± 161	3088 ± 201.6	3695 ± 344.4	4363 ± 226	6603 ± 399.6[e]
HCl	1828 ± 290.8	2815 ± 378	3589 ± 529	3240 ± 302.4	4854 ± 69.6	4573 ± 518	4472 ± 512	4487 ± 363.6
			Dobutamine (μg/kg/min)					
			5	10	20	40	80	160
Control	2564 ± 176.8	1948 ± 94.2	1330 ± 110.8	1235 ± 77.6	1079 ± 92.2	1035 ± 97.8	923 ± 153	1065 ± 282
LAC	1612 ± 104.2	1674 ± 147.6	1794 ± 229	1713 ± 198.8	1619 ± 247.2	1576 ± 300.4	1408 ± 266.8	1305 ± 271
HCl	2685 ± 250.6	3595 ± 426.8	3621 ± 595.4	3052 ± 412.2	2788 ± 402.2	2481 ± 335.8	1937 ± 237.2	1677 ± 193.6

[a] Values are means ± SEM and represnt SVR changes (dyne-sec/cm^5). Pre-m, Before acid infusion; Post-m, after acid infusion.
[b] Significant difference compared to control group ($p < 0.01$).
[c] Significant difference compared to LAC group ($p < 0.01$).
[d] Significant difference compared to LAC group ($p < 0.05$).
[e] Significant difference compared to control group ($p < 0.05$).

IV. Discussion

Metabolic acidosis is common during cardiovascular dysfunction. The effect of acidosis on cardiovascular function has been extensively studied, and the results show that the hemodynamic effects of some drugs during normal acid–base and severe acidotic conditions are different (2,3,7). The results of our study support these published findings. In the present study there was, in general, a decrease in the responsiveness to the infusion of endogenous catecholamines in acidotic animals compared to subjects with normal pH.

Depressed response to intravenous catecholamines in humans during acidosis was reported by Campbell *et al.* as early as 1958 (8). More recent studies showed that the correction of acidemia using $NaHCO_3$ did not improve the hemodynamics in critically ill patients who had metabolic acidosis associated with hypoxia (9,10). Furthermore, clinical and experimental studies suggest that high doses of $NaHCO_2$ may be ineffective or even detrimental to the brain and cardiovascular system (9). The trend has been shifting from the use of alkalinizers to treat metabolic acidosis to the application of high doses of catecholamines (9,11). Epinephrine remains the drug of choice for cardiac arrest and severe, refractory hypotension, and it has been suggested that the most effective dose may be higher than those currently recommended during cardiopulmonary resuscitation (CPR) (12). However, the optimal doses of catecholamine for circulatory support in metabolic acidosis are still controversial. Currently recommended doses of the drugs used in this study for CPR by the American Heart Association are as follows: EPI, 1–4 μg/min; NE, 2–12 μg/min; and DOB, 2.5–20 μg/kg/min. Interestingly, the large logarithmically incremental doses of EPI and NE used in this study were unable to overcome the depressant effects of acidosis created in the dogs. However, there have been reports of successful use of high-dose EPI (2,13).

There are several possible mechanisms for the deterioring cardiovascular responses to catecholamine stimulation. First, myocardial intracellular acidosis from accumulation of LAC may contribute directly to the impairment of myocardial function. The low pH and/or the high concentration of LAC may cause a decrease in responsiveness or affinity of adrenergic receptors to agonist. Some data suggest that elevated LAC by itself may have negative effects on myocardial and brain cells separate from low pH (7). Second, the prolonged sympatho-adrenal stimulation that results from acidosis, may result in depletion of endogenous catecholamines and energy substances and progressive deterioration of myocardial contractility. Third, the pulmonary and systemic vasoconstriction induced by acidosis and the infusions of EPI or NE will increase afterload to the myocardium. This increase in afterload can further compound the problem of an already depressed heart. Finally, there is evidence that a high partial pressure of CO_2 in the myocardium may adversely affect cardiac contractility and may correlate well with further decreases in myocardial pH, intramural ST segment changes, anaerobic generation of adenosine triphosphate (ATP), and histological injury (9,11). It is likely that the decreased response to catecholamines in the presence of severe acidosis represents some combination of the above factors.

Although the response to endogenous catecholamines was depressed in the acidotic animal, we observed a more normal dose–response relationship with DOB on the cardiovascular parameters measured. Kosugi and Tajimi also found that DOB produced increases in CI and SVI during severe lactic acidosis as well as during normal pH (14).

The increase in CI by the infusion of DOB was likely the result of increased HR and decreased SVR, reducing cardiac afterload during normal acid–base status and severe acidosis. Kosugi and Tajimi (14) and our results suggest that DOB might be more useful than EPI and NE in improving cardiac output and therefore oxygen delivery during severe lactic acidosis. However, we recognize that our normovolemic acidotic model is not the same as hypovolemic or hypoxic acidosis and may lead to pharmacological and clinical differences.

Marked increases in pulmonary artery pressure resulted from infusion of LAC or HCl. Additional increase in MPAP occurred with infusion of each drug in the presence of acidosis. This was associated with a decrease in arterial oxygen saturation. There are reports that acidosis created by LAC infusion produced hypoxic-induced increases in pulmonary arterial pressure or pulmonary vascular resistance (15,16). The increase in pulmonary vascular resistance caused by the infusion of HCl may be mediated by thromboxane A_2 or prostacyclin synthesis (15).

We postulated that there is a hemodynamic difference between acidosis produced by LAC or HCl, and some of our data tend to support this hypothesis. Pulmonary vasoconstriction, as reflected by the mean pulmonary artery pressure, was consistently higher following LAC infusion than HCl infusion, especially during the infusion of EPI and NE ($p < 0.06$). Likewise, cardiac index was consistently lower following LAC infusion in comparison with HCl infusion during EPI infusion ($p < 0.06$). However, cardiac index following HCl acidosis and dobutamine infusion was consistently lower than control and lactic acid infusion ($p < 0.05$). We do not have a good explanation for this interaction between dobutamine and lactic acid, versus that between dobutamine and HCl.

In summary, our data suggest that, in intact anesthetized dogs, acute lactic or HCl acidosis exerts an increase in MPAP, which is further increased following the infusion of EPI, NE, and DOB. In contrast, cardiac index is decreased following lactic acidosis and the infusion of EPI and NE, but increased following DOB infusion.

References

1. Oliva, P. B. (1970). Lactic acidosis. *Am. J. Med.* **48,** 209–225.
2. Jaffe, A. S. (1989). New and paradoxes, acidosis and cardiopulmonary resuscitation. *Circulation* **80,** 1079–1083.
3. Zhou, H. Z., Malhotra, D., and Shapiro, J. I. (1991). Contractile dysfunction during metabolic acidosis: Role of impaired energy metabolism. *Am. J. Physiol.* **261,** H1481–H1486.
4. Paradis, N. A., Goetting, M. G., and Rivers, E. P. (1990). High-dose epinephrine therapy and return of spontaneous circulation during human pseudo-electromechanical dissociation. *Ann. Emerg. Med.* **19,** 491–492.
5. Yakaitis, R. W., Thomas, J. D., and Mahaffey, J. E. (1975). Influence of pH and hypoxia on the success of defibrillation. *Crit. Care Med.* **13,** 139–142.
6. Miller, R. D. (1986). "Anesthesia," pp. 1165–1197. Churchill Livingstone, London.
7. Wildenthal, K., Mierzwiak, D. S., Myers, R. W., and Mitchell, J. H. (1968). Effects of acute lactic acidosis on left ventricular performance. *Am. J. Physiol.* **214,** 1352–1359.
8. Campbell, G.S., Hale, D. B., Crisp, N. W., Weil, M. H., and Brown, E. B. (1958). Depressed response to intravenous sympathomimetic agents in humans during acidosis. *Dis. Chest* **33,** 18–22.
9. Graf, H., Leach, W., and Arieff, A. I. (1985). Evidence for a detrimental effect of sodium bicarbonate therapy in hypoxic lactic acidosis. *Science* **227,** 754–756.
10. Kette, F., Weil, M. H., and Gazmuri, R. J. (1991). Buffer solutions may compromise cardiac resuscitation by reducing coronary perfusion pressure. *J. Am. Med. Assoc.* **266,** 2121–2126.
11. Arieff, A. J. (1990). Indications for use of bicarbonate in patients with metabolic acidosis. *Br. J. Anaesth.* **67,** 165–177.
12. "Textbook of Advanced Cardiac Life Support," 2nd ed., pp. 98, 116 and 118. American Heart Association.
13. Callaham, M. L. (1990). High-dose epinephrine therapy and other advances in treating cardiac arrest. *West. J. Med.* **152,** 697–703.
14. Kosugi, I., and Tajimi, K. (1985). Effects of dopamine and dobutamine on hemodynamics and plasma catecholamine levels during severe lactic acid acidosis. *Circ. Shock* **17,** 95–102.
15. Lejeune, P., Brimioulle, S., Leeman, M., Hallemans, R., Melot, C., and Naeije, R. (1990). Enhancement of hypoxic pulmonary vasoconstriction by metabolic acidosis in dogs. *Anesthesiology* **73,** 256–264.
16. Leoppky, J. A., Svotto, P., Riedel, C. E., Roach, R. C., and Chick, T. W. (1992). Effects of acid–base status on acute hypoxic pulmonary vasoconstriction and gas exchange. *J. Appl. Physiol.* **72,** 1787–1797.

Role of Oxygen Free Radicals and Lipid Peroxidation in Cerebral Reperfusion Injury

Laurel E. Moore and Richard J. Traystman
Department of Anesthesiology and Critical Care Medicine
Johns Hopkins Medical Institutions
Baltimore, Maryland 21287

I. Introduction

Despite a national emphasis on the reduction of cardiovascular risk factors in the last decade, stroke remains a major cause of morbidity and mortality in the United States. Although prevention remains the key to reducing these figures, measures aimed at reducing infarct size and resulting morbidity remain significant areas of research. At least three problems hinder our attempts to reduce the mortality and morbidity associated with stroke. First, although the public has been widely informed as to the early symptoms of myocardial infarction resulting in early aggressive therapy, cerebral ischemia is less well understood by the public. This may be due, in part, to the limited clinical techniques currently available to minimize ischemic injury to the brain. Second, whereas return of blood flow is necessary for the salvage of ischemic neurons, this reperfusion in itself is associated with direct and indirect injury to tissue, and the brain may be particularly susceptible to reperfusion injury. Finally, reperfusion injury is multifactorial in its genesis, and protection against injury is unfortunately not a matter of blocking parallel mechanisms. This article reviews the known pathophysiology of reperfusion injury and outlines some current areas of research which attempt to therapeutically minimize cerebral ischemia and reperfusion injury.

Advances in Pharmacology, Volume 31

II. Free Radicals

A free radical is simply a molecule with an unpaired electron. The unpaired electron makes the molecule a reactive species as it attempts to achieve electrical neutrality. Although free radicals have long been felt to be involved in cerebral ischemic injury (1), this has proved difficult to support because of problems in documenting the presence of free radicals *in vivo*. The decrease in endogenous free radical scavengers associated with ischemia (1) supports the concept of free radical production during ischemia but does not prove their existence. Several techniques have been developed, however, which do support their presence during ischemia and reperfusion (2–7). Although each technique is associated with problems *in vivo*, through these techniques free radical formation during reperfusion has been demonstrated.

There are several possible mechanisms for free radical formation during ischemia and reperfusion. In the absence of oxygen to serve as a terminal electron acceptor, the electron transport chain within mitochondria becomes highly reduced (8). In this reduced state oxygen radical formation may result, particularly when oxygen is resupplied (9). Furthermore, during ischemia the release of excitatory amino acids can stimulate *N*-methyl-D-aspartate (NMDA) receptors within brain to produce nitric oxide (NO) (10). There is evidence to support increased NO production during focal ischemia (11). NO with its unpaired electron is able to initiate free radical chain reactions in the presence of superoxide anion (12). The potential roles of NMDA receptor blockade and NO synthase inhibition in reducing ischemic injury are discussed below.

Cerebral ischemia is associated with the rapid failure of adenosine triphosphate (ATP)-dependent ionic pumps. Whereas under normal conditions these pumps maintain normal intracellular and extracellular ionic gradients and therefore membrane potentials, during ischemia their failure causes the rapid efflux of K^+ and influx of Na^+, Cl^-, and Ca^{2+} (13). Ca^{2+} influx may be particularly damaging during ischemia because it appears to be involved with lipid membrane breakdown. Ca^{2+} influx may activate phospholipase C with resultant breakdown of cell membrane lipids and release of free fatty acids (14). In an ATP-deficient environment, phospholipase A, an adenosine 3′,5′-cyclic monophosphate (cAMP)-dependent enzyme, may also cause the breakdown of phospholipid in cell membranes (15).

The significance of these events is supported by the fact that ischemia produces a rapid increase in free fatty acids. The concentrations of fatty acids and arachidonic acid correlate with the duration of ischemia (16). Furthermore, accumulation of arachidonic acid is greatest in regions most sensitive to the effects of ischemia and reperfusion (17). Arachidonic acid directly intercalates into cell membranes and alters the packing of lipid molecules within these lipid membranes (18). Arachidonic acid also stimulates the cyclooxygenase and lipoxygenase pathways with resultant increases in thromboxane, prostaglandins (19), and superoxide anion (3). Prostaglandins may be responsible for decreased vascular reactivity postischemia (20),whereas oxygen radicals can produce brain injury directly by lipid peroxidation (21) or indirectly by altering vascular reactivity (22). Of potential clinical relevance, the postischemic accumulation of arachidonic acid (19) and superoxide anion (3) as well as postischemic neurological injury (23) can be reduced by pretreatment with cyclooxygenase inhibitors.

During ischemia, in addition to the rapid increase in arachidonic acid, there is also an increase in interstitial adenosine (24) and hypoxanthine (25). These result from the breakdown of brain nucleotides (e.g., ATP) in the setting of oxygen deprivation. This adenosine can be further degraded to inosine and hypoxanthine which then become substrates for the xanthine oxidase pathway. Whereas under normal conditions brain xanthine oxidase activity is low, Ca^{2+} influx during ischemia activates proteases which convert xanthine dehydrogenase to xanthine oxidase (26). In the presence of oxygen (e.g., with reperfusion), metabolism via the xanthine oxidase pathway produces oxygen radicals which are then available to begin chain reactions. Supporting the importance of purine metabolism as a source for radical production, the administration of allopurinol, a xanthine oxidase inhibitor, improves the survival of gerbils exposed to focal ischemia (27). Allopurinol, however, has hydroxyl radical scavenging effects (28) so the precise mechanism of this improved outcome following focal ischemia remains uncertain.

III. Mechanisms of Brain Injury

Lipid peroxidation appears to be a major mechanism by which radicals produce brain injury. Because of the high concentration of polyunsaturated fats in brain, this organ may be particularly susceptible to insult. Lipid peroxidation may be worsened in the presence of a high inspired oxygen concentration (29) or lactic acidosis (30) as occurs with hyperglycemia.

In the presence of oxygen (e.g., reperfusion), these peroxidation reactions continue to propagate within brain once initiated. For example, superoxide anion production may in turn cause the production of hydroxyl radicals, a much more unstable species, by Fenton chemistry. Beckman *et al.* have suggested that a likely mechanism for injury to endothelium involves the production of peroxynitrite by the reaction of NO and superoxide (12). Peroxynitrite anion may be as reactive as hydroxyl radical but is not dependent on the presence of iron as with Fenton chemistry (12). Within brain radicals appear to impair endothelial cell functions which maintain homeostasis of water and electrolytes (31). Furthermore, oxidative mechanisms appear to be involved in synaptic damage within the brain (32).

IV. Potential Therapeutic Agents

To be useful, potential therapeutic agents need to have access to the site at which reperfusion injury is occurring. Agents which are able to cross the blood–brain barrier may have the greatest effectiveness in cerebral protection (33). Interestingly, however, large molecules such as polyethylene glycol-conjugated superoxide dismutase which have poor access to the brain (34) also reduce infarct volume during focal ischemia (35), suggesting an important role of endothelium in reperfusion injury. This discussion of potential therapies aimed at reducing neural injury from ischemia focuses primarily on free radical scavengers, agents which reduce lipid peroxidation, and inhibitors of NO synthesis.

Several endogenous free radical scavengers or antioxidants have been tested for their efficacy in reducing neurological, biochemical, metabolic, or histological injury from cerebral ischemia. α-Tocopherol, an endogenous antioxidant, reduces lipid peroxidation when administered prior to ischemia (36). Similarly, rats fed a diet with excess α-tocopherol have reduced lipid peroxidation in response to ischemia compared to rats fed a diet deficient in α-tocopherol (37).

Deferoxamine, an iron-chelating agent, presumably acts to decrease the amount of free iron available during reperfusion for hydroxyl radical formation. We have demonstrated that a reduced intracellular pH, but not bicarbonate concentration, impairs somatosensory evoked potential recovery following 12 min of complete ischemia (38). One effect of acidosis may be increased disassociation of free iron from proteins such as ferritin or transferrin (30). Hurn *et al.* have demonstrated that treatment with deferoxamine prior to incomplete global ischemia improves metabolic recovery and postischemic hypoperfusion in the setting of hyperglycemic acidosis (39). Futhermore, when deferoxamine is conjugated to a high molecular weight starch, preventing significant entry to brain parenchyma, the beneficial effect of deferoxamine was eliminated supporting the idea that the protective action of deferoxamine occurs within brain parenchyma and not the microvasculature (39). LY178002, a potent inhibitor of iron-dependent lipid peroxidation, improves electrophysiological recovery following global ischemia in dogs. (40). The precise mechanism by which the drug provides neuroprotection is uncertain. These findings support the hypothesis that iron-dependent free radical formation is a significant mechanism of brain injury postischemia, particularly in the setting of acidosis.

Superoxide dismutase (SOD) and catalase are endogenous free radical scavengers which may act to scavenge tonically produced free radicals under normal conditions. Under ischemic conditions, however, endogenous concentrations may be insufficient. SOD catalyzes the conversion of superoxide anion to hydrogen peroxide. Catalase coverts hydrogen peroxide to water. Deferoxamine inhibits the conversion of hydrogen peroxide to hydroxyl radical by chelating free iron. In large doses SOD has been demonstrated to improve neurological recovery from transient spinal cord ischemia (41). Unfortunately SOD has a serum half-life of 8 min owing to rapid renal clearance (42), and its large molecular weight (32,000) prevents significant access to brain (43). These factors may explain the limited effectiveness of intravenous SOD in improving cerebral recovery postischemia (44).

Superoxide dismutase can be conjugated with polyethylene glycol (PEG–SOD), which increases its circulatory half-life to 40 hr in rats (45). Although administration of PEG–SOD does not increase brain SOD activity *in vitro* (34), *in vitro* PEG–SOD significantly increases the access of SOD to endothelial cells in culture (46). PEG–SOD alone (35) or in combination with PEG–catalase (47) reduces infarct size from focal ischemia.When administered prior to ischemia, PEG–SOD increases blood flow to ischemic regions during focal ischemia produced by unilateral middle cerebral artery and bilateral carotid artery occlusion (35). Interestingly, PEG–SOD does not improve postischemic hyperemia (48) or delayed hypoperfusion (49) following global ischemia. These findings may support a vascular effect of PEG–SOD which permits improved collateral flow during focal ischemia but may not improve cerbral function in the setting of global ischemia. We have demonstrated, however, improvement in postischemic hypercapnic reactivity following PEG–SOD before or after global ischemia (50). This finding also supports a vascular effect of PEG–SOD such as decreased radical production with resultant improved prostaglandin synthesis. In humans, a recently completed Phase II trial investigating PEG–SOD in severely head injured patients (51) demonstrated that 5000 or 10,000 U/kg i.v. bolus reduced the mean percentage of time intracranial pressure was greater than 20 mmHg within the first 5 days postinjury, and that 10,000 U/kg reduced the number of patients remaining in a persistent vegetative state at 3 and 6 months postinjury. Unlike PEG–SOD, liposomal entrapment of SOD has been successful at increasing brain SOD levels within 1 hr of administration in rats (52).

The 21-aminosteroids are a group of agents which lack classic steroidal effects (and side effects) but are active *in vitro* in preventing lipid peroxidation. The most fully studied among the 21-aminosteroids is tirilizad mesylate (U74006F). Tirilizad has such diverse effects as lipid peroxyl radical scavenging (53) and α-tocopherol preservation during ischemia (54). Tirilizad may be neuroprotective in a variety of experimental paradigms. For example, the drug is associated with improved survival and decreased histological injury after transient focal ischemia in gerbils (55). We have found tirilizad to improve mean recovery time of intracellular pH, inorganic phosphate, and somatosensory evoked potentials when administered before (56) or after (57) incomplete global ischemia. With complete global ischemia (cerebrospinal fluid infusion), however, tirilizad had no effect on these same metabolic and neurophysiological measures (58). The reason for this significant difference between incomplete and complete global ischemia may represent structural injuries such as white matter shearing from compression, but it raises the issue that mechanisms of neural injury other than lipid peroxidation may be more important in the setting of complete global ischemia than incomplete ischemia.

Although the results of tirilizad in global ischemia have been mixed, the drug may have other beneficial effects which may be of clinical relevance when administered for the purpose of cerebral protection. Sterz *et al.* reported that whereas tirilizad had no effect on delayed hypoperfusion or cerebral oxygen consumption after 12.5 min of cardiac arrest, those dogs receiving tirilizad were easier to resuscitate (59). Improved survival in dogs treated with tirilizad 24 hr after 10 min of normothermic ventricular fibrillation may be the result of an improved neurological condition (60). However, the improved survival could also result from protective effects of the drug on the cardiovascular or renal systems (61). No human studies evaluating tirilizad have been completed, but tirilizad may be effective in such clinical settings as acute cord injury where high-dose methylprednisolone has been shown to be effective in reducing injury (62).

Ischemia causes an increase in extracellular excitatory amino acids (63) which in turn can stimulate NMDA receptors to produce NO (10). NO may then interact with superoxide anion to produce hydroxyl radicals (12). Therefore, potential means to prevent NO production and subsequent radical proliferation include decreasing the release of excitatory amino acids, blockade of NMDA receptors, and inhibition of NO synthesis. Even mild hypothermia has been demonstrated to decrease the release of excitatory amino acids during ischemia (64). Baldwin *et al.* demonstrated that mild hypothermia during reperfusion reduces intracranial pressure and facilitates recovery of somatosensory evoked potentials following 20 min of complete global ischemia (65). A decreased accumulation of excitatory amino acids is probably only one of several mechanisms by which hypothermia is protective in this setting.

More specific to the role of excitatory amino acids and NMDA receptors, preliminary data from our laboratory demonstrates that NPC17742, a competitive NMDA receptor antagonist, reduces both caudate and hemispheric infarct volume when administered just prior to reperfusion following 1 hr of left middle cerebral artery occlusion. This is presumably due to a decrease in NO production. NMDA receptor stimulation, however, may cause several effects during ischemia including receptor-mediated opening of Ca^{2+} channels permitting Ca^{2+} influx. MK-801, a known NMDA receptor antagonist, ameliorates brain injury after prolonged focal ischemia (66). This neuroprotection is enhanced when MK-801 is administered in combination with nimodipine, a Ca^{2+} channel blocker. (67). Therefore NMDA receptor blockade may be neuroprotective by more than one mechanism.

The hypothesis that neuroprotection by NMDA receptor blockade may in part occur via decreased NO production is supported by evidence that NO synthase inhibition is also protective in transient (68) and permanent (69) focal ischemia. We have shown that administration of N^G-nitro-L-arginine methyl ester (L-NAME), a NO synthase inhibitor, reduces caudate infarct volume following 1 hr of left middle cerebral artery occlusion (68). This protective effect was reversed by administration of L-arginine, supporting NO synthesis as the mechanism of protection rather than possible flow redistribution or muscarinic effects known to be associated with L-NAME (70). Futhermore, the protective effect of L-NAME was equivalent whether the drug was adminstered before ischemia or just prior to reperfusion, which suggests that NO production during reperfusion is important in brain injury.

V. Conclusion

In conclusion, as our understanding of the mechanisms of cerebral reperfusion injury increases, our ability to selectively affect neurological outcome and survival improves. Unfortunately, these mechanisms are highly complex and closely interrelated so actual clinical application remains distant. Over the past years innumerable agents have been demonstrated to be neuroprotective in one animal model or another, or in cell culture. Most of these agents do not afford complete protection, and it seems clear that no single neuroprotective agent is going to be found which will ameliorate the effects of ischemia-reperfusion injury. More likely, since there are multiple mechanisms of injury (i.e., calcium channels, excitatory amino acids, free radicals, lipid peroxidation) a neuroprotective "cocktail" which utilizes a combination of drugs, or sequential treatment using several different pharmacological agents, may in the end be appropriate therapy. In anesthesiology we have the rare opportunity to predict and monitor cerebral ischemia (e.g., cerebral aneurysm or aortic arch surgery) and therefore may be among the first clinicians to affect neurological outcome pharmacologically once these agents become clinically applicable.

References

1. Flamm, E. S., Demopoulos, H. B., Seligman, M. L., Poser, R. G., and Ransohoff, J. (1978). Free radicals in cerebral ischemia. *Stroke* **9,** 445–447.
2. Kontos, H. A., Wei, E. P., Ellis, E. F., Jenkins, L. W., Povlishock, J. T., Rowe, G. T., and Hess, M. L. (1985). Appearance of superoxide anion radical in cerebral extracellular space during increased prostaglandin synthesis in cats. *Circ. Res.* **57,** 142–151.
3. Armstead, W. M., Mirro, R., Busija, D. W., and Leffler, C. W. (1988). Postischemic generation of superoxide anion by newborn pig brain. *Am. J. Physiol.* **255,** H401–H403.
4. Imaizumi, S., Kayama, T., and Suzuki, J. (1984). Chemiluminescence in hypoxic brain—the first report. Correlation between energy metabolism and free radical reaction. *Stroke* **15,** 1061–1065.
5. Lai, E. K., Crossley, C., Sridhar, R., Misra, H. P., Janzen, E. G., and McCay, P. B. (1986). *In vivo* spin trapping of free radicals generated in brain, spleen, and liver during gamma radiation of mice. *Arch. Biochem. Biophys.* **244,** 156–160.
6. Tomonaga, T., Imaizumi, S., Yoshimoto, T., Suzuki, J., and Fujita, Y. (1985).(Protective effect of radical scavengers on cerebral infarction: Experimental study utilizing the spin trapping method of ESR). *No to Shinkei* **37,** 555–560.
7. Cao, W., Carney, J. M., Duchon, A., Floyd, R. A., and Chevion, M. (1988). Oxygen free radical involvement in ischemia and reperfusion injury to brain. *Neurosci. Lett.* **88,** 233–238.
8. Demopoulos, H. B., Flamm, E. S., Pietronigro, D. D., and Seligman, M. L. (1980). The free radical pathology and the microcirculation in the major central nervous system disorders. *Acta Physiol. Scand. Suppl.* **492,** 1–119.
9. Cino, M., and Del Maestro, R. F. (1989). Generation of hydrogen peroxide by brain mitochondria: The effect of reoxygenation following postdecapitative ischemia. *Arch. Biochem. Biophys.* **269,** 623–638.
10. Bredt, D. S., and Snyder, S. H. (1989). Nitric oxide mediates glutamate-linked enhancement of cGMP levels in the cerebellum. *Proc. Natl. Acad. Sci. U.S.A.* **86,** 9030–9033.
11. Malinski, T., Bailey, F., Zhang, Z. G., and Chopp, M. (1993). Nitric oxide measured by a porphyrinic microsensor in rat brain after transient middle cerbral artery occlusion. *J. Cereb. Blood Flow Metab.* **13,** 355–358.
12. Beckman, J. S., Beckman, T. W., Chen, J., Marshall, P. A., and Freeman, B. A. (1990). Apparent hydroxyl radical production by peroxynitrite: Implications for endothelial injury from nitric oxide and superoxide. *Proc. Natl. Acad. Sci. U.S.A.* **87,** 1620–1624.
13. Siesjo, B. K. (1984). Cerebral circulation and metabolism. *J. Neurosurg.* **60,** 883–908.

14. Wieloch, T., and Siesjo, B. K. (1982). Ischemic brain injury: The importance of calcium, lipolytic activities, and free fatty acids. *Pathol. Biol.* **30,** 269–277.
15. Edgar, A. D., Strosznajder, J., and Horrocks, L. A. (1982). Activation of ethanolamine phospholipase A2 in brain during ischemia. *J. Neurochem.* **39,** 1111–1116.
16. Shiu, G. K., and Nemoto, E. M. (1981). Barbiturate attenuation of brain free fatty acid liberation during global ischemia. *J. Neurochem.* **37,** 1448–1456.
17. Westerberg, E., Deshpande, J. K., and Wieloch, T. (1987). Regional differences in arachidonic acid release in rat hippocampal CA1 and CA3 regions during cerebral ischemia. *J. Cereb. Blood Flow Metab.* **7,** 189–192.
18. Klausner, R. D., Kleinfeld, A. M., Hoover, R. L., and Karnovsky, M. J. (1980). Lipid domains in membranes. Evidence derived from structural perturbations induced by free fatty acids and lifetime heterogeneity analysis. *J. Biol. Chem.* **255,** 1286–1295.
19. Gaudet, R. J., Alam, L., and Levine, L. (1980). Accumulation of cyclooxygenase products of arachidonic acid metabolism in gerbil brain during reperfusion after bilateral common carotid artery occlusion. *J. Neurochem.* **35,** 653–658.
20. Kontos, H. A., Wei, E. P., Povlishock, J. T., and Christman, C. W. (1984). Oxygen radicals mediate the cerebral arteriolar dilation from arachidonate and bradykinin in cats. *Circ. Res.* **55,** 295–303.
21. Yoshida, S., Inoh, S., Asano, T., Sano, K., Kubota, M., Shimazaki, H., and Ueta, N. (1980). Effect of transient ischemia on free fatty acids and phospholipids in the gerbil brain. Lipid peroxidation as a possible cause of postischemic injury. *J. Neurosurg.* **53,** 323–331.
22. Wei, E. P., Christman, C. W., Kontos, H. A., and Povlishock, J. T. (1985). Effects of oxygen radicals on cerebral arterioles. *Am. J. Physiol.* **248,** H157–H162.
23. Chan, P. H., and Fishman, R. A. (1980). Transient formation of superoxide radicals in polyunsaturated fatty acid-induced brain swelling. *J. Neurochem.* **35,** 1004–1007.
24. Van Wylen, D. G., Park, T. S., Rubio, R., and Berne, R. M. (1986). Increases in cerebral interstitial fluid adenosine concentration during hypoxia, local potassium infusion, and ischemia. *J. Cereb. Blood Flow Metab.* **6,** 522–528.
25. Nihei, H., Kanemitsu, H., Tamura, A., Oka, H., and Sano, K. (1989). Cerebral uric acid, xanthine, and hypoxanthine after ischemia: The effect of allopurinol. *Neurosurgery* **25,** 613–617.
26. Kinuta, Y., Kimura, M., Ishikawa, Y., Itokawa, M., and Kikuchi, H. (1989). Changes in xanthine oxidase in ischemic rat brain. *J. Neurosurg.* **71,** 417–420.
27. Beckman, J. S., Campbell, G. A., Hannan, C. J., Jr., Karfias, C. S., and Freeman, B. A. (1986). Involvement of superoxide and xanthine oxidase with death due to cerebral ischemia-induced seizures in gerbils. *In* "Superoxide and Superoxide Dismutase in Chemistry, Biology, and Medicine" (G. Rotilio, ed.), pp. 602–607. Elsevier, New York.
28. Moorhouse, P. C., Grootveld, M., Halliwell, B., Quinlan, J. G., and Gutteridge, J. M. (1987). Allopurinol and oxypurinol are hydroxyl radical scavengers. *FEBS Lett.* **213,** 23–28.
29. Mickel, H. S., Vaishnav, Y. N., Kempski, O., von Lubitz, D., Weiss, J. F., and Feuerstein, G. (1987). Breathing 100% oxygen after global brain ischemia in Mongolian gerbils results in increased lipid peroxidation and increased mortality. *Stroke* **18,** 426–430.
30. Rehncrona, S., Hauge, H. N., and Siesjo, B. K. (1989). Enhancement of iron-catalyzed free radical formation by acidosis in brain homogenates: Differences in effect by lactic acid and CO_2. *J. Cereb. Blood Flow Metab.* **9,** 65–70.
31. Lo, W. D., and Betz, A. L. (1986). Oxygen free-radical reduction of brain capillary rubidium uptake. *J. Neurochem.* **46,** 394–398.

32. Pellmar, T. C., and Neel, K. L. (1989). Oxidative damage in the guinea pig hippocampal slice. *Free Radical Biol. Med.* **6,** 467–472.
33. Traystman, R. J., Kirsch, J. R., and Koehler, R. C. (1991). Oxygen radical mechanisms of brain injury following ischemia and reperfusion. *J. Appl. Physiol.* **71,** 1185–1195.
34. Haun, S. E., Kirsch, J. R., Helfaer, M. A., Kubos, K. L., and Traystman, R. J. (1991). Polyethylene glycol-conjugated superoxide dismutase fails to augment brain superoxide dismutase activity in piglets. *Stroke* **22,** 655–659.
35. Matsumiya, N., Koehler, R. C., Kirsch, J. R., and Traystman, R. J. (1991). Conjugated superoxide dismutase reduces extent of caudate injury after transient focal ischemia in cats. *Stroke* **22,** 1193–1200.
36. Yamamoto, M., Shima, T., Uozumi, T., Sogabe, T., Yamada, K., and Kawasaki, T. (1983). A possible role of lipid peroxidation in cellular damages caused by cerebral ischemia and the protective effect of α-tocopherol administration. *Stroke* **14,** 977–982.
37. Yoshida, S., Busto, R., Santiso, M., and Ginsberg, M. D. (1984) Brain lipid peroxidation induced by postischemic reoxygenation *in vitro:* Effect of vitamin E. *J. Cereb. Blood Flow Metab.* **4,** 466–469.
38. Maruki, Y., Koehler, R. C., Eleff, S. M., and Traystman, R. J. (1993). Intracellular pH during reperfusion influences evoked potential recovery after complete cerebral. *Stroke* **24,** 697–703.
39. Hurn, P. D., Koehler, R. C., Blizzard, K. K., and Traystman, R. J. (1993). Metabolic injury associated with low end-ischemic brain bicarbonate is reduced by desferoxamine. *J. Cereb. Blood Flow Metab.* **13,** S32 (abstract).
40. Toung, T., Kirsch, J., Koehler, R., and Traystman, R. (1993). LY178002 improves brain electrophysiological activity after transient global ischemia in dogs. *Anesthesiology* **79,** A754 (abstract).
41. Lim, K. H., Connolly, M., Rose, D., Siegman, E., Jacobowitz, I., Acinapura, A., and Cunningham, J. N., Jr. (1986). Prevention of reperfusion injury of the ischemic spinal cord: Use of recombinant superoxide dismutase. *Ann. Thorac. Surg.* **42,** 282–286.
42. Turrens, J. F., Crapo, J. D., and Freeman, B. A. (1984). Protection against oxygen toxicity by intravenous injection of liposome-entrapped catalase and superoxide dismutase. *J. Clin. Invest.* **73,** 87–95.
43. Petkau, A., Chelack, W. S., Kelly, K., Barefoot, C., and Monasterski, L. (1976). Tissue distribution of bovine 125-I-superoxide dismutase in mice. *Res. Commun. Chem. Pathol. Pharmacol.* **15,** 641–654.
44. Forsman, M., Fleischer, J. E., Milde, J. H., Steen, P. A., and Michenfelder, J. D. (1988). Superoxide dismutase and catalase failed to improve neurologic outcome after complete cerebral ischemia in the dog. *Acta Anaesthesiol. Scand.* **32,** 152–155.
45. White, C. W., Jackson, J. H., Abuchowski, A., Kazo, G. M., Mimmack, R. F, Berger, E. M., Freeman, B. A., McCord, J. M., and Repine, J. E. (1989). Polyethylene glycol-attached antioxidant enzymes decrease pulmonary oxygen toxicity in rats. *J. Appl. Physiol.* **66,** 584–590.
46. Beckman, J. S., Minor, R. L., Jr., White, C. W., Repine, J. E., Rosen, G. M., and Freeman, B. A. (1988). Superoxide dismutase and catalase conjugated to polyethylene glycol increases endothelial enzyme activity and oxidant resistance. *J. Biol. Chem.* **263,** 6884–6892.
47. Liu, T. H., Beckman, J. S., Freeman, B. A., Hogan, E. L., and Hsu, C. Y. (1989). Polyethylene glycol-conjugated superoxide dismutase and catalase reduce ischemic brain injury. *Am. J. Physiol.* **256,** H589–H593.
48. Helfaer, M. A., Kirsch, J. R., Haun, S. E., Moore, L. E., and Traystman, R. J. (1991). Polyethylene glycol conjugated superoxide dismutase fails to blunt post-ischemic reactive hyperemia. *Am. J. Physiol.* **261,** H548–H553.

49. Schurer, L., Grogaard, B., Gerdin, B., and Arfors, K. E. (1990). Superoxide dismutase does not prevent delayed hypoperfusion after incomplete cerebral ischemia in the rat. *Acta Neurochir.* **103,** 163–170.
50. Kirsch, J. R., Helfaer, M. A., Haun, S. E., Koehler, R. C., and Traystman, R. J. (1993). Polyethylene glycol conjugated superoxide dismutase improves recovery of post-ischemic hypercapnic cerebral blood flow in piglets. *Pediatr. Res.* In press.
51. Muizelaar, J. P., Marmarou, A., Young, H. F., Choi, S. C., Wolf, A., Schneider, R. L., and Kontos, H. A. (1993). Improving the outcome of severe head injury with the oxygen radical scavenger polyethylene glycol-conjugated superoxide dismutase: A phase II trial. *J. Neurosurg.* **78,** 375–382.
52. Imaizumi, S., Woolworth, V., Fishman, R. A., and Chan, P. H. (1990). Liposome-entrapped superoxide dismutase reduces cerebral infarction in cerebral ischemia in rats. *Stroke* **21,** 1312–1317.
53. Braughler, J. M., and Pregenzer, J. F. (1989). The 21-aminosteroid inhibitors of lipid peroxidation: Reactions with lipid peroxyl and phenoxy radicals. *Free Radical Biol. Med.* **7,** 125–130.
54. Hall, E. D., Andrus, P. K., and Yonkers, P. A. (1993). Brain hydroxyl radical generation in acute experimental head injury. *J. Neurochem.* **60,** 588–594.
55. Hall, E. D., Pazara, K. E., and Braughler, J. M. (1988). 21-Aminosteroid lipid peroxidation inhibitor U74006F protects against cerebral ischemia in gerbils. *Stroke* **19,** 997–1002.
56. Maruki, Y., Koehler, R. C., Kirsch, J. R., Blizzard, K. K., and Traystman, R.J. (1993). Effect of the 21-aminosteroid tirilazad on cerebral pH and somatosensory evoked potential after incomplete ischemia. *Stroke* In press.
57. Kim, H., Koehler, R. C., Kirsch, J. R., Blizzard, K. K., and Traystman, R. J.(1993). Dose-dependent improvement in metabolic recovery from hyperglycemic ischemia with tirilazad post-treatment. *J. Cereb. Blood Flow Metab.* **13** (Suppl. 1), S682 (Abstract).
58. Helfaer, M. A., Kirsch, J. R., Hurn, P. D., Blizzard, K. K., Koehler, R. C., and Traystman, R. J. (1992). Tirilazad mesylate does not improve early cerebral metabolic recovery following compression ischemia in dogs. *Stroke* **23,** 1479–1486.
59. Sterz, F., Safar, P., Johnson, D. W., Oku, K.-I., and Tisherman, S. A. (1991). Effects of U74006F on multifocal cerebral blood flow and metabolism after cardiac arrest in dogs. *Stroke* **22,** 889–895.
60. Natale, J. E., Schott, R. J., Hall, E. D., Braughler, J. M., and D'Alecy, L.G. (1988). Effect of the aminosteroid U74006F after cardiopulmonary arrest in dogs. *Stroke* **19,** 1371–1378.
61. Podrazik, R. M., Obedian, R. S., Remick, D. G., Zelenock, G. B., and D'Alecy, L. G. (1989). Attenuation of structural and functional damage from acute renal ischemia by the 21-aminosteroid U74006F in rats. *Curr. Surg.* **46,** 287–292.
62. Bracken, M. B., Shepard, M. J., Collins, W. F., Holford, T. R., Young, W., Baskin, D. S., Eisenberg, H. M., Flamm, E., Leo-Summers, L., Maroon, J., Marshall, L. F., Perot, P. L., Piepmeier, J., Sonntag, V. K. H., Wagner, F. C., Wilberger, J. E., and Winn, H. R. (1990). A randomized, controlled trial of methylprednisolone or naloxone in the treatment of acute spinal-cord injury. Results of the second national acute spinal cord injury study. *N. Engl. J. Med.* **322,** 1405–1411.
63. Benveniste, H., Drejer, J., Schousboe, A., and Diemer, N. H. (1984). Elevation of the extracellular concentrations of glutamate and aspartate in rat hippocampus during transient cerebral ischemia monitored by intracerebral microdialysis. *J. Neurochem.* **43,** 1369–1374.
64. Busto, R., Globus, M. Y., Dietrich, W. D., Martinez, E., Valdes, I., and Ginsberg, M. D. (1989). Effect of mild hypothermia on ischemia-induced release of neurotransmitters and free fatty acids in rat brain. *Stroke* **20,** 904–910.

65. Baldwin, W. A., Kirsch, J. R., Hurn, P. D., Toung, W. S. P., and Traystman, R. J. (1991). Hypothermic cerebral reperfusion and recovery from ischemia. *Am. J. Physiol.* **261,** H774–H781.
66. Park, C. K., Nehls, D. G., Graham, D. I., Teasdale, G. M., and McCulloch, J.(1988). Focal cerebral ischaemia in the cat: Treatment with glutamate antagonist MK-801 after induction of ischaemia. *J. Cereb. Blood Flow Metab.* **8,** 757–762.
67. Uematsu, D., Araki, N., Greenberg, J. H., Sladky, J., and Reivich, M. (1991). Combined therapy with MK-801 and nimodipine for protection of ischemic brain damage. *Neurology* **41,** 88–94.
68. Nishikawa, T., Kirsch, J. R., Koehler, R. C., Miyabe, M., and Traystman, R. J. (1993). Nitric oxide synthase inhibition reduces caudate injury following transient focal ischemia in cats. *Stroke* submitted.
69. Nishikawa, T., Kirsch, J. R., Koehler, R. C., Bredt, D. S., Snyder, S. H., and Traystman, R. J. (1993). Effect of nitric oxide synthase inhibition on cerebral blood flow and injury volume during focal ischemia in cats. *Stroke* in press.
70. Buxton, I. L. O., Cheek, D. J., Eckman, D., Westfall, D. P., Sanders, K. M.,and Keef, K. D. (1993). N^{G}-Nitro-L-arginine methyl ester and other alkyl esters of arginine are muscarinic receptor antagonists. *Circ. Res* **72,** 387–395.

Effects of Volatile Anesthetics on Cerebrocortical Laser Doppler Flow: Hyperemia, Autoregulation, Carbon Dioxide Response, Flow Oscillations, and Role of Nitric Oxide

Antal G. Hudetz, Joseph G. Lee, Jeremy J. Smith, Zeljko J. Bosnjak, and John P. Kampine

Departments of Anesthesiology and Physiology
The Medical College of Wisconsin
Milwaukee, Wisconsin 53226

I. Introduction

Volatile anesthetics cause dose-dependent increases in cerebral blood flow (1–5). Halothane increases cerebral blood flow more than isoflurane owing to the greater depressant effect of isoflurane on cerebral metabolic requirement (2). In addition, halothane and isoflurane exert regionally selective effects on cerebral blood flow, with halothane causing a greater increase in blood flow in the neocortex and isoflurane having more pronounced effects on subcortical structures (6). Autoregulation of cerebral blood flow is believed to be maintained during anesthesia with isoflurane but not with halothane (4,5,37). However, this has not been established well in a regionally specific manner in the brain and, therefore, requires further study.

The mechanism underlying cerebral vasodilator effects of volatile anesthetics is not entirely clear. Nitric oxide (NO) or endothelium-derived

Advances in Pharmacology, Volume 31

relaxing factor (EDRF) has been suggested as a possible mediator for halothane-induced cerebral vasodilation (8). In the brain NO is produced by vascular endothelial cells, neurons (9), and astrocytes (10). NO from either source could stimulate guanylate cyclase in vascular smooth muscle cells resulting in an increase in cyclic GMP and consequent vasodilatation. This has been shown in isolated cerebral arteries (11) and *in vivo* in the mouse cerebral cortex (12). Vasodilator prostanoids (PGI_2, PGE_2, or PGD_2) may contribute in synergism with NO to the vasodilatation in the microcirculation (8,13,14). However, the involvement of these mediators in halothane-induced increases in cerebrocortical red cell flow has not been investigated.

Laser Doppler flowmetry is a novel and increasingly popular technique that can be used to monitor, in a highly localized manner, microvascular perfusion of various tissues, in particular, that of the brain (15–17). LDF has the important advantage over many other flow measurement techniques that it allows continuous monitoring of microvascular perfusion and can be used in a minimally invasive manner in the rat cerebral cortex (16). Laser Doppler flowmetry monitors nondirectional flow of red blood cells in the microcirculation. Because the cerebral capillary hematocrit is lower than the systemic arterial hematocrit (36), it cannot be inferred that the influences of anesthetics on red blood cell flow mirror those on total or regional cerebral blood flow. For example, a fraction of red blood cells may be shunted away from the exchange capillaries by arterio-venous anastomoses during vasodilatation (18). As the red blood cell is the major oxygen carrier in the blood, the distinction between red blood cell flow and bulk blood flow in the brain is an important one. No studies have to date compared the effects of halothane and isoflurane on red blood cell flow and flow autoregulation in the brain microcirculation.

Laser Doppler flowmetry reveals characteristic spontaneous oscillations in tissue perfusion which are assumed to be related to spontaneous arterial vasomotion. Oscillations in microvascular blood flow (flowmotion) are general characteristics of the microcirculation in several tissues. In the brain, they occur mainly in the frequency range of 4–12 cycles per minute (16). The physiological significance of flow oscillations is that they appear to be closely associated with autoregulation of cerebral blood flow. Both vasomotion and "flowmotion" depend strongly on arterial pressure and arteriolar tone such that they are enhanced during hypotension and during vasoconstriction. It is thought that flow oscillations are abolished by barbiturates and local anesthetics, although the confounding, blood pressure depressant effects of barbiturates have not been accounted for. There is no information on how volatile anesthetics influence spontaneous flow oscillations in the cerebral cortex and whether anesthetic influences on flow oscillations depend on the involvement of nitric oxide as a mediator.

In the present studies we compared the direct effects of halothane and isoflurane on laser Doppler flow (LDF), flow autoregulation, response to carbon dioxide, and spontaneous flow oscillations as measured by laser Doppler flowmetry in the parietal cerebral cortex of the rat. We also investigated whether NO is involved in the LDF response to halothane and in spontaneous oscillations of LDF.

II. Measurement of Laser Doppler Flow

Experimental procedures and protocols were approved by the Institutional Animal Care Committee. Adult male Sprague-Dawley rats (250–400 g body weight) were anesthetized with intraperitoneal pentobarbital (65 mg/kg body weight). One or both femoral arteries and veins were cannulated for the measurement of blood pressure, performance of hemorrhage, and drug administration. The rats were tracheotomized, intubated, and ventilated with 30% oxygen in nitrogen plus 0–5% CO_2 in hypercapnic experiments. The inspired gases were mixed using a three-tube flowmeter. Body temperature was maintained at 37 ± 0.2°C with a water-circulated heating pad. When ventilated, the rats were anesthetized with 0.5–1.0 minimum alveolar concentration (MAC) of either halothane or isoflurane or a pentobarbital infusion (0.1 mg/kg/hr). Inspired and expired oxygen, carbon dioxide, and volatile anesthetic concentrations were monitored continuously using a POET II gas analyzer (Criticare Systems, Inc., Milwaukee, WI).

The head of the rat was placed in a stereotaxic apparatus. The scalp and connective tissue were removed in a 1 cm^2 area. A burr hole of 1–2 mm in diameter was drilled in the cranium, over the right parietal cerebral cortex, utilizing a low-speed air drill with the aid of a stereomicroscope. The burr hole was drilled approximately 3 mm posterior to the bregma and 3 mm to the right of the midline. The burr hole was made as deep as possible without penetrating the skull, as previously described (16). Red blood cell flow was monitored utilizing a Perimed PF3 laser Doppler flowmeter and a small flow probe (PF316, Dental Probe) with a tip diameter of 1 mm. The probe was lowered into the bottom of the cranial burr hole using a micromanipulator. Care was taken to position the probe over a tissue area devoid of large blood vessels visible through the thinned bone. Once established, the probe position was not altered for the duration of the experimental protocol. The experimental setup is shown schematically in Fig. 1.

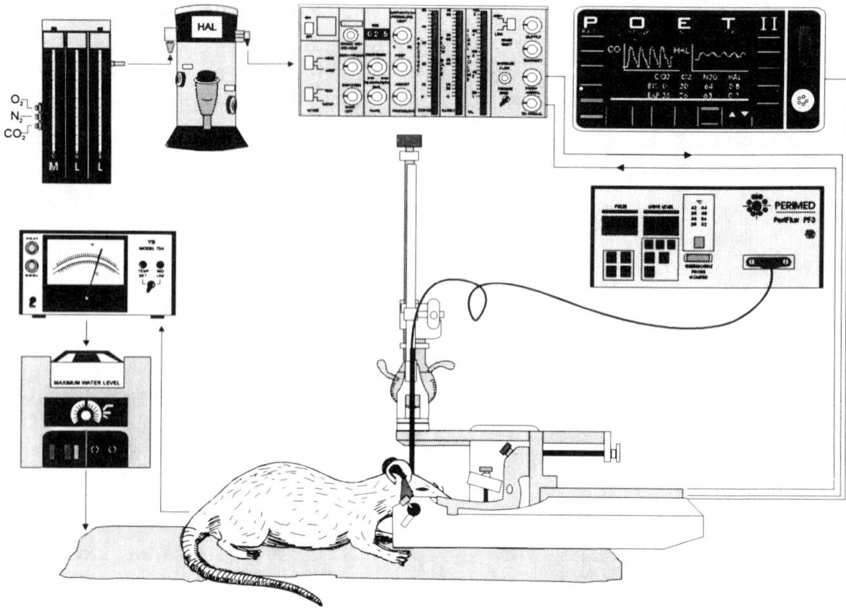

Fig. 1 Schematic illustration of the experimental setup used to study microcirculation of the rat cerebral cortex using laser Doppler flowmetry. The head of the anesthetized, artificially ventilated animal is secured in a stereotaxic frame, and cortical laser Doppler flow is monitored transcranially after the bone is thinned to a translucent plate.

Each LDF recording was averaged over a 5-min period. Laser Doppler flow response was, in all protocols, expressed as percent change from control. A control (baseline) LDF recording was taken at commencement of the experiment, at steady-state conditions U either pentobarbital infusion or at 0.5 or 1.0 MAC volatile anesthetic, depending on the experimental protocol, and at 1 hr following surgical procedures. Data were expressed as means ± SEM. Each mean represents data from 5 to 8 rats. When studying autoregulation, the slopes of the upper, nearly linear portion of the LDF versus mean arterial pressure (MAP) curves were determined by linear regression. The autoregulation coefficient was defined as the percent LDF change from baseline per millimeter Hg change in MAP between 140 and 60 mmHg. CO_2 reactivity was defined as the percentage LDF change per millimeter Hg change in end-tidal CO_2 ($etCO_2$) from the hypocapnic to the hypercapnic level. Cerebrocortical vascular resistance (CVR) was estimated as MAP/LDF where LDF was expressed as the percentage of control. Statistical comparisons were made using one-way analysis of variance (ANOVA) and Student Neuman–Keuls test for same anesthetic comparisons and two-way ANOVA and Duncan's test for comparison of different anesthetic effects. A p value below 0.05 was considered significant.

III. Resting Flow

To study the direct dose-dependent effects of volatile anesthetics on cerebrocortical red cell flow, the animals received 0.5, 1, 1.5, and 2 MAC of either halothane or isoflurane. Each animal received only one volatile anesthetic. After the stabilization hour, baseline LDF was recorded for 5 min and the anesthetic concentration was increased to 1 MAC. If, at 1 MAC, MAP decreased, the α_1-adrenoreceptor agonist phenylephrine (0.5–5 μg/kg/min) was infused to return arterial blood pressure to control levels. After stabilization at 1 MAC, LDF was again recorded, and the recordings were repeated at 1.5 and 2 MAC. End-tidal CO_2 was maintained at 35–37 mmHg in all experiments.

Control MAP in the halothane group was 115 $\pm$ 7 mmHg compared with 125 $\pm$ 5 mmHg in the isoflurane group (no difference). Control LDF in the halothane and isoflurane groups at 0.5 MAC volatile anesthetic were 159 $\pm$ 8 and 158 $\pm$ 12 perfusion units, respectively. In the halothane group LDF increased significantly at each increment of MAC ($p < 0.05$). Similarly the increase in LDF was significant ($p < 0.05$) at each 0.5 MAC increase in isoflurane concentration. When the effects of halothane and isoflurane were compared, isoflurane caused a greater LDF increase at 1.5 and 2 MAC than did halothane at the same anesthetic concentrations ($p < 0.01$) (Fig. 2).

The present results obtained with LDF appear to be at variance with earlier studies which suggested that halothane causes greater vasodilatation in cerebral cortex than does isoflurane. An explanation of this apparent contradiction may be that laser Doppler flowmetry measures red blood cell flow in a small area of cortical tissue. Maekawa *et al.* (19) demonstrated that isoflurane did not increase local cerebrocortical blood flow in highly discrete regions (200 μm in diameter) until isoflurane concentrations exceeded 1 MAC. Above 1 MAC, isoflurane caused heterogeneous changes in local cerebrocortical blood flow and metabolism. It is possible that isoflurane has a preferential dilating effect on smaller vessels (vessel diameter <10–15 μm) than does halothane in the cerebral cortex. Gelman *et al.* (20) demonstrated a tendency for a larger increase in cerebral blood flow, determined with 9-μm microspheres, during isoflurane than during halothane anesthesia. Entrapment of 15-μm but not 9-μm microspheres is complete in most organs (21); however, blood flow determined with

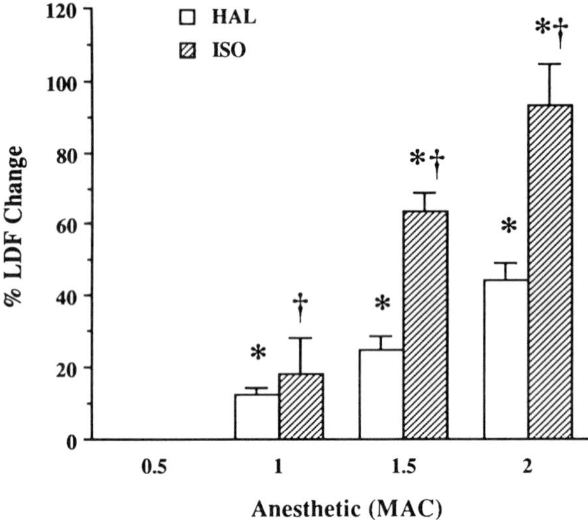

Fig. 2 Comparison of the direct effects of halothane (HAL) and isoflurane (ISO) on laser Doppler flow (LDF). During increases in anesthetic concentration, mean arterial blood pressure was maintained constant with intravenous phenylephrine infusion (0.5–5 μg/kg/min). LDF changes are expressed as the percent change from control LDF (0%) at 0.5 MAC volatile anesthetic. Data are expressed as means ± SEM. *Significant difference ($p < 0.05$) versus adjacent lower MAC. †Significant difference ($p < 0.05$) versus equi-MAC halothane.

9-μm spheres reflects nutritive blood flow (22). Microvessel dilatation to isoflurane may result in increased capillary hematocrit and, consequently, a larger increase in red blood cell flow as measured by LDF than whole blood flow. This possibility needs to be examined with independent measurements of small vessel hematocrit and/or capillary red cell flow (23).

IV. Autoregulation

In the autoregulation experiments, after a stabilization period of 1 hr at 0.5 MAC halothane or isoflurane or pentobarbital infusion (0.1 mg/kg/hr), MAP was increased, over 5 min, to 140 mmHg with intravenous phenylephrine (0.5–5 μg/kg/min). Because of the differences in initial MAP from experiment to experiment, all autoregulation protocols were commenced at this MAP level. Baseline LDF was recorded and anesthesia was maintained at either the original anesthetic concentration or at 1.5 MAC halothane or 1.5 MAC isoflurane. LDF was then recorded at MAP levels of 140, 120, 100, 80, 60, and 40 mmHg. Decreases in arterial blood pressure were achieved initially be reducing phenylephrine infusion rates to zero, and then by hemorrhage via the left femoral artery following systemic heparinization of the animals. The preparation was allowed to stabilize for 3–5 min at each blood pressure level prior to LDF recordings. End-tidal CO_2 was maintained at 35–37 mmHg in all experiments.

There were no statistically significant differences in baseline LDF during pentobarbital infusion, 0.5 MAC halothane, and 0.5 MAC isoflurane at 176 ± 28, 211 ± 15, and 190 ± 10 perfusion units, respectively. Figure 3 illustrates that between mean arterial blood pressures of 60 and 140 mmHg, autoregulation of LDF was present at all anesthetic concentrations used. However, LDF at 1.5 MAC was greater at all corresponding blood pressures than LDF at 0.5 MAC in the MAP range 60–140 mmHg ($p < 0.05$). Autoregulation coefficients (% LDF change/mmHg MAP change) were between 0.42 ± 0.07 at 0.5 MAC isoflurane and 0.20 ± 0.05 at 1.5 MAC

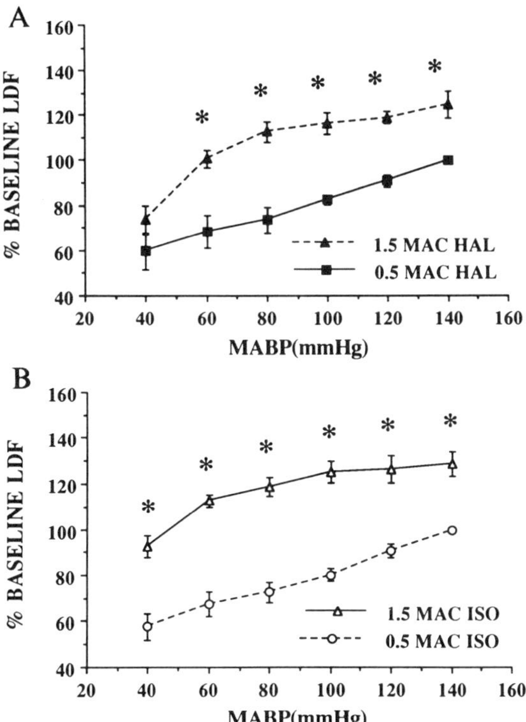

Fig. 3 Effects of 0.5 and 1.5 MAC halothane (HAL, A) and isoflurane (ISO, B) on autoregulation of laser Doppler flow (LDF) to changes in mean arterial blood pressure (MABP). LDF changes are expressed as the percent change from baseline control (100%) LDF at 0.5 MAC and a MABP of 140 mmHg. Data are expressed as means ± SEM. *Significant difference ($p < 0.05$) versus 0.5 MAC at the same MABP. From (17), with permission.

isoflurane. Autoregulation of LDF appeared to be less attenuated at higher MAC, reaching significance ($p < 0.05$) for 1.5 MAC isoflurane versus 0.5 MAC isoflurane.

The finding that the LDF–MAP curve tended to be flatter at 1.5 MAC for both halothane and isoflurane was unexpected. Previous studies have demonstrated that relative changes in LDF signal obtained from cerebral tissues correlate well with blood flow measured by radioactive microspheres (24) or the hydrogen clearance technique (15,25). It cannot be excluded, however, that the sensitivity of the laser Doppler flowmeter is somewhat decreased at high flow velocity or high tissue hematocrit as a result of increased probability of multiple scattering of the laser light. Alternatively, the relatively flat LDF–MAP relationship at high arterial pressures may reflect "false" autoregulation (26) caused by swelling of brain tissue and the compression of blood vessels against the dura and the skull as the anesthetic concentration is increased. It is also possible that the increased cerebral volume associated with higher concentrations of volatile anesthetics (7) resulted in stagnation of capillary and venular flow, compared to overall brain blood flow, thus minimizing LDF increases at 1.5 MAC halothane and isoflurane. Another possible explanation is that, at high arterial blood pressure and anesthetic concentration, red blood cell flow is shunted away from the microcirculation while plasma flow is not.

Phenylephrine infusion is unlikely to interfere with our results as Sokrab and Johansson (27) showed that of three adrenergic agonists, only epinephrine, but not phenylephrine, increases local cerebral blood flow. In the presence of volatile anesthetic, phenylephrine infused to increase mean arterial blood pressure does not result in cerebral vasoconstriction (28,29).

The reason for the apparent "better" autoregulation of laser Doppler flow in the cerebral cortex at higher (1.5 versus 0.5 MAC) volatile anesthetic concentrations must await further clarification.

V. Hypocapnia and Hypercapnia

To study the dose-dependent effects of volatile anesthetics on the cerebrocortical flow response to CO_2, the first group of animals were allowed to stabilize for 1 hr on pentobarbital infusion (0.1 mg/kg/hr) or at 0.5 MAC halothane or isoflurane and baseline LDF was measured at normocapnia ($etCO_2$ 35–37 mmHg). Hyperventilation to $etCO_2$ of 22 mmHg was then performed and another LDF measurement taken after 5 min. Hyperventilation was discontinued, normocapnia reestablished, and a control measurement performed 5 min later. Then 5% CO_2 was added to the inspired gas mixture to yield an $etCO_2$ of 65–67 mmHg and another measurement performed, again after 5 min. In the experiments performed at 1.5 MAC, the preparation was allowed to stabilize for 1 hr at 0.5 MAC halothane or isoflurane, and baseline LDF was measured at 0.5 MAC anesthetic at normocapnia ($etCO_2$ 35–37 mmHg). Anesthetic concentration was then increased to 1.5 MAC and CO_2 reactivity measured at hypo-, normo-, and hypercapnia as described above. The α_1-adrenoreceptor agonist phenylephrine (0.5–5 μg/kg/min) was administered intravenously, as required, to maintain MAP at control levels (i.e., MAP at 0.5 MAC at normocapnia).

Figure 4 shows that LDF was linearly related to $etCO_2$ at all anesthetic concentrations. The CO_2 reactivity (% change in LDF per mmHg change in $etCO_2$ from hypo- to hypercapnia) was greater ($p < 0.05$) at 1.5 MAC than at 0.5 MAC for halothane but not for isoflurane. The CO_2 reactivity during pentobarbital (1.78 ± 0.19%/mmHg) was similar to that at 0.5 MAC isoflurane (2.28 ± 0.22%/mmHg), and both were greater ($p < 0.05$) than that observed at 0.5 MAC halothane (1.19 ± 0.14%/mmHg). The CO_2

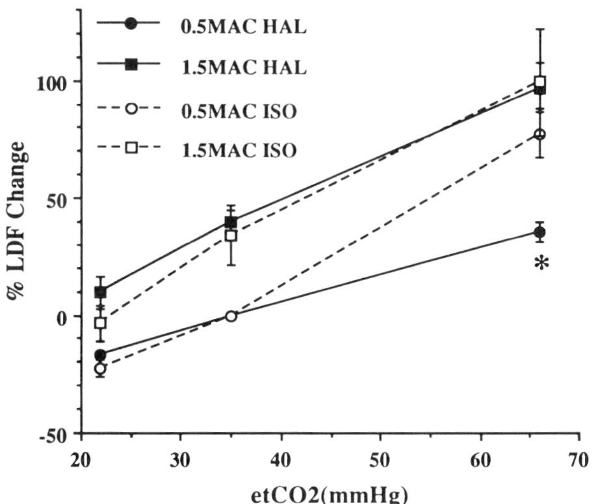

Fig. 4 Comparison of the effects of 0.5 and 1.5 MAC halothane (HAL) and isoflurane (ISO) on LDF at hypo-, normo-, and hypercapnia ($etCO_2$ of 22, 36, and 66 mmHg). LDF changes are expressed as the percent change from control LDF at 0.5 MAC, at normocapnia ($etCO_2$ 36 mmHg). Data are expressed as means ± SEM. The CO_2 curve is shifted upward at 1.5 MAC compared to 0.5 MAC for both halothane and isoflurane, that is, LDF is greater at 1.5 MAC than at 0.5 MAC at each $etCO_2$ ($p < 0.01$). CO_2 reactivity (% LDF change/mmHg $etCO_2$ change) is greater at 1.5 MAC halothane (1.99 ± 0.25) than at 0.5 MAC (1.19 ± 0.14). *Lower than all other groups at the same $etCO_2$ ($p < 0.01$).

reactivities at 1.5 MAC halothane and 1.5 MAC isoflurane were not significantly different (1.99 ± 0.25 and 2.67 ± 0.35%/mmHg, respectively).

These results suggest that changes in arterial p_{CO_2} cause qualitatively similar changes in red cell flow in the cerbrocortical microcirculation as they do on whole blood flow in the brain. The magnitude of the CO_2 response as observed in this study was somewhat smaller than those obtained with laser Doppler flowmetry by others (15,30), whereas Cucchiara *et al.* (3) demonstrated CO_2 reactivities of similar magnitude to ours with halothane and isoflurane. However, most previous studies which utilized LDF to assess the cerbrovascular CO_2 response used an open cranial window preparation in which the dura mater was opened and reflected. We found in preliminary experiments that full craniotomy, even without cutting and removing the dura, resulted in serious and lasting hyperemia of the cerebral cortex as indicated by baseline LDF perfusion units significantly higher than those obtained in the thinned skull preparation as used in the present study. We believe that tissue trauma associated with durectomy could have resulted in enhanced baseline flow and proportionally larger CO_2 reactivity. Transcranial monitoring of LDF as practiced in this and our previous studies (16,17) overcomes this problem.

Some difference between the responses in regional blood flow and laser Doppler flow is possible because the measurement of LDF is highly local. In our studies, we used a small flow probe in which the separation of light-emitting and light-receiving fibers was 0.5 mm. The small fiber separation results in a shallow depth of measurement, of the order of a few hundred micrometers. If the response of intracerebral blood vessels to CO_2 is not uniform across various cortical layers, then the surface measurements of red cell flow using LDF may not fully reflect blood flow measurements from deep cortex. Despite these potential differences, the present results demonstrate that the known CO_2 response of cerebral blood flow is closely reflected by the similar response of LDF.

VI. Role of Nitric Oxide

The experiments investigating the role of nitric oxide were carried out to determine if the halothane-induced increase in cerebrocortical red blood cell flow may be mediated by NO alone or in synergism with vasodilator cyclooxygenase products. After the surgical setup time of 1.5 hr postinduction, all rats achieved a steady-state baseline with 1 MAC halothane for 1 hr. Indomethacin (2 mg/kg) was infused intravenously over 5 min followed 15 min later by the infusion of the nitric oxide synthase (NOS) inhibitor N^{ω}-nitro-L-arginine methyl ester (L-NAME, 20 mg/kg i.v.) (31).

A separate group of animals received L-NAME only. Thirty minutes after L-NAME infusion, the inspired halothane concentration was increased to 1.7 MAC. During this period MAP was maintained at the pretreatment pressure using an infusion of phenylephrine (0.5–5 μg/kg/min). This pressure level was chosen because of the difficulties encountered in maintaining pressure at the increased level induced by L-NAME.

Infusion of L-NAME alone increased MAP by 32% and reduced baseline LDF by 28 ± 3% ($p < 0.001$) (Fig. 5). These effects were reversed by L-arginine (200 mg/kg i.v.). In the other experimental group, infusion of indomethacin decreased baseline LDF by 19 ± 2% ($p < 0.001$) and had no effect on MAP. Subsequent infusion of L-NAME produced an additional decrease in flow of 28 ± 3% ($p < 0.001$) while raising MAP by 16 ± 10 mmHg ($p < 0.05$) suggesting an additive effect of NO and vasodilator cyclooxygenase products (e.g., PGI_2) in maintaining resting cortical flow. On raising inspired halothane from 1.0 to 1.7 MAC, LDF increased by 17 ± 5% in the L-NAME-treated animals. Following combined treatment with indomethacin plus L-NAME, the LDF response to halothane was attenuated but still significant at 5 ± 3% ($p < 0.05$).

It should be noted that in these experiments MAP was allowed to fall, during increases in halothane, to the corresponding pretreatment level.

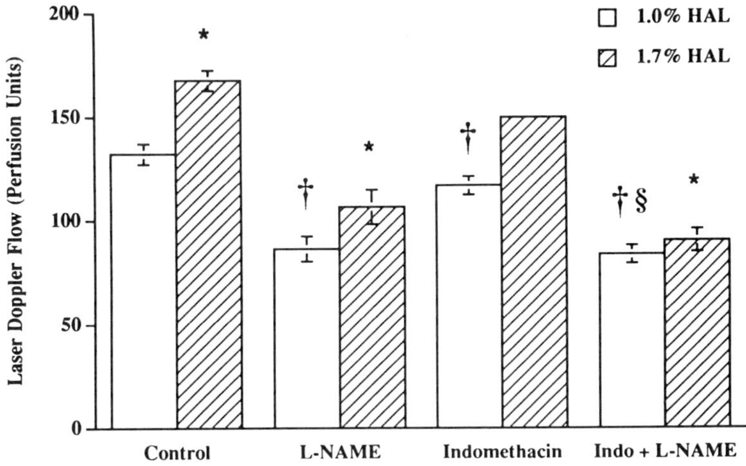

Fig. 5 Effect of NO synthase inhibition on halothane (HAL)-induced increases in cerebrocortical laser Doppler flow (LDF) with and without indomethacin pretreatment. Baseline LDF is reduced to a similar level by L-NAME and by combined tratment with indomethacin plus L-NAME. After the combined tratment, halothane-induced hyperemia is largely attenuated but still differs significantly from zero. *Significant difference ($p < 0.5$) compared to 1.0% halothane of the same treatment group. †Significant difference ($p < 0.05$) compared to control. §Significant difference ($p < 0.5$) compared to indomethacin only treated group.

To account for the possible confounding effect of changes in MAP concomitant with those in flow, the halothane-induced decrease in cerebrovascular resistance (CVR) was also evaluated in each experimental group. We found that on raising halothane from 1.0 to 1.7 MAC, CVR decreased by 30 ± 3% in the L-NAME-treated group and by 19 ± 3% in the group treated with indomethacin plus L-NAME. This attenuation of the CVR response to halothane after combined treatment with the enzyme inhibitors was significant but clearly smaller than that observed in LDF response.

These results suggest that both nitric oxide and cyclooxygenase products have a role in the maintenance of resting cerebrocortical LDF. The reduced vasodilator response to halothane after cyclooxygenase and NOS inhibition may have several interpretations. The most straightforward interpretation is that the endothelium-derived mediators of the cerebrovascular effect of halothane were partially removed by the combination of indomethacin and L-NAME. However, it is also possible that the greatly increased vascular smooth muscle (VSM) tone secondary to infusion of indomethacin plus L-NAME attenuated the response of VSM to NO and PGI_2. It should be noted in this regard that baseline CVR increased significantly more after indomethacin plus L-NAME (108 ± 9%) than after L-NAME alone (66 ± 5%). Third, the drug combination could have obtunded autoregulation of flow to decreasing MAP and resulted in an apparent attenuation of halothane-induced vasodilatation. Finally, the attenuation of halothane-induced vasodilatation following simultaneous blockade of NOS and cyclooxygenase could also be due to the diminution of a shear-related vasodilator component as a result of low posttreatment baseline flow and/or the removal of endothelium-derived relaxing factors NO and PGI_2. However, as the CVR data show, even the simultaneous blockade of endothelial NOS and cyclooxygenase did not completely eliminate the hyperemic response to halothane as high as 1.7MAC.

In conclusion, the results suggest that nitric oxide alone is not an obligatory mediator of the cerebrocortical laser Doppler flow response to halothane in the rat. Apparently, other vasodilator mechanisms (either endothelium dependent or independent) are able to maintain the cerebrovascular responsiveness after the inhibition of NOS and cyclooxygenase.

VII. Spontaneous Flow Oscillations

In the spontaneous flow oscillation experiments, anesthesia was maintained by intravenous pentobarbital infusion at 3.3 mg/hr and an LDF baseline was established at about 60 min. As in the previous studies, L-

NAME was infused intravenously (20 mg/kg) over a period of 10 min. Thirty minutes after the administration of L-NAME the concentration of inspired halothane was raised from 0 to a level between 0.4and 0.9% and a new steady-state level of LDF was established. Halothane was then turned off, and, after normal conditions were reestablished, the flow response to CO_2 was tested by elevating the concentration of CO_2 in the inspired air to 5%. In a second group of animals equilibration was achieved first at 0.5% halothane anesthesia. Thirty minutes later hyperventilation was performed by increasing the rate of ventilation to achieve an $etCO_2$ level of 22 mmHg for 6 to 12 min. Normal $etCO_2$ (35 mmHg) was then reestablished. Next, hypercapnia was produced by inhalation of 5% CO_2. Arterial blood pressure was maintained at the preceding, normocapnic level by the controlled intravenous infusion of phenylephrine (0.5–5 μg/kg/min).

Before the administration of L-NAME spontaneous variations in flow were absent. After L-NAME infusion, flow oscillations readily appeared or were greatly enhanced. Figure 6 illustrates spontaneous cerebrocortical LDF oscillations as an example. The frequeny of flow oscillations was remarkably stable, except during hypercapnia (see below). At times, however, the amplitude of oscillations showed large variations, and in those cases it was difficult to establish the primary frequency of the oscillations.

Figure 7 illustrates that halothane inhalation rapidly suspended the LDF oscillations in the presence of L-NAME. Hypercapnia had a similar effect. Sometimes the oscillations were not completely abolished by moderate

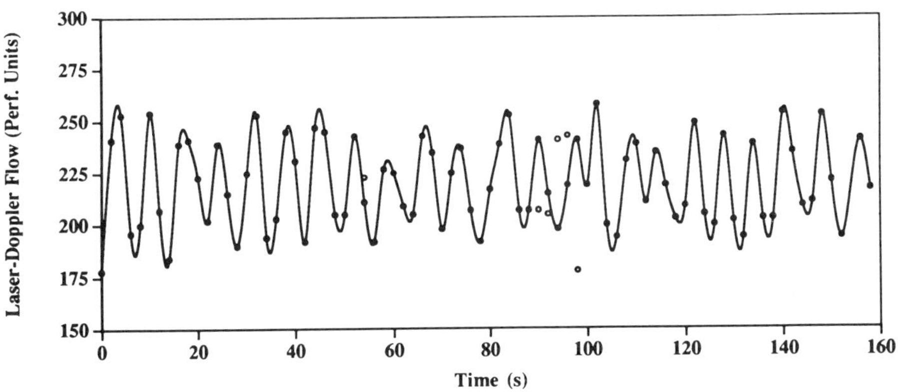

Fig. 6 Example of spontaneous cerebrocortical laser Doppler flow oscillations as observed after infusion of 20 mg/kg L-NAME. To better illustrate the fundamental oscillatory pattern, the graph was constructed by sampling the continuous LDF signal at 0.5 Hz and interpolating the data using a spline function curve.

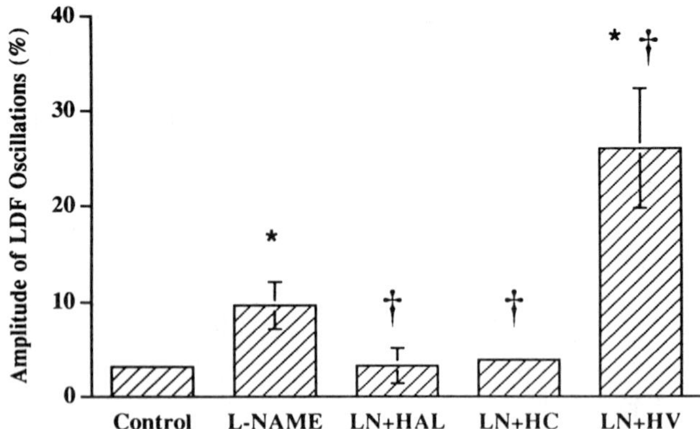

Fig. 7 Effects of L-NAME (20 mg/kg i.v.) and post L-NAME halothane (LN + HAL), hypercapnia (LN + HC), and hyperventilation (LN + HV) on the amplitude of spontaneous flow oscillations in cerebral cortex. The control value was obtained during pentobarbital anesthesia. Halothane was administered at 1 MAC. Peak-to-peak oscillation amplitude is expressed as the percentage of mean LDF. Means ± SD are shown. *Significant difference ($p < 0.05$) from control. †Significant difference ($p < 0.05$) from L-NAME.

hypercapnia, but marked slowing of the oscillation frequency was observed. It is notable that the inhibitory effect of hypercapnia on LDF oscillations occurred even after administration of L-NAME, suggesting that NOS inhibition did not eliminate this effect of CO_2. Likewise, hyperventilation resulted in a further significant increase in the oscillation amplitude after treatment with L-NAME, which was not a pressure-related effect since MAP was well maintained.

In our previous experiments, in which 1.0% halothane anesthesia was used, flow oscillations were rarely seen. We suspected that anesthesia interfered with the oscillations as suggested before (32). The present study demonstrates that the inhalational anesthetic halothane in concentrations as low as 0.4% may suspend the LDF oscillations in the cerebral cortex. Stadnicka *et al.* (33) have shown that, in order of their potency, halothane, isoflurane, and enflurance suppressed the amplitude and frequency of norepinephrine-induced oscillatory contractions of mesenteric veins before as well as after administration of L-NAME. This effect of volatile anesthetics is in contrast to the effect of sodium pentobarbital which appears to enhance the oscillations (16), although this may be associated with the barbiturate's hypotensive effect. This difference in the effects of anesthetics on flow oscillations is consistent with the fact that barbiturates are cerebral vasoconstrictors whereas volatile anesthetics are known cerebral vasodilators.

The latter results further suggest that nitric oxide is not required for the generation of cerebrocortical flow oscillations, a result that is now supported by studies of Dirnagl *et al.* (34) and Morita-Tsuzuki *et al.* (35). Hypocapnia enhances and both hypercapnia and halothane attenuate the flow oscillations after inhibition of nitric oxide synthase, suggesting that the action of neither of these depends exclusively on NO as a mediator.

VIII. Conclusions

Halothane and isoflurane produce dose-dependent increases in red blood cell flow in the rat cerebral cortex with isoflurane having a greater effect. Halothane but not isoflurane potentiates hypercapnic hyperemia of red cell flow. Neither anesthetic agent abolishes autoregulation as assessed by laser Doppler flowmetry. Nitric oxide does not appear to be an obligatory mediator of halothane-induced hyperemia but may act in concert with other endothelium-derived relaxing factors. Inhibition of NOS enhances spontaneous oscillations in cortical LDF. The oscillations are abolished or strongly attenuated by halothane and hypercapnia and augmented by hypocapnia, most likely via an NO-independent or partially NO-dependent pathway.

Acknowledgments

The technical help of James Wood and the secretairal help of Anita Tredeau are greatly appreciated. This work was supported by the VA Merit Review 7793-02P, Anesthesiology Research Training Grant GM 08377, and National Science Foundation Grant BCS-9001425.

References

1. Cucchiara, R. F., Theye, R. A., and Michenfelder, J. D. (1974). The effects of isoflurane on canine cerebral metabolism and blood flow. *Anesthesiology* **40,** 571–574.
2. Drummond, J. D., Todd, M. M., Scheller, M. S., and Shapiro, H. M. (1986). A comparison of the direct cerebral vasodilating potencies of halothane and isoflurane in the New Zealand White rabbit. *Anesthesiology* **65,** 462–467.
3. McPherson, R. W., and Traystman, R. J. (1988). Effects of isoflurane on cerebral autoregulation in dogs. *Anesthesiology* **69,** 493–499.
4. Miletich, D. J., Ivankovich, A. D., and Albrecht, R. F. (1976). Absence of autoregulation of cerebral blood flow during halothane and enflurane anesthesia. *Anesth. Analg. (N. Y.)* **55,** 100–109.
5. Todd, M. M., and Drummond, J. C. (1984). A comparison of the cerebrovascular and metabolic effect of halothane and isoflurane in the cat. *Anesthesiology* **60,** 276–282.
6. Hansen, T. D., Warner, D. S., Todd, M. M., Vust, L. J., and Trawick, D. S. (1988). Distribution of cerebral blood flow during halothane versus isoflurane anesthesia in rats. *Anesthesiology* **69,** 332–337
7. Drummond, J. C., Todd, M. M., Toutant, S. M., and Shapiro, H. M. (1983). Brain

surface protrusion during enflurane, halothane and isoflurane anesthesia in cats. *Anesthesiology* **59,** 288–293.

8. Koenig, H. M., Pelligrino, D. A., and Albrecht, R. F. (1993). Halothane vasodilation and nitric oxide in rat pial vessels. *J. Neurosurg. Anesthesiol.* **5,** 264–271.
9. Bredt, D. S., Hwang, P. M., and Synder, S. H. (1990). Localization of nitric oxide synthase indicating a neural role for nitric oxide. *Nature (London)* **347,** 768–770.
10. Murphy, S., Minor, R. L., Jr., Welk, G., and Harrison, D. G. (1990). Evidence for an astrocyte-derived vasorelaxing factor with properties similar to nitric oxide. *J. Neurochem.* **55,** 349–351.
11. Eskinder, H., Hillard, C. J., Flynn, N., Bosnjak, Z. J., and Kampine, J. P. (1992). Role of guanylate cyclase–cGMP systems in halothane-induced vasodilation in canine cerebral arteries. *Anesthesiology* **77,** 482–487.
12. Nahrwold, M. L., Lust, W. D., and Passaneau, J. V. (1977). Halothane-induced alterations of cyclic nucleotide concentrations in three regions of the mouse nervous system. *Anesthesiology* **47,** 423–427.
13. Jensen, N. F., Todd, M. M., Kramer, D. J., Leonard, P. A., and Warner, D. S. (1992). A comparison of the vasodilating effects of halothane and isoflurane on the isolated rabbit basilar artery with and without intact endothelium. *Anesthesiology* **76,** 624–634.
14. Muldoon, S. M., Hart, J. L., Bowen, K. A., and Freas, W. (1988). Attenuation of endothelium mediated vasodilation by halothane. *Anesthesiology* **68,** 31–37.
15. Haberl, R. L., Heiser, M. L., Marmarou, A., and Ellis, E. F. (1989). Laser-Doppler assessment of brain microcirculation: Effect of systemic alterations. *Am. J. Physiol.* **25,** H1247–H1254.
16. Hudetz, A. G., Roman, R. J., and Harder, D. R. (1992). Spontaneous flow oscillations in the cerebral cortex during acute changes in mean arterial pressure. *J. Cereb. Blood Flow Metab.* **12,** 491–499.
17. Lee, J. G., Hudetz, A. G., Smith, J. J., Hillard, C. J., Bosnjak, Z. J., and Kampine, J. P. (1994). The effects of halothane and isoflurane on cerebrocortical microcirculation and autoregulation as assessed by laser-doppler flowmetry. *Anesth. Analg. (N.Y.)* **79,** 58–65.
18. Pries, A. R., Fritzsche, A., Ley, K., and Gaehtgens, P. (1992). Redistribution of red blood cell flow in microcirculatory networks by hemodilution. *Circ. Res.* **70,** 1113–1121.
19. Maekawa, T., Tommasino, C., Shapiro, H. M., Keifer-Goodman, J., and Kohlenberger, R. W. (1986). Local cerebral blood flow and glucose utilization during isoflurane anesthesia in the rat. *Anesthesiology* **65,** 144–151.
20. Gelman, S., Fowler, K. C., and Smith, L. R. (1984). Regional blood flow during isoflurane and halothane anesthesia. *Anesth. Analg. (N. Y.)* **63,** 557–565.
21. Fan, F. C., Schuessler, G. B., Chen, R. Y. Z., and Chien, S. (1979). Determinations of blood flow and shunting of 9 and 15 μm spheres in regional beds. *Am. J. Physiol.* **237,** H25–H33.
22. Dinda, P. K., Buel, M. G., Da Costa, L. R., and Beck, I. T. (1983). simultaneous estimation of arteriolar, capillary, and shunt blood flow of the gut mucosa. *Am. J. Physiol.* **245,** G29–G37.
23. Hudetz, A. G., Weigle, G. M., Fenoy, F. J., and Roman, R. J. (1992). Use of fluorescently labeled erythrocytes and digital cross-correlation for the measruement of flow velocity in the cerebral microcirculation. *Microvasc. Res.* **43,** 334–341.
24. Eyre, J. A., Essex, T. J., Flecknell, P. A., Bartholomeow, P. H., and Sinclair, J. I. (1988). A comparison of measurements of cerebral blood flow in the rabbit using laser Doppler spectroscopy and radionuclide labelled microspheres. *Clin. Phys. Physiol. Meas.* **9,** 65–74.

25. Skarphedinsson, J. O., Harding, H., and Thoren, P. (1988). Repeated measurements of cerebral blood flow in rats. Comparisons between the hydrogen clearance method and laser Doppler flowmetry. *Acta Physiol. Scand.* **134,** 133–142.
26. Miller, J. D., Garibi, J., North, J. B., and Teasdale, G. M. (1975). False autoregulation after cold injury to the cerebral cortex. *In* "Cerebral Circulation and Metabolism" (T. W. Langfitt, *et al.,* eds.), pp. 95–98. Springer-Verlag, New York.
27. Sokrab, T. E. O., and Johansson, B. B. (1989). Regional cerebral blood flow in acute hypertension induced by adrenaline, noradrenaline and phenylephrine in the conscious rat. *Acta Physiol. Scand.* **137,** 101–105.
28. Mutch, W. A. C., Malo, L. A., and Ringaert, K. R. A. (1989). Phenylephrine increases regional cerebral blood flow following hemorrhage during isoflurane–oxygen anesthesia. *Anesthesiology* **70,** 276–279.
29. Mutch, W. A. C., Patel, P. M., and Ruta, T. S. (1990). A comparison of the cerebral pressure–flow relationship for halothane and isoflurane at haemodynamically equivalent end-tidal concentrations in the rabbit. *Can. J. Anaesth.* **37,** 223–230.
30. Iadecola, C. (1992). Does nitric oxide mediate the increases in cerebral blood flow elicited by hypercapnia? *Proc. Natl. Acad. Sci. U.S.A.* **89,** 3913–3916.
31. McPherson, R. W., Kirsch, J. R., Moore, L. E., and Traystman, R. J. (1993). N^{ω}-Nitro-L-arginine methyl ester prevents cerebral hyperemia by inhaled anesthetics in dogs. *Anesth. Analg. (N. Y.)* **77,** 891–897.
32. Hundley, W. G., Renaldo, G. J., Levasseur, J. E., and Kontos, H. A. (1988). Vasomotion in cerebral microcirculation of awake rabbits. *Am. J. Physiol.* **254,** H67–H71.
33. Stadnicka, A., Flynn, N. M., Bosnjak, Z. J., and Kampine, J. P. (1993). Enflurane, halothane, and isoflurane attenuate contractile responses to exogenous and endogenous norepinephrine in isolated small mesenteric veins of the rabbit. *Anesthesiology* **78,** 326–334.
34. Dirnagl, U., Lindauer, U., and Villringer, A. (1993). Nitric oxide synthase blockade enhances vasomotion in the cerebral microcirculation of anesthetized rats. *Microvasc. Res.* **45,** 318–323.
35. Morita-Tsuzuki, Y., Bouskela, E., and Hardebo, J. E. (1993). Effects of nitric oxide synthesis blockade and angiotension II on blood flow and spontaneous vasomotion in the rat cerebral microcirculation. *Acta Physiol. Scand.* **148,** 449–454.
36. Cremer, J. E., and Seville, M. P. (1983). Regional brain blood flow, blood volume and hematocrit values in the adult rat. *J. Cereb. Blood Flow Metab.* **3,** 254–256.
37. Drummond, J. C., Todd, M. M., and Shapiro, H. M. (1983). Cerebral blood flow autoregulation in the cat during anesthesia with halothane and isoflurane. *Anesthesiology* **59,** A305.

Cerebral Blood Flow during Isovolemic Hemodilution: Mechanistic Observations

Michael M. Todd

Department of Anesthesia
University of Iowa College of Medicine
Iowa City, Iowa 52242

I. Introduction

Isovolemic hemodilution has long been known to result in an increase in cerebral blood flow (CBF). However, the mechanisms underlying this apparently simple event remain the subject of debate and investigation. Specifically (and in simpleminded terms), there are two major possibilities: (1)The increase in CBF is an active, vasodilatory process, occuring as a compensatory response to a reduction in arterial O_2 content, and is hence similar to that seen during hypoxia. (2) The increase in CBF is a passive response to a reduction in whole blood viscosity. Because this increase in flow can occur without a change in vascular diameter, no active vasodilation is required. At the risk of overstating the case, most published data seem to support the first option, with very little data in existence to suggest that changes in viscosity are ever the sole cause of anemic CBF changes in the normal brain. However, there are data suggesting that viscosity plays some role (1), although the magnitude of this contribution is unclear.

We have conducted four experiments intended to provide some insight into the mechanisms by which hemodilution alters CBF in the normal brain. These experiments are summarized below.

Advances in Pharmacology, Volume 31

II. Influence of Hemodilution on Cerebral Blood Volume

If the CBF increase produced by hemodilution were due solely to changes in viscosity, that is, did not involve any active vasodilation, then the flow increase might not be accompanied by any increase in total cerebral blood volume (CBV). If vasodilation did occur, CBV should rise. We therefore simultaneously measured both CBF and CBV during progressive hemodilution. To do this, we developed a triple-label tracer technique to allow the measurement of CBF, cerebral red cell volume (CRCV), CPV, total CBV (=CRCV + CPV), and tissue hematocrit (=CRCV/CBV) (2). ^{99m}Tc was used to label red cells. [^{14}C] dextran was used as a plasma marker, and [^{3}H] nicotine was used to measure CBF via the indicator fractionation method. After tracer administration and appropriate circulation times, the brain was rapidly (~500 msec) heat fixed using focused high-energy (8–10 kW) microwave irradiation. After fixation ^{99m}Tc radioactivity in tissue and blood were measured in a well counter which is "blind" to both ^{3}H and ^{14}C activity. The samples were then allowed to sit for 5 days while the ^{99m}Tc decayed ($t_{1/2}$ 6.1 hr). ^{3}H and ^{14}C beta activity were then determined in a liquid scintillation counter. Experiments validating this method have been published (3,4).

CBF, CPV, and CRCV were measured in rats subjected to isovolemic hemodilution, and total CBV and cerebral hematocrit were calculated. The results [published in Todd *et al.* (5)] demonstrated that the expected increase in CBF was accompanied by an increase in CBV (Fig. 1), indicating the occurrence of at least some active cerebrovasodilation. Tissue hematocrit was also lower than arterial hematocrit, a phenomenon explained by the Fåhreus–Lindgvist effect (6,7). More importantly, tissue hematocrit decreased more rapidly during hemodilution than did arterial hematocrit.

There are a number of interesting implications of this work, other than simply finding that CBV rises during hemodilution. First, the magnitude of the CBV increase seems to be in excess of that which might be expected from simple arteriolar dilation. For example, the slope of the regression line plotting CBF versus CBV was much steeper for these animals than for normal animals subjected to changes in P_aCO_2 [data taken from Todd *et al.* (4)]. The implication is that at least a portion of the CBV increase is occurring distal to the arterioles that provide the most vascular resistance. Second, tissue hematocrit appeared to fall "faster" than did large vessel hematocrit. This finding has been confirmed by others in mesentary (8), and raises the possibility that tissue O_2 availability/delivery may not be adequately predicted by cerebral arterial oxygen content (CaO_2) calculations based on large vessel hematocrit. Because calculated red cell transit

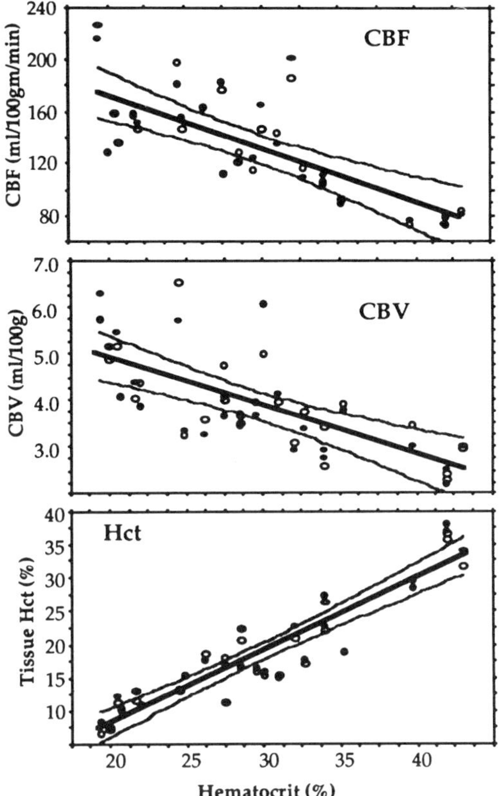

Fig. 1 CBF, CBV, and tissue hematocrit (Hct) during hemodilution in rats. Each data point represents one hemisphere from one animal. However, regression lines and confidence intervals are for both hemispheres combined. [From Todd *et al.* (5).]

time has also decreased significantly, it is possible that O_2 unloading from red cells might be time limited. Both of these factors raise the possibility that marked hemodilution may compromise tissue O_2 availability more than we have heretofore thought.

III. Influence of Focal Cortical Brain Lesion on Regional Cerebral Blood Flow Response to Hemodilution

If the CBF response to hemodilution is an active compensatory response to changing arterial O_2 content, then it is possible that interventions known to abolish hypoxic cerebrovasodilation might also abolish anemic vasodilation. Perhaps the most relevant intervention known to abolish hypoxic vasoresponsiveness is brain injury (9,10). We therefore measured CBF during hemodilution in pentobarbital-anesthetized rabbits that had previously (2 hr earlier) been subjected to a small (~8 mm diameter) freeze injury of the right posterior parietal cortex. Unlesioned rabbits served as controls. Isovolemic hemodilution (to a minimum hematocrit of ~12%) was achieved in three steps (hematocrits of 28, 19, and 12%). CBF was measured with radioactive microspheres in multiple cortical and subcorti-

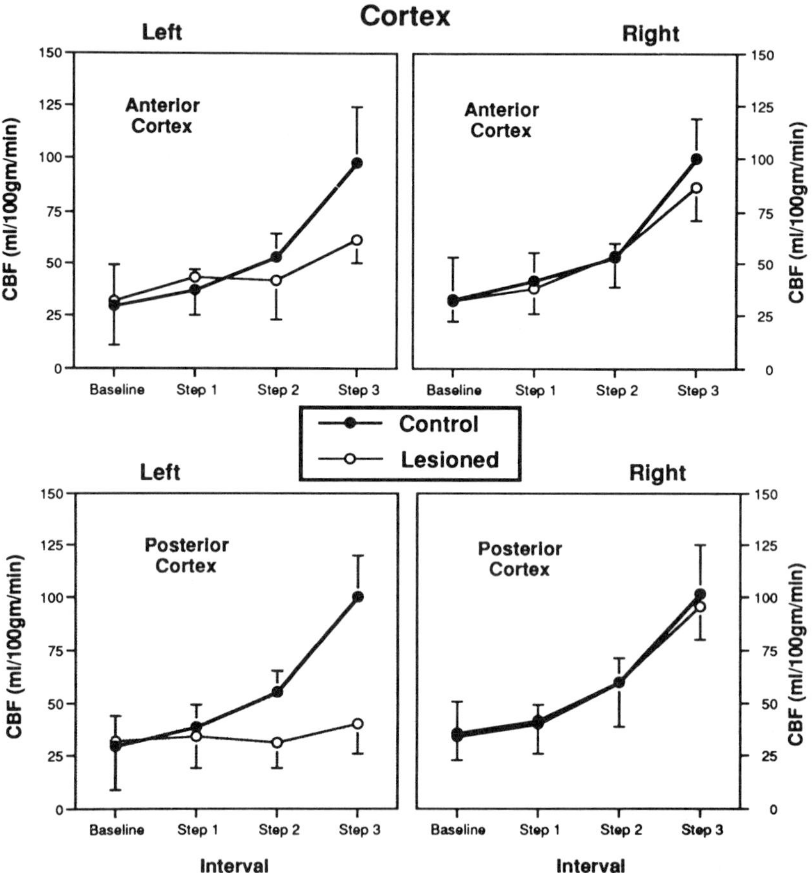

Fig. 2 Regional cortical CBF responses to three-step isovolemic hemodilution in normal and freeze-lesioned rabbits. The lesion was continued to the left posterior cortex and was approximately 8 mm in diameter. No visible injury and no Evans blue staining was observed in the left anterior portion.

cal regions, specifically those both near to and distant from the injury focus.

The results are summarized in Fig. 2. There are no differences in the flow patterns seen in the hemisphere contralateral to the lesion (right). However, a very different pattern was seen for cortical CBF from anterior and posterior portions of the lesioned hemisphere, with a marked attenuation of the CBF response to hemodilution seen in lesioned animals. Of particular interest is the attenuation seen in the anterior portion of the hemisphere, which is distant from the lesion focus and which contains no visibly damaged tissue (as assessed by Evans blue staining). In addition, no abnormalities were seen in subcortical samples, indicating that this was confined to the cortical mantle.

These results again support the concept that the CBF responses to hemodilution are active and can be abolished over a wide area by a small focal lesion. If viscosity were an important factor, we would expect to see normal CBF changes at least in ipsilateral brain regions that were clearly uninjured (i.e., in which there should be no physical vascular disruption). In addition, the attenuated flow responses occur with a pattern similar to that produced by spreading depression (11), although spreading depression (SD) is typically associated with a post-SD hypoperfusion (which was not seen in these animals). This may be of mechanistic importance, since SD is dependent on sequential activation of the *N*-methyl-D-aspartate (NMDA) receptor (12). Furthermore, the link between the NMDA receptor and NO synthesis is believed to play some role in the flow changes produced by SD (13). From a clinical standpoint, this finding provides experimental support for the observation by Miller *et al.* that traumatized patients with admission hematocrits below 30% have a poorer outcome than nonanemic individuals with otherwise similar injuries (14).

IV. Comparison of Cerebrovascular and Metabolic Changes Produced by Hypoxia and Hemodilution

One problem with understanding the mechanisms by which hemodilution influences organ blood flow is the fact that both CaO_2 and viscosity change as hematocrit falls. To separate these fators, an experiment should ideally attempt to manipulate these variables independently. One approach is to compare the CBF responses to hypoxia and hemodilution, both of which reduce CaO_2 but only one of which alters viscosity.

Two groups have directly compared the CBF effects of hypoxia and hemodilution (15,16). Both reported that reductions in CaO_2 produced by either intervention resulted in similar CBF increases, suggesting that viscosity played little role [although Korosue and Heros (16) noted that hemodilution increased CBF after focal cerebral occlusion, whereas hypoxia did not, perhaps implicating viscosity-related mechanisms in that situation]. However, minor limitations in experimental design (e.g., the number of animals studied and variability in CaO_2) make careful statistical comparisons between hypoxia and anemia difficult.

We measured CBF (using radioactive microspheres), cerebral O_2 delivery (DO_2 = CBF × CaO_2), cerebral venous oxygen content (CvO_2, with blood obtained from the exposed confluence of sinuses via a 25-gauge needle mounted on a micromanipulator), oxygen extraction ratio (OER), and $CMRO_2$ in pentobarbital-anesthetized rabbits. The animals were subjected to controlled reductions in CaO_2 produced by hypoxia or hemodilution. In each rabbit, CaO_2 was sequentially decreased from control values (~19 ml O_2/dl) to 13, 9, and 6 ml O_2/dl, either by reducing p_{aO_2} (to a minimum of ~26 mmHg) or by isovolemic hemodilution (to a minimum hematocrit of 14%).

If viscosity were an important determinant of anemic CBF, we would expect to see higher CBF values at any given CaO_2 in the rabbits subjected to hemodilution. Surprisingly, CBF at the lowest two CaO_2 values was greater in hypoxic animals, and analysis of covariance (ANCOVA) demonstrated significantly different CaO_2 versus CBF response slopes. These data are shown in Fig. 3. Also, although cerebral DO_2 was well maintained in hypoxic animals, it fell significantly with hemodilution, from 7.95 ± 2.92 to 5.08 ± 1.10 ml O_2/100 g/min (mean ± SD). Conversely, OER rose to a greater degree with hemodilution (from 0.43 to 0.68) than during hypoxia (from 0.46 to 0.54). $CMRO_2$ was well maintained.

There are two possible explanations for these results. The first is that during severe reductions in CaO_2, oxygen is more "available" to brain in hemodiluted animals than during hypoxia. This hypothesis is supported by somewhat higher venous p_{O_2} values seen in anemic animals at the lowest CaO_2 (i.e., 27 ± 3 mmHg as compared with 18 ± 2 mmHg during hypoxia), despite lower CBF and DO_2 values in the diluted rabbits. One could argue that the brain does not need to increase CBF as much during hemodilution because O_2 can be more readily extracted. This contradicts the hypothesis which was generated by our CBF/CBV experiments (see above). There are two possible explanations for the improved O_2 availability. First, Pries *et al.* noted that the distribution of red cells was more uniform throughout the microvasculature during hemodilution (8). Second, hemoglobin is an important buffer in blood, and hemodilution reduces this buffering capacity. This is manifested by lower cerebral venous pH

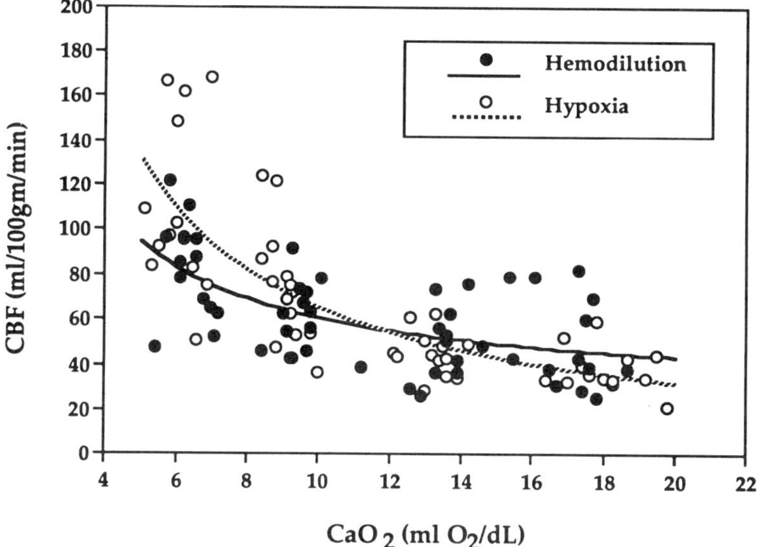

Fig. 3 Scattergrams and regression lines [CBF = $m(1/CaO_2) + b$] for forebrain CBF versus CaO_2 produced by either hypoxia or hemodilution. The slopes of the two regression lines are significantly different.

and higher p_{CO_2} levels (Table I), both of which shift the oxyhemoglobin desaturation curve to the right (i.e., P50 will rise). This should enhance O_2 unloading. Because hemoglobin content is unaltered in hypoxemic blood, this mechanism is not available.

Table I

Reduction of Buffering Capacity of Blood with Hemodilution[a]

Condition	Venous p_{CO_2} (mmHg)	Venous pH
Baseline (CaO_2 ~19)		
Hypoxia	48.2 ± 3.3	7.34 ± 0.05
Hemodilution	46.8 ± 4.1	7.33 ± 0.05
Reduced O_2 (CaO_2 ~6)		
Hypoxia	41.8 ± 4.5	7.33 ± 0.06
Hemodilution	45.6 ± 5.3[b]	7.27 ± 0.04[b]

[a] Values are means ± SD.

[b] Significantly different ($p < 0.05$) compared to baseline.

There is a second possibility. Pial vessel studies indicate that hemodilution produces vasoconstriction (17,18). If true, might not "anemic vasoconstriction" act to limit the CBF increase that would otherwise be seen? There is a potential mechanism for this. Hemodilution reduces whole blood viscosity, which will in turn reduce shear stress on the endothelium. Shear stress is known to influence the synthesis and/or release of a number of vasodilatory autocoids, including nitric oxide and prostacyclin (19–23). As viscosity drops, the release of these compounds will decrease as well. Hemodilution might thus lead to a progressive attenuation of intrinsic (background?) vasodilation, which would tend to counteract CBF increases driven by a falling CaO_2. There is precedent for this concept, since Kaiser and Sparks reported that *in vitro* acetylcholine-induced vasodilation (which is endothelial dependent) was blunted by saline perfusion (24).

V. Role of Nitric Oxide in Cerebral Blood Flow Response to Hemodilution

As a follow-up to the previous experiment, and as a first step in examining the role of NO in the CBF response to hemodilution, we subjected either hypoxic or anemic pentobarbital-anesthetized rabbits (CaO_2 ~6 ml O_2/dl

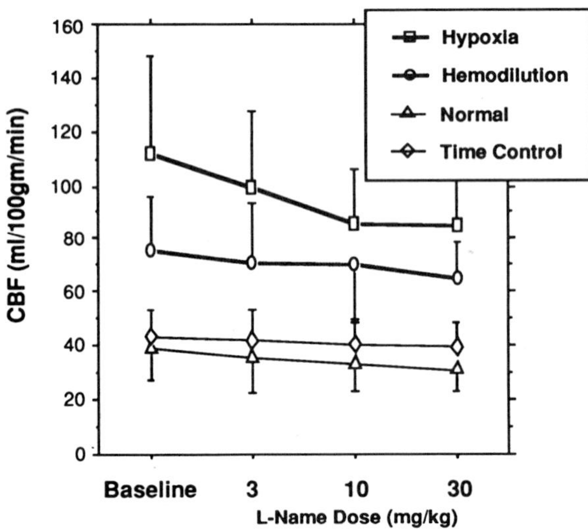

Fig. 4 CBF responses to L-NAME in normal, hemodiluted, and hypoxic animals. Time control animals received no L-NAME. The slopes of the lines are significantly different by repeated measures analysis of variance ($A \times B$).

in each group) or normoxic/nonanemic controls to progressively greater doses of N^{w}-nitro-L-arginine methyl ester (L-NAME), an inhibitor of NO synthesis (3, 10, and 30 mg/kg), with CBF being measured after each dose. The results are shown in Fig. 4. In both control and hemodiluted animals, the changes induced by L-NAME were indistinquishable from those in untreated time controls. However, a much greater reduction in CBF was seen in hypoxic animals.

Although these experiments are preliminary, they suggest that NO may be playing a different role in hypoxic versus diluted animals. Other experiments are underway to examine the role of adenosine as well as the interactions between NO and prostacyclin in these different flow responses. For example, this work indicates that whereas treatment of hypoxic animals with progressively increasing doses of 8-phenyltheophylline decreases CBF, it has little or no effect in diluted animals.

VI. Discussion

These experiments, when combined with those of others, suggest that arterial O_2 content is indeed the most important determinant of CBF change during hemodilution, at least in normal animals. The observed increase in CBF is clearly associated with an increase in CBV, indicating some active vasodilation, and this vasodilator response can be attenuated by a focal brain lesion. This attentuation is present not only in directly damaged tissue, but also in other ipsilateral cortical regions. We can speculate that this may be related to induced spreading depression, with consequent alterations in the function of NMDA receptor and perhaps NO production. However, in spite of these conclusions, a careful comparison of the CBF responses to matched hypoxic and anemic reductions in CaO_2 does reveal significant differences although the differences are in a direction opposite of what could be explained by the usual "low viscosity" effects. In fact, the findings raise the possibility that hemodilution-induced viscosity reductions may actually act as a vasoconstrictor, possibly by attentuating shear stress-induced release of endothelially produced vasodilator compounds.

References

1. Hudak, M. L., Koehler, R. C., Rosenberg, A. A., Traystman, R. J., and Jones, M. D. (1986). Effect of hematocrit on cerebral blood flow. *Am. J. Physiol.* **251,** H63–H70.
2. Cremer, J. E., and Seville, M. P. (1983). Regional brain blood flow, blood volume and haematocrit values in the adult rat. *J. Cereb. Blood Flow Metab.* **3,** 254–256.
3. Todd, M. M., Weeks, J. B., and Warner, D. S. (1993). The influence of intravascular

volume expansion on cerebral blood flow and blood volume in normal rats. *Anesthesiology* **78,** 945–953.
4. Todd, M. M., Weeks, J. B., and Warner, D. S. (1993). Microwave fixation for the determination of cerebral blood volume in rats. *J. Cereb. Blood Flow Metab.* **13,** 328–336.
5. Todd, M. M., Weeks, J. B., and Warner, D. S. (1992). Cerebral blood flow, blood volume, and brain tissue hematocrit during isovolemic hemodilution with hetastarch in the rat. *Am. J. Physiol.* **263,** H75–H82.
6. Barbee, J. H., and Cokelet, G. R. (1971). The Fåhraeus effect. *Microvasc. Res.* **3,** 6–16.
7. Lammertsma, A. A., Brooks, D. J., Beaney, R. P., Turton, D. R., Kensett, M. J., Heather, J. D., Marshall, J., and Jones, T. (1984). *In vivo* measurement of regional cerebral haematocrit using positron emission tomography. *J. Cereb. Blood Flow Metab.* **4,** 317–322.
8. Pries, A. R., Fritzsche, A., Ley, K., and Gaehtgens, P. (1992). Redistribution of red blood cell flow in microcirculatory networks by hemodilution. *Circ. Res.* **70,** 1113–1121.
9. Feustel, P. J., and Nelson, L. R. (1987). Cortical blood flow regulation during hypoxemia in experimental head injury. *J. Surg. Res.* **43,** 86–93.
10. Lewelt, W., Jenkins, L. W., and Miller, J. D. (1982). Effects of experimental fluid-percussion injury of the brain on cerebrovascular reactivity to hypoxia and to hypercapnia. *J. Neurosurg.* **56,** 332–338.
11. Lauritzen, M. (1984). Long-lasting reduction of cortical blood flow of the rat brain after spreading depression with preserved autoregulation and impaired CO_2 response. *J. Cereb. Blood Flow Metab.* **4,** 546–554.
12. Lauritzen, M., and Hansen, A. J. (1992). The effect of glutamate receptor blockade on anoxic depolarization and cortical spreading depression. *J. Cereb. Blood Flow Metab.* **12,** 223–229.
13. Goadsby, P. J., Kaube, H., and Hoskin, K. L. (1992). Nitric oxide synthesis couples cerebral blood flow and metabolism. *Brain Res.* **595,** 167–170.
14. Miller, J. D., Butterworth, J. F., Gudeman, S. K., Faulkner, J. E., Choi, S. C., Selhorst, J. B., Harbison, J. W., Lutz, H. A., Young, H. F., and Becker, D. P. (1981). Further experience in the management of severe head trauma. *J. Neurosurg.* **54,** 289–299.
15. Jones, M. D., Traystman, R. J., Simmons, M. A., and Molteni, R. A. (1981). Effects of changes in arterial O_2 content on cerebral blood flow in the lamb. *Am. J. Physiol.* **240,** H209–H215.
16. Korosue, K., and Heros, R. C. (1992). Mechanism of cerebral blood flow augmentation by hemodilution in rabbits. *Stroke* **23,** 1487–1493.
17. Hudak, M. L., Jones, M. D., Popel, A. S., Koehler, R. C., Traystman, R. J., and Zeger, S. L. (1989). Hemodilution causes size-dependent constriction of pial arterioles in the cat. *Am. J. Physiol.* **257,** H912–H917.
18. Muizelaar, J. P., Bouma, G. J., Levasseur, J. E., and Kontos, H. A. (1992). Effect of hematocrit variations on cerebral blood flow and basilar artery diameter *in vivo. Am. J. Physiol.* **262,** H949–H954.
19. Buga, G. M., Gold, M. E., Fukuto, J. M., and Ignarro, L. J. (1991). Shear stress-induced release of nitric oxide from endothelial cells grown on beads. *Hypertension* **17,** 187–193.
20. Busse, R., Pohl, U., and Luckhoff, A. (1989). Mechanisms controlling the production of endothelial autacoids. *Z. Kardiol.* **6,** 64–69.
21. Nerum, R. M., Harrison, D. G., Taylor, W. R., and Alexander, R. W. (1993). Hemodynamics and vascular endothelial biology. *J. Cardiovas. Pharmacol.* **21**(Suppl. 1), S6–10.

22. Pohl, U., Herlan, K., Huang, A., and Bassenge, E. (1991). EDRF-mediated shear-induced dilation opposes myogenic vasoconstriction in small rabbit arteries. *Am. J. Physiol.* **261,** H2016–H2023.
23. Rubanyi, G. M., Freay, A. D., Kauser, K., Johns, A., and Harder, D. R., (1990). Mechanoreception by the endothelium: Mediators and mechanisms of pressure- and flow-induced vascular responses. *Blood Vessels* **27,** 246–257.
24. Kaiser, L., and Sparks, H. V. J. (1987). Effect of hemodilution on endothelium-dependent vasodilation in the *in vivo* canine femoral artery. *Circ. Shock* **23,** 107–118.

Cerebral Physiology during Cardiopulmonary Bypass: Pulsatile versus Nonpulsatile Flow

Brad Hindman
Department of Anesthesiology
University of Iowa College of Medicine
Iowa City, Iowa 52242

I. Introduction

There is considerable neurological morbidity associated with cardiac surgery, with associated increases in perioperative mortality, length of intensive care unit (ICU) and hospital stay (1), and, in 2–4% of patients, long-term neurological and cognitive disability (2,3). The majority of neurological injuries associated with cardiac surgery are believed to be the result of focal or multifocal cerebral ischemic insults. The extent of infarction resulting from a focal ischemic insult is critically dependent on the physiological conditions present during the first few hours after the inciting event (4). It is for this reason that the conduct of cardiopulmonary bypass (CPB) is likely to have an important influence on neurological outcome. A long-standing question in CPB management is whether conventional nonpulsatile CPB may exacerbate (or possibly even initiate) neurological injury occurring in the course of cardiac surgery and whether pulsatile CPB might improve cerebral perfusion and/or neurological outcome.

There are several mechanistic theories as to how pulsatile perfusion might produce differential circulatory effects as compared to equivalent mean flow achieved with nonpulsatile perfusion. The extent to which these mechanisms apply to the cerebral circulation, if at all, is virtually

Advances in Pharmacology, Volume 31

unknown. Baroreceptor reflexes, which depend on arterial pulse frequency, pulse pressure, and *dP*/*dt* (5–7), play an important role in systemic hemodynamic control. In the absence of pulsatile flow, reflex mechanisms mediated via the carotid sinus, cardiovascular control center, and sympathetic nervous system, (7) increase arterial and venous smooth muscle tone (7,8). In this way systemic vascular resistance is increased and blood volume distribution altered by nonpulsatile perfusion. Some authors propose the greater peak hydraulic energy of pulsatile flow may recruit capillary beds that would otherwise be closed under nonpulsatile conditions (9,10). By so doing, pulsatile perfusion could decrease vascular resistance, increase the uniformity of tissue perfusion (11), and possibly improve substrate delivery (9). It has been discovered that the greater peak sheer stress of pulsatile flow promotes release of endothelial-derived vasodilators such as prostacyclin (12) and nitric oxide (13). In the absence of pulsatile flow, a decrease in the local concentration of these substances may increase organ vascular resistance.

Animal studies uniformly indicate that pulsatile flow favorably influences cerebral perfusion. In 1971 Matsumoto *et al.*, using a microscope focused on the brain surface and conjunctiva of normothermic dogs, compared microvascular effects of pulsatile and nonpulsatile perfusion (14). They found that nonpulsatile perfusion resulted in capillary collapse, intravascular sludging, and marked venodilation. Pulsatile perfusion, on the other hand, was free of these effects. In 1972 Sanderson *et al.* found dogs undergoing 1 to 3 hr of normothermic pulsatile CPB were without subsequent neuropathological changes, whereas animals undergoing nonpulsatile CPB had a high incidence of diffuse changes (15). Capillary beds collapsed during nonpulsatile perfusion, and a large proportion of ischemic neuronal cell changes were distributed in arterial boundary zones, suggesting cerebral perfusion at the arteriolar or capillary level was impaired. In 1981, Mori *et al.* reported that nonpulsatile perfusion resulted in increased cerebral anaerobic metabolism during systemic cooling in dogs, whereas pulsatile perfusion did not (16). In 1985 Dernevik *et al.*, studying normothermic dogs, found cerebral blood flow (CBF) to be significantly greater (~18%) during pulsatile CPB as compared to nonpulsatile CPB (17). In this study, the brain arterial–venous oxygen content difference did not differ between pulsatile and nonpulsatile flow, suggesting that the cerebral metabolic rate for oxygen (CMR_{O_2}) was also approximately 18% greater during pulsatile flow. In 1986 Tranmer *et al.* also reported CBF to be approximately 18% greater with pulsatile perfusion in normothermic dogs (18). After baseline CBF measurements, one hemisphere was made ischemic by unilateral occlusion of the internal and middle cerebral arteries. CBF in the ischemic hemisphere was significantly greater with pulsatile as compared to nonpulsatile perfusion. This last study is especially provocative as it suggests pulsatile perfusion might indeed better support the acutely injured brain.

It is on the above theoretical and experimental bases that some cardiac surgery groups routinely employ pulsatile CPB (19) or, alternatively, use it in patients considered at high risk of neurological complications (20). Nevertheless, there is, to date, no clinical evidence of superior neurological outcome with pulsatile perfusion. In a study of 312 patients undergoing coronary artery bypass grafting (CABG), Shaw *et al.* found no difference in neurological outcome between patients undergoing pulsatile or nonpulsatile perfusion (21). Unfortunately, perfusion technique was not randomized, and criteria of assigning patients to pulsatile versus nonpulsatile perfusion were not stated. Therefore, it is possible that selection bias may have masked neurological outcome differences between perfusion techniques. In a much smaller study, Henze *et al.* randomized patients undergoing CABG ($n = 22$) to pulsatile versus nonpulsatile CPB (22). In contrast to the animal studies cited above, CBF during bypass did not differ between groups. Neither were there neurological outcome differences between groups. However, the sample size of the study is probably too small relative to the occurrence rate (e.g., 1 case of arm paresis in each group) to provide sufficient statistical power by which to conclude confidently that pulsatile perfusion is without neurological benefit.

Thus, despite consistently positive animal studies, clinical studies do not support the notion of improved cerebral perfusion and neurological outcome with pulsatile perfusion. On what basis might these differences exist? Existing animal studies can be criticized in several respects. First, all the studies referred to above were conducted with dogs. In contrast to humans, dogs have extensive collateralization between the internal and external carotid systems, such that CBF is mainly derived from the external carotid. It is possible that improved brain perfusion with pulsatile flow in dogs is due to increases in extracranial blood flow, increases which might not occur or which might be irrelevant in humans. Second, early animal studies [e.g., Matsumoto *et al.* (14) and Sanderson *et al.* (15)] used perfusion techniques/devices which are not comparable to current technologies. Injury patterns observed with nonpulsatile flow in the early 1970s might not be observed now. For example, Laursen *et al.* perfused pigs with nonpulsatile bypass for 3 hr at normothermia and observed no histopathological changes (23). Third, all experimental studies showing improved perfusion with pulsatile flow were performed at normothermia. This is in contrast to the routine clinical practice, where CPB is performed primarily with moderate hypothermia. Hypothermia is very likely to influence aspects of vascular control thought to mediate the effects of pulsatility: vascular smooth muscle reactivity, blood rheology (sheer stress), endothelial synthesis of vasoregulators, and the autonomic nervous system. Thus, perhaps differences in cerebral perfusion between pulsatile and nonpulsatile flow observed at normothermia might be attenuated or even eliminated at hypothermia. The latter possibility is supported by the findings of Henze *et al* (22) (see above). In addition to the limitations of the human studies previously discussed, a final and major limitation is their failure to document and/or quantiate the degree of pulsatility achieved. In neither study referred to above were pulsatile waveforms described, shown, or quantitated. There is no way to judge whether the degree of pulsatility achieved was sufficient to truly represent pulsatile flow. In other words, was pulsatile flow "pulsatile enough?"

This leads to a fundamental physiological question, namely, Are there essential aspects or determinants of pulsatile flow that distinguish it from nonpulsatile flow? If so, what are they? One of the few studies to begin to address this issue is that of Grossi *et al.* (24). Dogs were placed on normothermic pulsatile CPB. Pulse waveform shape and frequency were varied over a range of values. These investigators observed no difference in systemic lactate production between pulsatile and nonpulsatile CPB until a critical degree of pulsatility, defined by these authors as the "pulsatility index," was achieved. The authors found both pulse shape and pulse frequency to influence lactate production. Thus, in any study of pulsatile perfusion it appears necessary first to assess whether pulse characteristics do, in fact, adequately represent pulsatile perfusion in the system being tested. However, because the essential characteristics and mechanisms distinguishing pulsatile from nonpulsatile flow are still largely unknown, choice of any single parameter (pulse pressure, dP/dt, systolic/diastolic ratio, pulsatility index, peak power) is arbitrary.

II. Pulsatile versus Nonpulsatile Bypass

With these considerations in mind our laboratory has conducted an experiment to answer the following question: Do CBF and CMR_{O_2} differ between pulsatile and nonpulsatile cardiopulmonary bypass at 27°C? We present our preliminary findings. We used a modification of our previously described rabbit model of cardiopulmonary bypass (25). Rabbits provide a good model of human cerebrovascular physiology because the rabbit brain is supplied exclusively by the internal carotid and vertebral arteries in a pattern similar to humans (26). Rabbit CBF and CMR_{O_2}, cerebrovascular responses to changes in P_{aCO_2} and arterial pressure (autoregulation), closely approximate human values. The pulsatile perfusion system (Medical Engineering Consultants, Los Angeles, CA) consisted of two pumping chambers in series. Each chamber has an inlet and outlet valve to ensure unidirectional flow. The first chamber receives blood from the venous reservoir and pumps it through the membrane oxygenator. The second chamber receives blood from the oxygenator and pumps it to the animal. Pump ejection period (msec), pulse frequency (pulse/min), and stroke volume (ml) are independently variable. Carotid arterial pressure was recorded from a solid state micromanometer placed via the facial artery. Cerebral blood flow was measured via the microsphere technique. CMR_{O_2} was calculated via the Fick equation, using cerebral venous blood from the confluens sinnum. The experiment was conducted in two parts.

III. Influence of Arterial Pressure Waveform

In Part A of the experiment, we sought to determine whether CBF and CMR_{O_2} were influenced by the shape and/or frequency of the arterial pressure waveform. Animals anesthetized with fentanyl and diazepam were randomized to one of three groups based on pulse ejection period: 100 msec (group A-1, $n = 8$), 120 msec (group A-2, $n = 8$), and 140 msec (group A-3, $n = 8$). Once a stable brain temperature of 27°C was achieved on CPB, each animal underwent three consecutive 20-min periods at each of three pulse frequencies (150, 200, 250 pulses/min; these pulse rates are physiologically appropriate for the rabbit) in random order. Systemic flow was maintained constant throughout the experiment at 80 ml/kg/min via adjustment of pump stroke volume. CBF and CMR_{O_2} were determined at the end of each 20-min perfusion period. Thus, in total, CBF and CMR_{O_2} were determined at nine different combinations of ejection period and pulse rate.

At one end of the spectrum our goal was to produce an arterial pressure waveform appearing virtually indistinguishable from a normal rabbit arterial pressure waveform (including a dicrotic notch). This corresponds to group A-1 animals with a pulse rate of 150 pulses/min (see Fig. 1). It is doubtful that in the clinical setting an equivalent waveform could be produced. This is because the relatively small aortic cannulas used in human CPB result in marked damping of pulsatile waveforms (9,11,24). Nevertheless, we wanted to test for CBF and CMR_{O_2} dependency using the best possible waveform. At the other end of the spectrum we produced an arterial pressure waveform that had a relatively small pulse pressure, small dP/dt, and a less "physiological" appearance. This corresponds to group A-3 animals with a pulse rate of 250 pulses/min (see Fig. 1).

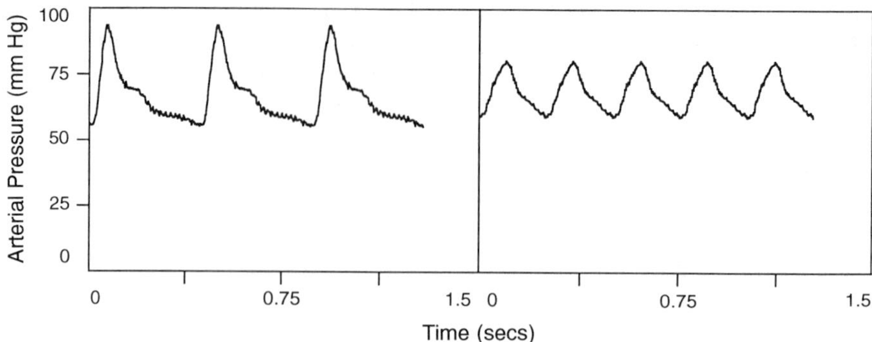

Fig. 1 Group-averaged carotid arterial pressure waveforms. (Left) Group A-1 (n = 8) at pulse rate of 150 pulses/min. (Right) Group A-3 (n = 8) at pulse rate of 250 pulses/min.

Some authors advocate use of energy indices to quantitate pulsatile flow [e.g., "energy equivalent pressure" (9) or "pulsatility index" (24)]. This requires measurements of arterial flow waveforms as opposed to, or in addition to, pressure waveforms. We saw no advantage to that approach. First, it is arterial pressure, not flow, that is monitored clinically. Second, energy indices do not always distinguish pulsatile from nonpulsatile flow, in that, in some cases, nonpulsatile flow can have greater mean hydraulic power (9). Third, Grossi *et al.* observed that both pulse shape and pulse frequency were important in distinguishing pulsatile and nonpulsatile flow, not energy indices per se (24). Despite the intended variation in pulse rate and ejection period, mean arterial pressure was equivalent among groups and frequencies (~66 mmHg). Using multivariate analysis of variance, we found CBF (~29 ml/100g/min) and CMR_{O_2} (~1.8 ml/100 g/min) to be independent of group (ejection period), pulse frequency, and order of determination. Thus, as far as the cerebral circulation was concerned, all nine arterial waveforms were physiologically equivalent. We concluded, therefore, that any combination of ejection period (100–140 msec) and pulse frequency (150–250) provided an equivalent and, very likely, adequate representation of pulsatile perfusion.

In the second part of the experiment (Part B), we compared CBF and CMR_{O_2} between animals randomized to undergo either pulsatile bypass (n = 8; ejection period 120 msec, rate 250 pulses/min) or nonpulsatile bypass (n = 8) at 27°C. Nonpulsatile bypass was maintained with a centrifugal pump, thereby eliminating even the small oscillations in arterial pressure and flow obtained with standard roller pumps. Systemic flow was kept constant at 100 ml/kg/min. In contrast to Part A, anesthesia was maintained with isoflurane (inspired concentration 1% v/v, at 26°C). This was done to avoid the issue of potential differences in fentanyl and diazepam metabolism between pulsatile and nonpulstile flow. CBF and CMR_{O_2} measurements were made at 60 min of bypass duration. In the pulsatile group, arterial pressure was approximately 87/54 mmHg. Mean arterial pressure (~67 mmHg) was equivalent between groups (see Fig. 2). Using unpaired *t*-tests, we detected no significant difference in CBF (~31 ml/100g/min) nor CMR_{O_2} (~1.7 ml/100 g/min) between pulsatile and nonpulsatile groups.

Thus, in our rabbit model of cardiopulmonary bypass at 27°C, we did not observe any significant difference in brain blood flow nor oxygen metabolism between pulsatile and nonpulsatile perfusion. Although this finding is somewhat at odds with normothermic dog studies (see above), it is in agreement with the only human study to make a similar comparison [Henze *et al.* (22)]. Thus, in our model, there is no evidence that pulsatility is important in determining the adequacy of cerebral perfusion during hypothermic CPB.

Why might pulsatility be unimportant to CBF and CMR_{O_2} at 27°C? First, baroreceptor responses are likely to be highly attenuated and/or eliminated by the combined effects of anesthesia (27,28) and hypothermia. Furthermore, it appears, even at normothermia, that baroreceptor responses and the sympathetic nevous system activity have little sustained effect on CBF and CMR_{O_2} (29). Second, evidence indicates that cerebrovascular smooth muscle tone and responsiveness to vasoregulators is altered by hypothermia. Using an isolated rat cerebral arteriole preparation, Ogura *et al.* found that hypothermia directly induced vasodilation (30). At 20°C vasoconstrictive and vasodilatory responses to acute changes in pH, as well as vasoconstrictive responses to $PGF_{2\alpha}$, were abolished. Thus, as temperature is reduced during cardiopulmonary bypass, CBF may be

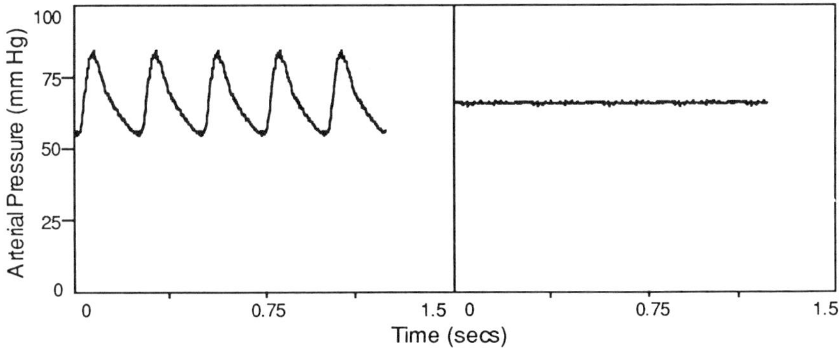

Fig. 2 Group-averaged carotid arterial pressure waveforms. (Left) Pulsatile bypass (n = 8) at pulse rate of 250 pulses/min. (Right) Nonpulsatile bypass (n = 8).

determined less by "normal" determinants of cerebrovascular tone and more by the effect of hypothermia on cerebrovascular smooth muscle. Finally, the synthesis of many vasoregulators thought to be influenced by arterial pulsation (e.g., nitric oxide) may also be inhibited by hypothermia.

IV. Summary

In summary, neurological injury continues to be a significant source of perioperative morbidity and mortality in cardiac surgical procedures. To reduce the incidence and/or severity of these complications, a better understanding of the cerebral physiology of CPB is needed. Specifically, it is important to determine how the conduct of CPB either contributes to or modifies the response of the brain to neurological insults. Our research indicates that, in the uninjured brain, nonpulsatile perfusion per se does not appear disadvantageous in terms of brain blood flow or oxygen metabolism at 27°C. Conversely, pulsatile perfusion does not appear to confer any special benefits. Whether pulsatility might influence neurological outcome in the presence of an ischemic insult remains to be determined.

References

1. Tuman, K. J., McCarthy, R. J., Najafi, H., and Ivankovich, A. D. (1992). Differential effects of advanced age on neurologic and cardiac risks of coronary artery operations. *J. Thorac. Cardiovasc. Surg.* **104,** 1510–1517.
2. Shaw, P. J., Bates, D., Cartlidge, N. E. F., Heaviside, D., French, J. M., Julian, D. G., and Shaw, D. A. (1986). Neurological complications of coronary artery bypass graft surgery: Six month follow-up study. *Br. Med. J.* **293,** 165–167.
3. Shaw, P. J., Bates, D., Cartlidge, N. E. F., French, J. M., Heaviside, D., Julian, D. G., and Shaw, D. A. (1987). Long-term intellectual dysfunction following coronary artery bypass graft surgery: A six month follow-up study. *Q. J. Med.* **62,** 259–268.
4. Siesjo, B. K. (1992). Pathophysiology and treatment of focal cerebral ischemia. Part I: Pathophysiology. *J. Neurosurg.* **77,** 169–184.
5. Angell James, J. E. (1971). The effects of altering mean pressure, pulse pressure and pulse frequency on the impulse activity in barorecptor fibres from the aortic arch and right subclavian artery in the rabbit. *J. Physiol.* (*London*) **214,** 65–88.
6. Chapleau, M. W., and Abboud, F. M. (1989). Determinants of sensitization of carotid baroreceptors by pulsatile pressure in dogs. *Circ. Res.* **65,** 566–577.
7. Chapleau, M. W., Hajduczok, G., and Abboud, F. M. (1989). Pulsatile activation of baroreceptors causes central facilitation of baroreflex. *Am. J. Physiol.* **256,** H1735–H1741.
8. Minami, K., Vyska, K., and Korfer, R. (1992). Role of the carotid sinus in response of integrated venous system to pulsatile and nonpulsatile perfusion. *J. Thorac. Cardiovasc. Surg.* **104,** 1639–1646.
9. Shepard, R. B., and Kirklin, J. W. (1969). Relation of pulsatile flow to oxygen consumption and other variables during cardiopulmonary bypass. *J. Thorac. Cardiovasc. Surg.* **58,** 694–702.

10. Raj, J. U., Kaapa, P., and Anderson, J. (1992). Effect of pulsatile flow on microvascular resistance in adult rabbit lungs. *J. Appl. Physiol.* **72,** 73–81.
11. Williams, G. D., Seifen, A. B., Lawson, N. W., Norton, J. B., Readinger, R. I., Dungan, T. W., Callaway, J. K., and Campbell, G. S. (1979). Pulsatile perfusion versus conventional high-flow nonpulsatile perfusion for rapid core cooling and rewarming of infants for circulatory arrest in cardiac operation. *J. Thorac. Cardiovasc. Surg.* **78,** 667–677.
12. Watkins, W. D., Peterson, M. B., Kong, D. L., Kono, K., Buckley, M. J., Levine, F. H., and Philbin, D. M. (1982). Thromboxane and prostacyclin changes during cardipulmonary bypass with and without pulsatile flow. *J. Thorac. Cardiovasc. Surg.* **84,** 250–256.
13. Hutcheson, I. R., and Griffith, T. M. (1991). Release of endothelium-derived relaxing factor is modulated both by frequency and amplitude of pulsatile flow. *Am. J. Physiol.* **261,** H257–H262.
14. Matsumoto, T., Wolferth, C. C., and Perlman, M. H. (1971). Effects of pulsatile and nonpulsatile perfusion upon cerebral and conjunctival microcirculation in dogs. *Am. Surg.* **37,** 61–64.
15. Sanderson, J. M., Wright, G., and Sims, F. W. (1972). Brain damage in dogs immediately following pulsatile and non-pulsatile blood flow in extracorporeal circulation. *Thorax* **27,** 275–286.
16. Mori, A., Sono, J., Nakashima, M., Minami, K., and Okada, Y. (1980). Application of pulsatile cardiopulmonary bypass for profound hypothermia in cardiac surgery. *Jpn. Circ. J.* **45,** 315–322.
17. Dernevik, L., Arvidsson, S., and William-Olsson, G. (1985). Cerebral perfusion in dogs during pulsatile and nonpulsatile extracorporeal circulation. *J. Cardiovasc. Surg.* **26,** 32–35.
18. Tranmer, B. I., Gross, C. E., Kindt, G. W., and Adey, G. R. (1986). Pulsatile versus nonpulsatile blood flow in the treatment of acute cerebral ischemia. *Neurosurgery* **19,** 724–731.
19. Blauth, C. I., Smith, P. L., Arnold, J. V., Jagoe, J. R., Wootton, R., and Taylor, K. M. (1990). Influence of oxygenator type on the prevalence and extent of microembolic retinal ischemia during cardiopulmonary bypass. Assessment by digital image analysis. *J. Thorac. Cardiovasc. Surg.* **99,** 61–69.
20. Kuroda, Y., Uchimoto, R., Kaieda, R., Shinkura, R., Shinohara, K., Miyamoto, S., Oshita, S., and Takeshita, H. (1993). Central nervous system complications after cardiac surgery: A comparison between coronary artery bypass grafting and valve surgery. *Anesth. Analg. (N. Y.)* **76,** 222–227.
21. Shaw, P. J., Bates, D., Cartlidge, N. E. F., French, J. M., Heaviside, D. Julian, D. G., and Shaw, D. A. (1989). An analysis of factors predisposing to neurological injury in patients undergoing coronary bypass operations. *Q. J. Med.* **72,** 633–646.
22. Henze, T., Stephan, H., and Sonntag, H. (1990). Cerebral dysfunction following extracorporal circulation for aortocoronary bypass surgery: No differences in neuropsychological outcome after pulsatile versus nonpulsatile flow. *Thorac. Cardiovasc. Surg.* **38,** 65–68.
23. Laursen, H., Bodker, A., Andersen, K., Waaben, J., and Husum, B. (1986). Brain oedema and blood–brain barrier permeability in pulsatile and nonpulsatile cardiopulmonary bypass. *Scand. J. Thorac. Cardiovasc. Surg.* **20,** 161–166.
24. Grossi, E. A., Connolly, M. W., Krieger, K. H., Nathan, I. M., Hunter, C. E., Colvin, S. B., Baumann, F. G., and Spencer, F. C. (1985). Quantification of pulsatile flow during cardiopulmonary bypass to permit direct comparison of the effectiveness of various types of "pulsatile" and "nonpulsatile" flow. *Surgery* **98,** 547–554.

25. Hindman, B. J., Dexter, F., Cutkomp, J., Smith, T., Todd, M. M., and Tinker, J. H. (1992). Cerebral blood flow and metabolism do not decrease at stable brain temperature during cardiopulmonary bypass in rabbits. *Anesthesiology* **77,** 342–350.
26. Scremin, O. U., Sonnenschein, R. R., and Rubtinstein, E. H. (1982). Cerebrovascular anatomy and blood flow measurements in the rabbit. *J. Cereb. Blood Flow Metab.* **2,** 55–66.
27. McCallum, J. B., Stekiel, T. A., Bosnjak, Z. J., and Kampine, J. P. (1993). Does isoflurane alter mesenteric venous capacitance in the intact rabbit? *Anesth. Analg. (N. Y.)* **76,** 1095–1105.
28. Taneyama, C., Goto, H., Kohno, N., Benson, K. T., Sasao, J., and Arakawa, K. (1993). Effects of fentanyl, diazepam, and the combination of both on arterial baroflex and sympathetic nerve activity in intact and baro-denervated dogs. *Anesth. Analg. (N. Y.)* **77,** 44–48.
29. Heistad, D. D., and Kontos, H. A. (1983). Cerebral circulation. *In* "Handbook of Physiology, Section 2: The Cardiovascular System, Volume III. Peripheral Circulation and Organ Blood Flow, Part 1" (J. T. Shephard and F. M. Abbound, vol. eds.), pp. 137–182. American Physiological Society, Bethesda, Maryland.
30. Ogura, K., Takayasu, M., and Dacey, R. G. (1991). Effects of hypothermia and hyperthermia on the reactivity of rat intracerebral arterioles *in vitro*. *J. Neurosurg.* **75,** 433–439.

Anesthetic Actions on Cardiovascular Control Mechanisms in the Central Nervous System

William T. Schmeling and Neil E. Farber
Departments of Anesthesiology and Pharmacology and Toxicology
The Medical College of Wisconsin
and Zablocki Veterans Affairs Medical Center
Milwaukee, Wisconsin 53226

I. Introduction

The disruption of normal cardiovascular regulation by volatile anesthetics is well documented. Modulation of circulatory hemodynamics may be accomplished via direct anesthetic actions in the periphery, including inotropic and chronotropic actions on the heart (1), a direct vascular smooth muscle dilatation (2,3), changes in sympathetic and parasympathetic tone to the heart and peripheral vasculature (4–13), and alterations in circulating catecholamine levels (5,6,10,11). Alternatively, volatile anesthetics may modulate central neural control mechanisms or produce a degree of ganglionic blockade (9–11,14) to influence cardiovascular homeostasis (5,6,10,11). The extent to which volatile anesthetics alter cardiovascular control systems within the central nervous system (CNS) versus their peripheral actions has been the subject of controversy for over 30 years. Price *et al.* (5) found that halothane decreased medullary stimulation-induced pressor and depressor responses in vagotomized, decerebrate dogs, implying an effect on regulatory mechanisms within the CNS. In contrast, using a variant of the isolated, supported head technique, Wang *et al.* (13) demonstrated that in pentobarbital-anesthetized

Advances in Pharmacology, Volume 31

dogs there were minimal reductions in reflex vascular responses to medullary stimulation when halothane was delivered to the head perfusion but marked changes when halothane was given caudally to the body only. These findings suggested a peripheral rather than central site of action in regulating cardiovascular responses.

Both halothane and isoflurane produce dose-dependent decreases in mean systemic arterial pressure, which are typically described to be the results of differential effects by the anesthetics on cardiac output and total peripheral resistance, although alteration in the heart rate may be more variable (10,11). However, the relative effects of these anesthetics on central neural control mechanisms modulating cardiovascular homeostasis have not been well delineated. Evidence indicates that halothane may produce decreases in systemic arterial pressure predominantly by a depression in cardiac output, whereas isoflurane administration results primarily in a reduction in vascular resistance (10). Additionally, halothane has been shown to depress cardiac performance to a greater extent than isoflurane, perhaps owing to a decrease in afterload by isoflurane (10,15–19). Halothane and isoflurane have been found to attenuate the baroreceptor reflex, with halothane suppressing the reflex to a greater extent than isoflurane (10,20–23). The mechanisms by which the volatile anesthetic agents suppress the baroreceptor reflex have been shown to include both peripheral and central sites of actions (7,10,12,20–23). However, the relative contribution of the CNS to the cardiovascular modulatory effects of the volatile agents remains unclear.

Previous studies have demonstrated a significant action of the volatile anesthetics on alterations in systemic arterial pressure produced by stimulation at multiple cardiovascular control centers in the neuraxis (24–26). However, such pressor responses induced by central neuronal stimulation might be altered by volatile anesthetic action at peripheral sites, including autonomic ganglia, or by disrupting end organ sensitivity (10,11,24,25). The purpose of the first part of these experiments was to investigate the action of halothane on evoked potentials at intermediolateral (ILC) and intermediomedial thoracic spinal sites evoked from previously identified CNS pressor sites. These studies were conducted in order to delineate specific volatile anesthetic actions on cardiovascular control mechanisms totally within the CNS by examining anesthetic action on these potentials at ILC sites within the thoracic spinal cord, which represent the sites of the preganglionic sympathetic fibers to the heart (27–32). These evoked potentials have been hypothesized to represent a monosynaptic short latency projection in combination with a longer latency component representing polysynaptic projections traversing bulbar cardiovascular centers (26–29). Such evoked potentials are modulated by baroreceptor activity and are subject to alteration by pharmacological intervention (26). Potentials were evoked from documented pressor sites in the CNS, as identified above. These sites have been previously identified as having extensive neuronal projections directly and polysynaptically to the ILC, but the effects of the volatile anesthetics on such potentials have not been previously investigated.

The purpose of the second part of the present investigation was to examine the actions of the volatile anesthetic isoflurane on CNS-mediated pressor and depressor responses in dogs. Sites within the ventrolateral hypothalamus, reticular formation, and medial medulla were selected based on previous evidence documenting their extensive involvement in CNS-mediated cardiovascular control mechanisms (24,27–29). The pressor results are compared to similar previous studies from our laboratory, utilizing similar techniques in cats, which were chronically instrumented, midcollicularly transected, or anesthetized with one of several baseline anesthetics. Some of the data presented here have been adapted from previously published investigations (24,33).

II. Studies Performed in Cats

Twenty-four male and female cats (1.9–3.5 kg) were utilized in the experiments. Studies were also conducted in nineteen male and female mongrel dogs, with weights of 11–29 kg. All animals were fasted overnight prior to experimentation. Fluid replacement was accomplished with normal saline and was continued for the duration of each experiment. Studies were generally performed at the same time each day. Ventilatory and anesthetic gas concentrations were continuously monitored by mass spectrometry (Marquette Electronics, Inc., Milwaukee, WI).

A. Preparation for Acute Studies

Seventeen animals were anesthetized by an intraperitoneal injection of α-chloralose (50–60 mg/kg) and urethane (500–600 mg/kg). Seven animals were anesthetized by an intraperitoneal injection of sodium pentobarbital (30–35 mg/kg). Anesthesia was supplemented by intravenous administration throughout the experiment, as required. In all animals, the right femoral artery and left cephalic vein were cannulated for measurement of arterial pressure and for fluid and drug administration, respectively. After the establishment of an anesthetic state, animals were endotracheally intubated and placed on positive pressure ventilation with 100% oxygen using a Narco ventimeter ventilator and a semiclosed circuit. Arterial blood samples were determined for measurements of blood gas tensions (Radiometer ABL2, Coppenhagen, Denmark), and ventilation was adjusted and/or sodium bicarbonate administered in order to maintain arterial gas tensions within normal limits.

Animals were subsequently placed in a stereotaxic and iliac crest hip pin restraint, a midline incision was utilized to expose the skull, and the galea, periosteum and temporalis muscle were reflected laterally. Small burr holes were made through the calverium with a fiber-optic, high-speed, air-turbine drill (Midwest Dental Products, Des Plaines, IL), and Kopf micromanipulators (David Kopf Instruments, Tujunga, CA) were utilized to place stereotaxically a 23-gauge coaxial, Formvar-insulated, stimulating electrode at target sites in the right ventrolateral hypothalamus [Horsely-Clark anterior (A) 10.0 mm, lateral (L) 2.5 mm, depth (D) −4.0 mm], left mesencephalic reticular formation (A 2.0 mm, L 2.0 mm, D −1.0 mm), and the right medulla (posterior 11.0 mm, lateral 4.5 mm, depth 5.0 mm). The sites were determined using the atlases of Bleir (34), Berman (35), and Jasper and Ajmone-Marsan (36). The stimulus for pressor responses was a 10-sec train of 0.05 and 0.1 msec square pulses delivered at 100 Hz through a constant current unit using digital stimulators (WPI Electronics, New Haven, CT). Current intensity was monitored and adjusted continuously during stimulation presentation by means of an optically isolated current monitor. This was utilized as additional assurance that, even if electrode impedance varied during the course of an experiment, current density remained the same as at selected control values.

After electrode placement, a "threshold" current which elicited a small but consistent increase in pressure was established. A random stimulus sequence of threshold and 2 and 4 times threshold current was then accomplished in order to generate a graded series of pressor responses at each CNS site. During pressor site stimulation, systemic pulsatile and mean arterial pressure and heart rate were continuously monitored and recorded on a Grass polygraph. All hemodynamic and stimulation parameters were constantly recorded on a FM tape recorder (Vetter Model 300A), for subsequent analog-to-digital conversion and computer-assisted analysis (see below). Core body temperature was monitored by a rectal thermocouple and maintained during all experiments by application of heating pads and/or heat lamps. Humidity was not controlled.

B. Preparation for Studies of Intermediolateral Cell Column Evoked Potentials

In those cats which exhibited reproducible responses to stimulation of any cardiovascular pressor site, a midline incision over the first to fifth thoracic vertebrae was subsequently accomplished. Muscle and fascia were dissected and the fifth and first vertebral dorsal processes clamped and immobilized. A laminectomy of the second to fourth thoracic vertebrae was accomplished, skin flaps anchored, and the dura incised. A pool of warm mineral oil was maintained over the spinal cord. The CNS stimulating electrodes in the hypothalamic, reticular formation, and medullary cardiovascular pressor sites (see above) were then utilized to deliver single square wave pulses of 0.1–0.3 msec duration, 0.1–1.0 mA, 0.3–0.5 Hz. A hydraulic microdrive or Burleigh Inchworm (Burleigh Instruments, Fishers, NY) stepping motor was utilized to advance carefully tungsten microelectrodes, with a tip diameter of 5–25 μm, into the region of the intermediolateral cell column at T2–T3 until an evoked field potential was recorded. Previous evidence has indicated that the highest ratio of presynaptic neurons to sympathetic white rami lie at this level of the cord (30–32,37). Utilizing Bak amplifiers (Bak Instruments, Rockville, MD) (1000–10,000 times gain, 15–3000 Hz), electrical activity, spontaneous and evoked, in the intermediolateral cell column was displayed on Tektronix oscilloscopes (Tektronics, Pittsfield, MA) and the evoked activity visualized using stimulus-linked delayed sweeps. After establishing baseline potentials evoked by stimulation of each CNS cardiovascular center previously demonstrated to evoke a consistent pressor response (see above), halothane was administered at varying concentrations in a random fashion. After equilibration of end-tidal concentrations of halothane, stimulation of pressor sites was repeated and the evoked potentials recorded and averaged, as above. After recording potentials at each end-tidal concentration, halothane was discontinued, and a 5- to 10-sec train of 0.05 to 01 msec square wave pulses at 100 Hz was delivered to the recording electrode in the intermediolateral cell column to elicit an increase in sympathetic outflow and produce a pressor response.

All data were continuously recorded on tape and/or digitized on-line during each experiment. Data were digitized at 200 Hz for analysis of the hemodynamic parameters. All evoked potentials were digitized at 2 to 64 kHz utilizing a triggered analog-to-digital acquisition routine. Six to 24 separate evoked potentials were averaged. Potentials recorded from the intermediolateral cell column were averaged using computer-assisted averaging techniques and software specifically developed for this purpose. All data were digitized utilizing a Metrabyte (Kiethley Instruments, Taunton, MA) Dash-16 A/D converter interfaced to an IBM 486 clone microcomputer.

III. Studies in Mongrel Dogs

Nineteen fasting mongrel dogs were anesthetized with intravenous α-chloralose (50–60 mg/kg) and urethane (500–600 mg/kg), endotracheally intubated, and positive pressure ventilated as above. Fluid replacement with normal saline was accomplished, and fluid maintenance was continued at 3 ml/kg/hr for the duration of each experiment. Anesthesia was supplemented during each experiment as required. The femoral artery and vein were cannulated, and in several animals a contralateral femoral arterial flow probe (Transonic Instruments, Ithaca, NY) was placed. Animals were placed in a stereotaxic restraint, and, as above, 23-gauge coaxial stimulating electrodes were placed in the right ventrolateral hypothalamus (A 20.0 mm, L 3.0 mm, *D* 5.0 mm), left reticular formation in the vicinity of the nucleus gigantocellularis and nucleus reticularis pontis caudalis [posterior (P) 4.0 mm, L 4.0 mm, D − 11.0 mm], and right medial medulla (P 12.0 mm, L 5.0 mm, D − 5.0 mm) according to the atlas of Lim *et al.* (38). Stimuli (0.1 msec duration, 10-sec train, 100 Hz) were delivered to each electrode in order to elicit a threshold and 2 and 4 times threshold response of pressor or depressor alterations in systemic arterial pressure via a constant current unit. The variation in cranial anatomy between dogs is greater than that of cats, and thus placement of electrodes in specific CNS sites is somewhat more difficult. Several of the animals exhibited depressor responses associated with electrical stimulation of CNS sites. If depressor responses were elicited, the electrode placement was not changed, but rather the effects of halothane on the depressor response was evaluated.

After control responses were obtained, varying concentrations of isoflurane [1.5, 2.5, and 3.0% end-tidal] were adminstered in oxygen. Between 30 and 60 min was allowed after each concentration change for equilibration, and the stimulation sequence was repeated. Arterial blood samples were obtained for determination of arterial gas tensions, and ventilation was adjusted and/or sodium bicarbonate was administered to maintain normal gas tensions throughout each experiment.

At the conclusion of all experiments, all animals were sacrificed with a lethal dose of pentobarbital, and stimulation sites in the hypothalamus, reticular formation, medulla and intermediolateral cell column (ILC) were marked by passing a 5-mA DC current between the tip of each electrode and neutral ground for approximately 15 sec. The brain and/or spinal cord was perfused *in situ* with saline followed by 10% (v/v) formalin–saline solution and was removed from the skull. The brain was then placed in 10% formalin–saline solution containing 0.5% sodium ferrocyanide to complete fixation and to develop Prussian Blue marks at the electrode tip sites. After 48 hr of fixation, the brains were blocked, frozen, sectioned for histological examination and electrode sites photographed and documented utilizing techniques previously described (24,26,33,39).

Pressor or depressor responses were calculated as the change in systolic arterial pressure from baseline measurements taken immediately prior to stimulation of each CNS site. Pressor or depressor responses before and during anesthesia were analyzed with a repeated measures analysis of variance. Pairwise comparisons of interventions were performed with contrasts derived from the repeated measures analysis using the Bonferroni modification of the *t*-test. All statistical analyses were performed using SAS system software on a VAX mainframe computer. All changes were considered significant when the probability (p) value was less than 0.05. All data are expressed as the mean ± standard error of the mean (SEM).

IV. Central Nervous System Pressor Site Responses in Cats and Intermediolateral Cell Column Evoked Potential

Previous investigations utilizing techniques similiar to those used in the present study have specifically identified the actions of the volatile anesthetics halothane (33) and isoflurane (24) on a 10-sec train stimulation (0.1 msec, 100 Hz) of pressor sites in the hypothalamus, mesencephalic reticular formation, and medulla. These studies demonstrated that both anesthetics significantly, dose-dependently inhibited the evoked arterial pressor responses in chronically instrumented cats (Figs. 1–5). Additionally, the presence (pentobarbital or chloralose–urethane) or absence (midcollicular transection) of a baseline anesthetic significantly altered the disruptive effects of halothane on the arterial pressor responses. Specifically, cardiovascular control centers in the medulla were significantly more resistant to halothane inhibition than those of hypothalamic or mesencephalic reticular formation sites. Additionally, baseline anesthesia with chloralose–urethane resulted in significantly greater preservation of pressor responses at all sites, as compared to pentobarbital. Finally, midcollicularly transected animals exhibited a greater susceptibility to halothane-induced suppression of pressor responses at medullary levels than those animals which received a baseline anesthetic.

Single-pulse stimulation of the same CNS vasomotor sites in the present experiments resulted in consistent evoked potentials in the intermediolateral cell column (Figs. 6–8). Typically, these potentials were composed of a field component in association with discrete unit spiking. The administration of graded concentrations of halothane markedly attenuated the peak-to-peak amplitudes of evoked potentials which occurred following electrical stimulation of CNS sites.

Typical results obtained in two experiments are illustrated in Figs. 6 and 7. Figure 6 illustrates the effect of 2.0% halothane on the individual

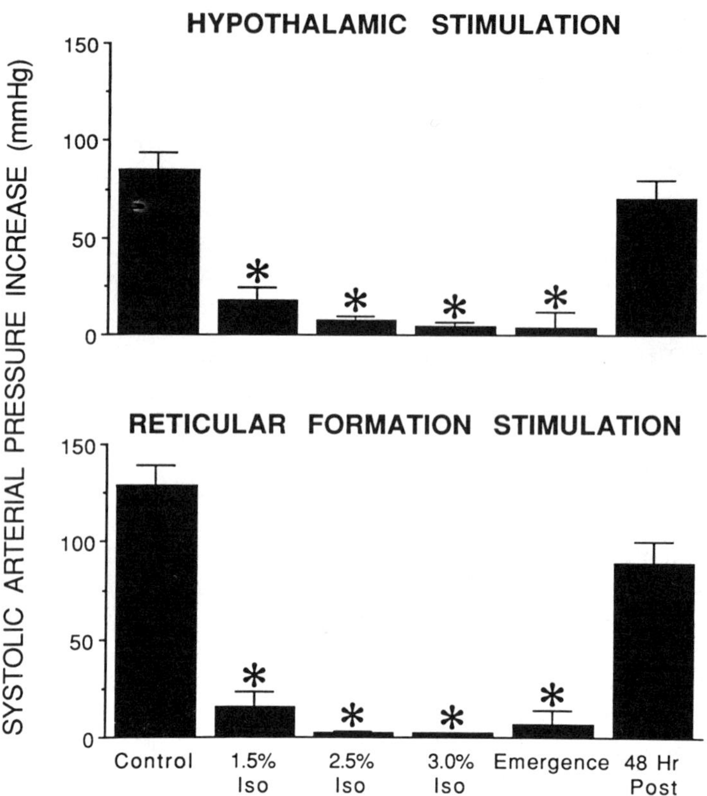

Fig. 1 Effect of graded levels of isoflurane (1.5, 2.5, and 3.0% Iso) on systolic arterial pressure pressor responses to 4 times threshold stimulation of sites in the hypothalamus and reticular formation in chronically instrumented cats. Data are means ± SEM. *Significantly different from conscious control pressor responses. [From Poterack *et al.* (24), with permission.]

and averaged evoked potentials obtained from stimulation of pressor sites within the reticular formation and medulla of a pentobarbital-anesthetized animal. Figure 7 shows simliar ILC evoked potentials recorded after stimulation of hypothalamic and medullary areas in an animal anesthetized with chloralose–urethane.

In all experiments there was a dose-dependent attenuation of the peak-to-peak amplitudes of the evoked potentials produced by varying concentrations of halothane. However, consistent with previous data regarding the differential effects of halothane on systemic pressor responses produced by stimulation at these sites, the potentials evoked from medullary levels appeared to be more resistant. Additionally, potentials were less

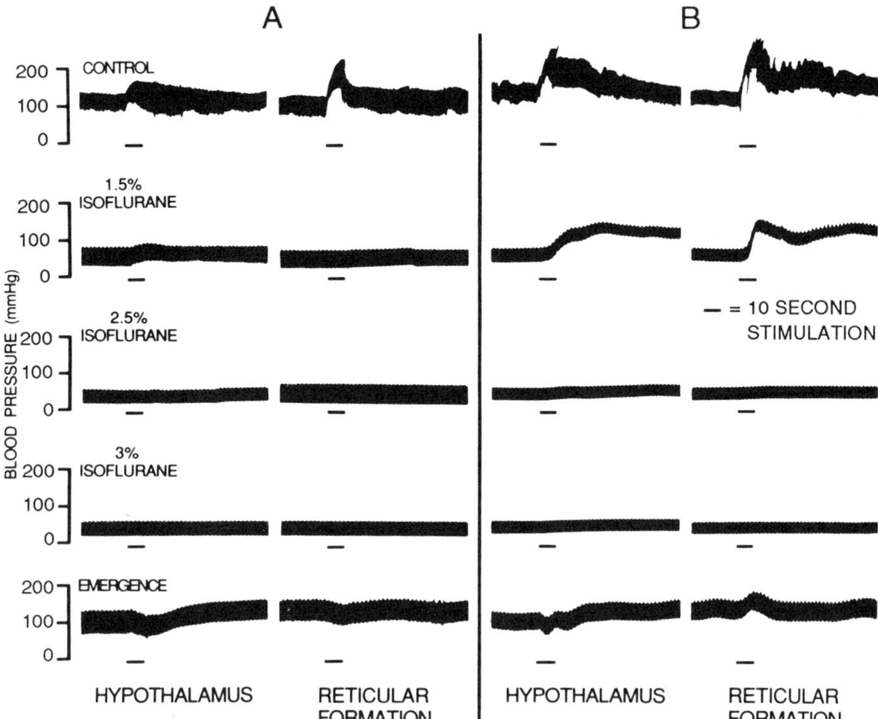

Fig. 2 Effect of graded concentrations of isoflurane on systemic arterial pressure responses to stimulation of hypothalamic and reticular formation sites in two different chronically instrumented cats. Isoflurane significantly attenuated the pressor respones at high concentrations. Note also the appearance of a depressor response to hypothalamic stimulation (cats A and B) and reticular formation site stimulation (cat A) during emergence. [From Poterack *et al.* (24), with permission.]

disrupted by halothane anesthesia in chloralose–urethane-anesthetized animals than in animals anesthetized with pentobarbital. Figure 8 represents both the evoked pressor responses and associated ILC-evoked potentials from hypothalamic and medullary sites in a chloralose–urethane-anesthetized animal.

V. Anesthetic Responses in Dogs

The administration of isoflurane to dogs resulted in graded decreases in systemic hemodynamic parameters, including systolic (167 ± 6 to 73 ± 5 mmHg at 3.0% isoflurane) and diastolic (103 ± 5 to 40 ± 3 mmHg at

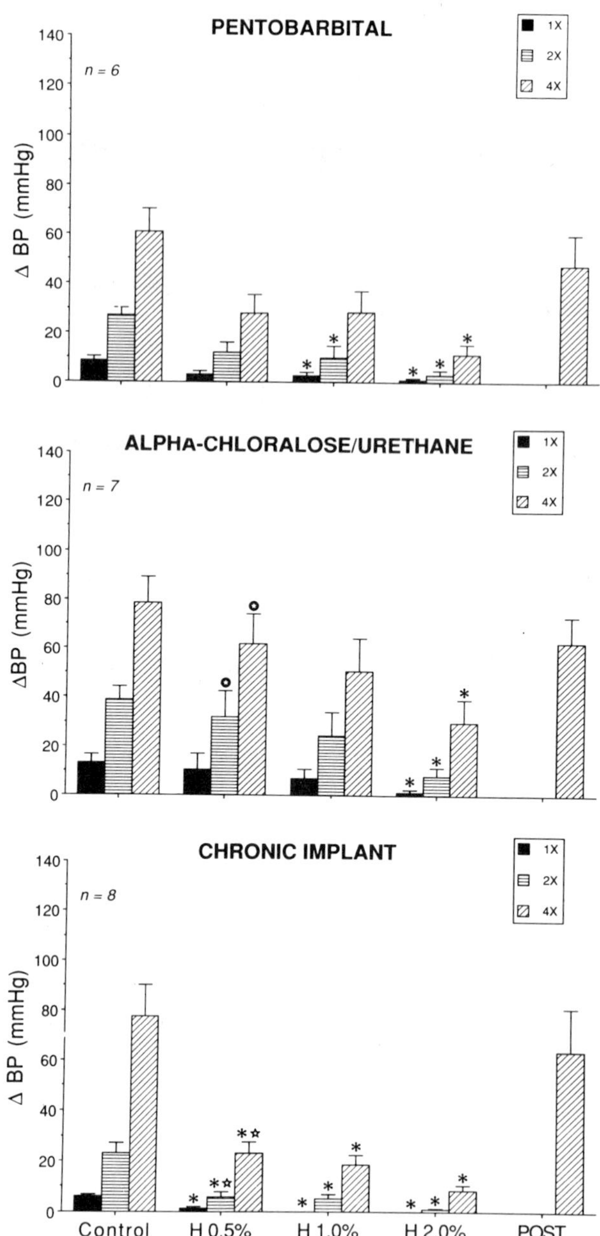

Fig. 3 Effect of graded doses of halothane (H) on systolic arterial pressure increases to stimulation (1×, threshold; 2× and 4×, 2 or 4 times threshold current) of hypothalamic pressor sites in cats chronically implanted or anesthetized with chloralose–urethane or pentobarbital. All data are means ± SEM. *Significantly different from control response. ☆Significantly different from chloralose–urethane animals. ✪Significantly different from chronically implanted animals.

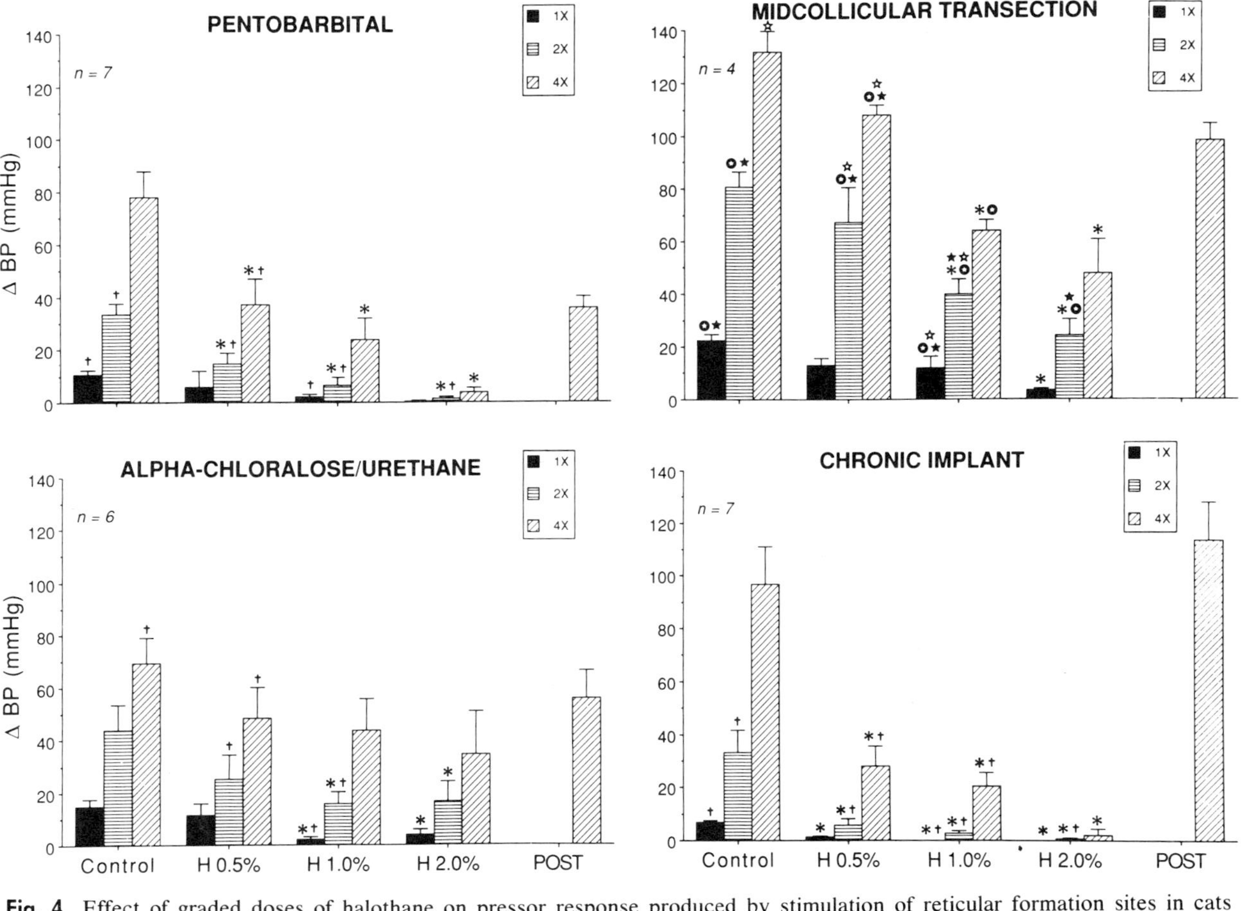

Fig. 4 Effect of graded doses of halothane on pressor response produced by stimulation of reticular formation sites in cats chronically implanted, anesthetized with either chloralose–urethane or pentobarbital, or midcollicularly transected. †Significantly different from midcollicularly transected animals. ★Significantly different from pentobarbital-anesthetized animals. All other symbols are the same as in Fig. 3.

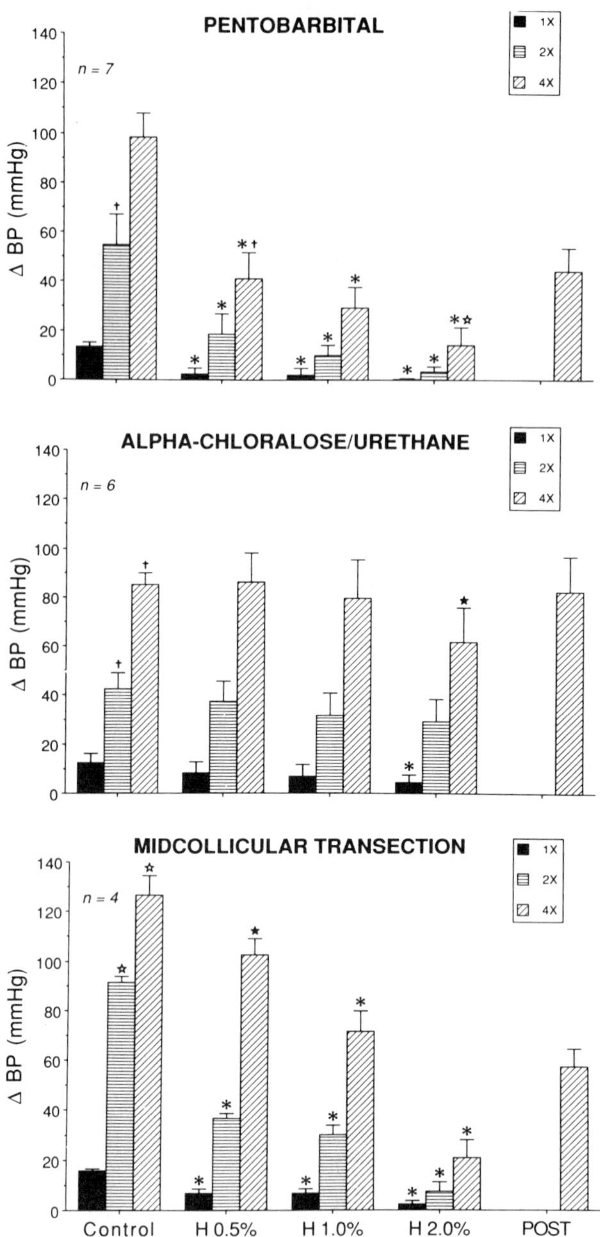

Fig. 5 Effect of graded doses of halothane on pressor responses produced by stimulation of medullary sites in animals anesthetized with chloralose–urethane or pentobarbital or transected at midcollicular levels. All symbols are the same as in Figs. 3 and 4.

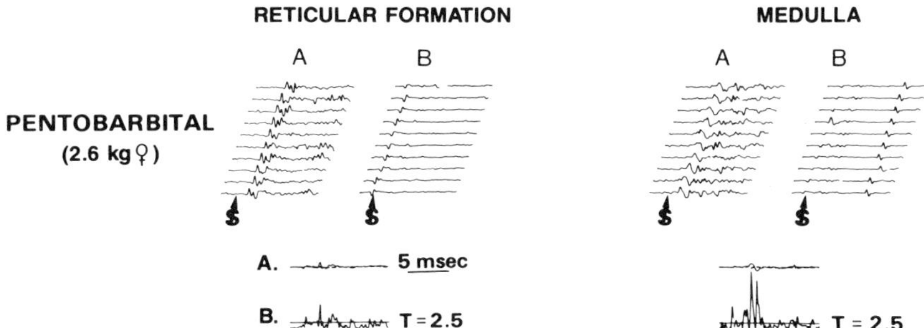

Fig. 6 Individual and averaged evoked potentials recorded in the intermediolateral cell (ILC) column to single-pulse stimulation (0.2 msec, 0.3 Hz, 1.0mA) of previously identified pressor loci in the reticular formation and medulla of a pentobarbital-anesthetized cat. Ten individual potentials, before (A) and after 2% halothane (B), are illustrated at top. Shown below are the averaged evoked potentials (n = 20) of the individual potentials. The solid line in each case represents the control averaged evoked potential prior to 2% halothane (dashed line). Below the averaged potentials is a point-by-point analysis (B) for significance (t = 2.5) of both the control and posthalothane averaged evoked potentials. Halothane significantly attenuated the amplitude of the evoked potential with a slight increase in latency. Medullary sites appeared more resistant than sites within the ventrolateral hypothalamus.

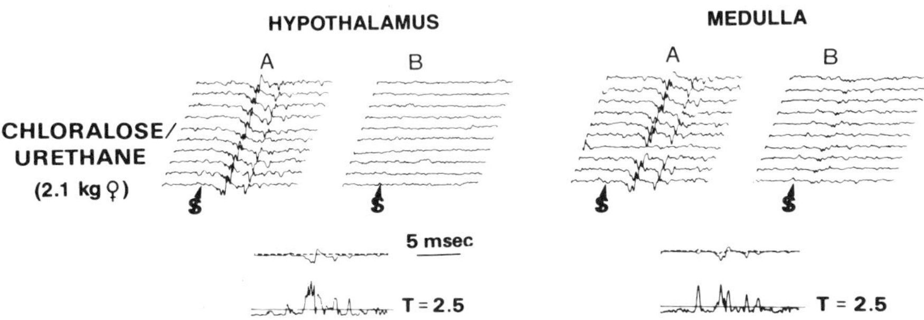

Fig. 7 Individual and averaged evoked potentials recorded in the intermediolateral cell to single-pulse stimulation (0.3 msec, 0.5 Hz, 1.0 mA) of previously identified pressor loci in the hypothalamus and medulla of a cat anesthetized with chloralose–urethane. Individual evoked potentials are illustrated at top, before (A) and after 2.0% halothane (B). The average of 20 individual potentials (bottom), before (solid line) and after 2% halothane (dashed line), are also illustrated. Halothane resulted in an attenuation of the evoked potentials recorded in the ILC after either hypothalamic or medullary stimulation. In a similar fashion to cats anesthetized with pentobarbital (Fig. 6), halothane reduced the amplitude of potentials evoked from both reticular and hypothalamic levels significantly more than at medullary sites. However, the overall suppression of amplitude and increase in latency were less than in pentobarbital–anesthetized animals.

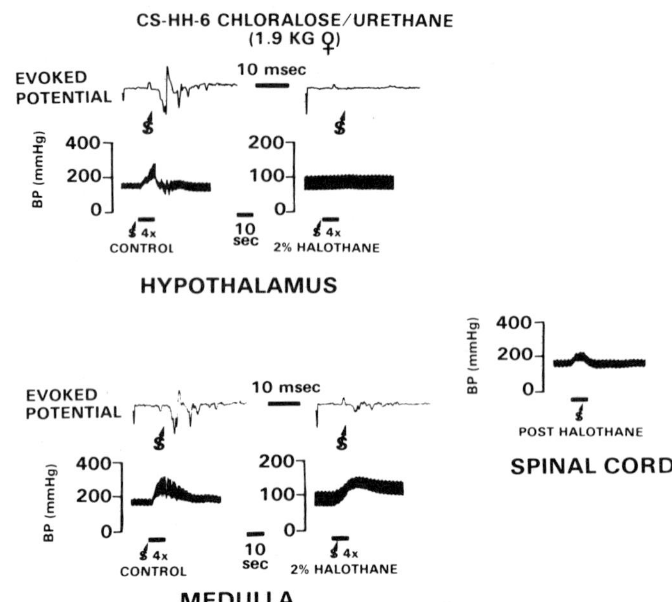

Fig. 8 Average evoked potential recorded at the ILC produced by single-pulse stimulation (0.3 msec, 0.3 Hz, 1.0 mA) of hypothalamic and medullary loci. Below each averaged evoked potential, before and after 2% halothane, is shown the associated systemic arterial pressor response to train stimulation (100 Hz, 10 sec, 1.0 mA) of each site. Note the difference in blood pressure scales and the attenuation of pressor responses and the associated evoked potentials. Note also that the evoked potential from medullary sites as well as the associated pressor responses appeared more resistant to the effects of halothane. The effect of a 10-sec (100 Hz, 200 mA) train stimulation of the ILC recording electrode and the elicited pressor response are also shown.

Table I

Effect of Isoflurane on Systemic Hemodynamic Parameters[a]

Parameter	Control	Isoflurane 1.5%	Isoflurane 2.5%	Isoflurane 3.0%	Post isoflurane
HR (beats/min)	185 ± 8	167 ± 8	144 ± 8[b]	135 ± 9[b]	161 ± 9[b]
SBP (mmHg)	168 ± 6	136 ± 5[b]	106 ± 6[b]	73 ± 5[b]	153 ± 5
DBP (mmHg)	103 ± 5	82 ± 4[b]	63 ± 3[b]	40 ± 4[b]	92 ± 4
MBP (mmHg)	125 ± 5	100 ± 4[b]	77 ± 4[b]	51 ± 4[b]	113 ± 4
HR × SBP (bpm × mmHg)	309 ± 7	228 ± 5[b]	154 ± 14[b]	103 ± 12[b]	247 ± 16[b]
RU	1.37 ± 0.15	1.02 ± 0.21	0.97 ± 0.13	1.00 ± 0.19	1.54 ± 0.31

[a] All data are means ± SEM. HR, Heart rate; SBP, systolic arterial blood pressure; DBP, diastolic arterial blood pressure; MAP, mean arterial blood pressure; RU, resistance units.

[b] Significantly different from control ($p < 0.05$).

3.0% isoflurane) arterial pressures. Isoflurane also decreased heart rate (185 ± 9 to 135 ± 9 beats/min at 3.0% isoflurane) (Table I). These changes occurred similarly in all dogs which exhibited pressor or depressor responses to CNS stimulation. In addition, as expected, isoflurane administration resulted in concentration-dependent decreases in hind limb vascular resistance (1.37 ± 0.15 to 1.00 ± 0.19 resistance units at 3.0% isoflurane) (Table I).

Stimulation of hypothalamic, reticular formation, and classic medullary vasoactive sites resulted in either graded pressor or depressor responses (Figs. 9–12). Isoflurane at the lowest concentration (1.5%) resulted in a significant inhibition of the pressor responses evoked at each CNS site. Of particular interest was the finding that low concentrations (1.5%) of isoflurane resulted in a frequent conversion of the previous pressor responses to depressor responses. These findings are consistent with previous results (24) (Fig. 2) from our laboratory in chronically instrumented cats utilizing similar stimulation sites and parameters. However, in contrast, higher concentrations of isoflurane were required to significantly attenuate depressor responses evoked at each CNS site. Stimulation of

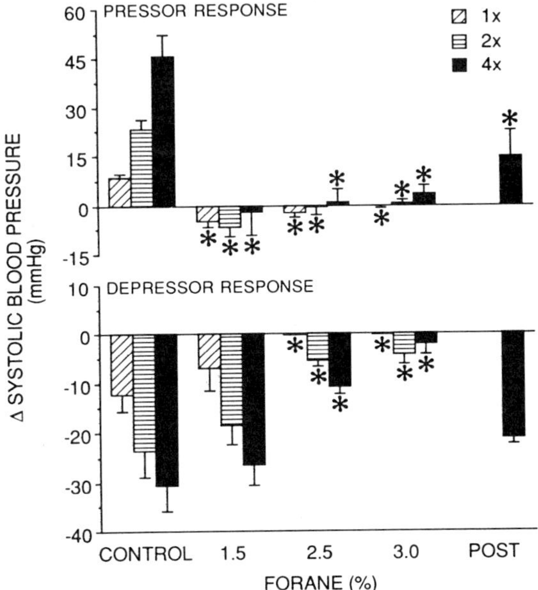

Fig. 9 Effect of isoflurane (1.5, 2.5, and 3.0%) on graded pressor and depressor responses to stimulation (1×, threshold; 2× and 4×, 2 and 4 times threshold) of sites within the ventrolateral hypothalamus of dogs. *Significantly different from control.

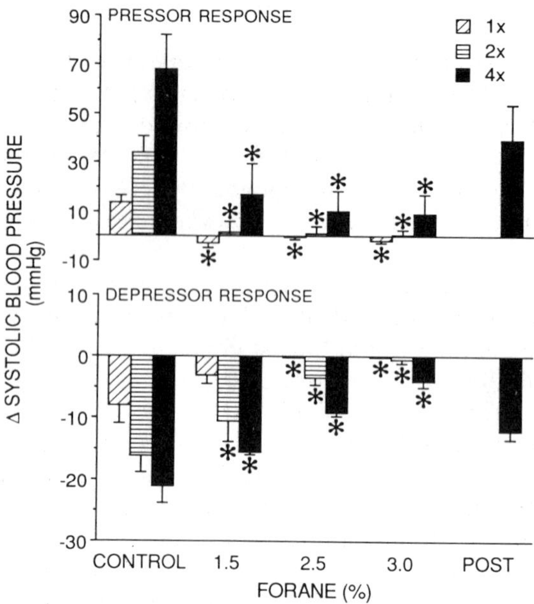

Fig. 10 Effect of isoflurane (1.5, 2.5, and 3.0%) on graded pressor and depressor responses to stimulation (1×, threshold; 2× and 4×, 2 and 4 times threshold) of sites within the reticular formation of dogs. *Significantly different from control.

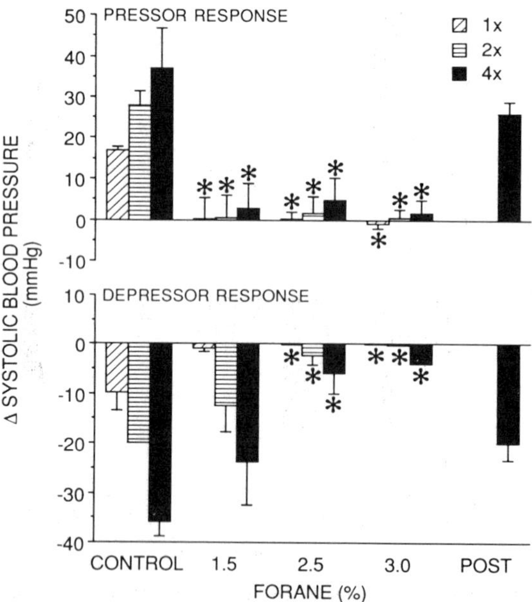

Fig. 11 Effect of isoflurane (1.5, 2.5, and 3.0%) on graded pressor and depressor responses to stimulation (1×, threshold; 2× and 4×, 2 and 4 times threshold) of sites within the medulla of dogs. *Significantly different from control.

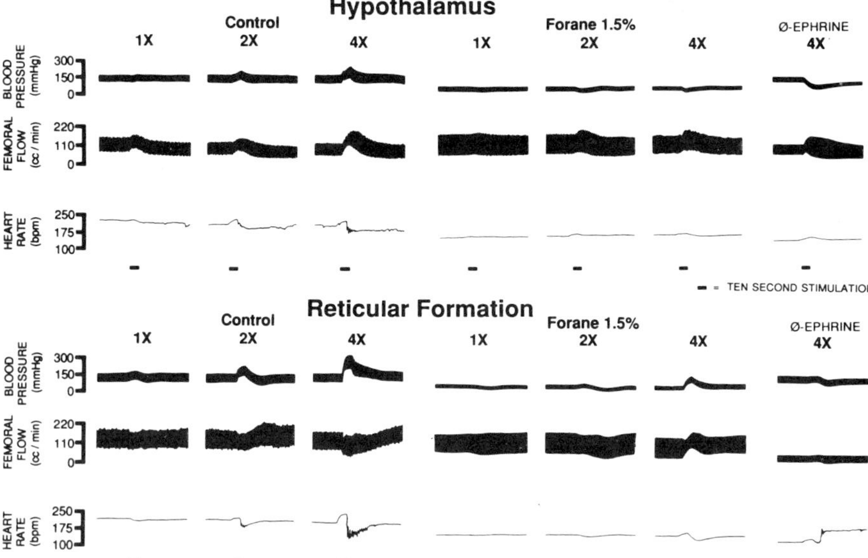

Fig. 12 Typical experiment illustrating the effect of 1.5% isoflurane on threshold (1×) and 2 (2×) and 4 (4×) times threshold stimulation of pressor sites in the hypothalamus and reticular formation. Note the paradoxical conversion of the pressor response to a depressor response associated with a significant blunting of heart rate change elicited by the stimulation. Also illustrated is the restoration of systemic arterial pressure by phenylephrine infusion during isoflurane administration. Stimulation during this restoration of systemic arterial pressure was associated with an increase in the paradoxical depressor response at hypothalamic sites and a conversion of a small residual pressor response at reticular formation levels to a depressor response.

the ventrolateral hypothalamic and reticular formation sites was associated with graded increases in heart rate which were also attenuated by isoflurane in a concentration-dependent manner. Medullary stimulation resulted in large variations in heart rate, and thus no obvious effects by isoflurane were observed.

Stimulation of cardiovascular reticular formation and medullary sites in animals in which pressor responses were elicited evoked associated increases in calculated hind limb peripheral resistance. The administration of isoflurane significantly attenuated this effect (Fig. 12). Significantly, restoration of systemic arterial pressure with phenylephrine in selected experiments did not change the pressor to depressor conversion produced by isoflurane, but rather it appeared to augment the depressor response produced by stimulation of the CNS cardiovascular sites (Fig. 12).

VI. Histological Documentation of Electrode Sites

Subsequent histology demonstrated that hypothalamic, reticular formation, and medullary sites were all located within 1.0–1.5 mm of the target coordinates in all three axes. Additionally, all electrodes placed within the ILC were associated with consistent evoked field potentials, whereas those located outside had been associated with inconsistent or absent potentials. Histological placement of all sites was similar to previous studies investigating anesthetic action at these levels of the neuraxis (24,26,33).

VII. Discussion

The purpose of the studies presented in this article was to delineate the effects of halothane and isoflurane on vasomotor and preganglionic sympathetic neuronal responses evoked by electrical stimulation of cardiovascular control sites within the CNS. Previous results (24,33) have suggested that graded concentrations of anesthetics may differentially attenuate CNS-induced pressor responses following activation of several cardiovascular control sites, including diencephalic sites in the ventrolateral hypothalamus, the mesencephalic reticular formation, and medial medulla. These studies also suggested that the ability of halothane to modulate cardiovascular homeostasis appears to be, in part, dependent on an intact suprabulbar system (33). In these studies, midcollicular transection resulted in a greater resistance to the effects of halothane at the reticular formation and an enhanced sensitivity at the medullary level, which was otherwise markedly resistant to the effects of halothane. In addition, work from our laboratory has shown that the inhibitory actions of halothane on CNS pressor responses are also dependent on the use and type of a baseline anesthetic agent, in that pressor responses evoked in cats anesthetized with pentobarbital were more susceptible to the effects of halothane than were those in cats anesthetized with chloralose–urethane (33).

It is well known that general anesthetics interfere with autonomic responses and central mechanisms regulating the cardiovascular system (10–13,20–24,39). The administration of urethane or Nembutal has been shown to blunt or even reverse the pressor effects resulting from norepinephrine injections into the nucleus tractus solitarius (40). Chloralose and the combination of chloralose and urethane generally remain the most popular baseline anesthetics used in cardiovascular or neuroscience laboratory investigations, whereas pentobarbital has been regarded as having marked depressive effects on the CNS and CNS-mediated responses

(41–46). In contrast, we have previously demonstrated that the pressor responses in cats anesthetized with chloralose–urethane or pentobarbital were similar to those elicited in conscious, chronically instrumented cats (33). However, the stimulation current intensities required to elicit a threshold response tended to be greater in pentobarbital-anesthetized cats. Although the administration of halothane attenuated the evoked pressor responses in anesthetized cats similar to the effects in chronically instrumented animals, cats anesthetized with pentobarbital were significantly more susceptible to the effects of halothane as compared to those anesthetized with chloralose–urethane. In other words, after administration of chloralose–urethane, the attenuation of CNS-mediated pressor responses by halothane was less pronounced. Chloralose has been postulated to preserve certain forms of reflex behavior better than other anesthetics, and it has been reported to produce the least amount of depression of the baroreceptor reflex (42,45,47). Additionally, urethane may result in profound anesthesia with minimal effects on circulatory dynamics (48). Therefore, animals under barbiturate anesthesia might be expected to be more susceptible to halothane-induced disruption of cardiovascular homeostasis mediated within the central nervous system.

The role of various hypothalamic, reticular formation, and medullary areas in cardiovascular function is well documented (27,40,49–62). Previous evidence has revealed extensive afferent and efferent cardiopulmonary connections between the medial reticular formation and more rostral and caudal cardiovascular centers (28). Hypothalamic and reticular formation modulation of sympathetic outflow may partially utilize polysynaptic paths traversing bulbar cardiovascular sites (26–29,39,62). Peiss and Manning (43) have alluded to the role of the reticular formation as a "central internuncial system." It is well documented that the profuse arborization and multisynaptic potentials of reticular neurons provide a substantial substrate where anesthetic actions may occur (12,41,61,63,64). However, few previous studies have been directed toward the action of anesthetic agents on reticular modulation of cardiovascular function. Barbiturates have been previously demonstrated to attenuate differentially pressor responses, with caudal sites more resistant than rostral cardiovascular areas (43). Such differential effects have not been thoroughly examined for other anesthetics. The present results, along with those of previous investigations, suggest important differences between halothane and isoflurane in terms of the degree and sites of disruption on cardiovascular control mechanisms within the central nervous system.

The present investigations extend previous studies in cats (24) across species to dogs. Isoflurane altered central neuronal pools involved in vasomotor control in a selective fashion, directly supporting the previous evidence in cats. We have provided evidence suggesting that those neuronal elements contributing to the maintenance of neurogenic tone are more susceptible to the inhibitory effects of isoflurane than those mediating active vasodilatation, as vasopressor responses were more attenuated by the anesthetic than were vasodepressor responses. In fact, administration of isoflurane, in contrast to halothane, frequently resulted in conversion of pressor responses to depressor responses. Vasomotor area stimulation-induced changes in heart rate were also significantly affected by the administration of isoflurane.

Residual central neurogenic elements are still capable of producing active cardiovascular responses even at concentrations equal to 1.5 MAC isoflurane, concentrations which abolish cardiovascular responses to somatic stimuli (W. T. Schmeling and N. E. Farber, unpublished observations, 1994). These findings would suggest residual CNS modulation of cardiovascular homeostasis during inhalational anesthesia even at concentrations previously believed to block adrenergic responses (MACbar) (49). The hypotension produced by isoflurane administration, presumably by a combination of direct vasodilatation, attenuated sympathetic tone, and negative chronotropic and inotropic actions, resulted in diminished changes in hind limb blood flow associated with CNS stimulation. This patterning may contribute to the unique cardiovascular responses observed clinically with isoflurane.

In addition, we have shown that the stimulation of specific pressor sites within the CNS results in a consistent, reproducible evoked potential in the intermediolateral cell column, the sites of the sympathetic preganglionic cell bodies. Moreover, halothane administration markedly attenuates these evoked potentials elicited by vasomotor site stimulation, with the medulla being most resistant to the effects of halothane as compared to the reticular formation and hypothalamus. The present investigation is in agreement with our previous findings regarding antagonism of halothane effects by chloralose–urethane as compared to pentobarbital. We have shown that the effects of halothane on ILC evoked responses elicited by stimulation of CNS vasomotor areas were more augmented by pretreatment with pentobarbital and not chloralose–urethane, supporting a differential anesthetic action on CNS cardiovascular control mechanisms.

Central modulation and regulation of the circulatory system represent an area of intensive research. Several vasomotor regulatory sites have been described including sites within the hypothalamus, reticular formation, and medulla. These regions, and the major primary afferent terminations of the baroreceptors, including the nucleus tractus solitarius, have extensive interconnections regulating cardiovascular homeostasis. Early studies (65) suggested that the firing of sympathetic neurons was directly proportional to the intensity and frequency (over an approximately 30 to 120 Hz range) of stimulation in hypothalamic cardiovascular centers. Other early work suggested that a decrease in the stimulation frequency in the hypothalamus might result in a reversal of a previous pressor response to a depressor response. Additional data (65,66) suggested that buffer nerves at medullary levels might modulate the evoked alteration in sympathetic neurogenic tone to the vasculature. Evidence exists that several descending pathways modulate hypothalamic control of the cardiovascular system, including projections medial to the dorsal motor nucleus of the vagus, the medial reticular formation, and the nucleus ambiguus and ventrolateral to the mesencephalic tegmentum and pons, with both direct and indirect connections to the spinal cord. Multiple previous studies utilizing degeneration and electrolytic lesioning techniques and horseradish peroxidase have supported such direct projection to the preganglionic sympathetic neurons in the intermediomedial and ILC cell colums (26–29,39,50,62,67,68). Descending pathways from CNS cardiovascular centers include a lateral path, which traverses bulbar and spinal areas and terminates in the intermediolateral nuclei of the spinal cord, and a medial neuronal path, which terminates both in brain stem areas (such as the nucleus tractus solitarius) and in the spinal cord. Both pressor and depressor modulators of the cardiovascular system have been found to exist within the CNS, and a variety of neuronal pathways mediating these responses have been described.

Many studies have shown extensive modulation of bulbar cardiovascular activity from higher CNS sites. Evidence has also indicated that the hypothalamus, reticular formation, and major primary afferent terminations of the baroreceptors (including the nucleus tractus solitarius) have extensive interconnections modulating cardiovascular homeostasis.

It has been postulated that vasopressor outflow from the rostral CNS is organized into at least two pathways, one of which is inhibited by baroreceptor stimulation and the other which is largely independent from baroreceptor input (62,68,69). Previous evidence has shown that the benzodiazepines influence evoked supramedullary cardiovascular responses but are not hypotensive and do not affect medullary pressor responses or alter responses to bilateral carotid occlusion (39). In contrast, agents like clonidine and Δ^9-tetrahydocannabinol have been postulated to exert their hypotensive actions, at least partially, through brain stem sites of action (26,39,70). It may be postulated that agents like the volatile anesthetics which appear to have effects similar to both of the above types of agents have major sites of action at multiple CNS levels.

The disruption of normal cardiovascular function by all the volatile anesthetics is well known and well appreciated clinically. The mechanisms by which these agents modulate circulatory alterations have been attributed to direct effects on vasculature, myocardium, sympathetic and parasympathetic branches of the autonomic nervous system, and the CNS. The mechanism(s) by which volatile anesthetics may disrupt CNS modulation of the cardiovascular system is as yet unknown. However, the attenuation of CNS stimulation-induced pressor responses by halothane and isoflurane may involve an action at one or several locations along the sympathoexcitatory pathway, a disruption of the central integration of afferent neuronal information, or the generation of appropriate responses. Alternatively, these agents may affect CNS pressor responses by altering the responsiveness of pre- and or postganglionic neurons. Although the differential disruption of CNS cardiovascular control has been described for the volatile agents, α-chloralose, urethane, and barbiturates, the relative contributions of peripheral versus central mechanisms are still somewhat unclear. Conflicting reports exist in the literature regarding CNS versus peripheral actions of the volatile anesthetics in modulating circulatory function (10,11).

Work from our laboratory and others has previously demonstrated that inhalational anesthetics alter single-unit neuronal firing in CNS sites subserving cardiovascular and thermoregulatory functions (24,71). In the present study the finding that halothane blunts evoked potentials in the intermediolateral cell column which occur in response to CNS stimulation provides important evidence that the attenuation of CNS-mediated pressor responses by halothane is mediated, at least in part, via disruptions of CNS processing mechanisms or signal generation and supports evidence of a differential alteration at various CNS levels.

In summary, marked pressor responses were elicited from three specific CNS sites in the hypothalamus, reticular formation, and medulla. Graded levels of halothane or isoflurane significantly attenuated these hemodynamic responses in several different animal models and species. The degree of disruption by the volatile anesthetics is dependent on several variables including the presence or absence, and type, of baseline anesthetic used; the presence of an intact suprabulbar system; the specific CNS vasomotor area stimulated; and, indeed, the volatile anesthetic agent utilized. There appear to be differential effects of halothane and isoflurane in the disruption of CNS cardiovascular control mechanisms. These differential effects may, in part, contribute to dissimilar circulatory effects, which are observed clinically with these two agents.

These results along with those previously reported from our laboratory on halothane and isoflurane effects on CNS-mediated cardiovascular control in chronic and acute cats and acute dogs allows us to postulate that halothane is acting in the CNS to modulate pressor responses. More specifically, based on the alterations in responses in midcollicularly transected animals, we would postulate that these anesthetic agents may be disrupting CNS-induced pressor responses predominantly at the pontomedullary (bulbar) level. Potential sites of action might include the nucleus tractus solitarius, tractus solitarius, medullary tegmental fields, or secondary projections to caudal and ventrolateral medulla (A1 and C1 areas, respectively) which are intimately involved in baroreceptor modulation of the cardiovascular system.

Acknowledgments

Supported in part by U.S. Public Health Service Grants HL 361444, National Institutes of Health Anesthesiology Research Training Grant GM 36144, NIH Anesthesiology Research Training Grant GM 08377, and Medical Research Funds from the Department of Veterans Affairs.

References

1. Bosnjak, Z., and Kampine, J. P. (1983). Effects of halothane, enflurane, and isoflurane on the SA node. *Anesthesiology* **58,** 314–321.
2. Altura, B. M., Altura, B. T., Carella, A., Turlapaty, P. D. M. V., and Weinberg, J. (1980). Vascular smooth muscle and general anesthetics. *Fed. Proc.* **39,** 1584–1591.
3. Bosnjak, Z. (1993). Ion channels in vacular smooth muscle. *Anesthesiology* **79,** 1392–1401.
4. Alper, M. H., Fleisch, J. H., and Flacke, W. (1969). The effects of halothane on the responses of cardiac sympathetic ganglia to various stimulants. *Anesthesiology* **31,** 429–463.
5. Price, H. L., Linde, H. W., and Morse, H. T. (1963). Central nervous actions of halothane affecting the systemic circulation. *Anesthesiology* **24,** 770–778.
6. Price, H. L., Price, M. L., and Morse, H. T. (1965). Effects of cyclopropane, halothane, and procaine on the vasomotor center of the dog. *Anesthesiology* **26,** 55–60.
7. Seagard, J. L., Hopp, F. A., Bosnjak, Z. J., Elegbe, E. O., and Kampine, J. P. (1983). Extent and mechanism of halothane sensitization of the carotid sinus baroreceptors. *Anesthesiology* **58,** 432–437.
8. Skovsted, P., Price, M. L., Price, H. L. (1970). The effects of short-acting barbiturates on arterial pressure, preganglionic sympathetic activity and barostatic reflexes. *Anesthesiology* **33,** 10–18.
9. Biscoe, T. J., and Millar, R. A. (1966). The effect of cyclopropane, halothane and ether on sympathetic ganglionic transmission. *Br. J. Anaesth.* **38,** 3–12.
10. Seagard, J. L., Bosnjak, Z. J., Hopp, F. A., Jr., and Kampine, J. P. (1985). Cardiovascular effects of general anesthesia. *In* "Effects of Anesthesia" (B. G. Covino, H. A. Fozzard, K. Rehder, and G. Strichartz, eds.), pp. 149–177. American Physiological Society, Bethesda, Maryland.
11. Schmeling, W. T., Bosnjak, Z. J., and Kampine, J. P. (1990). Anesthesia and the autonomic nervous system. *Semin. Anesth.* **9**(4), 223–231.
12. Seagard, J. L., Hopp, F. A., Donegan, J. H., Kalbfleisch, J. H., and Kampine, J. P. (1982). Halothane and the carotid sinus reflex: Evidence for multiple sites of action. *Anesthesiology* **57,** 191–202.

13. Wang, H., Epstein, R. A., Markee, S. J., and Bartelstone, H. J. (1968). The effects of halothane on peripheral and central vasomotor control mechanisms of the dog. *Anesthesiology* **29,** 877–886.
14. Bosnjak, Z. J., Seagard, J. L., Wu, A., and Kampine, J. P. (1982). The effects of halothane on sympathetic ganglionic transmission. *Anesthesiology* **57,** 473–479.
15. Pagel, P. S., Kampine J. P., Schmeling, W. T., and Warltier, D. C. (1990). Comparison of end systolic pressure length relations and preload recruitable stroke work as indices of myocardial contractility in the conscious and anesthetized chronically instrumented dog. *Anesthesiology* **73,** 278–290.
16. Pagel, P. S., Kampine, J. P., Schmeling, W. T., and Warltier, D. C. (1991). Influence of volatile anesthetics on myocardial contractility *in vivo:* Desflurane versus isoflurane. *Anesthesiology* **74,** 900–907.
17. Pagel, P. S., Kampine, J. P., Schmeling, W. T., and Warltier, D. C. (1991). Comparison of the systemic and coronary hemodynamic actions of desflurane, isoflurane, halothane and enflurane in chronically instrumented dogs. *Anesthesiology* **74,** 539–551.
18. Pagel, P. S., Kampine, J. P., Schmeling, W. T., and Warltier, D. C. (1991). Alteration of left ventricular diastolic function by desflurane, isoflurane and halothane in the chronically instrumented dog with autonomic nervous system blockade. *Anesthesiology* **74,** 1103–1114.
19. Kersten, J., Pagel, P. S., Tessmer, J. P., Roerig, D. L., Schmeling, W. T., and Warltier, D. C. (1993). Dexmedetomidine alters the hemodynamic effects of desflurane and isoflurane in chronically instrumented dogs. *Anesthesiology* **79,** 1022–1032.
20. Devcic, A., Schmeling, W. T., Kampine, J. P., and Warltier, D. C. (1994). Oral dexmedetomidine preserves baroreceptor function and decreases anesthetic requirements of halothane-anesthetized dogs. *Anesthesiology* **81.**
21. Kotrly, K. J., Ebert, T. J., Vucins, E., Igler, F. O., Barney, J. A., and Kampine, K. P. (1984). Baroreceptor reflex control of heart rate during isoflurane anesthesia in humans. *Anesthesiology* **60,** 173–179.
22. Skovsted, P., Price, M. L., and Price, H. L. (1969). The effects of halothane on arterial pressure, preganglionic sympathetic activity and barostatic reflexes. *Anesthesiology* **31,** 507–514.
23. Skovsted, P., and Sapthavichaikul, S. (1977). The effects of isoflurane on arterial pressure, pulse rate, autonomic nervous activity, and barostatic reflexes. *Can. Anaesth. Soc. J.* **24,** 304–314.
24. Poterack, K. A., Kampine, J. P., and Schmeling, W. T. (1991). Effects of isoflurane, midazolam, and etomidate on cardiovascular responses to stimulation of central nervous system pressor sites in chronically instrumented cats. *Anesth. Analg.* **73,** 64–75.
25. Bloor, B. C., and Shcmeling, W. T. (1993). Cardiovascular effects of alpha$_2$-adrenoceptors. *Anaesth. Pharmacol. Rev.* **1,** 246–263.
26. Schmeling, W. T., Hosko, M. J., and Hardman, H. F. (1981). Potentials evoked in the intermediolateral column by hypothalamic stimulation-suppression by Δ^9-tetrahydrocannabinol. *Life Sci.* **29,** 673–680.
27. Calaresu, F. R., and Ciriello, J. (1979). Electrophysiology of the hypothalamus in relation to central regulation of the cardiovascular system. *In* "Brain and Hypertension" (P. Meyer and H. Schmitt, eds.), pp. 129–136.
28. Korner, P. I. (1971). Integrative neural cardiovascular control. *Physiol. Rev.* **51,** 321–367.
29. Ciriello, J., and Calaresu, F. R. (1977). Descending hypothalamic pathways with cardiovascular function in the cat: A silver impregnation study. *Exp. Neurol.* **57,** 561–580.
30. Cummings, J. F. (1969). Thoracolumbar preganglionic neurons and adrenal innervation in the dog. *Acta Anat.* **73,** 27–37.

31. Henry, J. L., and Calaresu, F. R. (1974). Responses of single units in the intermediolateral nucleus to stimulation of cardioregulatory nuclei in the cat. *Brain Res.* **77,** 314–319.
32. Henry, J. L., and Calaresu, F. R. (1974). Pathways from medullary nuclei to spinal cardioacceleratory neurons in the cat. *Exp. Brain Res.* **20,** 305–314.
33. Farber, N. F., Samso, E., Kampine, J. P. and Schmeling, W. T. (1994). The influence of halothane on cardiovascular responses in the neuraxis of cats. *Anesthesiology* submitted.
34. Bleier, R. (1961). "The Hypothalamus of the Cat," pp. 58–62. Johns Hopkins Press, Baltimore, Maryland.
35. Berman, A. "The Brain Stem of the Cat," pp. 14–16 and 48–50. Univ. of Wisconsin Press, Madison.
36. Jasper, H. H., and Ajmone-Marsan, C. (1954). "A Stereotaxic Atlas of the Diencephalon of the Cat." National Research Council of Canada, Ottawa.
37. Calaresu, F. R., and Thomas, M. R. (1975). Electrophysiological connections in the brain stem involved in cardiovascular regulation. *Brain Res.* **87,** 335–338.
38. Lim, R. K. S., Chan-Nao, L., and Moffitt, R. L. (1960). "A Stereotaxic Atlas of the Dog's Brain." Thomas, Springfield, Illinois.
39. Antonaccio, M. J. (1977). Neuropharmacology of central mechanisms governing the circulation. *In* "Cardiovascular Pharmacology" (M. Antonaccio, ed.), pp. 131–165. Raven, New York.
40. Vlahakos, D., Gavras, I., and Gavras, H. (1985). Alpha-adrenoceptor agonists applied in the area of the nucleus tractus solitarii in the rat: Effect of anesthesia on cardiovascular responses. *Brain Res.* **347,** 372–375.
41. Shimoji, K., Fujioka, H., and Ebata, T. (1984). Anesthetics block excitation with various effects on inhibition in MRF neurons. *Brain Res.* **295,** 190–193.
42. Greisheimer, E. M. (1965). The circulatory effects of anesthetics. *In* "Handbook of Physiology, Circulation," Sec. 2, Vol. 3, Chapt. 70, pp. 2477–2510. American Physiological Society, Washington, D.C.
43. Peiss, C. W., and Manning, J. W. (1964). Effects of sodium pentobarbital on electrical and reflex activation of the cardiovascular system. *Circ. Res.* **14,** 228–235.
44. Ngai, S. H., and Bolme, P. (1966). Effects of anesthetics on circulatory regulatory mechanisms in the dog. *J. Pharmacol. Exp. Ther.* **153,** 495–504.
45. Cox, H. R., and Bagshaw, R. J. (1979). Influence of anesthesia on the response to carotid hypotension in dogs. *Am. J. Physiol.* **237,** H424–H432.
46. Florez, J., and Borison, H. L. (1969). Effects on central depressant drugs on respiratory regulation in the decerebrate cat. *Respir. Physiol.* **6,** 318–329.
47. Borison, H. L. (1978). Central nervous system respiratory depressants–Anesthetics, hypnotics, sedatives and other respiratory depresants. *Pharmacol. Ther.* **3,** 377–395.
48. Sollman, T. (1949). "Manual of Pharmacology," 7th Ed. Saunders, Philadelphia, Pennsylvania.
49. Farber, N. E., Samso, E., Schwabe, D., and Schmeling, W. T. (1994). Dexmedetomidine modulates cardiovascular responses to stimulation of central nervous system pressor sites in cats anesthetized with halothane. *Anesth. Analg.* **78,**(2S), S204.
50. Barron, K. W., and Heesch, C. M. (1990). Cardiovascular effects of posterior hypothalamic stimulation in baroreflex-denervated rats. *Am. J. Physiol.* **259,** H720–H727.
51. Kannan, H., Hayashida, Y., and Yamashita, H. (1989). Increase in sympathetic outflow by paraventricular nucleus stimulation in awake rats. *Am. J. Physiol.* **256,** R1325–R1330.
52. Miura, M., and Reis, D. J. (1972). The role of the solitary and paramedian reticular nuclei in mediating cardiovascular reflex responses from carotid baro- and chemoreceptor. *J. Physiol. (London)* **223,** 525–548.
53. Cochrane, K. L., and Nathan, M. A. (1993). Cardiovascular effects of lesions of the

rostral ventrolateral medulla and the nucleus reticularis parvocellularis in rats. *J. Auton. Nerv. Syst.* **43,** 69–82.
54. Chai, C. Y., Lin, R. H., Lin, A. M. Y., Pan, C. M., Lee, E. H. Y., and Kuo, J. S. (1988). Pressor responses from electrical or glutamate stimulations of the dorsal or ventrolateral medulla. *Am. J. Physiol.* **255,** R709–R717.
55. Owsjannikow, P. (1871). Die fonischen und reflextorischen centren der gefass nerven. *K sachs Ges der Wiss Mathematischenphysische Klasse Ber* **23,** 135–147.
56. Karplus, J. P., and Kreidl, A. (1909). Gehirn and sympathicus. I Zwischenhinbasis und halssympathicus. *Pfluegers Arch. Physiol.* **129,** 138–144.
57. Reis, D. J., Granata, A. R., Joh, T. H., Ross, C. A., Ruggiero, D. A., and Park, D. H. (1984). Brain stem catecholamine mechanisms in tonic and reflex control of blood pressure. *Hypertension* **6**(Suppl. 2), 7–15.
58. Reis, D. J., Morrison, S., and Ruggiero, D. A. (1988). The C1 area of the brainstem in tonic and reflex control of blood pressure. State of the art lecture. *Hypertension* **11**(Suppl. 1), 8–13.
59. Yamashita, H., Kannan, H., Kasai, M., and Osaka, T. (1987). Decrease in blood pressure by stimulation of the rat hypothalamic paraventricular nucleus with L-glutamate or weak current. *J. Auton. Nerv. Syst.* **19,** 229–234.
60. Miura, M., and Reis, D. J. (1969). Termination and secondary projections of carotid sinus nerve in the cat brain stem. *Am. J. Physiol.* **217,** 142–153.
61. Syka, J., Papelar, J., and Radil-Weiss, T. (1975). Influence of increasing doses of pentobarbital on the mesencephalic reticular formation in rats. Spontaneous firing of neuronal pairs and activity evoked by polarization. *Brain Res.* **61,** 151–155.
62. Gebber, G. L. and Klevans, L. R. (1972). Central nervous system modulation of cardiovascular reflexes. *Fed. Proc.* **31,** 1245–1252.
63. Scheibel, M. E., and Scheibel, A. B. (1958). Structural substrates for integrative patterns in the reticular brain stem core. *In* "Henry Ford Hospital International Symposium on Reticular Formation of the Brain" (H. H. Jasper, L. D. Proctor, K. L. R. S. Knighton, W. C. Noshay, and R. T. Costello, eds), Chapt. 2, p. 40. Little, Brown, Boston.
64. Bard, P. (1960). Anatomical organization of the central nervous system in relation to control of the heart and blood vessels. *Phys. Rev.* **40,** 3–26.
65. Pitts, R. F., Larabee, M. G., and Bronk, D. W. (1941). An analysis of hypothalamic cardiovascular control. *Am. J. Physiol.* **134,** 359–383.
66. Pitts, R. F., and Bronk, D. W. (1941). Excitability cycle of the hypothalamas–sympathetic–neuron system. *Am. J. Physiol.* **134,** 504–522.
67. Coote, J. H. (1988). The organization of cardiovascular neurons in the spinal cord. *Rev. Physiol. Biochem. Pharmacol.* **110,** 148–285.
68. Loewy, A. D. (1990). Central autonomic pathways. *In* "Central Regulation of Autonomic Functions" (A. D. Loewy and K. M. Spyer, eds.), pp. 88–103. Oxford Univ. Press, New York.
69. Gebber, G. L., Taylor, D. G., and Weaver, L. C. (1973). Electrophysiological studies on organization of central vasopressor pathways. *Am. J. Physiol.* **224,** 470–481.
70. Hosko, M. J., Schmeling, W. T., and Hardman, H. F. (1982). Δ^9-Tetrahydrocannibinol: Site of action for autonomic effects. *In* "Cannabinoids," Academic Press, New York.
71. Poterack, K. A., Kampine, J. P., and Schmeling, W. T. (1991). The effect of halothane on thermosensitive neurons in the preoptic region of the anterior hypothalamus in acutely instrumented cats. *Anesthesiology* **75,** 625–633.

Index

Contents of Previous Volumes

Volume 23

Volume 24

Volume 25

Volume 26

Volume 27

Volume 28

Volume 29A

Volume 29B

Volume 30

ISBN 0-12-032932-8